NATURAL TOXINS 2

Structure, Mechanism of Action, and Detection

ADVANCES IN EXPERIMENTAL MEDICINE AND BIOLOGY

Recent Volumes in this Series

Volume 389
INTRACELLULAR PROTEIN CATABOLISM
Edited by Koichi Suzuki and Judith S. Bond

Volume 390
ANTIMICROBIAL RESISTANCE: A Crisis in Health Care
Edited by Donald L. Jungkind, Joel E. Mortensen, Henry S. Fraimow, and Gary B. Calandra

Volume 391
NATURAL TOXINS 2: Structure, Mechanism of Action, and Detection
Edited by Bal Ram Singh and Anthony T. Tu

Volume 392
FUMONISINS IN FOOD
Edited by Lauren S. Jackson, Jonathan W. DeVries, and Lloyd B. Bullerman

Volume 393
MODELING AND CONTROL OF VENTILATION
Edited by Stephen J. G. Semple, Lewis Adams, and Brian J. Whipp

Volume 394
ANTIVIRAL CHEMOTHERAPY 4: New Directions for Clinical Application and Research
Edited by John Mills, Paul A. Volberding, and Lawrence Corey

Volume 395
OXYTOCIN: Cellular and Molecular Approaches in Medicine and Research
Edited by Richard Ivell and John A. Russell

Volume 396
RECENT ADVANCES IN CELLULAR AND MOLECULAR ASPECTS OF ANGIOTENSIN RECEPTORS
Edited by Mohan K. Raizada, M. Ian Phillips, and Colin Sumners

Volume 397
NOVEL STRATEGIES IN THE DESIGN AND PRODUCTION OF VACCINES
Edited by Sara Cohen and Avigdor Shafferman

A Continuation Order Plan is available for this series. A continuation order will bring delivery of each new volume immediately upon publication. Volumes are billed only upon actual shipment. For further information please contact the publisher.

NATURAL TOXINS 2

Structure, Mechanism of Action, and Detection

Edited by

Bal Ram Singh

University of Massachusetts, Dartmouth
North Dartmouth, Massachusetts

and

Anthony T. Tu

Colorado State University
Fort Collins, Colorado

PLENUM PRESS • NEW YORK AND LONDON

Library of Congress Cataloging in Publication Data

Natural toxins 2: structure, mechanism of action, and detection / edited by Bal Ram Singh and Anthony T. Tu.
p. cm.—(Advances in experimental medicine and biology; v. 391)
"Proceedings of an American Chemical Society Symposium on Natural Toxins, held April 2–7, 1995, in Anaheim, California"—T.p. verso.
Includes bibliographical references and index.
ISBN-13:978-1-4613-8016-0
1. Toxins—Congresses. I. Singh, B. R. (Bal Ram), 1958– . II. Tu, Anthony T., 1930– III. American Chemical Society. IV. Series.
QP631.N383 1996 96-3810
615.9'5—dc20 CIP

Proceedings of an American Chemical Society Symposium on Natural Toxins, held April 2 – 7, 1995, in Anaheim, California

ISBN-13:978-1-4613-8016-0 e-ISBN-13: 978-1-4613-0361-9
DOI: 10.1007/978-1-4613-0361-9

Softcover reprint of the hardcover 1st edition 1996

A Division of Plenum Publishing Corporation
233 Spring Street, New York, N. Y. 10013

10 9 8 7 6 5 4 3 2 1

PREFACE

From beach encounters, aquaculture perils, and processed-food poisoning to snake bites and biological warfare, natural toxins seem never to be far from the public's sight. A better understanding of toxins in terms of their origin, structure, structure–function relationships, mechanism of action, and detection and diagnosis is of utmost importance to human and animal food safety, nutrition, and health. In addition, it is now clear that many of the toxins can be used as scientific tools to explore the molecular mechanism of several biological processes, be it a mechanism involved in the function of membrane channels, exocytosis, or cytotoxicity. Several of the natural toxins have also been approved as therapeutic drugs, which has made them of interest to several pharmaceutical companies. For example, botulinum neurotoxins, which have been used in studies in the field of neurobiology, have also been used directly as therapeutic drugs against several neuromuscular diseases, such as strabismus and blepherospasm. Toxins in combination with modern biotechnological approaches are also being investigated for their potential use against certain deadly medical problems. For example, a combination of plant toxin ricin and antibodies is being developed for the treatment of tumors. The great potential of natural toxins has attracted scientists of varying backgrounds—pure chemists to cancer biologists—to the study of fundamental aspects of the actions of these toxins.

A symposium on natural toxins was held at the 209th National Meeting of the American Chemical Society in April 1995 in Anaheim, California. This was the second such symposium on natural toxins. The first symposium was held in 1990 at the 200th National Meeting of the American Chemical Society, Washington, D.C. In the last decade, research on toxins has targeted not only ways to prevent or reverse their harmful effects, but also studies of these same effects as a means of understanding other biomedical processes. Consequently, research on toxins is gaining momentum, and a symposium at a five-year interval was quite appropriate. The latest symposium on natural toxins was intended to gather researchers with varying background to share not only their latest results but also their expertise to enrich one another's research efforts. This is obvious from the wide range of topics covered during the symposium. The objective of the symposium was to have a focused discussion on the toxins of various origins (animal, plant, marine, and microbial) to learn their common and contrasting modes of action.

In contrast to the first symposium-based book on natural toxins, topics in this book are not arranged based on their origin. The book is divided into five sections. The first section provides an overview of toxins from all the four major classes based on origin. The other sections are arranged to reflect common themes of our understanding of these toxins. In the past several years, as more detailed information on the structure and the mechanism of action of toxins is emerging, it is becoming increasingly clear that several toxins, irrespective of their origin, seem to have common structural features and a common strategy of action.

Therefore, there is a tremendous opportunity to derive useful common information from toxins of different origins.

The second section contains articles on the origin and structure–function aspects of toxins. An understanding of the structure of toxins and the relationship between structure and function has always been a fascinating topic. However, recent advances in the tools available for the structure determination as well as for structure alteration are moving the field of toxin research at a much faster pace than in the past. This is obviously reflected in the number of chapters in this section of the book.

The mechanism of action of many toxins is unique in terms of the target tissues and the nature of the damage caused. However, being external agents, almost all toxins encounter similar physiological barriers before they reach their cellular targets. Many toxins act on cell membranes as their main target, others cross the cell membrane to interfere with cellular metabolism. An understanding of the mechanism of action of toxins provides opportunities to design and develop antidotes for toxins and their harmful effects. The section on the mechanism of action reflects this versatility.

Toxins are acutely active external agents that interfere with some of the most important biochemical systems. A positive aspect of the existence of toxins is their application to the study of fundamental biochemistry, which has provided critical information that would not be available were toxins not used. In some cases, research on the mechanism of toxin action has led to discoveries in other fields of biological systems, use of bacterial toxins to understand signal transduction is one such example. Snake venoms are increasingly being used to investigate membrane structure–function relationships. Toxins are also being used or being developed as therapeutic drugs. This topic was discussed at the symposium by some of the leading scientists in field, and their articles in the book provide some prime examples of toxin use in the scientific and medical advances.

Detection and diagnosis are the core of prevention and treatment against toxins. Toxins that are active at trace concentrations and present in the environment or in food supply must be detected and removed. While the old methodology of animal use for bioassay is still commonly used, modern instrumentation and biotechnology are providing new avenues in the field of detection and diagnosis of toxins. The section on the detection, diagnostics, and therapy contains articles discussing some of these cutting-edge methods by leading experts in the field.

We would like to thank David J. Armstrong, Program Chairman, Agricultural and Food Chemistry Division, American Chemical Society, for his strong encouragement and help in organizing the symposium and providing partial funds. Cynthia Musinan was extremely helpful in the dispersal of funds. Financial support from the Food Safety Division of the United States Department of Agriculture (Grant 94-37201-1158), U.S. Army Medical Research Institute of Infectious Diseases (Grant DAMD17-95-1-5009), and List Biological Laboratories, Inc., are gratefully acknowledged. Financial and administrative support from the University of Massachusetts, Dartmouth were invaluable in the organization of the symposium and in the preparation of the book. We would also like to acknowledge Patricia M. Vann and Danielle McPhail, Plenum Press, for their persistent help in editing and producing this volume.

It is our hope that the book will provide a broad current knowledge on toxins. We believe that these chapters will provide a basis for the future research on toxins that will open new frontiers in the field.

Bal Ram Singh
Anthony T. Tu

CONTENTS

OVERVIEW OF TOXINS

ORIGIN, STRUCTURE AND FUNCTION

MECHANISM OF ACTION

SCIENTIFIC AND MEDICAL TOOLS

DETECTION, DIAGNOSTICS AND THERAPY

NATURAL TOXINS 2

Structure, Mechanism of Action, and Detection

MARINE NATURAL PRODUCTS

Diversity in Molecular Structure and Bioactivity

Paul J. Scheuer

University of Hawaii at Manoa
Honolulu, Hawaii 96822-2275

INTRODUCTION

The difference between a harmful toxin and a beneficial drug is largely a matter of dose. Historically, toxins provided important leads towards drug discovery, as their purification could be followed by a readily available and unambiguous bioassay - death of a small rodent. Modern sensitive enzyme-based assays have made all natural products - organic compounds isolated from living organisms - a potential source for drug discovery. The oceans of the earth which cover 70% of its surface and are the habitat of a rich and diverse fauna and flora, much of it still to be discovered and described, offer a prime resource for new molecular structures with a broad spectrum of bioactivities.

However, the road from recognition of a desirable bioactivity to an FDA-approved commercial drug is long, expensive, and beset with obstacles, which include access to starting materials, patent protection, competitive alternatives, size of potential market. Marine natural product research is only about twenty years old - too young to have produced a new morphine or a better penicillin. But the research has already generated enough promising leads to engender cautious optimism for the future.

SOME PROMISING LEADS

The enzyme topoisomerase II has been shown to be a useful indicator for predicting activity against certain lung cancers. Among marine secondary metabolites which have been isolated in my laboratory are two structurally diverse compounds which possess that activity.

Elenic acid (Figure 1) was isolated from an Indonesian sponge, an undescribed species of *Plakinastrella*.[1] Its structure, a C_{22} fatty acid with a phenol terminus, is remarkably unspectacular. With only two functionalized and widely separated termini the compound can easily be manipulated synthetically for discovery of its active moiety. Its topoisomerase II activity, 0.1 μg/mL, compares favorably with currently used clinical agents, as e.g. doxorubicin or etoposide.[2]

Natural Toxins II, Edited by B. R. Singh and A. T. Tu
Plenum Press, New York, 1996

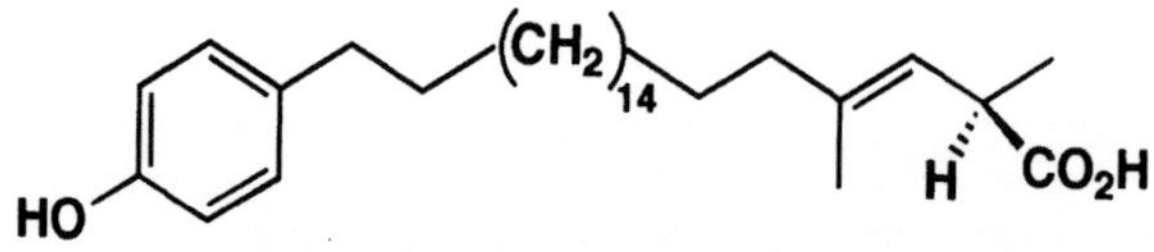

Figure 1. Elenic acid.

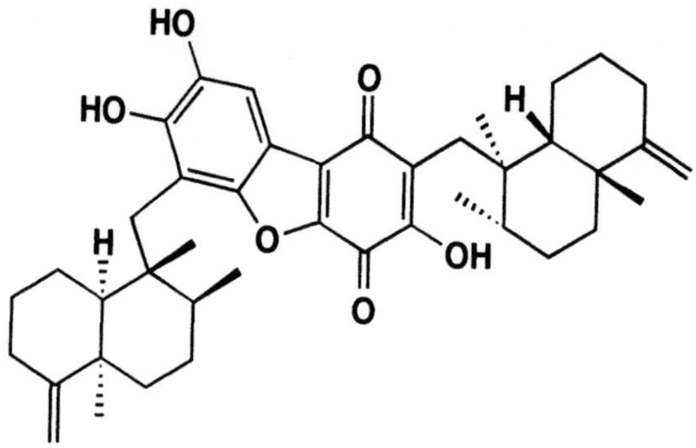

Figure 2. Popolohuanone.

The second compound, popolohuanone E (Figure 2), displays topoisomerase II activity of the same order of magnitude, but possesses a far more complex molecular structure.[3] It was isolated from a Micronesian sponge, *Dysidea* sp. Its structure is made up of two nearly identical fragments that are joined through an additional furan ring. The monomeric building blocks are frequently encountered sponge metabolites: a sesquiterpene (usually a drimane) joined to a C_6 oxygenated benzenoid or quinoid ring. These C_{21} precursors are exemplified by ilimaquinone (Figure 3),[4] which exhibits anti-HIV activity[5] and has been shown to inhibit protein transport.[6] Many variants of the highly reactive benzenoid part of these molecules have been discovered over the years. They constitute a rich resource for structure-activity studies.

Chemotaxonomic considerations have provided valuable guidelines in the study of bioactive and/or toxic constituents of terrestrial flowering plants. When examining, say, a

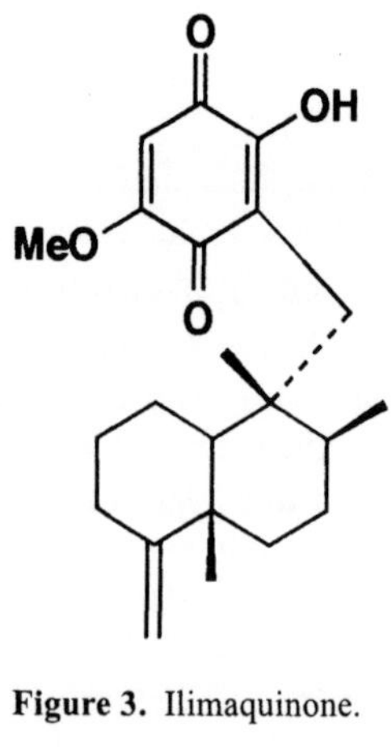

Figure 3. Ilimaquinone.

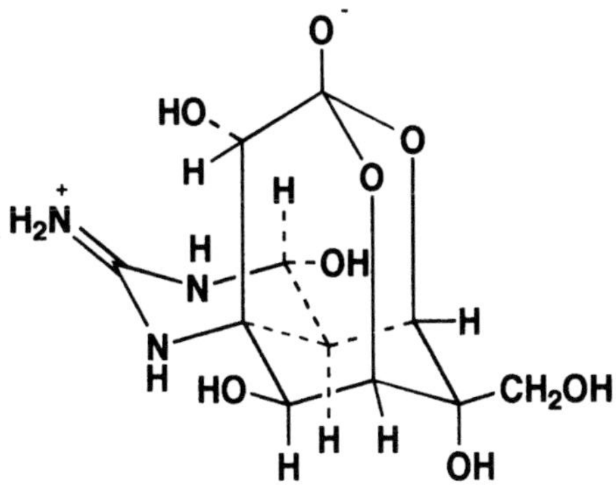

Figure 4. Tetrodotoxin.

new species of *Rauwolfia*, one can be reasonably certain to find indole alkaloids. Or, if one studies cyanogenic glycosides, one would concentrate on certain genera in the family Rosaceae. The greater complexity of the marine environment with its large floating communities of plants and animals makes such an approach much more tenuous. Widespread occurrence of epibiotic and symbiotic organisms introduce an additional element of uncertainty. At times, the first clue that an undiscovered and unsuspected agent is responsible for the biosynthesis of a marine natural product comes through isolation of identical compounds from phyletically distant taxa. A well-known example is the puffer fish toxin, tetrodotoxin (Figure 4),[7] which was also discovered in a California amphibian.[8] More than twenty years later and after many failed attempts to study the biosynthesis of tetrodotoxin in laboratory culture of the host organisms it could be shown that bacteria of two separate genera, *Pseudomonas* sp.[9] and *Vibrio* sp.,[10] are primary producers of the toxin.

In contrast to the deliberate attempt to discover the biosynthetic agent of tetrodotoxin the revelation of the origin of okadaic acid (Figure 5) was serendipitous. The acid was isolated from a black sponge, *Halichondria okadai*, as halichondrine A by researchers at the Fujisawa Pharmaceutical Company in Japan. Disappointing bioassay results in an antitumor screen prompted Fujisawa to drop the project. We were fortunate to inherit it and with Professor Jon Clardy's help were able to determine the molecular structure.[11,12] We also renamed it so as to reflect on its chemical nature. Simultaneously, Schmitz and coworkers at the University of Oklahoma isolated the acid from a Caribbean sponge, *H. melanodocia*. The only slight suspicion that the sponge was perhaps not the primary producer of okadaic acid came from the low yield in which the compound could be isolated from the animal. Soon after the structure of okadaic acid was published, the compound was discovered in Professor Yasumoto's laboratory as a constituent of the dinoflagellate *Prorocentrum lima*.[13] The organism was being cultured as part of a program to screen microalgae as potential

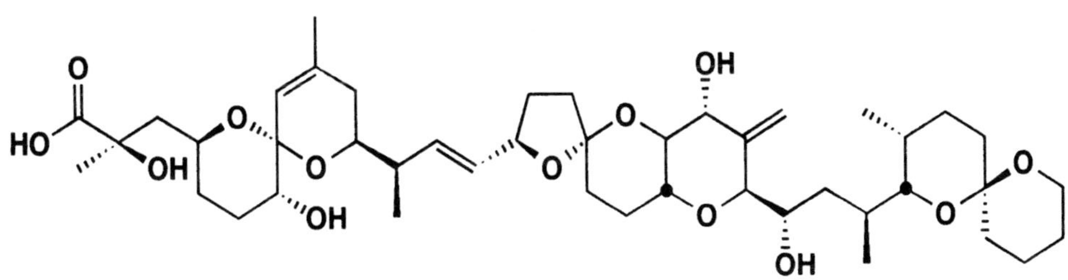

Figure 5. Okadaic acid.

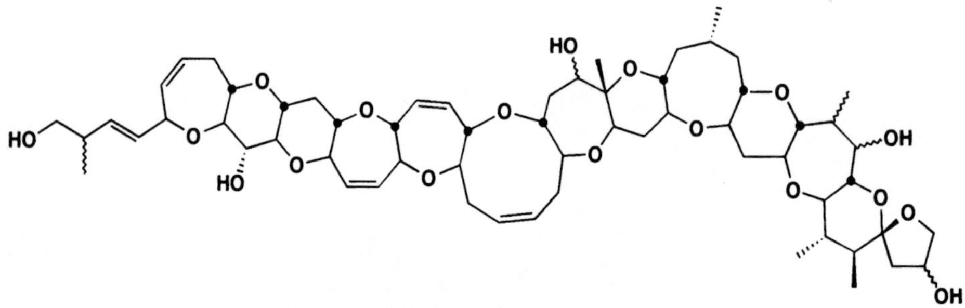

Figure 6. Ciguatoxin.

contributors to ciguatera fish poisoning.[14] This proved to be a most fortunate coincidence, as parallel behavior of ciguatoxin (Figure 6) and okadaic acid on thin-layer chromatograms provided the first vital clue that the two compounds were structurally related and belong to the group of polyethers, that includes the brevetoxins, which are the red tide toxins in the Gulf of Mexico and are constituents of the dinoflagellate *Gymnodinium breve*.

This important contribution to the study of ciguatera fish poisoning proved to be only the first important contribution of okadaic acid to biomedical science. A second role was discovered by Fujiki and coworkers,[15] who have studied the mechanism of carcinogenesis. Cancers are believed to be generated in two distinct stages, by an initiating and a promoting agent. Okadaic acid proved to be a powerful promoter that was structurally unrelated to the well-known phorbol esters. Yet another discovery, the property of okadaic acid to inhibit

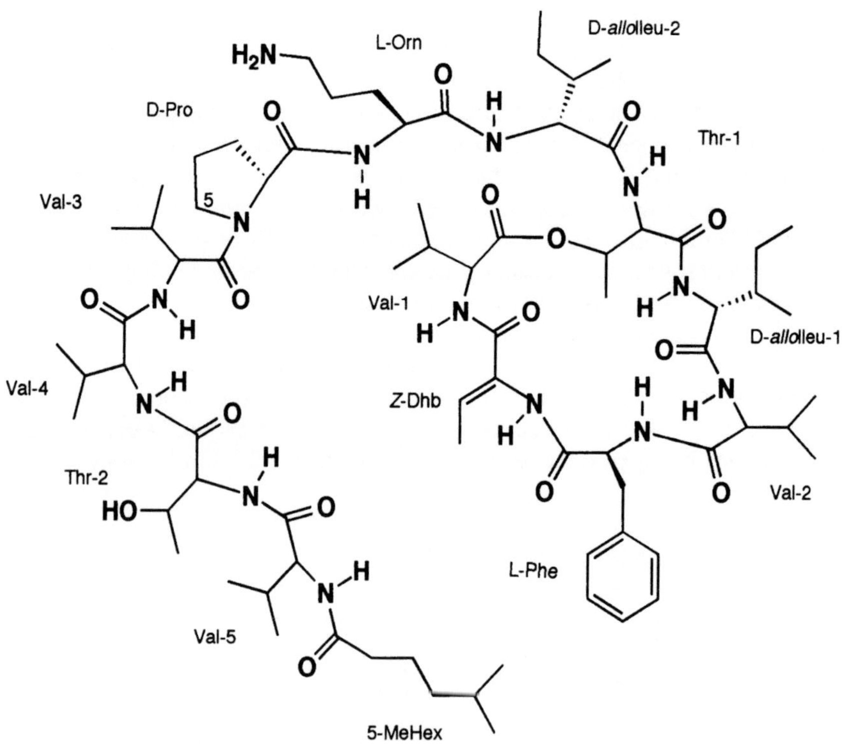

Figure 7. Kahalalide F.

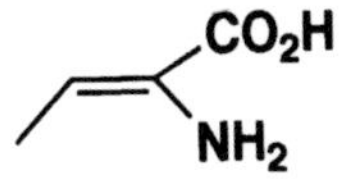

Figure 8. (Z)-2-Aminobut-2-enoic acid.

phosphatases 1 and 2A selectively[16] has made this compound an important tool in the study of protein transport and cellular regulation.[17]

The foregoing examples of bioactive marine metabolites have been fortuitous discoveries or have resulted from random screening of large numbers of organisms. Ecological phenomena, especially on tropical coral reefs, where competition for space and food is intense, constitute an excellent source of organic compounds possessing unique structural features and desirable pharmacological properties.[18]

Mollusks that lack an external shell would be subject to predation were it not for the fact that many of these animals have elaborated chemical defenses, which most often are acquired through the diet, but occasionally are the result of biosynthesis. A dietary link was first suspected in 1965 between a sea hare [19] and a red alga[20] on which it grazes. Many similar relationships have been established between other herbivorous mollusks and their algal diet, as well as between carnivores and their prey, which include sponges, corals, tunicates and other sessile invertebrates.

Sacoglossan mollusks feed on green algae and ingest functioning chloroplasts which retain photosynethetic capability.[21] *Elysia rufescens* is a sacroglossan, which is described from Hawaii, but no algal food source is mentioned.[22] We have collected this animal in early spring at Ka'alawai Beach near Black Point, O'ahu and the green alga, *Bryopsis pennata*, on which it feeds. From the sacoglossan we were able to isolate six cyclic depsipeptides ranging from a C_{31} tripeptide to a C_{75} tridecapeptide,[23] all of them composed largely of

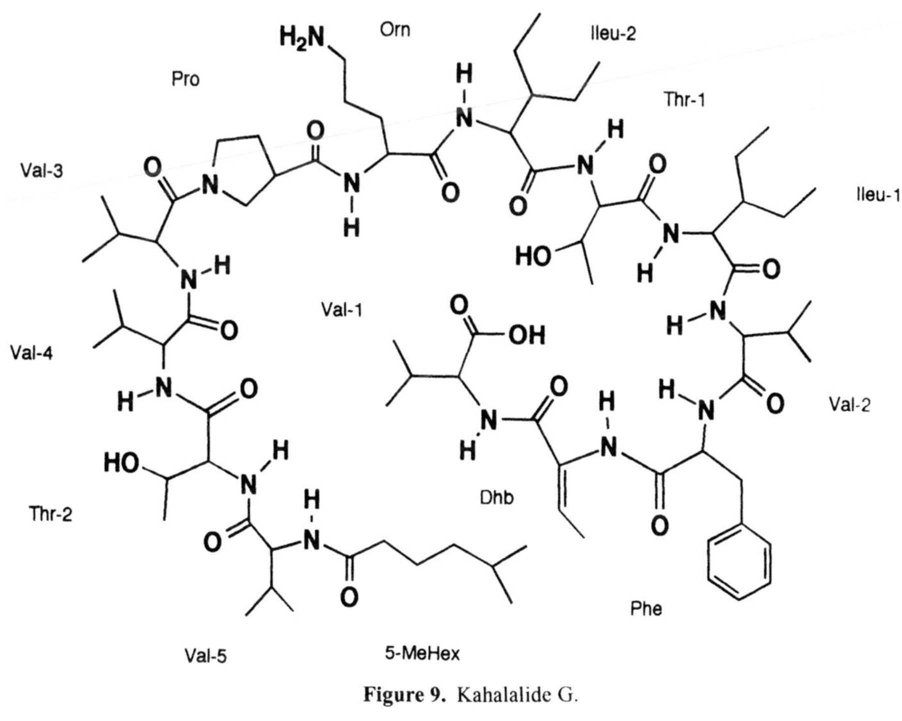

Figure 9. Kahalalide G.

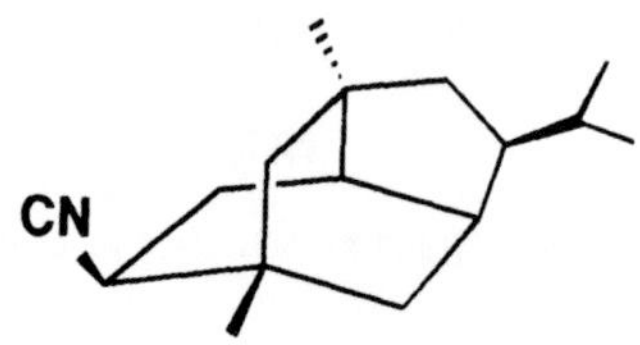

Figure 10. 9-Isocyanopupukeanane.

common amino acids. Only the largest, kahalalide F (Figure 7), $C_{75}H_{124}N_{14}O_{16}$, includes a single rare amino acid, (*Z*)-2-amino-but-2-enoic acid (Figure 8). It is also the only one with significant selective antitumor activity.[24] The fatty acid moieties of the six depsipeptides range in size and complexity from butanoic to 9-methyl-3-hydroxy decanoic acid. Four of the six are *iso*-acids. The alga, *B. pennata*, contains only two principal peptides - kahalalide F and its acyclic analog, kahalalide G (Figure 9). It is tempting to speculate that the acyclic kahalalide G, which is absent in the mollusk, is the precursor of the cyclized kahalalide F.

Sea hares, as was mentioned above, were among the earliest marine invertebrates to be studied by natural product chemists. Many of them live near shore, often are relatively large and conspicuous, move slowly if at all - reasons which make the animals an attractive target for collection. Besides, sea hares have long had a reputation of being toxic and, as chemists would soon discover, conveniently accumulate sequestered chemicals in their socalled digestive glands. Conversely, the carnivorous nudibranchs, while often spectacularly beautiful, were not investigated until much later. The first report[25] was triggered by an observation[26] that a nudibranch, *Phyllidia varicosa*, was ichthyotoxic and that its toxic secretion was a small heat-stable molecule with an unpleasant unidentifiable odor. Attempts to identify this substance were stymied, when it was discovered that the secretion rapidly declined in captivity and that the food source was unknown. It was only through accidental observation of *P. varicosa* preying on a sponge, *Ciocalypta* sp. (originally misidentified as *Hymeniacidon* sp.), that sufficient amounts of the active substance could be isolated and purified for structural elucidation. The identity of the *Phyllidia* secretion, 9-isocyanopupukeanane (Figure 10) was a gratifying outcome, indeed: a compound with a new tricyclic sesquiterpene skeleton bearing the rare isocyano function.

Many years later we isolated[27] from a Micronesian sponge, probably a *Ciocalypta* sp., axisonitrile-3 (Figure 11), which proved to be selectively active against solid tumors. Axisonitrile-3 was a known compound that had been described earlier by Fattorusso and coworkers.[28]

CONCLUSION

Diversity in molecular structure is readily apparent by even a small selection of bioactive marine natural products. Diversity in biological action is much more difficult to

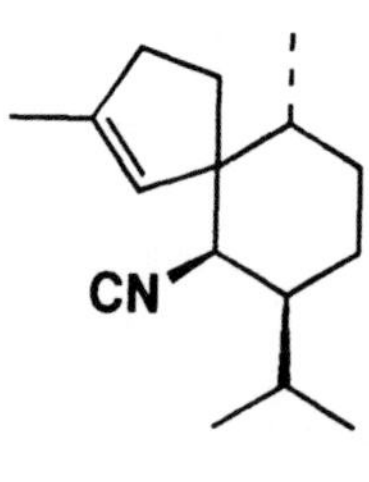

Figure 11. Axisonitrile-3.

demonstrate since widely accessible bioassay systems are often disease-oriented and hence driven by available funding: cancer research is better supported than, say, antimalaria studies. This situation, however, is likely to improve as underlying enzyme mechanisms replace organism-based bioassays.

ACKNOWLEDGMENTS

I am grateful to many devoted coworkers, who are not cited. I should like to thank the National Science Foundation, the Sea Grant College Program, and PharmaMar, S.A. for their continuing financial support.

REFERENCES

1. Juagdan, E. G., Kalidindi, R. S., Scheuer, P. J., and Kelly-Borges, M. 1995. Elenic acid, an inhibitor of topoisomerase II, from a sponge, *Plakinastrella* sp., *Tetrahedron Lett.*, 36:3977-3980.
2. Kasahara, K., Fujiwara, Y., Sugimoto, Y., Nishio, K., Tamura, T., Matsuda, T., Saijo, N. 1992. Determinants of response to the DNA topoisomerase II inhibitors doxorubicin and etoposide in human lung cancer cell lines, *J. Natl. Cancer Inst.* 84:113-118.
3. Carney, J. R. and Scheuer, P. J. 1993. Popolohuanone E, a topoisomerase-II inhibitor with selective lung tumor cytotoxicity from the Pohnpei sponge, *Dysidea* sp., *Tetrahedron Lett.* 34:3727-3730.
4. Luibrand, R. T., Erdman, T. R., Vollmer, J. J., Scheuer, P. J., Finer, J. and Clardy, J. 1979. Ilimaquinone, a sesquiterpenoid quinone from a marine sponge, *Tetrahedron* 35:609-612.
5. Loya, S., Tal, R., Kashman, Y. and Hizi, A. 1990. Ilimaquinone, a selective inhibitor of the RN_{ase} H activity of human immunodeficiency virus type I reverse transcriptase, *Antimicrob. Agents Chemother.* 34:2009-2012.
6. Takizawa, P. A., Yucel, J. K., Veit, B., Faulkner, D. J.; Deerinck, T., Soto, G., Ellisman, M. and Malhotra, V. 1993. Complete vesiculation of Golgi membranes and inhibition of protein transport by a novel sea sponge metabolite, ilimaquinone, *Cell* 73:1079-1090.
7. Woodward, R. B. 1964. Structure of Tetrodotoxin, *Pure Appl. Chem.* 9:49-74
8. Mosher, H. S., Fuhrman, F. A., Buchwald, H. D. and Fischer, H. G. 1964. Tarichatoxin-tetrodotoxin: A potent neurotoxin, *Science* 144:1100-1110.
9. Yasumoto, T., Yasumura, D., Yotsu, M., Michishita, T., Endo, A. and Kotaki, Y. 1986. Bacterial production of tetrodotoxin and anhydrotetrodotoxin by *Pseudomonas* sp., *Agric. Biol. Chem.* 50:793-795.
10. Noguchi, T., Jeon, J.-K., Arakawa, O., Sugita, H., Deguchi, Y., Shida, Y. and Hashimoto, K. 1986. Occurrence of tetrodotoxin and anhydrotetrodotoxin in *Vibrios* sp. isolated from the intestines of a xanthid crab, *Atergatis floridus*, *J. Biochem.* 99:311-314.
11. Tachibana, K. 1980. *Structural studies on marine toxins*. Ph.D. Dissertation, University of Hawaii, Honolulu, HI.
12. Tachibana, K., Scheuer, P. J., Tsukitani, Y., Kikuchi, H., vanEngen, D., Clardy, J., Gopichand, Y. and Schmitz, F. J. 1981. Okadaic acid, a cytotoxic polyether from sponges of the genus *Halichondria*, *J. Am. Chem. Soc.* 103:2469-2471.
13. Murakami, Y., Oshima, Y. and Yasumoto, T. 1982. The identification of okadaic acid as a toxic component of a marine dinoflagellate *Prorocentrum lima*, *Bull. Jap. Soc. Sci. Fish.*, 48:69-72.
14. Scheuer, P. J. 1994. Ciguatera and its offshoots: Encounters en route to a molecular structure, *Tetrahedron*, 50:3-18.
15. Suganuma, M., Fujiki, H., Suguri, H., Yoshizawa, S., Hirota, M., Nakayasu, M., Ojika, M., Wakamatsu, K., Yamada, K. and Sugimura, T. 1988. Okadaic acid: An additional non-phorbol-12-tetradecanoate-13-acetate-type tumor promoter, *Proc. Natl. Acad. Sci., U.S.A.* 85:1768-1771.
16. Haystead, T. A. J., Sim, A. T. R., Carling, D., Honnor, R. C., Tsukitani, Y., Cohen P. and Hardie, D. G. 1989. Effects of the tumor promoter okadaic acid on intracellular protein phosphorylation and metabolism. *Nature*, 337:78-81.
17. Cohen, P., Holmes, C. F. B. and Tsukitani, Y. 1990. Okadaic acid: a new probe for the study of cellular regulation, *Trends Biochem. Sci.* 15:98-102.
18. Scheuer, P. J. 1990. Some marine ecological phenomena: Chemical basis and biomedical potential, *Science* 248:173-177.

19. Yamamura, S. and Hirata, Y. 1963. Structure of aplysin and aplysinol, naturally occurring bromo-compounds, *Tetrahedron* 19:1485-1496.
20. Irie, T., Yasunari, Y., Suzuki, T., Imei, N., Kurosawa, E. and Masamune, T. 1965. A new sesquiterpene hydrocarbon from *Laurencia glandulifera.*
21. Ireland, C. and Scheuer, P.J. 1979. Photosynthetic marine mollusks. In vivo ^{14}C incorporation into metabolites of the Sacoglossan *Placobranchus ocellatus*, *Science* 203:922-923.
22. Kay, A. E. 1979. *Hawaiian Marine Shells*. Bishop Museum Press, Honolulu, HI, p. 454.
23. Hamann, M. T. 1992. *Biologically active constituents of some marine invertebrates*. Ph.D. Dissertation, University of Hawaii, Honolulu, HI.
24. Hamann, M. T. and Scheuer, P. J. 1993. Kahalalide F, a bioactive depsipeptide from the sacoglossan mollusk *Elysia rufescens* and the green alga *Bryopsis* sp. *J. Am. Chem. Soc.*, 115:5825-5826.
25. Burreson, B. J., Scheuer, P. J., Finer, J. and Clardy, J. 1975. 9-Isocyanopupukeanane, a marine invertebrate allomone, *J. Am. Chem. Soc.* 97:4763-4764.
26. Johannes, R. E. 1963. A poison-secreting nudibranch (Mollusca: Opisthobranchia), *Veliger* 5:104-105.
27. Pham, A. T., Ichiba, T., Yoshida, W. Y., Scheuer, P. J., Uchida, T., Tanaka, J. and Higa, T. 1991. Two marine sesquiterpene thiocyanates, *Tetrahedron Lett.*, 32:4843-4846.
28. DiBlasio, B., Fattorusso, E., Magno, S., Mayol, L., Pedone, C., Santocroce, C and Sica, D. 1976. Axisonitrile-3, axisothiocyanate-3, and axamide-3. Sesquiterpenes with a novel spiro[4,5] decane skeleton from the sponge *Axinella cannabina*, *Tetrahedron* 32:473-478.

2

PLANT TOXINS

The Essences of Diversity and a Challenge to Research

Gary D. Manners

Western Regional Research Center
Agricultural Research Service
United States Department of Agriculture
800 Buchanan St.
Albany, California 94710

INTRODUCTION

Historically, toxins from plants are associated with murder, assassination and suicide (Mann, 1992). The deaths of several famous historical figures have directly or indirectly involved toxins from poisonous plants. Socrates' forced suicide involved a toxic alkaloid from poison hemlock (*Conium maculatum)*, while the many victims of Livia (wife of Emperor Augustus) and Agrippina (wife of Claudius) succumbed to the toxic tropane alkaloids of deadly nightshade (*Atropa belladona*). Cleopatra is reported to have tested the extracts of several poisonous plants including henbane (*Hyoscyamus niger),* deadly nightshade and nux-vomica (*Strychnos nux-vomica*) on her slaves before she chose the asp.

In the middle ages, proficiency in poisoning increased as did witchcraft and sorcery. Henbane, deadly nightshade and mandrake (*Mandragora officinarium*) were common ingredients in witches brew. Aconite from monkshead (*Chrondrodendron tomentosum*) became the poison of choice by the "professional poisoners" who devised ingenious methods (poison lipsticks, poison rings, etc) to accomplish their mission. Poison yew (*Taxus baccata*), jimson weed (*Datura stramonium*) and foxglove (*Digitalis purpurea*) also were used.

Today, the use of toxic plants for intentional poisonings are rare. In fact, accidental poisonings (exclusive of drug overdose) of humans as the result of natural toxins from plants or animals accounted for only 51 deaths among the 6,043 deaths reported for poisonings by all types of substances in 1988 (National Center for Health Statistics, 1991). While the impact of poisonous substances in plants on humans has been reduced in significance, the same can not be said for domestic animals. It is estimated (Nielsen and James, 1992) that the total annual death attributable to poisonous plants exceeds 250,000 animals per year for cattle and for sheep in the United States. These losses translated to an estimated direct economic loss in excess of 240 million dollars in 1989.

The historic examples of acute poisoning in humans and the losses of domestic animals to plant toxins are consistent with the usual concept that a "poisonous" plant is one

Natural Toxins II, Edited by B. R. Singh and A. T. Tu
Plenum Press, New York, 1996

capable of poisoning man or domestic animals. This conception is supported by a predominance of information in the literature about the effect of plant toxins on mammals. However, the "poisons" from plants which are responsible for human and domestic animal poisonings constitute only a small minority of the "toxins" in plants which have no effect on humans or domestic animals, but are important in the relationship of plants to other plants and to insects through acute, chronic and sequestered modes of action.

While the effect of plant toxins on mammals is the most visible and the most severe economically, plant toxin impacts in plant-plant and plant-insect relationships are also extensive and economically significant. This presentation will be a summary overview of the role of plant toxins in all three of these relationships. The reader can gain access to greater in-depth information about each of the relationships by referring to the published reviews and individual citations in the text.

Areas involving plant toxins which will not be included in this presentation are: contact dermatitis toxins, psychic toxins, carcinogens, inorganic accumulated toxins and toxins of fungal origin. In accord with their economic prominence, more emphasis will be placed on the role of plant toxins involved in plant-animal relationships. In particular, a research program investigating cattle losses to poisonous larkspur will be described as an example of how a research program directed to solving a toxic plant problem is formulated and executed, and how the research results can be applied.

ORIGINS OF PLANT TOXINS

Plant toxins occur in plants as secondary metabolites. Unlike the amino acids and sugars, the function of the majority of secondary metabolites in the evolution of plants is unclear. It can be argued that the plant toxins constitute a subcategory of secondary compounds in plants which contribute to their survival adaptations. Evidence that physiologically active secondary compounds provide a defense mechanism for the plant is mostly unproved. However, anecdotal evidence relating predation resistance of plant species and the toxicity of chemicals obtained from these plants to their predators is compelling evidence in favor of the evolution and existence of natural chemical defense mechanisms in plants. It has been suggested (Bell, 1985) that the evidence of the defensive role of chemicals in plants could best be obtained from a study of the organisms that prey upon them. If the successful predator has adopted a specific method to circumvent the toxic nature of a plant, it would be a strong argument for the role of the secondary compound as a defensive constituent in the plant. This adaptation by predators does occur and will be described in this presentation. Such evidence offers rebuttal to the argument that secondary compounds exist as "evolutionary noise" with no defined role in plants.

Secondary metabolites which are identified as plant toxins can be categorized into seven generalized categories: alkaloids; glycosides; organic acids; alcohols; resins and resinoids (terpenoids including benzenoid and phenolics); proteinaceous compounds; and mineral toxins (inorganic compounds). These categories comprise a broad and diverse range of chemical characteristics for plant toxins. Their role in the interactions between plants and plants, plants and insects, and plants and animals is equally broad and diverse.

The breadth of the categorization of the plant toxins speaks to the diverse nature of these compounds. Examples representative of these categories will be included in the following discussion of the interaction of plant toxins with plants, insects and mammals.

PLANT TOXINS IN PLANT-PLANT INTERACTIONS

The phenomenon of secondary metabolites of plants acting to affect the growth of other plants through some manner of inhibitory biochemical mechanism is termed *allelopathy,* and the chemical constituents involved in alleopathy are termed allelochemicals. Although the term "allelopathy" was coined by Molish in 1937, major research efforts in the discipline did not take place until the 1970s. The expansion of research has correspondingly significantly increased the literature over the last two decades and several books (Rice, 1984; Thompson, 1985; Putnam and Tang, 1986; Waller, 1987, Inderjit et al., 1995) have been written detailing the subject. This brief survey of allelopathy will provide examples of a few alleochemicals among a broad array of plant secondary metabolites which have been identified as toxins to other plants.

Allelochemicals acting directly as phytotoxins affect various aspects of plant growth and metabolism subacutely. Their effects may affect physical functions (e.g. membrane permeability, stomatal movement) or biochemical processes (e.g. specific enzyme activity, protein synthesis) through a direct mode of action. The allelochemical interference in physical and biochemical plant processes is generally considered to involve a combination of phytotoxins. The allelopathic interaction of plants is complicated by factors affecting the production of allelochemicals and their release into the environment, their absorption and translocation in the receptor organism, concentration at the site of action, and by factors that determine their effectiveness after their release from the producing organism (Einhellig, 1987).

Plant allelochemicals can also act upon plants and plant process by indirect means. Alteration of soil properties and organisms (eg. microbes, nematodes, insects) in soil are examples of indirect allelochemical actions. This overview will only describe allelochemicals which act directly on plants.

Natural Allelochemicals

While virtually all plants contain chemical constituents which have allelopathic potential, only those chemical constituents that are released to the surrounding environment can be considered potential allelochemicals. A potential allelochemical can only be designated as an allelopathic agent after it has undergone a rigorous proof of demonstrated phytotoxic activity. The chemical nature of recognized allelochemicals is diverse, and almost every class of secondary plant metabolite has been implicated in allelopathy.

Phenolics. Simple phenolics, benzoic acid derivatives and flavanoids are frequently associated with allelopathy, and numerous investigations have associated them with phytotoxic activity (Rice, 1984; Thompson, 1985; Putnam and Tang, 1986). These compounds have been shown to be inhibitory to ATP generation and electron transport in chloroplasts and mitochondria (Moreland and Novitzky, 1987). Flavanoids were found to have the highest inhibitory action and the phenolics were the lowest. In seedling growth experiments, the flavanoid tricin [**1**], isolated from quack grass, has been shown to be inhibitory (Weston et al., 1987), and large amounts of the C-glucosylflavonoids vitexin and isovitexin [**2**,**3**], detected as exudates from the seed coat of mung bean, were found to be inhibitory to seedling germination and growth (Tang and Zhang, 1986). The common allelochemicals salicylic acid, *p*-hydroxybenzoic acid, hydroquinone and umbelliferone were effective in suppressing the growth of several weeds when applied as a spray (Shettel and Balke, 1983). In a study based upon field observations of the allelopathic action of the diminutive, noncompetitive plant small everlasting *(Antennaria microphylla)* to the invasive noxious weed leafy spurge

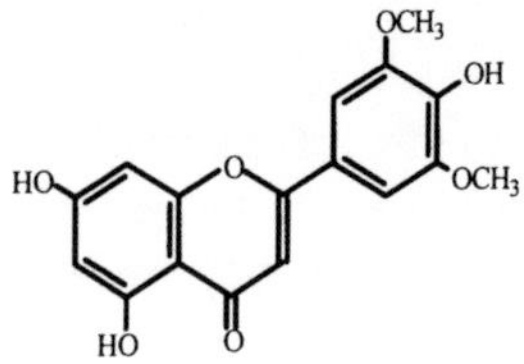

1: tricin

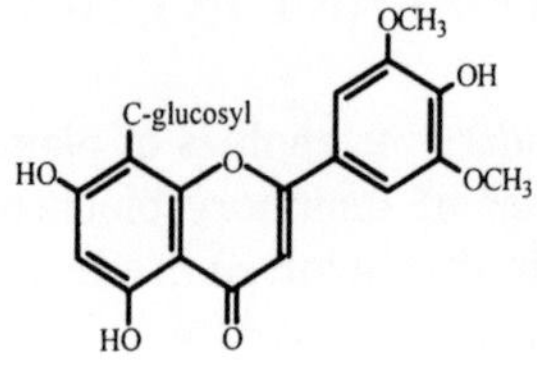

2: vitexin

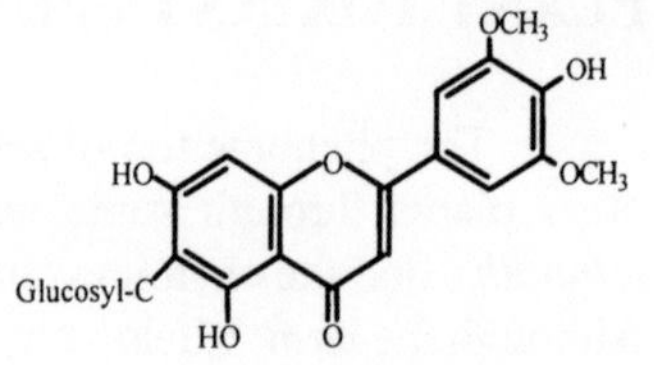

3: isovitexin

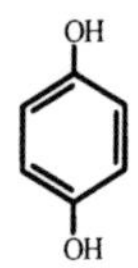

4: hydroquinone

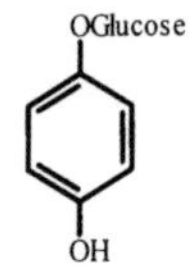

5: arbutin

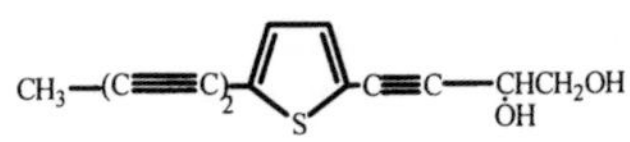

6: polyacetylene
(*Centaurea repens*)

Illustration 1. Structures [**1-6**].

(*Euphorbia esula*), small everlasting was shown to be allelopathic to the leafy spurge in cell culture. Hydroquinone [**4**], occurring in the small everlasting cells was determined to be phytotoxic to the spurge cells (Hogan and Manners, 1990). Cultured cells of both plants were found to be capabable of detoxifying hydroquinone to the nontoxic glycoside arbutin [**5**]. However, small everlasting displayed a higher efficiency in detoxification of the phenolic than did leafy spurge. In the circumstance of a chronic exposure of both plants to hydroquinone, the differential in detoxification ability between the two plants was considered as an important factor in providing an allelopathic advantage for the noncompetitive small everlasting (Hogan and Manners, 1991).

Polyacetylenes. This class of compounds is commonly found to occur among the Compositae plant family, among whose plant species several weeds have been alleged to be allelopathic. The knapweeds (*Centaurea* sp.) are invasive noxious weeds whose competitive advantage has been attributed to allelopathy. Field and greenhouse studies with diffuse knapweed (*Centaurea diffusa*) failed to associate allelopathic action with phenolic acids or sesquiterpene lactones (Muir et al., 1987). However, Stevens (Stevens, 1986) found one polyacetylene [**6**], among five polyacetylenes isolated and characterized from Russian knapweed (*Centaurea repens*), to be strongly phytotoxic in a series of bioassays. The polyacetylene was also found to be present at 4 to 5 ppm in soil surrounding the roots of the plant during the growing season.

The foregoing are summary examples of allelopathic research results from field, greenhouse, and micro level research. A majority of allelopathic research is conducted at the macro (field) level. Major challenges for allelopathic research exist at both micro and macro levels. At the micro level, researchers in allelopathy must gain more information about the microbiological alterations of allelochemicals in the soil interface between the allelopathic plant and the target plant. Substantially greater information about the biochemical processes of phytotoxicity in target species is necessary to significantly advance the concept of allelochemical models for new herbicides.

At the macro level, researchers in allelopathy must increase the validation of allelochemicals through field testing. Integrated allelochemical/herbicide weed control systems require increased attention as a viable alternative to existing herbicide based weed control protocols. In a manipulated agricultural system, crop management schemes must be further exploited to take advantage of the beneficial aspects of allelopathic plants. Expanded research in cover-crop residues, crop co-planting, and crop rotation involving allelopathic crop plants is necessary and will add credibility to the economic potential of allelopathy in agroecosystems (Einhellig and Leather, 1988; Inderjit et al., 1995).

PLANT TOXINS IN PLANT-INSECT INTERACTIONS

Insects as herbivores occupy a paramount position in nature. On a comparative biomass scale, the total biomass of insects in tropical forests is considered to be seven times that of vertebrates (Holden, 1989). Yet even with a high potential to severely impact plant species to satisfy nutritional needs, they are estimated to consume only 10 percent of the natural ecosystem and roughly 13 percent of the cultivated plants. Clearly, plants must possess inherent defensive mechanisms which provide resistance to insect predation. The naturally occurring chemical constituents in plants are recognized to be a primary factor in the defense of plants against insect attack.

The chemical defense of plants is observed to act against insects at two different levels: as a direct toxin (insecticidal) and as an indirect toxin (antifeedant or growth inhibitor). In a contradictory sense, plant phytochemicals are also involved in another interrelationship with insects (sequestration) in which the chemical toxins of the plant are utilized by the insect to provide it with a defense against its predators.

Plant Insecticides

The utilization of natural insecticides and synthetic insecticides based upon natural insecticides has been an interest of research for more than 50 years. In that period a wide variety of phytochemicals acutely toxic to insects have been identified. This discussion will be limited to three of these plant based insecticides. A review of the literature for the identified natural insecticides, antifeedants and growth regulators found in plants, except for the pyrethroids and insect hormones, has recently appeared (Addor, 1995).

Pyrethrins. Although phytochemicals from a wide variety of chemical families (alkaloids, terpenes, furanocoumarins, polyacetylenes, unsaturated amides, and others) have been shown to possess insecticidal properties, only one group, the pyrethrins, has achieved major commercial success. The recognized insecticidal activity of pyrethrum (dried and powdered flowers of *Tanacetum coccineum* (Persian insect flower) and *Tanacetum cinerari-*

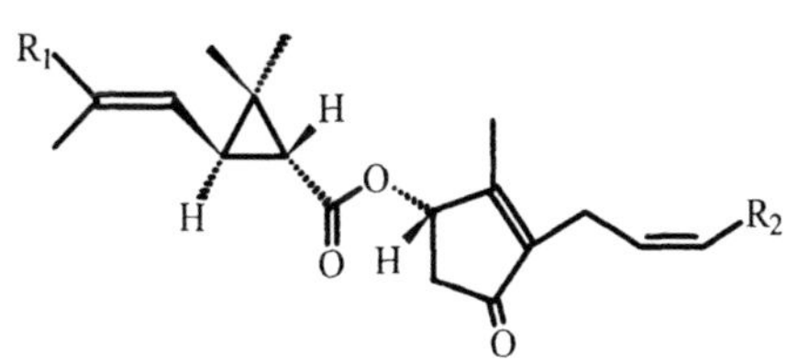

	7: pyrethrin I	**8:** pyrethrin II	**9:** jasmolin I	**10:** jasmolin II	**11:** cinerin I	**12:** cinerin II
R1=	CH_3	CH_3O_2C	CH_3	CH_3O_2C	CH_3	CH_3O_2C
R2=	$CH{=}CH_2$	$CH{=}CH_2$	CH_2CH_3	CH_2CH_3	CH_3	CH_3

Illustration 2. Structures [**7-12**].

ifolium (Dalmatian insect flower), two daisylike herbaceous perennials of Compositae, can be traced back to the 19th century. The insect powder gained wider commercial application in the early 20th century (Casida and Quistad, 1994). Six closely related pyrethrins (**7-12**) with insecticidal properties have been isolated from pyrethrum extracts. Synthetic analogs of the natural pyrethrins became available between 1940 and 1970 and were used in household insecticides. In the early 1970's synthetic pyrethrins (pyrethroids) with improved photo-stability increased the commercial application of these compounds to allow their use in field applications (Elliott et al., 1973; Naumann, 1990a, Naumann, 1990b). Today the pyrethroids are the second largest class of insecticides.

The pyrethrins possess the most desirable combination of high biological activity coupled with low mammalian toxicity. The synthetic pyrethroids have been improved over the pyrethrins to achieve a thousand fold increase in activity with even lower mammalian toxicity (Naumann, 1990a; Henrick, 1995). The pyrethrins and pyrethroids are considered to act on almost all parts of the nervous system of insects in their primary mode of action. They are speculated to act stereospecifically to block sodium channels in excitable nerve membranes in several parts of the insect nervous system to rapidly produce loss of coordinated movement, periods of convulsive activity and ultimate paralysis (Ruigt, 1985). The pyrethrins which produce repetitive discharges in the peripheral and central nervous system are classified as Type I pyrethroids. Type II pyrethroids (most modern synthetics) block the action potentials without inducing repetitive after discharges (Soderlund and Bloomquist, 1989; Henrick, 1995).

The development of the pyrethroids from the pyrethrins serves as an example of the most successful development of a commercially important synthetic insecticide from a natural precursor. The commercial pyrethroids are among the most active insecticides known and they control a wider range of insect and acarid pests at lower application rates than most other insecticides. Their high toxicity to insects and low mammalian toxicity make these compounds a role model for all present and future insecticides.

Unsaturated Amides. In 1971, a group of pungent compounds (isobutylamides) obtained from members of the families Compositae and Rutaceae were described by Jacobsen (Jacobsen, 1971) to have limited use as insecticides. The compounds were characterized as unstable with a marked tendency to polymerize. Examination of extracts of the fruit of the black pepper *Piper nigrum,* using the aduzki bean weevil as a bioassay, resulted in the isolation of three isobutylamides (**13-15**) which displayed insecticidal activity (when combined equimolarly) to the pyrethrins (Miyakado et al., 1983, 1989). Based upon the observations of Miyakado and earlier reported test results for synthetic analogs of the compounds described by Jacobsen, Elliot and co-workers defined a basic amide structure to

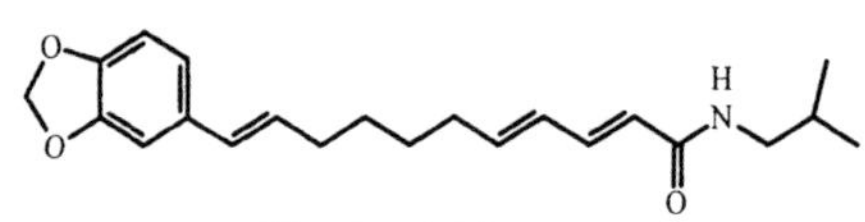

13: pipericide

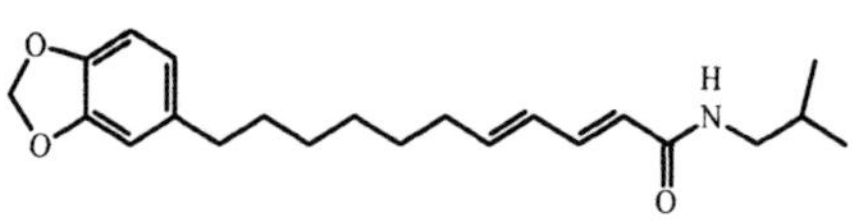

14: dihydropipericide

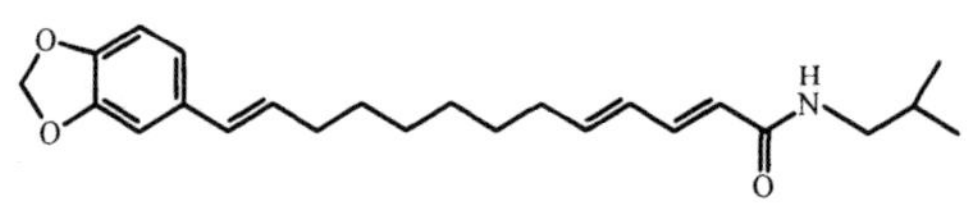

15: guineesine

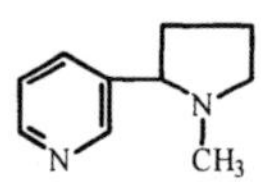

16: nicotine

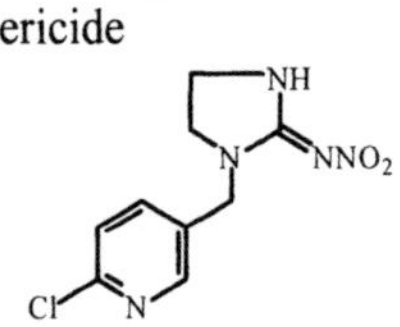

17: imidachloprid

Illustration 3. Structures [**13-17**].

search for in Piperaceae. Their search produced 172 amides of which 28 were reported to have insecticidal activity (Elliott et al., 1987).

Since the work of Elliott and Miyakado, several synthetic analogs of the insecticidal isobutylamides have been synthesized (Addor, 1995). Increased information about structure/activity relationships among these compounds has been compiled with some synthetic analogs displaying higher insecticidal activity than the pyrethrins. However, none of the analogs have provided the necessary combination of potency, breadth of action and stability to warrant commercialization.

Nicotine. The use of the alkaloid nicotine [**16**] as an insecticide has been known for at least two centuries. Its use in field applications was superseded by the pyrethroids which are of significantly lower toxicity to mammals than nicotine (LD_{50} 3-188 mg/kg orally) (Coats, 1994). Nicotine and the major related alkaloids nornicotine and anabasine (structures not shown) are believed to act by blocking acetylcholine receptors (Corbett et al., 1984).

Most of the synthetic changes to the nicotine nucleus have not resulted in increased insecticidal potency. Increases in the chain length of the *N*-acyl group produced a selective increase in activity to the tobacco hornworm but the compounds were not as effective against the tobacco budworm and the cabbage looper (Severson et al., 1988). More recently, the synthesis of an *N*-nitroimino analog of nicotine (imidacloprid [**17**]) has produced a compound for commercialization which is effective against a wide range of insects with low mammalian toxicity (Elbert et al., 1990). These recent advances in nicotinoid analog effectiveness as insecticides produce some hope that the untapped historical insecticidal character of nicotine can lead to commercial insecticides.

Plant Antifeedants and Growth Regulators

Plant terpenoids and benzopyrans represent the major source of insect antifeedants and growth regulators. Unlike the plant insecticides, which act on insects as direct toxins affecting neurological and neuro-muscular functions, these phytochemicals act as indirect toxins affecting the hormonal systems in insects. The consequence of their effects is to interrupt insect metamorphosis, reduce growth, reduce fecundity, be anti-ovipositional, deter feeding and reduce general fitness. This discussion will consider two groups of compounds which are recognized as insect growth regulators and antifeedants.

Precocenes. The precocenes (**18,19**) were first isolated from extracts of *Ageratum houstonianum,* a commonly cultivated ornamental bedding plant by, Bowers et al. (Bowers et al., 1976). These benzopyrans (chromenes) act as chemical antagonists of juvenile hormones (antijuvenile hormone agents) in insects to produce precocious metamorphosis. Specifically, when *Oncopeltus fasciatus* nymphs were confined with purified **18** and **19,** premature metamorphosis occurred and the surviving adults were sterile. Among adult females, gonads failed to develop and eggs were not produced (Bowers, 1982). The generally accepted mode of action of the precocenes is the epoxidation of the chromene double bond by enzymes in the insect corpora allata to form highly reactive epoxides or quinone methides which can alkylate and destroy cellular macromolecules with a consequent irreversible cellular obliteration (Staal, 1986). The precocenes are considered unusual in their ability to penetrate an intact insect, migrate to and penetrate the copora allata and act as a competitive substrate for natural oxidizing enzymes which produce the lethal ingredients responsible for chemical ablation of the corpora allata in a number of insect taxa.

Other natural precocenes and synthetic analogs have been prepared and biologically evaluated (Isman, 1989). Synthetic chromanes showed significantly less activity than the parent chromenes (precocenes) while the acetylchromenes showed antijuvenile hormone

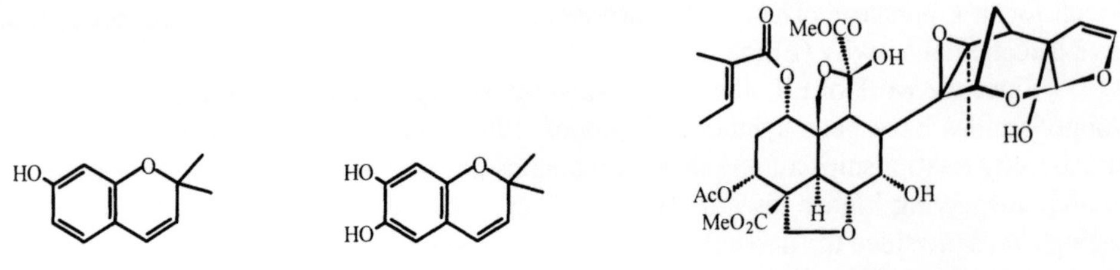

18: precocene 1 **19:** precocene 2 **20:** azadirachtin

Illustration 4. Structures [**18-20**].

activity comparable to the precocenes. Even though the precocenes have been shown to be potent antijuvenile hormone agents, their lack of selectivity among insects and a relatively low activity (compared to other insecticides) have severely limited their commercialization.

Azadirachtins. In the last 20 years a great deal of research effort by entomologists and phytochemists has been directed to the identification and biological evaluation of chemical constituents from the neem tree (*Azadirachta indica*) in relation to the recognized insecticidal properties of its fruits and leaves (Schmutterer, 1990). A large number of tetranortriterpenes have been isolated from this tree and other related members of the Meliaceae family. A recent review reports information about the chemistry of the melicane tetranortriterpenoids obtained from these plants (Kraus et al., 1993).

Azadirachtin [**20**] was the first tetranorditerpenoid from the neem tree to be associated with insect antifeedant activity and growth-retardation. Evidence supporting the ability of **20** to interfere with the neuroendocrine control of metamorphosis in susceptible insects has been summarized (Rembold, 1989a, 1989b) and the effect of the compound on insects has been reviewed (Schmutterer, 1990). A formulated neem seed extract has been developed for commercial purposes (Margosan-O) (Larson, 1989). The formulation contains about 3000 PPM azadirachtin and has been marketed for insect control in greenhouses, nurseries, forests and homes.

Neem-based products are considered broad-spectrum against most orders of insects. Although at high concentrations the azadirachtins can act as a toxic insecticide, in most applications allowed concentrations produce only antifeedant effects. The neem-based products are considered to be of low toxicity to mammals.

Plant Toxin Sequestration by Insects

The foregoing discussion of plant-insect interactions has focused on natural phytochemicals which may function in the defense of plants to predation by insects. While the toxicity of the phytochemicals cited is clearly documented, their precise role in the actual defensive mechanism is not as easily documented. The concept of plant chemicals acting in the defense of the plants that produce them seems undeniable. However, proof of their co-evolutionary role in providing protection to plants from insect predation is not definitive. In contrast, the utilization of plant chemicals by insects for protection against predation by vertebrates has been observed and verified. The most studied example of this plant phytochemical-insect-vertebrate relationship is the sequestration of poisonous cardiac glycosides (cardinolides) of milkweed (*Asclepias* spp.) by larvae of the monarch butterfly (Parsons, 1965; Reichstein et al., 1968) and the subsequent vomit-response occurring with birds that prey on the monarch caterpillar, pupa and adult (Brower et al., 1968; Brower, 1969). Sequestration of cardenolides for defense against vertebrate predators was demonstrated for a number of insects that feed on members of Asclepiadaceae (Duffey and Scudder, 1972, 1974; von Euw et al., 1967). Insects were found to have variable abilities to sequester cardenolides and they were also found to be capable of sequestering poisonous pyrrolizidine alkaloids (Rothschild and Marsh, 1978). An excellent review of cardenolide occurrence in Asclepiadaceae was compiled by Seiber et al. in 1983 (Seiber et al., 1983) and more information about plant-derived allomones is included in a review by Whitman (Whitman, 1988).

Calotropin [**21**] is representative of some of the cardenolides found in Asclepiadaceae. These cardenolides are distinguishable from the digitalis cardiac glycosides (i.e. digitoxigenin [**22**] (aglycone)) by the configuration of the A:B ring junction (5α (trans-A/B) vs 5β (cis-A/B)). The digitalis cardiac glycosides are not found in Asclepiadaceae. Calotropin is an example of the milkweed cardenolides which involve C-2 and C-3 hydroxyl groups in a cyclic bridge to a single sugar moiety. This unusual cyclic bonding is a major factor in the high resistance of the cardenolides to acid hydrolysis. This is not a characteristic of the digitalis cardiac glycosides (Seiber et al., 1983; references therein)

There are several research observations which support the utilization of cardenolides in milkweed by monarch butterflies for purposes of defense. First, the monarch larvae feed exclusively on species of *Asclepias* through all developmental stages (Urquhart, 1960). Second, the insects are capable of selective storage of some cardenolides among different members of Asclepiadaceae (Roeske et al., 1976; Brower et al., 1982). Finally, metabolic conversion to specific types of cardenolides occurs in the monarch (Seiber et al., 1980). These data and the emetic response ellicited by the cardenolides in vertebrates provide compelling evidence on behalf of the utilization of phytochemicals by insects for the purpose of defense against predators.

Significant research challenges remain in the area of plant-insect interactions. The successful commercialization of the pyrethroids, and the potential of the azadiractins illustrate the importance of continued survey of plant phytochemicals as insect control agents. The increased resistance of the public to persistent pesticides accentuates the need for new pesticides with low mammalian toxicity and high biodegradability. As illustrated by the pyrethrins, natural phytotoxins have the potential to serve as models for the synthesis of bioactive compounds which will meet the current environmental criteria. The insect-growth-inhibitory steroidal saccharide esters [**23**] recently isolated from *Physalis peruviana* (Ellinger et al., 1994) are an example of new possibilities in exploiting plant-insect interactions.

PLANT TOXINS IN PLANT-ANIMAL INTERACTIONS

Toxins in plants have their greatest economic impact in the relationship between plants and animals. A direct economic loss of 245 million dollars has been calculated for

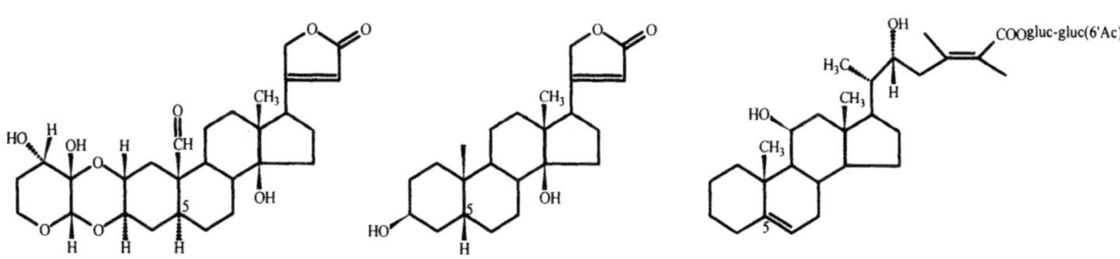

21: calotropin **22:** digitoxigenin **23:** steroidal saccharide ester (*Physalis peruviana*)

Illustration 5. Structures [**21-23**].

poisoning deaths of cattle and sheep in the United States in 1989 (Nielsen and James, 1992). This loss represents an immediate loss to ranchers, but it does not include indirect costs associated with lost lamb and calf crops, poison plant eradication costs, veterinary fees, fencing and herding costs to deny access of livestock to poisonous plants, and the loss of usable grazing lands. When these indirect costs are included the total economic loss were estimated to be in excess of 400 million dollars in 1989 (Nielsen and James, 1992). Allowing for increases in livestock prices, the 1995 costs probably approach 500 million dollars in the United States and 1 billion dollars on a world wide basis.

The impact of plant toxins can be catastrophic with the scores of animals killed in a single acute poisoning episode or it can be insidious with losses occurring through abortions, birth defects, chronic internal organ damage (liver and gastrointestinal) and photosensitization. Often the insidious loss can be a greater problem to a rancher than the catastrophic loss since the time between the exposure to the toxin and the result of its toxicity will often separate the victim from the source and thereby remove or reduce the ability to correlate effect with cause. Such is not the case in a catastrophic event. Additionally, in an open range situation an insidious event can often lead to the loss of small groups of cattle or sheep without the rancher having knowledge of the event. Regardless of the method of the loss or the impact, a secondary plant metabolite, probably produced by the plant as a part of a defensive mechanism against native predation, has successfully eliminated a predator with no co-evolutionary relationship.

As in the case of the plant toxins involved in plant-plant and plant-insect interactions, the variety of toxins involved in plant-animal relationships are numerous and represent a broad spectrum of chemical types. Several books which include comprehensive reviews of the variety of plant toxins in plant animal relationships are available (Keeler and Tu, 1983; Cheeke, 1985; Cheeke, 1989a; Cheeke, 1989b; Keeler and Tu, 1991). In this discussion, I will again provide examples of representative toxins which will be described according to their site of action.

Cardiovascular and Pulmonary Toxins

Plant toxins which affect contractions of the heart muscle or metabolic respiration processes are associated with acute poisoning in mammals. Two groups of glycosides (cyanogenic glycosides and cardiac glycosides) are present in plants commonly available to grazing animals and are responsible for livestock losses.

Cyanogenic Glycosides. About 60 cyanogenic glycosides have been found in a wide variety of plant families (Seigler, 1992) and their toxic threat to man and animals has been reviewed (Poulton, 1983; Poulton, 1988a; Poulton, 1988b; Poulton, 1990; Nahrstedt, 1993). Amygdalin [**24**] is one of the most common cyanogenic glycosides and is found in the seeds of domestic fruits of the Roseacea family i.e. cherry, apple, peach, apricot and pear, and is the "active ingredient" in the controversial laetrile treatments of cancer in the late 1970s. Prunasin [**25**] was described by Kingsbury (Kingsbury, 1964) to be associated with the poisoning of livestock eating the leaves of several species of wild cherry (*Prunus virginiana*). This compound has been described as the primary source of toxicity in arrowgrass (*Triglochin maritima*) and Western chokecherry (*Prunus virginiana* var. *demissa*)(James et al., 1980) which are commonly available to grazing animals in the Western United States.

The cyanogenic glycosides are, without exception, β-glucosides of cyanohydrins. While they are bitter, they are not acutely toxic to higher animals. In the presence of the degrading enzymes (β-glucosidases in the same plant), they undergo cyanogenesis to liberate hydrogen cyanide (HCN). The acute toxicity of HCN stems from its extreme affinity for cytochrome oxidase. A very low concentration of HCN (33μM) is capable of completely

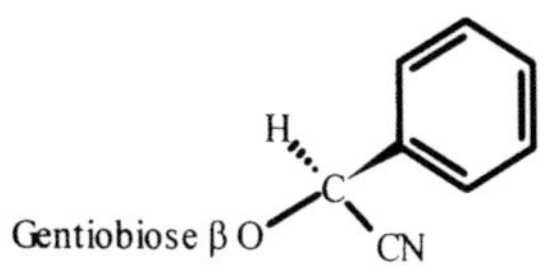

24: amygdalin

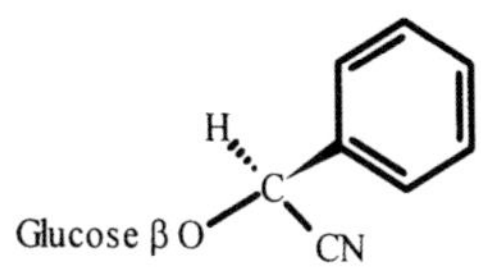

25: prunasin

Illustration 6. Structures [**24,25**].

blocking electron transfer through the mitochrondrial electron transport chain, thereby preventing the utilization of oxygen by the cell. The HCN effectively poisons by cellular asphyxiation. Acute toxicity depends primarily upon rate of plant ingestion, level of β-glucosidase and animal size and condition. Death can occur in a matter of a few minutes.

Successful antidotes for acute cyanide poisoning are available. Sodium thiosulfate, a sulfur donor, can detoxify HCN by promoting its conversion to thiocyanate when catalyzed by the enzyme rhodanase (Baumeister et al., 1975; Egekeze and Oehme, 1980); however, the detoxification effect is slow. Another procedure utilizes cobalt compounds (hydroxocobalamine) to bind directly with cyanide to form physiologically inactive cyanocobalamine (Offterdinger and Weger, 1969).

Teratogenic Toxins

Natural teratogens found in plants and fungi, synthesized teratogens, viral diseases and genetic defects are considered the sources of congenital deformities in humans and animals. The identification of teratogens responsible for genetic defects has been closely associated with epidemic episodes of congenital deformities in livestock and the evaluation of hazardous constituents of drugs, food or other commercial items. The variety of naturally occurring teratogens is broad and plant-derived and fungal-derived teratogens have been reviewed (Keeler, 1983; Hood and Szczech, 1983).

A large proportion of the plant-derived teratogens are alkaloids. In most cases these teratogenic alkaloids are also acute toxins. Those alkaloids that are teratogenic to livestock are included in a recent review by Roitman and Panter (Roitman and Panter, 1995). This discussion will include information and examples from only two of the 14 classes of alkaloids (steroidal and quinolizidine) which contain teratogens. The reader is referred to the aforementioned reviews to locate information about other alkaloid teratogens.

Quinolizidine Alkaloids. The "crooked calf" syndrome which can afflict rangeland cattle in the western United States was attributed to the maternal consumption of several *Lupinius* species after other possible factors (genetic, viral) had been eliminated (Kingsbury, 1964; Finnell et al., 1991; Keeler, 1983; Shupe et al., 1967, 1968). The syndrome is clinically manifested in skeletally deformed calves which are born at full term and are otherwise in good health. In mild cases, the skeletal abnormalities appear as bowed front legs. Severe cases reveal twisted front legs which can not be extended, and occasionally the spine, neck and hind legs are also affected (Kingsbury, 1964). Animals with mild affliction are able to stand. However, in severe cases the animals are unable to stand even with assistance (Finnell et al., 1991).

The "crooked calf" syndrome was reproduced through feeding experiments with lupine species and alkaloid extracts of the lupines, and the quinolizidine alkaloid anagyrine [**26**] was established as the most likely causative agent (Keeler, 1976). All attempts to

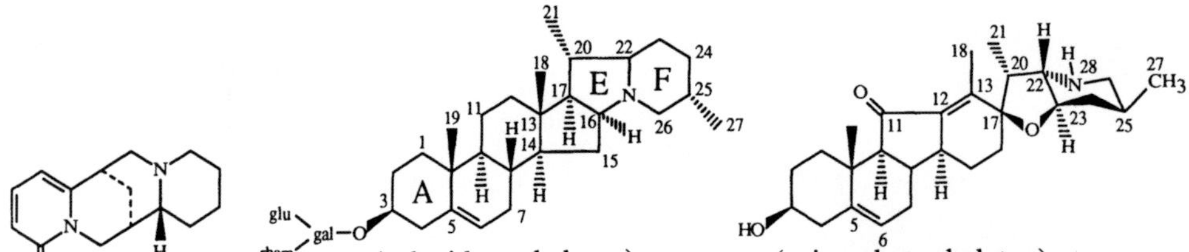

(solanidane skeleton) (spirosolane skeleton)

26: (-)-anagyrine **27:** solanine **28:** jervine

Illustration 7. Structures [**26-28**].

reproduce the syndrome in animals other than cattle (sheep and goats) have been unsuccessful (Panter, 1993). It has been speculated that the susceptibility of cattle to the teratogenic effects of the lupines could be associated with the ruminal conversion of anagyrine [**26**] to teratogenic piperidine alkaloids (Keeler and Panter, 1989). However, no evidence of the conversion of anagyrine to piperidine alkaloids was detected when blood alkaloids of cattle, sheep and goats fed *Lupinus caudatus* were analyzed using gas chromatography-mass spectrometry methodology (Gardner and Panter, 1993). In addition, an acute toxic myopathic condition induced in cattle by quinolizidine alkaloids (including **26)** of lupines (Baker and Keeler, 1991) could not be detected in sheep or goats. The animal-specific toxic and teratogenic activity of quinolizidine alkaloids remains unexplained.

Steroidal Alkaloids. Steroidal alkaloids occur primarily in the Liliaceae and Solanaceae plant families. All of these are acute toxins: Liliaceae: death camas (*Zigadenus* spp.), false hellebores (*Veratrum* spp.); Solanaceae: tomato (*Lycopersicon* spp.), potato, nightshade and eggplant (*Solanum* spp.). The toxicity is attributed to glycoalkaloids, and more than 1100 are known, with considerable variation in their carbohydrate and alkaloid structural moieties (Ripperger and Schreiber, 1981). The alkaloids are normally divided into solanum (i.e. solanine [**27**]) and veratrum (i.e. jervine [**28**]) structural groups. Field observations and plant-feeding experiments with death camas (*Zygadenus* sp.) show the veratrum alkaloids to be fast acting pulmonary toxins at low concentrations, with victims showing pulmonary congestion and subcutaneous hemorrhaging in the thoracic region (Panter et al., 1987). The high toxicity and the similarity in appearance of death camas to grass has led to major poisoning episodes (James et al., 1980; Kingsbury, 1964; Panter et al., 1987). Clinical manifestations of poisoning by various species of *Solanum* are consistent with the solanum alkaloids acting as gastrointestinal toxins. Post mortem examinations of animals succumbing to toxic *Solanum* reveal gastric and intestinal hemorrhage, mucosal ulceration and necrosis (Baker et al., 1988). The solanum alkaloids have also been shown to be potent toxins at low concentrations (Baker et al., 1989) and to be the cause of severe livestock poisonings (Kingsbury, 1964).

Prior to the 1960s, congenital birth defects affecting up to 25% of sheep flocks were observed in the western United States (Kingsbury, 1964). Cranio-facial deformities, popularly called "monkey face" lamb, commonly appeared and were attributed to genetic origins. Mild cases of the affliction were characterized by a poorly developed jaw and a dished-in facial appearance. Severe cases were marked by gross malformations of the face, including no eyes or nose, a single eye, a single fused eye with two corneas and underdeveloped cerebrum, and infrequently, a skin-covered proboscis above a cycloptic single eye (Binns et al., 1962). Lambs carried to full term were usually born alive but died soon after birth because of an inability to eat or breathe. Lambs carried past term died *in utero* and caused the dam to die shortly thereafter (Kingsbury, 1964).

Various feeding studies with *Veratrum californicum* (western false hellebore) in pregnant sheep reproduced the "monkey face" lamb syndrome and established reproductive and toxic factors involved in the syndrome (see references in Keeler, 1983; Roitman and Panter, 1995). It was found that the craniofacial defects could only be induced if plant material was consumed on the 14th day of gestation, and that certain bone and cartilage defects were induced during the 28th-33rd day. The veratum steroidal alkaloids were established as the causative agents and pure alkaloids, including jervine [**28**], could also produce the malformations. The malformations were also induced in cattle and sheep in feeding trial utilizing *Veratrum californicum*. Other mammals (rabbits, rats, mice, hamsters) and chicks also experienced congenital malformations when exposed to pure veratrum alkaloids.

The structural similarity between the solanum and veratrum alkaloids precipitated the examination of the teratogenic potential of the human food plants tomato, eggplant, and potato (Keeler, 1983; Gaffield et al., 1992). These studies confirmed that the solanum alkaloids, both as free alkaloids and as glycosides, could induce congenital malformations. Structural-activity studies for both the solanum and veratrum alkaloids established that the spatial orientation of the nitrogen atom above the plane of the steroid nucleus (Keeler et al., 1976; Brown and Keeler, 1978) and the existence of a 5,6 double bond (Gaffield and Keeler, 1993,1994) were important to maximize teratogenicity among the steroidal alkaloids. In a comparison of relative teratogenicity of 13 veratrum and solanum alkaloids, jervine was found to be the most teratogenic. The teratogenicity of jervine was reduced in half when the C-12,C-13 bond was saturated and was reduced even further when both the C-12,C13 and C-5,C-6 double bonds were saturated (Gaffield and Keeler, 1994). Similar saturation of double bond in the solanum alkaloids also reduced teratogenicity.

The research information on the toxicity and teratogenicity of the veratrum and solanum alkaloids enabled alterations in range management techniques two decades ago to reduce or eliminate many of the problems associated with grazing plants containing these toxins. However, the biochemical mechanisms of teratogenicity and toxicity induced by these alkaloids is still not clearly understood. Information about these mechanism will provide important insights into the physiological responses induced by these toxins.

Hepatotoxins and Abortifacients

Plant toxins which act as hepatotoxins or abortifacients do not act as acute toxins. Their toxicity is of a more subtle, insidious nature with few outward manifestations until shortly before death (hepatotoxins) or abortion (abortifacients). In the case of range animals, the site and degree of exposure of a group of animals will often go unnoticed until an animal succumbs or aborts leaving the rancher with little information about either the toxic source or the location. The mode of action of these toxins dictates their consideration as toxins to humans, particularly in relation to their possible inclusion in herbal medicines and natural home remedies.

Pyrrolyzidine Alkaloids. More than 250 pyrrolyzidine alkaloids have been reported to occur in members of the Boraginaceae (many genera), Leguminosae (particularly *Crotalaria* spp.) and Compositae (primarily *Senecio* spp.) plant families all over the world and their chemistry, toxicology and pathology has been reviewed (Peterson and Culvenor, 1983; Mattocks, 1986; Roitman and Panter, 1995). Livestock which graze plants containing these alkaloids are exposed to potential chronic doses of lethal liver toxins. Animal deaths attributed to pyrrolyzidine toxicosis occur as a result of a severely dysfunctional liver (Bull et al., 1968). The victims generally succumb well after exposure to the toxic plants without clinical signs until few days before death. Field studies and feeding experiments show horses

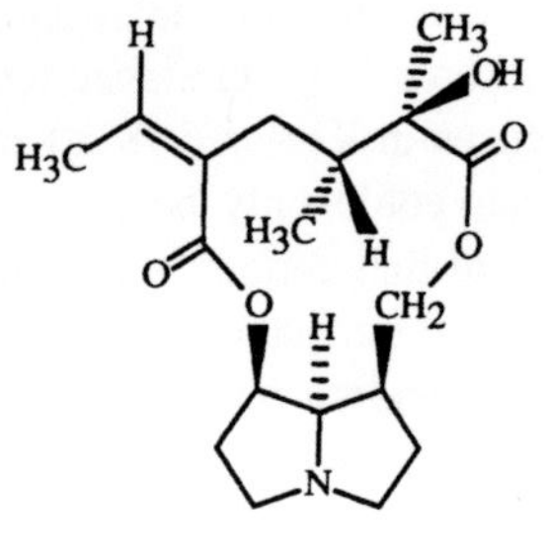

29: platyphylline

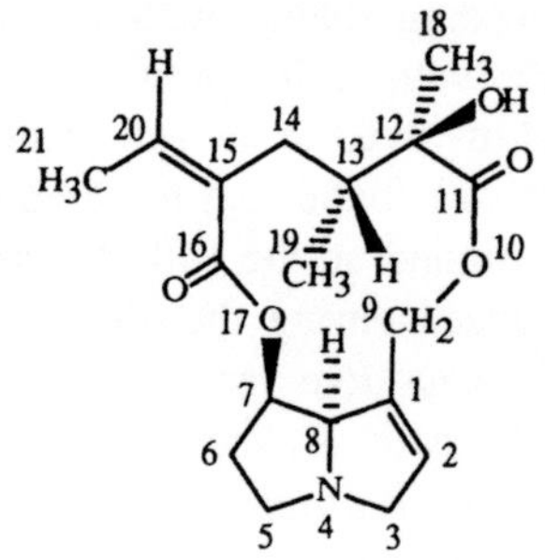

30: senecionine

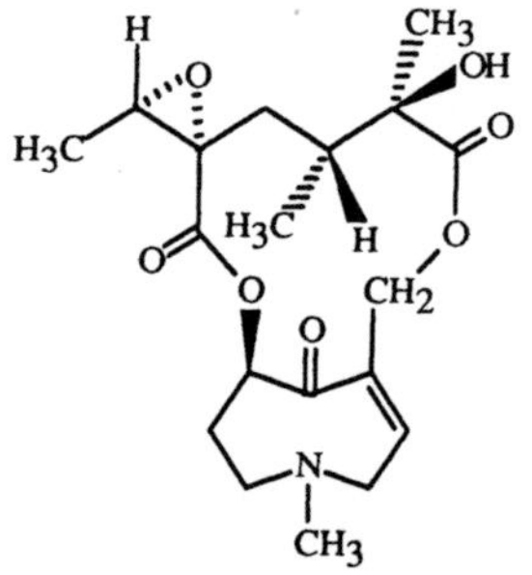

31: otosenine

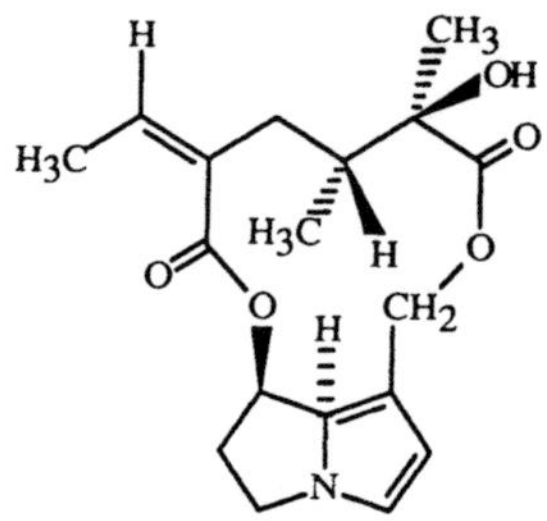

32: pyrrole ester alkaloid

Illustration 8. Structures [**29-32**].

and cattle to be the major livestock species to be affected by pyrrolizidines. However, sheep and pigs also are susceptible (Bull et al., 1956, 1968). In a ranking of susceptibility to poisoning by *Senecio jacobea*, Hooper (Hooper, 1978) ranked pigs as the 200 times more susceptible to pyrrolizidines than sheep and goats and 14 times more susceptible than cattle and horses. The species differences were attributed to the differences among the animals to produce pyrroles in the liver (Shull et al., 1976).

Structurally, the pyrrolizidine alkaloids are composed of a necine base (two fused five-membered rings with a nitrogen at one of the vertices) and one or more branched carboxylic acids attached as esters to one or two of the necine hydroxyl groups. The esterified branched mono-and di-carboxylic acids are unique to the pyrrolizidine alkaloids. The alkaloids are represented by three major structural groups: saturated bases (i.e. platyphylline [**29**]), 1,2-unsaturated necines (i.e. senecionine [**30**]) and *seco* necines (i.e. (otosenine [**31**]). The N-oxides of these alkaloids frequently co-occur with the free alkaloids (Roitman and Panter, 1995).

The pyrrolizidine alkaloids themselves are not considered toxic until they are transformed by liver microsomes to pyrrolic alkaloid esters [**32**]. The pyrrolic alkaloid esters act as alkylating agents capable of alkylating and cross-linking vital cell constituents (Bull et al., 1968; Mattocks 1986, 1992). The saturated necines (i.e. **29**) do not form pyrrolic esters and are therefore not hepatotoxic. The toxicity of the pyrollic metabolites depends upon their stability and reactivity. Highly reactive pyrroles attack the liver at the site of their formation, while less reactive pyrroles may migrate to other organs. Naturally occurring bacteria have been isolated from the rumen of sheep which can alter pyrrolizidine alkaloids so they do not form pyrollic ester, but are easily excreted.

The prevention of pyrrolizidine alkaloid toxicosis in livestock relies primarily on denying animals access during times when the alkaloid content is high. Biocontrol insects are also considered as a viable option where large population of toxic plants exist. The ability of some bacteria to detoxify these alkaloids has suggested the potential of intra-ruminal detoxification by inoculation of the bacteria (Peterson and Culvenor, 1983)

Outbreaks of pyrrolizidine poisonings in humans have occurred and have generally been associated with contaminated grain used for meal or bread. The pyrrolizidines have also been detected in "comfrey" teas (Roitman, 1981) commonly available at herbal medicine and health food stores. The possibility of the occurrence of pyrrolizidine residues in meat or milk of livestock consuming toxic plants exists, but no toxic responses in humans from this source has been reported (Peterson and Culvenor, 1983).

Diterpene Acids. The needles of ponderosa pine (*Pinus ponderosa*) have long been recognized to cause abortions in cattle in late stages of pregnancy (Kingsbury, 1964; James et al., 1977). The calves born in the last trimester generally are alive and their survival depends upon the stage of gestation and post natal care. Cows have also been reported to die from pine needle toxicosis (James et al., 1977).

Until recently, attempts to establish the nature of the toxic principle in pine needles responsible for inducing abortions in cattle have been generally unsuccessful. The lack of success was related to the lack of an adequate bioassay technique to adequately establish abortifacient extracts in pine needles. Feeding trials utilizing pregnant cows (James et al., 1994) was established as the proper methodology to determine which solvent extracts of ponderosa pine needles contained abortifacient active materials. Utilizing this bioassay, the labdane diterpene acid, isocupressic acid [**33**], was characterized as the abortifacient in pine needles (Gardner et al., 1994) and was found to be active at a twice-daily dose rate of 100mg/kg. It was determined that the consumption of about 3 kg of needles by a cow would provide the isocupressic acid dose required to initiate an abortion. Other labdane diterpene acids, occurring in much lower amounts in the needles, were also found to be active abortifacients.

A subsequent examination of the source of acute toxicity of the pine needles to cattle has shown that abietane or rosin acids present in the needles are potent nephrotoxins and neurotoxins. Feeding studies utilizing abietane diterpene acids (i.e. abietic acid [**34**]) obtained from pine tips (new spring and summer growth) were found to induce clinical signs consistent with kidney failure and the dosed cows developed progressive paralysis. Histological examination of tissues from euthanized cattle intoxicated by pine needles revealed nephrotic and neurologic lesions attributed to the abietane diterpene acids (Stegelmeir et al., 1995).

Neurotoxins

Interference with the ability of neuronal cells to generate, conduct and transmit nervous impulses through the body can result in severe neuro-physiological events which

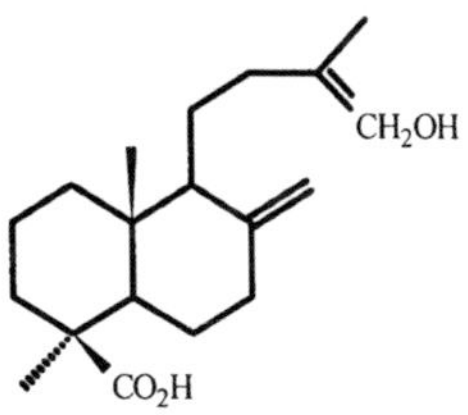

33: isocupressic acid

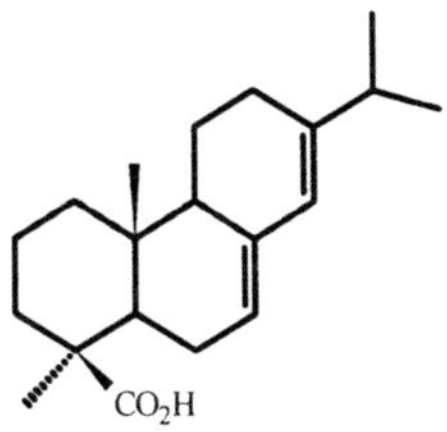

34: abietic acid

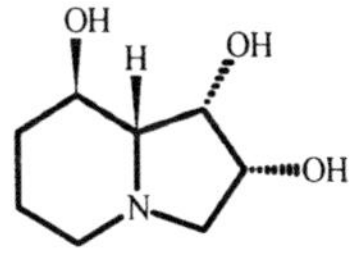

35: swainsonine

Illustration 9. Structures [**33-35**].

commonly lead to death in mammals. A wide range of plant toxins are capable of perturbing neuronal functions by: 1) altering axon physical geometries; 2) interfering with nerve impulse generation and transmission by altering intercellular ion flux; 3) blocking, inhibiting or competing with synaptic impulse transmission and/or neurotransmitters; and 4) the disturbance of synaptic connectivity through excessive accumulation of lysosomal catabolite (Huxtable, 1994). This discussion will describe the neurotoxic effects of the glycosidase-inhibiting indolizidine alklaloids and the acetylcholine-inhibiting action of norditerpenoid alkaloids. The case of the norditerpenoid alkaloids will be utilized to describe the initiation, planning and execution of a research program directed to solving a toxic plant problem.

Indolizidine Alkaloids. Two persistent, yet heretofore unresolved, poisonous plant problems affecting live stock in Australia ("peastruck" syndrome) and the United States (locoism) have recently been ascribed to gylcosidase inhibition by polyhydroxyindolizidine alkaloids, in particular swainsonine [**35**]. The physiological activity of these compounds has created substantial interest and their chemistry, physiological and neuro-physiological activity have been reviewed (Colegate et al., 1991; James et al., 1989; Molyneux and James, 1991a; Molyneux, 1993a).

The "peastruck" syndrome and locoism have long been associated with the consumption by livestock of *Swainsona* spp. (Darling pea) in Australia (Everist, 1981) and *Oxytropis* and *Astragalus* spp. (loco weed) in the United States (Kingsbury, 1964). Intoxication manifests itself in affected animals through trembling of head and limbs, slight staggering and carrying the head abnormally high in sheep and low in cattle and horses. As the disease progresses, clumsiness increases and fallen animals are unable to regain their feet. Sheep have been observed to jump high over small or imaginary objects and cattle and horses can become excitable to the point of unmanageability.

Clinical symptoms are consistent with neurological abnormalities in affected animals. Histological evaluation of tissue from severely poisoned animals reveals extensive vacuolation, ulceration of the abomasum with accompanying edma, and the occurrence of axonal spheroids in many areas of the brain. Degradation and fragmentation of Golgi cells in the brain is also observed (Everist, 1981, Kingsbury, 1964). The neurological disorders can be accompanied by emaciation, reproductive problems and congestive right heart failure (Molyneux and James, 1991b). Horses are considered more susceptible to the disease than are sheep and cattle.

The comparability of the histopathological abnormalities observed for *Swainsona* poisoned cattle to those of the genetic disorder α-mannosidosis led to the elucidation of the cause of the "peastruck" syndrome (Colegate et al., 1991). A bioassay-guided examination of *Swainsona* extracts led to the isolation and identification of swainsonine [**35**] as the toxic agent. The characterization of **35** in *Swainsona* led investigators of the *Oxytropis* and *Astragalus* locoweed species to its characterization as the elusive locoing toxin (Molyneux and James, 1991a).

Histological, and chemical information from *Swainsona* poisonings established that swainsonine inhibited the enzymes required for the processing of oligosaccharides (Dorling et al., 1978, 1993). The isolation of a mannose-rich oligosaccharide from lesions in the lymph nodes of sheep fed *Swainsona cavescens* which was not present in the plant material, confirmed its synthesis in the animal's tissue (Dorling et al., 1993). The lysosomal accumulation of the mannose-rich oligosaccharides is consistent with reduced neuronal cell longevity and a corresponding increase in neurologic impairment in *Swainsona* and locoweed-poisoned animals.

The specific inhibition of α-mannosidases by swainsonine has produced a great deal of interest in the compound . It has been used as a probe in glycoprotein biochemical processes, glycoprotein synthesis, modification and storage, genetic glycoprotein disorders, cancer cell proliferation, alterations of T-lymphocyte responses and immunomodulatory effects (James et al, 1989). The ability of the indolizidine alkaloid to inhibit metastasis of cancerous tumors has spurred investigation of its chemotherapeutic potential (Olden et al., 1991).

Norditerpenoid Alkaloids: Attacking a Toxic Plant Problem

The norditerpenoid alkaloids are complex, multi-cyclic C_{19} diterpene alkaloids which are highly substituted with hydroxyl, methoxyl and ester groups. These alkaloids occur in members of the Ranunculaceae family (primarily in *Aconitum* spp. and *Delphinium* spp.) and many have been determined to be highly toxic to humans and animals. They occur as two skeletal structure types [**36**, **37**] . Three structural sub-types are associated with skeletal type **36**, based upon C-7 substitution patterns and C-8, C15 unsaturation: aconitine type (i.e. aconitine [**38**]); lycoctonine type (i.e. lycoctonine [**36**]) and pyrodelphinine type (i.e. pyrodelphinine [**39**]). Skeletal type **37** is designated the heteratisine type (i.e. heteratisine [**37**]) norditerpenoid alkaloids (Pelletier et al., 1984). Only a small number of pyrodelphinine and heteratisine type norditerpenoid alkaloids have been characterized and there is very limited toxicity data available for them (Benn and Jacyno, 1983). More than 300 aconitine and lycoctonine norditerpenoid alkaloids have been characterized (Pelletier et al., 1984, 1991). *Aconitum* spp. contain almost exclusively aconitine type alkaloids while *Delphinium* spp. contain primarily lycoctonine type with a few aconitine type alkaloids.

Several aconitine and lycoctonine norditerpenoid alkaloids have been determined to be potent neurotoxins in mammals (Benn and Jacyno, 1983). Aconitine [**38**] is considered the principal toxin in *Aconitum* species (commonly called wolfbane and monkshood) used for poison arrows and spears employed in the hunting of wild game and in warfare and murder since the time of ancient Greece (Benn and Jacyno, 1983; Mann, 1992). *Aconitum napellus* is considered the most dangerous plant in Great Britain and is reported to be lethal

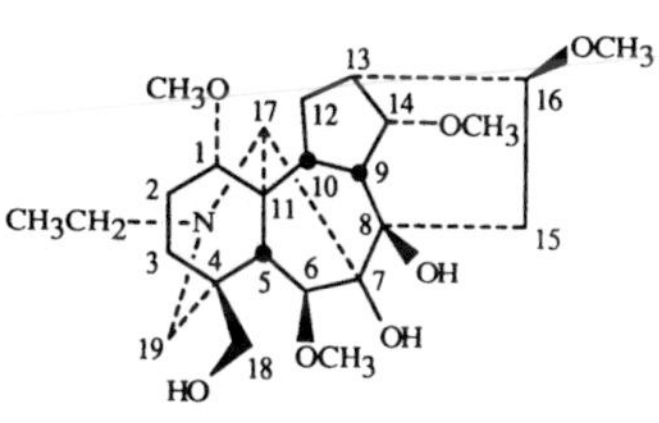

36: lycoctonine

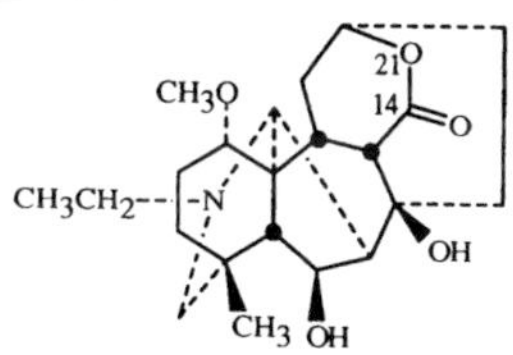

37: heteratisine

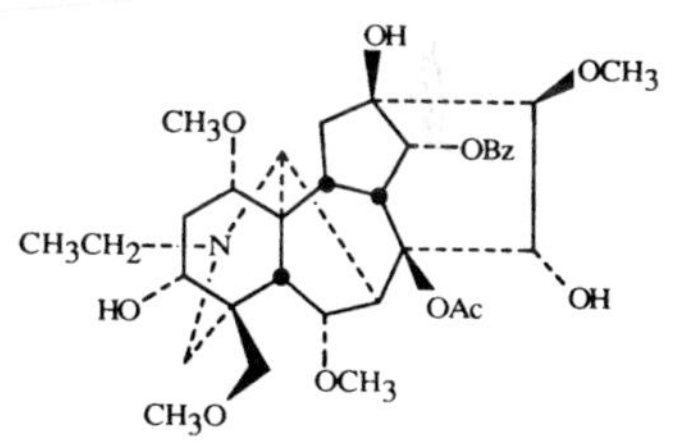

38: aconitine

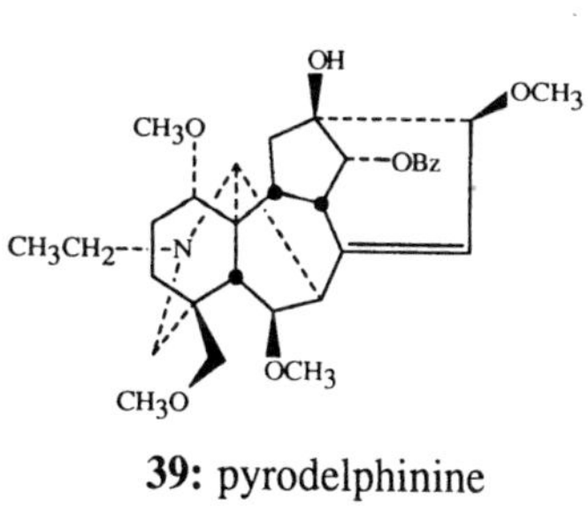

39: pyrodelphinine

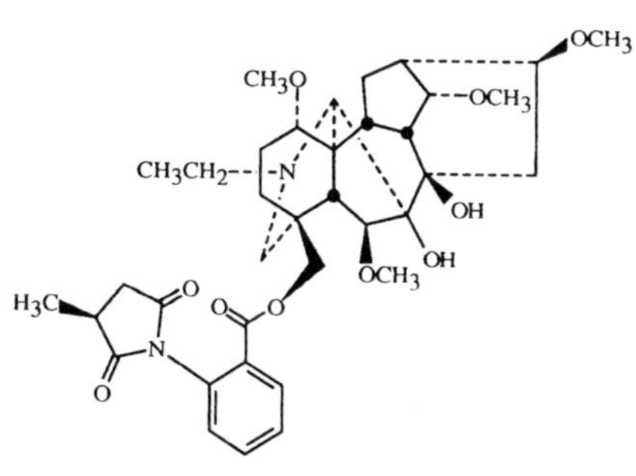

40: methyllycaconitine

Illustration 10. Structures [**36-40**].

to a horse which consumes only 0.075% of its weight of fresh root material (Kingsbury, 1964). In both animals and humans, clinical signs of poisoning by *Aconitum* is manifested swiftly, symptoms are intense and death often occurs within a few hours. Humans display nausea, vertigo, tingling skin, impaired speech and vision, anxiety and oppressive chest pain. In livestock, frothy salivation, constant swallowing, bloating and belching are commonly observed (Kingsbury, 1964). Electrophysiological studies have attributed the toxicity of aconitine to its ability to block neuronal cell membrane sodium channels, thereby causing prolonged depolarization of neuronal excitation (Catterall, 1980)

The toxicity of norditerpenoid alkaloids in *Delphinium* spp. has primarily been associated with livestock poisonings. *Delphinium* species (larkspurs) are common on livestock grazing ranges of the western United States and Canada. The plants are considered the major cause of livestock losses to poisoning on those rangelands (Nielsen and Ralphs, 1988). The larkspurs are designated as "low larkspur" or "tall larkspur" based upon habitat and height at maturity. The tall larkspurs are found at higher elevations in wet habitat and have a spring to fall growing season. The low larkspurs occur at lower elevation in drier habitat and have a spring to mid-summer growing season. Some intermediate height larkspurs are also found at both elevations and in both wet and dry habitat. Consumption of toxic larkspurs leads to clinical signs associated with neuromuscular toxicosis , including: (initial) uneasiness; elevated respiration rate; rapid pulse; stiff gait; spread leg stance; (intermediate) animal collapses front legs first to rest on sternum; (advanced) head goes to ground; or animal falls on side (Olsen, 1978). The animals may rise and fall several times, accompanied by constipation, bloat and regurgitation. Death is usually the result of respiratory paralysis and/or asphyxiation caused by aspiration of vomitation.

Norditerpene alkaloids with *N*-(methylsuccimido)anthranilic acid esterification of the C18 ethanolic group of lycoctonine (i.e. methyllycaconitine [**40**]) have been established as the primary toxin in the low larkspurs (Benn and Jacyno, 1983). When the *N*-(methylsuccinimido)anthranoyllycoctonine (MSAL) norditerpenoid alkaloid methyllycaconitine (MLA) [**40**], obtained from a low larkspur, was administered to a calf, clinical symptoms associated with larkspur poisoning were observed (Nation et al., 1982). In a neuromuscular *in vitro* test system, MLA was confirmed to be highly toxic with a curare-like action associated with the postsynaptic blockage of nicotinic cholinergic receptors (Dozortseva-Kubanova, 1959, Aiyar et al., 1979). Experiments with cultured fetal rat hippocampal tissues showed MLA to act as nicotinic acetylcholine receptor (nAChR) antagonist to block postsynaptic neurotransmission at the neuromuscular junction (Alkondon et al., 1992). The alkaloid was also found to be a potent competitor of α-bungarotoxin binding to nAChR in insects in the nanomolar range (Jennings et al., 1986), and that the binding was reversible (Alkondon et al., 1992).

The Larkspur Problem

Although information about the toxicity and the toxin of a low larkspur was available in the mid-1980's, similar information about the toxicity or toxin of the tall larkspurs was not available. The tall larkspurs constitute the largest poisonous plant threat to ranchers in the Western U.S. Hundreds of cattle are lost to tall larkspur poisoning each year and the losses represent a significant economic loss to ranchers. In response to need to reduce these losses, a collaborative study involving range scientists, an animal scientist, a veterinary scientist and a natural products chemist was begun by the Agricultural Research Service in the late 1980s. In the context of this presentation, this collaborative research program serves as a good example of the planning and execution of research directed to solving a toxic plant problem.

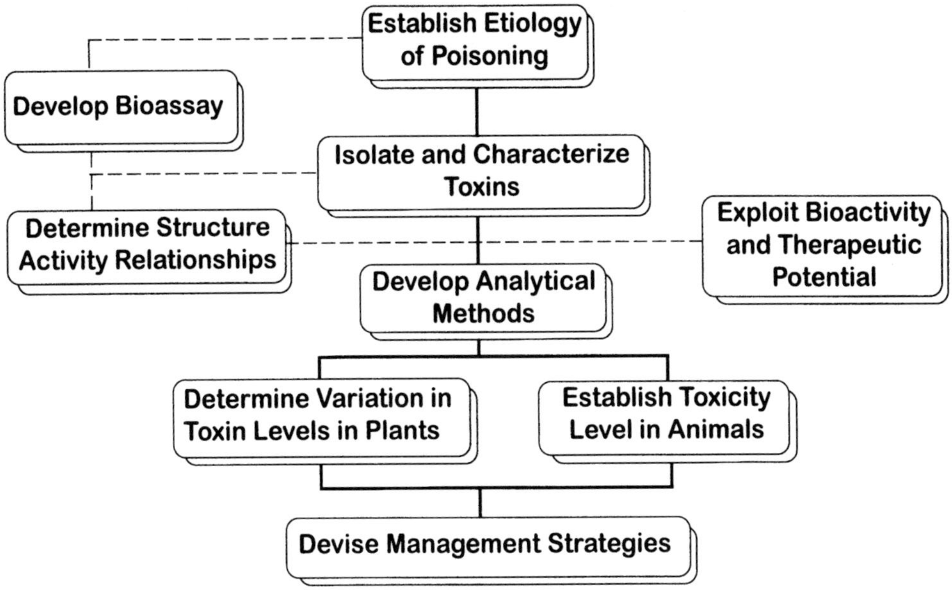

Figure 1. Research protocol for attacking a poisonous plant problem.

Larkspur Problem: Research Approach. Plant toxin problems which affect domestic livestock, such as the larkspur problem, generally are directed toward a research objective of reducing or eliminating the existing threat of the poisonous plant. The information obtained for a plant toxin(s) in a plant poisonous to animals can be applied in two ways to satisfy the research objective: 1) devising preventive range management schemes and 2) developing detoxification methods. Since the lethality of a plant toxin is directly related to dose, information about the variations in the level of the toxin in the plant in relation to its growth stages and its availability to animals can provide the basis for range management schemes to alleviate losses. Additionally, information about structure/activity relationships and about the physiological mode of action of the toxin in the animal are important guides to developing antidotes or intervention mechanisms to be used in the animal.

The inherent variability of plants and the secondary metabolites that they contain dictates flexibility and innovation in the isolation of specific plant toxins. The development and utilization of a bioassay system to assess the biological activity of toxin containing plant extracts and individual toxins is a necessity and should involve the affected animal if it is feasible. The research scheme depicted in Figure 1 is a modified version of a research approach proposed by Molyneux (Molyneux, 1993b). This research approach has been effective applied in determining the toxins involved in many poisonous plant problems involving livestock, most recently including abortions caused by pine needles (Gardner et al., 1994) and locoism (Colegate et al., 1991). The approach has also been applied to the investigation of the larkspur problem.

In the examination of the toxic tall larkspurs, several adjustments of the research approach were necessary to develop an efficient research plan. First, the etiology of tall larkspur poisoning is well established and several species of tall larkspur are clearly recognized as threats to livestock. Furthermore, research information exists on the isolation of toxic components from both *Aconitum* and *Delphinium* species and on the toxicity, structure/activity relationships and mode of action of the alkaloids obtained from toxic larkspurs (see previous discussion). Less information is available about toxins present in the tall larkspurs and their quantification. The extreme toxicity of the larkspurs severely limited the use of large animals as a bioassay system for the verification of toxins in these *Delphiniums,* thereby dictating the development of a suitable alternate bioassay. This information generated a research plan which focused on the isolation and characterization of toxins through the modification of existing methodology, the development of a new analytical method for quantifying toxins in the larkspurs, the development of a small animal bioassay, and the accumulation of toxic materials for large animal testing and structure/activity evaluation.

Working within the framework of the research plan and in awareness of the research objective, the collaborative research team began the examination of the toxic tall larkspurs. Large collections of tall larkspur plants (*Delphinium barbeyi, Delphinium occidentale* and *Delphinium glaucescens*) were made ***at sites of poisoning episodes as close to the time of the episode as possible.*** Species identity was verified and voucher specimens were deposited. Plant toxicity was verified in feed experiments and plant material was prepared for chemical extraction. Alkaloid fractions and individual alkaloids were obtained through the modification of a reported method (Pelletier et al., 1981) and individual alkaloids were toxicologically evaluated in a mouse bioassay (Manners et al., 1993). A gas liquid chromatographic (GLC) analytical technique was developed to detect 7,8 methylenedioxylycoctonine (MDL) norditerpenoid alkaloids in larkspur extracts (Manners and Ralphs, 1989) and a high performance liquid chromatographic (HPLC) analytical method was developed to quantify MSAL norditerpenoid alkaloids (Manners and Pfister, 1993). The MSAL norditerpenoid alkaloids were established as the major toxins in the tall larkspurs and the HPLC analysis method was applied to extracts of field samples of tall larkspurs to determine their phenological variation relative to range management recommendations (Manners et al., 1993; Pfister et al., 1994a). Toxic MSAL norditerpenoid alkaloids obtained from the tall larkspurs and synthetic analogs were utilized in a study of structure/activity relationships (Manners et al., 1995), in the evaluation of MSAL norditerpenoid alkaloids structural changes on binding affinity to lizard muscle nicotinic acetylcholine receptors (Dobelis et al., 1993), and for the evaluation of physostigmine as an antidote to larkspur intoxication (Pfister et al., 1994b).

Larkspur Problem: Research Results. The results of these investigations provide important information about the toxic nature of the tall larkspurs and range management scheme which can reduce livestock exposure. The primary toxin in these plants is the MSAL norditerpenoid alkaloid methyllycaconatine (MLA) [**40**] which can occur in toxic specimens well in excess of 1% of the dry plant weight. The alkaloid has an LD_{50} in mice of 7.5 mg/kg (i.v.) and based upon an estimated oral toxic dose of 6 mg/kg in cattle, an average cow could consume a toxic dose if roughly 1/3 of its food requirement consisted of tall larkspur containing 0.2% MLA. Sites with MLA in excess of 1% could provide a toxic dose to a cow through the consumption of a few plants. The occurrence of the toxin was determined to be highest in the reproductive tissues in the early growth stages of the plant and to decrease to levels which were acceptable for grazing in the early summer. Cows intoxicated through the consumption of larkspur were effectively revived through the intravenous administration of physostigmine, thereby providing a larkspur toxicosis intervention method.

The toxicological evaluation of seven of the nine MSAL norditerpenoid alkaloids reported to occur in *Delphinium* species and seven other related naturally occurring or synthetic norditerpenoid alkaloids provides specific information about structural features which are important to their toxicological activity. It was determined that the tertiary alkaloid nitrogen and the anthranilic acid esterification of the lycoctonine C-18 ethanolic group changes the basic nontoxic lycoctonine to a moderately toxic compound. The addition of the *N*-methylsuccimido to the anthranilic acid residue to form MLA increases the toxicity by a factor of three. Changes in the substitution of the oxygen at C-14 of MLA also change the toxicity. Removal of the C-14 methyl group to form the hydroxyl or esterification of the oxygen as an acetate increases the toxicity in mice by 2 to 3 more times than that of MLA. Derivatization of the vicinal hydroxyl groups at C-7 and C-8 or opening of the *N*-methylsuccimido group does not, however, lower the toxicity relative to MLA. These observations are consistent with observed differences of binding affinity to lizard muscle nAChR sites and other information which describes MLA as an antagonist of acetylcholine neurotransmitter binding sites (see earlier discussion) and they provide a platform to expand research directed to developing intervention by detoxification of the MSAL norditerpenoid alkaloids.

This summary of research results for the investigation of tall larkspur toxicity demonstrate the value of a multi-disciplinary team effort in attacking a toxic plant problem. Since these problems involve a complicated interaction of plants, animals and toxins, a research team whose members combine their individual expertize in each aspect of the plant-animal-toxin interaction can more efficiently proceed to accomplishing the research objective. Similar effective interdisciplinary collaboration can be seen in other toxic plant studies reported in this presentation.

CONCLUSION

It has been the purpose of this presentation to survey the interrelationships that exist between toxic plants, their neighbors and their predators. It is hopeful that the reader will grasp some sense of the wide diversity that exists among natural toxins in plants, the wide diversity of their toxic actions and the scope of research effort necessary to identify them and characterize their physiological activity. A great deal of work has been accomplished in isolating and identifying these compounds, but a great deal of work needs to be done in determining how they act. It is important to note that a large number of today's drugs come from natural sources and that it is often only a matter of dilution that can change a toxin to a therapeutic agent.

REFERENCES

Addor, R. W. 1995. Insecticides. In: *Agrochemical From Natural Products*, Godfrey, C. R. A., ed. New York, Marcel Dekker. pp.1-62.

Alkondon, M., Pereira, E. F. R., Wonnacott, S., and Albuquerque, E. X. 1992. Blockage of Nicotinic Currents in Hippocampal Neurons Defines Methyllycaconitine as a Potent and Specific Receptor Antagonist. *Molecular Pharmacology* 41: 802-808.

Baker, D. C., and Keeler, R. F. 1991. Myopathy in Cattle Caused by *Thermopsis montana*. In: *Toxicology of Plant and Fungal Compounds-Handbook of Natural Toxins*, Vol.6, Keeler, R. F., and Tu, A. T., eds. New York, Marcel Dekker. pp.61-82.

Baker, D. C., Keeler, R. F., and Gaffield, W. 1989. Pathology in Hamsters Administered Solanum Plant Species That Contain Steroidal Alkaloids. *Toxicon* 27: 1331-1337.

Baker, D. C., Keeler, R. F., and Gaffield, W. 1988. Mechanism of Death in Hamsters Gavaged Potato Sprout Material. *Toxicological Pathology* 16: 333-339.

Baumeister, R. G. H., Schievelbein, H., and Zickgraf-Rudel, G. 1975. Toxicological and Clinical Aspects of Cyanide Metabolism. *Drug Research* 25: 1056-1064.

Bell, E. A. 1985. The Biological Significance of Secondary Metabolites in Plants. In: *Plant Toxicology*, Seawright, A. A., Hegarty, M. P. J. L. F., and Keeler, R. F., eds. Yeerongpilly, Queensland. Queensland Dept. of Primary Industries, Animal Research Institute. pp.50-56.

Benn, M. H., and Jacyno, J. 1983. The Toxicology and Pharmacology of Diterpenoid Alkaloids. In: *Alkaloids: Chemical and Biological Perspectives*, Vol.1, Pelletier, S. W., ed. New York, Wiley & Sons. pp.153-210.

Binns, W., James, L. F., Shupe, J. L., and Thacker, E. J. 1962. Cyclopian-Type Malformations in Lambs. *Archives of Environmental Health* 5: 106-108.

Bowers, W. S. 1982. Toxicology of the Precocenes. In: *Insecticide Mode of Action*, Coats, J. R., ed. New York, Academic Press. pp.403-427.

Bowers, W. S., Ohta, T., Cleere, J. S., and Marsella, P. A. 1976. Discovery of Insect Anti-Juvenile Hormones in Plants. *Science* 193: 542-547.

Brower, L. P., Seiber, J. N., Nelsen, C. J., Lynch, S. P., and Tuskes, P. M. 1982. Plant Determined Variation in the Cardenolide Content and Emetic Potency of Monarch Butterflies, *Danaus plexippus*, Reared on the Milkweed *Asclepias eriocarpa* in California. *Journal of Chemical Ecology* 8: 579-633.

Brower, L. P. 1969. Ecological Chemistry. *Scientific American* 220: 22-29.

Brower, L. P., Ryerson, W. N., Coppinger, L. L., and Glazier, S. C. 1968. Ecological Chemistry and the Palatability Spectrum. *Science* 188: 19-25.

Brown, D., and Keeler, R. F. 1978. Structure-Activity Relation of Steroid Teratogens: 3. Solanidan-epimers. *Journal of Agricultural and Food Chemistry* 26: 566-569.

Bull, L. B., Culvenor, C. C. J., and Dick, A. T. 1968. *The Pyrrolizidine Alkaloids: Their Chemistry, Pathogenicity and Other Biological Properties*. New York, North Holland Publishing Co., John Wiley and Sons. 293pp.

Bull, L. B., Culvenor, C. C., Keast, J. C., and Edgar, G. 1956. An Experimental Investigation of the Hepatotoxic and of Other Effects on Sheep of the Consumption of *Heliotropium europaeum* L.: Heliotrope Poisoning of Sheep. *Australian Journal of Agricultural Research* 7: 281-332.

Casida, J. E., and Quistad, G. B., eds. 1994. *Pyrethrum Flowers: Production, Chemistry, Toxicology and Uses*. New York, University Press. pp93-115.

Catterall, W. A. 1980. Neurotoxins That Act on Voltage-Sensitive Sodium Channels in Excitable Membranes. *Annual Reviews in Pharmacology and Toxicology* 20: 15-43.

Cheeke, P. R., ed. 1989a. *Toxicants of Plant Origin*. Vol.1. *Alkaloids*. Boca Raton, FL, CRC Press. 335pp.

Cheeke, P. R., ed. 1989b. *Toxicants of Plant Origin*. Vol.2 *Glycosides*. Boca Raton, FL, CRC Press. 277pp.

Cheeke, P. R., ed. 1985. *Natural Toxins in Feeds and Poisonous Plants*. Boca Raton, FL, CRC Press. 287pp.

Coats, J. R. 1994. Risks From Natural Versus Synthetic Insecticides. In: *Annual Review of Entomology*, Vol.39, Mittler, T. E., Radovsky, F. J., and Resh, V. H., eds. Palo Alto, Annual Reviews Inc. pp.489-515.

Colegate, S. M., Dorling, P. R., and Huxtable, P. R. 1991. Swainsonine: a Toxic Indolizidine Alkaloid from the Australian Swainsona species. In: *Toxicology of Plant and Fungal Compounds-Handbook of Natural Toxins*, Vol.6, Keeler, R. F., and Tu, A. T., eds. New York, Marcel Dekker. pp.159-189.

Corbett, J. R. 1984. Insects Acting Elsewhere in the Nervous System.. In: *The Biochemical Mode of Action of Pesticides*, Corbett, J. R., Wright, K., and Baillie, A.C. eds. London, Academic Press. pp.141-178.

Dobelis, P., Madl, J. E., Manners, G. D., Pfister, J. A., and Walrond, J. P. 1993. Antagonism of Nicotinic Receptors by Delphinium Alkaloids. *Neuroscience Abstracts* 631: 12.

Dorling, P. R., Colegate, S. R., and Huxtable, C. R. 1993. Plants Affecting Livestock: an Approach to Toxin Isolation. In: *Bioactive Natural Products*, Colegate, S. R., and Molyneux, R. J. eds. Boca Raton, FL, CRC Press. pp.481-506.

Dorling, P. R., Huxtable, C. R., and Vogel, P. 1978. Lysosomal Storage in *Swainsonia* spp. Toxicosis. *Neuropathology and Applied Neurobiology* 4: 285-295.

Dozortseva-Kubanova, P. M. 1959. Pharmacology of the Alkaloid Methyllycaconitine (Mellictine). *Farmakologiya i Toksikologiya* (Moscow) 22: 30-33.

Duffey, S. S., and Scudder, G. G. E. 1974. Cardiac glycosides in *Oncopeltus fasciatus* (Dallas) Hemiptera:Lygaeidae. I. The Uptake and Distribution of Natural Cardenolides in the Body. *Canadian Journal of Zoology* 52: 283-290.

Duffey, S. S., and Scudder, G. G. E. 1972. Cardiac Glycosides in North American Asclepiadacae, a Basis for Unpalatability in Brightly Coloured Hemiptera and Coleoptera. *Journal of Insect Physiology* 18: 63-68.

Egekeze, J. O., and Oehme, F. W. 1980. Cyanides and Their Toxicity: a Literature Review. *The Veterinary Quarterly* 2: 104-114.

Einhellig, F. A., and Leather, G. R. 1988. Potentials for Exploiting Allelopathy to Enhance Crop Production. *Journal of Chemical Ecology* 14: 1831-1844.

Einhellig, F. A. 1987. Interactions Among Allelochemicals and Other Stress Factors of the Plant Environment. In: *Allelochemicals: Role in Agriculture and Forestry*, Waller, G. D., ed. ACS Symposium Series, 330. Washington, DC, American Chemical Society. pp.343-357.

Elbert, A., Overbeck, H., Iwaya, K., and Tsuboi, S.. 1990. Inidacloprid, a Novel Systemic Nitromethylene Analogue Insecticide for Crop Protection. In: *Brighton Crop Protection Conference - Pests and Diseases*, Vol.1, . Surrey, UK, The British Crop Protection Council. pp.21-39.

Ellinger, C. A., Haddon, W. F., Harden, L., Waiss Jr., A. C., and Wong, R. Y. 1994. Insect Inhibitory Steroidal Saccharide Esters from *Physalis peruviana*. *Journal of Natural Products* 57: 348-356.

Elliott, M., Farnham, A. W., Janes, N. F., Johnson, D. M., and and Pulman, D. A. 1987. Synthesis and Insecticidal Activity of Lipophilic Amides. Parts 1-4: *Pesticide Science* 18: 191-244.

Elliott M., Farnham, A. W., Janes, N. F. N. P. H., Pulman, D. A., and Stevenson, J. H. 1973. A Photostable Pyrethroid. *Nature* 246: 169-170.

Everist, S. L. 1981. *Poisonous Plants of Australia*. London: Angus and Robertson. 966 pp.

Finnell, R. H., Gay, C. C., and Abbott, L. C. 1991. Teratogenicity of Rangeland Lupinus: the Crooked Calf Disease. In: *Toxicology of Plant and Fungal Toxins -Handbook of Natural Toxins*, Vol.6, Keeler, R. F., and Tu, A. T., eds. New York, Marcel Dekker. pp.27-39.

Gaffield, W., and Keeler, R. F. 1994. Structure-Activity Relations of Teratogenic Natural Products. *Pure and Applied Chemistry* 66: 2407-2410.

Gaffield, W., and Keeler, R. F. 1993. Implication of C-5, C-6 Unsaturation as a Key Structural Factor in Steroidal Alkaloid-Induced Mammalian Teratogenesis. *Experientia* 49: 922-924.

Gaffield, W., Keeler, R. F., and Baker, D. C. 1992. Teratogenicity and Toxicity of Solanum Alkaloids. In: *Natural Toxins:Toxicology, Chemistry and Safety*, Keeler, R. F., Mandava, N. B., and Tu, A. T., eds. Ft. Collins, CO, Alken Inc. pp.18-34.

Gardner, D. R., Molyneux, R. J., James, L. F., Panter, K. E., and Stegelmeir, B. L. 1994. Ponderosa Pine Needle-Induced Abortion in Beef Cattle: Identification of Isocupressic Acid as the Principal Active Compound. *Journal of Agricultural and Food Chemistry* 42: 756-761.

Gardner, D. R., and Panter, K. E. 1993. Comparison of Blood Plasma Alkaloid Levels in Cattle, Sheep and Goats Fed Lupinus caudatus. *Journal of Natural Toxins* 2: 1-11.

Henrick, C. A. 1995. Pyrethroids. In: *Agrochemicals From Natural Products*, Godfrey, C. R. A., ed. New York, Marcel Deckker. pp.63-146.

Hogan, M. E., and Manners, G. D. 1991. Differential Allelochemical Detoxification Mechanism in Tissue Cultures of *Antenarria microphylla* and *Euphorbia esula*. *Journal of Chemical Ecology* 17: 167-174.

Hogan, M. E., and Manners, G. D. 1990. Allelopathy of Small Everlasting (*Antennaria microphylla)*: Phytotoxicity to Leafy Spruge (*Euphorbia esula)* in Tissue Culture. *Journal of Chemical Ecology* 16: 931-939.

Holden, C. 1989. Entomologists Wane as Insects Wax. *Science* 246: 754-756.

Hood, R. D., and Szczech, G. M. 1983. Teratogenicity of Fungal Toxins and Fungal-Produced Antimicrobial Agents. In: *Plant and Fungal Toxins-Handbook of Natural Toxins*, Vol.1, Keeler, R. F., and Tu, A. T., eds. New York, Marcel Dekker. pp.201-233.

Hooper, P. T. 1978. Pyrrolizidine Alkaloid Poisoning-Pathology With Particular Reference to Differences in Animal and Plant species. In: *Effects of Poisonous Plants on Livestock*, Keeler, R. F., Van Kamper, K. R., and James, L. F., eds. New York, Academic Press. pp.161-176.

Huxtable, C. R. 1994. Some Characteristics of Neurones Which Endow Vulnerability to Chemical Toxins. In: *Plant-Associated Toxins: Agricultural, Phytochemical and Ecological Aspects*, Colegate, S. M., and Dorling, P. R., eds. Wallingford, UK, CAB International. pp.357-362.

Inderjit, S., Dahshini, K. M. N., and Einhellig, F. A., eds. 1995. *Allelopathy: Organisms, Processes and Applications*, . ACS Symposium Series, 582. Washington, DC, American Chemical Society. 381pp.

Isman, M. B. 1989. Toxicity and Fate of Acetylchromenes in Pest Insects. In: *Insecticides of Plant Origin*, Vol 1.ACS Symposium Series No. 387, Arnason, J. T., Philogene, B. J. R., and Morand, P., eds. Washington, DC, American Chemical Society. pp.44-58.

Jacobson, M. 1971. The Unsaturated Isobutylamides. In: *Naturally Occurring Insecticides*, Jacobson, M., and Crosby, D. G., eds. New York, Marcel Dekker. pp.137-149.

James, L. F., Molyneux, R. J., Panter, K. E., Gardner, D. R., and Stegelmeir, B. L. 1994. Effect of Feeding Ponderosa Pine Needle Extracts and Their Residues to Pregnant Cattle. *Cornell Veterinarian* 84: 33-39.

James, L. F., Elbein, A. D., Molyneux, R. J., and Warren, C. D.. 1989. *Swainsonine and Related Glycosidase Inhibitors*. Ames, Iowa: Iowa State Press. 504 pp.

James, L. F., Keeler, R. F., Johnson, A. E., Williams, M. C., Cronin, E. H., and Olsen, J. D. 1980. *Plants Poisonous to Livestock in the Western United States*. Agriculture Information Bulletin, 415. Washington, DC, U.S. Department of Agriculture.

James, L. F., Call, J. W., and Stevenson, A. H. 1977. Experimentally Induced Pine Needle Abortion in Range Cattle. *Cornell Veterinarian* 67: 294-299.

Jennings, K. R., Brown, D. G., and Wright Jr., D. P. 1986. Methyllycaconitine, a Naturally Occurring Insecticide with a High Affinity for the Insect Cholinergic Receptor. *Experientia* (Basel) 35: 611-613.

Keeler, R. F., and Tu, A. T., eds. 1991. *Toxicology of Plant and Fungal Compounds-Handbook of Natural Toxins*, Vol.6, . New York, Marcel Dekker.

Keeler, R. F., and Panter, K. E. 1989. Piperidine Alkaloid Composition and Relation to Crooked Calf Disease-Inducing Potential of *Lupinus formosus*. *Teratology* 40: 423-432.

Keeler, R. F. 1983. Naturally Occurring Teratogens in Plants. In: *Plant and Fungal Toxins-Handbook of Natural Toxins*, Vol.1, Keeler, R. F., and Tu. A.T., eds. New York, Marcel Dekker. pp.161-199.

Keeler, R. F., and Tu, A. T., eds. 1983. *Plant and Fungal Toxins-Handbook of Natural Toxins*, Vol.1, . New York, Marcel Dekker. 934pp.

Keeler, R. F. 1976. Lupin Alkaloids for Teratogenic and Nonteratogenic Lupin: III. Identification of Anagyrine as the Probable Teratogen by Feeding Trials. *Journal of Toxicology and Environmental Health* 1: 887-898.

Keeler, R. F., Young, S., and Brown D. 1976. Spina Bifida, Exencephaly, and Cranial Bleb Produced by the Solanum Alkaloid Solasodine. *Research Communications in Chemistry, Pathology and Pharmacology* 13: 723-730.

Kingsbury, J. M. 1964. *Poisonous Plants of the United States and Canada*. Englewood Cliffs, N.J.: Prentice-Hall. 626pp.

Kraus, W., Bokel, M., Schwinger, M., Bogler, B., Soellner, R., Wendisch, D., Steffens, R., and Wachendorf, U. 1993. The Chemistry of Azadirachtin and Other Insecticidal Constituents of Meliaceae. In: *Phytochemistry and Agriculture*, van Beek, T. A., and Breteler, H., eds. Oxford, Clarendon Press. pp.18-39.

Larson, R. O. 1989. The Commercialization of Neem. In: *Focus on Phytochemical Pesticides*, Volume 1, *The Neem Tree*, Jacobson, M., ed. Boca Raton, FL, CRC Press. pp.155-167.

Mann, J. 1992. Murder, Magic and Medicine. New York: Oxford University Press. 232pp.

Manners, G. D., Panter, K. E., and Pelletier, S. W. 1995. Structure-Activity Relationships of Norditerpenoid Alkaloids Occurring in Toxic Larkspur (*Delphinium*) Species. *Journal of Natural Products* 58:863-869.

Manners, G. D., Panter, K. E., Ralphs, M. H., Pfister, J. A., Olsen, J. D., and James, L. F. 1993. Toxicity and Chemical Phenology of Norditerpenoid Alkaloids in the Tall Larkspurs (*Delphinium)* Species. *Journal of Agricultural and Food Chemistry* 41: 96-100.

Manners, G. D., and Pfister, J. A. 1993. Normal Phase Liquid Chromatographic Analysis of Toxic Norditerpenoid Alkaloids. *Phytochemical Analysis* 4: 14-18.

Manners, G. D., and Ralphs, M. H. 1989. Capillary Gas Chromatographic Analysis of *Delphinium* Alkaloids. *Journal of Chromatography* 466: 427-432.

Mattocks, A. R. 1992. Recent Advances in Pyrrolizidine Research. In: *Poisonous Plants*, James, L. F., Keeler, R. F., Bailey Jr. E.M., Cheeke, P. R., and Hegarty, M. P., eds. Ames, Iowa, Iowa State University Press. pp.161-168.

Mattocks, A. R. 1986. *Chemistry and Toxicology of Pyrrolizidine Alkaloids*. London: Academic Press-Harcourt Bruce. 393pp.

Miyakado, M., Nakayama, I., and Ohno, N. 1989. Insecticidal Unsaturated Isobutylamides. In: *Insecticides of Plant Origin*, Aranason, J. T., Philogene, B. J., and Morand, P., eds. ACS Symposium Series, 387. Washington, DC, American Chemical Society. pp.173-187.

Miyakado, M., Nakayama, I., Ohno, N., and Yoshioka, H. 1983. Structure, Chemistry and Actions of the Piperaceae Amides: New insecticidal Constituents Isolated from the Pepper Plant. In: *Natural Products for Innovative Pest Management*, Whitehead, D. L., and Bowers, W. S., eds. Oxford, Pergamon Press. pp.369-384.

Molyneux, R. F. 1993a. Water Soluble Alkaloids: Cryptic Bioactive Natural Products. In: *Bioactive Natural Products - Detection, Isolation and Structure Determination*, Colegate, S. M., and Molyneux, R. J., eds. Boca Raton, FL, CRC Press. pp.59-74.

Molyneux, R. J. 1993b. Toxic Range Plants and Their Constituents. In: *Phytochemistry and Agriculture*, van Beek, T. A., and Breteler, H., eds. Oxford, Clarendon Press. pp.151-170.

Molyneux, R. J., and James, L. F. 1991a. Swainsonine, the Locoweed Toxin: Analysis and Distribution. In: *Toxicology of Plant and Fungal Compounds-Handbook of Natural Toxins*, Vol.6, Keeler, R. F., and Tu, A. T., eds. New York, Marcel Dekker. pp.191-214.

Molyneux, R. J., and James, L. F. 1991b. Swainsonine, Slaframine and Castanospermine. In: *Mycotoxins and Phytoalexins*, Sharma, R., and Salunkhe, D. K., eds. Boca Raton, FL, CRC Press. pp.637-656.

Moreland, D. E., and Novitzky, W. P. 1987. Effects of Phenolic Acids, Coumarins and Flavanoids on Isolated Chloroplasts and Mitochondria. In: *Allelochemicals: Role in Agriculture and Forestry*, Waller, G. R., ed. ACS Symposium Series, 330. Washington, DC, American Chemical Society. pp.247-261.

Muir, A. D., Majak, W., Balza, F., and Towers, G. H. N. 1987. A Search for the Allelopathic Agents in Diffuse Knapweed. In: *Allelochemical: Role in Agriculture and Forestry*, Waller, G. R., ed. ACS Synposium Series, 330. Washington, DC, American Chemical Society. pp.239-246.

Nahrstedt, A. 1993. Cyanogenesis and Foodplants. In: *Phytochemistry and Agriculture*, van Beek, T. A., and Breteler, H., eds. Oxford, Clarendon Press. pp.107-129.

Nation, P. N., Benn, M. H., Roth, S. H., and Wilkens, J. L. 1982. Clinical Signs and Studies of the Site of Action of Purified Larkspur Alkaloid, Methyllycaconitine, Administered Parenterally to Calves. *Canadian Veterinary Journal* 23: 264-266.

National Center For Health Statistics. 1991. *Vital Statistics of the United States, 1988, Mortality*. Part B. p. 2. Washington, DC, U.S. Public Health Service.

Naumann, K. 1990a. *Chemistry of Plant Protection.* Volume 4. *Synthetic Pyrethroid Insecticides: Structures and Properties*. Berlin: Springer-Verlag. 340pp.

Naumann, K. 1990b. *Chemistry of Plant Protection. Volume 5. Synthetic Pyrethroid Insecticides: Chemistry and Patents*. Berlin: Springer-Verlag. 340pp.

Nielsen, D. B., and James, L. F. 1992. The Economic Impacts of Livestock Poisonings by Plants. In: *Poisonous Plants*, James, L. F., Keeler, R. F., Bailey Jr., E. M., Cheeke, P. R., and Hegarty, M. P., eds. Ames, Iowa, Iowa State University Press. pp.3-10.

Nielsen, D. B., and Ralphs, M. H. 1988. Larkspur, Economic Considerations. In: *The Ecology and Economic Impact of Poisonous Plants on Livestock Production*, James, L. F., Ralphs, M. H., and Nielsen, D. B., eds. Boulder, CO, Westview Press. pp.119-129.

Offterdinger, H., and Weger, N. 1969. Circulation and Respiration in Cyanide Poisoning and its Therapy with Ferrihemoglobin and Cobalt Containing Agents. *Archives of Experimental Pathology and Pharmacology* 264: 289-294.

Olden, K., White, S. L., Newton, S. A., Molyneux, R. J., and Humphries, M. J. 1991. Antineoplastic Potential and Other Possible Uses of Swainsonine and Related Compounds. In: *Toxicology of Plant and Fungal Compounds - Handbook of Natural Toxins*, Vol.6, Keeler, R. F., Tu, A. T., eds. New York, Marcel Dekker. pp.575-588.

Olsen, J. D. 1978. Tall Larkspur Poisoning in Cattle and Sheep. *Journal of the American Veterinary Medical Association* 173: 762-765.

Panter, K. E. 1993. Ultrasound Imaging: a Bioassay Technique to Monitor Fetotoxity of Natural Toxicants and Teratogens. In: *Bioactive Natural Products: Detection, Isolation and Structural Determination*, Colegate, S. M., and Molyneux, R. J., eds. Boca Raton, CRC Press. pp.465-480.

Panter, K. E., Ralphs, M. H., Smart, R. A., and Duelke, B. 1987. Death Camas Poisoning in Sheep: a Case Report. *Veterinary and Human Toxicology* 29: 45-48.

Parsons, J. A. 1965. A Digitalis-like Toxin in the Monarch Butterfly *Danaus plexippus* L. *Journal of Physiology* 178: 290-304.

Pelletier, S. W., and Joshi, B. S. 1991. Carbon-13 and Proton NMR Shift Assignments and Physical Constants of Norditerpenoid Alkaloids. In: *Alkaloids: Chemical and Biological Perspectives*, Vol.7, Pelletier, S. W., ed. New York, Springer-Verlag. pp.297-564.

Pelletier, S. W., Mody, N. V., Joshi, B. S., and Schramm, L. C. 1984. Carbon-13 and Proton NMR Shift Assignments and Physical Constants of C19-Diterpenoid Alkaloids. In: *Alkaloids: Chemical and Biological Perspectives*, Vol.2, Pelletier, S. W., ed. New York, Wiley & Sons. pp.205-462.

Pelletier, S. W., Dailey Jr., O. D., Mody, N. V., and Olsen, J. D. 1981. Isolation and Structure Elucidation of the Alkaloids of *Delphinium glaucescens* Rybd. *Journal of Organic Chemistry* 46: 3284-3293.

Peterson, J. E., and Culvenor, C. C. J. 1983. Hepatotoxic Pyrrolizidine Alkaloids. In: *Plant and Fungal Toxins-Handbook of Natural Toxins*, Keeler, R. F., and Tu, A. T., eds. New York, Marcel Dekker. pp.637-671.

Pfister, J. A., Manners, G. D., Gardner, D. R., and Ralphs, M. H. 1994a. Toxic Alkaloid Levels in Tall Larkspur (*Delphinium barbeyi*) in Western Colorado. *Journal of Range Management* 47: 355-358.

Pfister, J. A., Panter, K. E., Manners, G. D., and Cheney, C. D. 1994b. Reversal of Tall Larkspur (*Delphinium barbeyi*) Poisoning in Cattle With Physostigmine. *Veterinary and Human Toxicology* 36: 511-514.

Poulton, J. E. 1990. Cyanogenesis in Plants. *Plant Physiology* 94: 401-411.

Poulton, J. E. 1988a. Localization and Catabolism of Cyanogenic Glycosides. In: *Cyanide Compounds in Biology*, Vol.140, CIBA Foundation, eds. Chichester, Wiley. 261pp.

Poulton, J. E. 1988b. Toxic Compounds in Plant Foodstuffs:Cyanogens. *Food Proteins* 381-401.

Poulton, J. E. 1983. Cyanogenic Compounds in Plants and Their Toxic Effects. In: *Plant and Fungal Toxins-Handbook of Natural Toxins*, Vol.1, Keeler, R. F., and Tu, A. T., eds. New York, Marcel Dekker. pp.117-157.

Putnam, A. R., T. S. Tang, and eds. 1986. *The Science of Allelopathy*. New York: John Wiley and Sons.317pp.

Reichstein, T., von Euw, J., Parsons, J. A., and Rothschild, M. 1968. Heart Poisons in the Monarch Butterfly. *Science* 161: 861-866.

Rembold, H. 1989a. In: *Focus on Phytochemical Pesticides*, Volume 1, *The Neem Tree*, Jacobsen, M., ed. Boca Raton, FL, CRC Press. pp.19-34.

Rembold, H. 1989b. Azadirachtins: Their Structure and Mode of Action. In: *Insecticides of Plant Origin*, Arnason, J. T., Philogene, B. F. R., and Morand, P., eds. ACS Symposium Series, 387. Washington, DC, American Chemical Society. pp.150-163.

Rice, E. L. 1984. *Allelopathy*. New York: Academic Press. 357pp.

Ripperger, H., and Schreiber, K. 1981. Solanum Steroid Alkaloids. In: *The Alkaloids-Chemistry and Physiology*, Vol.19, Manske, R. H. F., ed. pp.81-192 New York, Academic Press.

Roeske, C. N., Seiber, J. N., Brower, L. P., and Moffitt, C. M. 1976. Milkweed Cardenolides and Their Comparative Processing by Monarch Butterflies *Danaus plexippus* L. *Recent Advances in Phytochemistry* 10: 93-167.

Roitman, J. N., and Panter, K. E. 1995. Livestock Poisoning Caused by Plant Alkaloids. In: *Toxic Action of Marine and Terrestial Alkaloids*, Blum, M.S. ed. Fort Collins, CA, Alaken, Inc. pp 53-124.

Roitman, J. N. 1981. Comfrey and Liver Damage. *The Lancet* ii: 944.

Rothschild, M., and Marsh, C. D. 1978. Some Peculiar Aspects of Danaid/Plant Relationships. *Entomological Experimentation and Application* 24: 437-450.

Ruigt, G. S. F. 1985. Pyrethroids. In: *Conprehensive Insect Physiology, Biochemistry and Pharmacology*, Vol.12, Dekut, G. A., and Gilbert, L. I., eds. Oxford, UK, Pergamon. pp.183-262.

Schmutterer, H. 1990. Properties and Potential of Natural Pesticides from the Neem Tree, *Azadirachta indica*. In: *Annual Review of Entomology*, Vol.35, Mittler, T. E., Radovsky, F. J., and Resh, V. H., eds. Palo Alto, Annual Reviews Inc. pp.271-297.

Seiber, J. N., Lee, S. M., and Benson, J. M. 1983. Cardiac Glycosides (Cardenolides) in Species of Asclepias (Asclepiadaceae). In: *Plant and Fungal Toxins-Handbook of Natural Toxins*, Vol.1, Keeler, R. F., and Tu, A. T., eds. New York, Marcel Dekker. pp.43-83.

Seiber, J. N., Nelson, C. J., and Benson, J. M. 1980. Pharmacodynamics of Some Individual Milkweed Cardenolides Fed to Larvae of the Monarch Butterfly. *Journal of Chemical Ecology* 8: 321-339.

Seigler, D. S. Cyanide and Cyanogenic Glycosides.1991. In: *Herbivores:Their Interactions with Secondary Plant Metabolites. The Chemical Participants*, Vol.1, Rosenthal, G. A., and Berenbaum, M. R., eds. San Diego, Academic Press. pp.35-77.

Severson, R. F., Arrendale, R. F., Cutler, H. G., Jones, D., Sisson, V. A., and Stephenson, M. G. 1988. Chemistry and Biological Activity of Acylnornicotines from *Nicotiana repandae.* In: *Biologically Active Natural Products, Potential Use in Agriculture*, Cutler, H. G., edACS Symposium Series, 335. Washington, DC, American Chemical Society. . pp.335-362.

Shettel, N., and Balke, N. E. 1983. Plant Growth Response to Several Allelopathic Chemicals. *Weed Science* 31: 293-298.

Shull, L. R., Buchmaster, G. W., and Cheeke, P. R. 1976. Factors Influencing Pyrrolizidine (Senecio) Alkaloid Metabolism: Species, Liver Sulfhydryls and Rumen Fermentation. *Journal of Animal Science* 43: 1247-1253.

Shupe, J. L., Binns, W., James, L. F., and Keeler, R. F. 1968. A Congenital Deformity in Calves Induced by the Maternal Consumption of Lupine. *Australian Journal of Agricultural Research* 19: 335-340.

Shupe, J. L., Binns, W., James, L. F., and Keeler, R. F. 1967. Lupine, a Cause of Crooked Calf Disease. *Journal of the American Veterinary Medical Association* 43: 1247-1253.

Soderlund, D. M., and Bloomquist, J. R. 1989. Neurotoxic Actions of Pyrethroid Insecticides. In: *Annual Reviews of Entomology*, Vol.34, Mittler, T. E., Radovsky, F. J., and Resh, V. H., ed. Palo Alto, USA, Annual Reviews Inc. pp.77-96.

Staal, G. B. 1986. Anti Juvenile Hormone Agents. In: *Annual Review of Entomology*, Vol.31, Mittler, T. E., Radovsky, F. J., and Resh, V. H., eds. pp.391-429 Palo Alto, Annual Reviews Inc.

Stegelmeir, B. L., Gardner, D. R., James, L. F., Panter, K. E., and Molyneux, R. J. 1995. The Toxic and Abortifacient Effects of Ponderosa Pine. *Veterinary Pathology* (in press).

Stevens, K. L. 1986. Allelopathic Polyacetylenes From *Centaurea repens* (Russian Knapweed). *Journal of Chemical Ecology* 12: 1205-1211.

Tang, T. S., and Zhang, B. 1986. Qualitative and Quantitative Determination of the Allelochemical Sphere of Germinating Mung Bean. In: *The Science of Allelopathy*, Putnam, A. R., and Tang, C. S., eds. New York, John Wiley and Sons. pp.229-242.

Thompson, A. C., ed. 1985. *The Chemistry of Allelopathy*, . ACS Symposium Series, 286. Washington, DC, American Chemical Society. 470pp.

von Euw, J., Fishelson, L., Parsons J.A., Reichstein, T., and Rothschild, M. 1967. Cardenolides (Heart Poisons) in a Grasshopper Feeding on Milkweeds. *Nature* (Lond.) 214: 35-39.

Waller, G. R., ed. 1987. *Allelochemical: Role in Agriculture and Forestry*, . ACS Symposium Series, 330. Washington, DC, American Chemical Society. 660pp.

Weston, L. A., Burke, B. A., and Putnam, A. R. 1987. Isolation Characterization and Activity of Phytotoxic Compounds from Quackgrass (*Agropyron repens* (L.)Beauv.). *Journal of Chemical Ecology* 13: 403-421.

Whitman, D. W. 1988. Allelochemical Interactions Among Plants, Herbivores and Their Predators. In: *Novel Aspects of Insect-Plant Interactions*, Barbosa, P., and Letourneau, D. K., eds. pp.11-64 New York, Wiley-Interscience.

3

OVERVIEW OF SNAKE VENOM CHEMISTRY

Anthony T. Tu

Department of Biochemistry and Molecular Biology
Colorado State University
Fort Collins, Colorado 80523-1870

1. INTRODUCTION

Snake venom is not composed of single compounds but is a complex mmixture of proteins. It is not known exactly how many proteins are present in a venom, but it probably consists of fifty to sixty components. Each protein possesses its own biological activity. A further complication is that composition and activity of a snake venom vary among the species. In general, the closer the phylogenetic relationship of the snakes, the more similar are the venom properties and composition.

A snakebite causes all types of toxic action because a snake venom contains a variety of toxic components. As mentioned earlier, a snake venom is a mixture of different proteins containing both toxic and nontoxic components.

Even for toxic components, the potency of toxicity is not the same. Usually, neurotoxins are most powerful. Therefore, a snakebite due to a snake whose venom contains neurotoxins exerts more toxic action and becomes dangerous to a victim. Cobra, krait, and sea snake venoms contain potent neurotoxins, and a victim quickly develops muscle paralysis that leads to death when the poisoning is severe. On the other hand, most common rattlesnake venoms contain few neurotoxins. Instead, they contain potent tissue damaging toxins such as hemorrhagic and myonecrotic toxins. Thus, rattlesnake bite often accompanies hemorrhage and the muscle damage of tissues and organs with possible dysfunction or complete loss of a body part such as fingers and toes. However, rattle snakes from South America and Mojave rattlesnake are extremely poisonous because their venoms contain potent presynaptic neurotoxins.

2. TOXIC COMPONENTS

Some snake venoms are more toxic than others because the type and amount of toxic components present in their venoms are different. Usually neurotoxins are most powerful. Cardiotoxins, cytotoxins, hemorrhagic toxins, and myotoxins are also highly toxic but less toxic than neurotoxins.

Natural Toxins II, Edited by B. R. Singh and A. T. Tu
Plenum Press, New York, 1996

2.1. Neurotoxins

There are several types of neurotoxins and their structures; the site of action and the mechanism are not identical. Snake neurotoxins are peripheral neurotoxins, rather than centrally neurotoxic; apparently they do not pass through the blood-brain barrier.

A. Presynaptic Neurotoxins. The presynaptic neurotoxins are also called β-toxins in contrast to postsynaptic or α-toxins. When a β-toxin is added to the neuromuscular preparation, the muscle contraction starts without stimulation of the nerve axon. β-Toxin usually does not affect the depolarization of the muscle itself or have a binding ability to the acetylcholine receptor. It is thus clear that the β-toxin somehow affects the presynaptic end of the nerve and initiates the release of acetylcholine and then eventually stops the release.

This can be clearly seen by observing the change in the miniature endplate potential (MEPP). The MEPP is a very small potential, observed in the neuromuscular junction, that is due to the natural leakage of acetylcholine from the vesicle. When β-toxin is applied, the MEPP frequently decreases first (5-10 min), then suddenly increases (for several hours). Finally, the frequency decreases until it becomes zero.

There are several types of presynaptic toxins. They are structurally distinct among themselves. However, there is one common property, and that is the possession of phospholipase A activity. Phospholipase A is one of the common enzymes found in various snake venoms and animal tissues. However, not all phospholipases A are toxic. The toxic phospholipase A is usually a basic protein. There is as yet no satisfactory explanation for this. Tsai et al. (1987) found that the basic amino acids tended to cluster near the surface region at the NH_2-terminal side in basic phospholipase A. This may have something to do with toxicity.

One type of presynaptic toxin is composed of two subunits bound together. The basic subunit has phospholipase A activity, whereas the acidic subunit has no enzymic activity. Crotoxin is the first presynaptic toxin isolated from snake venom and is also one of the most well-studied presynaptic toxins.

The role of the acidic subunit A is to guide the toxin to a specific site, then the basic subunit B functions as a presynaptic toxin (Hendon and Tu, 1979). Each subunit alone is relatively nontoxic, but combined, the toxicity is greatly enhanced (Trivedi et al., 1989). The undissociated crotoxin itself shows phospholipase A activity, indicating the active site of subunit B is not masked by subunit A (Radvanyi and Bon, 1984). Besides neurotoxicity, subunit B also has hemolytic activity. Subunit B attaches to many parts of the erythrocyte membranes (Jeng et al., 1978) and also on the postsynaptic membrane (Bon et al., 1979), in addition to the presynaptic binding site.

From a structural viewpoint, both subunits A and B are in the isoforms (Aird et al., 1985; 1986; Faure et al., 1991). The difference lies in the length of the polypeptide chains. This suggests that the isoforms originate from posttranslational proteolytic cleavage.

Mojave toxin from *Crotalus scutulatus* is also structurally similar to crotoxin and is composed of two subunits (Bieber et al., 1975; Cate and Bieber, 1978). There are considerable amino acid sequence homologies between the two toxins (Aird et al., 1990). The amino acid sequence and spectroscopic properties of subunits of crotoxin are similar to other presynaptic phospholipases A from snake venoms (Aird et al., 1989a).

Crotoxin's neurotoxic action is very similar to β-bungarotoxin (β-btx), but has some difference. For instance, crotoxin and its subunit B have a postsynaptic effect, whereas β-btx has no such activity (Bon et al., 1979).

The acidic and basic subunit types presynaptic toxins are fairly common in neurotoxic snake venoms. For instance, such toxins have been isolated from the venoms of *C. viridis concolor* (Aird et al., 1989b) and *C. durissus collilineatus* (Lennon and Kaiser, 1990). The amino acid sequence of the basic subunit also has considerable homology to other snake venom phospholipases A.

Although most studies of presynaptic toxins are focused on the nerve ending of the neuromuscular junction or synasptosomes, the toxin may have broader biological action. For instance, Mojave toxin inhibits calcium channel dihydropyridine receptor binding in rat brain (Valdes et al., 1989).

The second type of presynaptic neurotoxin from a chemical viewpoint is that two polypeptide chains are connected by a disulfide bond. The most typical toxin of this type is β-btx. It consists of two chains: The A chain has 120 amino acids, with a relative molecular mass (M_r) of 13,500, and the B chain has 60 amino acid residues with an M_r of 7000. The amino acid sequence of the A chain is similar to the phospholipase A sequence and, in fact, the A chain does possess phospholipase A activity. For presynasptic activity, phospholipase A activity is essential. For instance, when Ca^{2+} is replaced with Sr^{2+}, phospholipase A activity and presynaptic activity are both lost. When histidine residues of the A chain are modified chemically, phospholipase A activity is lost, as well as presynasptic toxic activity.

In β-btx, the A chain is the one that has phospholipase A activity: The nomenclature is somewhat opposite that of crotoxin or Mojave toxin; in crotoxin or Mojave toxin, subunit B is the one showing phospholipase A activity. Oxidation of methionine at the 6 and 8 positions lowered the toxicity without affecting antigenicity. Moreover, the NH_2-terminal region of the A chain plays a crucial role in maintaining functional activity (Chang and Yang, 1988).

Heterodimeric β-btx was examined by x-ray diffraction. The crystals are monoclinic, space group C2 with unit cell parameters of a = 176.5 Å, b = 39.3 Å, c = 92.7 Å, and β = 114.8°. These heterodimers appear to be associated as two crystallographically distinct $(AB)_4$ tetramers, each having dihedral D2 symmetry. The two are positioned with equivalent molecular two-fold axes, coincident with crystallographic cyads, but rotated by 55° relative to one another (Kwong et al., 1989).

The exact mechanism of β-btx is not yet known. But it may be that phospholipase A creates a hole in the nerve end membrane and Ca^{2+} flows to the cytoplasm. As a result, vesicles containing the nerve transmitter acetylcholine discharge it. The nucleotide sequence encoding β-btx A_2-chain has been determined (Danse and Garnier, 1990).

Because the action of a presynaptic toxin, β-btx, is to start the initial burst of acetylcholine followed by the stop of acetylcholine release, it eventually causes paralysis of the muscle. The mechanism does not involve the hydrolysis of acetylcholine; therefore, it is reasonable that anticholinesterase does not overcome β-btx's effect.

The third type of presynaptic toxin is a tertiary complex of three polypeptide chains. Taipoxin from the venom of the Australian snake, taipan, has three subunits, α, β, and γ, with an M_r of 46,000. The number of amino acid residues present in the subunits is 120, 120, and 135, respectively. The α-chain is basic and has phospholipase A activity.

The fourth type is a quaternary complex of four polypeptide chains. Textilotoxin, isolated from *Pseudonaja textilis*, consists of A, B, C, and D subunits. Subunit D consists of two identical covalently linked polypeptide chains (Tyler et al., 1987; Pearson et al., 1991).

Tryptophan residues in subunits A, B, and D are relatively exposed to solvent, whereas subunit C exhibits no fluorescence. Probably subunit C does not contain trypotophan (Aird et al., 1989b).

The last type is a single polypeptide chain presynaptic neurotoxin. An example of this is notexin, from the venom of *Notechis scutatus scutatus*, consisting of 119 amino acid residues, with seven disulfide bonds. It has an $M_r$of 13,400. Notexin has isotoxins; they differ in only one amino acid residue among the two isotoxins (Chwetzoff et al., 1990). The three-dimensional structure of notexin was determined by crystallography. The core of the protein is very similar to other phospholipases A. The difference, however, exists mainly in the area of residues 56-80 and 85-89 (Westerlund et al., 1992).

From all the presynaptic toxins examined, one sees that they possess phospholipase A activity; but the reverse is not true. There are many proteins with phospholipase A activity, and not all of them are toxic; only those with basic phospholipase A are toxic, and only some of them are presynaptic neurotoxins.

Not every presynaptic toxin is identical in relation to the release of acetylcholine from the presynaptic site. With β-btx, there is an initial burst of acetylcholine, but eventually the release is stopped. Even though toxins may behave like β-btx, the length of time for acetylcholine release is different for each toxin. Some presynaptic toxins do not release acetylcholine from the beginning and simply stop the release. In such an event, the depolarization wave never reaches the muscle, and the muscle is paralyzed.

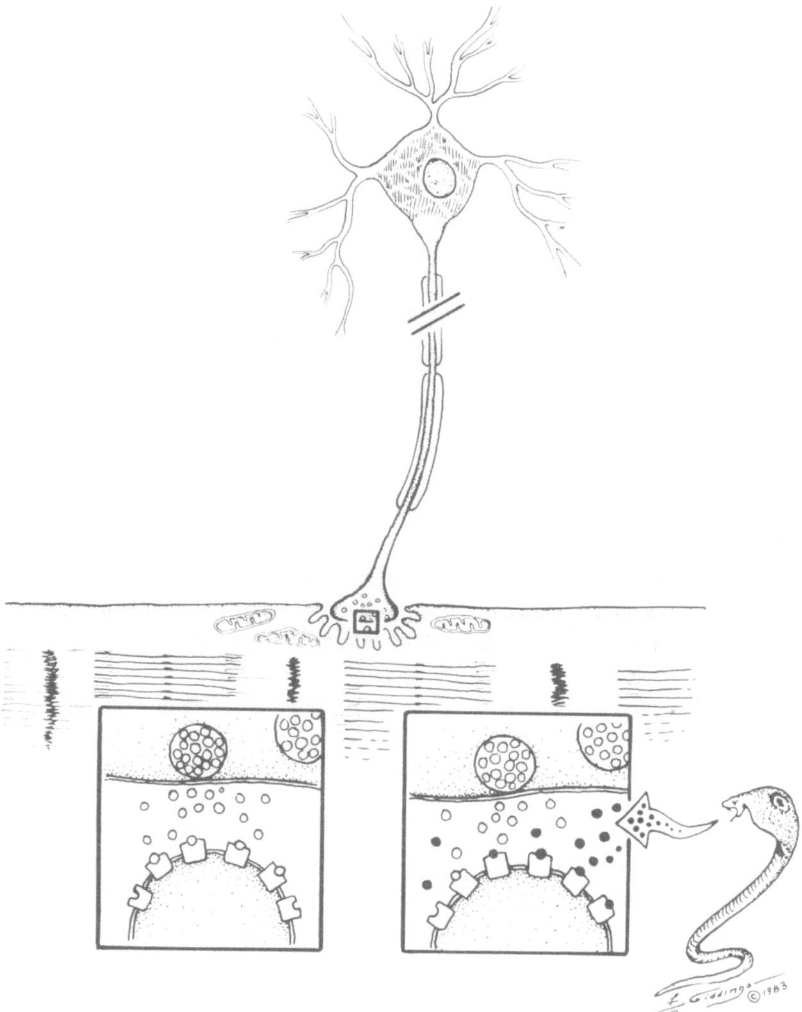

Figure 1. A diagram showing nerve transmission across the neuromuscular junction. Left: Normal transmission by acetylcholine (open circle). Right: Blockage of acetylcholine receptor by postsynaptic neurotoxin (solid circle).

B. Postsynaptic Neurotoxins. Postsynaptic neurotoxins are commonly found in the venoms of Hydrophiidae and Elapidae. The toxins affect the neuromuscular junction at the postsynaptic site by combining with acetylcholine receptor (AChR). This is diagrammatically shown in Fig. 1. These neurotoxins act on the muscle side, rather than the nerve side. The solabeled *postsynaptic neurotoxins* are really the toxins affecting the particular site of the muscle and should not have been designated neurotoxins. Actually, postsynaptic neurotoxins bind to the acetylcholine receptor in the muscle that is to receive the neurotransmitter acetylcholine. On the other hand, the attachment of acetylcholine to the acetylcholine receptor is considered a part of the nerve-transmitter mechanism. From this functional viewpoint it is not unreasonable to call them postsynaptic neurotoxins because of their activity. Therefore, the paralysis of the muscle by postsynaptic neurotoxin poisoning is essentially due to the formation of an acetylcholine receptor-neurotoxin complex. One should realize that usually a snake venom contains multiple numbers of neurotoxins. *Bungarus multicinctus* venom is well known as the source of α- and β-btx, but the venom contains many other neurotoxins. For instance, toxin F, which also blocks neuronal nicotinic receptors, has been isolated (Loring et al., 1986). Venom from a similar snake, *B. fasciatus*, also contains various neurotoxins (Liu et al., 1989a,b). Sea snake, *Acalytophis peronii*, venom also contains major and minor neurotoxins. The only difference between the major and minor postsynaptic toxins is in the 43rd residue. The major toxin at this position contains glutamine, whereas the minor toxin contains glutamic acid (Mori and Tu, 1988a,b).

Before further discussing the action of postsynaptic neurotoxins, it would be useful to review normal nerve transmission very briefly. When a normal nerve impulse (depolarization wave) passes through the axon and reaches the end of that axon, the calcium ion concentration is increased and the neurotransmitter, acetylcholine (ACh), is suddenly released from the vesicle at the end of the nerve. Acetylcholine moves across the synaptic crevice and reaches the acetylcholine receptor in the muscle. The AChR is composed of five subunits, α_2,βγδ. When two molecules of acetylcholine attach to the α-subunits, the AChR

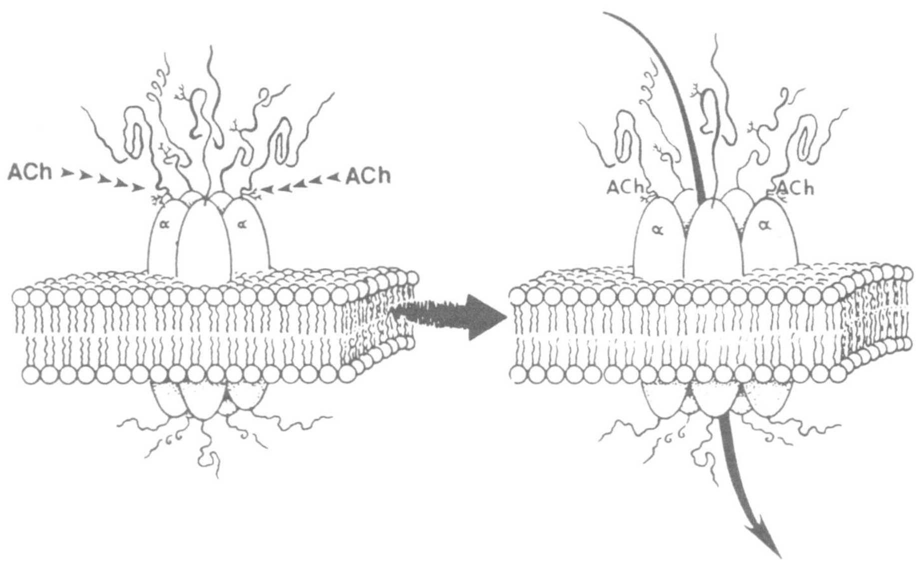

Figure 2. As two moles of acetylcholine (ACh) attach to two α-subunits of acetylcholine receptor (AChR), the pore opens to form an ion channel in the membrane that allows ions to pass through the channel. This is the role of AChR in muscle depolarization.

changes configuration and becomes an open ion channel, permitting certain ions to pass through (Fig. 2). By this mechanism, the depolarization wave reaches the muscle and is further propagated through the muscle plasma membranes, T-tubules, and sarcoplasmic

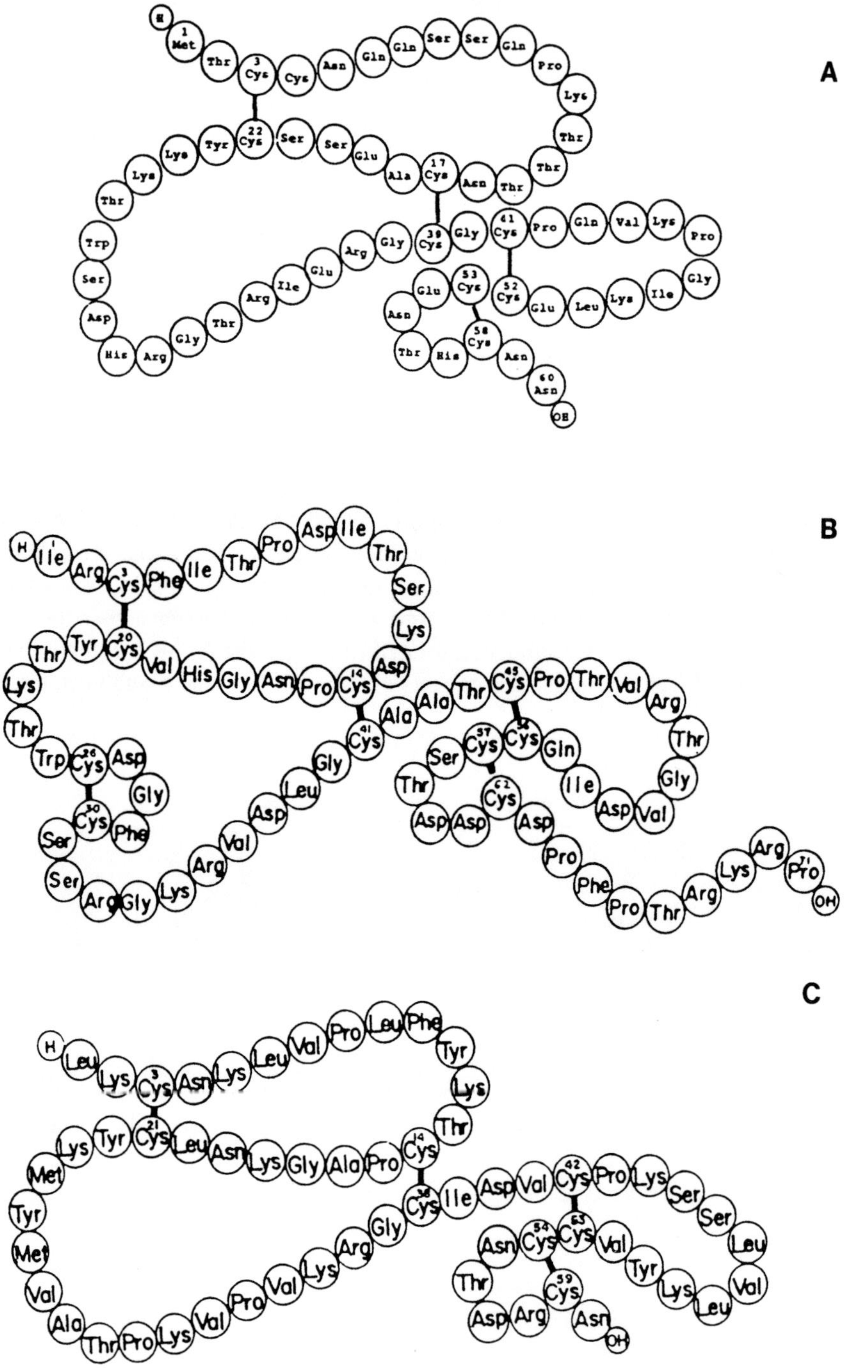

Figure 3. Examples of neurotoxins (A, C) and a cardiotoxin (B). (A) Primary structure of lapemis toxin, a short-chain postsynaptic neurotoxin. (B) Cardiotoxin from *Naja naja* venom. (C) Toxin B from *Naja naja* venom.

reticulum (SR). The SR has a very high concentration of calcium ion. When the depolarization wave reaches the SR, the calcium ion suddenly leaks out of the SR into the myoplasm, causing the myofilaments to contract. As soon as the muscle is relaxed, the calcium ion moves back into the SR.

The structure of postsynaptic neurotoxins is well studied. There are actually two types of these neurotoxins (Fig. 3A,C). One type has four disulfide bonds (called type I or short-chain neurotoxins). The short-chain neurotoxin has one or two amino acids at segment 8, whereas the long-chain neurotoxins have a longer segment 8 (see Fig. 3). Another difference is that there is only one amino acid within segment 5 of the short-chain neurotoxin, whereas the long-chain neurotoxin has three amino residues within the segment (see Fig. 3).

Both short- and long-chain neurotoxins have the same biological activity; namely, to bind to AChR, but there is some difference in chemical properties. It was well documented that the invariant tryptophan residue in short-chain neurotoxin is essential, because the chemical modification of this residue caused the loss of neurotoxicity (51-53). However, the modification of a tryptophan residue in α-btx, which is a long-chain neurotoxin, did not appreciably change the toxicity (Chang et al., 1990).

Most neurotoxins isolated from Australian Elapidae venoms were reported as presynaptic neurotoxins, but a postsynaptic one was isolated from *Acanthophis antarcticus* (Australian death adder) (Sheumack et al., 1990).

One interesting aspect from a structural viewpoint is that the two types of postsynaptic neurotoxins are very similar to Elapidae venom cardiotoxins (see Fig. 3B). Cardiotoxins stop the heartbeat when they make contact with the heart. Cardiotoxins have four disulfide bonds and a very short segment 8. In this manner, they are similar to short-chain neurotoxins. Although the similarity in disulfide bonds and the peptide backbone is remarkable for cardiotoxins and postsynaptic neurotoxins, there are considerable differences between them in amino acid composition and sequences. Cardiotoxins do not bind to the AChR, whereas there is strong binding between the neurotoxins and the AChR. The hydrophilic index of cardiotoxins shows them to be quite hydrophobic molecules, whereas the neurotoxins are quite hydrophilic molecules. Cardiotoxins are more general toxins, affecting cell membranes, whereas neurotoxins are specific toxins, binding to acetylcholine receptors.

Neurotoxins are relatively small-sized proteins, but they contain four or five disulfide bonds. Thus, they have a compact structure and, molecularly, they are very stable.

Postsynaptic neurotoxins are composed mainly of an antiparallel β-sheet and a β-turn structure, with only a small amount of α-helical structure (Yu et al., 1975; Tu, 1990; Betzel et al., 1991; LeGoas et al., 1992; Yu et al., 1990; Tu et al., 1976). The toxin is comprised of three loops, A, B, and C (Fig. 4). Loop B is considered most important, and it is believed that this loop is attached to the acetylcholine-binding site of the AChR. Loop B is also the antigenic determinant.

The amino acid sequences of over 100 postsynaptic neurotoxins have been determined by many investigators; therefore, we will not discuss the sequence of all the toxins. However, one should be aware of the incorrect sequence of α-btx, as originally reported earlier (Mebs et al., 1971). The correct primary structure of α-bungarotoxin was later established (Ohta et al., 1987). The original paper (Mebs et al., 1971) reported the sequences of Ile-Pro-Ser (9-11), His-Pro (67-68), and Arg-Gln (71-72). However, these sequences are incorrect, and the correct sequences have now been established as Ser-Pro-Ile (9-11), Pro-His (67-68), and Gln-Arg (71-72) by Ohta et al. (1987).

The primary structure of postsynaptic toxins is unique to snakes, and there are no homologies with the toxins of scorpions, spiders, or bees. However, there is an interesting report that significant homologous sequences to snake postsynaptic neurotoxins are found in visna virus and HIV-I *tat* proteins (Gourdou et al., 1990).

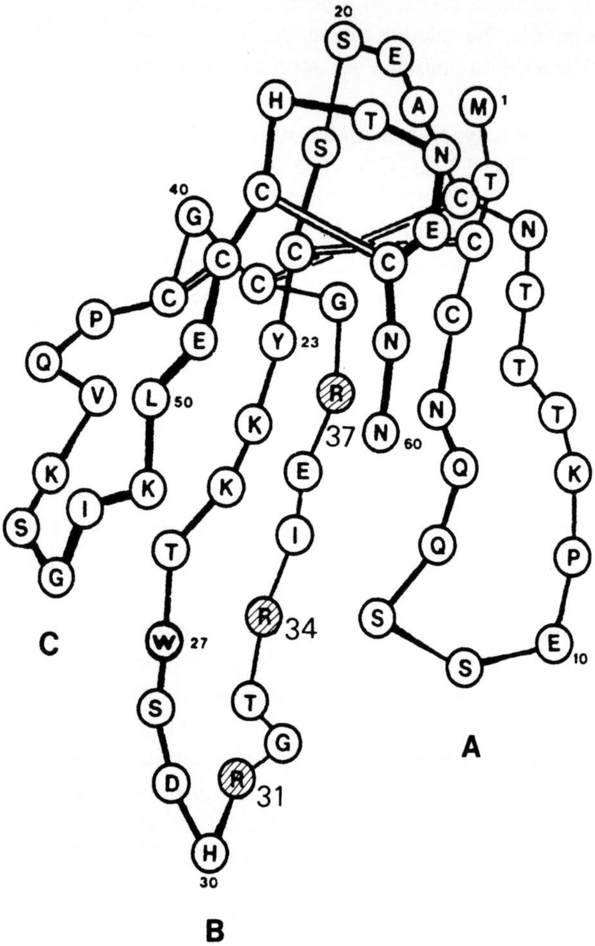

Figure 4. Chemical structure of lapemis toxin showing three main loops A, B, and C.

Snake venoms also contain nonneurotoxic proteins, with structures very similar to a postsynaptic neurotoxin. For instance, mambia is a platelet aggregation inhibitor isolated from the venom of *Dendroaspis jamesonii*. It has 59-amino acid residues, with four disulfide bonds and a high homology to postsynaptic neurotoxins (McDowell et al., 1992).

Although postsynaptic neurotoxins are small polypeptides with an M_r of about 6,800, they are antigenic. However, by conjugating neurotoxin to a protein with a higher M_r, antigeneity can be further enhanced (Sunthornandh et al., 1992). There is a toxic fusion protein in snake venoms. Ducanal et al. (1989) constructed a recombinant expression plasmid encoding a protein A-neurotoxin fusion protein in *Escherichia coli*. The median lethal dose (LD_{50}) values of the fused toxin and native toxin are 130 and 20 nmol/kg mouse, respectively.

The AChR is a pentamer that is comprised of five subunits (two α, one each of β, γ, and δ), and two of them are identical (see Fig. 3). The presence of four different subunits can readily be seen in electrophoresis after reduction (Fig. 5). The receptor is a ligand (acetylcholine)-gated channel protein, allowing ions to pass through when activated (see Fig. 2). The ligand, acetylcholine, attaches to the α-subunits. Since there are two α-subunits, the stoichiometry of ligand-receptor interaction is 2 mol of acetylcholine per receptor. Postsy-

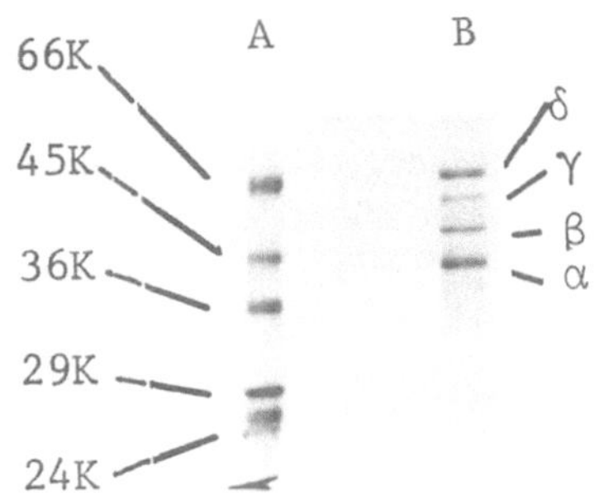

Figure 5. Four subunits from AChR, α, β, γ, and δ (B). A is standard molecular weight marker.

naptic toxins attach to the same sites as acetylcholine; however, the AChR receptor fails to form a channel (Fig. 6).

Each subunit is a glycoprotein; however, it is not yet clear just what role the polysaccharide, which is present in each subunit, plays. There are several types of polysaccharides in each subunit. One of them is shown here:

```
M(α1-2) - M(α1-6)\
                   > M(α1-6)\
          M(α1-3)/           > M(β1-4)-NAc G (β1-4)-NAcG
M(α1-2) - M(α1-2) - M(α1-3) /
```

where M is mannose and NAcG is *N*-acetylglucosamine (Nomoto et al., 1986).

The toxin attachment site in the α-subunit is not simply a single amino acid residue; many sites are involved in the toxin binding.

Normally, the interaction of AChR and a postsynaptic neurotoxin is studied by using a radiolabeled neurotoxin. However, a simple, nonradioactive, but sensitive, method was

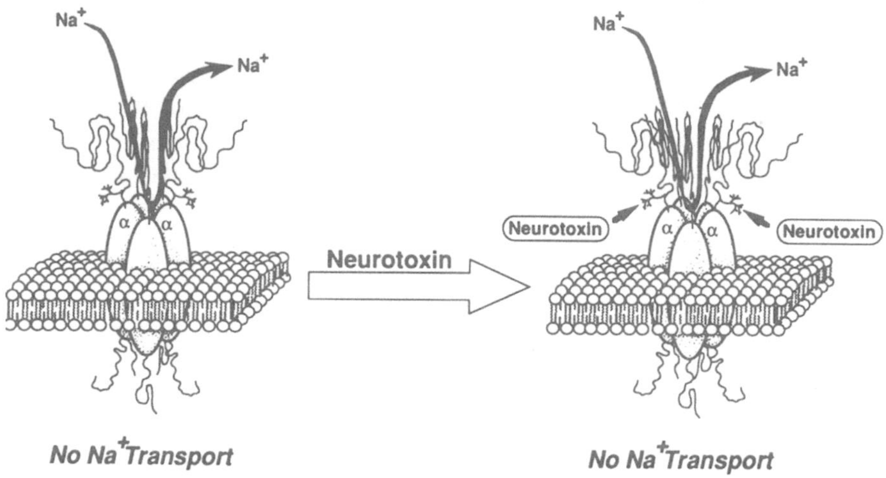

Figure 6. Attachment of a neurotoxin to the same site as that of acetylcholine causes the AChR to fail in forming an ion channel in the membrane.

developed by Nomoto et al. (1992), who used horseradish peroxidase (HRP) conjugated neurotoxin.

The acetylcholine receptor and a neurotoxin form a noncovalent bond-type complex. The most important question is what portion of a neurotoxin is really involved in the receptor binding. Is it a particular residue, or are several residues involved? Figure 4 is a two-dimensional structure of a postsynaptic neurotoxin—lapemis toxin from *Lapemis hardwickii*—that is based on the x-ray diffraction study of another similar toxin.

From studies of chemical modification of amino acid residues studied by many investigators, it was shown that the ones located in loop B are essential for neurotoxicity. For instance, the Arg-31, Arg-34, Trp-37, Tyr-23, Lys-24, and Lys-25 are known to be related to neurotoxicity. It is logical to assume that loop B is most likely to bind to the AChR.

To clarify this problem, synthetic peptides identical with A, B, and C loops were made and their ability to bind to the acetylcholine receptor was studied (Miller and Tu, 1991).

Peptide synthesis:

```
         10        20        30        40         50        60
MTCCNQQSSQPKTTTNCAESSCYKKTWSDHRGTRIERGCGCPVQKPGIKLECCHTNECNN  Lapemis toxin
  CCNQQWWQPKTTTNC                                             Peptide A1
  |             |     YKKTWSDHRGTRIERG                        Peptide B2
  |__S______S___|    CYKKTWSDHRGTRIERGC                       Peptide B1
                     |                |  CPQVKPGIKLEC         Peptide C1
                     |_____S_____S____|  |          |
                                         |__S____S__|

EACDFGHIKLMNPQRSTVWY                                        Nonsense peptide
```

Only the peptide identical with the central loop B bound to the acetylcholine receptor, whereas the other peptides had no detectable binding. The disulfide bond is essential for binding. When the central loop peptide was reduced and alkylated, the binding ability was lost. This finding suggested that the central loop plays a dominant role in the toxin's ability to bind the receptor.

Hydrophilicity analysis of lapemis toxin showed that the central loop is the most hydrophilic region. Since the ligand-binding region of the acetylcholine receptor is in the outside of the cell membranes, it is also hydrophilic. Therefore, it is also logical that the most hydrophilic portion of neurotoxin binds to the acetylcholine receptor-ligand-binding site, which is also hydrophilic.

The antigenic determinant is located at loop B, which is also the acetylcholine receptor-binding region (Kase et al., 1989).

Most AChR studies were done using skeletal muscle or torpedo tissues. The acetylcholine receptor concentration in the brain is very small, but it is present. Recently, the AChR in the brain has been actively studied using snake postsynaptic neurotoxins. Some of these are rather typical neurotoxins that bind to both skeletal muscles and the brain, and some of them are specific to the brain AChR. Since a brain α-subunit of AChR binds to α-btx, there must be a similarity between the toxin-binding site for the brain AChR and the muscle AChR (McLane et al., 1990; Scheidler et al., 1990).

C. Potassium Channel-Binding Neurotoxins. The potassium channel plays an important role in the repolarization process in nerve transmission and is less well known than the sodium channels in the nerve. The The K^+ channel is composed of membrane protein and has six transmembrane helical regions. Both the NH_2- and COOH-terminal chains are located inside the membrane. The first snake toxin found to bind K^+ is dendrotoxin. This toxin is a potent convulsant and facilitates transmitter release by inhibition of voltage-sensitive K^+ channels (Weller et al., 1985; Penner et al., 1986; Harvey and Karlsson, 1980; Black et al., 1988; Benoit and Dubois, 1986).

Table 1.

Venom	Name	Identical with
Dendroaspis angusticeps	α-DaTX	Dendrotoxin, $C_{13}S_2C_3$
	β-DaTX	New toxin
	γ-DaTX	New toxin
	δ-DaTX	$C_{13}S_1C_3$
D. polylepis polylepis	DTX_14	Toxin I
Bungaris multicinctus	β-btx	β-Bungarotoxin

Dendrotoxins are more suitable for study of the K^+ channels than β-btx because they lack the intrinsic phospholipase A activity (Moczydlowski et al., 1988). Dendrotoxin induces repetitive firing in rat visceral sensory neurons by inhibiting a slowly inactivating outward K^+ current (Stansfeld et al., 1986).

Dendrotoxin (DTX) has an M_r of 7000 (Busch et al., 1971) and strongly binds to synaptic plasma membranes of rat or chick brain (Benishin et al., 1988). The receptor has a high M_r of 405,000-465,000 (Black et al., 1986). Rhem and Lazdunski (1988) also isolated the K^+ channel proteins that bind to DTX I. The purified material has three bands of M_r 76,000-80,000, 38,000, and 35,000 in polyacrylamide gel electrophoresis (PAGE). By using neuraminidase and glycopeptidase, K^+ channel proteins that bind to DTX, β-btx, and MCD were reduced to 65,000 Da. This indicates that a peptide core of the K^+ channel protein that binds to the toxins is about 65,000 Da (Rehm, 1989). β-Bungarotoxin, normally considered to be a presynaptic neurotoxin affecting the nerve ending, is also a K^+ channel blocker (Peterson et al., 1986; Schmidt et al., 1988). There are considerable sequence homologies between β-bungarotoxin and dendroapsis venom toxins. The K^+ channel inhibitory action of β-btx is independent of its phospholipase A activity (Rowan and Harvey, 1988). Then, one may wonder whether many other presynaptic snake toxins have any K^+ channel-blocking activity. This question has not yet been answered because few other presynaptic toxins have been examined for K^+ channel-blocking activity. However, there is evidence that other snake presynaptic toxins may also be K^+ channel blockers. Alvarez and Garcia-Sancho (1989), using crude venoms of *Notechis scutulatus*, *Oxyuranus scutulatus*, and *Vipera russelli*, found that they did inhibit K^+ channels. The first two venoms are known to contain potent presynaptic toxins. One should, however, notice that Alvarez and Garcia-Sancho used K^+ channels of red cells, whereas most other studies were done on synaptosomes. Anderson and Harvey (1988) used other tissues, such as diaphragm and the nerve-muscle preparation, and observed the same inhibition as in synaptosomes studied by many other workers.

D. Antiacetylcholinesterase Neurotoxins. The fourth type of neurotoxin is the one that binds to acetylcholinesterase (Rodriquez-Ithurralde et al., 1981; Cervenansky et al., 1991). When acetylcholinesterase is not functioning, acetylcholine (after binding to the acetylcholine receptor) cannot be hydrolyzed; consequently, normal nerve transmission is impaired. Acetylcholinesterase action of *D. angusticeps* venom was first reported by Rodriguez-Ithurralde et al. (1983).

Antiacetylcholinesterase-type neurotoxins have so far only been isolated from African mambas (*Dendroaspis*). The names of the snake venoms from which anticholinesterase-type toxin was isolated are shown in Table 2.

Anticholinesterase-type neurotoxin has 57-60 amino acids in a single polypeptide chain, cross-linked by three disulfide bonds. The two-dimensional structure of fasciculin 2 from dendroaspis venom is shown in Fig. 15. Fasciculin 2 is identical with toxin F_7 isolated by Viljoen and Botes (1973). Similarly, toxins C and D from *D. polylepis polylepis* venom

Table 2.

Venom	Toxin	Reference
Dendroaspis angusticeps	F_7	Lin et al., 1987
D. polylepis polylepis	C	Lin et al., 1987
D. angusticeps	Fasciculin	Dajas et al., 1987

are also related to acetyl-cholinesterase-type neurotoxin (Joubert and Strydom, 1978; Karlsson et al., 1984). Although anticholinesterase neurotoxins are structurally similar to postsynaptic-type neurotoxins and cardiotoxins, they differ immunologically.

The crystalline structure of fasciculin 2 indicates that the toxin is structurally related to both cardiotoxin and α-neurotoxins (le Du et al., 1989). The crystals are tetragonal, with unit cell dimensions of a = 48.9 Å and c = 82.0 Å and with the space group of P41212 or P43212. There are 16 molecules in the unit cell. Fasciculin 1 was also examined by x-ray crystallography. The unit cell values for fasciculin 1 are a = 40.4 Å and c = 81.1 Å, with the space group P4(1)2(1)2 or P4(3)2(1)2. It is estimated that there is one molecule in the asymmetric unit (Ménez and Ducruix, 1990).

The toxin binds to acetylcholinesterase and renders acetylcholine unhydrolyzed. This causes continuous excitement of the muscle. The inhibition of acetylcholinesterase is seen not only in vitro, but also in vivo. For instance, 80% of the acetylcholinesterase activity in the locus coeruleus was inhibited by the injection of fasciculin 2 in rats (Abo et al., 1989). The inhibition of the enzyme by fasciculin is longlasting, and a 74% inhibition five days after injection was observed (Quillfeldt et al., 1990).

By inhibiting acetylcholinesterase, fasciculin increased the amplitude and time course of the endplate potential (Lee et al., 1985). Fasciculin also increased the amplitude of the miniature endplate potential (Cervenansky et al., 1991).

Acetylcholinesterase enveloped in an artificial liposome can also bind to fasciculin (Puu and Koch, 1990).

Because of the inhibition of acetylcholinesterase, dendrotoxins or other facilitatory toxins enhance the release of acetylcholine. Thus, dendrotoxins and fasciculins have synergistic action that enhances the lethality.

Fasciculin 2 has no presynaptic action on transmitter release or on postsynaptic receptor-blocking action; the main action is on anticholinesterase (Anderson et al., 1985). There was no significant change in dopamine or serotonin concentration in rats after fasciculin 2 injection (Bolioli et al., 1989).

2.2. Cardiotoxins

Cardiotoxins are also quite toxic components present in cobra venoms. They have four disulfide bonds and the position of the four disulfide bonds is almost identical to neurotoxins. Yet, from a functional viewpoint, they are quite different. From close examination of the amino acid sequence of cardiotoxin, it is quite different from neurotoxins.

A detailed NMR study of cardiotoxin indicated that there is a major structural difference that exists in all loops between cardiotoxins and postsynaptic neurotoxins (Bhaskaran et al., 1994). The difference in the secondary structure makes cardiotoxins bind to heart membranes, whereas neurotoxins do not bind.

```
                 5         10        15        20        25        30        35        40
Myotoxin I    Y K R C H K K E G H C F P K T V I C L P P S S D F C K M D C R W K W K C C I I G S V N

Myotoxin II   Y K R C H K K G G H C F P K E K I C T P P S S D F G K M D C R W K W K C C K K G S V N
              Y K R C H K K G G H C F P K T V I C L P P S S D F G K M D C R W R W K C C K K G S V N

Myotoxin a    Y K Q C H K K G G H C F P K E E I C I P P S S D L G K M D C R W K W K C C K K G S G

Peptide C     Y K R C H K K G G H C F P K T V I C L P P S S D F G K M D C R W K W K C C K K G S V N

Crotamine     Y L Q C H K K G G H C F P K E K I C L P P S S K F G K M D C R W R W K C C K K G S G
```

Figure 7. Primary structure of myotoxin *a* and related peptides.

2.3. Tissue Damaging Toxins

A large number of hemorrhagic toxins were isolated, and they were found to be zinc containing proteases (Bjarnason and Tu, 1978; Hite et al., 1992). Hemorrhagic toxins cause hemorrhage by direct action on the capillary endothelium after initial dilation of the endoplasmic reticulum.

Snake venom proteases fall into two broad categories, serine proteases such as thrombin-like enzymes (Itoh et al., 1987), and zinc metalloproteases such as hemorrhagic toxins. Among zinc proteases, some are hemorrhagic and some are nonhemorrhagic. Proteolysis specificity of two groups of zinc proteases is very similar and it is not possible to conclude which bond of proteolysis is responsible for hemorrhagic action. Amino acid sequence comparison of hemorrhagic and nonhemorrhagic zinc proteases reveals that they are very similar. So it is not possible to determine which portion of protein causes hemorrhagic action. Further study is needed to differentiate the two types of zinc proteases; one is hemorrhagic and the other is nonhemorrhagic. Another prominent effect of snake venom on tissue damage is the destruction of the muscle. So far three components were identified to be myotoxic. One is myotoxin *a* and its analogues. The second one is some phospholipase A_2, and the third one is some toxins possess both hemorrhagic and myotoxic actions. Among them, myotoxin *a* is most extensively studied (Fig. 7). The amino acid sequences of myotoxin *a* and related compounds are shown here in figure 7.

Myotoxin *a* binds to Ca^{++}-ATPase in the sarcoplasmic reticulum (SR) and inhibits SR Ca^{++} loading activity (Volpe et al., 1986; Utaisincharoen et al., 1991).

Phospholipase A_2 has many biological actions such as presynaptic action, indirect hemolytic action, and myonecrotic action. Strangely, not all phospholipase A_2 possesses such toxic actions. Usually the enzyme that exerts such actions is basic phospholipase A_2 (Hawgood and Bon, 1991).

The third type of myonecrotic toxins is the type of toxins that possesses both hemorrhagic and myonecrotic actions. Examples are viriditoxin (Fabiano and Tu, 1981), HTb from *Crotalus atrox* (Ownby et al., 1978), and a hemorrhagic factor from *Vipera aspis aspis* (Komori and Sugihara, 1988).

2.4. Hemolytic Factors

Hemolysis is the rupture of the erythrocyte membrane, with the hemoglobin present inside the cell diffusing to the surrounding media.

Snake venoms contain direct and indirect hemolytic factors, although there is not always a very clear distinction between them. The direct hemolytic factor itself can hemolyze the red cell. Indirect lytic factor lyses the red cell very slowly; however, its lytic action can be greatly accelerated by the addition of phosphatidylcholine. Indirect hemolytic factor is identified as phospholipase A_2. Since the proportion of these two factors differs from venom to venom, the rate of hemolysis varies, depending on the venom. Hemolytic action depends on a variety of factors including the species of animal from which the erythrocyte originated. Venoms of Elapidae, such as those from cobras, usually contain both direct and indirect factors. Hemolysis is, no doubt, one of the toxic effects that snake venoms produce, but it is not the main lethal factor.

The direct hemolytic factor is a highly basic polypeptide with a molecular weight of 7000 daltons. Usually its activity is synergistically increased with phospholipase A_2. The main function of the direct hemolytic factor would appear to be to disrupt the matrix of the cell membrane organization in such a way that venom phospholipase A_2 can attack membrane phosphatidylcholine more effectively. Lysophosphatidylcholine produced as a result of phospholipase A_2 hydrolysis further enhances the hemolytic action. Many other snake venom components such as cardiotoxin, myotoxin, and crotamine also induce hemolysis.

3. NON-TOXIC COMPONENTS

A snake venom also contains many nontoxic components including many enzymes. Only a few of them are discussed here.

3.1. Cobra Venom Factor (CVF)

There is a special way whereby the complement system can be activated without an antigen-antibody interaction. A factor that induces this alternative pathway can be isolated from the venoms of cobras and other Elapidae.

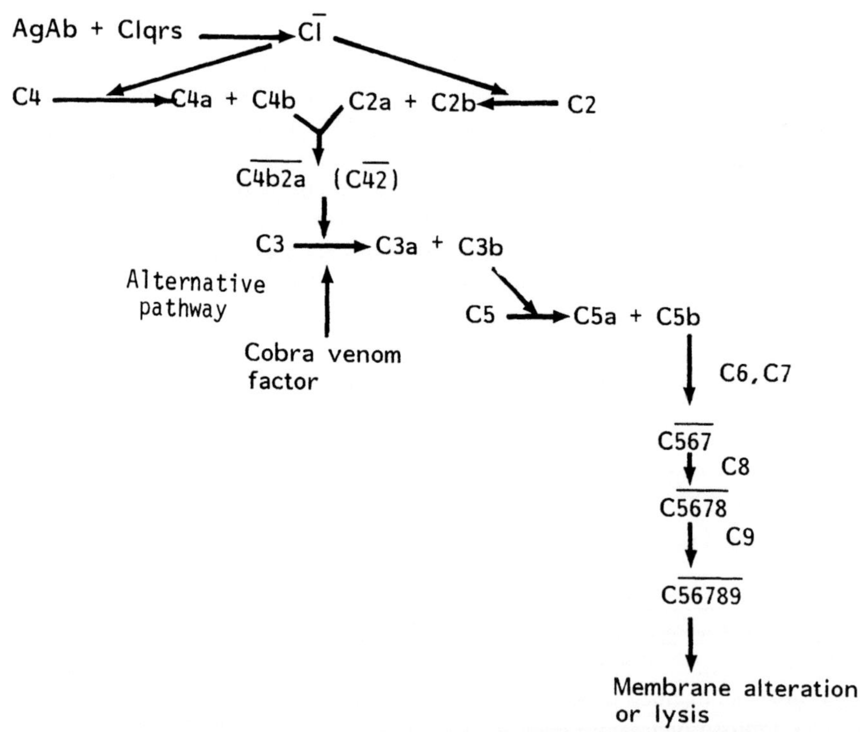

Figure 8. The site of cobra venom factor (CVF) in complementary system.

Antibody-independent activation of the complement system starting with C3 can be achieved by means of CVF (cobra venom factor), which interacts with a serum cofactor. The CVF-serum cofactor complex acts on C3 and activates the terminal complement sequence (Fig. 8). The result may be cytolysis if bacteria or erythrocytes are involved, or the activation of a noncytolic mechanism such as the release of histamine from mast cells, platelets, or leukocytes, enhanced phagocytosis, contraction of smooth muscles, the aggregation and fusion of platelets, or the promotion of blood coagulation.

CVF has been used in the investigation of the role of complement in various immunologic reactions. The anticomplement action of the CVF has potential use in the prevention of transplanted organ rejection. CVF as an immunosuppressor has been investigated by a number of workers; their results showed increased survival time of transplanted or grafted tissues.

3.2. Nerve Growth Factor (NGF)

The presence of NGF in snake venoms is quite common. NGF promotes the growth of peripheral sympathetic neurons and also promotes the development of some sensory neurons during early development.

NGF isolated from *Naja naja* venom has a high degree of amino acid sequence homology to mouse NGF. NGF isolated from mouse salivary gland has three subunits—α, β, and γ—of which the biologically active subunit is the β-unit. However, snake venom NGF does not have αβγ complexes. The similarity between the amino acid sequences of NGF from two sources suggests that the active sites of NGF are highly conserved.

The presence of this factor in snake venoms is common. It was isolated from venoms of several Elapidae, Viperidae, and Crotalidae. NGF is a glycoprotein and has no lethal activity.

3.3. Autopharmacological Action

Upon envenomation, substances are released from the tissues that were not originally present in the venoms, including bradykinin, histamine, serotonin, and ATP. Such liberated substances probably have pronounced effects on victims, especially on the vascular system. Hypotension and the pooling of blood in the vascular system are due to severe vasodilation known as "bradykinin shock" and are prominent in severe poisoning. The autopharmacological substances also increase capillary permeability, which induces tissue edema, anemia, and rapid decrease of circulating plasma volume. Snakebites by Crotalidae and Viperidae produce intense pain, which is probably due to the kinins and biogenic amines released.

Bradykinin is a strong hypotensive agent because of its vasodilating action and is a potent substance for increasing capillary permeability. Snake venoms contain enzymes that release bradykinin from its plasma globulin precursor bradykininogen.

```
kininogen
    |
    |  bradykinin-releasing enzyme
    ↓
Arg-Pro-Pro-Gly-Phe-Ser-Pro-Phe-Arg (bradykinin)
```

Lysyl-bradykinin (kallidin), which has an additional lysine residue at the *N*-terminal atoms can also be released. This kinin-releasing activity is present in all venoms of Crotalidae and Viperidae.

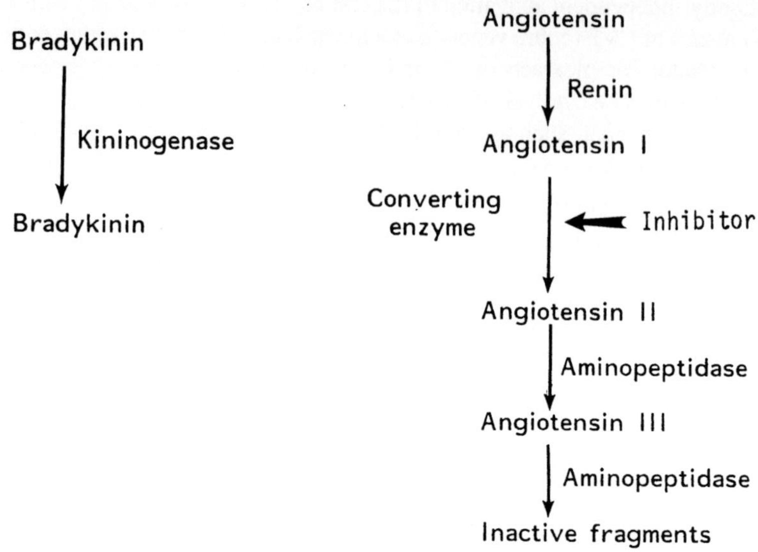

Figure 9. The site of angiotensin converting enzyme inhibitor.

3.4. Angiotensin-Converting Enzyme Inhibitor (ACEI)

The inhibitor is also called bradykinin-potentiating factor because it enhances the hypotensive activity of bradykinin (Fig. 9).

Peptides structurally similar to ACEI have also been isolated from snake venoms and have been shown to have inhibitory action on angiotensin-converting enzyme. Converting enzyme liberates His-Leu from angiotensin I to form angiotensin II, which is a powerful hypertensive activity.

3.5. Effect on Blood Coagulation

Snake venoms exert profound effects on the blood coagulation system; some accelerate the process itself and others retard it. Frequently, snake venoms are divided into two types, coagulant (procoagulant) and anticoagulant, but this is an oversimplification. It is not unusual to find one venom containing both coagulant and anticoagulant factors, and sometimes a venom exerts coagulant or anticoagulant effects depending on the concentration used.

When fibrinolytic activity is stronger than coagulant activity, a fibrin clot is not observed, since true fibrin is never actually produced. When coagulant action is stronger than fibrinolytic action, a fibrin clot may actually form. The relative proportions of anticoagulant and coagulant fractions in a venom determine whether a venom is more anticoagulant or more procoagulant. Venoms of Elapidae such as the cobras and mambas are usually anticoagulant, whereas those of Crotalidae and Viperidae are more complex and often contain both coagulant and anticoagulant factors, which can be separated by fractionation.

Malayan pit viper (*Agkistrodon rhodostoma*) is a good example of so-called coagulant venom, since it converts fibrinogen to a form of fibrin. However, the clot formed is different from an ordinary fibrin clot and actually consists of microclots, which are either removed from the plasma and rapidly lysed by the subsequent extensive fibrinolytic reaction of tissues or removed by the reticuloendothelial system. The result is a prolonged defibrinogenated state during which blood is incoagulable. In this respect, the venom can be described as anticoagulant. This is why one simply cannot state with precision that the Malayan pit viper venom is coagulant or anticoagulant. Both statements are correct, yet neither describes the true actions of the venom accurately.

A. Thrombin-Like Enzymes. Because the high "coagulant" activity of Malayan pit viper venom, the thrombin-like enzyme arvin (ancrod) was isolated from this venom. Arvin has been extensively studied because it is used as a therapeutic drug for diseases such as thrombosis or myocardial infarction, which involve fibrinogen.

Although arvin is frequently called a thrombin-like enzyme, it is not identical to thrombin. The notable difference is that arvin does not give peptide B as a product when fibrin is formed. Fibrinogen consists of a subunit that contains three polypeptide chains, Aα, Bβ, and γ. When thrombin hydrolyzes the *N*-terminal arginyl-glycine bonds of the Aα and Bβ chains, fibrinopeptides A and B are released. The remaining portion of fibrinogen polymerizes to form fibrin, which is a blood clot.

$$\text{fibrinogen} \xrightarrow{\text{thrombin}} \text{fibrin} + \text{fibrinopeptide A} + \text{fibrinopeptide B}$$

$$\text{fibrinogen} \xrightarrow{\text{arvin}} \text{fibrin} + \text{fibrinopeptide A}$$

Fibrin clots formed with thrombin differ in their chemical composition and physical properties from those formed by arvin. Instead of a highly polymerized cross-linked clot of normal fibrin, the fibrin produced with arvin is composed of short polymers that are rapidly dispersed. However, defibrinogenation is a one-shot effect, not a continuing process as long as the liver is functioning properly. The normal half-life of fibrinogen is three days; thus it takes the liver three days to restore the fibrinogen level to 50% of normal after the defibrinogen effect has ceased.

Thrombin-like enzymes are relatively common, are isolated from many other pit viper and viper venoms, and are called by different names such as reptilase, batroxobin, crotalase, or are simply called "thrombin-like enzyme." Unlike thrombin, these "thrombin-like enzymes" affect factor XIII in a variable way; ancrod activates the factor very slowly, whereas reptilase does so more quickly.

Some snake venoms, especially from Crotalidae and Viperidae, produce fibrin clots very similar to those produced by thrombin. Such clot formation induces thrombosis in any part of the body. Vasospasm and microcirculation obstruction by fibrin clots contribute to decreased renal blood flow and subsequent decreased urine outflow. These effects are generally transient, since the clots are lysed rapidly and metabolized *in vivo*.

B. Anticoagulation Enzymes. Several anticoagulant factors were isolated from snake venoms. They have fibrinolytic activities. So far there is no enzyme that hydrolyzes only fibrinogen or fibrin. All venom proteolytic enzymes hydrolyze both fibrinogen and fibrin. Unlike thrombin, snake venom fibrinogenolytic and fibrinolytic enzymes are non-specific proteases hydrolyzing both substrates. Therefore, it is not surprising that they do not produce a fibrin clot from fibrinogen.

3.6. Effect on Platelets

Snake venoms have a diverse effect on platelets. Some promote platelet aggregation and some inhibit the aggregation. Certain crotalid venoms contain both factors. By fractionation of crude venom, one can separate these two factors (Ouyang et al., 1979). The venoms that have the platelet aggregation factor can cause an *in vivo* decrease in platelet count on envenomation.

4. ENZYMES

Most venoms contain a variety of enzymes and most are hydrolytic, with the notable exception of 1-amino acid oxidase, which causes the oxidative deamination of amino acid. The variety of enzymes present in snake venoms is summarized here.

4.1. Hydrolytic Enzymes

There are many hydrolytic enzymes in snake venoms. Among them phospholipase A_2 is commonly present and is one of the enzymes most extensively studied.

A. Phospholipase A_2 (Phospholipase A, Lecithinase). Its specificity is directed toward "the site of fatty acids" rather than "the type of fatty acids." Some venoms contain multiforms of phospholipase A_2 (isozyme) that have different molecular weights (11,000-15,000 d), isoelectric points, and immunological properties.

Phospholipase A_2 has diverse biological effects; it disrupts the electron transport chain and the integrity of the mitochonrial structure. It increases the membrane permeability of the nerve axon and breaks down brain synaptic vesicles. Lysolecithin formed by the action of venom phospholipase A_2 is actually responsible for the disruption of red cell membranes, hence hemolysis.

The enzyme itself will cause mild myonecrosis, and the necrotic activity can be greatly enhanced by the addition of phosphatidylcholine. Apparently phospholipase A_2 itself is an indirect agent, producing lysophosphatidylcholine, which is the direct agent. Phospholipase A_2 itself can be separated from the main toxic fraction; however, many presynaptic toxins such as β-bungarotoxin and notexin have weak phospholipase A_2 activity.

B. Phosphodiesterase. Snake venoms commonly contain enzymes that hydrolyze phosphodiester bonds. There are two types of phosphodiesterase; one is an exonuclease and the second is an endonuclease.

Exonuclease. Exonuclease removes successive mononucleotide units from the polynucleotide chain in stepwise fashion, starting from the 3-end.

Snake venom exonucleases can hydrolyze almost any type of polynucleotide of any chain length. The type of bases, linkages, and sugars do not have much effect on the rate of hydrolysis. Thus, the enzyme hydrolyzes RNA, DNA, synthetic polynucleotides, and native or denatured DNA.

Frequently, the presence of ATPase in snake venoms is described. However, it is clear that the hydrolysis of ATP is due to the action of an exonuclease rather than the separate enzyme, ATPase.

The ATPase of most biological systems hydrolyzes the α-phosphate producing ADP + P_i. When ATP is used as a substrate, snake venom exonuclease hydrolyzes it to AMP and PP (pyrophosphate). However, when ATP is mixed with snake venom, it is hydrolyzed to adenosine, P_i and PP because snake venom contains an additional enzyme, 5'-nucleotidase. This can be summarized as follows:

$$\text{ATP} \xrightarrow{\text{exonuclease}} \text{PP}+ \text{5'-AMP}$$

$$\text{5'-AMP} \xrightarrow{\text{5'-nucleotidease}} \text{adenosine} + \text{P}_i$$

Endonuclease. Snake venoms do possess endopolynucleotidase activities; thus they can hydrolyze RNA and DNA, producing oligonucleotide fragments. The important question is whether snake venom contains specific RNase or DNase like pancrease, or whether the hydrolysis is due to a nonspecific endonuclease that has not been isolated in pure form.

C. Phosphomonoesterase. Snake venoms contain nonspecific as well as specific phosphomonoesterases. Frequently phosphomonoesterase is referred to as a phosphatase, depending on its optimum pH, by which the enzyme is designated as an acid phosphatase or an alkaline phosphatase.

Nonspecific Phosphomonoesterase. The presence of phosphomonoesterase in snake venom is common; this enzyme hydrolyzes a variety of compounds containing phosphomonoester bond. Some venoms contain both acid and alkaline phosphatase, whereas others contain only one type.

Phosphomonoesterase can be separated from the main toxic fractions. Because of the lack of any important biological effect caused by this enzyme in venom action, very few investigators have attempted to isolate it.

Specific Phosphomonoesterase. A specific phosphomonoesterase that can be found commonly in snake venoms is 5'-nucleotidase, which hydrolyzes a variety of 5'-nucleotide phosphates. Thus the enzyme does not hydrolyze 2'- or 3'-phosphates of a guanosine, adenosine, cytidine, uridine, or adenosine 2',3'-cyclic phosphates. Nucleoside diphosphates of triphosphates are also not hydrolyzed by 5'-nucleotidase.

Toxicity of 5'-nucleotidase isolated from *Bungarus fasciatus* venom has an LD_{50} value of greater than 50 μg/g in mice. Neurotoxin isolated from the same venom has a low LD_{50} value, 0.04 μg/g. Thus, the enzyme does not participate in the major lethal action of snake venoms.

D. Acetylcholinesterase. Acetylcholinesterase is commonly present in the venoms of Elapidae and Hydrophiidae but not in those of Viperidae and Crotalidae. Since acetylcholinesterase is involved in nerve transmission, it was thought at one time that venom acetylcholinesterase was responsible for the neurotoxic action. However, it has now been proven that the enzyme can be separated from the main toxic factor.

E. Proteolytic Enzymes. Certain snake venoms are rich in proteolytic enzymes, especially those of Crotalidae and Viperidae, which have strong endopeptidases. Elapidae venoms usually do not contain endopeptidase but are rich in di- and tripeptidases. Some snake venoms contain very specific proteases, cleaving only specific peptide bonds.

Endopeptidases. At one time, venom endopeptidase was considered to be identical to pancreatic trypsin. A number of venom proteases were isolated and their specificities were investigated. Unlike trypsin, the sites of hydrolysis of venom proteases are more varied and cannot be generalized. Some venom proteases are metalloproteins that contain calcium and/or zinc ions.

Specific Proteases. There are many varieties of special venom proteases. For instance, some venoms contain the bradykinin-releasing factor (kininogenase), which hydrolyzes two specific peptide bonds in bradykininogen to release bradykinin. Snake venoms are also known to have a pronounced effect on the blood coagulation system. The factors affecting blood coagulation are also proteases. The special proteases involved in blood coagulation and bradykinin release were also discussed previously. Venom collagenase hydrolyzes collagen but not other proteins; it is believed that venom collagenolytic enzyme is a true collagenase. Collagenase activity is strongest in the venom of Crotalidae and less potent in the venom of Viperidae. Most Elapidae venoms contain no or minimal collagenase activity.

Elastin is a yellow scleroprotein present in elastic tissues. Venom elastase hydrolyzes elastic fibers of the rat aorta and the synthetic substrate for elastase. The enzyme has not yet been isolated in pure form.

F. Arginine Esterase and Other Esterases. Arginine esters are convenient substrates for trypsin assay because the enzyme can be measured spectrophotometrically. Like trypsin, some snake venoms hydrolyze this substrate. Therefore, it was thought at one time that snake venoms contained trypsin or trypsin-like enzymes; however, venom proteases and trypsin have different sites of proteolysis. The protease and arginine ester hydrolase were eventually separated into different fractions. Arginine esterase activities are usually associated with very specific proteases such as bradykinin-releasing enzyme and with blood coagulation activities. Not all snake venoms hydrolyze arginine esters. Usually arginine ester hydrolases are present in the venoms of Crotalidae and Viperidae, but they are not found in the venoms of Elapidae and Hydrophiidae. Thus, the enzyme distribution in snake venoms has taxonomic significance at the family level.

G. Hyaluronidase. Hyaluronidase is commonly present in venoms of Elapidae, Viperidae, and Crotalidae. Hyaluronic acid, a mucopolysaccharide present in skin, connective tissue, and bone joints, serves to promote intercellular adhesion or acts as a lubricant. Chemically, hyaluronic acid consists of repeating units of the type $(\text{-N-G})_n$ where N is *N*-acetyl-*d*-glucosamine and G is d-glucuronic acid. The glycosidic linkages N to G and G to N are β-1,4- and β-1,3-respectively. Snake venom hyaluronidase hydrolyzes the glucosaminidic bond between C_1 of the glucosamide unit and C_4 of glucuronic acid.

The enzyme is frequently referred to as the "spreading factor" because hydrolysis of hyaluronic acid facilitates toxin diffusion into the tissues of the victim. Hyaluronidase itself can be separated from the toxic fraction, hence is not a main factor. Despite the importance of this enzyme in snake venom action, it has not been isolated in pure form.

H. NAD Nucleosidase. NAD nucleosidase, or sometimes simply called NADase, is present in some snake venoms. It hydrolyzes the nicotinamide *N*-ribosidic linkage of NAD. The products of this reaction are nicotinamide and adenosine diphosphate ribose.

The enzyme NADase should not be confused with nucleotide pyrophosphatase. The products of nucleotide pyrophosphatase are nicotinamide mononucleotide and 5'-AMP. In snake venoms, there is no separate nucleotide pyrosphosphatase, but the action is due to snake venom exonuclease (phosphodiesterase).

4.2. Nonhydrolytic Enzymes

Most venom enzymes are hydrolytic; 1-amino acid oxidase is a notable exception.

A. L-Amino Acid Oxidase. Snake venom amino acid oxidase is different from microorganism (bacteria and fungi) enzymes in that the former acts of levorotary-amino acids, and the latter work on dextrorotary-amino acids. The enzyme converts free amino acids into α-keto acids.

The enzyme contains FAD as a prosthetic group and there are many isozymes with different isoelectric points. 1-Amino acid oxidase is largely responsible for the yellow color of snake venoms and does not cause any major toxic actions.

5. ENZYME INHIBITORS

A number of enzyme inhibitors have been isolated from snake venoms. Specific enzymatic activity for certain venoms is found to be low because of the presence of corresponding enzyme inhibitors.

Those for phospholipase A_2, acetylcholinesterase, angiotensin-converting enzyme, and proteinase have thus far been reported and the effects attributed to specific protein-protein interactions.

Angiotensin-converting enzyme inhibitor is identical to the bradykinin-potentiating factor. The kidney is involved in homeostatic control of arterial blood pressure by releasing renin which produces angiotensin I. Angiotensin I is converted into angiotensin II by angiotensin-converting enzyme. The inhibitor acts on the enzyme; thus the formation of angiotensin II is inhibited, resulting in the inactivation of central pressor activities.

6. USE OF SNAKE VENOM COMPONENTS

6.1. Source of Enzymes

A snake venom is a rich source of enzymes. Many enzymes are purified from snake venoms and sold commercially. These include phosphodiesterase, L-amino acid oxidase, 5'-nucleotidase, thrombin-like enzymes (ancrod, atroxin, crotalase, reptilase), and thrombocytin.

6.2. Medical and Scientific Uses

A. Postsynaptic Neurotoxins. Postsynaptic neurotoxins are an indispensable tool in neurochemistry as they are a good antagonist to acetylcholine, a nerve transmitter. By using a neurotoxin, one can determine the number of acetylcholine receptors (AChR) in a given neuromuscular junction. Using this technique, it was found that myasthenia gravis patients have much less AChR than normal persons.

Myasthenia gravis is an autoimmune disease in which the antibody keeps binding to its own AChR and destroys AChR. The consequence is muscle paralysis eventually leading to the patient's death.

B. Cobra Venom Factor (CVF). Because CVF activates an anti-complementary factor C_3 without the presence of antibody, the injection of CVF to a person will activate the anticomplementary system and reduce the factors. This property is extensively used in organ transplantation. A person pretreated with CVF has a better chance of receiving an organ without rejection.

C. Nerve Growth Factor (NGF). NGF stimulates the growth of the sympathetic chain ganglia *in vivo* and a halo-like outgrowth of nerve fibers from embryonic sensory ganglia cultured *in vitro*. Alzheimer's disease is atrophy of nerve cells in the brain. There is considerable attention among medical scientists that NGF may help the regrowth of atrophied brain cells.

D. Thrombin-Like Enzymes. Ancrod (arvin), a very specific snake venom component, has been investigated extensively for its anticoagulant action. Ancrod is a very specific protease whose action has some similarity to that of thrombin and occurs in the terminal sequence of a complex blood coagulation mechanism. Ancrod hydrolyzes only the Aα chain of fibrinogen and produces a polymer of the type $(\alpha B\beta\gamma)_n$ rather than the $(\alpha\beta\gamma)_n$ type normal fibrin clot. The microclot produced by ancrod from fibrinogen is readily hydrolyzed by plasmin that was activated from tissue plasminogen. This results in a defibrination effect. This property is extensively used in the treatment of a patient who has suffered from myocardial infarction.

E. Fibrinolytic Enzymes. Some snake venom proteolytic enzymes can dissolve blood clots, fibrin, without causing hemorrhage. Unlike tissue plasminogen activator (tPA) which liberates plasmin from plasminogen, venom fibrinolytic enzymes hydrolyze fibrin directly. These enzymes have potential therapeutic use in dissolving thrombi. Many such enzymes have been isolated (Retzios and Markland, 1988; Willis and Tu, 1988; Siigur and Siigur, 1991).

REFERENCES

Abo, V., Viera, L., Silveira, R., and Dajas, F. (1989). Effects of local inhibition of locus caeruleus acetylcholinesterase by fasciculin in rats. *Neurosci. Lett.* 98:253-257.

Aird, S. D., Kaiser, I. I., Lewis, R. V., and Kruggel, W. G. (1985). Rattlesnake presynaptic neurotoxins: Primary structure and evolutionary origin of the acidic subunit. *Biochemistry* 24:7054-7058.

Aird, S. D., Kaiser, I. I., Lewis, R. V., and Kruggel, W. G. (1986). A complete amino acid sequence for the basic subunit of crotoxin. *Arch. Biochem. Biophys.* 249:296-300.

Aird, S. D., Middaugh, C. R., and Kaiser, I. I. (1989b). Spectroscopic characterization of textilotoxin, a presynaptic neurotoxin from the venom of the Australian eastern brown snake (*Pseudonaja t. textilis*). *Biochim. Biophys. Acta* 997:219-223.

Aird, S. D., Steadman, B. L., Middaugh, C. R., and Kaiser, I. I. (1989a). Comparative spectroscopic studies of four crotoxin homologies and their subunits. *Biochim. Biophys. Acta* 997:211-218.

Aird, S. D., Yates, J. R. III, Martino, P. A., Shabanowitz, J., Hunt, D. F., and Kaiser, I. I. (1990). The amino acid sequence of the acidic subunit B-chain of crotoxin. *Biochim. Biophys. Acta* 1040:217-224.

Alvarez, J., and Garcia-Sancho, J. (1989). Inhibition of red cell Ca^{2+}-dependent K^+ channels by snake venoms. *Biochim. Biophys. Acta* 980:134-138.

Anderson, A. J., and Harvey, A. L. (1988). Effects of the potassium channel blocking dendrotoxins on acetylcholine release and motor nerve terminal activity. *Br. J. Pharmacol.* 93:215-221.

Anderson, A. J., Harvey, A. L., and Mbugua, P. M. (1985). Effects of fasciculin 2, an anticholinesterase polypeptide from green mamba venom, on neuromuscular transmission in mouse diaphragm preparations. *Neurosci. Lett.* 54:123-128.

Benishin, C. G., Sorensen, R. G., Brown, W. E., Krueger, B. K., and Blaustein, M. P. (1988). Four polypeptide components of green mamba venom selectively block certain potassium channels in rat brain synaptosomes. *Mol. Pharmacol.* 34:152-159.

Benoit, E., and Dubois, J. M. (1986). Toxin I from the snake *Dendroaspis polylepis polylepis*: A highly specific blocker of one type of potassium channel in myelinated nerve fiber. *Brain Res.* 377:374-377.

Betzel, C., Lange, G., Pal, G.-P., Wilson, K. S., Maelicke, A., and Saenger, W. (1991). The refined crystal structure of α-cobratoxin from *Naja naja siamensis* at 2.4-Å resolution. *J. Biol. Chem.* 266:21530-21536.

Bhaskaran, R., Huang, C. C., Chang, D. K., and Yu, C. (1994). *J. Mol. Biol.* 235:1291-1306.

Bieber, A. L., Tu, T., and Tu, A. T. (1975). Studies of an acidic cardiotoxin isolated from the venom of Mojave rattlesnake (*Crotalus scutulatus*). *Biochim. Biophys. Acta* 400:178-188.

Bjarnason, J. B., and Tu, A. T. (1978). Hemorrhagic toxins from western diamondback rattlesnake (*Crotalus atrox*) venoms: Isolation and characterization of five toxins and the role of zinc in hemorrhagic toxin *e*. *Biochemistry* 17:3395-3404.

Black, A. R., Breeze, A. L., Othman, I. B., and Dolly, J. O. (1986). Involvement of neuronal acceptors for dendrotoxin in its convulsive action in rat brain. *Biochem. J.* 237:397-404.

Black, A. R., Donegan, C. M., Denny, B. J., and Dolly, J. O. (1988). Solubilization and physical characterization of acceptors for dendrotoxin and β-bungarotoxin from synaptic membranes of rat brain. *Biochemistry* 27:6814-6820.

Bolioli, B., Castello, M. E., Jerusalinsky, D., Rubinstein, M., Medina, J., and Dajas, F. (1989). Acetylcholinesterase inhibition by fasciculin in rats. *Brain Res.* 504:1-6.

Bon, C., Changeux, J. P., Jeng, T. W., and Fraenkel-Conrat, H. (1979). Postsynaptic effects of crotoxin and of its isolated subunits. *Eur. J. Biochem.* 99:471-481.

Busch, A. E., Kavanaugh, M. P., Osborne, P. B., North, R. A., and Adelman, J. P. (1991). Identification of amino acid residues involved in dendrotoxin block of rat voltage-dependent potassium channels. *Mol. Pharmacol.* 40:572-576.

Cate, R. L., and Bieber, A. L. (1978). Purification and characterization of Mojave (*Crotalus scutulatus scutulatus*) toxin and its subunits. *Arch. Biochem. Biophys.* 189:397-408.

Cervenansky, C., Dajas, F., Harvey, A. L., and Karlsson, E. (1991). Fasciculins, anticholinesterase toxins from mamba vneoms: Biochemistry and pharmacology. In *Snake Venoms* (A. L. Harvey, ed.), Pergamon Press, New York, pp. 303-321.

Chang, C. C., Kawata, Y., Sakiyama, F., and Hayashi, K. (1990). The role of an invariant tryptophan residue in α-bungarotoxin and cobrotoxin. Investigation of active derivatives with the invariant tryptophan replaced by kynurenine. *Eur. J. Biochem.* 193:567-572.

Chang, L. S., and Yang, C. C. (1988). Role of the N-terminal region of the A chain in β_1-bungarotoxin from the venom of *Bungarus multicinctus* (Taiwan-banded krait). *J. Protein Chem.* 7:713-726.

Chwetzoff, S., Mollier, P., Bouet, F., Rowan, E. G., Harvey, A. L., and Ménez, A. (1990). On the purification of notexin. Isolation of a single amino acid variant from the venom of *Notechis scutatus scutatus*. *FEBS Lett.* 261:226-230.

Dajas, F., Bolioli, B., Castello, M. E., and Silveira, R. (1987). Rat striatal acetylcholinesterase inhibition by fasciculin (a polypeptide from green mamba snake venom). *Neurosci. Lett.* 77:87-91.

Danse, J.-M., and Garnier, J.-M. (1990). cDNA deduced aminoacid sequences of two novel kappa-neurotoxins from *Bungarus multicinctus*. *Nucleic Acids Res.* 18:1050.

Ducancel, F., Boulain, J.-C., Trémeau, O., and Ménez, A. (1989). Direct expression of *E. coli* of a functionally active protein A - snake toxin fusion protein. *Protein Eng.* 3:139-143.

Fabiano, R. J., and Tu, A. T. (1981). Purification and biochemical study of viriditoxin, tissue damaging toxin, from prairie rattlesnake venom. *Biochemistry* 20:21-27.

Faure, G., Guillaume, J. L., Camoin, L., Saliou, B., and Bon, C. (1991). Multiplicity of acidic subunit isoforms of crotoxin, the phospholipase A_2 neurotoxin from *Crotalus durissus terrificus* venom, results from post translational modifications. *Biochemistry* 30:8074-8083.

Gourdou, I., Mabrouk, K., Harkiss, G., Marchot, P., Watt, N., Hery, F., and Vigne, R. (1990). Neurotoxicity in mice due to cysteine-rich parts of visna virus and HIV-1 *tat* proteins. *C. R. Acad. Sci. III* 311:149-155.

Harvey, A. L., and Karlsson, E. (1980). Dendrotoxin from the venom of the green mamba, *Dendroaspis angusticeps*. A neurotoxin that enhances acetylcholine release of neuromuscular junctions. *Naunyn Schmiedeberg's Arch. Pharmacol.* 312:1-6.

Hawgood, B., and Bon, C. (1991). Snake venom presynaptic toxins. *Handbook Natural Toxins* 5:3-52.

Hendon, R. A., and Tu, A. T. (1979). The role of crotoxin subunits in tropical rattlesnake neurotoxic action. *Biochim. Biophys. Acta* 578:243-252.

Hite, L. A., Shannon, J. D., Bjarnason, J. B., and Fox, J. W. (1992). Sequence of a cDNA clone encoding the zinc metalloproteinase hemorrhagic toxin *e* from *Crotalus atrox*: Evidence for signal, zymogen, and disintegrin-like structure. *Biochemistry* 31:6203-6211.

Itoh, N., Tanaka, N., Mihashi, S., and Yamashina, I. (1987). Molecular cloning and sequence analysis of cDNA for batroxobin, a thrombin-like snake venom enzyme. *J. Biol. Chem.* 262:3132-3135.

Jeng, T. W., Hendon, R. A., and Fraenkel-Conrat, H. (1978). Search for the relationship among the hemolytic, phospholipolytic and neurotoxic activities of snake venoms. *Proc. Natl. Acad. Sci. USA* 75:600-604.

Joubert, F. J., and Strydom, D. J. (1978). Snake venoms: The amino acid sequences of trypsin inhibitor E of *Dendroaspis polylepis polylepis* (black mamba) venom. *Eur. J. Biochem.* 87:191-198.

Karlsson, E. D., Mbugua, P., and Rodriquez-Ithurralde, D. (1984). Fasciculins, anticholinesterase toxins from the venom of green mamba *Dendroaspis angusticeps*. *Pharmacol. Ther.* 30:259-276.

Kase, R., Kitagawa, H., Hayashi, K., Tanoue, K., and Inagaki, F. (1989). Neutralizing monoclonal antibody specific for α-bungarotoxin. Preparation and characterization of the antibody, and localization of antigenic region of α-bungarotoxin. *FEBS Lett.* 254:106-110.

Komori, Y., and Sugihara, H. (1988). Biological study of muscle degenerating hemorrhagic factor from the venom of *Vipera aspis aspis* (aspic viper). *Int. J. Biochem.* 20:1417-1423.

Kwong, P. D., Hendrickson, W. A., and Sigler, P. B. (1989). beta- Bungarotoxin. Preparation and characterization of crystals suitable for structural analysis. *J. Biol. Chem.* 264:19349- 19353.

le Du, M. H., Marchot, P., Bougis, P. E., and Fontecilla-Camps, J. C. (1989). Crystals of fasciculin 2 from green mamba snake venom. Preparation and preliminary x-ray analysis. *J. Biol. Chem.* 264:21401-21402.

Lee, C. Y., Tsai, M. C., Tsaur, M.-L., Lin, W.-W., Carlsson, F. H. H., and Joubert, F. J. (1985). Pharmacological study on angusticeps-type toxins from mamba snake venoms. *J. Pharmacol. Exp. Ther.* 233:491-498.

Le Goas, R., La Plante, S. R., Delsac, M.-A., Guittet, E., Robin, M., Charpentier, I., and Lallemand, J.-Y. (1992). α-Cobratoxin: Proton NMR assignments and solution structure. *Biochemistry* 31:4867-4875.

Lennon, B. W., and Kaiser, I. I. (1990). Isolation of a crotoxin- like protein from the venom of a South American rattlesnake (*Crotalus durissus collilineatus*). *Comp. Biochem. Physiol.* 97B:695-699.

Lin, W. W., Lee, C. Y., Carlsson, F. H. H., and Joubert, F. J. (1987). Anticholinesterase activity of angusticeps-type toxins and protease inhibitor homologues from mamba venoms. *Acta Pacific J. Pharmacol.* 2:79-85.

Liu, C. S., Hsiao, P. W., Chang, C.-S., Tzeng, M. C., and Lo, T. B. (1989a). Unusual amino acid sequence of fasciatoxin, a weak reversibly acting neurotoxin in the venom of the banded krait, *Bungarus fasciatus*. *J. Biochem.* (*Tokyo*) 259:153-158.

Liu, C.-S., Chen, J.-P., Chang, C.-S., and Lo, T.-B. (1989b). Amino acid sequence of a short chain neurotoxin from the venom of banded krait. *J. Biochem.* 105:93-97.

Loring, R. H., Andrews, D., Lane, W., and Zigmond, R. E. (1986). Amino acid sequence of toxin F, a snake venom toxin that blocks neuronal nicotinic receptors. *Brain Res.* 385:30-37.

McDowell, R. S., Dennis, M. S., Louie, A., Shuster, M., Mulkerrin, M. G., and Lazarus, R. A. (1992). Mambin, a potent glyco-protein IIb-IIIa antagonist and platelet aggregation inhibitor structurally related to the short neurotoxins. *Biochemistry* 31:4766-4772.

McLane, K. E., Wu, X., and Conti-Tronconi, B. M. (1990). Identification of a brain acetylcholine receptor α subunit able to bind α-bungarotoxin. *J. Biol. Chem.* 265:9816-9824.

Mebs, D., Narita, K., Iwanaga, S., Samejima, Y., and Lee, C. Y. (1971). Amino acid sequence of α-bungarotoxin from the venom of *Bungarus multicinctus*. *Biochem. Biophys. Res. Commun.* 44:711-716.

Ménez, R., and Ducruix, A. (1990). Preliminary x-ray analysis of crystals of fasciculin 1, a potent acetylcholinesterase inhibitor from green mamba venom. *J. Mol. Biol.* 216:233-234.

Miller, R. A., and Tu, A. T. (1991). Structure-function relationship of lapemis toxin: A synthetic approach. *Arch. Biochem. Biophys.* 291:69-75.

Moczydlowski, E., Lucchesi, K., and Ravindran, A. (1988). An emerging pharmacology of peptide toxins targeted against potassium channels. *J. Membr. Biol.* 105:95-111.

Mori, N., and Tu, A. T. (1988a). Isolation and primary structure of the major toxin from sea snake, *Acalytophis peronii*, venom. *Arch. Biochem. Biophys.* 260:10-17.

Mori, N., and Tu, A. T. (1988b). Amino acid sequence of the minor neurotoxin from *Acalytophis peronii* venom. *Biol. Chem. Hoppe-Seyler* 369:521-526.

Nomoto, H., Nagaki, Y., Ishikawa, M., Shoji, H., and Hayashi, K. (1992). Solid-phase neurotoxin binding assay for nicotinic acetylcholine receptor—changes of the binding ability of the receptor with various treatments. *J. Nat. Toxicol.* 1:33-44.

Nomoto, H., Takahashi, N., Nagaki, Y., Endo, S., Arata, Y., and Hayashi, K. (1986). Carbohydrate structure of acetylcholine receptor from *Torpedo californica* and distribution of oligo-saccharides among the subunits. *Eur. J. Biochem.* 157:233-242.

Ohta, M., Ohta, K., Nishitani, H., and Hayashi, K. (1987). Primary structure of α-bungarotoxin: Six amino acid residues differ from the previously reported sequence. *FEBS Lett.* 222:79-82.

Ougang, C., Teng, C. M., and Huang, T. F. (1987). Characterization of snake venom principles affecting blood coagulation and platelet aggregation. *Asia Pacific J. Pharmacol.* 2:169-179.

Ownby, C. L., Bjarnason, J., and Tu, A. T. (1978). Hemorrhagic toxins from rattlesnake (*Crotalus atrox*) venom: Pathogenesis of hemorrhage induced by three purified toxins. *Am. J. Pathol.* 93:201-218.

Pearson, J. A., Tyler, M. I., Retson, K. V., and Howden, M. E. (1991). Studies on the subunit structure of textilotoxin, a potent presynaptic neurotoxin from the venom of the Australian common brown snake

(*Pseudonaja textilis*). 2. Amino acid sequence and toxicity studies of subunit D. *Biochim. Biophys. Acta* 1077:147-150.

Penner, R., Petersen, M., Pierau, F. K., and Dreyer, F. (1986). Dendrotoxin: A selective blocker of a non-inactivating potassium current in guinea-pig dorsal root ganglion neurons. *Pflugers Arch.* 407:365-369.

Peterson, M., Penner, R., Pierau, F. K., and Dreyer, F. (1986). β-Bungarotoxin inhibits a noninactivating potassium current in guinea pig dorsal root ganglion neurons. *Neurosci. Lett.* 68:141-145.

Puu, G., and Koch, M. (1990). Comparison of kinetic parameters for acetylcholine, soman, ketamine and fasciculin towards acetyl-cholinesterase in liposomes and in solution. *Biochem. Pharmacol.* 40:2209-2214.

Quillfeldt, J., Raskovsky, S., Dalmaz, C., Dias, M., Huang, C., Netto, C. A., Schneider, F., Izquierdo, I., Medina, J. H., and Silveira, R. (1990). Bilateral injection of fasciculin into the amygdala of rats: Effects on two avoidance tasks, acetylcholinesterase activity, and cholinergic muscarinic receptors. *Pharmacol. Biochem. Behav.* 37:439-444.

Radvanyi, F., and Bon, C. (1984). Investigations on the mechanism of action of crotoxin. *J. Physiol.* (*Paris*) 79:327-333.

Retzios, A. D., and Markland, F. S. (1988). A direct-acting fibrinolytic enzyme from the venom of *Agkistrodon contortrix contortrix*: Effects on various components of the human blood coagulation and fibrinolysis system. *Thrombosis Res.* 52:541

Rhem, H. (1989). Enzymatic deglycosylation of the dendrotoxin-binding protein. *FEBS Lett.* 247:28-30.

Rhem, H., and Lazdunski, M. (1988). Purification and subunit structure of a putative K^+-channel protein identified by its binding properties for dendrotoxin I. *Proc. Natl. Acad. Sci. USA* 85:4919-4923.

Rodriguez-Ithurralde, D., Silveira, R., Barbeito, L., and Dajas, F. (1983). Fasciculin, powerful antichoinesterase polypeptide from *Dendroaspis angusticeps* venom. *Neurochem. Int.* 5:267-274.

Rodriguez-Ithurralde, D., Silveira, R., and Dajas, F. (1981). Gel chromatography and anticholinesterase activity of *Dendroaspis angusticeps* venom. *Braz. J. Med. Sci.* 14:394.

Rowan, E. G., and Harvey, A. L. (1988). Potassium channel blocking actions of betabungarotoxin and related toxin on mouse and frog motor nerve terminals. *Br. J. Pharmacol.* 94:839-847.

Scheidler, A., Kaulen, P., Brüning, G., and Erber, J. (1990). Quantitative autographic localization of [^{125}I]α-bungarotoxin binding sites in the honeybee brain. *Brain Res.* 534:332-335.

Schmidt, R. R., Betz, H., and Rehm, H. (1988). Inhibition of β-bungarotoxin binding to brain membranes by mast cell degranulating peptide, toxin I, and ethylene glycol bis(β-aminoethyl ether)-N,N,N',N'-tetraacetic acid. *Biochemistry* 27:963-967.

Sheumack, D. D., Spence, I., Tyler, M. I., and Howden, M. E. H. (1990). The complete amino acid sequence of a postsynaptic neurotoxin isolated from the venom of the Australian death adder snake *Acanthophis antarcticus*. *Comp. Biochem. Physiol.* 95B:45-50.

Siigur, E., and Siigur, J. (1991). Purification and characterization of lebetase, a fibrinolytic enzyme from *Vipera lebetina* (snake) venom. *Biochem. Biophys. Acta* 1074:223-229.

Stansfeld, C. E., Marsh, S. J., Halliwell, J. V., and Brown, D. A. (1986). 4-Aminopyridine and dendrotoxin induce repetitive firing in rat visceral sensory neurons by blocking slowly inactivating outward current. *Neurosci. Lett.* 64:299-304.

Sunthornandh, P., Matangkasombut, P., and Ratanabanangkoon, K. (1992). Preparation, characterization and immunogenicity of various polymers and conjugates of elapid postsynaptic neurotoxins. *Mol. Immunol.* 29:501-510.

Trivedi, A., Kaiser, I. I., Tanaka, M., and Simpson, L. L. (1989). Pharmacologic experiments on the interaction between crotoxin and the mammalian neuromuscular junction. *J. Pharmacol. Exp. Ther.* 251:490-496.

Tsai, I. H., Liu, H. C., and Chang, T. (1987). Toxicity domain in presynaptic toxic phospholipase A_2 of snake venom. *Biochim. Biophys. Acta* 916:94-99.

Tu, A. T. (1990). Neurotoxins from sea snake and other vertebrate venoms. In *Marine Toxins: Origin, Structure, and Molecular Pharmacology* (S. Hall and G. Strinchartz, eds.). *ACS Symp. Ser.* 418:336-346.

Tu, A. T., and Hong, B. S. (1971). Purification and chemical studies of a toxin from the venom of *Lapemis hardwickii* (Hardwick's sea snake). *J. Biol. Chem.* 246:2772-2779.

Tu, A. T., Hong, B. S., and Solie, T. N. (1971). Characterization and chemical modifications of toxins isolated from the venoms of the sea snake *Laticauda semifasciata*, from Philippines. *Biochemistry* 10:1295-1304.

Tu, A. T., Jo, B. H., and Yu, N. (1976). Laser Raman spectroscopy of venom neurotoxins: Conformation. *Int. J. Peptide Res.* 8:337-343.

Tu, A. T., and Toom, P. M. (1971). Isolation and characterization of the toxic components of *Enhydrina schistosa* (common sea snake) venom. *J. Biol. Chem.* 246:1012-1016.

Tyler, M. I., Barnett, D., Nicholson, P., Spence, I., and Howden, M. E. H. (1987). Studies on the subunit structure of textilotoxin, a potent neurotoxin from the venom of the Australian common brown snake (*Pseudonaja textilis*). *Biochim. Biophys. Acta* 915:210-216.

Utaisincharoen, P., Baker, B., and Tu, A. T. (1991). Binding of myotoxin *a* to sarcoplasmic reticulum Ca^{++}-ATPase: A structural study. *Biochemistry* 30:8211-8216.

Valdes, J. J., Thompson, R. G., Wolff, V. L., Menking, D. E., Rael, E.D., and Chambers, J. P. (1989). Inhibition of calcium channel dihydrophyridine receptor binding by purified Mojave toxin. *Neurotoxicol. Teratol.* 11:129-133.

Viljoen, C. C., and Botes, D. P. (1973). The purification and amino acid sequence of toxin F_{vii} from *Dendroaspis angusticeps* venom. *J. Biol. Chem.* 248:4915-4919.

Weller, U., Bernhardt, U., Siemen, D., Dreyer, F., Vogel, W., and Habermann, E. (1985). Electrophysiological and neurobiochemical evidence for the blockade of potassium channel by dendrotoxin. *Naunyn-Schmiedeberg's Arch. Pharmacol.* 330:77-83.

Westerlund, B., Nordlund, P., Uhlin, U., Eaker, D., and Eklund, H. (1992). The three-dimensional structure of notexin, a presynaptic neurotoxic phospholipase A_2 at 2.0 Å resolution. *FEBS Lett.* 301:159-164.

Willis, T. W., and Tu, A. T. (1988). Purification and biochemical characterization of atroxase, a nonhemorrhagic fibrinolytic protease from western diamondback rattlesnake venom. *Biochemistry* 27:4769-4777.

Yu, C., Lee, C.-S., Chuang, L.-C., Shei, Y.-R., and Wang, C. Y. (1990). Two-dimensional NMR studies and secondary structure of cobrotoxin in aqueous solution. *Eur. J. Biochem.* 193:789-799.

Yu, N., Lin, T., and Tu, A. T. (1975). Laser Raman scattering of neurotoxins isolated from the venoms of sea snakes *Lapemis hardwickii schistosa*. *J. Biol. Chem.* 250:1782-1785.

4

CRITICAL ASPECTS OF BACTERIAL PROTEIN TOXINS

Bal Ram Singh

Department of Chemistry
University of Massachusetts Dartmouth
North Dartmouth, MA 02747

INTRODUCTION

Bacterial protein toxins could be classified as the most dangerous group of agents faced by the human population as well as by other groups of biological systems. Consideration of the potency of toxins makes them dreadful. For example, based on the lowest minimal lethal dose consideration, botulinum neurotoxin produced by the anaerobic bacteria *Clostridium botulinum*, tops the list of known toxins of both proteinaceous and non-proteinaceous nature with a mouse lethal dose of 0.3 ng, which on molar basis is 300, 2 million, 5 million and 100 million fold lower than diphtheria, cholera, cobrotoxin, and sodium cyanide, respectively (Middlebrook, 1989). In addition, there are other important factors contributing to the dangerous nature of bacterial toxins. One such factor is the ubiquitous nature of bacteria that makes toxins' access to an unlimited target hosts easily permissible. Second, the invisible nature of bacteria to naked eye leaves only the option of preventive measures rather than avoidance, unlike in cases of most animal and plant toxins. Lastly, the short duplication time of bacteria provides them with ample opportunities to go through genetic evolution to produce bacterial strains adapted for the adversarial environmental conditions. Such trait perhaps also allows them to develop resistance to antibiotics. Short duplication time also helps bacteria to spread extensively and faster to cover a wide area of target population.

The purpose of toxin production by bacteria is a question that is not easy to settle. A dogmatic view most widely accepted in the absence of any evidence otherwise relates to three main features: First, it is a simple case of parasitism that provides bacteria with a forceful tool to invade its host with a degree of likely success. Second, faced with physiological need of every organism for survival, the toxin provides bacteria with a weapon to survive in adverse conditions of either other bacterial competition such as the case of colicins produced by a strain of *E. coli*, or against the odd of being destroyed by immune system of the host as in the case of most toxins for the animals and humans. Finally, these factors of parasitism and survival lead to the goal of proliferation and reproduction, and the ultimate goal of every living being. Despite the acceptance of the above dogmatic view on the possible

Natural Toxins II, Edited by B. R. Singh and A. T. Tu
Plenum Press, New York, 1996

purpose of the production of toxins, it is still an open question. Could toxins be produced by a sheer chance of genetic alteration or metabolic byproduct? Do the toxins play any critical physiological role(s) in the life of bacterial cell itself? Adequate experimental data are not available to answer these questions; in most cases they have not been attempted. It might be the last frontier of the bacterial protein toxin research, partly because the most pressing issues of bacterial protein toxins relate to the harmful effects of such toxins to other organisms, and most recently to the use of the specific potency of many toxins for therapeutic purposes.

Although bacterial toxins are produced by a variety of bacterial strains varying in morphology and physiology, most bacterial toxins can be classified basically in two categories based on their site of action. One, a group of protein toxins with extracellular target of action. This group includes cytolysins that target plasma membrane as the site of their action. Two, a group of toxins that have intracellular target of action, and includes toxins that range from being cytotoxins such as diphtheria to neurotoxins such as botulinum and tetanus.

Among the toxins acting on the plasma membrane are membrane pore forming hemolysins such as *Staphylococcus aureus* α-toxin (Walker et al., 1992), aerolysins produced by certain *Aeromonas* species (Buckley, 1992; Hirono et al., 1992), colicins produced by *E. coli* (Cramer et al., 1990; Pattus et al., 1990), thiol-activated cholesterol-binding toxins such as streptolysin O produced by *Streptococcus pyogenes* (Alouf et al., 1991; Sekiya et al., 1993), and several others that will not be listed in detail in this chapter (interested readers are referred to a detail account by Menestrina et al., 1994). Another group of toxins acting on plasma membrane are *Staphylococcus aureus* δ-lysin and streptolysin S from *Streptococcus pyogenes* (Bernheimer and Rudy, 1986; Freer et al., 1984; Thiaudiere et al., 1991), which act as detergents on the membrane in their mode of toxic action. Additionally, there are several toxins that have enzymatic activities to act on membrane components resulting in the cytolysis. These enzymatic activities include phospholipase A2 of *Vibrio parahemolyticus* (Shinoda et al., 1991), phospholipase C of α-toxin from *Clostridium perfringens* (Mollby, 1978), sphingomyelinase of *Bacillus cereus* (Tomita et al., 1991), phospholipase D of *Bacillus subtilis* (Fehrenbach and Jurgens, 1991), and cholesterol oxidase of *Brevibacters spp.* (Fehrenbach and Jurgens, 1991).

Another group of toxins that are so far known not to act beyond plasma membrane are staphylococcal enterotoxins and heat shock syndrome toxins produced *by Staphylococcus aureus*. These toxins act as superantigens (Marrack and Kapler, 1990), and have binding sites for major histocompatibility complex II on macrophage cell membrane as well as for the T-cell antigen receptor. Upon binding with T-cells, these toxins evoke massive release of cytokines which become harmful in such a large amount.

In case of toxins that have intracellular targets for their toxic action, they must first cross the membrane barrier to reach their targets. These toxins have a common macrostructure that reflects their common mode of action. Toxins with intracellular targets are made of two types of subunits, B and E. The B subunit generally helps the toxin anchor to the cell surface through receptor and membrane lipid groups, and assists in the translocation of the E subunit inside the cell. E subunit generally possesses specific enzymatic activity for modification of an intracellular target. This group of toxins includes some of the most commonly known toxins such as diphtheria, cholera, pertussis, tetanus and botulinum. Even though the bacteria producing these toxins are widely varied, the enzymatic activities of some of these toxins are same. For example, diphtheria, cholera and pertussis, and variety of other toxins have ADP-ribosyltransferase activity (Althaus and Richter, 1987; Ward, 1987) although substrates recognized by these toxins are totally different. Botulinum and tetanus neurotoxins have protease activity and recognize various components of neuronal exocytosis machinery (Montecucco and Schiavo, 1993; 1994).

Irrespective of the type of bacterial toxin, a common feature remains the membrane, either as the site of action or a barrier to cross. Membrane being a non-polar barrier to toxins which are water soluble, interaction between toxins and membranes have remained a very intriguing problem for researchers. Toxins such as staphylococcal enterotoxins perhaps only interact with the surface of the membrane, and such process is physically compatible given the presence of protein receptors and lipid polar head groups. However, for other toxins which either form pores in membranes to cause toxicity (e.g., hemolysins) or form pores to translocate themselves to reach the intracellular target, question remains as to how a water soluble protein integrates itself into a non-polar lipid bilayer. Whereas there are individual traits attributable to individual toxic proteins, and a general feature of these toxins appears to be the existence of transmembrane and amphiphilic structure in the polypeptide chains (Singh and Be, 1992). Additionally, the polypeptide structures of toxins are flexible for conformational adaptation once in contact with the surface of membrane or are sensitive to environmental conditions such as low pH of the endosomes. Nature of conformational changes in toxins include not only secondary and tertiary structural foldings but also quaternary structure in several cases.

TOXINS WITH PLASMA MEMBRANE AS TARGET OF ACTION

As an example of the interaction of a toxin that targets plasma membrane for its toxic action, features of staphylococcal α-toxin are briefly discussed to elaborate the relevant points. Staphylococcal α-toxin is hemolytic toxin that is secreted as a single water soluble 33 kDa protein (Jiang et al., 1991; Walker et al., 1992).

Staphylococcal α-toxin is a 293 amino acid protein that seems to consist of two domains (N- and C-domains) that are separated by a glycine-rich loop (residues 119-143) (Gray and Kahoe, 1984). The loop is surface accessible in the monomeric form but becomes inaccessible in the hexameric form (Tobkes et al., 1985). The toxin is a β-sheet dominated protein with 5 % α-helix, 57 % β-sheets, 12% β-turns and 28 % random coils (Tobkes et al., 1985).

The toxin causes lysis of erythrocytes by damaging their membranes in the following three steps:

1. Binding of the native α-toxin to the cell membrane. Binding to cells is assumed to involve a protein receptor on susceptible cells (Hildebrand et al., 1991) although no such receptor has been isolated and characterized.
2. Conformational changes in the protein is followed by oligomerization of the toxin to form amphiphilic hexameric complex generating a transmembrane channel. The channel activity results in the leakage of ions. Conformational changes are believed to occur upon interaction with the receptor and membrane lipids. Such structural change does not seem to be significant at the secondary structure level (Table 1; Tobkes et al., 1985). It has been suggested that tertiary structural alteration is significantly involved in the interaction of the toxin with membrane (Forti and Menestrina, 1989; Tobkes et al., 1985; Walker et al., 1992). There has not been a direct analysis of the tertiary structure of the toxin to demonstrate the point. However, based on results of differential proteolytic susceptibility (Tobkes et al., 1985; Walker et al., 1992), it is indirectly shown that the tertiary structure of the toxin changes. The structural changes partly turn the protein inside out that results in the formation of hexamer structure-a amphiphilic structure that has its hydrophilic domain lining the channel for the leakage of ions.
3. Osmotic shock resulting from the leakage of ions through the pore formed by the toxin that leads to the lysis of the cell.

Table 1. Secondary structure contents of staphylococcal α-toxin (Tobkes et al., 1985)

Toxin	α-helix(%)	β-sheet(%)	β-turn(%)	random coil(%)
Monomer	5	57	11	27
Hexamer	7	52	16	27
Monomer in deoxycholate	5	55	12	28
Hexamer in deoxycholate	10	62	9	19

While oligomeric structure of toxin is believed to be formed on the membrane surface, it can also be demonstrated *in vitro* under several conditions including in the presence of deoxycholate (Bhakdi et al., 1981). Deoxycholate, however, does not seem to change the back bone secondary structure (Table 1). Deletion mutant studies (Walker et al., 1992) have suggested that removal of 2 to 22 amino acid residues from N-terminus of the toxin results in the loss of lytic activity although it still forms hexameric structure both on red blood cell membrane and in deoxycholate. Mutants with missing 3 or 5 amino acids from the C-terminus fail to form significant amount of the hexameric structure, but retain hemolytic activity, albeit very low. These results in combination with proteolytic accessibility of different mutant compared with the wild type toxin have lead to the following model (Fig. 1) for the molecular mode of staphylococcal α-toxin action. According to this model, staphylococcal α-toxin first binds to the target cells as a monomer, which is transformed into a non-lytic oligomeric intermediate before formation of the lytic pore. While the C-terminal residues are apparently important for the formation of oligomers, N-terminal residues seem to be critical for the formation of the lytic pore of the oligomer.

TOXINS WITH INTRACELLULAR TARGET OF ACTION

Among the toxins with intracellular targets, we will focus mostly on the botulinum neurotoxins although some references will be made to tetanus neurotoxin because of the structural and functional relationships between tetanus and botulinum neurotoxins. While

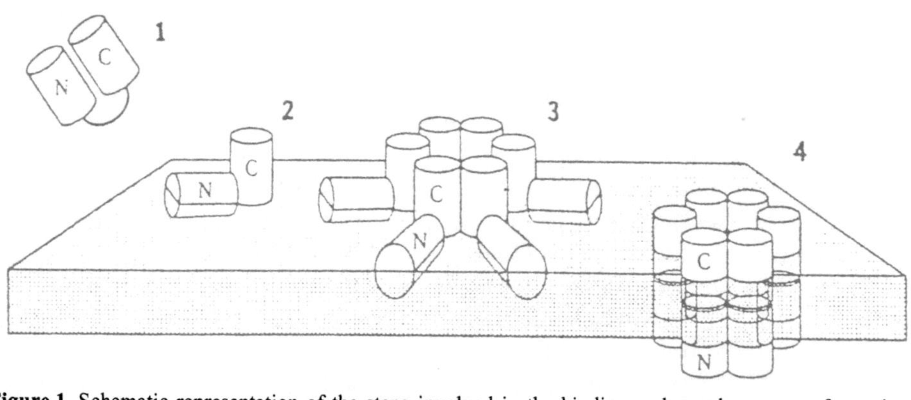

Figure 1. Schematic representation of the steps involved in the binding and membrane pore formation by *Staphylococcus aureus* α-toxin (adapted from Walker et al., 1992). (1) Monomeric form of water soluble α-toxin consists of a N-terminal domain and a C-terminal domain that are separated by a glycine-rich loop. (2) The toxin binds to membrane in its monomeric form. (3) A non-lytic oligomer is formed consisting of up to six subunits. (4) The subunits then penetrate further to form the lytic pore.

greater research progress has been made with other toxins in this group such as diphtheria, cholera, etc., botulinum and tetanus neurotoxins provide a unique set of proteins with respect to their mode of action in terms of their potency, specificity, involvement of metals in their biological function, and the intracellular targets of their action.

Similarities between Botulinum and Tetanus Neurotoxins and Implications

Botulinum and tetanus neurotoxins have several common features including over 30% sequence homology (Binz et al., 1990a; Eisel et al., 1986; Fairweather and Lyness, 1986; Whelan et al., 1992b), and antagonistic effects of heterologous protein fragments (Simpson, 1984). Both the neurotoxins have similar secondary structural contents (Singh et al., 1990), and both have the so called 'catalytic domain' on the C-terminal domains of their respective light chains (Singh, 1990). Tetanus is capable of causing botulism symptoms at high concentrations (Matsuda et al., 1982), and blocks the release of acetylcholine from the presynaptic membranes just like botulinum neurotoxins. Therefore, it is tempting to assume a similar mechanism of action for botulinum and tetanus neurotoxins at the molecular level. Indeed, experimental studies have revealed many similarities between the two neurotoxins including the membrane channel formation by the N-terminal fragments of their respective heavy chains and blockage of the neurotransmitter release from cultured cells (Bittner et al., 1989) and identical cleavage site of type B botulinum and tetanus neurotoxins' proteolytic activity against synaptobrevin-2 (Montecucco and Schiavo, 1993). Thus, the results related to the botulinum neurotoxins are relevant to the tetanus neurotoxin, and vice versa.

Disease and the Organism

Botulism, the deadly food poisoning disease is caused by the growth of various strains of *Clostridium botulinum* in food. The organism produces a large polypeptide (neurotoxin) which is the most toxic protein known to the human kind. Seven serotypes of botulinum neurotoxins produced by different strains of *C. botulinum* have been characterized, and serotypes A, B and E are known to cause botulism in humans. Ingestion of food contaminated with the neurotoxin causes flaccid muscle paralysis that can result in patients' death. Wound botulism has also been reported where the organism can grow in the wounds, and produces the neurotoxin that causes paralysis.

Infant botulism has been observed in babies of 3-35 weeks where ingestion of *C. botulinum* spores leads to the organisms' growth and production of the neurotoxin in the intestine (Miduara and Arnon, 1976; Pickett et al., 1976). Recently, non-botulinum clostridial species, such as *Clostridium butyricum* and *C. barati* have been found to be responsible for human botulism cases (McCrosckey et al., 1988; Arnon, 1992). In addition, botulism has been held responsible for several sudden infant death syndrome (SIDS) cases (Arnon, 1992).

The Neurotoxin

Botulinum neurotoxins (seven serotypes, A-G) are relatively large water soluble proteins (150 kDa) produced by the *Clostridium botulinum*. Each protein has two polypeptide chains (a 100 kDa heavy chain and a 50 kDa light chain) linked through a disulfide bond (Fig. 2). In the proposed mode of action of botulinum and tetanus neurotoxins (Simpson, 1986, 1989), the C-terminal half of the heavy chain binds to the nerve membrane leading to internalization of the neurotoxin in the nerve cell through endocytosis. Subsequently, the pH of the endosome is lowered causing the heavy chain to get integrated in the membrane

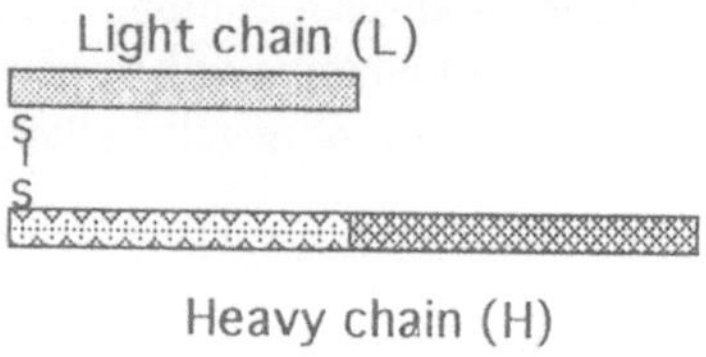

Figure 2. Schematic diagram of botulinum neurotoxin showing its light and heavy chains. The two different domains of the heavy chain shaded with different patterns indicate the N-terminal and C-terminal halves (about 50 kDa each). These two domains are believed to play different functional roles during the intoxication process. The light chain has been shown to contain the toxic site.

through the N-terminal half of the heavy chain for the membrane channel formation. The whole neurotoxin or a part (light chain) of it is translocated through this channel into the cytoplasm of the nerve cell where it blocks the neurotransmitter release.

The primary sequences of types B, C1, D, E, F and G have already been published (Binz et al., 1990a; 1990b; 1990c; Campbell et al., 1993; Hauser et al., 1990; East et al., 1992; Thompson et al., 1990; Whelan et al., 1992a; 1992b). While knowledge of primary structures has been very useful for the comparison of botulinum and tetanus neurotoxins, they had not provided any obvious clues to its mechanism of toxic action until recently. Several reports have appeared (Wright et al., 1992; Schiavo et al., 1992a, 1992b; 1992c; 1993a; 1993b; Fujii et al., 1992) which strongly suggest botulinum and tetanus neurotoxins to be zinc proteases. One significant observation made from the sequence data had been the existence of 3 histidine residues on a short stretch of the light chain that may be involved in the toxic site of the neurotoxin (Binz et al., 1990a). Recent studies have provided evidence that botulinum and tetanus neurotoxins bind to zinc not only in a stoichiometric ratio (Wright et al., 1992; Schiavo et al., 1992c) but that zinc-protease activity is part of the toxic activity of the neurotoxins (Schiavo et al., 1992a). Because of the structural and functional similarities between botulinum and tetanus neurotoxins, results of botulinum or tetanus are commonly used to understand each other's mode of action (Simpson, 1989a).

The neurotoxin appears to have a highly ordered polypeptide folding with predominantly ß sheet structure (approximately 50%) although a significant amount of α-helical structures (20-30%) is present (Table 2; Singh and DasGupta, 1989a; 1989b; 1990). Similar back bone secondary structure has been observed for the tetanus neurotoxin also (Singh et al., 1990). An observation made during the structural analysis of the neurotoxin and its toxoid was that the secondary structural content remained unchanged during the toxoiding (Singh and DasGupta, 1989c). This is interesting because toxoiding with formaldehyde only destroys the toxicity while retaining the immunogenic properties, indicating that the toxic site is perhaps not just made of a sequential site (contiguous stretch of a polypeptide segment), but rather a site that is composed of several polypeptide segments brought together by tertiary structural folding. Accordingly, tertiary structure was found to change considerably upon toxoiding (Singh and DasGupta, 1989c).

Mode of Action

The mode of action of the botulinum neurotoxin is not well understood at the molecular level. Based on some experimental evidence and analogies with other dichain

Table 2. Secondary structure estimation of botulinum and tetanus neurotoxins as analyzed by circular dichroism under physiological pH conditions (Singh et al., 1990; Singh and DasGupta, 1989b)

Neurotoxin	α-helix(%)	β-sheets(%)	β-turns(%)	Random coils(%)
Type A botulinum	21	44	5	30
Tetanus	20	51	0	30

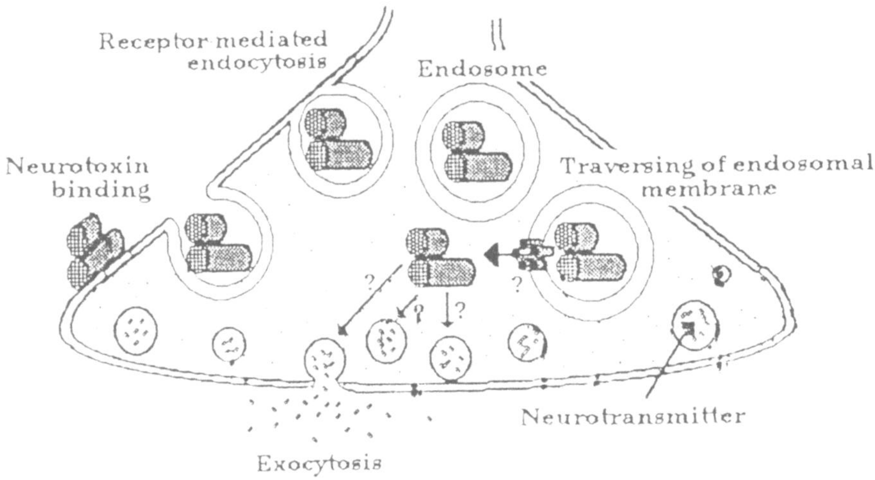

Figure 3. Schematic representation of the steps involved in the binding, internalization and intracellular activity of the botulinum neurotoxins. The steps with question mark are intensive areas of current research.

toxins such as diphtheria, cholera and *Pseudomonas* exotoxin A, a working model has been proposed (Simpson, 1981; 1986). Three major steps (Fig. 3) involved are:

(i) Extracellular Step involves the binding of the neurotoxin to presynaptic membranes through the C-terminal half of the heavy chain. The C-terminal half of the heavy chain binds to the gangliosides on the presynaptic membrane (Simpson, 1981; 1986). According to the double receptor model (Montecucco, 1986), the C-terminal domain of the neurotoxin binds first with the ganglioside, altering the protein structure and making it compatible for binding with a protein receptor. Structural changes in botulinum and tetanus neurotoxins upon binding with gangliosides (or detergent) were experimentally observed (Lazarovici et al., 1987; Singh et al., 1991). These changes could also be involved in exposing transmembrane and surface-seeking peptide domains. Recently, several reports have appeared suggesting existence of different protein receptors for different serotypes of botulinum neurotoxin and of tetanus neurotoxin (Li and Singh, 1995; Nishiki et al., 1993; 1994; Schengrund et al., 1992; Schiavo et al., 1991). However, no confirmed involvement of these putative receptors has been demonstrated in the neurotoxin-mediated toxicity of the neuronal cells.

(ii) Internalization and Translocation. Upon binding, the neurotoxin is internalized through endocytosis (Fig. 3). Inside the cell, the pH of the endosome is lowered to 4-5, which leads to the formation of a membrane channel by the N-terminal half of the heavy chain. This channel helps translocate the whole or a part of the neurotoxin into the cytoplasm.

Botulinum and tetanus neurotoxins, which are extensively soluble in water, are known to form efficient membrane channels at a low pH in artificial membranes (Hoch et al. 1985; Boquet and Duflot, 1982). Membrane channel formation by a water soluble protein is an intriguing phenomenon, because for water solubility, hydrophobic domains are needed on the surface of a protein whereas for membrane channel formation, adequate hydrophobic segments will be required for the interaction with non-polar membrane bilayer. A major question to be answered is how are the polypeptides integrated in the lipid bilayer. Are the hydrophobic segments of these neurotoxins "hidden" in aqueous medium which get exposed at a low pH to interact with membrane? A screening of primary sequences of tetanus neurotoxin revealed one segment (H-650-681) in the heavy chain and one segment in the light chain (L-223-253) with sufficient hydrophobicity and a sequence length capable of spanning the lipid bilayer (Eisel et al., 1986). Similar observations were noted for type A botulinum neurotoxin, except no hydrophobic segment was observed in the light chain (Thompson et al., 1990). Hydrophobic protein segments can only indicate membrane interacting domains. A membrane interacting segment does not necessarily form a "channel" by itself, especially to translocate large proteins such as the neurotoxin or its light chain. Because of this function, it is likely that the channel will probably have an amphiphilic structure to interact with the non-polar lipid bilayer on one surface and with hydrophilic groups of the protein on the other. Membrane channel formed by water soluble proteins is a phenomenon which is not well understood. Examples are diphtheria, cholera, *Pseudomonas* exotoxin A, botulinum and tetanus (Jiang et al., 1989; Jinno et al., 1989; Papini et al., 1987a, 1987b). In order to remain water soluble, these proteins must contain adequate hydrophilic domains on the exterior of the protein. However, for membrane channel formation, hydrophobic segments of proteins are required to interact with the non-polar lipid bilayer.

To explain adequate interaction of botulinum and tetanus neurotoxin with lipid bilayer, we analyzed hydrophobicity and hydrophobic moment characteristics (Be et al., 1994; Doyle and Singh, 1993; Singh and Be, 1992) of both neurotoxins. The results indicate several polypeptide segments with properties that can characterize them as integral membrane segments or as surface-seeking segments (amphiphilic) (Fig. 4), which are compatible for interaction with lipid bilayer. Surface and membrane segments were identified in both light and heavy chains of the neurotoxin, suggesting a likely interaction of both the chains with the lipid bilayer.

A surface-seeking peptide domain has low average hydrophobicity because of the presence of the hydrophilic amino acid residues, but it has relatively high hydrophobic

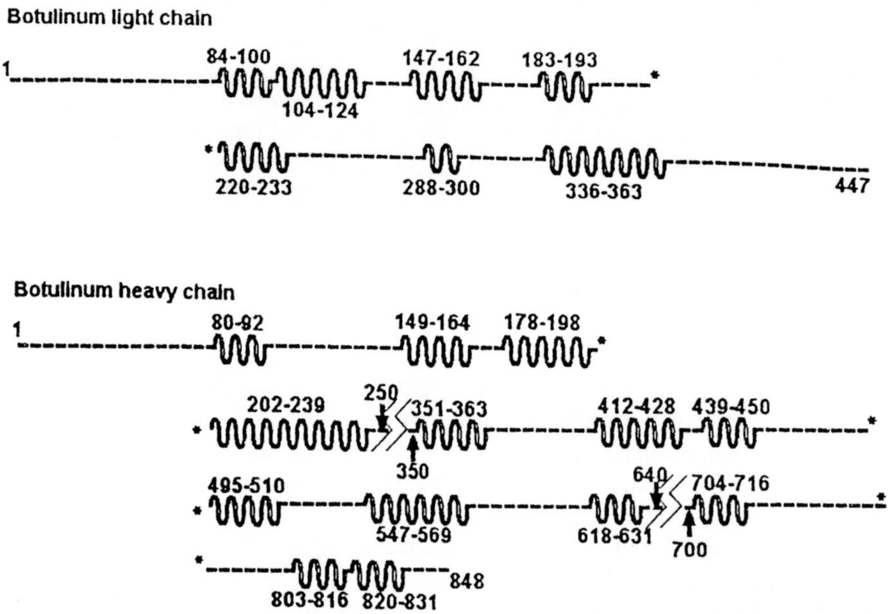

Figure 4. Cartoon drawing of the light and heavy chains of type A botulinum neurotoxin indicating amphiphilic (open helices) and transmembrane (solid helices) peptide segments. The numbers correspond to the amino acid sequence numbers in respective chains (see Fig. 2).

moment, i.e. amphiphilicity due to the topographical location of hydrophobic and hydrophilic residues. Therefore, a surface-seeking domain could interact with hydrophobic groups on one surface and with hydrophilic groups on the other. The presence of transmembrane and surface-seeking domains in the light chains of botulinum and tetanus neurotoxins is puzzling because light chain by itself has not been shown to have any membrane channel activity (Boquet and Duflot, 1982; Hoch et al., 1985), except for a recent report for the release of calcein (Kamata and Kozaki, 1994; F.-N. Fu and B. R. Singh, unpublished results). The role of transmembrane and surface-seeking segments of the light chain could be in (i) the interaction of light chain with the heavy chain; and/or in (ii) the direct interaction of light chain with lipid bilayer during its translocation. Light and heavy chains of type A botulinum neurotoxin have significant interactions with each other (Singh and DasGupta, 1989a). The role of hydrophobic surfaces of light chain in its translocation can be envisaged in the "cleft" model proposed by Montecucco and co-workers (Montecucco et al., 1989). According to this model, the heavy chain protects the hydrophilic groups of the light chain by combining with them and leaving the hydrophobic surface of the light chain to interact with the lipid bilayer. Experimental results of photo-crosslinking techniques have suggested light chains can enter the lipid bilayer at low pH (Montecucco et al., 1989).

Respective light and heavy chains of botulinum and tetanus neurotoxins have substantial similarities in the location and characteristics of membrane compatible peptide segments (Be et al., 1994), which is consistent with their common mode of biological activities. Based on our results, it may be possible to explain membrane channel formation by water soluble proteins such as botulinum or tetanus neurotoxins because they have sufficient numbers of transmembrane or amphiphilic (surface-seeking) polypeptide segments. It should also be pointed out that hydrophobicity calculations did not reveal any transmembrane segment in the botulinum light chain (Thompson et al., 1990). However, hydrophobic moment analysis (Be et al., 1994) resulted in the first observation of potential membrane interacting domains of botulinum light chain. The presence of transmembrane and amphiphilic segments is also consistent with experimental results which suggest direct light chain interaction with lipid bilayers (Montecucco, 1986). Analysis of hydrophobicity alone had revealed at least one segment in tetanus heavy chain (H-203-234) which although may be capable of traversing the membrane bilayer, but it may not be adequate to form a channel for the translocation of the light chain for two reasons: (i) the segment is too small to form one side of the whole channel structure; and (ii) it has strong hydrophobic residues that may not allow its interaction with the light chain for translocation of the latter. A channel activity for the translocation of light chain can be partly rationalized under the above conditions if we assume that the interaction between light and heavy chains are through hydrophobic domains, and that the neurotoxins exist as oligomers (at least a dimer) which has been suggested at least for the botulinum neurotoxin (Shone et al., 1985; Ledoux et al., 1994). In an oligomeric structure two or more heavy chains can be envisaged to form the channel, and their respective light chains can pass through the hydrophobic region of the heavy chain on one side and with the other light chain on the other.

A second explanation of membrane channel formation by botulinum and tetanus neurotoxins is as follows: since hydrophobic moment calculations have helped identify several membrane interacting hydrophobic domains both in light and heavy chains of both neurotoxins, more than one segment of the heavy chains can be allowed to integrate into the membrane bilayer. Because some of these segments are amphiphilic in nature, interaction with hydrophilic domains of light chain is possible during translocation of the latter. This should also allow interaction of the light chain directly with lipid bilayer which will be consistent with the photo-labeling experiments that suggest direct light chain interaction with lipid bilayer (Montecucco et al., 1988).

It is also possible that a combination of both modes suggested above is involved in the formation of membrane channel for the translocation botulinum neurotoxin. The main assumptions of the hypothesis are: (i) The presence of amphiphilic and transmembrane segments in the light and heavy chains as predicted by the hydrophobic moment analysis. (ii) Existence of botulinum neurotoxin in oligomeric form.

Tetanus neurotoxin has been extensively analyzed in the past (Robinson and Hash, 1982) for its molecular size using analytical ultracentrifugation, and it was reported that purified tetanus neurotoxin exists as a monomer (125-145 kDa), although "aggregation" was observed at certain concentrations. Moreover, to avoid "aggregation", Robinson and co-workers (Robinson and Hash, 1982) used their toxin preparations "immediately" after purification on a preparative scale electrophoresis which was carried out in 4 M urea. Urea is likely to facilitate disaggregatation of tetanus neurotoxin into the monomeric form.

Botulinum neurotoxin type A is also reported to form "aggregation" based on native gel electrophoresis band of 450-600 KDa (Shone et al., 1985; Ledoux et al., 1994). Whether or not the "aggregated" form is a natural form of the neurotoxins remains to be confirmed. We have carried out native gel electrophoresis of type A botulinum and tetanus neurotoxins (Ledoux et al., 1994), which indicates that the tetanus neurotoxin exists primarily as a dimer although a small fraction exists as trimer, whereas type A botulinum neurotoxin exists in the form of at least a trimer. Type A botulinum neurotoxin band is spread over a range of 450-600 kDa, but most of it appears as a trimer. This observation is consistent with those of shone et al. (Shone et al., 1985). Crystal forms of type A botulinum neurotoxin reported recently are also believed to exist as a dimer (Stevens et al., 1991). We have further confirmed the existent of oligomeric type A botulinum neurotoxin by chemical cross-linking experiments (Ledoux et al., 1994).

Chemical cross-linking experiments with both BS^3 and glutaraldehyde suggested the presence of oligomeric forms of type A botulinum neurotoxin corresponding to molecular weights of 249 and 366 kDa (Ledoux et al., 1994). Although both of these species do not correspond to exact dimer or trimer molecular weight, these are obviously distinct molecular species. A molecular weight species higher than a trimer was also observed at the top of the gel in the form of sharp bands (Ledoux et al., 1994).

An intriguing observation was made with a careful analysis of the amino acid sequences of the heavy and light chains of tetanus and botulinum neurotoxins (Singh et al., 1994). It suggested possible sites of association between monomeric neurotoxin molecules in the formation of oligomers. Specifically, leucine zipper-like structures have been located in the heavy chain of the tetanus neurotoxin, and in the heavy and light chains of the type A neurotoxin. The ideal leucine zipper possesses a leucine residue every seventh residue of a polypeptide chain having a helical structure. Aberrations from the ideal could include, for example, the substitution of leucine for a different residue, resulting in a leucine zipper-like structure. Possible structures of this type have been identified in the type A neurotoxin heavy and light chains, as well as in the tetanus heavy chain (Fig. 5). The type A botulinum heavy chain exhibits such a structure with leucine and isoleucine residues. The two exceptions (noted in the table) are residues which offset the "scheme" of the zipper structure by one residue. The substitution of an isoleucine residue for a leucine residue does not seem unreasonable, since the relative bulk of the side chains of each residue is comparable, and would allow the "packing" of these leucine-like structures to form a "zipper." An interesting possibility appears in the light chain of the type A neurotoxin, where a leucine zipper-like structure with a hydrophobic residue every seventh residue, with the exception of residue 270 (Aspartate) and 291 (Lysine). When these regions (in separate neurotoxin molecules) are placed in proximity and arranged in a parallel fashion (C-terminus to N-terminus, C-terminus to N-terminus), the positively-charged lysines and negatively-charged aspartate residues will repel. However, if these regions are placed in anti-parallel fashion (C-terminus

Type A (heavy chain)

IPYGVKRLEDFDASLKDALLKYIYDNRGTLIGQVDRLKDKVNNTL

801 808 815 823 830 837 845

Tetanus (heavy chain)

IPYAPEYKSNAASTIEIHNIDDNTIYQYL LANKWVFITITNDRLSSANLYI

536 543 550 557 564 1027 1034 1041 1048

Type A (light chain)

LEVDTNPLLGAGKFATDPAVTLAHELIHAGHRLYGIAINPNRVFKVNTNAYYEMSGL

200 207 214 221 228 235 242 249 256

EVSFEELRTFGGHDAKFIDSLQENEFRLYYYNKFK

263 270 277 284 291

LFEFYKLLCVRGIITSKTDSL

422 428 435 442

Figure 5. Primary amino acid sequence of type A botulinum neurotoxin and tetanus neurotoxin showing possible leucine zipper-like structures. Residues with asterisk(*) mark offset the zipper scheme by one residue.

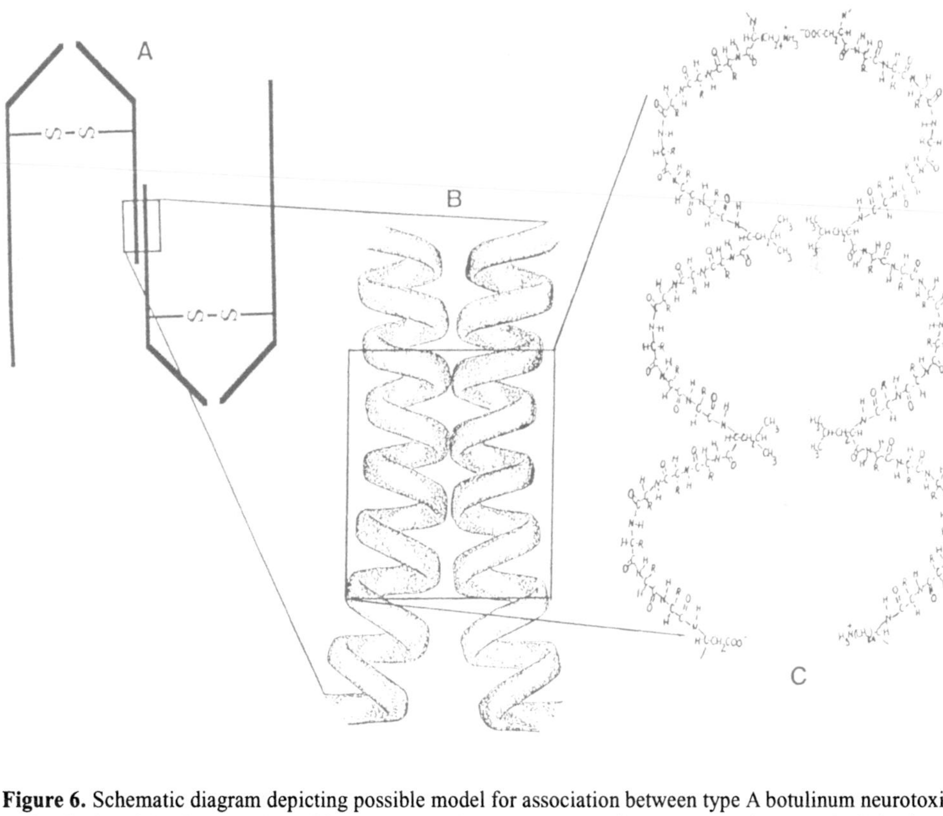

Figure 6. Schematic diagram depicting possible model for association between type A botulinum neurotoxin molecules involving leucine-zipper like structure. (A) Representation of an association between the light chains of the monomeric neurotoxin molecules for a dimer formation. (B) Representation of a helical structure which is assumed for the leucine-like structure. (C) Depiction of amino acid residues which may be in favorable contact. The amino acid sequence corresponds to the residues 270 to 291 on the light chain of type A botulinum neurotoxin (Fig. 5). The two sequences are represented antiparallel to depict favorable ionic contacts.

Table 3. Change in the secondary structures of type A botulinum and tetanus neurotoxins with lowering of pH (Singh et al., 1990; 1991; Singh and DasGupta, 1989b)

pH	α-helix(%)	β-sheets(%)	β-turns(%)	random coils(%)
Botulinum				
7.2	21	44	5	30
5.5	29	45-49	0	22-26
Tetanus				
7.0	22	51	0	27
5.5	25	47	0	28

to N-terminus, N-terminus to C-terminus), the lysine and the aspartate residues could provide a favorable coulombic contact (Fig. 6). In addition to this structure, the light chain of type A also has a significantly long stretch of zipper-like structure with leucine and isoleucine between residues 200-291 (Fig. 5). The heavy chains of the tetanus and botulinum neurotoxins also seem to possess leucine zipper-like structures (Fig. 5). Type A botulinum heavy chain possesses such a structure between heavy chain residues 353 and 397, whereas tetanus heavy chain has them between heavy chain residues 536 and 564, and 1027 and 1048 (Fig. 5). The first possible leucine zipper-like region of tetanus heavy chain (residues 536-564, Fig. 5) possesses lysine and aspartate residues at position 543 and 557, respectively. When the polypeptides are placed antiparallel, the lysine and aspartate residue may come in favorable ionic contact. The tetanus heavy chain also has another sequence of leucine and isoleucine residues (residues 1027-1048, Fig. 5) with the potential to form a leucine zipper structure. These data suggested leucine zipper-like structures are possible sites of association among monomeric forms of the neurotoxin molecules in the formation of oligomeric forms of the neurotoxin. Since more than one of these regions exists per neurotoxin molecule, it is possible that these regions act together in causing an association between neurotoxin molecules.

The significance of the observations presented in this report pertain not only to the function of botulinum and tetanus neurotoxins at the molecular level, but also to many other water-soluble proteins such as colicin and other bacterial toxins (diphtheria, cholera, pertussis, etc.) that are believed to form membrane channels as part of their mode to gain entry into their target cells. In the case of botulinum and tetanus neurotoxins, the neurotoxins are known to have a common mode of action (Simpson, 1989a) which may have a basis in the common observation of the existence of oligomeric forms. There are significant differences between botulinum and tetanus neurotoxins in terms of primary sites and degree of toxicity which, in part, could be the result of differential oligomeric species observed for the neurotoxins.

If botulinum and tetanus neurotoxins exist as dimer or trimer/tetramer in aqueous solutions (Fig. 7), what could be the physiological role of such structures? The observation could be relevant to explain the behavior of botulinum neurotoxin with mouse phrenic hemidiaphragm at different concentrations (Bandyopadhyay, 1987; Maisey et al., 1988). Because both the neurotoxins exist in more than one oligomeric form, it is possible that these oligomeric forms are in equilibrium with each other (Fig. 7), and this equilibrium could be altered in different conditions such as in low pH and upon interaction with membranes. A model of oligomeric channel is shown in Figure 8 assuming a trimeric form of botulinum neurotoxin.

Low pH is apparently required for the strong channel formation activity of botulinum and tetanus neurotoxins and their heavy chains (Boquet and Duflot, 1982; Hoch et al., 1985). A pH of 5.0 or lower induces the channel formation. Two possible effects of a low pH are

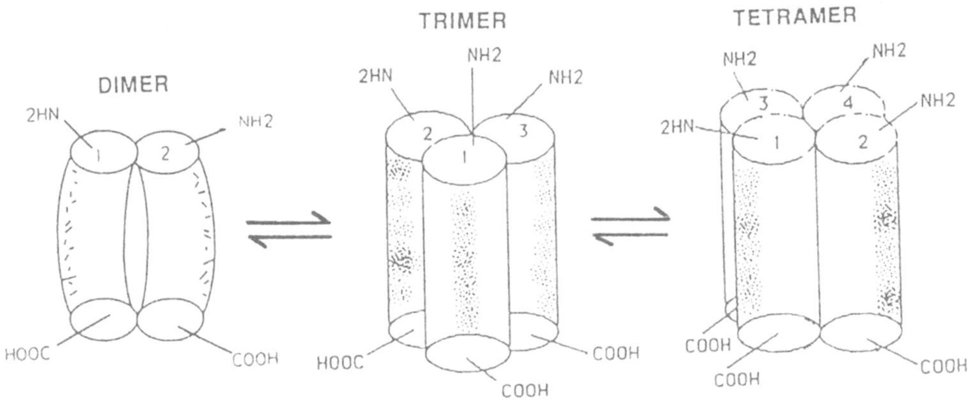

Figure 7. Schematic representation of oligomeric structure of type A botulinum and tetanus neurotoxins. Based on results in Ledoux et al. (1994) and modified after Singh (1993), it is assumed that botulinum neurotoxin exists as trimer and tetramer whereas tetanus neurotoxin exists as dimer and trimer. The arrows indicate possible interconversion between two oligomeric form. The shaded areas indicate the location of amphiphilic/transmembrane region of the monomeric units.

(i) neutralization of negative charges of the protein and/or (ii) conformational changes in the polypeptide folding which allow the integration of the neurotoxin into the membrane bilayer. It has been observed that low pH increases the surface hydrophobicity of tetanus neurotoxin as demonstrated by a labeled Triton-X-100 binding (Duflot and Boquet, 1982). Similar observations were made for other dichain toxins such as *Pseudomonas* exotoxin A (Idziorek et al., 1990). Based on our calculations, the potential membrane interacting domains of the neurotoxins are predicted at pH 7.0 as the hydrophobicity scale used is based on partition

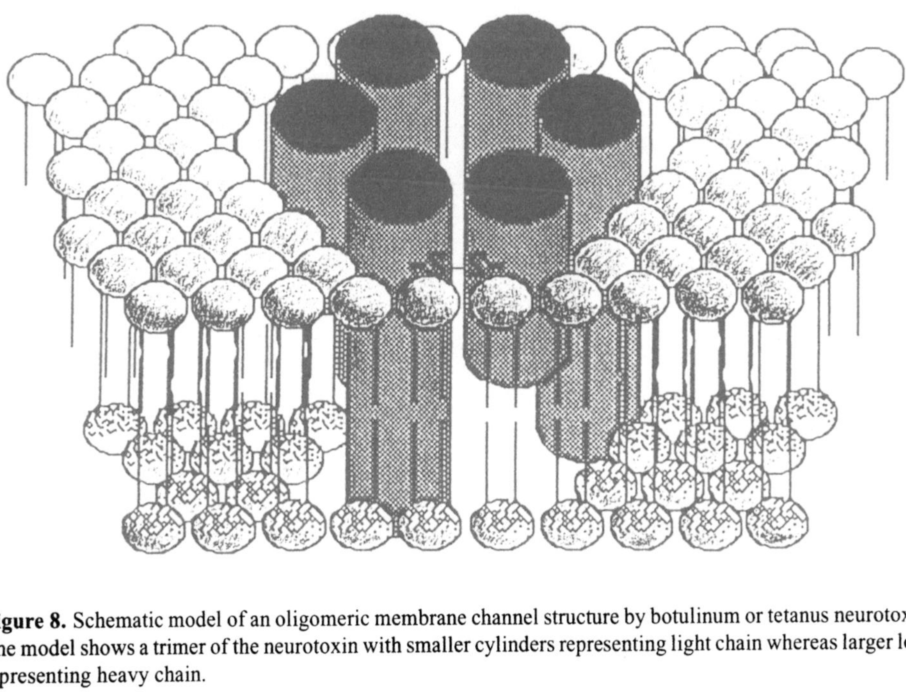

Figure 8. Schematic model of an oligomeric membrane channel structure by botulinum or tetanus neurotoxin. The model shows a trimer of the neurotoxin with smaller cylinders representing light chain whereas larger lobe representing heavy chain.

coefficients of amino acids between a non-polar solvent and water (Fauchere and Pliska, 1983). Therefore, it is likely that topographical changes are introduced by the low pH which could expose the membrane compatible domains of the neurotoxin. FT-IR studies of the tetanus neurotoxin did not suggest a significant gross conformational change with pH (Singh et al., 1990). However, there was a clear indication of conformational changes in certain segments of the polypeptide. Secondary structure of type A botulinum neurotoxin, on the other hand, seems to change significantly upon lowering the pH from 7.2 to 5.5 (Table 3; Singh et al., 1994). This observation will be consistent with the presence of membrane interacting domains even at pH 7.0. The low pH could perhaps refold the protein at the tertiary structural level without altering much of the secondary structure. Trp fluorescence quenching experiments have suggested a significant alteration in the tertiary structure of tetanus neurotoxin at low pH (B. R. Singh, unpublished results). Also, similar indications were observed from the differential scanning calorimetric analysis of type A botulinum at pH 7.0 and pH 4.0 (unpublished results).

(iii) Intracellular Step of the botulinum neurotoxin action is not well understood. It is known that the light chain subunit of the neurotoxin, once inside the nerve cells, is sufficient to block the neurotransmitter release (Bittner et al., 1989; Lomneth et al., 1991; Mochida et al., 1989; Stecher et al., 1989). The neurotoxin does not affect either biosynthesis or packaging of the acetylcholine in the nerve cell (cf. Simpson, 1989b). Recent experimental evidence has suggested that botulinum and tetanus neurotoxins are zinc-proteases, and the substrates for the different serotypes of botulinum neurotoxins and tetanus neurotoxin are the constitutive components of the secretory machinery (see Huttner, 1993). The substrate for the protease activity of the type B botulinum and tetanus neurotoxins has been identified as synaptobrevin-2, a 19 kDa protein present on synaptic vesicles (Schiavo et al., 1992a). It has now been established that synaptobrevein-2 also acts as intracellular substrate for the proteolytic activity of botulinum neurotoxin types D, F and G (Schiavo et al., 1993a; 1993b; Schiavo and Montecucco, 1994) although site of cleavage on the synaptobrevin-2 for each neurotoxin is different. Intracellular target for the proteolytic activity of types A and E has

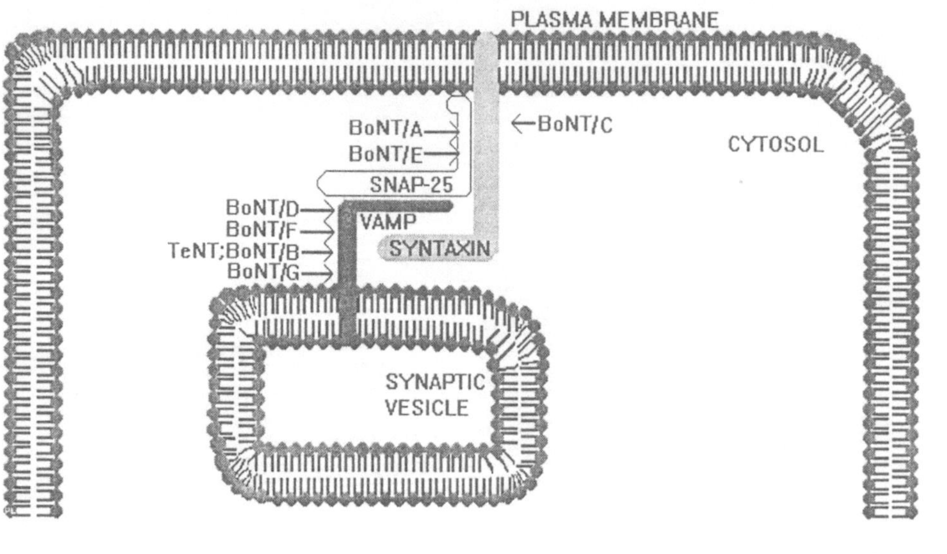

Figure 9. Schematic diagram of synaptic vesicle and presynaptic plasma membrane showing target substrates of the proteolytic activity of botulinum and tetanus neurotoxins.

been as SNAP-25 (a presynaptic plasma membrane protein) (Blasi et al., 1993a; Schiavo et al., 1993a). Again, the site of cleavage for the two neurotoxin is different on SNAP-25 (Schiavo et al., 1993c). Finally, the intracellular target for the type C1 botulinum neurotoxin has been identified as syntaxin (Blasi et al., 1993b). The targets identified, so far, as the substrates of botulinum and tetanus neurotoxins are the constitutive components of the secretory machinery (Huttner, 1993; Menestrina et al., 1994; Montecucco and Schiavo, 1993; Fig. 9) rather than any of the regulatory components.

An alternative mechanism of the involvement of the transglutaminase catalyzed immobilization of the synaptic vesicles have been proposed for tetanus-induced blockage of the neurotransmitter release (Facchiano et al., 1993). In the past, attempts have been made to demonstrate ADP-ribosyl transferase activity in botulinum neurotoxins (cf. Singh, 1990) without any success. The hypothesis of ADP-ribosyl transferase activity of botulinum neurotoxins derives from analogies with diphtheria, cholera and botulinum C2 toxins, all with similar extracellular mode of action (Simpson, 1989b).

The most recent experimental evidence (Fujii et al., 1992; Schiavo et al., 1992a; Wright et al., 1992; Blasi et al., 1993; Montecucco and Schiavo, 1993) clearly suggests that botulinum neurotoxins are metal-binding proteins, and the most likely metal ligand is zinc, although iron has also been suggested to bind to botulinum neurotoxins (Bhattacharyya and Sugiyama, 1989). The question remains as to what is the role of the metal binding to the neurotoxin and how is this protease activity related to the toxic activity of the neurotoxin? Based on current data, it appears that the toxic site of the neurotoxin consists of not only the binding site of the metal. This is likely because the metal binding site (HEXXH, where X is any amino acid residue) has been conserved in all the serotypes of the botulinum neurotoxins, but botulinum neurotoxins have different substrates for proteolysis (see Huttner 1993). Discovery of different substrates of different serotypes of botulinum neurotoxins poses further question for molecular basis of the substrate specificity of botulinum neurotoxins with nearly identical zinc-protease motif (Table 4).

There are two possible ways in which neurotoxins may be providing the observed differential substrate specificity. (i) The three dimensional structure of the light chain (the toxic subunit) of each neurotoxin inherently differ significantly. (ii) The metal binding to the light chain introduces differential conformation for the binding of each neurotoxin to a specific substrate. The toxic site of botulinum neurotoxin is likely to consist of more than one polypeptide domain, and metal binding may be causing the formation of the native toxic site. It is possible that while the metal binding sites are conserved, the other peptide segments of the toxic domain are not conserved among all the serotypes of botulinum neurotoxins, and the toxic site of the neurotoxin results from a specific folding of each neurotoxin polypeptide initiated by the metal binding.

A recent study (Foran et al., 1994) has provided some insight into the enzymatic properties of the protease activity of type B botulinum and tetanus neurotoxins. Several

Table 4. Primary amino acid sequence in the vicinity of zinc-protease motifs, substrates and cleavage sites of botulinum and tetanus neurotoxins

Neurotoxin	Amino Acid Squence	Substrate	Peptide Bond Cleavage
Botulinum type A	D P A V T L A H E L I H A G H R N Y G I	Snap-25	NQ-RA
Botulinum type B	D P A L I L M H E L I H V L H G L Y G I	Snaptobrevin	SQ-FE
Botulinum type C1	D P I L I L M H E L N H A M H N Y G I	Syntaxin	?
Botulinum type D	D P V I A L M H E L T H S L H Q Y G I	Snaptobrevin	QK-LS
Botulinum type E	D P A L T L M H E L I H S L H G L Y G A	Snap-25	DR-LM
Botulinum type F	D P A I S L A H E L I H A L H G L Y G A	Snaptobrevin	DQ-K-L
Botulinum type G	D P A L T L M H E L I H V L H G L Y G	Snaptobrevin	SA-A-K
Tetanus	D P A L L L M H E L I H V L H G L Y G	Snaptobrevin	SQ-FE

Table 5. Kinetic properties of the protease activity of type B botulinum and tetanus neurotoxins with 62 amino acid long fragment of synaptobrevin (Foran et al., 1994)

Neurotoxin	Km (mM)	Vmax (nM s^{-1})	Kcat (s^{-1})
Tetanus	2.4	55	2.7
Botulinum	0.65	62	12.3

observations make the protease activity of these neurotoxins unique. First, known inhibitors of metalloproteases such as captopril and dithiothreitol were largely ineffective on the protease activity at their normal concentrations. Second, unusually large size of the substrate is required. A peptide fragment consisting of the cleavage site of the substrate, synaptobrevin, did not inhibit the protease activity of the neurotoxins, suggesting that perhaps substrate is recognized by multiple contacts between the substrate and the neurotoxin. Third, proteolytic activity of the neurotoxin was not inhibited at 1 mM Zn^{2+} ion concentration whereas other metalloproteases such as thermolysin are reduced at submicromolar concentrations of Zn^{2+} (Pangburn and Walsh, 1975). In addition, the kinetic studies of type B botulinum and tetanus neurotoxin suggested unusually large Km values (Table 5).

Treatment of the neurotoxin with metal chelators such as ethylenediamine tetraacetic acid (EDTA) inactivate most (80%) of the enzymatic activity, and large part (80%) of the lost activity can be restored with the addition of Zn^{2+} ions activity (Foran et al., 1994). Similar loss of toxic activity of type A botulinum neurotoxin was recently observed in our studies with the neurotransmitter release studies in PC12 cells (F. -N. Fu, R. Lomneth and B. R. Singh, unpublished data). However, the loss of activity was not restored with the replenishment with Zn^{2+} ions. The loss of activity at least in part seems to be caused by the change in the polypeptide folding (Fu and Singh, 1995).

It is also possible that substrates other than those already identified exist in the neuronal cells that are modified by botulinum neurotoxins either proteolytically or non-proteolytically. Two main reasons for such a possibility are as follows:

a. Irreversible impairment of the exocytosis by proteolytic degradation of constitutive components of secretory machinery does not explain Ca^{2+}-induced reversibility of the neurotransmitter release from type A botulinum and tetanus neurotoxins (Dreyer, 1989; Simpson, 1989b; Ashton et al., 1993).
b. Biochemical task of degradation of a constitutive component (e.g., synaptobrevin) on about 500,000 vesicles present in each nerve ending of each neuro-muscular junction (Tauc and Poulain, 1993) within a few minutes (mice death occurs within 30 min of injection of the neurotoxin) is perhaps very large even for an enzyme. This may be especially true for botulinum and tetanus neurotoxins as their enzymatic activity seems to be poor (Table 5; Foran et al., 1994). Furthermore, there are two zones of the vesicles within the neuronal cells. About 200 vesicles are in active zones at a time, and even if the neurotoxin can act upon that population for proteolysis, it may act on another target to simultaneously block the movement of the vesicles within the cell.

In addition, based on the electrophysiological data, the vesicular mode itself has been questioned for the exocytosis of neurotransmitters (Tauc and Poulain, 1993). Therefore, it is very important to identify other possible targets for the neurotoxins in addition to those already identified. It is likely that the neurotoxins act on the regulatory and constitutive components of the secretory machinery simultaneously. An evidence of this possibility is a report that tetanus neurotoxin activates transglutaminase activity which could cross-link vesicle synapsin 1 to actin to block the vesicular movement (Facchiano et al., 1993).

Interestingly, in almost all the reports coming out of main research group that initially demonstrated the proteolytic activity of the neurotoxin on certain neuronal substrates, it has been suggested that other substrates are likely to exist (Menestrina et al., 1994; Montecucco and Schiavo, 1993; 1994).

CONCLUDING REMARKS

Bacterial protein toxins although of varied origins to cover many classes of bacteria seem to have a common functional mode for their biological activity. Because of such commonality in function, their structural evolution points to one critical direction, which involves their ability to interact with membranes. All the known toxins are water soluble with potential for interaction with membrane. The water solubility permits their transportation to target cells through the body fluid. Initial toxin interaction with membrane involves polar head groups of lipid bialyer as well as specific association with protein receptors. In many cases, the initial association with membrane 'primes' toxins for their ultimate integration into the non-polar lipid bilayer.

Detailed molecular features as well as events leading to the integration of toxins into lipid bilayers is not clear at this point, and is a subject of intense current research. While the available knowledge of the structure of a handful toxins such as cholera, *Psudomonas* exotoxin A and porin, some common factors are visible, no toxic motif(s) has been identified yet. Availability of three dimensional structure of an adequate number of toxins is likely to reveal interesting structural features. Another area of needed research is the characterization of receptors for toxins. In most cases, cell surface receptors hold the key of toxin entry and potency. Receptor knowledge will not only help develop antidotes against toxins, but also could also help our understanding of toxins themselves.

Toxins that have intracellular targets are perhaps the most complex group of proteins. Such toxins are generally multi-subunit toxins. Although the subunits have their defined function, they still seem to have a well coordinated series of steps among themselves to express toxicity. Interaction among the toxin subunits and between toxin and its target are the areas of current and future research.

Bacterial toxin research is obviously warranted for public safety. However, their is also a silver lining about the toxins. Some of the toxins are currently being used as therapeutic drugs (Scott, 1989) while others are being projected to be altered for their use against such diseases as cancer. Thus, there is more than curiosity involved when it comes to research with protein toxins, especially bacterial protein toxins.

ACKNOWLEDGMENTS

The work was in part supported by National Institute of Neurological Disorders and Stroke (NINDS, NIH, Grant NS33440). The author would like to thank David Ledoux and Paul Lindo for drawing some of the figures.

REFERENCES

Alouf, J. E., Knoll, H. and Kohler, W., 1991, The family of mitogenic, shock-inducing and superantigenic toxins of Staphylococci and Streptococci. In: *Sourcebook of Bacterial Protein Toxins* (Alouf, J. E. and Freer, J. H., eds.), Academic Press, London. pp. 367-414.

Althaus, F. R. and Richter, C., 1987, *ADP-Ribosylation of Proteins*, Springer-Verlag, Berlin. pp. 131-182.

Arnon, S. S., 1992, Infant botulism. In: *Textbook of Pediatric Infectious Diseases* (R. D. Feigen and J. D. Cherry, eds.), 3rd edition, WB Saunders, Philadelphia. pp. 1095-1102.

Ashton, A. C., de Paiva, A. M., Poulain, B., Tauc, L. and Dolly, J. O., 1993, Factors underlying the characteristic inhibition of the neuronal release of transmitters by tetanus and various botulinum neurotoxins. In: *Botulinum and Tetanus Neurotoxins: Neurotransmission and Biomedical Aspects* (B. R. DasGupta, ed.), Plenum Press, New York. pp. 191-213.

Bandyopadhyay, S., Clark, A. W., DasGupta, B. R. and Sathyamoorthy, V. 1987, Role of the heavy and light chains of botulinum toxin in neuromuscular paralysis. *J. Biol. Chem.* 262: 2660-2663.

Be, X., Fu, F. -N. and Singh, B. R., 1994, Hydrophobic moment analysis of amino acid sequences of botulinum and tetanus neurotoxins to identify functional domains. *J. Natural Toxins* 3: 49-68.

Bernheimer, A. W. and Rudy, B., 1986, Interactions between membranes and cytolytic peptides. *Biochim. Biophys. Acta* 864: 123-141.

Bhakdi, S, Fussle, R and Tranum-Jensen, J., 1981, Staphylococcal α-toxin: oligomerization of hydrophilic monomers to form amphiphilic hexamers through contact with deoxycholate detergent micelles. *Proc. Natl. Acad. Sci. U.S.A.* 78: 5475-5479.

Bhattacharyya, S. D. and Sugiyama, H., 1989, Inactivation of botulinum and tetanus toxins by chelators. *Infect. Immun.* 57: 3053-3057.

Binz, T., Eisel, U., Kurazono, H., Binschek, T., Bigalke, H. and Niemann, H., 1990a, Tetanus and botulinum A toxins: Comparison of sequences and microinjection of tetanus toxin A subunit-specific mRNA into bovine chromaffin cells. In: *Bacterial Protein Toxins* (R. Rappuoli et al., eds.). Gustav Fisher Verlag, Stuttgart. pp. 97-101.

Binz, T., Kurazono, H., Popoff, M. W., Frevert, J., Wernars, K. and Niemann, H., 1990b, The complete sequence of botulinum neurotoxin type A and comparison with other clostridial neurotoxins. *J. Biol. Chem.* 265: 9153-9158.

Binz, T., Kurazono, H., Popoff, M. R., Eklund, M. W., Sakaguchi, G., Kozaki, S. Krieglstein, K., Henschen, A, Gill, D. M. and Niemann, H., 1990c, Nucleotide sequence of the gene encoding the sequence of Clostridium botulinum. *Nucleic Acids Res.* 18: 5556.

Bittner, M. A., Habig, W. H. and Holz, R. W., 1989, Isolated light chain of tetanus toxin inhibits exocytosis: Studies in digitonin permeabilized cells. *J. Neurochem.* 53: 966-968.

Blasi, J., Chapman, E. R., Link, E., Binz, T., Yamasaki, S., De Camilli, P., Sudhof, T. C., Niemann and Jahn, R., 1993a, Botulinum neurotoxin A selectively cleaves the synaptic protein SNAP-25. *Nature* 365: 160-163.

Blasi, J., Chapman, E. R., Yamasaki, S., Binz, T., Niemann H. and Jahn, R., 1993b, Botulinum neurotoxin C blocks nurotransmitter by means of cleaving HPC-1/syntaxin. *EMBO J.* 12: 4821-4828.

Boquet, P. and Duflot, E., 1982, Tetanus toxin fragment forms channels in lipid vesicles at low pH. *Proc. Natl. Acad. Sci. USA* 79: 7614-7618.

Buckley, J. T., 1992, Crossing three membranes. Channel formation by aerolysin. *FEBS Lett.* 307: 30-33.

Campbell, K., Collins, M. D. and East, A. K., 1993, Nucleotide sequence of the gene coding for Clostridium botulinum (Clostridium argentinense) type G neurotoxin: geneological comparison with other clostridial neurotoxins. *Biochim. Biophys. Acta* 1216: 487-491.

Cramer, W. A., Cohen, F. S., Merril, A. R. and Song, H. Y., 1990, Structure and dynamics of the colicin E1 channel. *Mol. Microbiol.* 4: 519-526.

Cull-Candy, S. G., Lundh, H. and Thesleff, S., 1976, Effect of botulinum toxinon neuromuscular transmission in rat. *J. Physiol.* 260: 177-203.

Cutler, D., 1992, Progress by poisoning. *Nature* 359, 773.

Doyle, J. and Singh, B. R., 1992, Molecular basis of of low pH-dependent membrane translocation of botulinum and tetanus neurotoxins. *In: Botulinum and Tetanus Neurotoxins: Neurotransmission and Biomedical Aspects (B. R. DasGupta, ed.),* Plenum Press. 231-235.

Dreyer, F., 1989, Peripheral actions of tetanus toxin. In: *Botulinum Neurotoxin and Tetanus Toxin* (L. L. Simpson, ed.), Academic Press, San Diego. pp. 179-202.

East, A. K., Richardson, P. T., Allaway, D., Collins, M.D., Roberts, T. A. and Thompson, D., 1992, Sequence of the gene encoding type F neurotoxin of Clostridium botulinum. *FEMS Microbiol. Lett.* 96: 225-230.

Eisel, U., Jayausch, W., Goretzki, K., Henschen, A., Engels, J., Weller, U., Hude, M., Habermann, E. And Niemann, H., 1986, Tetanus toxin: primary structure, expression in E. coli and homology with botulinum toxins. *EMBO J.* 5: 2495-2502.

Facchiano, F. and Luini, A., 1992, Tetanus toxin potentially stimulates tissue transglutaminase. A possible mechanism of neurotoxicity. *J. Biol. Chem.* 267: 13267-13271.

Facchiano, F., Valtorta, F., Benfenati, F. and Luini, A., 1993, The transglutaminase hypothesis for the action of tetanus toxin. *Trends Biochem. Sci.* 18: 327-329.

Fauchere, J. L. and Pliska, V., 1983, Hydrophobic parameters of amino acid side chains from the partitioning of N-acetyl-amino acid amides. *Eur. J. Med. Chem.Ther.* 18: 369-375.

Fairweather, N. F. and Lyness, V. A., 1986, The complete nucleotide sequence of tetanus toxin. *Nucleic Acids Res.* 14: 7809-7812.

Fehrenbach, F. J. and Jergens D., 1991, Comparative membrane-active (lytic) processes. In: *Sourcebook of Bacterial Protein Toxins* (J. E. Alouf and J. H. Freer, eds.), Academic Press, London. pp. 187-213.

Foran, P., Shone, C. C. and Dolly, J. O., 1994, Differences in the protease activities of tetanus and botulinum B toxins revealed by the cleavage of vesicle-associated membrane protein and various sized fragments. *Biochemistry* 33: 15365-15374.

Freer, J. H., Birbeck, T. H., and Bhakoo, M. (1984) Interaction of staphylococcal δ-lysin with phospholipid monolayers and bilayers: a short course. In: *Bacterial Protein Toxins* (J. H. Freer, ed.), Academic Press, London. pp. 181-189.

Fu, F. -N. and Singh, B. R., 1995, Role of zinc binding in the toxic structure of botulinum neurotoxin. *Protein Sci.* 7:110.

Fujii, N., Kimura, K., Yokosawa, N., Tsuzuki., K and Oguma, K., 1992, A zinc-protease domain in botulinum and tetanus neurotoxins. *Toxicon* 30: 1486-1488.

Gray, G. and Kehoe, M., 1984, Primary sequence of the alpha-toxin gene from Staphylococcus aureus. *Infect. Immun.* 46: 615-618.

Hauser, D., Eklund, M. W., Kurazono, H., Binz, T., Niemann, H., Gill, D. M., Boquet, P. and Popoff, M. R., 1990, Nucleotide sequence of Clostridium botulinum C1 neurotoxin. *Nucleic Acids Res.* 18: 4924.

Hildebrand, A., Pohl, M. and Bhakdi, S., 1991, *Staphylococcus aureus* α-toxin, dual mechanism of binding to target cells. *J. Biol. Chem.* 266: 17195-17200.

Hirono, I., Aoki, T., Asao, T. and Kozaki, S., 1992, Nucleotide sequences and characterization of haemolysin genes from *Aeromonas hydrophila* and *Aeromonas sobria*. *Microbial Pathogenesis* 13: 433-446.

Hoch, D. H., Romero-Mira, M., Ehlrich, B. E., Finkelstein, A., DasGupta, B. R. and Simpson, L. L., 1985, Channels formed by botulinum, tetanus and diphtheria toxins in planer lipid bilayers: relevance to translocation of proteins across membranes. *Proc. Natl. Acad. Sci. USA* 82: 1692-1696.

Huttner, W. B., 1993, Snappy exocytoxins. *Nature* 365: 104-105.

Idziorck, T., FitzGerald, D. and Pastan, I., 1990, Low pH-induces changes in*Pseudomonas* exotoxin and its domains: increased binding of Triton X-114. *Infect. Immun.* 58: 1415-1420.

Jiang, J. X., Abrams, F. S. and London, E., 1991, Folding changes in membrane-inserted diphtheria toxin that may play important roles in its translocation. *Biochemistry* 30: 3857-3864.

Jiang, G., Solow, R. and Hu, V. W., 1989, Characterization of diphtheria toxin-induced lesions in liposomal membranes: an evaluation of the relationship between toxin insertion and "channel" formation. *J. Biol. Chem.* 264: 13424-13429.

Jinno, Y., Ogata, M., Chaudhary, V. K. and Willingham, M. C., 1989, Domain II mutants of *Psedomonas* exotoxin deficient in translocation. *J. Biol. Chem.* 264: 15953-15959.

Kamata, Y. and Kozaki, S., 1994, The light chain of botulinum neurotoxin forms channels in lipid membrane. *Biochem. Biophys. Res. Commun.* 205: 751-757.

Lazarovici, P., Yanai, P. and Yavin, E., 1987, Molecular interactions between micellar polysialogangliosides and affinity purified tetanotoxins in aqueous solution. *J. Biol. Chem.* 262: 2645-2651.

Ledoux, D. N., Be, X. and Singh, B. R., 1994, Quaternary structure of botulinum and tetanus neurotoxins as probed by chemical cross-linking and native gel electrophoresis. submitted to *Toxicon* 32: 1095-1104.

Li, L. and Singh, B. R., 1995, Isolation and characterization of a protein receptor for type E botulinum neurotoxin. *Protein Sci.* 7: 94.

Lomneth, R., Martin, T. F. J. and DasGupta, B. R., 1991, Botulinum neurotoxin light chain inhibits norepinephrine secretion in PC12 cells at an intracellular membranous or cytoskeletal site. *J. Neurochem.* 57: 1413-1421.

Marrack, P. and Kappler, J., 1990, The staphylococcal enterotoxins and their relatives. *Science* 248: 705-711.

Maisey, E. A., Wadsworth, J. D. F., Poulain, B., Shone, C. C., Melling, J., Gibbs, P., Tauc, L. and Dolly, J. O., 1988, Involvement of constituent chains of botulinum neurotoxins A and B in the blockade of neurotransmitter release. *Eur. J. Biochem.* 177: 683-691.

Matsuda, M., Sugimoto, N., Ozutsumi, K. and Hirai, T., 1982, Acute botulinum-like intoxication of tetanus neurotoxin in mice. *Biochem. Biophys. Res. Commun.* 104: 799-805.

McCroskey, L. M., Hatheway, C. L., Fenicia, L., Pasolini, B. and Aureli, P., 1986, Characterization of an organism that produces type E botulinum toxin but which resembles Clostridium butyricum from the feces of an infant with type E botulism. *J. Clin. Microbiol.* 23: 201-202.

McInnes, C. and Dolly, J. O., 1990, Ca^{2+}-dependent noradrenaline release from permeabilized PC12 cells is blocked by botulinum neurotoxin a and its light chain. *FEBS Lett.* 261: 323-326.

Menestrina, G., Schiavo, G. and Montecucco, C., 1994, Molecular mechanism of action of bacterial protein toxins. *Molec. Aspects Med.* 15: 79-193.

Middlebrook, J., 1989, Cell surface receptors for protein toxins. In: Botulinum Neurotoxin and Tetanus Toxin (L. L. Simpson, ed.), Academic Press, San Diego. pp. 95-119.

Midura, T. F. and Arnon, S. S., 1976, Infant botulism: identification of *Clostridium botulinum* and its toxins in faeces. *Lancet* 2: 934-936.

Mochida, S., Poulain, B., Weller, U., Habermann, E. and Tauc, L., 1989, Light chain of tetanus toxin intracellularly inhibits acetylcholine release at neuro-muscular synapses, and its internalization is mediated by heavy chain. *FEBS Lett.* 253: 47-51.

Mollby, R., 1978, Bacterial phospholipases. In: *Bacterial Toxins and Membranes* (J. Jeljaszewics, and T. Wadstrom, eds.), Academic Press, London. pp. 367-424.

Montecucco, C., 1986, How do tetanus and botulinum toxins bind to neuronal membrane? *Trends in Biochem. Sci.* 11: 314-317.

Montecucco, C. and Schiavo, G., 1993, Tetanus and botulism neurotoxins: a new group of zinc proteases. *Trends Biochem. Sci.* 18: 324-327.

Montecucco, C. and Schiavo, G., 1994, Mechanism of action of tetanus and botulinum neurotoxins. *Mol. Microbiol.* 13: 1-8

Montecucco, C., Schiavo, G. and DasGupta, B. R., 1989, Effect of pH on the interaction of botulinum neurotoxins A, B and E with the liposomes.*Biochem. J.* 259: 47-53.

Montecucco, C., Schiavo, G., Gao, Z., Blaustein, R., Boquet, P and DasGupta, B. R., 1988, Interaction of botulinum and tetanus toxins with the lipid bilayer surface. *Biochem. J.* 251: 379-383.

Nishiki, T., Kamata, Y., Nemoto, Y., Omori, Y., Ito, T., Takahashi, M. and Kozaki, S., 1994, Identification of protein receptor for *Clostridium botulinum* type B neurotoxin in rat brain synaptosomes. *J. Biol. Chem.* 269: 10498-10503.

Nishiki, T., Ogasawara, J., Kamata, Y. and Kozaki, S., 1993, Solubilization and characterization of the acceptor for *Clostridium botulinum* type B neurotoxin from rat brain synaptic membranes. *Biochim. Biophys. Acta* 1158: 333-338.

Pangburn, M. K. and Walsh, K. A., 1975, Thermolysin and neutral proteases: Mechanistic considerations. *Biochemistry* 14: 4050-4054.

Papini, E., Colonna, R., Cusinato, F., Montecucco, C., Tomasi, M. and Rappuoli, R., 1987a, Lipid interaction of diphtheria toxin and mutants with altered fragment B: (1) Liposome aggregation and fusion. *Eur. J Biochem. 169*, 629-635.

Papini, E., Schiavo, G., Tomasi, M., Colombatti, M., Rappuoli, R. and Montecucco, C., 1987b, Lipid interaction of diphtheria toxin and mutants with altered fragment B: (2) Hydrophobic photolabelling and cell intoxication. *Eur. J. Biochem.* 169: 637-644.

Pattus, F., Massotte, D., Wilmsen, H. U., Lakey, J., Tsernoglou, D., Tucker, A. and Parker, M. W., 1990, Colicins: prokaryotic killer pores. *Experientia 46*: 180-192.

Pickett, J., Berg, B., Chaplin, E., and Brunstetter-Shafer, M. A., 1976, Syndrome of botulism in infancy: clinical and electrophysiologic study. *N. Engl. J. Med.* 295: 770-772.

Robinson, J. P. and Hash, J. H., 1982, A review of the molecular structure of tetanus neurotoxin. *Mol. Cell. Biochem.* 48: 33-44.

Schengrund, C.-L., Ringler, N. J. and DasGupta, B. R., 1992, Adherence of botulinum and tetanus neurotoxins to synaptosomal proteins. *Brain Res. Bull.* 29: 917-924.

Schiavo, G., Benfenati, F., Poulain, B., Rossetto, O., Polverinao de Laureto, P., DasGupta, B. R. and Montecucco, C., 1992a, Tetanus and botulinum-B neurotoxins block neurotransmitter release by proteolytic cleavage of synatobrevin. *Nature* 359: 832-835.

Schiavo, G., Malizio, C., Trimble, W. S., de Laureto, P. P., Milan, G., Sugiyama, H., Johnson, E. and Montecucco, C., 1994, Botulinum G neurotoxin cleaves VAMP/synaptobrevin at a single Ala-Ala peptide bond. *J. Biol. Chem.* 269: 20213-20216.

Schiavo, G., Poulain, B., Rossetto, O., Benfenati, F., Tauc, L. and Montecucco, C., 1992b, Tetanus is a zinc protein and its inhibition of neurotransmitter release and proteolytic activity depend on zinc. *The EMBO J.* 11: 3577-3583.

Schiavo, G., Rossetto, O., Ferrari, G. and Montecucco, C., 1991, Tetanus toxin receptor. Specific cross-linking of tetanus toxin to a protein of NGF-differentiated PC12 cells. *FEBS Lett.* 290: 227-230.

Schiavo, G., Rossetto, O., Santucci, A., DasGupta, B. R. and Montecucco, C., 1992c, Botulinum neurotoxins are zinc proteins. *J. Biol. Chem.* 267: 23479-23483.

Schiavo, G., Rosetto, O., Catsicas, S., Polverino De Lareto, P., DasGupta, B. R., Benfenati, F. and Montecucco, C., 1993a, Identification of the nerve-terminal targets of botulinum neurotoxin serotypes A, D and E. *J. Biol. Chem.* 268: 23784-23787.

Schiavo, G., Shone, C. C., Rossetto, O., Alexander, F. C. G. and Montecucco, C., 1993b, Botulinum neurotoxin serotype F is a zinc endopeptidase specific for VAMP/synaptobrevin. *J. Biol. Chem.* 268: 11516-11519.

Schiavo, G., Santucci, A., DasGupta, B. R., Mehta, P. P., Jontes, J., Benfenati, F., Wilson, M. C. and Montecucco, C., 1993c, Botulinum neurotoxins serotypes A and E cleave SNAP-25 at distinct COOH-terminal peptide bonds. *FEBS Lett.* 335: 99-103.

Scott, A. B., 1989, Clostridial toxins as therapeutic agents. In: *Botulinum Neurotoxin and Tetanus Toxin* (L. L. Simpson, ed.), Academic Press, San Diego. pp. 399-412.

Sekiya, K., Satoh, R., Danbara, H. and Futaesaku, Y., 1993, A ring-shaped structure with crown formed by streptolysin-O on the erythrocyte membrane. *J. Bacteriol.* 175: 5953-5961.

Shinoda, S., Matsuoka, H., Tsuchie, T., Miyoshi, S. I., Yamamoto, S., Taniguchi, H. and Mizuguchi, Y., 1991, Purification and characterization of a lecithin-dependent haemolysin from *Escherichia coli* transformed by a *Vibrio parahemolyticus* gene. *J. Gen. Microbiol.* 137: 2705-2711.

Shone, C., Wilton-Smith, P., Appleton, N., Hambleton, P., Modi, N., Gately, S., and Melling, J., 1985, Monoclonal antibody-based immunoassay for type A Clostridium botulinum toxin is comparable to mouse bioassay. *Appl. Environment. Miocrobiol.* 50: 63-67.

Simpson, L. L., 1984, The binding fragment from tetanus antagonizes the neuromuscular blocking actions of botulinum toxin. *J. Pharmacol. Exptl. Ther.* 229: 182-187.

Simpson, L. L., 1986, Molecular pharmacology of botulinum toxin and tetanus toxin. *Ann. Rev. Pharmacol. Toxicol.* 26: 427-453.

Simpson, L. L. (ed.), 1989a, *Botulinum toxin and Tetanus Toxin*. Academic Press, San Diego.

Simpson, L. L., 1989b, Peripheral actions of the botulinum toxins In: *Botulinum Neurotoxin and Tetanus Toxin* (L. L. Simpson, ed.), Academic Press, San Diego. pp. 153-178.

Singh, B. R., 1990, Identification of specific domains in botulinum and tetanus neurotoxins. *Toxicon* 28: 992-996.

Singh, B. R., 1993, Structure-function relationship of botulinum and tetanus neurotoxins. In: *Botulinum and Tetanus Neurotoxins: Neurotransmission and Biomedical Aspects (B. R. DasGupta, ed.)*, Plenum Press. pp. 377-392.

Singh, B. R. and Be, X., 1992, Use of hydrophobic moment to analyze membrane interacting domains of botulinum, tetanus and other toxins. In: *Techniques in Protein Chemistry III* (R. H. Angeletti, ed.), Academic Press, Orlando. pp. 373-383.

Singh, B. R. and DasGupta, B. R., 1989a, Structure of heavy and light chain subunits of type A botulinum toxin analyzed by circular dichroism and fluorescence measurements. *Mol. Cell. Biochem.* 85: 67-73.

Singh, B. R. and DasGupta, B. R., 1989b, Molecular topography and secondary structure comparisons of botulinum toxin types A, B and E. *Mol. Cell. Biochem.* 86: 87-95.

Singh, B. R. and DasGupta, B. R., 1989c, Molecular differences between type A botulinum toxin and its toxoid. *Toxicon* 27: 403-410.

Singh, B. R. and DasGupta, B. R., 1989d, Changes in the molecular topography of the light and heavy chains of type A botulinum neurotoxin following their separation. *Biophysical Chem.* 34: 259-267.

Singh, B. R. and DasGupta, B. R., 1990, Conformational changes associated with the nicking and activation of botulinum neurotoxin type E. *Biophysical Chem.* 38: 123-130.

Singh, B. R., Fuller, M. P. and DasGupta, B. R., 1991, Botulinum neurotoxin type A: Structure and interaction with micellar concentration of SDS determined by FT-IR spectroscopy. *J. Protein Chem.* 10: 637-649.

Singh, B. R., Fuller, M. P. and Schiavo, G., 1990, Molecular structure of tetanus neurotoxin as revealed by Fourier transform infrared and circular dichroic spectroscopy. *Biophysical Chem. 36*, 155-166.

Singh, B. R., Ledoux, D. N. and Fu, F.-N., 1994, An analysis of the protein structure of botulinum and tetanus neurotoxins to understand molecular basis of membrane channel formation. In: *Advances in Venom and Toxin Research* (Tan, N. H., Oo, S. L., Thambyrajah, V. and Azila, N., eds.), Malaysian Society on Toxinology, Kuala Lumpur. pp. 103-108.

Stecher, B., Weller, U., Habermann, E., Gratzl, M. and Ahnert-Hilger, G., 1989, The light chain but not the heavy chain of botulinum A toxin inhibits exocytosis from permeabilized adrenal chromaffin cells. *FEBS Lett.* 255: 391-394.

Stevens, R. C., Evenson, M. L., Tepp, W. and DasGupta, B. R., 1991, Crystallization and preliminary x-ray analysis of botulinum neurotoxin type A. *J. Mol. Biol.* 222: 887-880.

Tauc, L. and Poulain, B., 1993, Ca^{2+} dependent evoked quantal neurotransmitter release does not necessarily involve exocytosis of synaptic vesicles. In: *Botulinum and Tetanus Neurotoxins. Neurotransmission and Biomedical Aspects* (B. R. DasGupta, Ed.), Plenum Press, New York. pp. 71-86.

Thiaudire, E. Siffert, O., Talbot, J. -C., Board, J. Alouf, J. E. and Dufourcq, J., 1991, The amphiphilic-helix concept. Consequences on the structure of staphylococcal δ-toxin in solution and bound to lipid bilayers. *Eur. J. Biochem.* 195: 203-213.

Thompson, D. E., Brehm, J. K., Oultram, J. D., Swinfield, T. J., Shone, C. C., Atkinson, T., Melling, J. and Minton, N. P., 1990, The complete amino acid sequence of the Clostridium botulinum type A neurotoxin deduced by nucleotide sequence analysis of the encoding gene. *Eur. J. Biochem.* 189: 73-81.

Tobkes, N., Wallace, B. A. And Bayley, H., 1985, Secondary structure and assembly mechanism of an oligomeric channel protein. *Biochemistry* 24: 1915-1920.

Tomita, M., Taguchi, R. and Ikezawa, H., 1991, Sphingomyelinase of Bacillus cereus as a bacterial haemolysin. *J. Toxicol. Toxin Rev. 10*: 169-207.

Walker, B., Krishnasastry, M., Zorn, L and Bayley, H., 1992, Assembly of the oligomeric membrane pore formed by staphylococcal a-hemolysin examined by truncation mutagenesis. *J. Biol. Chem.* 267: 21782-21786.

Ward, W. H. J., 1987, Diphtheria toxin: a novel cytocidal enzyme. *Trends Biochem. Sci.* 12, 28-31.

Whelan, S. M., Elmore, M. J., Bodsworth, N. J.,Atkinson, T. and Minton, N. P., 1992a, The complete amino acid sequence of the Clostridium botulinum type E neurotoxin, deduced by neucleotide sequence analysis of the encoding gene. *Eur. J. Biochem.* 204: 657-667.

Whelan, S. M., Elmore, M. J., Bodsworth, N. J., Brehm, J. K., Atkinson, T. and Minton, N. P., 1992b, Molecular cloning of Clostridium botulinum structural gene encoding the type B neurotoxin and determination of its entire nucleotide sequence. *Appl. Environ. Microbiol.* 58: 2345-2354.

Wright, J. F., Pernollet, M., Reboul, A., Aude, C. and Colomb, M. G., 1992, Identification and partial characterization of a low affinity metal binding site in the light chain of tetanus toxin. *J. Biol. Chem.* 267: 9053-9058.

5

STRUCTURE AND FUNCTION OF COBRA NEUROTOXIN

C. C. Yang

Institute of Life Sciences
National Tsing Hua University
Hsinchu, Taiwan 30043, Republic of China

1. INTRODUCTION

Snake neurotoxins are the main toxic proteins of cobra, krait, tiger snake and sea snake venoms which block neuromuscular transmission and cause animals death of respiratory paralysis. Snake neurotoxins are classified into two distinct types, postsynaptic and presynaptic neurotoxins, in relation to the neuromuscular junction. Postsynaptic neurotoxins bind specifically to nicotinic acetylcholine receptor (AChR) at the motor endplate and produce a nondepolarizing block of neuromuscular transmission. Presynaptic neurotoxins block the release of acetylcholine from the presynaptic motor nerve terminals.

Cobrotoxin, isolated in a crystalline state (Fig. 1) from the venom of Taiwan cobra *Naja naja atra*, binds specifically to the AChR on the postsynaptic membrane and thus blocks neuromuscular transmission. Cobrotoxin was isolated by means of ammonium sulfate fractionation followed by repeated chromatography on CM-cellulose (Yang, 1964, 1965), but now by a single step chromatography on SP-Sephadex C-25 column (Yang, *et al.*, 1981). Cobrotoxin has lethal toxicity 7 times greater than that of the original venom and was proved to be the main toxic protein in cobra venom (Chang and Yang, 1969). It is a small basic protein consisting of a single peptide chain of 62 amino acid residues, crosslinked by four disulfide bonds. The complete amino acid sequence and the position of disulfide bonds in cobrotoxin have been established (Fig. 2) (Yang *et al.*, 1969, 1970).

2. STRUCTURE OF POSTSYNAPTIC NEUROTOXIN

To date, more than 120 toxins with neurotoxic activity have been isolated in pure state from elapid and hydrophid (sea-snake) venoms. Over 100 highly homologous postsynaptic neurotoxins belonging to two distinct size groups, short and long neurotoxins, have been sequenced (Yang, 1974, 1984; Mebs, 1988; Endo and Tamiya, 1991). Short neurotoxins contain 60-62 amino acid residues with four disulfide bonds, and long neurotoxins comprise

Natural Toxins II, Edited by B. R. Singh and A. T. Tu
Plenum Press, New York, 1996

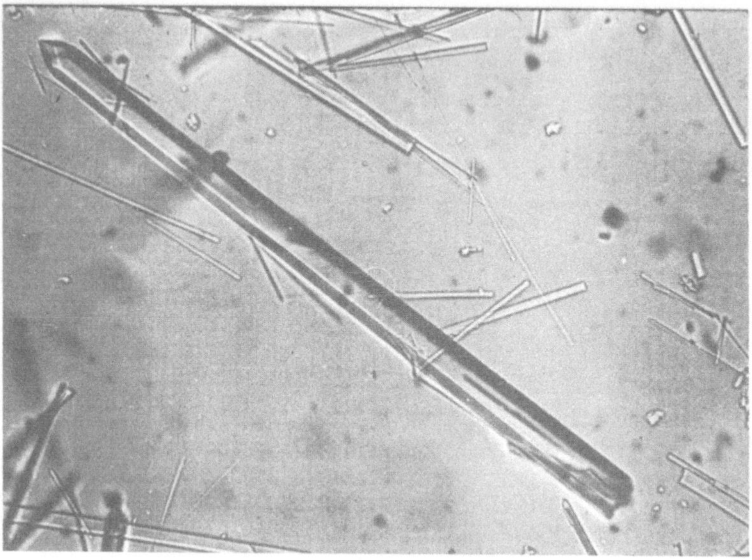

Figure 1. Crystalline cobrotoxin (150 X).

71-74 residues with five disulfide bonds. Cobrotoxin (structural type, 62-4; 7 kD) is a short neurotoxin and α-bungarotoxin (74-5; 8 kD) is a long neurotoxin.

Short Neurotoxins: 60-62 amino acids with four disulfide bonds

Cobrotoxin	(structural type, 62-4)
Erabutoxins	(62-4)
N. nigricollis toxin α	(61-4)

Long Neurotoxins: 71-74 amino acids with five disulfide bonds

α-Bungarotoxin	(74-5)
Siamensis toxin 3	(71-5)
N. nivea toxin α	(71-5)

The pairing of four disulfide bonds is similar in both short and long neurotoxins (Fig. 2). The "extra" disulfide bond in the long neurotoxins, at 29-33, pinches off a short pentapeptide section in the second disulfide loop, thereby shortening the loop to about the same length as in the short neurotoxins. The extra disulfide bond is exposed at the surface of the toxin molecule and can be selectively reduced without affecting binding affinity (Martin *et al.*, 1983). Long neurotoxins extend seven amino acid residues beyond the carboxyl terminus of the short toxins (Fig. 2). The removal of the C-terminal four or five residues of α-bungarotoxin by mild tryptic digestion (Wu *et al.*, 1983) or carboxypeptidase P (Endo *et al.*, 1987) induces no global conformational change in the molecule but affects only a limited region. Therefore, the C-terminal tail of long neurotoxin appears to be unimportant in maintaining the specific polypeptide chain folding.

The homology and common mode of action of postsynaptic neurotoxins imply similar three-dimensional structure. X-ray crystallography has provided models for erabutoxin b (structural type, 62-4) (Low *et al.*, 1976; Tsernoglou and Petsko, 1976; Low and Corfield, 1986), cobratoxin (71-5) (Walkinshaw *et al.*, 1980), and α-bungarotoxin (74-5)

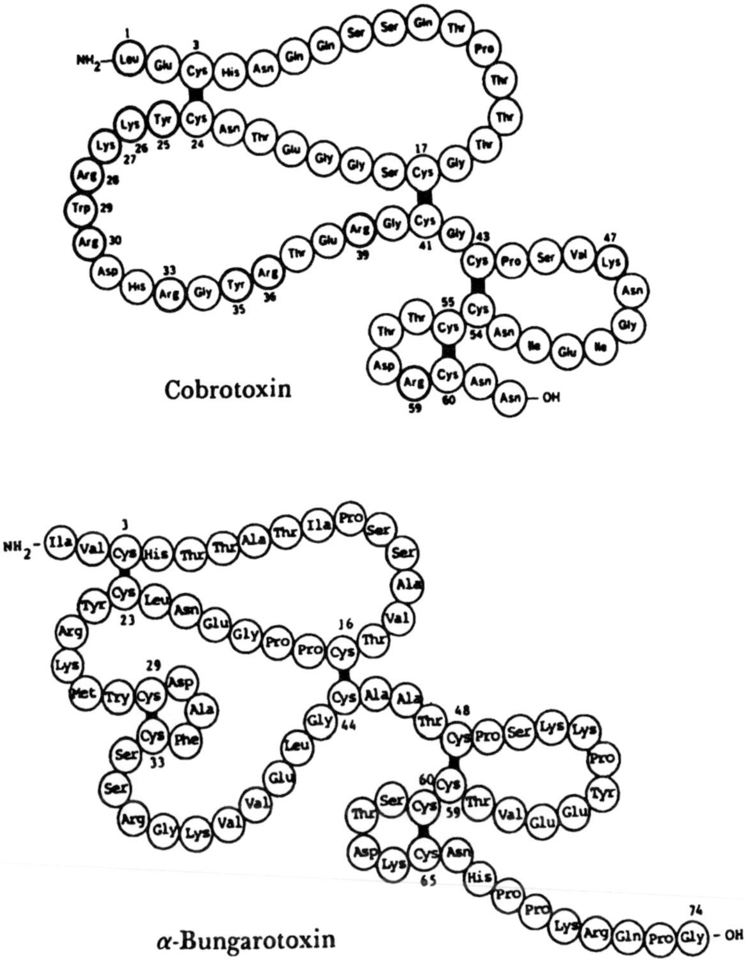

Figure 2. Structure of cobrotoxin and α-bungarotoxin showing the amino acid sequence and position of disulfide bonds.

(Love and Stroud, 1986). All postsynaptic neurotoxins are similar in their overall folding, but differ in details such as the extent of secondary structure and the position of an invariant side-chain. Recently, the NMR three dimensional structure of cobrotoxin in solution has been determined (Yu *et al.*, 1993). The mean solution structure was compared with the X-ray crystal structure of homologous protein erabutoxin b which has been solved to a resolution of 1.4 Å (Low and Corifield, 1986) (Fig. 3). This yielded information that both structures resemble each other except at the exposed loops and surface regions, where the solution structure reveals the higher flexibility in its conformation.

3. STRUCTURE AND FUNCTION OF COBROTOXIN

It is significant that cobrotoxin has only 62 amino acid residues, it contains four disulfide bonds and is devoid of free sulfhydryl groups. In order to examine the relationship between disulfide bonds and biological activity, we dissolved cobrotoxin in 8 M urea solution, and added the reducing agent β-mercaptoethanol to split disulfide bonds. On

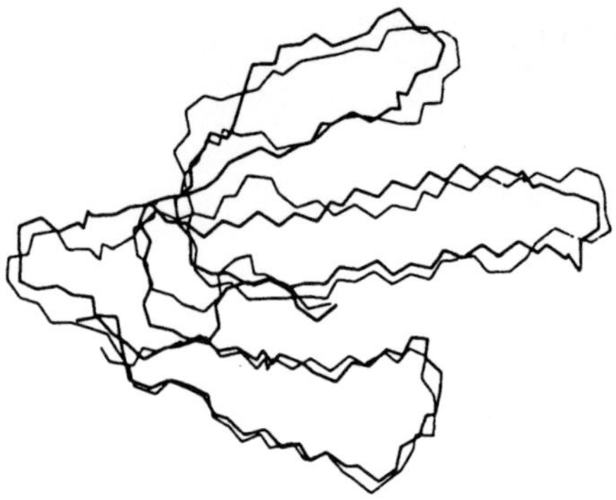

Figure 3. Superposition of the backbone atoms of the NMR 3-D structure of cobrotoxin in solution (thin line) and the X-ray crystal structure of erabutoxin b (thick line).

reduction with β-mercaptoethanol, cobrotoxin displays eight sulfhydryl groups. The toxin lost its lethal toxicity and antigenic specificity, and conformational change occurred concurrently (Yang, 1967). Reduced cobrotoxin, however, on reoxidation, yielded a biologically active product showing essentially the same ORD (Fig. 4) and IR spectra as the native toxin. This fact indicates that the integrity of the disulfide bonds in cobrotoxin is essential for its biological functions.

Cobrotoxin contains two tyrosine residues at positions 25 and 35. Spectrophotometric titration of cobrotoxin (Chang *et al.*, 1971a) showed that one tyrosyl group was titrated freely with a normal apparent pK 9.65, whereas the other ionized slowly only after irreversible conformational changes had occurred at pH above 11.3. The reaction of cobrotoxin with tetranitromethane resulted in selective nitration of the Tyr-35 without altering the biological activity and conformation of the toxin (Chang *et al.*, 1971a). However, when the invariant Tyr-25 that appears to be "buried" in the molecule was modified in the presence of 5 M guanidine-HCl, the toxin lost its biological activity, and the CD spectrum changed greatly (Fig. 5). It is noteworthy that the tyrosine residue at position 25 is common to all snake neurotoxins; this result, together with those above, indicates that for Tyr-25 to remain intact is important to maintain the active conformation of the toxin.

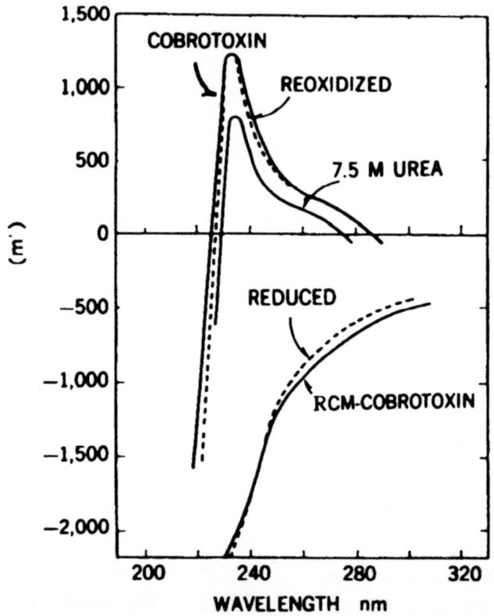

Figure 4. Optical rotatory dispersion (ORD) of cobrotoxin, reduced cobrotoxin and reoxidized cobrotoxin in phosphate buffer, pH 5.9, ionic strength 0.1.

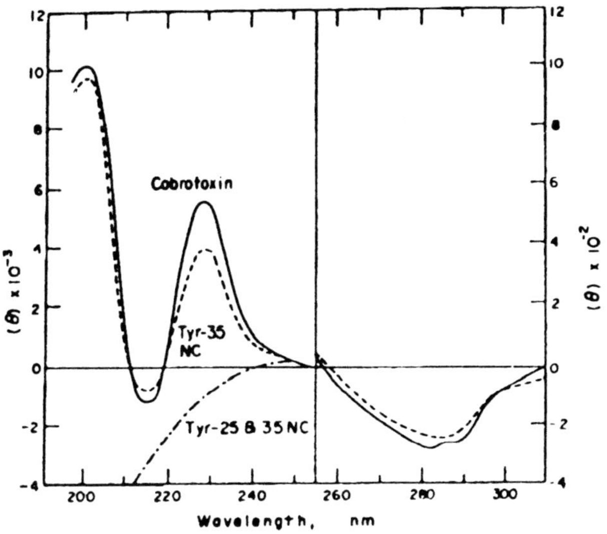

Figure 5. CD curves of cobrotoxin and its nitrated derivatives in phosphate buffer, pH 7.0, ionic strength 0.1.

It seems necessary to distinguish between two types of important groups. The structurally important groups of one type are required for a molecule to attain its active conformation. Disulfide bonds and tyrosine residue at position 25 are structurally important to maintain the active conformation of the toxin. The second type is functionally important groups. For the postsynaptic neurotoxins, groups directly involved in binding the toxins to an AChR of the muscle motor endplate thus prevent transmission across the cholinergic synapse.

4. CHEMICAL MODIFICATION OF COBROTOXIN

Cobrotoxin is a basic protein having four free amino groups and six arginine residues at positions 28, 30, 33, 36, 39 and 59 in sequence (Fig. 2). The important roles of cationic groups in cobrotoxin in relation to the lethal toxicity and antigenic specificity were investigated by selective and stepwise chemical modification of arginine and lysine residues.

Modification of arginine residues was conducted with phenylglyoxal at various pH (Yang *et al.*, 1974). Reaction of cobrotoxin with phenylglyoxal at pH 8.0 resulted in an almost complete loss of lethal toxicity and four of the six arginine residues were modified. However, the rate of inactivation was decreased significantly when the pH of the reaction was decreased. Only one arginine residue was modified when the reaction was carried out at pH 6.0 and the modified derivative retained full biological activity.

From the map of peptides produced by chymotryptic digestion of the modified derivative, we identified the arginine residue modified at pH 6.0 to be Arg-28. Arg-28 and Arg-33 were two of the six arginine residues modified by reaction at pH 6.7 (Table 1); this greatly decreased the lethal toxicity but the antigenic specificity was not altered significantly. The lethality was lost almost completely and the antigenic activity decreased about 30% when an additional arginine residue at position 30 was modified at pH 7.5. These results indicate that the cationic groups contributed by the guanidino groups of arginine residues are functionally important (Yang *et al.*, 1974).

Table 1. Stepwise modification of arginine residues in cobrotoxin with phenylglyoxal at varying pH

	Arg-residues modified	Lethality (%)	Antigenic activity (%)
Cobrotoxin	None	100	100
at pH 6.0	Arg-28	100	98
pH 6.7	Arg-28 & 33	22.6	94
pH 7.5	Arg-28, 30 & 33	3.1	70
pH 8.0	Arg-28, 30, 33 & 36	1.6	34

The status of free amino groups in cobrotoxin was investigated by stepwise modification with trinitrobenzene sulfonate (TNBS) (Chang *et al.*, 1971b). When cobrotoxin was allowed to react with TNBS in 1.1-fold molar excess, the most reactive ε-amino group of Lys-27 was selectively trinitrophenylated without altering the activity of the toxin. However, complete loss of lethality was observed when Lys-27 and Lys-47 were modified with TNBS in 2.2-fold molar excess indicates that the positive charge contributed by the ε-amino group of Lys-47 is functionally important for the biological activity of cobrotoxin.

Guanidination of cobrotoxin with O-methylisourea made essentially all three lysine residues convert to homoarginine with no effect on lethal toxicity or antigenic specificity (Chang *et al.*, 1971b). This result indicates that the effect of trinitrophenylation on activity of the toxin differs from that of guanidination. An explanation is that trinitrophenylation with TNBS converts the positively charged ú‘-amino group into a neutral state, whereas guanidination yields a substituted group that maintains the positive charge.

Selective and stepwise chemical modifications of cobrotoxin indicate that at least two cationic groups, ε-amino group of Lys-47 and guanidino group of Arg-33, both of which are common to all known postsynaptic snake neurotoxins, held at a certain critical distance in the molecule are functionally important for its neuromuscular blocking activity.

5. NEUROTOXIN-AChR INTERACTIONS

Cobrotoxin binds specifically to the AChR on the postsynaptic membrane in a competitive manner with cholinergic agonists thus to block neuromuscular transmissions. AChR are complex transmembrane proteins formed by five homologous subunits in a stoichiometry of $\alpha_2\beta\gamma\delta$. Cobrotoxin binds reversibly to the acetylcholine-binding sites present in two α-subunits.

In order to probe the structure of the ACh-binding sites on the native, membrane-bound AChR, Dennis *et al.* (1988) employed a photoaffinity reagent [^{3}H] DDF for these sites. The labeled amino acids are located within three distinct regions of the large amino-terminal hydrophilic domain of the α-subunit primary structure. These findings are in accord with models proposed for the transmembrane topology of the α-chain that assign the amino-terminal segment α1-210 to the synaptic cleft.

Atassi *et al.* (1991) using synthetic overlapping peptides of the extracellular parts of the α-chain of *Torpedo* and human AChR localized the full profile of the toxin-binding regions on receptors. The main binding activity for cobrotoxin resided within region α122-138 and a lower binding activity was present in the overlap α23-60 and in peptide α194-210. The region α182-198 showed little or no binding to cobrotoxin was unexpected, in view of the fact that this constituted a major binding region to long neurotoxin, α-bungarotoxin (Mulac-Jericevic and Atassi, 1987). The quantitative radiometric titration also showed that the long neurotoxin bound to both peptides α122-138 and α182-198, but short neurotoxin did not bind to peptide α182-198, where peptide α122-138 was fully capable of binding both toxins. These results indicate that the region within residues α122-138 of both *Torpedo* and human AChR constitutes a universal binding site for long and short neurotoxins on AChR. Take into considerations of binding affinity of various synthetic peptides derived from the AChR sequence and the three-dimensional structure of cobrotoxin constructed from high-resolution NMR studies, the α-subunit residues 126-136 can be fitted onto the central loop of cobrotoxin on the basis of energy minimization that satisfies binding data and structural constraints.

Based on relative binding affinities of peptide-peptide interactions, the position of AChR peptides relative to the α-bungarotoxin molecule was assessed (Ruan *et al.*, 1990). The receptor peptides were docked onto the appropriate regions of α-bungarotoxin and a 3-D model was constructed of the binding-site cavity for the toxin on human AChR. The cavity appears to be conical, 30.5 Å in depth, involving several receptor regions that make contact with the α-bungarotoxin loop regions. Recently, Basus *et al.* (1993) reported the 2-D NMR characterization of the complex formed between a synthetic peptide α185-196 and α-bungarotoxin to determine the three-dimensional structure of the complex. The structural study reveals a receptor binding cleft created between the N-terminal loop 1 and the middle loop 2 of α-bungarotoxin and suggests that α-bungarotoxin undergoes a conformational change upon peptide binding.

Short and long neurotoxins have very similar dissociation constants with AChR (10^{-9}-10^{-11} M) and LD_{50} values (50-150 μg/kg for mice). They differ chiefly in their rates of association and dissociation from the receptor. Long neurotoxins generally associate and dissociate much more slowly. With rat phrenic-nerve preparations, binding of the short neurotoxins, cobrotoxin and erabutoxin b, were slowly reversible, whereas that of α-bungarotoxin was not. Binding of cobrotoxin to the sciatic-nerve sartorius-muscle preparation of the frog was reversible, but that of α-bungarotoxin was irreversible. It may explain that the differences in reversibility between long and short neurotoxins are due to the inability of short neurotoxins to bind to the contact region within residues α182-198 of AChR. Thus the participation of region α182-198 in the neurotoxin binding may explain the differences in reversibility between long and short neurotoxins.

Chemical modifications of individual amino acid residues demonstrated that no single amino acid is responsible for the full biological activity of neurotoxins. The specific tight binding of neurotoxins to AChR is thought to be achieved through interactions with multiple points attachment (Yang, 1987, 1988, 1992). According to Fig. 6, four disulfide bonds and Tyr-25, which is buried in the molecule, form a central core to maintain and stabilize the active conformation of the toxin, and from the core three loops are extended. The clustering of disulfides near one end of the neurotoxin allows considerable flexibility in the extended loops which contain many functionally important residues lying on the concave surface of the molecule.

The loops constitute two antiparallel β-sheets, one double and one triple stranded β sheets (Yu *et al.*, 1993) (Fig. 6). The double stranded antiparallel β sheet is found to be on the N-terminal side with the strand 1 and strand 2 linked by turn 1 segment. The triple stranded antiparallel β sheet is formed among the strand 3 (residues 25-30), strand 4 (36-40) and strand 5 (51-56), in which strand 3 lies in between the others. Strands 3 and 4 are connected by trun 2, the residues involved are 31-34. This turn is noted to have higher mobility because of its high exposure to solvent. An eight residue long segment (residues 42-49) in loop 3 is found to have

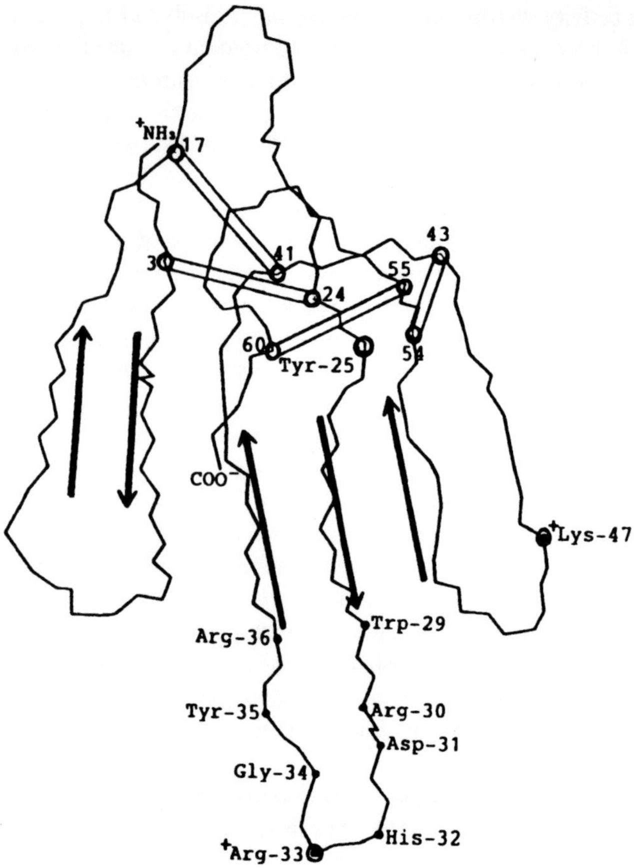

Figure 6. Perspective drawing of the course of the polypeptide chain of cobrotoxin. Arrows indicate the direction of the antiparallel double- and triple-stranded β-sheets.

higher conformational freedom because the entire segment is exposed to solvent. The higher mobility observed for turn 2 and for the exposed segment in loop 3 may be due to the presence of cationic reactive groups, Arg-33 and Lys-47, respectively, which are functionally responsible for the neuromuscular blocking activity of the protein (Yang, 1993).

6. CLONING AND EXPRESSION OF cDNA ENCODING COBROTOXIN

Total RNA was isolated by acid guanidinium thiocyanate-phenol-chloroform extraction method (Chomczynski and Sacchi, 1987) from venom glands of Taiwan cobra *Naja naja atra*. Polymerase chain reaction (PCR) was employed to amplify cDNAs constructed from the total RNA to facilitate the cloning and sequencing of cobrotoxin gene (Maniatis *et al.*, 1989). The PCR products were then subcloned into pUC19 vector and transformed into *E. coli* strain DH5αF'. Plasmids that purified from the positive clones were prepared for nucleotide sequencing by dideoxynucleotide chain-termination method (Sanger *et al.*, 1977). The plasmid-derived clone was used as probe to hybridize λGEM-4 cDNA library. Sequencing several clones containing about 0.5 kb DNA inserts contained a complete and unambiguous full-length reading frame of 249 base pairs covering a precursor of cobrotoxin gene (Fig.7) with a deduced mature protein sequence of 62 amino acids which are identical to the amino acid sequence of cobrotoxin (Yang *et al.*, 1969) and a 21 amino-acid segment of signal peptide.

GGCTGCAGTGGCAAG					ATG	AAA	ACT	CTG	CTG	CTG	ACC	TTG
					M	K	T	L	L	L	T	L
					-21							
CTG	GTG	GTG	ACA	ATC	GTG	TGC	CTG	GAC	TTA	GGA	TAC	ACC
L	V	V	T	I	V	C	L	D	L	G	Y	T
			-10									-1
CTG	GAA	TGT	CAC	AAC	CAA	CAA	TCA	TCG	CAA	ACT	CCA	ACC
L	E	C	H	N	Q	Q	S	S	Q	T	P	T
1									10			
ACT	ACA	GGT	TGT	TCA	GGT	GGG	GAG	ACC	AAT	TGC	TAT	AAA
T	T	G	C	S	G	G	E	T	N	C	Y	K
						20						
AAG	CGT	TGG	CGT	GAT	CAC	CGT	GGA	TAT	AGA	ACC	GAG	AGG
K	R	W	R	D	H	R	G	Y	R	T	E	R
			30									
GGA	TGT	GGT	TGC	CCT	TCA	GTG	AAG	AAC	GGC	ATT	GAA	ATT
G	C	G	C	P	S	V	K	N	G	I	E	I
40										50		
AAC	TGT	TGC	ACA	ACA	GAC	AGA	TGC	AAC	AAT	TAG	CTCTCCG	
N	C	C	T	T	D	R	C	N	N	end		
							60					

AGTGGCTAATTCCTTGAGTTTGCTCTCATCATCATGACATCTTGAAATT
ATAGCTAGCTTAGCACCAGAGCACTCCTCTGCTTGACACTCACACTCT
ACTAGCCGAGACCCGAGCGAAAAAAAAAAAAAAAAAAAAAAAAA

Figure 7. The cDNA nucleotide sequence and the amino acid sequence of cobrotoxin. The 21 residues long signal peptide is underlined.

All snake neurotoxins present the same cDNA organization with a short 5' noncoding region, followed by an open reading frame and a long 3' noncoding region. In all cases, the open reading frame codes for a 21 residue signal peptide which is either identical or highly conserved to all precursors of neurotoxins. For example, the deduced amino acid sequence of the signal peptide (underlined) of cobrotoxin (Fig. 8) differs from those of the sea-snake neurotoxins erabutoxins a and b (Tamiya *et al.*, 1985; Obara *et al.*, 1989), *Aipysurus laevis* toxins b and d (Ducancel *et al.*, 1990) by two nucleotide base pairs at the gene level and one amino acid residue i.e., percent identity is about 95% which is higher than that between their respective mature toxins (56%). As many as 11 nonpolar residues including 7 Leu, 3 Val and 1 Ile are present among the 21 amino acids found in the signal peptide, which presumably favorize the translocation of the neosynthesized 62 residue cobrotoxin through the endoplasmic reticulum (Blobel and Dobberstein, 1975).

Expression of cobrotoxin in *E. coli* vector generated a polypeptide which can cross-react with the antiserum against the native cobrotoxin from the same cobra venom though with a much lower activity. This genetic approach offers the capacity to produce large amounts of a protein by microorganism and the possibility of engineering a protein by site-specific mutagenesis. Detailed structural and site-specific mutational studies on the cDNA clones of cobrotoxin may complement our previous chemical modifications of the functional role of some amino acid residues in neurotoxins and lead to insight into the modes of action for these biologically active molecules.

```
                                       1         10        20
Cobrotoxin     MKTLLLTLLVVTIVCLDLGYT LECHNQQSSQTPTTTGCSGG
Erabutoxin a   ·····V··············· RI·F·H····PQ··KT··P·

                       30        40        50        60
               ETNCYKKRWRDHRGYRTERGCGCPSVKNGIEINCCTTDRCNN
               ·SS··N·Q·S·F··TII·······T··P··KLS··ESEV···
```

Figure 8. Comparison of the amino acid sequences of cobrotoxin and erabutoxin a. A dot (·) represents the same amino acid residue as shown for cobrotoxin. The singal sequence is underlined. The signal peptides of the precursors of toxins from sea-snakes, *L. semifasciata* and *A. laevis* are identical in terms of amino acid sequence and the corresponding nucleotide sequences only differ from each other by a single silent mutation.

7. CONCLUSION AND PERSPECTIVES

In our laboratory we first isolated the major lethal protein (termed Cobrotoxin) of non-enzymatic nature from the venom of Taiwan cobra (*Naja naja atra*) in 1964 and subsequently purified and crystallized the protein. The primary structure and the disulfide linkages with various efforts by chemical modification and immunological methods in elucidation of the structure-function relationship of this important venom neurotoxin have since been accomplished. Structure-activity correlations have been drawn from chemical modification carried out on both pre- and post-synaptic neurotoxins. With recent advances in DNA recombination and protein engineering, we feel that the time is now ripe to apply these techniques to the isolation and characterization of the genes encoding these toxins. Detailed structural and site-specific mutational studies on the cDNA clones of neurotoxins of both types may complement our previous chemical modifications of the functional role of some amino acid residues in neurotoxins and lead to insight into the modes of action for these biologically active molecules.

ACKNOWLEDGEMENTS

This work was supported by the National Science Council, Republic of China. I express my sincere appreciation to my colleagues for their contributions to this study.

REFERENCES

Basus, V.J., Song, G. and Hawrot, E. (1993) NMR solution structure of an α-bungarotoxin/nicotinic receptor peptide complex. *Biochemistry* **32**: 12290-12298.

Blobel, G. and Dobberstein, B. (1975) Transfer of proteins across membrans. *J. Cell Biol.* **67**: 835-851.

Chang, C. C. and Yang, C. C. (1969) Immunochemical studies on cobrotoxin. *J. Immunol.* **102**: 1437-1444.

Chang, C. C., Yang, C. C., Hamaguchi, K., Nakai, K. and Hayashi, K. (1971a) Studies on the status of tyrosyl residues in cobrotoxin. *Biochim. Biophys. Acta* **236**: 164-173.

Chang, C. C., Yang, C. C., Nakai, K. and Hayashi, K. (1971b) Studies on the status of free amino and carboxyl groups in cobrotoxin. *Biochim. Biophys. Acta* **251**: 334-344.

Chomczynski, P. and Sacchi, N. (1987) Single-step method of RNA isolation by acid guanidinium thiocyanate-phenol-chloroform extraction. *Anal. Biochem.* **162**: 156-159.

Dennis, M., Giraudat, J., Kotzyba-Hibert, F., Goeldner, M., Hirth, C., Chang, J.Y., Lazure, C., Chretien, M. and Changeux, J.P. (1988) Amino acids of the *Torpedo marmorata* acetylcholine receptor α subunit labeled by a photoaffinity ligand for the acetylcholine binding site. *Biochemistry* **27**: 2346-2357.

Ducancel, F., Guignery-Frelat, G., Boulain, J.C. and Menez, A. (1990) Nucleotide sequence and structure analysis of cDNA encoding short-chain neurotoxins from venom glands of sea snake (*Aipysurus laevis*). *Toxicon* **28**: 119-123.

Endo, T. and Tamiya, N. (1991) Structure-function relationships of postsynaptic neurotoxins from snake venoms. in "*Snake Toxins*" (Harvey, A.L., ed.) Pergamon Press, New York, pp. 165-222.

Endo T., Oya, M., Tamiya, N. and Hayashi, K. (1987) Role of C-terminal tail of long neurotoxins from snake venoms in molecular conformation and acetylcholine receptor binding: Protein nuclear magnetic resonance and competition binding studies. *Biochemistry* **26**: 4592-4598.

Love, R. A. and Stroud, R. M. (1986) The crystal structure of α-bungarotoxin at 2.5 Å resolution: Relation to solution structure and binding to acetylcholine receptor. *Protein Eng.* **1**: 37-46.

Low, B. W. and Corfield, P. W. R. (1986) Erabutoxin b: Structure-function relationships following initial protein refinement at 0.140-nm resolution. *Eur. J. Biochem.* **161**: 579-587.

Low, B. W., Preston, H. S., Sato, A., Rosen, L. S., Searle, F. E., Rudko, A. D. and Richardson, J. S. (1976) Three dimensional structure of erabutoxin b neurotoxic protein: Inhibitor of acetylcholine receptor. *Proc. Natl. Acad. Sci. USA* **73**: 2991-2994.

Maniatis, T., Fritsch, E.F. and Sambrook, J. (1989) *Molecular Cloning, A Laboratory Manual.* Cold Spring Harbor Laboratory Press, Cold Spring Harbor, N.Y.

Martin, B. M., Chibber, B. A. and Maelicke, A. (1983) The sites of neurotoxicity in α-cobratoxin. *J. Biol. Chem.* **258**: 8714-8722.

Mebs, D. (1988) Snake venom toxins: Structural aspects. in "*Neurotoxins in Neurochemistry*" (Dolly, J. O., ed.) Ellis Horwood, Chichester, England, pp. 3-12.

Mulac-Jericevic, B. and Atassi, M.Z. (1987) Profile of the α-bungarotoxin-binding regions on the extracellular part of the α-chain of *Torpedo california* acetylcholine receptor. *Biochem. J.* **248**: 847-852.

Obara, K., Fuse, N., Tsuchiya, T., Nonomura, Y., Menez, A. and Tamiya, T. (1989) Sequence analysis of a cDNA encoding a erabutoxin b from the sea-snake *Laticauda semifasciata. Nucleic Acids Res.* **17**, 10490.

Ruan, K.H., Spurlino, J., Quiocho, F.A. and Atassi, M.Z. (1990) Acetylcholine receptor-α-bungarotoxin interactions: Determination of the region-to-region contacts by peptide-peptide interactions and molecular modeling of the receptor cavity. *Proc. Natl. Acad. Sci. USA* **87**: 6156-6160.

Ruan, K.H., Stiles, B.G. and Atassi, M.Z. (1991) The short-neurotoxin-binding regions on the α-chain of human and *Torpedo californica* acetylcholine receptors. *Biochem. J.* 274: 849-854.

Sanger, F., Nicklen, S. and Coulson, A.R. (1977) DNA sequencing with chain-terminating inhibitors, *Proc. Natl. Acad. Sci. USA* **74**: 5463-5467.

Tamiya, N., Lamouroux, A., Julien, J.F., Grima, B., Mallet, J., Fromageot, P. and Menez, A. (1985) Cloning and sequence analysis of the cDNA encoding a snake neurotoxin precursor. *Biochimie* **67**: 185-189.

Tsernoglou, D. and Petsko, G. A. (1976) The crystal structure of a postsynaptic neurotoxin from sea snake at 2.2 Å resolution. *FEBS Lett.* **68**: 1-4.

Walkinshaw, M. D., Saenger, W. and Maelicke, A. (1980) Three-dimensional structure of the long neurotoxin from cobra venom. *Proc. Natl. Acad. Sci. USA* **77**: 2400-2404.

Wu, S. H., Chen, C. J., Tseng, M. J. and Wang, K. T. (1983) The modification of α-bungarotoxin by digestion with trypsin. *Arch. Biochem. Biophys.* **227**: 111-117.

Yang, C. C. (1964) Purification of toxic proteins from cobra venom. *J. Formosan Med. Assoc.* **63**: 325-331.

Yang, C. C. (1965) Crystallization and properties of cobrotoxin from Formosan cobra venom. *J. Biol. Chem.* **240**: 1616-1618.

Yang, C. C. (1967) The disulfide bonds of cobrotoxin and their relationship to lethality. *Biochim. Biophys. Acta* **133**: 346-355.

Yang, C. C. (1974) Chemistry and evolution of toxins in snake venoms. *Toxicon* **12**: 1-43.

Yang, C. C. (1984) Snake neurotoxin. *The Snake* **16**: 90-103.

Yang, C. C. (1987) Cobrotoxin structure and function. in *"Pharmacology"* (Rand, M. J. and Raper, C., eds.) Elsevier Biomedical Division, Amsterdam, pp. 871-874.

Yang, C.C. (1988) Chemistry and immunochemistry of cobrotoxin. *J. Chin. Biochem. Soc.* **17**: 20-31.

Yang, C.C. (1992) Chemistry of snake neurotoxins and future perspective. *J. Chin. Chem. Soc.* **39**: 731-740.

Yang, C.C. (1993) Structure and function of snake neurotoxin. in *"Recent Advances in Molecular and Biochemical Research on Protein"* (Wei, Y.H., Chen, C.S. and Su, J.C., eds.) World Scientific, Singapore, pp. 105-118.

Yang, C. C., Chang, C. C. and Liou, I. F. (1974) Studies on the status of arginine residues in cobrotoxin. *Biochim. Biophys. Acta* **365**: 1-14.

Yang, C. C., King, K. and Sun, T. P. (1981) Chemical modification of lysine and histidine residues in phospholipase A2 from the venom of *Naja naja atra* (Taiwan cobra). *Toxicon* **19**: 645-659.

Yang, C. C., Yang, H. J. and Chiu, R. H. C. (1970) The position of disulfide bonds in cobrotoxin. *Biochim. Biophys. Acta* **214**: 355-363.

Yang, C. C., Yang, H. J. and Huang, J. S. (1969) The amino acid sequence of cobrotoxin. *Biochim. Biophys. Acta* **188**: 65-77.

Yu, C., Bhaskaran, R., Chuang, L. C. and Yang, C. C. (1993) Solution conformation of cobrotoxin: A nuclear magnetic resonance and hybrid distance geometry-dynamical simulated annealing study. *Biochemistry* **32**: 2131-2136.

6

STRUCTURE AND FUNCTION OF COBRA VENOM FACTOR, THE COMPLEMENT-ACTIVATING PROTEIN IN COBRA VENOM

Carl-Wilhelm Vogel,[1] Reinhard Bredehorst,[1] David C. Fritzinger,[2] Thomas Grunwald,[1] Patrick Ziegelmüller,[1] and Michael A. Kock[1]

[1] Department of Biochemistry and Molecular Biology
University of Hamburg, Martin-Luther-King-Pl. 6
20146 Hamburg, Germany
[2] SRA Technologies, Inc.
Rockville, Maryland 20850

INTRODUCTION

Cobra Venom Factor (CVF) is an unusual venom component known to be present in the venom of the cobra species *Naja, Ophiophagus*, and *Hemachatus* of the *Elapidae* family (1). CVF is not a toxin in the classical sense. As a matter of fact, the purified molecule is not toxic. It specifically interacts with components of the serum complement system, leading to complement activation which in turn leads to the consumption of complement activity.

THE COMPLEMENT SYSTEM

The complement system is an effector system of the immune system. It consists of approximately 20 plasma proteins and several receptors and regulatory proteins in the membranes of various blood cells and cells of the immune system. Complement activation occurs through two pathways called the classical pathway and the alternative pathway, respectively. The classical pathway of complement is typically activated by an antigen-antibody complex, whereas alternative pathway activation is typically antibody-independent. The process of complement activation through either pathway is a cascade mechanism where an activated protease acts on the zymogen of another protease, thereby leading to significant amplification of the initiation signal. At the step of C5 activation, the two pathways merge into the common final pathway of membrane attack (Fig. 1).

Natural Toxins II, Edited by B. R. Singh and A. T. Tu
Plenum Press, New York, 1996

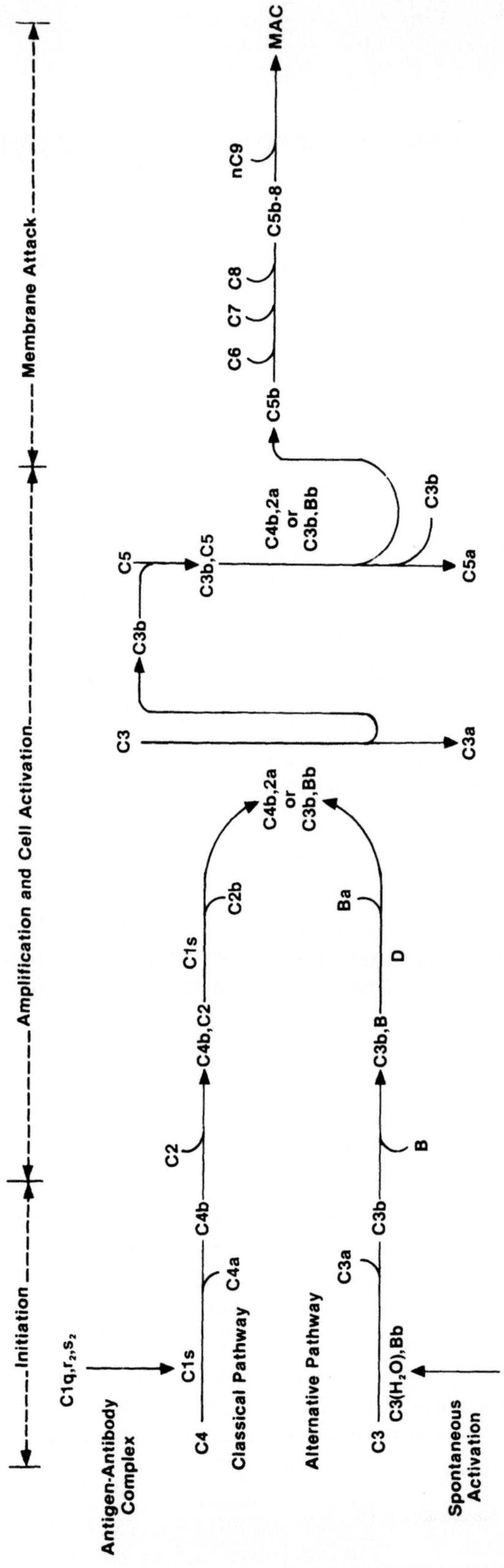

Figure 1. Pathways of complement activation (taken from ref. 1).

The biological functions of complement activation are mediated by the various activation products. The most impressive biological action is the physical destruction of a foreign cell by the so-called membrane attack complex (MAC), which is a high molecular weight complex consisting of the terminal complement components. Multiples of this membrane attack complex insert into the target membrane causing multiple channel formation, physical destruction of the membrane architecture, and eventually cell death. Other biological functions of complement activation include the initiation of the inflammatory response through the anaphylatoxins C3a and C5a, which are released during the activation process from the complement components C3 and C5, respectively. The depostion of C3b onto particles or cells is a process called opsonization which leads to subsequent phagocytosis of these particles by macrophages. Lastly, some complement activation peptides have been shown to exhibit an immunoregulatory function. Several reviews exist on the biochemistry and physiology of complement and its role in the pathogenesis of diseases (2-4).

STRUCTURE OF CVF

Cobra venom factor is a glycoprotein with a molecular mass of ~149,000 Da. It consists of three disulfide-linked chains with molecular masses of ~68,500 Da (α-chain), ~48,500 Da (β-chain), and ~32,000 Da (γ-chain) (Fig. 2). The γ-chain shows size heterogeneity which is most likely due to differential processing at the *C*-terminus (5,6). From circular dichroism spectroscropy the secondary structure of CVF was determined to consist of 11% α-helix, 47% ß-sheet, 18% β-turn, and 20% remainder (7). High resolution transmission electron microscropy revealed a somewhat irregular and elongated ellipsoid structure of the molecule with dimensions of 137 Å by 82 Å (7,8).

CVF is a glycoprotein with a carbohydrate content of ~7.4% (w/w). The oligosaccharide portion of CVF consists of three *N*-linked oligosaccharide chains of the complex type, of which two are in the α-chain and one is in the β-chain (6,9,10). The major oligosaccharide could be identified as a symmetric fucosylated biantennary complex type chain with an unusual α-galactosyl residue at its nonreducing end (10,11) (Fig. 3).

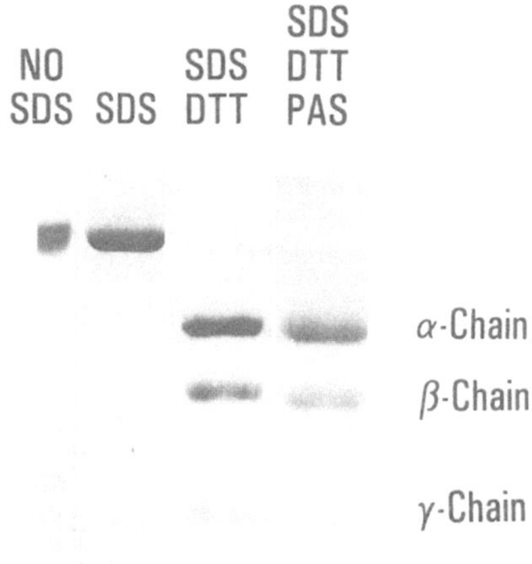

Figure 2. Polyacrylamide gel electrophoresis of purified CVF. Shown is the purified protein under nonreducing conditions in the absence and presence of SDS. Reducing conditions reveal the three-chain structure of CVF, the size heterogeneity of the γ-chain, and the absence of carbohydrate in the γ-chain (taken from ref. 6).

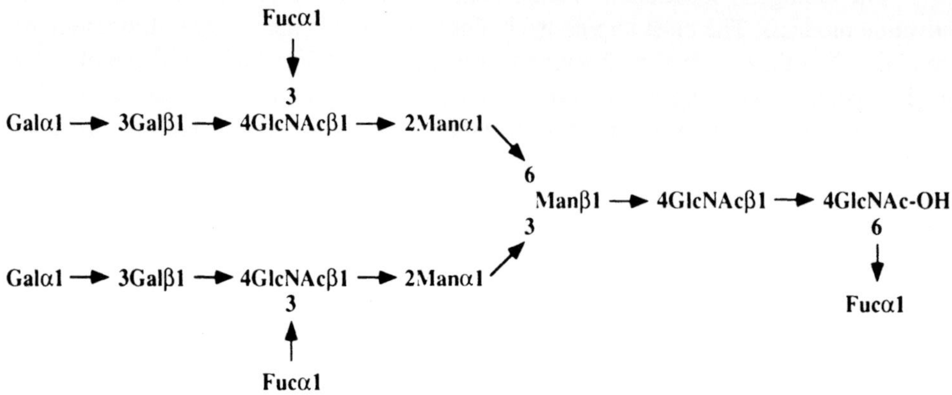

Figure 3. Structure of the major oligosaccharide of CVF.

MOLECULAR INTERACTION OF CVF WITH COMPLEMENT AND FUNCTIONAL HOMOLOGY WITH THE THIRD COMPONENT OF COMPLEMENT

When CVF is added to human or mammalian serum it activates complement and leads to complement consumption. In serum, CVF binds to factor B of the alternative pathway. When factor B is in complex with CVF, factor B is cleaved by factor D into Ba, the activation peptide that is released, and Bb that remains bound to CVF. The bimolecular complex CVF,Bb is a C3 convertase which cleaves C3 (12-15) (Fig. 4).

In addition to its binding site for factor B, CVF has a binding site for C5 (16). C5, when bound to CVF, becomes susceptible to cleavage by the CVF,Bb enzyme. Accordingly, the CVF,Bb enzyme has not only C3-cleaving activity, but also C5-cleaving activity and is referred to as C3/C5 convertase.

The formation and function of the CVF-dependent C3/C5 convertase are analogous to the formation and function of the C3b-dependent C3 convertase that is formed during the activation of the alternative pathway. Like CVF, C3b, the activated form of complement component C3, binds factor B to form the complex C3b,B, which is also cleaved by factor D into the bimolecular enzyme C3b,Bb and Ba. The bimolecular complex C3b,Bb is the C3 convertase of the alternative pathway that activates C3 analogously to the C3 convertase formed with CVF. Like CVF, C3b also has a binding site for C5, to which C5 needs to be bound in order to be susceptible to cleavage by the C3b,Bb enzyme. Accordingly, the C3b,Bb enzyme is also called the C3/C5 convertase of the alternative pathway (2).

Formation:

$$\text{CVF} + \text{Factor B} \xrightarrow[\text{Mg}^{2+}]{\text{Factor D}} \text{CVF,Bb} + \text{Ba}$$

$$\text{C3b} + \text{Factor B} \xrightarrow[\text{Mg}^{2+}]{\text{Factor D}} \text{C3b,Bb} + \text{Ba}$$

Function:

$$\text{C3} + H_2O \xrightarrow{\text{C3b,Bb or CVF,Bb}} \text{C3a} + \text{C3b}$$

Figure 4. Formation and enzymatic activity of CVF-dependent and C3b-dependent C3 convertases.

Although the two C3/C5 convertases CVF,Bb and C3b,Bb are highly homologous enzymes, sharing the molecular architecture, the active site-bearing Bb subunit, and the substrate specificity, the two enzymes exhibit considerable functional differences. The CVF,Bb enzyme is physico-chemically far more stable than C3b,Bb, it is resistant to inactivation by the regulatory proteins factors H and I, it exhibits different kinetic properties, and it does not require additional C3b for C5 cleavage (1).

In addition to the functional homology, CVF and C3 have been shown to exhibit several structural similarities, including immunologic crossreactivity, amino acid composition, circular dichroism sprectra and secondary structure, electron microscopic ultrastructure, and *N*-terminal amino acid sequences (1). Nevertheless, significant structural differences exist between the two molecules. Whereas C3 is a two-chain molecule with an apparent molecular mass, dependent on the species, of 170 to 190 kDa (17,18), CVF is a three-chain molecule with an apparent molecular mass of 149 kDa (6) that resembles C3c, one of the physiologic activation products of C3 (1,7). Another significant structural difference between C3 and CVF lies in their glycosylation. The extent of glycosylation of C3 molecules is less than that of CVF. Furthermore, the number, size, structure, and location of the oligosaccharide chains varies from species to species (1,19-21). The structural homology between CVF and C3 was fully confirmed once the derived primary structure of CVF became available from the molecular cloning of the molecule (see below).

STRUCTURE OF CVF mRNA

The molecular cloning of CVF was accomplished by using a λgt11 cDNA library prepared from a poly-A$^+$ RNA from freshly obtained cobra venom glands (22). The screening of this expression library was performed with a monospecific polyclonal goat anti-CVF antiserum.

The CVF mRNA is >5948 nucleotides in length. It contains a single open reading frame of 4926 nucleotides, coding for a pre-pro-protein of 1642 amino acid residues (Fig. 5). The mRNA has a 5'-untranslated region of at least three nucleotides and a 3'-untranslated region of 999 nucleotides, and a poly-A tail of at least 20 nucleotides. The pre-pro-protein has a signal sequence of 22 residues with a core rich in hydrophobic amino acids. The signal sequence is followed by the 627 amino acid α-chain. Immediately following the *C*-terminus of the α-chain, there are four arginine residues and a 79 amino acid peptide resembling a C3a anaphylatoxin. The γ-chain begins at position 711. The exact position of the *C*-terminus of the γ-chain is still unknown and is apparently heterogeneous. Based on the size of the γ-chain on SDS PAGE and the lack of a free sulfhydryl group (7), the γ-chain must terminate in the immediate proximity to the *C*-terminus of what would be C3g in C3. Assuming the *C*-terminus of the γ-chain to correspond to the *C*-terminus of the C3g domain in human C3, the γ-chain would be 252 residues long with its *C*-terminus being at position 962. The γ-chain is followed by a 279 residues long "C3d" domain which is not present in the mature CVF protein. The "C3d" domain in pro-CVF is 38 residues shorter at its *C*-terminus compared to human C3d. The ß-chain of CVF begins at position 1242 and extends for 379 residues to the end of the open reading frame.

The structure of CVF mRNA shows that CVF is, like C3 molecules, encoded as a single-chain pre-pro-protein that is subsequently processed into the mature molecule. The structure of the pre-pro-CVF molecule confirms the proposed chain relationships between CVF and C3, namely that the CVF α-chain is homologous to the C3 β-chain and that the 989 amino acid precursor chain for the CVF γ- and β-chains is homologous to the C3 α-chain (Fig. 6). Four arginine residues are present at the chain junction in both pro-CVF and pro-C3 (Fig. 7).

```
                    Signal Sequence                                     -1|+1  N-terminus of α-chain
 -22   M  E  R  M  A  L  Y  L  V  A  A  L  L  I  G  F  P  G  S  S  H  G | A  L  Y  T  L  I  T     7
   1  CCCATGGAGAGGATGGCTCTCTATCTGGTGGCTGCTCTATTGATTGGTTTTCCAGGGTCTTCTCATGGGGCTCTCTACACCCTCATCACC    90

   8   P  A  V  L  R  T  D  T  E  E  Q  I  L  V  E  A  H  G  D  S  T  P  K  Q  L  D  I  F  V  H    37
  91  CCTGCTGTTTTGCGAACAGACACAGAAGAGCAAATTTTGGTGGAGGCCCATGGAGACAGTACTCCAAAACAGCTTGACATCTTTGTTCAT   180

  38   D  F  P  R  K  Q  K  T  L  F  Q  T  R  V  D  M  N  P  A  G  G  M  L  V  T  P  T  I  E  I    67
 181  GATTTTCCACGGAAGCAGAAAACCTTGTTCCAAACCAGAGTAGATATGAATCCAGCAGGAGGCATGCTTGTCACTCCAACTATAGAGATT   270

  68   P  A  K  E  V  S  T  D  S  R  Q  N  Q  Y  V  V  V  Q  V  T  G  P  Q  V  R  L  E  K  V  V    97
 271  CCAGCAAAAGAAGTGAGTACGGACTCCAGGCAAAATCAATATGTGGTTGTGCAAGTAACTGGTCCTCAAGTGAGATTGGAAAAGGTGGTT   360

  98   L  L  S  Y  Q  S  S  F  L  F  I  Q  T  D  K  G  I  Y  T  P  G  S  P  V  L  Y  R  V  F  S   127
 361  CTCCTTTCTTACCAGAGTAGCTTTCTGTTTATCCAGACAGATAAAGGCATCTATACACCAGGGTCTCCAGTACTCTATCGTGTTTTTTCT   450
            CHO         CHO
 128   M  D  H  N  T  S  K  M  N  K  T  V  I  V  E  F  Q  T  P  E  G  I  L  V  S  S  N  S  V  D   157
 451  ATGGATCACAACACAAGCAAGATGAACAAAACTGTGATTGTTGAGTTTCAGACTCCAGAAGGCATTCTTGTCAGTTCTAATTCAGTTGAC   540
                                                                                            CHO
 158   L  N  F  F  W  P  Y  N  L  P  D  L  V  S  L  G  T  W  R  I  V  A  K  Y  E  H  S  P  E  N   187
 541  CTAAACTTCTTCTGGCCTTACAATTTACCAGACCTTGTCAGTTTGGGGACTTGGAGGATTGTGGCCAAATATGAACATTCCCCAGAGAAT   630

 188   Y  T  A  Y  F  D  V  R  K  Y  V  L  P  S  F  E  V  R  L  Q  P  S  E  K  F  F  Y  I  D  G   217
 631  TATACTGCATATTTTGATGTCAGGAAATATGTGTTGCCAAGCTTTGAAGTCCGTCTGCAACCATCAGAGAAGTTTTTTTACATTGACGGC   720

 218   N  E  N  F  H  V  S  I  T  A  R  Y  L  Y  G  E  E  V  E  G  V  A  F  V  L  F  G  V  K  I   247
 721  AATGAAAATTTCCACGTGTCTATCACTGCAAGGTACTTGTATGGAGAGGAAGTGGAAGGTGTGGCCTTTGTCCTCTTTGGAGTGAAAATA   810

 248   D  D  A  K  K  S  I  P  D  S  L  T  R  I  P  I  I  D  G  D  G  K  A  T  L  K  R  D  T  F   277
 811  GATGATGCTAAAAAGAGTATTCCAGACTCACTCACGAGAATTCCGATTATTGATGGAGATGGGAAAGCAACACTAAAAAGAGATACATTC   900

 278   R  S  R  F  P  N  L  N  E  L  V  G  H  T  L  Y  A  S  V  T  V  M  T  E  S  G  S  D  M  V   307
 901  CGTTCTCGATTTCCAAATCTCAATGAGCTTGTTGGGCATACTCTGTATGCATCTGTAACAGTCATGACAGAATCAGGCAGTGATATGGTA   990

 308   V  T  E  Q  S  G  I  H  I  V  A  S  P  Y  Q  I  H  F  T  K  T  P  K  Y  F  K  P  G  M  P   337
 991  GTGACTGAGCAAAGCGGCATTCATATTGTGGCATCTCCCTATCAGATCCACTTCACAAAAACCCCCAAATATTTCAAGCCAGGAATGCCA  1080

 338   Y  E  L  T  V  Y  V  T  N  P  D  G  S  P  A  A  H  V  P  V  V  S  E  A  F  H  S  M  G  T   367
1081  TATGAACTGACGGTGTATGTTACCAACCCTGATGGCTCACCAGCTGCCCATGTGCCAGTGGTATCAGAGGCCTTTCATTCTATGGGAACC  1170

 368   T  L  S  D  G  T  A  K  L  I  L  N  I  P  L  N  A  Q  S  L  P  I  T  V  R  T  N  H  G  D   397
1171  ACTTTGAGTGATGGGACTGCTAAGCTCATCCTGAACATACCATTGAATGCTCAAAGCCTACCAATCACTGTTAGAACTAACCATGGAGAC  1260

 398   L  P  R  E  R  Q  A  T  K  S  M  T  A  I  A  Y  Q  T  Q  G  G  S  G  N  Y  L  H  V  A  I   427
1261  CTCCCAAGAGAACGCCAGGCAACAAAGTCCATGACAGCCATAGCCTACCAAACCCAGGGAGGATCTGGAAACTATCTTCATGTAGCCATT  1350

 428   T  S  T  E  I  K  P  G  D  N  L  P  V  N  F  N  V  K  G  N  A  N  S  L  K  Q  I  K  Y  F   457
1351  ACATCTACAGAGATTAAGCCCGGAGATAACTTACCTGTCAATTTCAATGTGAAGGGCAATGCAAATTCACTGAAGCAGATCAAATATTTC  1440

 458   T  Y  L  I  L  N  K  G  K  I  F  K  V  G  R  Q  P  R  R  D  G  Q  N  L  V  T  M  N  L  H   487
1441  ACATACCTCATATTGAATAAAGGGAAGATTTTCAAGGTTGGCAGGCAACCCAGGAGAGATGGGCAGAATCTGGTGACCATGAATCTGCAT  1530

 488   I  T  P  D  L  I  P  S  F  R  F  V  A  Y  Y  Q  V  G  N  N  E  I  V  A  D  S  V  W  V  D   517
1531  ATCACTCCAGATCTCATCCCTTCCTTCCGGTTTGTGGCTTACTACCAAGTGGGAAACAACGAAATTGTGGCTGATTCTGTCTGGGTGGAT  1620

 518   V  K  D  T  C  M  G  T  L  V  V  K  G  D  N  L  I  Q  M  P  G  A  A  M  K  I  K  L  E  G   547
1621  GTGAAGGATACCTGCATGGGAACGTTGGTTGTGAAAGGAGACAATCTAATACAAATGCCAGGAGCTGCAATGAAAATCAAATTGGAAGGG  1710

 548   D  P  G  A  R  V  G  L  V  A  V  D  K  A  V  Y  V  L  N  D  K  Y  K  I  S  Q  A  K  I  W   577
1711  GATCCAGGTGCTCGGGTTGGTCTTGTGGCTGTGGACAAAGCAGTATATGTTCTCAATGATAAATATAAGATTAGCCAAGCTAAGATATGG  1800

 578   D  T  I  E  K  S  D  F  G  C  T  A  G  S  G  Q  N  N  L  G  V  F  E  D  A  G  L  A  L  T   607
1801  GACACAATAGAAAAGAGTGACTTTGGCTGTACAGCTGGCAGTGGCCAGAATAATCTGGGTGTGTTTGAAGATGCTGGACTGGCTCTGACA  1890
                                                  C-terminus of α-chain |           | N-terminus of "C3a"
 608   T  S  T  N  L  N  T  K  Q  R  S  A  A  K  C  P  Q  P  A  N | R  R  R  R | S  S  V  L  L  L   637
1891  ACCAGCACTAATCTCAACACCAAACAGAGATCAGCTGCAAAGTGTCCTCAGCCTGCAAATCGGAGGCGTCGCAGTTCTGTTTTGCTGCTT  1980
         CHO
 638   D  S  N  A  S  K  A  A  E  F  Q  D  Q  D  L  R  K  C  C  E  D  V  M  H  E  N  P  M  G  Y   667
1981  GACAGCAACGCAAGCAAAGCGGCAGAATTTCAGGATCAAGACCTGCGTAAATGCTGTGAAGATGTCATGCATGAGAACCCCATGGGGTAC  2070

 668   T  C  E  K  R  A  K  Y  I  Q  E  G  D  A  C  K  A  A  F  L  E  C  C  R  Y  I  K  G  V  R   697
2071  ACTTGTGAAAAGCGTGCAAAATACATCCAGGAGGGAGATGCTTGTAAGGCTGCCTTCCTTGAATGCTGTCGCTACATCAAGGGGGTCCGA  2160
                  C-terminus of "C3a" | N-terminus of γ-chain
 698   D  E  N  Q  R  E  S  E  L  F  L  A  R | D  D  N  E  D  G  F  I  A  D  S  D  I  I  S  R  S   727
2161  GATGAAAACCAACGGGAGAGCGAGTTGTTTCTGGCAAGAGATGATAATGAAGATGGTTTCATAGCAGATAGTGATATCATCTCAAGGTCT  2250

 728   D  F  P  K  S  W  L  W  L  T  K  D  L  T  E  E  P  N  S  Q  G  I  S  S  K  T  M  S  F  Y   757
2251  GATTTCCCCAAGAGTTGGTTGTGGCTAACAAAGGACTTGACCGAGGAGCCTAACAGTCAAGGGATTTCAAGCAAGACAATGTCTTTTTAT  2340
```

Figure 5. cDNA and derived amino acid sequence of CVF. The signal sequence, the *N*- and *C*-termini of the chains, the potential glycosylation sites, and the thioester site are indicated. Amino acid residue numbering starts at the *N*-terminus of the pro-CVF molecule (*N*-terminus of the mature α-chain) (modified from ref. 22).

758 L R D S I T T W V V L A V S F T P T K G I C V A E P Y E I R 787
2341 CTGAGGGATTCCATCACAACCTGGGTGGTGCTGGCTGTAAGCTTTACACCCACCAAAGGGATCTGTGTGGCTGAACCTTATGAAATAAGA 2430

788 V M K V F F I D L Q M P Y S V V K N E Q V E I R A I L H N Y 817
2431 GTCATGAAAGTCTTCTTCATTGATCTTCAAATGCCATATTCAGTAGTGAAGAATGAGCAGGTGGAGATTCGAGCTATTCTGCACAACTAC 2520

818 V N E D I Y V R V E L L Y N P A F C S A S T K G Q R Y R Q Q 847
2521 GTTAACGAGGATATTTATGTGCGAGTGGAACTGTTATACAACCCAGCCTTCTGCAGTGCTTCCACAAAAGGACAAAGATACCGACAGCAG 2610

848 F P I K A L S S R A V P F V I V P L E Q G L H D V E I K A S 877
2611 TTCCCAATTAAAGCCCTGTCCTCCAGAGCAGTACCGTTTGTGATAGTCCCATTAGAGCAAGGATTGCATGATGTTGAGATTAAAGCAAGT 2700

878 V Q E A L W S D G V R K K L K V V P E G V Q K S I V T I V K 907
2701 GTCCAGGAAGCGTTGTGGTCAGACGGTGTGAGGAAGAAACTGAAAGTTGTACCTGAAGGGGTACAGAAATCCATTGTGACTATTGTTAAA 2790

908 L D P R A K G V G G T Q L E V I K A R K L D D R V P D T E I 937
2791 CTGGACCCAAGGGCAAAAGGAGTTGGTGGAACACAGCTAGAAGTGATCAAAGCCCGCAAATTAGATGACAGAGTGCCTGACACAGAAATT 2880

C-terminus of γ-chain(?)

938 E T K I I I Q G D P V A Q I I E N S I D G S K L N H L I I T 967
2881 GAAACCAAGATTATCATCCAAGGTGACCCTGTGGCTCAGATTATTGAAAACTCAATTGATGGAAGTAAACTCAACCATCTCATTATCACT 2970

Thioester Site

968 P S G C G E Q N M I R M A A P V I A T Y Y L D T T E Q W E T 997
2971 CCTTCTGGCTGTGGGGAGCAAAATATGATCCGCATGGCCGCACCAGTTATTGCCACCTACTACCTGGACACCACAGAGCAGTGGGAGACT 3060

998 L G I N R R T E A V N Q I V T G Y A Q Q M V Y K K A D H S Y 1027
3061 CTCGGCATAAATCGCAGGACTGAAGCTGTCAATCAGATCGTGACTGGTTATGCCCAGCAGATGGTGTACAAGAAAGCAGATCATTCCTAT 3150

1028 A A F T N R A S S S W L T A Y V V K V F A M A A K M V A G I 1057
3151 GCAGCATTTACAAACCGTGCATCTAGTTCTTGGCTAACAGCATATGTCGTAAAAGTCTTTGCCATGGCTGCCAAAATGGTAGCAGGCATT 3240

1058 S H E I I C G G V R W L I L N R Q Q P D G A F K E N A P V L 1087
3241 AGTCATGAAATCATTTGTGGAGGTGTGAGGTGGCTGATTCTGAACAGGCAACAACCAGATGGAGCGTTCAAAGAAAATGCCCCTGTACTT 3330

1088 S G T M Q G G I Q G A E E E V Y L T A F I L V A L L E S K T 1117
3331 TCTGGAACAATGCAGGGAGGAATTCAAGGTGCTGAAGAAGAAGTATATTTAACAGCTTTCATTCTGGTTGCGTTGTTGGAATCCAAAACA 3420

1118 I C N D Y V N S L D S S I K K A T N Y L L K K Y E K L Q R P 1147
3421 ATCTGCAATGACTATGTCAATAGTCTAGACAGCAGCATCAAGAAGGCCACAAATTATTTACTCAAAAAGTATGAGAAACTGCAAAGGCCT 3510

1148 Y T T A L T A Y A L A A A D Q L N D D R V L M A A S T G R D 1177
3511 TACACTACAGCCCTCACAGCCTATGCTTTGGCTGCTGCAGACCAACTCAATGATGACAGGGTACTCATGGCAGCATCAACAGGAAGGGAT 3600

1178 H W E E Y N A H T H N I E G T S Y A L L A L L K M K K F D Q 1207
3601 CATTGGGAAGAATACAATGCTCACACCCACAACATTGAAGGCACTTCCTATGCCTTGTTGGCCCTGCTGAAAATGAAGAAATTTGATCAA 3690

1208 T G P I V R W L T D Q N F Y G E T Y G Q T Q A T V M A F Q A 1237
3691 ACTGGTCCCATAGTCAGATGGCTGACAGATCAGAATTTTTATGGGGAAACATATGGACAAACCCAAGCAACAGTTATGGCATTTCAAGCT 3780

N-terminus of β-chain

1238 L A E Y E I Q M P T H K D L N L D I T I E L P D R E V P I R 1267
3781 CTTGCTGAATATGAGATTCAGATGCCTACCCATAAGGACTTAAACTTAGATATTACTATTGAACTGCCAGATCGAGAAGTACCTATAAGG 3870

1268 Y R I N Y E N A L L A R T V E T K L N Q D I T V T A S G D G 1297
3871 TACAGAATTAATTATGAAAATGCTCTCCTGGCTCGGACAGTAGAGACCAAACTCAACCAAGACATCACTGTGACAGCATCAGGTGATGGA 3960

CHO

1298 K A T M T I L T F Y N A Q L Q E K A N V C N K F H L N V S V 1327
3961 AAAGCAACAATGACCATTTTGACATTCTATAACGCACAGTTGCAGGAGAAGGCAAATGTTTGCAATAAATTTCATCTTAATGTTTCTGTT 4050

1328 E N I H L N A M G A K G A L M L K I C T R Y L G E V D S T M 1357
4051 GAAAACATCCACTTGAATGCAATGGGAGCCAAGGGAGCCCTCATGCTCAAGATCTGCACAAGGTATCTGGGAGAAGTTGATTCTACAATG 4140

1358 T I I D I S M L T G F L P D A E D L T R L S K G V D R Y I S 1387
4141 ACAATAATTGATATTTCTATGCTGACTGGTTTTCTCCCTGATGCTGAAGACCTTACAAGGCTTTCTAAAGGAGTGGACAGATACATCTCC 4230

1388 R Y E V D N N M A Q K V A V I I Y L N K V S H S E D E C L H 1417
4231 AGATATGAAGTTGACAATAATATGGCTCAGAAAGTAGCTGTTATCATTTACTTAAACAAGGTCTCCCACTCTGAAGATGAATGCCTGCAC 4320

1418 F K I L K H F E V G F I Q P G S V K V Y S Y Y N L D E K C T 1447
4321 TTTAAGATTCTCAAGCATTTTGAAGTTGGCTTCATTCAGCCAGGATCAGTCAAGGTGTACAGCTACTACAATCTAGATGAAAAATGTACC 4410

1448 K F Y H P D K G T G L L N K I C I G N V C R C A G E T C S S 1477
4411 AAGTTCTACCATCCAGATAAAGGAACAGGCCTTCTCAATAAGATATGTATTGGTAACGTTTGCCGATGTGCAGGAGAAACCTGTTCCTCG 4500

1478 L N H Q E R I D V P L Q I E K A C E T N V D Y V Y K T K L L 1507
4501 CTCAACCATCAGGAAAGGATTGATGTTCCATTACAAATTGAAAAAGCCTGCGAGACGAATGTGGATTATGTCTACAAAACCAAGCTGCTT 4590

1508 R I E E Q D G N D I Y V M D V L E V I K Q G T D E N P R A K 1537
4591 CGAATAGAAGAACAAGATGGTAATGATATCTATGTCATGGATGTTTTAGAAGTTATTAAACAAGGTACTGACGAAAATCCACGAGCAAAG 4680

1538 T H Q Y I S Q R K C Q E A L N L K V N D D Y L I W G S R S D 1567
4681 ACCCACCAGTACATAAGTCAAAGGAAATGCCAGGAGGCTCTGAATCTGAAGGTGAATGATGATTATCTGATCTGGGGTTCCAGGAGTGAC 4770

```
1568  L  L  P  T  K  D  K  I  S  Y  I  I  T  K  N  T  W  I  E  R  W  P  H  E  D  E  C  Q  E  E  1597
4771  CTGTTGCCCACGAAAGATAAAATTTCCTACATCATTACAAAGAACACATGGATTGAGAGATGGCCACATGAAGACGAATGTCAGGAAGAA 4860
                                                        C-terminus of β-chain |
1598  E  F  Q  K  L  C  D  D  F  A  Q  F  S  Y  T  L  T  E  F  G  C  P  T  |                     1620
4861  GAATTCCAAAAGTTGTGTGATGACTTTGCTCAGTTTAGCTACACATTGACTGAGTTTGGCTGCCCTACTTAAAAGTTCAGAAGAATCAAT 4950

4951  GATAGGAAGGAAATTCTCAGAAGACAGATTTTTGAGCCAATGCATATATGTTACTTTGCCTCTTGATCTTTTAGTTTTATGTCAATTTGC 5040

5041  TCTGTTATTTTCCCTTAAATTGTTTATACATAAAATAAATAATCGATTTCTTACTTTGATATGTTCTTGATTTTTAATAAACAATGGTGA 5130

5131  TTCATGATTATTTTTTTCTTCTTCTGATCCATCCAATATTTGAAGTGCTCTGAACAGAGCACTTATGGAGTAATGTTTAGTGATGGATG 5220

5221  AATAAGTTGGTGAGTCAATATTATCAGGCCCTATATACTCTTATGGAAGATCGATTTGTACCCAAAGAAACATAGATTGAAATGTGTTAC 5310

5311  TTTGAAAACAGAGGTTTCAGTTGTATATGTTTACACTTGGATACAATCTTAACTCTTAATAAACACTGATCTCAGAACATTTAACAGCTG 5400

5401  CTATTTAATAATGACAAAATATTCTTTGACTGCACCCACAGAAAACATTGCATTACATTAGAATGGGTTTTATCAGATGACTAAGTCTGC 5490

5491  TAGACTTGCCATCTGTCAAAATGTGCCTCTTCCCCAGCTCCAACTTTAAGGATAGTAACTAATAGATGTTCTCTCATTGGCTCCTGACAG 5580

5581  AGGTGTGGTAGCCACTGAGTTTCCCTGGATGACACTAGAAGCTGGCAGCACACTGCAGCCTGGTGGAGGGGCCTCTTTTGCTATCCCATG 5670

5671  AGCTTCTATTCATCCTCTTATCTGTTGGGATGGGGATGGGACGTCTCTGATTTTCCAGGTATACAGGTGATCTCATTTACTAACATCACC 5760

5761  ACTAACTTCAAGGATTGGTTGAGGGGTTATGCCAATGTGATTGAAGGTTTCACCCATGTGAATCTATTCTCCAATCCCAATGCTGTATCT 5850

5851  ATGCTGCTCATTTCTGCTTGTAAAAATGGTATAAAAAGAATAAACACTGCCCAGGCAGTCAGACATCTTTGGACACTGAAAAAAAAAAAA 5940

5941  AAAAAAAA                                                                                   5948
```

Figure 5. Concluded.

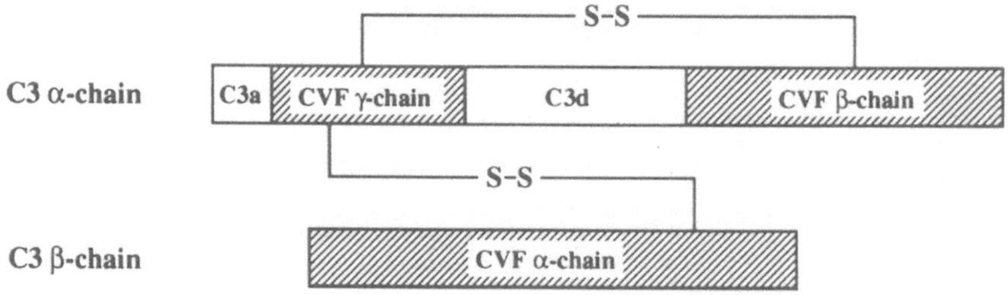

Figure 6. Schematic drawing showing the chain structures of C3 and CVF and their relationship.

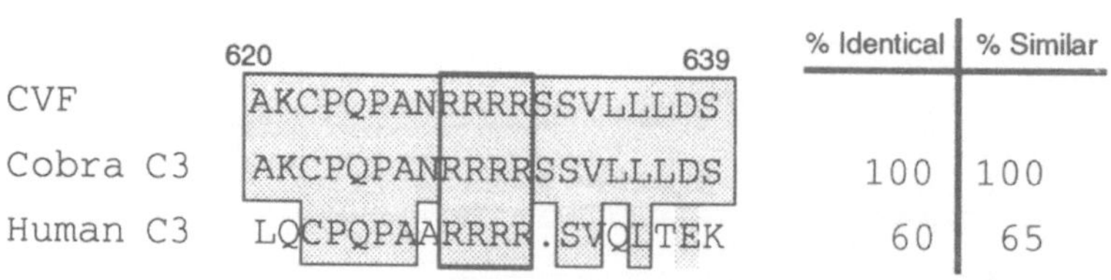

Figure 7. Sequence comparison of CVF, cobra C3, and human C3 at the chain junction. Comparisons were made with a sequence analysis program (23). Identical amino acid residues are boxed and shaded whereas conservative replacements are shaded only. Amino acid residue numbering is based on the CVF sequence as shown in Figure 5. The percent sequence identity and similarity with the CVF sequence is shown on the right. The four arginine residues in all three pre-pro-proteins are indicated (modified from ref. 22).

Table I. Codon frequency of CVF, cobra C3, and human C3

Amino Acid	Codon	CVF	Cobra C3	Human C3	Amino Acid	Codon	CVF	Cobra C3	Human C3
Gly	GGG	0.18	0.20	0.27	Trp	UGG	1.00	1.00	1.00
Gly	GGA	0.40	0.37	0.17	End	UGA	0.00	0.00	0.00
Gly	GGU	0.21	0.21	0.12	Cys	UGU	0.56	0.56	0.31
Gly	GGC	0.20	0.23	0.44	Cys	UGC	0.44	0.44	0.69
Glu	GAG	0.34	0.33	0.69	End	UAG	0.00	0.00	0.00
Glu	GAA	0.66	0.67	0.31	End	UAA	1.00	0.00	0.00
Asp	GAU	0.64	0.69	0.28	Tyr	UAU	0.57	0.57	0.18
Asp	GAC	0.36	0.31	0.72	Tyr	UAC	0.43	0.43	0.82
Val	GUG	0.39	0.42	0.50	Leu	UUG	0.22	0.20	0.10
Val	GUA	0.16	0.14	0.05	Leu	UUA	0.09	0.11	0.01
Val	GUU	0.26	0.22	0.12	Phe	UUU	0.59	0.59	0.22
Val	GUC	0.18	0.22	0.33	Phe	UUC	0.41	0.41	0.78
Ala	GCG	0.04	0.03	0.08	Ser	UCG	0.01	0.03	0.10
Ala	GCA	0.35	0.37	0.14	Ser	UCA	0.16	0.18	0.08
Ala	GCU	0.38	0.38	0.19	Ser	UCU	0.27	0.25	0.17
Ala	GCC	0.23	0.23	0.59	Ser	UCC	0.17	0.19	0.29
Arg	AGG	0.34	0.28	0.20	Arg	CGG	0.08	0.12	0.24
Arg	AGA	0.25	0.23	0.10	Arg	CGA	0.14	0.19	0.15
Ser	AGU	0.22	0.18	0.10	Arg	CGU	0.10	0.12	0.10
Ser	AGC	0.16	0.16	0.27	Arg	CGC	0.08	0.07	0.22
Lys	AAG	0.44	0.40	0.73	Gln	CAG	0.51	0.55	0.80
Lys	AAA	0.56	0.60	0.27	Gln	CAA	0.49	0.45	0.20
Asn	AAU	0.62	0.65	0.21	His	CAU	0.70	0.73	0.22
Asn	AAC	0.38	0.35	0.79	His	CAC	0.30	0.27	0.78
Met	AUG	1.00	1.00	1.00	Leu	CUG	0.29	0.29	0.49
Ile	AUA	0.16	0.16	0.10	Leu	CUA	0.07	0.07	0.08
Ile	AUU	0.52	0.56	0.15	Leu	CUU	0.13	0.14	0.05
Ile	AUC	0.32	0.28	0.75	Leu	CUC	0.21	0.19	0.27
Thr	ACG	0.05	0.05	0.15	Pro	CCG	0.03	0.04	0.14
Thr	ACA	0.43	0.40	0.15	Pro	CCA	0.55	0.52	0.22
Thr	ACU	0.28	0.31	0.12	Pro	CCU	0.29	0.32	0.19
Thr	ACC	0.23	0.25	0.58	Pro	CCC	0.13	0.12	0.45

The codon frequency is given as fractions of codons for a given amino acid. Data are for pre-pro-proteins. Data for human C3 are from (17). Data for cobra C3 are from (25).

The processing of the pro-CVF molecule includes, like that of pro-C3, the removal of the four arginine residues. The precursor chain is processed by removing the "C3a" and "C3d" domains thereby generating the CVF γ- and β-chains (compare Fig. 6). The enzymes involved in the processing of pro-CVF have not been identified although it has been suggested that the protease found in low concentrations in cobra venom which cleaves human C3 into a C3c-like molecule called C3o (24) is a processing enzyme for pro-CVF. These results further confirm that CVF with its three-chain structure resembles the two C3 degradation products C3c and C3o although the "C3d" domains missing in all three molecules differ in size and relative position (17,24). In this context, it is interesting to note that the *N*-terminal residue of the CVF γ-chain is an aspartic acid in contrast to the serine residue conserved in all C3s (compare Fig. 5) suggesting that the processing protease involved in the removal of the "C3a" domain from pro-CVF differs from the C3 convertase which cleaves C3a from C3 molecules.

Pro-CVF has five potential N-glycosylation sites of which are three in the α-chain (Asn 131, Asn 136, Asn 187), one in the "C3a" domain not present in the mature protein (Asn 640), and one in the β-chain (Asn 1324) (compare Fig. 5). Since the two potential glycosylation sites at Asn 131 and Asn 136 are so close together it is likely that for sterical reasons only one of the two will be glycosylated. Collectively, these results confirm that only the α- and β-chains of CVF are glycosylated and that there are two *N*-linked oligosaccharide chains in the α-chain and only one in the β-chain (6,10).

The GC composition of the open reading frame for pre-pro-CVF is 43.5% (for the whole cDNA: 42.4%). This is approximately the same as for cobra C3 (25), though more than 10% lower than found for mammalian C3s. Similarly, the codon usage for CVF closely resembles that of cobra C3 (Table I). The significance of these differences is not known and may reflect a different preference in codon usage in the species cobra since the GC percentage of cobra nerve growth factor (43.6%) and cobra acetylcholin receptor (44.2%) are similarly low (26,27).

Table II gives the amino acid composition of pre-pro-CVF, pro-CVF, CVF, and its three individual chains. Assuming the γ-chain to be made up of 252 amino acids, the mature CVF protein consists of 1258 amino acids, corresponding to a molecular mass of the protein of 141,454 Da. The protein molecular mass of the three individual chains is 69,502 Da (α-chain: 627 residues), 43,569 Da (ß-chain, 379 residues), and 28,383 Da (γ-chain, 252 residues), respectively.

The cDNA sequence and the derived protein sequence of CVF exhibit homology to the corresponding sequences of C3 from cobra and other species for which the full (cobra, human, mouse, rat, guinea pig) or partial (rabbit, *Xenopus laevis*) sequences have been reported. As is evident from Table III, the cDNA sequence of CVF is 93.3% identical to the cDNA sequence of cobra C3 and ~57% identical to C3 molecules from other vertebrates,

Table II. Derived amino acid composition of CVF

Residue	α-Chain	β-Chain	γ-Chain	Mature CVF	Pro-CVF	Pre-pro-CVF
Ala	37	17	13	67	110	113
Arg	23	15	12	50	70	71
Asn	34	23	8	65	81	81
Asp	35	27	18	80	97	97
Cys	3	13	2	18	27	27
Gln	27	16	12	55	75	75
Glu	27	29	17	73	98	99
Gly	42	17	12	71	91	94
His	13	10	2	25	32	33
Ile	37	30	24	91	109	110
Leu	50	35	18	103	135	139
Lys	39	29	19	87	107	107
Met	13	8	3	24	35	37
Phe	28	11	9	48	58	59
Pro	37	11	13	61	68	69
Ser	41	17	20	78	96	98
Thr	51	27	12	90	115	115
Trp	4	3	4	11	16	16
Tyr	26	18	7	51	71	72
Val	60	23	27	110	129	130
Total	627	379	252	1258	1620	1642
MW	69,502	43,569	28,383	141,454	182,186	184,502

The size of the γ-chain is an estimate, since the C-terminus is not known.

Table III. Sequence homology of CVF with C3 molecules

C3 Species	% Identity (Similarity) of Protein Sequence	% Identity of DNA Sequence
Cobra	84.7 (91.6)	93.3
Human	50.0 (69.1)	56.3
Mouse	51.7 (70.1)	57.3
Rat	51.0 (69.7)	57.8
Guinea Pig	51.3 (70.3)	56.1
Rabbit	48.9 (68.1)	56.8
Xenopus	47.6 (64.3)	56.6

Data are for pre-pro-proteins. The sequence comparisons to rabbit and *Xenopus* are with partial sequences.
Data for cobra C3 (25), human C3 (17), murine C3 (28,29), rat C3 (30), guinea pig C3 (31), rabbit C3 (32), and *Xenopus* C3 (33) were calculated using sequences as reported.

including human. At the protein level, CVF and cobra C3 exhibit a sequence identity of 84.7% and a sequence similarity of 91.6% if one allows for conservative substitutions. The protein sequence identity of CVF to the four fully sequenced mammalian C3s is ~51%, corresponding to a protein sequence similarity of ~70%. Protein sequence identity and similarity are ~3 - 5% lower for the partially sequenced portions of C3 from rabbit and *Xenopus*. The dot plot comparison of the CVF protein sequence with that of human C3 indicates that there are large stretches of homology between the two proteins, with the homology extending throughout the molecule (Fig. 8, right panel). The dot plot comparison of CVF with cobra C3 results in a continuous line (Fig. 8, left panel), confirming the extremely high degree of homology. The homology of CVF with cobra and human C3 is further evident by comparing the hydrophilicity/hydrophobicity plots as calculated from the amino acid sequences (Fig. 9). A careful analysis of the plots indicates that the pattern of

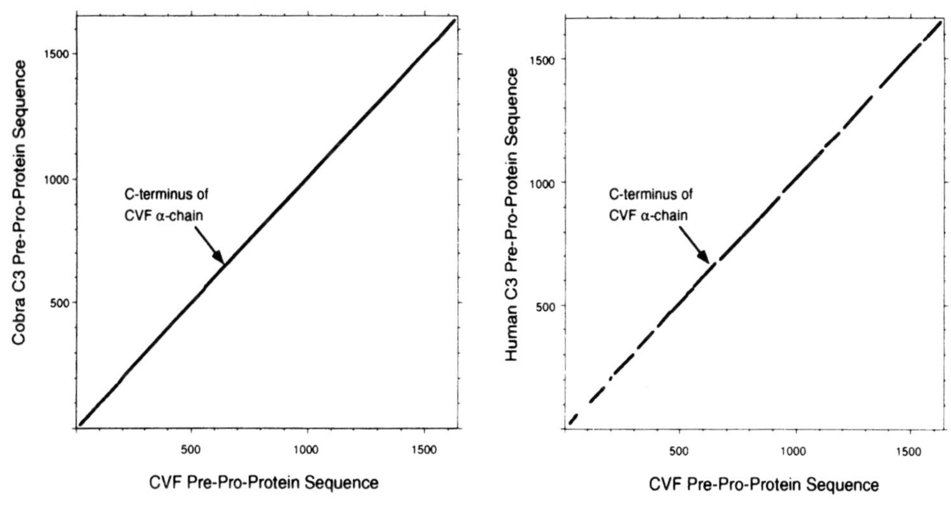

Figure 8. Dot plot comparisons of the CVF protein sequence to those of cobra C3 (left panel) and human C3 (right panel). The dot plots were generated using a sequence analysis program (34). In that program a moving window of 30 amino acid residues is compared, and where 21 or more residues are similar a dot is drawn on the graph.

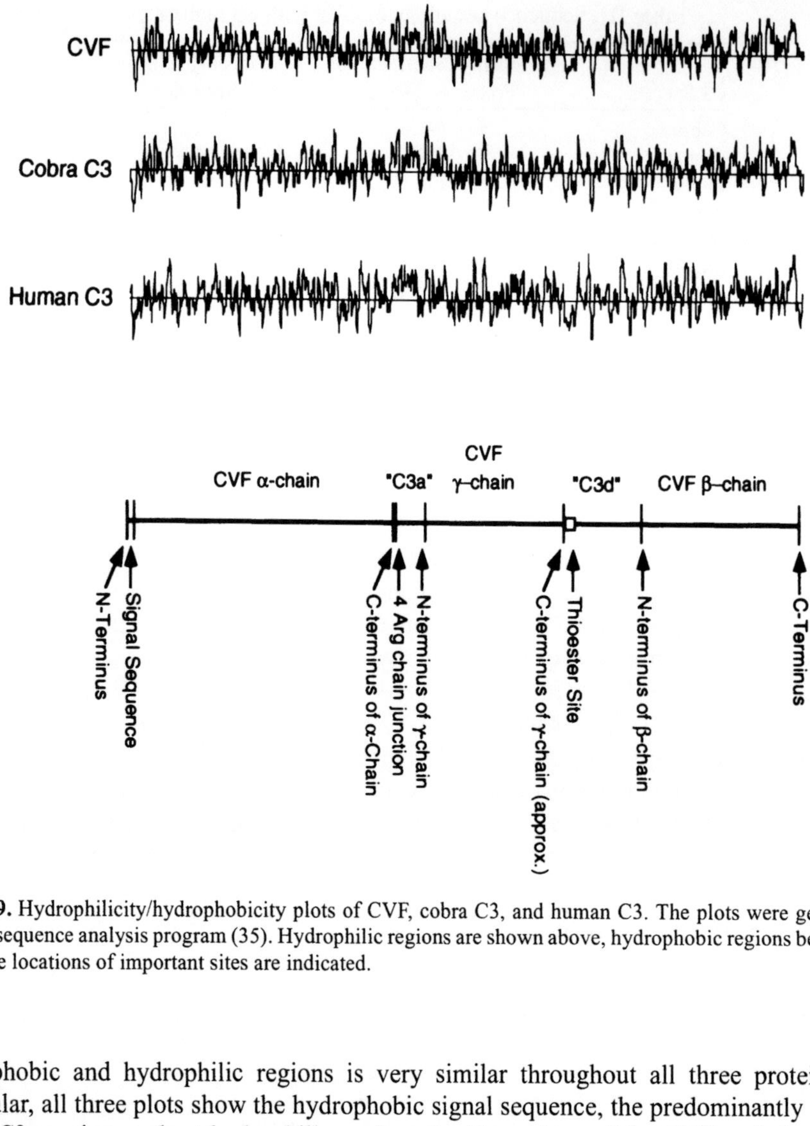

Figure 9. Hydrophilicity/hydrophobicity plots of CVF, cobra C3, and human C3. The plots were generated using a sequence analysis program (35). Hydrophilic regions are shown above, hydrophobic regions below the line. The locations of important sites are indicated.

hydrophobic and hydrophilic regions is very similar throughout all three proteins. In particular, all three plots show the hydrophobic signal sequence, the predominantly hydrophilic C3a region, a short hydrophilic peak at the *N*-terminus of the CVF γ-chain (corresponding to the *N*-terminus of the C3 α'-chain), a hydrophobic twin peak containing the thioester site, and a relatively wide hydrophilic peak just before the C-terminus.

Pro-CVF contains 27 cysteine residues of which are three in the α-chain, six in the "C3a" domain, two in the γ-chain, three in the "C3d" domain, and 13 in the β-chain. The total number and distribution of the cysteine residues in pro-CVF are identical to those in the C3 molecules of all other species, indicating that they are highly conserved. Based on the assignment of disulfide bonds in human C3 (36,37), the positions of the disulfide bonds in pro-CVF and mature CVF can be predicted (Fig. 10). Accordingly, the α-chain has one intra-chain disulfide bridge and one cysteine at position 522 involved in the formation of the inter-chain disulfide bridge with the γ-chain. The "C3a" domain has three intra-chain disulfide bridges. The γ-chain has only two cysteine residues, both involved in inter-chain disulfide bridges to the α-chain (residue 779) and β-chain (residue 835), respectively. The "C3d" domain contains one intra-chain disulfide bridge and the cysteine involved in the intramolecular thioester at position 971. The β-chain contains six intra-chain disulfide bridges and one cysteine (residue 1470) involved in the inter-chain disulfide bridge to the γ-chain.

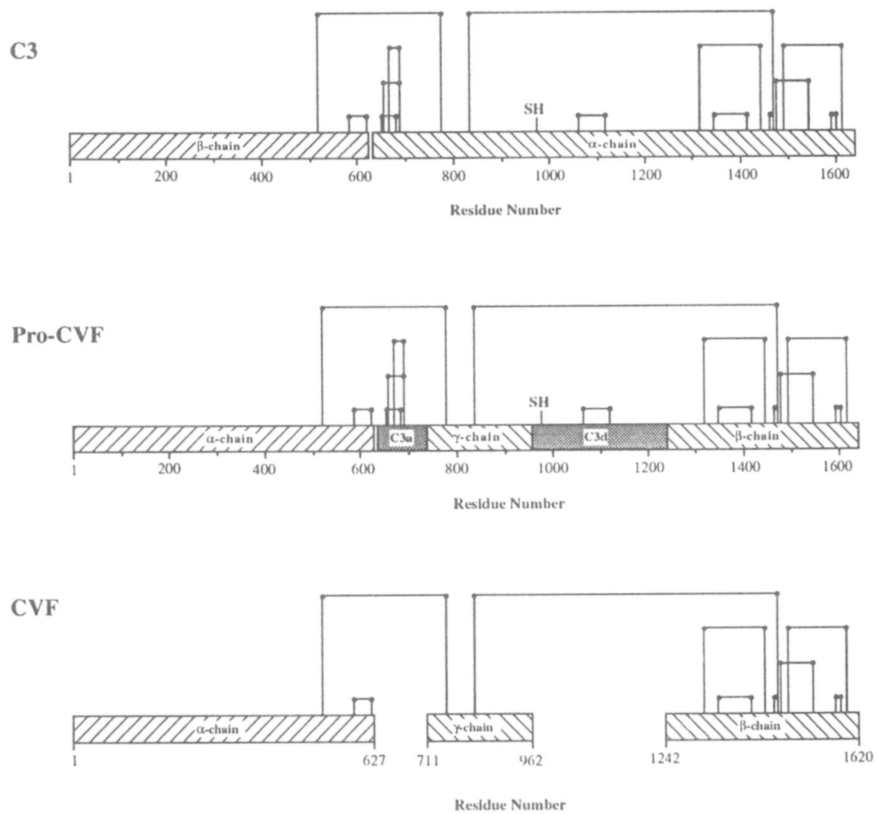

Figure 10. Schematic drawing showing the proposed positions of disulfide bonds in cobra C3, pro-CVF, and mature CVF. The data for C3 are taken from refs. 36,37. Proposed positions of disulfide bonds: Cys^{522}-Cys^{794}, Cys^{587}-Cys^{622}, Cys^{655}-Cys^{682}, Cys^{656}-Cys^{689}, Cys^{669}-Cys^{690}, Cys^{835}-Cys^{1470}, Cys^{1063}-Cys^{1119}, Cys^{1318}-Cys^{1446}, Cys^{1346}-Cys^{1415}, Cys^{1463}-Cys^{1468}, Cys^{1475}-Cys^{1547}, Cys^{1494}-Cys^{1618}, Cys^{1594}-Cys^{1603}. Position of the cysteine residue involved in the thioester: Cys^{971} (modified from ref. 22).

The mRNA for pro-CVF encodes for a thioester site homologous to C3 molecules including the cysteine and glutamine residues at positions 971 and 974, respectively (Fig. 11). Since the thioester site is in the "C3d" domain which is not present in the mature protein the

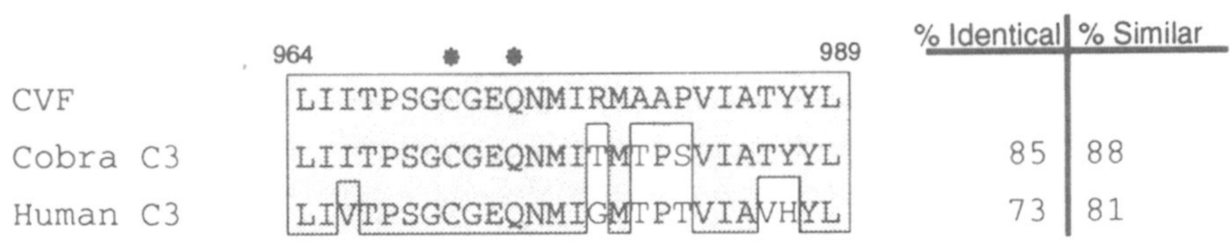

Figure 11. Sequence comparison of CVF, cobra C3, and human C3 at the thioester site. Comparisons were made with a sequence analysis program (23). Identical amino acid residues are boxed and shaded whereas conservative replacements are shaded only. Amino acid residue numbering is based on the CVF sequence as shown in Figure 5. The percent sequence identity and similarity with the CVF sequence is shown on the right. The cysteine and glutamine residues involved in the intramolecular thioester are identified by asterisks (modified from ref. 22).

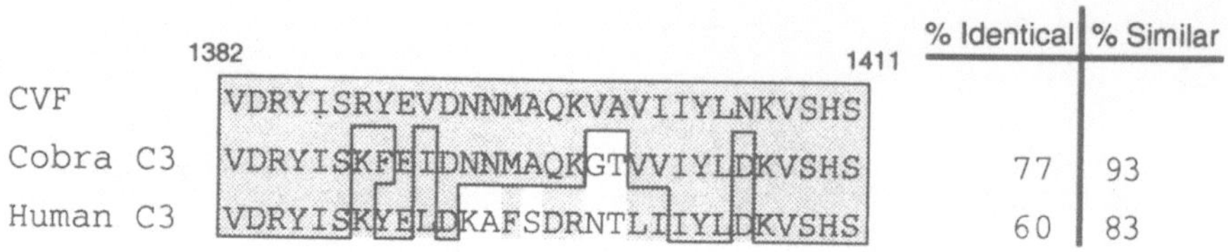

Figure 12. Sequence comparison of CVF, cobra C3, and human C3 at the proposed properdin binding site of human C3. Comparisons were made with a sequence analysis program (23). Identical amino acid residues are boxed and shaded whereas conserved replacements are shaded only. Amino acid residue numbering is based on the CVF sequence as shown in Figure 5. The percent sequence identity and similarity with the CVF sequence is shown on the right (modified from ref. 22).

conservation of the coding information for this structure does not seem to make sense. On the other hand, recent work has made it likely that the formation of the intramolecular thioester is a spontaneous non-enzymatic process not requiring an enzyme machinery (38). Accordingly, the formation of the thioester in pro-CVF in the cobra venom gland may not represent a fruitless energy expense, and the presence of the thioester sequence information may simply indicate that the CVF gene evolved rather recently by gene duplication from the C3 gene. This conclusion is consistent with the enormous overall homology of CVF and cobra C3 as well as the observation that pro-CVF contains another apparently conserved region not present in the mature protein, the anaphylatoxin ("C3a") domain.

With regard to the conservation of binding sites in the mature CVF protein for C3 ligands these are difficult to evaluate due to the high degree of sequence homology between CVF and cobra C3. A degree of homology higher than the overall homology between CVF and human C3 at a binding site may serve to predict its presence in CVF. Such is the case for the properdin binding site in the CVF β-chain (Fig.12) which exceeds with 60% identity and 83% similarity the overall sequence identity and similarity with human C3 (39). Whereas properdin is known to stabilize the rather labile C3b-dependent C3 convertase (40), the CVF-dependent C3 convertase is orders of magnitude more stable than the properdin-stabilized C3b-dependent convertase, making the significance of an apparently conserved properdin binding site in CVF obscure.

BIOLOGY OF COBRA VENOM FACTOR

One question that has never been properly answered is: What is the biological function of CVF in cobra venom? Cobra venom factor in purified form can be considered nontoxic. When introduced into the bloodstream of an animal, complement activation with consumption of complement component occurs (41). Intravascular complement activation may have some side effects, such as sequestration of neutrophils to the lungs with subsequent injury to lung tissue (42). However, unless massive amounts of CVF are administered i.v., the consequences of intravascular complement activation by CVF are insignificant compared with the toxic effects of the other venom components, particularly the neurotoxins and membrane toxins. It appears, therefore, that local complement activation at the site of venom injection into the prey animal may be the beneficial effect of CVF for the cobra. Local complement activation by CVF will release the anaphylatoxins C3a and C5a which, in turn, will cause the degranulation of mast cells and basophils, with subsequent increase of the vascular permeability. The increased vascular permeability locally at the site of venom injection may aid the toxic venom components to enter the bloodstream and to more quickly reach their site of action.

Whereas CVF has been able to activate the alternative pathway of complement in all vertebrate species tested, cobra serum is resistent to CVF. Incubation of cobra serum with CVF does not consume complement activity, and CVF assumes a different electrophoretic mobility upon incubation in cobra serum. Apparently, cobra serum has a protective mechanism against complement depletion by its own CVF, the biochemical nature of which is currently unknown but under investigation (43).

CURRENT USE OF COBRA VENOM FACTOR

The demonstration that CVF can be safely administered to laboratory animals with its only effect apparently being a temporary suppression of complement activity has made CVF a widely used experimental tool to study biological functions of complement by comparing normal with complement-depleted animals. Questions addressed include the role of complement in the generation of immune responses, in host defense against infectious agents, and in the pathophysiology of diseases. For a more detailed listing of investigations where CVF was used to deplete the complement activity of laboratory animals see ref. (1).

The property of the CVF,Bb enzyme to exhaustively activate complement has also been exploited for the selective killing of tumor cells by coupling of CVF to monoclonal antibodies with specificity for surface antigens of tumor cells. Antibody conjugates with CVF will target CVF to the cell surface, at which the CVF,Bb enzyme forms from complement factors B and D of the host complement system. The antibody-bound and, therefore, cell surface-bound CVF,Bb enzyme will continuously activate C3 and C5 and elicit complement-dependent target cell killing. The concept of coupling CVF to monoclonal antibodies has gained particular interest since it was learned from rather disappointing immunotherapeutic studies that almost all monoclonal antibodies against tumors may bind with good specificity, but that they do not mediate biological effector functions, such as complement activation. The coupling of CVF to monoclonal antibodies introduces a biological effector function, namely complement activation. Antibody conjugates with CVF have been shown to kill human melanoma cells (44), human lymphocytes and leukemia cells (45), and human neuroblastoma cells (46).

CONCLUSIONS AND OUTLOOK

CVF has been the focus of research for nearly a century. Much has been learned about its structure and function, and CVF has become and still remains a valuable research tool to study the biological functions of complement. After the molecule has now been cloned, current work is concentrating on recombinantly producing CVF which will require the proper posttranslational processing of the single-chain pro-CVF molecule to the three-chain active CVF.

Once the recombinant production of CVF is possible, the system can be used for a detailed structure/function analysis of the molecule. Of particular interest will be the determination of the molecular basis for the physical stability of the CVF-dependent C3 convertase and its resistance to the regulatory proteins Factors H and I, the major functional difference to the C3b-dependent C3 convertase. The high degree of homology with C3 may ultimately allow to produce a "human CVF", a derivative of human C3 with CVF-like functions. Such a "human CVF" can be expected to be significantly less immunogentic than CVF and may be a useful therapeutic agent to modulate complement activity, e.g. for the depletion of complement in the plasma of patients undergoing organ xenotransplantations (47) or with autoimmune diseases.

ACKNOWLEDGEMENTS

Work performed at Georgetown University was supported by NIH grants HL 29523, AI 26821, and CA 01039. Work performed at the University of Hamburg was supported by a research grant from Baxter Healthcare Corporation, McGaw Park, Ill., U.S.A. The authors wish to thank Ms. Leonie Bubenheim for expert typing of the manuscript.

REFERENCES

1. C.-W. Vogel: Cobra venom factor, the complement-activating protein of cobra venom. In *Handbook of Natural Toxins, Vol. 5, Reptile and Amphibian Venoms.* A. T. Tu, ed. Marcel Dekker, New York 1991, 147-188.
2. H. J. Müller-Eberhard. Molecular organization and function of the complement system. 1988. *Ann. Rev. Biochem.* 57, 321-347.
3. T. Kinoshita. Biology of complement: the overture. 1991. *Immunology Today* 12, 291-295.
4. B. P. Morgan, and M. J. Walport. Complement deficiency and disease. 1991. *Immunology Today* 12, 301-306.
5. G. Eggertsen, P. Lind, and J. Sjöquist: Molecular characterization of the complement activating protein in the venom of the Indian cobra (*Naja n. siamensis*). 1981. *Mol. Immunol.* 18, 125-133.
6. C.-W. Vogel, and H. J. Müller-Eberhard: Cobra venom factor: Improved method for purification and biochemical characterization. 1984. *J. Immunol. Meth.* 73, 203-220.
7. C.-W. Vogel, C. A. Smith, and H. J. Müller-Eberhard: Cobra venom factor: Structural homology with the third component of human complement. 1984. *J. Immunol.* 133, 3235-3241.
8. C. A. Smith, C.-W. Vogel, and H. J. Müller-Eberhard: Ultrastructure of cobra venom factor-dependent C3/C5 convertase and its zymogen, factor B of human complement. 1982. *J. Biol. Chem.* 257, 9879-9882.
9. A. H. Grier, M. Schultz, and C.-W. Vogel: Cobra venom factor and human C3 share carbohydrate antigenic determinants. 1987. *J. Immunol.* 139, 1245-1252.
10. D. C. Gowda, M. Schultz, R. Bredehorst, and C.-W. Vogel: Structure of the major oligosaccharide of cobra venom factor. 1992. *Mol. Immunol.* 29, 335-342.
11. D. C. Gowda, E. C. Petrella, T. T. Raj, R. Bredehorst, and C.-W. Vogel: Immunoreactivity and function of oligosaccharides in cobra venom factor. 1994. *J. Immunol.* 152, 2977-2986.
12. H. J. Müller-Eberhard, and K.-E. Fjellström: Isolation of the anticomplementary protein from cobra venom and its mode of action on C3. 1971. *J. Immunol.* 107, 1666-1672.
13. W. Vogt, L. Dieminger, R. Lynen, and G. Schmidt: Alternative pathway for the activation of complement in human serum. Formation and composition of the complex with cobra venom factor that cleaves the third component of complement. 1974. *Hoppe-Seyler's Z. Physiol. Chem.* 355, 171-183.
14. C.-W. Vogel, and H. J. Müller-Eberhard: The cobra venom factor-dependent C3 convertase of human complement. A kinetic and thermodynamic analysis of a protease acting on its natural high molecular weight substrate. 1982. *J. Biol. Chem.* 257, 8292-8299.
15. P. Hensley, M. C. O'Keefe, C. J. Spangler, J. C. Osborne, Jr., and C.-W. Vogel: The effects of metal ions and temperature on the interaction of cobra venom factor and human complement factor B. 1986. *J. Biol. Chem.* 261, 11038-11044.
16. I. Von Zabern, B. Hinsch, H. Przyklenk, H., G. Schmidt, and W. Vogt: Comparison of *Naja n. naja* and *Naja h. haje* cobra-venom factors: Correlation between binding affinity for the fifth component of complement and medication of its cleavage. 1980. *Immunology* 157, 499-514.
17. M. H. L. De Bruijn, and G. H. Fey: Human complement component C3: cDNA coding sequence and derived primary structure. 1985. *Proc. Natl. Acad. Sci. USA* 82, 708-712.
18. J. Alsenz, D. Avila, H. P. Huemer, I. Esparza, J. D. Becherer, T. Kinoshita, Y. Wang, S. Oppermann, and J. D. Lambris. Phylogeny of the third component of complement C3: Analysis of the conservation of human CR1, CR2, H, and B binding sites, ConA binding sites, and thiolester bond in the C3 from different species. 1992. *Dev. Comp. Immunol.* 16, 63-76.

19. S. Hase, N. Kikuchi, T. Ikenaka, and K. Inoue. Structures of sugar chains of the third component of human complement. 1985. *J. Biochem.* 98, 863-874.
20. S. Hirani, J. D. Lambris, and H. J. Müller-Eberhard. Structural analysis of the asparagine-linked oligosaccharides of human complement component C3. 1986. *Biochem. J.* 233, 613-616.
21. K. Miki, S. Ogata, Y. Misumi, and Y. Ikehara. Carbohydrate structures of the third component of rat complement. 1986. *Biochem. J.* 240, 6991-698.
22. D. C. Fritzinger, R. Bredehorst, and C.-W. Vogel. Molecular cloning and derived primary structure of cobra venom factor. 1994. *Proc. Natl. Acad. Sci. USA* 91, 12775-12779.
23. S. B. Needleman, and C. D. Wunsch: A general method applicable to the search for similarities in the amino acid sequences of two proteins. 1970. *J. Mol. Biol.* 48, 443-453.
24. M. C. O'Keefe, L. H. Caporale, and C.-W. Vogel: A novel cleavage product of human complement component C3 with structural and functional properties of cobra venom factor. 1988. *J. Biol. Chem.* 263, 12690-12697.
25. D. C. Fritzinger, E. C. Petrella, M. B. Connelly, R. Bredehorst, and C.-W. Vogel: Primary structure of cobra complement component C3. 1992. *J. Immunol.* 149, 3554-3562.
26. M. J. Selby, R. H. Edwards, and W. J. Rutter: Cobra nerve growth factor: Structure and evolutionary comparison. 1987. *J. Neuroscience Res.* 18, 293-298.
27. D. Neumann, D. Barchen, M. Horowitz, E. Kochva, and S. Fuchs: Snake acetylcholine receptor: Cloning of the domain containing the four extracellular cysteines of the alpha-subunit. 1989. *Proc. Natl. Acad. Sci. USA* 86, 7255-7259.
28. Å. Lundwall, R. A. Wetsel, H. Domdey, B. F. Tack, and G. H. Fey: Structure of murine complement component C3: I. Nucleotide sequence of cloned complementary and genomic DNA coding for the β-chain. 1984. *J. Biol. Chem.* 259, 13851-13856.
29. R. A. Wetsel, Å. Lundwall, F. Davidson, T. Gibson, B. F. Tack, and G. H. Fey: Structure of murine complement component C3: II. Nucleotide sequence of cloned complementary DNA coding for the α-chain. 1984. *J. Biol. Chem.* 259, 13857-13862.
30. Y. Misumi, M. Sohda, and Y. Ikehara: Nucleotide and deduced amino acid sequence of rat complement C3. 1990. *Nucleic Acids Res.* 18, 2178.
31. H. S. Auerbach, R. Burger, A. Dodds, and H. R. Colten: Molecular basis of complement C3 deficiency in guinea pigs. 1990. *J. Clin. Invest.* 86, 96-106.
32. M. Kusano, N. H. Choi, M. Tomita, K.-I. Mamamoto, S. Migita, T. Sekiya, and S. Nishimura: Nucleotide sequence of cDNA and derived amino acid sequence of rabbit complement component C3 α-chain. 1986. *Immunol. Inv.* 15, 365-378.
33. D. Grossberger, A. Marcuz, L. Du Pasquier, and J. D. Lambris: Conservation of structural and functional domains in complement component C3 of *Xenopus* and mammals. 1989. *Proc. Natl. Acad. Sci. USA* 86, 1323-1327.
34. J. V. Maizel, Jr., and R. P. Lenk: Enhanced graphic matrix analysis of nucleic acid and protein sequences. 1981. *Proc. Natl. Acad. Sci. USA* 78, 7665-7669.
35. T. P. Hopp, and K. R. Woods: Prediction of protein antigenic determinants from amino acid sequences. 1981. *Proc. Natl. Acad. Sci. USA* 78, 3824-3828.
36. R. Huber, H. Scholze, E. P. Paques, and J. Deisenhofer. Crystal structure analysis and molecular model of human C3a anaphylatoxin. 1980 *Hoppe-Seyler's Z. Physiol. Chem.* 361, 1389-1399.
37. K. Dolmer, and L. Sottrup-Jensen: Disulfide bridges in human complement component C3b. 1993. *FEBS Lett.* 315, 85-90.
38. M. K. Pangburn: Spontaneous reformation of the intramolecular thioester in complement protein C3 and low temperature capture of a conformational intermediate capable of reformation. 1992. *J. Biol. Chem.* 267, 8584-8590.
39. M. E. Daoudaki, J. D. Becherer, and J. D. Lambris: A 34-amino acid peptide of the third component of complement mediates properdin binding. 1988. *J. Immunol.* 140, 1577-1580.
40. D. T. Fearon, and K. F. Austen: Properdin: Binding to C3b and stabilization of the C3b-dependent C3 convertase. 1975. *J. Exp. Med.* 142, 856-863.
41. C. G. Cochrane, H. J. Müller-Eberhard, and B. S. Aikin: Depletion of plasma complement in vivo by a protein of cobra venom: Its effects on various immunologic reactions. 1970. *J. Immunol.* 105, 55-69.
42. G. O. Till, K. J. Johnson, R. Kunkel, and P. A. Ward. Intravascular activation of complement and acute lung injury. 1982. *J. Clin Invest.* 69, 1126-1135.
43. C.-W. Vogel, and H. J. Müller-Eberhard: The cobra complement system: I. The alternative pathway of activation. 1985. *Dev. Comp. Immunol.* 9, 311-325.

44. C.-W. Vogel, and H. J. Müller-Eberhard: Induction of immune cytolysis: tumor-cell killing by complement is initiated by covalent complex of monoclonal antibody and stable C3/C5 convertase. 1981. *Proc. Natl. Acad. Sci. USA* 78, 7707-7711.
45. B. Müller, and W. Müller-Ruchholtz. Covalent conjugates of monoclonal antibody and cobra venom factor mediate specific cytotoxicity via alternative pathway of human complement activation. *Leukemia Res.* 11, 461-468.
46. H. Juhl, E. C. Petrella, N.-K. V. Cheung, R. Bredehorst, and C.-W. Vogel: Complement killing of human neuroblastoma cells: A cytotoxic monoclonal antibody and its $F(ab')_2$-cobra venom factor conjugate are equally cytotoxic. 1990. *Mol. Immunol.* 27, 957-964.
47. J. R. Leventhal, A. P. Dalmasso, J. W. Cromwell, J. L. Platt, C. J. Manivel, R. M. Bolman III, and A. J. Matas: Prolongation of cardiac xenograft survival by depletion of complement. 1993. *Transplantation* 55, 857-866.

7

A CASE STUDY OF CARDIOTOXIN III FROM THE TAIWAN COBRA (*Naja naja atra*)

Solution Structure and Other Physical Properties

T. K. S. Kumar, C.-S. Lee, and C. Yu

Department of Chemistry
National Tsing Hua University
Hsinchu, Taiwan Republic of China

INTRODUCTION

Snake venoms are a mixture of many different toxins of which the cardiotoxins and the neurotoxins are the most toxic (1,2). The cardiotoxins and the neurotoxins (long and short), isolated from different snake venom species belonging to the *Elapidae* family show significant amount of homology (50-60%) in their primary amino acid sequences (3,4). More generally, cardiotoxins and neurotoxins belong structurally to the three-finger proteins, which also includes other proteins such as wheat germ agglutinin (5,6). The members of these toxin groups are small molecular weight proteins (6-7.5 Kda) with 60-75 amino acids (7). The primary structure homology between these two toxins groups includes similar location of the disulfide bridges (8). Despite the homology in their primary amino acid sequences, cardiotoxins and the neurotoxins, show drastically different biological activities. The target of neurotoxins is well identified as the nicotinic acetylcholine receptor at the post-synaptic level of the neuro-muscular junction (9,10). Even if a detailed mechanism is not known, it is clear that the function of neurotoxins is to block the receptor and thus prevent binding of acetylcholine receptor (11,12). By contrast, little is known about the mode of action of cardiotoxins (13). The cardiotoxins elicit depolarization and contraction of muscular cells, loss of excitability and depolarization of nervous cells (14). They are also known to prevent aggregation by lytic effect on platelets (15). Various types of cells like erythrocytes, epithelial cells and fetal lung cells are also known to lyse at higher concentrations of cardiotoxins (16-18). The mechanism of the lytic action of cardiotoxins is still controversial (19). Fluorescence studies reveal that the ability of cardiotoxins to insert themselves into negatively charged phospholipids in artificial membranes (20-22). The proteolytic digestion experiments indicate that the first 13 residues of cardiotoxins are in contact with the lipid membranes (23). However, it is difficult to extrapolate this result to an in vivo situation as it is known that the external layer of cell membranes are usually poor in such phospholipids (24). In the contrary, the possibility that cells sensitive to cardiotoxins could have localised

Natural Toxins II, Edited by B. R. Singh and A. T. Tu
Plenum Press, New York, 1996

clusters of negatively charged phospholipids which could act as targets, cannot be excluded. Cardiotoxins may themselves induce local phase separation, due to their positive charges, as demonstrated to occur in mixtures of anionic and zwitterionic phospholipids (25). The question of whether the binding to phospholipids is sufficient to cause lysis is open; as chemical modification of the ε-amino groups of all lysine residues prevents the lytic activity but not the insertion into the phospholipid vesicles (26). Interestingly, cross-linking experiments using radiolabelled cardiotoxins have shown that cardiotoxins could inhibit protein kinase C (27). It has been postulated that cardiotoxins might interact with a site on phosphatidylserine, rather than binding directly to protein kinase C (28). The protein kinase binding activity has been shown to be co-related with the hydrophobicity of the cardiotoxins (29). Unfortunately, it has not yet been possible to conclusively determine the structural determinants responsible for the array of biological activities of *Elapid* venom cardiotoxins. However, in recent years solution (30-34) and crystal (35,36) structures of cardiotoxins from various snake venom sources have been published and could be of substantial help in the understanding of structure-function relationship of cardiotoxins. In this chapter, we would like to share our understanding and knowledge on the structure and protein folding aspects of cardiotoxins from the Taiwan cobra, *Naja naja atra.*

CHEMISTRY AND STRUCTURE OF CARDOTOXIN III

The Taiwan cobra (*Naja naja atra*) is a rich source for the cardiotoxins (37). To date six isoforms have been isolated from this source (38). The isoforms are numbered on column chromatographic profiles (39). Of the six isoforms, cardiotoxin III (CTX III) is the major fraction. It constitutes 50 - 60% of the total dry weight of the cardiotoxin isoforms and hence the best characterized (of the toxin isoforms) from this source (*Naja naja atra*) (40).

CTX III is a single chain polypeptide chain of sixty amino acids with a molecular weight of 6700 Kda. It is a highly basic protein with an isoelectric point of over 10.0 (41). Like all other cardiotoxin isoforms, it has the four well conserved disulfide bridges and lacks in histidine residues. CTX III also lacks glutamic acid. The toxin is highly hydrophobic and its sequence shows a patterned distribution of the hydrophobic and hydrophilic amino acids (7). Eventhough, information from circular dichroism (42,43) and theoretical prediction (44,45) studies shows that the backbone of this toxin is predominantly a β-sheet; the determination of the three dimensional structure proved elusive for quite long. This is primarily because of difficulties in crystallizing the cardiotoxins due to the existence of extended hydrophobic zones on their surface, which impose severe constraints on the crystal packing (6). This impasse was circumvented by the use of 2D NMR techniques to elucidate the solution structure of these toxins. The solution structure of a number of cardiotoxins, including CTX III have been published recently (30-34).

```
1                                      20
 L K C N K L V P L F Y K T C P A G K N L

21                                     40
 C Y K M F M V A T P K V P V K R G C I D

41                                     60
 V C P K S S L L V K Y V C C N T D R C N
```

Figure 1. The amino acid sequence of CTX III.

Chemical modification studies directly executed on CTX III and the information deduced from the reports on other homologues cardiotoxins isolated from other sources have provided useful knowledge on the amino acid residue(s) important for the biological activity of this toxin (46-49). The highly conserved methionine (46) at position 24 and the two tyrosine residues at positions 22 and 51 are implicated in the toxic action exhibited by CTX III (50). Carboxymethylation of cysteine residues after reduction of the disulfide bridges is also known to render the toxin, non-toxic (51). In addition, quenching the positive charge(s) on the lysine residues by acetylation is also reported to severely impair the lytic activity of the toxin (52). Chemical modification of the crucial amino acid residues (involved in the biological activity) are known to indirectly perturb the 3D structure of the toxin, resulting in the loss of biological activity (53,54). It should be mentioned that cardiotoxins derive their potential as a toxin with a repertoire of biological activities, due to their unique three dimensional structure.

NMR AND SOLUTION STRUCTURE CALCULATION

Two dimensional Nuclear Magnetic Resonance (2D NMR) technique is facile and offers an easy handle to tackle problems involving structural elucidation of proteins in solution (55). Recently, we successfully employed this technique to study the solution structure (8,32,56) and protein folding (57) aspects of Cardiotoxin III from the Taiwan cobra.

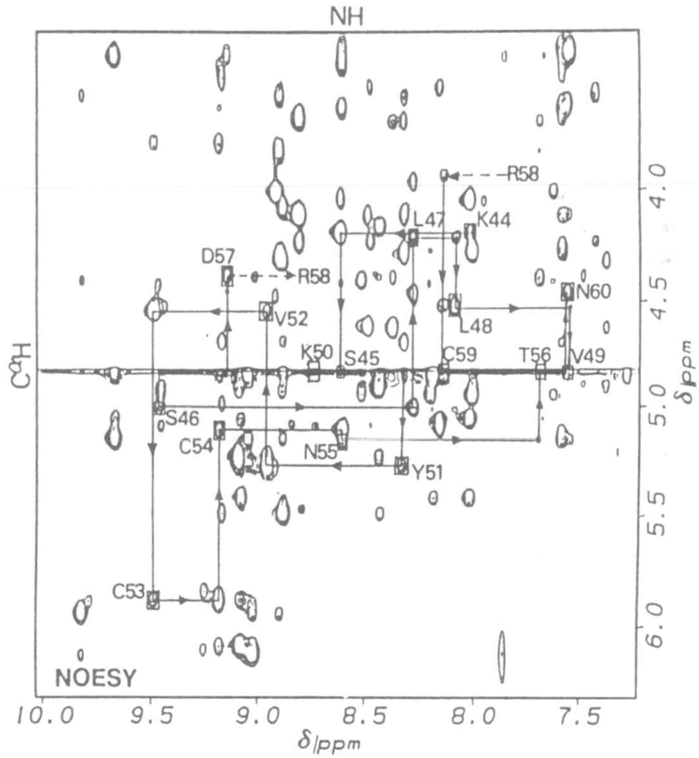

Figure 2. Part of the NOESY spectrum of CTX III recorded in H_2O at pH 3.0 and 20°C for a mixing period of 150ms. The cross peaks corresponding to the intraresidue NOE along the polypeptide backbone are squared and labelled. The sequential d $_{\alpha N}$ connectives are indicated for the sequence of residues 44 to 60.

Without getting into the technical details (for reasons of clarity to the non-specialised readers) of the technique, we only present the gist of the methodology involved.

In principle, 2D NMR structural determination of proteins in solution involves the linking of information derived from a combination of three complementary experiments – TOCSY (Total Correlated Spectroscopy), DQF-COSY (Double-Quantum Filtered Correlated Spectroscopy) and NOESY (Nuclear Overhauser Effect Spectroscopy). The TOCSY (58) spectra are used to identify spin systems of the amino acids in the protein. COSY (59,60) spectra yield complementary information to the TOCSY, which can be used to obtain the direct scalar connectives. The NOESY (61) spectra yield information on the through space relationship of protons in the protein. It should be noted that because of the vast abundance of hydrogen atoms in a protein molecule proton NMR is the preferred technique to determine the solution structure of a protein.

The information derived from the three spectra (TOCSY, DQF-COSY, NOESY) are coherently tied and the complete assignments of the observed NOEs' is carried out by the 'sequence-specific' (62) approach. This approach, though widely used, is tedious and more popular and rapid methods of structural determination, like the Main Chain Directed (MCD) approach are available to determine the secondary structural elements in the protein (63).

The data obtained from these experimental 2D NMR techniques is still raw (64). The experimental restraints, including the inter-proton distances and torsional angle structural constraints are then computed using the Distance Geometry (DG) algorithm (65-67). The structures obtained are further refined by subjecting them to a series of energy minimization protocols (using high speed computers) like the simulated annealing (SA) (65) or the restrained molecular dynamics procedures (68). Based on the considerations of NOE energy, constraint violation and root mean square deviation, an ensemble of structures are chosen as final possible structures for the protein (69).

THE SOLUTION STRUCTURE ANALYSIS

CTX III when subjected to 2D NMR techniques (as described above), followed by rigorous refinement procedures yielded a family of 20 structures (32). The structures were consistent with both experimental data and covalent geometry.

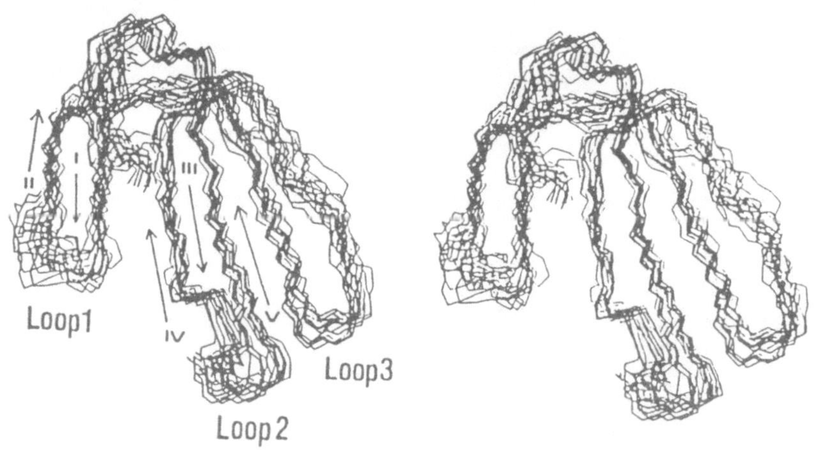

Figure 3. Stereoview of the least-squares superposition of the backbone atoms of the 20 selected solution structures of CTX III. Strands of the double and triple stranded antiparallel β-sheet that are linked by the 3 main loops are denoted by Roman numerals with the arrows indicating the direction of the backbone of the molecule.

The averaged structure of CTX III presents a picture of three fingers emerging from the palm of the hand. Hence CTX III can be branded as a 'three finger' protein. The secondary structure of CTX III is exclusively β-type, consisting of five β-strands protruding from the globular head or core and three main loops act as linkers of these strands. The core region is fortified by the cluster of disulfide bridges. The five β-strands primarily arrange themselves anti - parallely into double - stranded and triple-stranded β-sheets.The double-stranded β-sheet is formed at the N - terminal end of the molecule spanning residues 1 to 5 (STR-I, -Leu-Lys-Cys-Asn-Lys-) and 10 to14 (STR-II, -Phe-Tyr-Lys-Thr-Cys-). These two strands are interconnected by a loop (loop-1) formed from residues 6-9.The double-stranded β-sheet is bridged by hydrogen bonds formed between residues 3 and 12, 5 and 10, 12 and 3, and 14 and 1. STR-1 shows a backbone NH - O hydrogen bond between lys-5 and phe-10. The triple-stranded antiparallel β-sheet comprises of three strands (STR III, STR IV and STR V, respectively), juxtaposed in a step-wise manner. Strand III (STR III) and Strand IV (STR IV) comprise of residues, 20 -26 (STR III, - Leu-Cys-Tyr-Lys-Met-Phe-Met-) and 34-39 (STR IV, -Val-Lys-Arg-Gly-Cys-Ile-) respectively. The β-sheet secondary structure between these strands is characterised by a β-bulge formed by Pro-33 and Val-34. This bulge can be catalogued under G1 type β-bulge according to the classification of Richardson (70). STR III and STR IV are connected by a middle loop (loop-2), unlike loop-1, assumes an irregular conformation. No specific hydrogen bonding is seen among the residues comprising this loop. It should be mentioned that the solution structure of a homologous cardiotoxin, CTX I (isolated from the Taiwan cobra), suggests a tight turn between residues 26 and 29 that is similar to the copybook β1-type pattern (33). Strand V (STR V) is completed by an extended segment between 50-55 (-Lys-Tyr-Val-Cys-Cys-Asn-) that antiparallels residues 20 to 26 (STR III) to form the triple-stranded β-sheet, which shows a characteristic right handed twist. The double and triple strands are also connected by two disulfide bridges between residues 3-21 and 14-38. The disulfide bridge between 42-53 links STR V with the tail portion of STR IV forming the third loop (loop-3). Loop-3 makes a well defined type -1 turn between residues 46 (serine) and 49 (valine), except that the distance between Val-49 HN and Ser-46 CO is longer than expected. On the whole the triple stranded β-sheet is bridged by hydrogen bonds formed between residues 20 and 39, 21 and 54, 22 and 37, 23 and 52, 24 and 35, 25 and 50, 26 and 32, 39 and 20, 50 and 25, 52 and 23 and also 54 and 21. In addition to the 3 loops tethering the five β-sheet strands, there is also a C-terminal loop spanning residues 54 to 60. The loop is stabilised by he disulfide bridge, 52-59 . It is

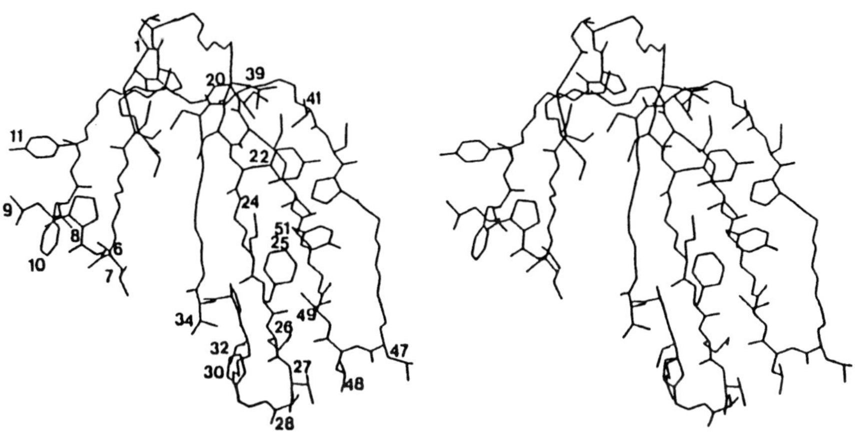

Figure 4. Stereo pair front view of 1 of the 20 solution conformers showing the side chains of the hydrophobic amino acid residues. Selected residues are numbered.

also firmly knotted to the rest of the molecule by the disulfide bridge 42-53 and by the hydrogen bonds, linking the C-terminal and the N-terminal ends of the molecule.

It is interesting to note that seven out of the eight cysteine residues that are present in CTX III, are part of one of the β-sheets or the C-terminal loop. The odd, Cys-42, however is the exception. The well conserved pattern of the disulfide linkage in the family of the cardiotoxins, probably could explain the structural stability of these class of toxins. A stereo view of the molecule possess a convex side and a concave side, the former being the side at which the polypeptide chain crosses over the β-sheet between the triple strand .The tail, C-terminal loop, is located on the concave side. Stretches of hydrophobic patches extends at the surface of the molecule. These hydrophobic regions are located in the loop-1 region between residues 6 and 11. A large chunk of such non-polar patches can also be visualised on STR III (spanning residues 47- 49). Majority of the side-chains of non -polar residues like Met-24, Met-26, Val-27, Ala-28, Leu-48 and Val-49 are on the concave side of the molecule, directed away from the C-terminal loop. However, some non-polar amino acids such as that of Phe-25, Val-32, Pro-33 and Val-52, point to the convex side of the molecule. Tyr-22 and Tyr-51 form the fringe of the hydrophobic patch. Most of the aromatic side-chains are located on the lower part of the molecule. The aromatic ring of Tyr-51 occupies a central position on the convex side of the triple-stranded β-sheet. The phenolic group of Tyr-51 is placed centrally on the surface of the protein, with Pro-43 located just above one face of the aromatic ring. It is interesting to note that the hydrophobic patches on the molecules are flanked by conserved short stretches of polar residues. Most of the hydrophobic side-chains project into the solvent with their polar groups.

It should be brought to notice that X-ray and NMR structures of eight cardiotoxins have been published so far (29-36,71). Most of the published structures agrees well with one another, barring, a few subtle variation(s). The available structural information, hopefully could facilitate the evolution of a consensus mode of action of these class of toxins.

FUNCTIONAL IMPLICATIONS

The three dimensional folding of the cardiotoxins appears to be well conserved. In fact Breckenridge and Dufton have compiled a list of 41 different cardiotoxin sequences from various cobra species (72). They found that 40 out of the 60 residues were highly conserved (> 90%).Therefore, structural variations must be restricted to subtle local and spatial rearrangements due to natural mutations. This also gives a clue that functional diversity of cardiotoxins may not be connected with major structural differences. Chemical modification studies on the toxin (CTX III) and other related cardiotoxins from other snake venom sources, have shown that invariant residues like lys-18, lys-23, lys-35 and tyr-22 are involved in the toxicity of the cardiotoxins (73). This shows that functionally important residues are concentrated in a definite area in the toxin comprising the β-sheets of STR II, III and IV. The hydrophobicity of loop-1 (comprising of residues 6 to 9) linking STR I and STR II is clearly a characteristic feature of the cardiotoxins, as such a hydrophobic patch seldom occurs in this region in the neurotoxins. This hydrophobic loop is always flanked by lys-12 in cardiotoxins. Chemical modification of this residue has been shown to impair the binding of cardiotoxins to phospholipid membrane vesicles (52). The lysine presumably helps the toxin to anchor on to the membrane vesicle *via* a charge-interaction with polar phosphate head of the phospholipids comprising the membrane. This interaction would then facilitate the facile insertion of the toxin whilst loop-1 into the hydrophobic part of the membrane (23). It appears that most of the lysis-related biological activity of cardiotoxins can be bestowed to the asymmetric distribution of hydrophobic and hydrophilic amino acids on the surface of the molecule (74,75). The concentration of disulfide bridges in the 'head' region of the molecule also appears to be important because reduction of disulfide bridge(s) results in total loss of toxicity of the molecule (76). The head region, probably is the "Achilles heal" of the molecule, as reduction of the disulfide bridge(s) in this region brings about a total collapse of the structure of the molecule (CTX III).

In the recent years, novel biological activities are being reported (28, 77-80). These activities are widely variant from one another. Hence, postulation of an unified mechanism for the mode of their action, explaining all the biological activities and consistent with their structure looks, atleast for the present, impossible.

STABILITY

The NMR solution structures of cardiotoxins reported thus far were carried out at room temperatures in the pH range of 3.0-4.0 (28-34). Cardiotoxins have a high content of charged amino acids in their sequence. It is found that charged amino acids in the toxin contribute significantly to the electrostatic interactions which in turn are involved in the maintenance of the three dimensional structure. As the state of ionization of the charged amino acids can be altered by varying the pH (81), we studied the effect of solvent pH variation on the structure of the molecule using circular dichroism (CD) studies (Yu *et al.*, unpublished results). This we believe could offer an interesting twist to the structure-activity relationship studies.

The Far UV CD spectrum (secondary structure region) of CTX III at different pH conditions shows that the 215 nm (an indicator of the β -sheet content) band is maximum at pH 6.0. The intensity of this Cotton band decreases as the pH of the solvent is either increased or decreased from pH 6.0. This finding raises an interesting question as to whether the NMR solution structures of cardiotoxins are stringently dependent on the pH of the solvent. Though we have not carried out systematic study, we feel that such a possibility evidently exists. Infact, differences in the NMR solution structures of α-cobratoxin (a neurotoxin) solved under different pH conditions is reported (82). We presume that subtle variations of this sort (in the structure) could explain the battery of drastically different biological activities reported for the same cardiotoxin molecule under different conditions.

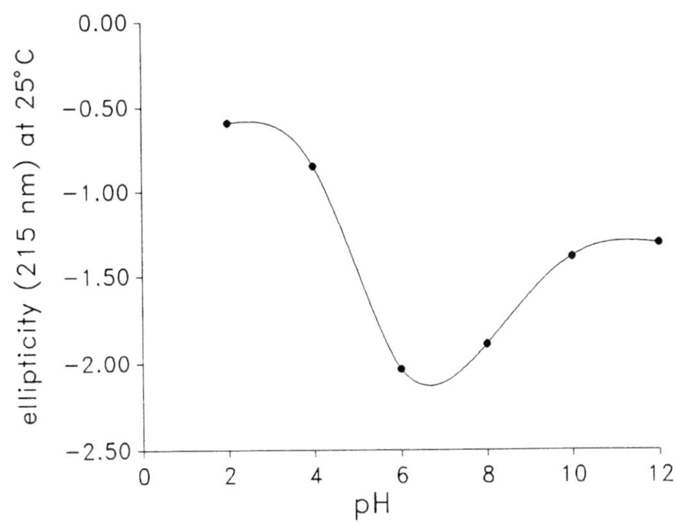

Figure 5. pH dependent ellipticity (at 215 nm) profile of CTX III.

PROTEIN FOLDING

Cardiotoxins, on account of their small size (60 amino acids) and interesting biological activities are fascinating molecules to work with. In short, it is 'dream' molecule for any protein chemist, as it is conducive to study their structure-function relationship. To a structural biologist, cardiotoxins are unique molecules as they are all β-sheet proteins. No segment of their backbone is in a helical conformation.

The problem of 'Protein folding' in the recent years has been a hot area of research. Intense research activity is centred in the area of identifying and characterising folding intermediates along the folding/unfolding pathway(s) of several proteins (83-85). Such a study would enable us to understand the different forces that come into play to direct a protein to its final native fold (86-88). Stable intermediate state(s) called the 'molten globule', have been identified along the folding/unfolding pathway of many proteins. A protein in its 'molten globule' state is believed to be compact and has pronounced secondary structure but no rigid tertiary structure (89-91). The 'molten globule' like intermediates are presently believed to be universal equilibrium protein folding intermediate(s) (92). However, this postulation has led to some controversy as 'molten globule' like stable intermediates have not been identified thus far along the folding/unfolding pathway of any all β-sheet protein (93). Studies in this direction were hampered primarily due to the scarcity of all β-sheet proteins. As CTX III is a small β-sheet protein and also as its complete solution structure was worked out by our group; we embarked on the task of identification of 'molten globule' like stable intermediates along the folding/unfolding pathway of cardiotoxin III from the Taiwan cobra. For obvious reasons, we used mild denaturant like 2,2,2- trifluoroethanol (TFE) at low pH (pH 2.0) for this study.

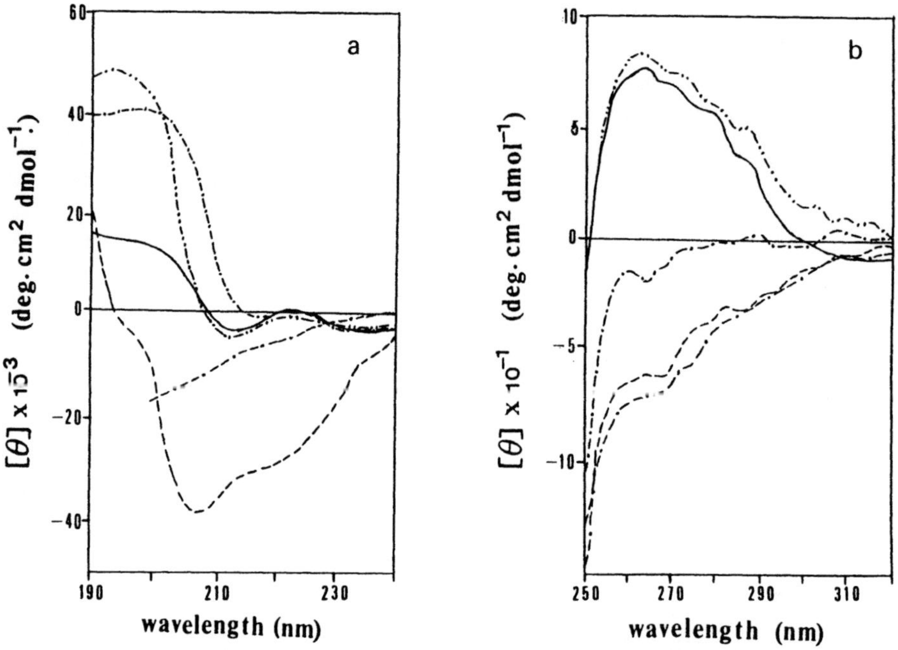

Figure 6. Far (a) and near (b) UV CD spectra of untreated (——) and treated CTX III. The alcohol concentrations used in the treatment of CTX III are 20% (— · · —), 50% (—· — · —), 80% (- - - -) and 80% TFE containing 2M urea (– – · – –). All the spectra were recorded at pH 2.5.

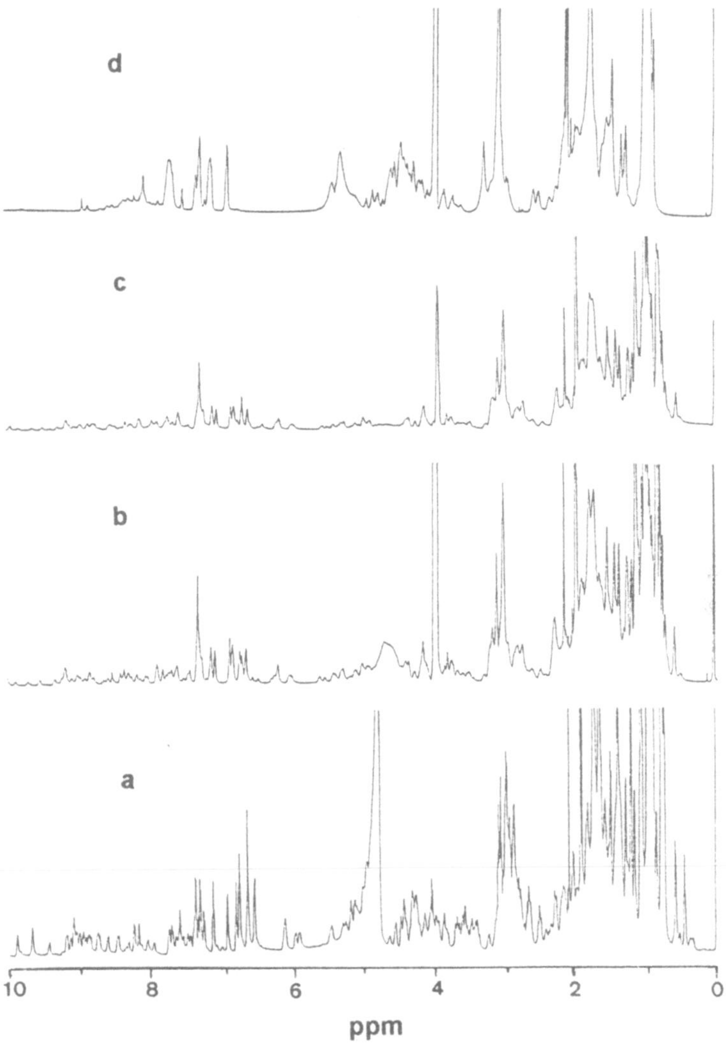

Figure 7. 1D- ^{1}H NMR spectra of CTX III in, a) native state (90% H_2O/10% D_2O); b) 50% TFE-d_3/H_2O (v/v); c) 80% TFE-d_3/H_2O(v/v); and d) 80% TFE- d_3 containing 2M urea-d_4.

The CTX folding problem was attacked by a combinational use of three different techniques, namely, NMR, CD and fluorescence spectroscopy. The process of unfolding is found to be totally reversible. Structural changes are evident at the TFE concentration of 20% (v/v). When the toxin is treated with 50% TFE concentration, the tertiary structural interaction are found to be lost. The Near UV CD spectrum (representing the tertiary structure) of this toxin at this concentration of TFE (50% v/v) shows a total loss of the positive ellipticity band centred at 275 nm. However, at this concentration of TFE (50% v/v), the secondary structure of the toxin is intact. This implies that CTX III gets into a 'molten globule' like state. This result also corroborates the results from the 1D NMR experiments in terms of the chemical shift dispersion and line broadening (94,95). It is interesting to observe that structural transitions continue to occur beyond 50 % TFE (v/v) and the transitions are only complete when the TFE concentration reaches 80% (v/v). Portions of

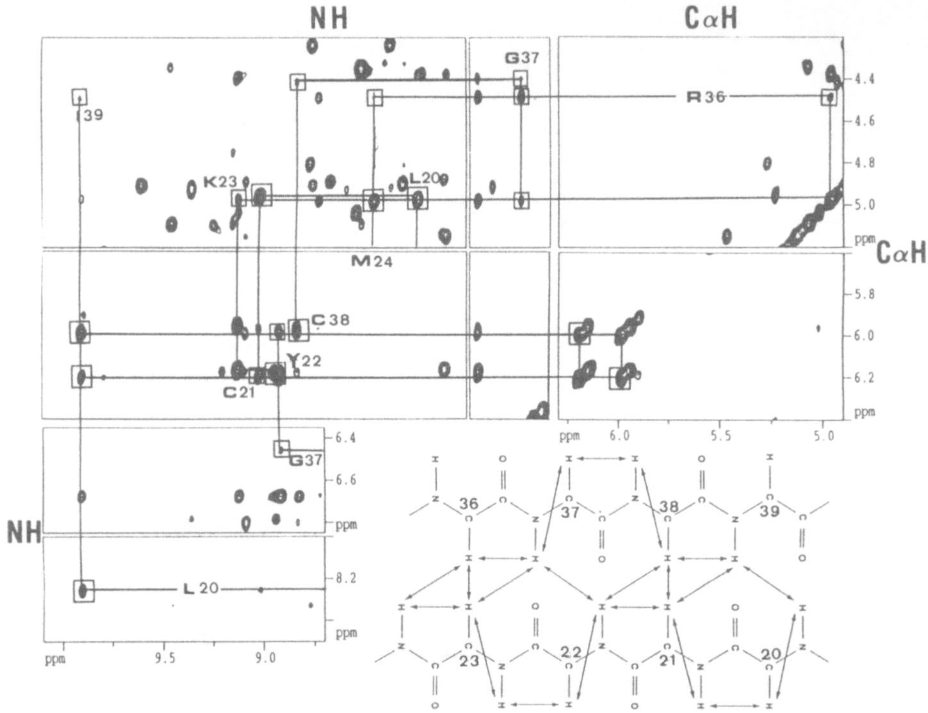

Figure 8. Expansion of NOESY spectrum for CTX III in 80% CTX III in 80% TFE-d_3/H_2O Twenty-one proton pairs are indicated with an arrow (in the lower-right of this figure) result in 21 NOE cross peaks labelled in a box. The NOE elements recognised and ordered by this MCD pattern clearly show that the secondary structure (β-sheet) remains intact - a feature characteristic of the 'molten globule' state.

the backbone of the toxin is found to get into a helix as evident from CD experiments. At this stage of unfolding of CTX III, all its tertiary structural contacts are lost.

The molecule in 80% (v/v) TFE is also characterized by its strong binding to hydrophobic dye like Anilino NaPTHalene Sulfonate (ANS). Such strong binding is indicative of the toxin being in a typical 'molten globule' like state (95). It appears from our results that different types of 'molten globule' like states can be trapped between 50% (v/v) and 80% (v/v) TFE concentrations. Addition of denaturant like urea at this stage (at 80% TFE) results in the complete denaturation of the toxin.

Two dimensional NMR experiments on the 'molten globule' state at 80% (v/v) TFE reveal that with the exception of a few, most of the original secondary structural contacts (as found in the native CTX III) are intact in the 'molten globule' state. These contacts could represent the probable helical segments formed in the toxin at 80% TFE. We believe that identification of the 'molten globule' state in an all β-sheet protein, such as CTX III, is an important landmark in the field of protein folding.

FUTURE PERSPECTIVES

In this chapter, only a case study of CTX III from the Taiwan cobra is presented. However, we believe that the lesson learnt herein, are generally applicable to all snake venom cardiotoxins. Though a lot of useful information is now available on CTX III and cardiotoxins from other venom sources, a lot remains to be explored. Unfortunately, site-directed mutants of cardiotoxins are still not available due to difficulties in cloning and expessing the cardiotoxin gene (96). Availability of such site-directed mutants in the future will not only offer better insight into the structure-function relationship in cardiotoxins but also pave the way for the construction of folding/unfolding pathways of the cardiotoxins. Ironically, despite their small size, cardiotoxins were not adopted as popular model molecules to understand protein folding. However, with the availability of complete NMR structural data, there is ample scope to carry out elaborate protein folding experiments on these toxins. Our group is presently engaged in the research involving the trapping of kinetic intermediates (97-99) along the folding pathway of CTX III employing the hydrogen/deuterium exchange using the rapid - quench technique (100-102). We hope to obtain useful data on the structural features of the kinetic intermediates.

Cardiotoxins, of late, have caught the attention of neurobiologists. Cardiotoxins have been shown to form ion-channels (3). Intense research activity focused on the understanding the structure of the cardiotoxin induced ion-channel(s) in the coming years can be expected. As more and more novel and fancied biological properties are being reported, it is imperative that there would be increased attention on snake venom cardiotoxins in the near future.

ACKNOWLEDGEMENTS

This work was supported by the Taiwan National Science Council grants (NSC84-2113-M007-016 & NSC84-2811-M007-010). This study was carried out on the 600 MHz NMR spectrometer at the Regional Instrumentation Center at Hsinchu, supported by the National Science Council, Taiwan (R.O.C). We thank the Academic Press, USA for permitting us to reproduce some of the figures from Refs. 32 & 57.

REFERENCES

1. Dufton, M. J., and Hider, R. C., 1991, in *Snake toxins*, Harvey,A.L., Ed., Pergamon Press, New York, pp 259-298.
2. Harvey, A.L.,1991,Cardiotoxins from snake venoms. In *Handbook of Natural toxins*, Tu, A.T., Ed., Marcel Dekker, New York, Vol 5,pp 85-106.
3. Dufton, M.J., and Hider, R.C., 1988, Structure and pharmacology of *Elapid* cytotoxins, *Pharmacol. Ther.* 36:1-40.
4. Harvey, A.L., 1985, Cardiotoxins from cobra venoms: possible mechanisms of action, *Toxicol. Toxin. Rev.* 4 : 41-69.
5. Menez, A., Botems, F., Roumestand, C., Gilquin, B., and Toma, F., 1992, Structural basis for functional diversity of animal toxins, *Proc. Roy. Soc. Edinburgh* 90B: 83-103.
6. Rees, B.,and Bilwes, A., 1993, Three-dimensional structures of neurotoxins and cardiotoxins, *Chem. Res. Toxicol.* 6: 385-406.
7. Dufton, M.J., and Hider, R.C., 1983, Conformational properties of the neurotoxins and cytotoxins isolated from *Elapid* snake toxin, *CRC Crit. Rev. Biochem.* 114: 113-114.
8. Yu, C., Bhaskaran, R., and Yang, C.C., 1994, Structures in solution of toxins from Taiwan cobra,*Naja naja atra*, derived from NMR spectra, *J.Toxin. Toxicol. Rev.* 13: 291-315.
9. Basus, V.J., Song, G., and Hawrot, E., 1993, NMR solution structure of an α-bungarotoxin/nicotinic peptide complex, *Biochemistry* 32: 12290-12298.
10. Conti-Tronconi, B.M., Tang. F., Diethelm, B.M., Wu, X., Tang, F., Beratazzon, T., Schroder, B., Reinhardt-Maeclicke, S., and Maelicke, A., 1991, α-bungarotoxin and the competing antibody Wf6 interact with different amino acids within the same cholinergic subsite, *Biochemistry* 30: 2575-2584.
11. Traztos, S.J., and Remoundos, M.S., 1990, Fine localisation of the major α -bungarotoxin binding-site to residues- α 189-195 of the torpedo acetylcholine receptor residues- 189, residues-190, and residues-195 are indispensible for binding, *J. Biol. Chem.* 265: 21462-21467.

12. Radding, W., Corfield, P.W.R., Levinson, L.S., Hashim, G.A., and Low, B.W., 1988, Alpha-toxin binding to acetylcholine receptor 179-191 peptides - Intrinsic fluorescence studies, *FEBS Lett.* 231: 212-216.
13. Fletcher, J.F., and Jiang, M.-S., 1993, Possible mechanisms of action of cobra snake venom cardiotoxins and bee venom mellitin, *Toxicon*, 31: 695-699.
14. Grognet, J.M., Menez, A., Drake, A., Hayashi, K., Morrison, I.E.G., and Hider, R.C., 1988, Circular dichroic spectra of *Elapid* cardiotoxins, *Eur. J. Biochem.* 172: 383-388.
15. Kini, R.M., and Evans, H.J., 1988, Mechanism of platelet effects of cardiotoxins from *Naja nigricollis crawshawii* (spitting cobra) snake venom, *Thromb. Res.* 52: 185-195.
16. Hinman, C.L., Lepisto, E., Stevens, R., Mongomery, I.N., Rauch, H.C., and Hudson, R.A., 1987, Effects of cardiotoxin D from *Naja Siamensis* snake venom upon murine splenic lymphocytes, *Toxicon* 25: 1011-1014.
17. Takechi, M., Tanaka, Y., and Hayashi, K., 1986, Binding of cardiotoxin analogue III from Formosan cobra to FL cells Le, *FEBS Lett.* 205:143-146.
18. Gatineau, E., Takeshi, M., Bowet, F., Mansuelle, P., Rochat, H., Harvey, A.L., Montenay-Garestier, T., and Menez, A., 1990, Delineation of the functional site of a snake venom cardiotoxin : preparation, structure, and function of monoacetylated derivatives, *Boichemistry* 29: 6480-6489.
19. Kini, R.M., and Evans, H.J., 1989, Role of cationic residues in cytolytic activity : modification from *Naja nigricollis* and correlation between cytolytic and anti-platelet effect, *Biochemistry* 28: 295-301.
20. Vincent, J.P., Balerna, M. and Lazdunski, M., 1978, Properties of association of cardiotoxin with lipid vesicles and natural membranes : a case study, *FEBS Lett.* 85: 103-108.
21. Defourcq, J., and Faucon, J.F., 1978, Specific binding of a cardiotoxin from *Naja mossambica mossambica* to charged phospholipids detected by intrinsic fluorescence, *Biochemistry* 17: 1170-1176.
22. Batenby, A.M., Bougis, P.E., Rochat, H., Verkleij, A.J., and de Kruijff,B., 1985, Penetration of a cardiotoxin into cardiolipin model membranes and its implications on lipid organisation, *Biochemistry* 24: 7101-7110.
23. Defourcq,J., Faucon,J.F., Bernard,E., Pezolot,M., Tessier,M., Bougis,P., Rietschoten,J., Delori,P., and Rochat,H., 1982, Structure-function relationships for the cardiotoxins interacting with phospholipids, *Toxicon* 20: 165-174.
24. Rothman, J.E., and Lenard, J., 1977, Membrane asymmetry, *Science* 195: 743-753.
25. Desormeaux, A., Laroche, G., Bougis, P.E., and Pezolot, M., 1992, Characterisation by infrared spectroscopy of the interaction of a cardiotoxin with phosphatidic acid and with binary mixtures of phosphatidic acid and phosphatidyl choline, *Biochemistry* 31: 12173-12182.
26. Bougis, P.E., Tessier, M., Rietschoten, J.V., Rochat, H., Faucon, J.F., and Dufourcq, J., 1983, Are interactions with phospholipids responsible for pharmacological activities of cardiotoxins ?, *Mol. Cell. Biochem.* 55: 49-64.
27. Knowden, M.J., Vitale, A.J., Trumble, M.J., Wesson, C.R., and Trumble, W.R., 1992, A bioassay for cobra cardiotoxin activity using semi-isolated cockroach heart, *Toxicon* 30: 295-301.
28. Chiou, S., Raynor, R.L., Zheng, B., Chambers, T.C., and Kuo, J.F., 1993, Cobra venom cardiotoxin (cytotoxin) isoforms and neurotoxin : Comparitive potency of protein kinase C inhibition and cancer cell cytotoxicity and modes of enzyme inhibition, *Biochemistry* 32: 2062-2067.
29. Kuo, J.F., Kem, W.R., Raynor, R.L., Mazzei, G.J., Schatzman, R.C., and Turner, R.C., 1983, Cobra polypeptide cytotoxin 1 and marine worm polypeptide cytotoxin A-IV are potent and selective inhibitors of phospholipid-sensitive Ca^{+2} dependent protein kinase, *FEBS Lett.* 153: 183-186.
30. O'Connell, J.F., Bougis, P.E., and Wuthrich, K., 1993, Determination of the NMR solution structure of cardiotoxin CTX IIb from *Naja mossambica mossambica. Eur. J. Biochem.* 213: 891-900.
31. Gilquin, B., Roumestand, C., Zinn-Justin, S., Menez, A., and Toma, F., 1993, Refined three-dimensional solution structure of snake cardiotoxin : analysis of the side-chain organization suggests the existence of a possible phospholipid binding site , *Biopolymers*, 33: 1659-1675.
32. Bhaskaran, R.,Huang, C.C., Chang, D.K., and Yu, C., 1994, Cardiotoxin III from the Taiwan cobra (*Naja naja atra*) : Determination of structure in solution and comparison with short neurotoxins, *J. Mol. Biol.* 235, 1291-1301.
33. Jahnke, W., Mier, D.F., Beress, L. and Kesseler, H., 1994, Structure of cobra cardiotoxin, CTX I, as derived from Nuclear Magnetic Resonance Spectroscopy and Distance geometry calculations, *J.Mol.Biol.*, 240, 445-458.
34. Bhaskaran, R., Haung, C.C., Tsai, Y.-C., Jayaraman, G., Chang, D.-K., and Yu, C., 1994, Cardiotoxin II from Taiwan cobra venom, *Naja naja atra*. Structure in solution and comparison among homologous cardiotoxins , *J. Biol. Chem.*, 269: 23500-23508.
35. Rees, B., Bilwes, A., Samama, J.P., and Moras, D., 1990, Cardiotoxin $V^{II}1$ 4 from *Naja mossambica mossambica* : the refined crystal structure, *J.Mol. Biol.* 214: 281-297.

36. Bilwes, A., Rees, B., Moras, D., Menez, R., and Menez, A., 1994, X-ray structure at 1.55 A° of Toxin γ, a cardiotoxin from *Naja nigricollis* venom : Crystal packing reveals a model for insertion into membranes, *J.Mol. Biol.* 239: 122-136.
37. Yang, C.C., 1974, Chemistry and evolution of toxins in snake venoms, *Toxicon* 12 : 1-43.
38. Chiou, S.H., Lee, B.S., and Yu, N.T., 1992, Structural analysis and comparison of cobrotoxin and cardiotoxins by near-IR Fourier transform Raman spectroscopy, *Biochem. Intl*, 274: 747-758.
39. Kumar, T.K.S., Radha, R., and Rambhav, S., What's in a name,*Toxicon*, 28: 135-138.
40. Yang, C.C., King, K., and Sun, T.P., 1981, Chemical modification of lysine and histidine residues in phospholipase A from the venom of *Naja naja atra* (Taiwan cobra), *Toxicon* 19: 645-659.
41. Hayashi, K., Takechi, M., Kaneda, N., and Sasaki, T., 1976, Amino acid sequence of cardiotoxin from the venom of *Naja naja atra, FEBS Lett.* 66: 210-214.
42. Drake, A.F., Dufton, M.J., and Hider, R.C., (1980), Circular dichroism of *Elapid* protein toxins, Eur. J. *Biochem.*, 105: 623-628.
43. Hung, M.C., and Chen, Y.H., 1977, Conformational stabilty of a snake cardiotoxin . *Int.J.Peptide.Protein Res.* 10, 277-289.
44. Dufton, M.J., and Hider, R.C., 1977, Snake toxin secondary structure predictions. *J.Mol.Biol.*, 115, 117-142.
45. Menez, A., Langlet, G., Tamiya, N., and Formagoet, P., 1978, Conformation of snake toxic polypeptides studied by a method of prediction and circular dichroism, *Biochemie*, 60, 505-513.
46. Carlsson, F.H.H., and Louw, A.I., 1978,The oxidation of methionine and its effect on the properties of cardiotoxin V^{II} 1 from *Naja melanoleuca* venom, *Biochim.Biophys.Acta*, 534, 325-329.
47. Gatineau, E., Toma, F., Montenay-Garestier, T., Takechi, M., Fromageot, P., and Menez, A., 1987, The role of tyrosine and tryptophan residues in the structure-activity relationships of a cardiotoxin from *Naja nigricollis* venom, *Biochemistry* 26: 8046-8055.
48. Fryklund, L., and Eaker, D., 1975, The complete structure of a cardiotoxin from the venom of *Naja nigricollis* (African black-necked spitting cobra)*Biochemistry*, 14: 2865-2871.
49. Menez, A., Gatineau, E., Roumestand, C., Harvey, A.L., Mouawad, L., Gilquin, B., and Toma, F., 1990, Do cardiotoxins possess a functional site - Structural and chemical modification studies reveal the functional site of the cardiotoxin from *Naja nigricollis, Biochemie* 72: 575-588.
50. Hung, M.C., Pau, Y.H., Cheng, K.L., and Chen, Y.H., 1978, The status of tyrosyl residues in Foromasn cobra cardiotoxin, *Biochim.Biophys.Acta* 535: 178-182.
51. Louw, A.I., and Visser, L., 1977, Kinetics of lysis by snake venom cardiotoxins, *Biochim.Biophys.Acta* 498: 143-153.
52. Surewicz, W.K., Stepanik, T.M., Szabo, A.G., Mantsch, H.H., 1988, Lipid induced changes in the secondary structure of snake venom cardiotoxins, *J. Biol. Chem.* 263: 786-790.
53. Roumestand, C., Gatineau, E., Gilquin, B., Menez, A., and Toma, F., 1990, in *Peptides: Chemistry, Structure and Biology*, Rivier,J.F. and Marshall,G.R., Eds. Escom, London, pp 622-624.
54. Chiou, S.H., Hung, C.C., Huang, H.C., Chen, S.T., Wang, K.T., and Yang, C.C., 1995, Cardiotoxins and cobrotoxins isolated from Taiwan cobra, *Biochem.Biophys.Res.Commun.*, 206: 22-32.
55. Wuthrich., K., 1990, Protein structure determination in solution by NMR spectroscopy, *J. Biol. Chem.* 265: 2059-2062.
56. Bhaskaran, R., Yu, C., and Yang, C.C., 1994, Solution structures and functional implications of the toxins from Taiwan cobra venom, *J. Protein Chem.* 13: 503-504.
57. Kumar, T.K.S., Jayaraman, G., Lee, C.S., Sivaraman, T., Lin, W.Y., and Yu, C., 1995, Identification of 'molten globule' state in an all β-sheet protein, *Biochem.Biophys.Res.Commun.*, (in press).
58. Bax, A., and Davies, D.G., 1985, MLEV-17 based two-dimensional homonuclear magnetisation transfer spectroscopy, *J.Magn.Reson.* 65: 355-360.
59. Rance, M., Sorensen, O.W., Bodenhaussen, G., Wagner, G., Ernst, R.R., and Wuthrich, K., 1983, Improved spectral resolution in COSY proton NMR spectra of proteins via double quantum filtering, *Biochem. Biophys. Res. Commun.* 117: 479-485.
60. Marion, D., and Wuthrich, K., 1983, Application of phase sensitive two-dimensional spectroscopy (COSY) for measurements of proton-proton spin-spin coupling constants in proteins, *Biochem. Biophys. Res. Commun.* 113: 967-974.
61. Kumar, A., Ernst, R.R., and Wuthrich, K., 1980, A two dimensional nuclear overhauser enhancement (2D NOE) experiment for elucidation of complete proton-proton cross-relaxation networks in biological macromolecules, *Biochem. Biophys. Res. Commun.* 95: 1-6.
62. Wuthrich, K., 1986, *NMR of Proteins and Nucleic acids*, Wiley, New York.
63. Englander, S.W., and Wand, A.J., 1987, Main chain directed strategy for the assignment of ^{1}H-NMR spectra of proteins, *Biochemistry* 26: 5953-5958.

64. Gippert, G.P., Yip, P.F., Wright, P.E., and Case, D.A., 1990, Computational methods for determining protein structures from NMR data, *Biochem. Pharmacol.* 40: 15-22.
65. Nilges, M., Clore, G.M., and Gronenborn, A.M., 1988, Determination of three-dimensional structures of proteins from interproton distance data by hybrid distance geometry - dynamical simulated annealing calculations, *FEBS Lett.* 229: 317-329.
66. Brunger, A.T., 1992, X-PLOR software manual version 3.0, Yale university, New Haven, CT.
67. Nilges, M., Clore, G.M., and Gronenborn, A.M., 1990, ^{1}H-NMR stereospecific assignments by conformational database searches, *Biopolymers* 29: 813-822.
68. Brooks, B.R., Bruccoleri, R.E., Olafsen, B.D., States, D.J., Swaminathan, S., and Karplus, M., 1983, CHARMM : a program for minimization and dynamics calculations, *J.Comput. Chem.* 4: 187-217.
69. Clore, G.M., Robien, M.A., and Gronenborn, A.M., 1993, Exploring the limits of precision of protein structures determined by Nuclear Magnetic Resonance spectroscopy, *J. Mol. Biol.* 231: 82-102.
70. Richardson, J.S., 1981, The anatomy and taxanomy of protein structure, *Adv. Protein. Chem.* 34: 167-339.
71. Singhal, A.K., Chien, K.Y., Wu, W.G., and Rule, G.S., 1993, Solution structure of cardiotoxin V from *Naja naja atra* , *Biochemistry* 32: 8036-8044.
72. Breckenridge, R., and Dufton, M.J., 1987, The structural evolution of cobra venom cytotoxins , *J. Mol. Evol.* 26: 274-283.
73. Karlsson, E., 1979, Chemistry of protein toxins in snake venoms, In *hand book of experimental pharmacology* (Lee, C.Y., Eds.) Springer Verlag, Berlin. Vol 52 pp 159-172.
74. Wise, B.C., Chou, C.H.J., Glass, D.B., Katoh, N., Kibler, R.F., Kuo, J.F., Raynor, R.F., Schatzman, R.C., and Turner, R.S., 1982, Phospholipid-sensitive Ca dependent protein kinase from heart . 2. Substrate specificity and inhibition by various agents, *J. Biol. Chem.* 257: 8489-8495.
75. Raynor, R.L., Bin, Z., and Kuo, J.F., 1991, Membrane interaction of amphiphilic polypeptides mastopran, mellitin, polymyxin B, and cardiotoxin - diffrential inhibition of protein kinase C, Ca^{+2}-calmodulin dependent protein kinase-II and synaptosomal membrane Na,K-ATPase and Na pump and differentiation of H 160 cells, *J. Biol. Chem.* 266: 2753-2758.
76. Lee, C.Y., 1972, Chemistry and pharmacology of polypeptide toxins in snake venoms, *Ann. Rev. Pharmacol.* 12: 265-293.
77. Chien, K.Y., Huang, W.N., Jean, J.H., and Wu, W.G., 1991, Fusion of spingomyelin vesicles induced by proteins from Taiwan cobra *(Naja naja atra)*, *J. Biol. Chem.* 266: 3232-3259.
78. Chien, K.Y., Chiang, C.M., Hseu, Y.U., Vyas, A.A., Rule, G.S., and Wu, W.G., 1994, Two distinct types of cardiotoxin as revealed by the structure and activity relationship of their interaction with zwitterionic phspholipid dispersions, *J. Biol. Chem.* 269: 14473-14483.
79. Ksenzhek, O.S., Gerod, V.S., Omel'Chenko, A.M., Semenov, S.N., Sotnichenkov, A.I., and Miroschinikov, A.I., 1978, Interaction of the cardiotoxin of the venom of the cobra *Naja naja oxiana* with phospholipid membrane model system, *Mol. Biol.* 12: 1057-1065.
80. Sandblom, J., and Diaz, E., 1984, Cardiotoxin-induced channels in lipid bilayer membranes, *Proc.Int.Union. Pure. Appl. Biophys* (8th Int. Congr.) pp-60.
81. Tanford, C., 1968, Protein denaturation, *Adv. Protein. Chem.* 23: 121-275.
82. Leroy, E., Mikou, A., Yinshan, Y., and Guittet, E., 1993, NMR solution structure of α-Cobratoxin at pH 3.2 and at pH 7.5, *J. Mol. Graphics*, 4: 65-70.
83. Barrick, D., Hughson, F.M., and Baldwin, R.L., 1994, Molecular mechanisms of acid denaturation - the role of histidine residues in the partial unfolding of apomyoglobin, *J. Mol. Biol.* 237: 588-601.
84. Hughson, F.M., Wright, P.E., and Baldwin, R.L., 1990, Structural characterisation of a partly folded apomyoglobin intermediate, *Science* 249: 1544-1548.
85. Kumar, T.K.S., Subbiah, V., Ramakrishna, T., and Pandit, M.W., 1994, Trichloacetic acid - induced unfolding of bovine pancreatic ribonuclease - existence of a molten globule like state, *J. Biol.Chem.* 269: 12620-12625.
86. Wuthrich, K., 1994, NMR assignments as a basis for structural characterisation of denatured states of globular proteins, *Curr. Opin. Struc. Biol.* 4: 93-99.
87. Radford, S.E., Dobson, C.M., and Evans, P.A., 1992, The folding of hen lysozyme involves partially structured intermediates and multiple pathways, *Nature* 358: 302-307.
88. Ballery, N., Desmadril, M., Minard, P., and Yon, J.M., 1993, Characterisation of an intermediate of phosphoglycerate kinase - chemical reactivity of genetically introduced cysteinyl residues during the folding process, *Biochemistry* 32: 669-678.
89. Ptitsyn, O.B., 1987, Protein folding - hypotheses and experiments, *J. Protein. Chem.* 6: 273-293.
90. Kuwajima, K., 1989, The molten globule state as a clue for understanding the folding cooperativity of globular protein structure, *Proteins:Stuc.Func.Genet.* 6: 87-103.

91. Goto, Y., Calciano, L.J., Fink, A.L., 1990, Mechanism of acid induced folding of proteins, *Proc. Natl. Acad. Sci.* (USA) 87:573-577.
92. Ptitsyn, O.B., and Uversky, V.N., 1994, The molten globule is a third thermodynamical state of protein molecules, *FEBS Lett.* 341: 15-18.
93. Jaenicke, R., 1991, Protein Folding: Local structures, Domains, Subunits and assemblies, *Biochemistry* 30: 3161-3169.
94. Ptitsyn, O.B., 1992, in *Protein Folding* (Creighton,T.E., Eds.) Freeman and Co. New York pp 243-300.
95. Jeng, M.F., and Englander, S.W., *1991,* Stable submolecular folding units in a compact form of Cyt-C, *J. Mol. Biol* 221: 1045-1061.
96. Marston, F.A.O., 1986, The purification of eucaryotic polypeptides, *Biochem. J.* 240: 1-12.
97. Goldenberg, D.P., 1992, Native and non-native intermediates in the BPTI folding pathway, *Trends Biochem. Sci.* 17: 257-261.
98. Weissman,J .S., and Kim, P.S., 1991, Reexamination of the folding pathway of BPTI - predominance of native intermediates, *Science* 253: 1386-1393.
99. Van Mierlo, C.P.M., Darby, N.J., Neuhaus, D., and Creighton, T.E., 1991, 2-Dimensional ^{1}H nuclear magnetic resonance of the (5-55) single disulfide folding intermediate of bovine pancreatic trypsin inhibitor, *J. Mol. Biol.* 222: 373-390.
100. Woodward, C.K., 1994, Hydrogen exchange rates and protein folding, *Curr. Opin. Struc. Biol.* 4: 112-116.
101. Briggs, M., and Roder, H., 1992, Early hydrogen-bonding events in the folding reaction of ubiquitin, *Proc. Natl. Acad. Sci.(USA)* 89: 2017-2021.
102. Udgaonker, J., and Baldwin, R.L., 1990, Early folding intermediate of ribonucease A, *Proc. Natl. Acad. Sci.(*USA) 87: 8197-8201.

8

THE STAPHYLOCOCCAL AND STREPTOCOCCAL PYROGENIC TOXIN FAMILY

Gregory A. Bohach,[1] Cynthia V. Stauffacher,[2] Douglas H. Ohlendorf,[3] Young-In Chi,[2] Gregory M. Vath,[4] and Patrick M. Schlievert[3*]

[1] Department of Microbiology, Molecular Biology, and Biochemistry
University of Idaho, Moscow, Idaho 83843
[2] Department of Biological Sciences
Purdue University, West Lafayette, Indiana 47907
[3] Department of Microbiology
[4] Department of Biochemistry
University of Minnesota, Minneapolis, Minnesota 55455

OVERVIEW OF CLASSIFICATION CRITERIA AND RELATEDNESS OF THE PYROGENIC TOXINS

The term "pyrogenic toxins" describes a group of protein exotoxins (Table 1) produced by organisms in two genera of bacteria, *Staphylococcus* and *Streptococcus*. Each pyrogenic toxin (PT) has unique biological activities, however, they often are grouped collectively into a single toxin family based on an ability that they share to induce multiple systemic effects (Bohach, et al., 1990). These shared activities result from the action of the toxins on the immune, cardiovascular, and other organ systems.

Classically, the PT family has included the enterotoxins and toxic shock syndrome toxin-1 (TSST-1) of *Staphylococcus aureus* and the pyrogenic exotoxins produced by *Streptococcus pyogenes* (Bergdoll, et al., 1973, Bergdoll, 1983, Bergdoll and Schlievert, 1984; Bohach, et al., 1990; Schlievert, et al., 1979; Schlievert, 1980). However, the PT family continues to expand as additional toxins are discovered. For example, several previously unrecognized PT-like exoproteins produced by *Streptococcus pyogenes* and other streptococci (groups B, C, F, and G) have been recently described (Hauser, et al., 1995; Iwasaki, et al., 1993; Mollick, et al., 1993; Reda, et al., 1994; Schlievert, et al., 1993). In addition,

* Phone and Fax numbers: G.A.B. (P) (208) 882-9663, (F) (208) 885-6518; C.V.S. (P) (317) 494-4937, (F) (317) 496-1189; D.H.O. (P) (612) 624-9471, (F) (612) 626-0623; Y.-I. C. (P) (317) 494-4937, (F) (317) 496-1189; G.M.V. (P) (612) 624-9471, (F) (612) 626-0623; P.M.S. (P) (612) 624-9471, (F) (612) 626-0623.

Natural Toxins II, Edited by B. R. Singh and A. T. Tu
Plenum Press, New York, 1996

Table 1. The pyrogenic toxin family

Organism	Toxin	Serotypes/variants
Staphylococcus aureus	Toxic shock syndrome toxins (TSST)	TSST-1
		TSST-ovine
	Staphylococcal enterotoxins (SE)	SEA
		SEB
		SEC1
		SEC2
		SEC3
		SEC-ovine
		SEC-bovine
		SED
		SEE
		SEG
Streptococcus pyogenes	Streptococcal pyrogenic exotoxins	SPEA
		SPEB
		SPEC
Streptococcus pyogenes	Mitogenic factor	MF (SPEF)
	Streptococcal superantigen	SSA
Group B, C, F, & G streptococci	SPEs	Undesignated

sequence and immunological variants of the classical PTs provide further diversity within the family (Lee, et al., 1992; Marr, et al., 1993).

As discussed below, sequence data obtained during the past ten years has revealed that several of the PTs share a significant level of amino acid sequence similarity. This relatedness has, in several instances, crossed the genus boundary between the staphylococci and streptococci and suggests that some of the toxins in this family have evolved from a common ancestral gene. In contrast, others in the toxin family share no significant amino acid identity and appear to have evolved separately. From this viewpoint it has become customary to divide the PT family into groups of toxins based on their relatedness at the primary structure level. One related group of PTs contains the staphylococcal enterotoxins (SE) serotypes SEB and SEC (Hovde, et al., 1990; Jones and Khan, 1986) plus two streptococcal toxins - type A streptococcal pyrogenic exotoxin (SPEA) and streptococcal superantigen (SSA) (Johnson, et al., 1986; Reda, et al., 1994; Weeks and Ferretti, 1986). Overall, primary sequences of the toxins within this group share at least 49% identity. In addition to its prototype SEA, the other group of related toxins contains, in order of decreasing sequence identity with SEA: SEE>SED> SPEC (Betley and Mekalanos, 1988; Couch, et al., 1988; Bayles and Iandolo, 1989; Goshorn and Schlievert, 1988). Three other PTs whose sequences are known, TSST-1, SPEB, and SPEF, appear to have only minimal or possibly no overall sequence relatedness to each other or to toxins in the two previously described groups (Blomster-Hautamaa, et al., 1986; Hauser and Schlievert, 1990; Iandolo, 1989; Iwasaki, et al., 1993). From a functional standpoint, this type of grouping should be interpreted with caution, as it does not take into consideration other potentially significant conserved structural features such as protein conformation. Furthermore, despite the lack of overall sequence relatedness, it is possible that some crucial short stretches of residues are conserved even among toxins that appear to be unrelated (discussed below and demonstrated in Figure 3).

Despite their collective diversity, PTs were initially grouped together based on one of their common features, the ability to induce a fever (Bohach, et al., 1990; Schlievert and Watson, 1978). A major contributing factor to pyrogenicity and several other important activities, is now known to be the potent T cell stimulating ability of the PTs through a mechanism known as superantigenicity (Marrack and Kappler, 1990; White, et al., 1989). The demonstration that all PTs exert their shared biological activities through this similar mechanism is consistent with sequence and structural information. The general picture that has developed is that even PTs with little or no overall sequence identity share certain conformational features that are probably critical for their shared activities. Evolutionary tendencies to conserve certain crucial structural features appear to have permitted some changes in amino acid composition, producing toxin variants that allow the organism to interact with the immune system receptors of several hosts. This review will consider the genetic and structural implications of PT biological activities and the role that shared and divergent features play in the evolution of the host-parasite interaction.

BIOLOGICAL PROPERTIES DEFINING THE PT FAMILY AND DISEASE ASSOCIATION

It is generally agreed that most of the shared biological properties of PTs are direct or indirect effects of the unique immunomodulatory role that superantigens have in the host. By definition, superantigens are endogenous or exogenous molecules that have the ability to stimulate proliferation of an exceptionally high percentage of the host T cells (Johnson, et al., 1992; Marrack and Kappler, 1990). This activity is attributed to their simultaneous interaction with Class II major histocompatibility complex (MHC) molecules (Fraser, 1989; Mollick, et al., 1989) and the T cell receptor (TCR), with receptor specificity determined by the Vβ region of the TCR (White, et al., 1989). The resulting cellular interactions are independent of the antigen specificity of the TCR. In fact, for the most part (discussed below), the interaction of superantigens with MHC Class II is not influenced by the nature of the peptide in the conventional binding groove of the receptor (Fraser, et al., 1993). Therefore, the interaction of TCR with superantigens via Vβ is less specific than its requirements for recognizing conventional antigens, making the probability of T cell stimulation orders of magnitude higher for superantigens. Other organisms, including viruses, parasites, fungi, and even mammalian cells, produce molecules with similar superantigenic properties. However, because more is known about their structure and function, PTs are considered to be the prototypes of microbial superantigens.

Although PT-induced T cell proliferation has been long recognized, its molecular and cellular mechanisms have only recently been defined with the discovery of the superantigen concept (Marrack and Kappler, 1990). This resulting superantigenic T cell stimulation and its requirement for antigen presenting cells explain the biological effects secondary to this large-scale cellular stimulation observed in vitro and in vivo. For example, activities such as massive cytokine induction (Fast, et al., 1988), pyrogenicity (Schlievert and Watson, 1978), and immunosuppression (Schlievert, 1993) are undoubtedly secondary effects of PT superantigenicity (Table 2). Thus, it seems obvious that superantigenicity plays a major role in staphylococcal and streptococcal disease. This is reflected in the strong correlation between symptoms of patients suffering staphylococcal and streptococcal toxic shock syndrome (TSS), which include fever, shock, and immunosuppression, and the biological properties in Table 2 (Bohach, et al., 1990). However, it is also likely that biological properties, other than those linked to superantigenicity, contribute to the disease. For example, some PTs have been shown to cross the blood brain barrier (Schlievert and Watson,

Table 2. Biological properties of pyrogenic toxins

Activity	Shared or unique?	Proposed mechanism(s)
Pyrogenicity	Shared	Direct action on fever center Superantigenicity
Enhancement of endotoxin shock	Shared	PT-LPS complex formation Cytotoxicity (hepatic, renal and immune)
cardiotoxicity	SPEs	Unknown
Immune cell stimulation	Shared	Superantigenicity
Cytokine release	Shared	Superantigenicity
Immunosuppresion	Shared	Superantigenicity
Lethality/shock	Shared	Superantigenicity PT-LPS complex formation
Direct capillary leak	Shared	Cytotoxicity
Emesis	SEs	Unknown
Hind limb paralysis	Group B SPE	Unknown

1978) and directly stimulate the fever center in the brain. Also, direct capillary leak and enhancement of endotoxin shock are well described biological phenomena that clearly could contribute to some of the symptoms in patients suffering from TSS (Lee, et al., 1991; Schlievert, 1982).

As a group, the PTs are probably best recognized for their clear role in the pathogenesis of TSS and the related toxigenic illness, scarlet fever. Both TSST-1 and various SEs have been confirmed for their role in the life-threatening symptoms of staphylococcal TSS as indicated by toxin analysis for *Staphylococcus aureus* clinical isolates published in numerous reports (Bohach, et al., 1990). Also, information obtained from extensive epidemiological surveys of the recent resurgence of severe invasive streptococcal diseases including streptococcal toxic shock syndrome (STSS) indicates a strong association with production of PTs, especially SPEA (Cone, et al., 1987; Demers, et al., 1993; Hauser, et al., 1989; Hoge, et al., 1993; Leggiadro, et al., 1993; Musser, et al., 1991; Stevens, et al., 1989; Wheeler, et al., 1991). Several of these studies have shown a significant but not exclusive association of SPEA with this disease. Other toxins associated with STSS include SPEB and SPEC, streptococcal superantigen, mitogenic factor, and SPEs produced by groups B, C, F, and G streptococci. Epidemiological evidence has been supported with the use of rabbit or mouse models using purified PTs in vivo alone, with endotoxin, or with chemically-sensitized animals (Arko, et al., 1984; DeAzavedo, et al., 1985; Miethke, et al., 1992; Schlievert, 1982; Stiles, et al., 1993).

There has been considerable debate regarding the importance of several unique biological properties of the PTs. Presently only the enterotoxigenicity (Dolman and Wilson, 1938) of SEs has a clear role in pathogenesis of any particular disease. Although they are superantigens, the SEs can be distinguished from other members of the PT family by their ability to induce emesis and diarrhea upon oral ingestion (Bergdoll, 1988) (Table 2). Therefore, SEs can cause either systemic effects leading to life-threatening shock or a self-limiting food poisoning, depending on the route of administration. They are well known for their ability to induce toxic-shock syndrome if the outcome of staphylococcal infection enables the toxin to enter the circulation (Bohach, et al., 1990). The resulting shock is due at least in part to the superantigen properties of the SEs, leading to massive induction of cytokines. In contrast, the pathogenesis of staphylococcal food poisoning is that of an intoxication in which preformed SEs are ingested. Consequently, the emetic reflex is stimulated through an action of the toxins in the gut. Although the cellular and molecular events leading to emesis are not well characterized, evidence has been provided for the role of inflammatory mediators (Boyle, et al., 1994; Scheuber, et al., 1987) as well as nervous impulse transmission to the brain via the vagus nerve (Bayliss, 1940; Clark, et al., 1962; Elwell, et al., 1975). Hence, they are directly responsible for the etiology of staphylococcal food poisoning. Likewise, the cardiotoxic properties of SPEs have been postulated to be involved in the pathogenesis of several streptococcal diseases such as scarlet fever, erysipelas, and STSS, in addition to the early events leading to rheumatic fever and guttate psoriasis (Schlievert, 1993). Finally, a recent association of several staphylococcal and streptococcal PTs with Kawasaki syndrome has been made by microbiological and immunological analysis of patients and could have significant implications for understanding pathogenesis of this disease (Abe, et al., 1992; Schlievert, 1993). Admittedly, the specific link between PT biological activities and these diseases have not been conclusively provided.

MOBILITY AND GENETIC REGULATION AS A MECHANISM TO ACHIEVE TOXIN DIVERSITY

The structural genes for most PTs have been cloned and sequenced (Bayles and Iandolo, 1989; Blomster-Hautamaa, et al., 1986; Betley and Mekalanos, 1988; Couch, et al., 1988; Goshorn and Schlievert, 1988; Hovde, et al., 1990; Iwasaki, et al., 1993; Jones and Khan, 1986; Johnson, et al., 1986; Hauser and Schlievert, 1990; Reda, et al., 1994; Weeks and Ferretti, 1986). The available evidence suggests that for some staphylococcal and streptococcal strains, there is a selective advantage to retain certain PT sequences. At the same time, considering the broad host range for both genera, it seems logical to predict that other selective pressures exist for these organisms to modify PT sequences. This diversity could allow them to produce toxins better capable of interacting with immune cells of the multiple hosts. The net result of these requirements is a large family of toxins with varying degrees of sequence identity, thus forming the basis for grouping the toxins according to their sequence similarities as described previously. Careful analysis of sequences of members of the PT family has provided several examples where diversity among the toxins could be due to gene duplication and/or homologous recombination. For example, SEC1 appears to have arisen from a recombination event involving the SEB gene and a structural gene encoding SEC2 or SEC3 (Couch and Betley, 1989). Although SEC1 is more similar to the other SEC subtypes, a stretch of residues near its N-terminus (residue numbers 16 through 26) is identical to the analogous region of SEB, but significantly different from the other SEC subtypes.

Additional minor variability probably results through point mutations, so that even within a particular PT, some sequence variability may occur. For instance, the most common form of mature SPEA contains 221 residues and has a molecular weight of 25,787. Recently, Nelson and coworkers (Nelson, et al., 1991) sequenced twenty examples of the *speA* gene from several M protein serotypes and found that most of the genes encode toxins that differ at most by a single amino acid. However, one allele, *speA4*, has 26 amino acid changes (89% similarity) compared to the most common form of the toxin. Similar variability has been documented for staphylococcal PTs, several of which display a significant degree of host specificity. The sequences of SEC toxins produced by strains of *Staphylococcus aureus* isolated from humans have slightly different sequences (>90% identity) compared to those produced by bovine (SEC-bovine) and ovine (SEC-ovine) isolates (Marr, et al., 1993). A similar phenomenon has been observed for TSST-1. The minor sequence differences result in significant differences in the ability of each of these toxins to stimulate T cells from several species of animals, again providing evidence of adaptability of the organism to its cadre of potential hosts (Lee, et al., 1992).

The belief most commonly held is that at least some of the PTs have arisen from a common ancestral gene which crossed the genus barrier and became stably introduced in both organisms. There has been considerable evidence in support of this proposal. The most convincing is the association of several PT structural genes on mobile elements (Iandolo, 1989). There is conclusive evidence that some staphylococcal and streptococcal PTs are encoded by structural genes located on bacteriophage genomes. Perhaps the best studied are the SPEs. Frobisher and Brown (1927) first showed that nontoxigenic streptococci could be converted to toxigenic strains by a filterable agent from scarlet fever isolates. It was later shown that infection with a lysogenic bacteriophage induced from strain T12gl causes a nontoxigenic strain, T253, to elaborate toxin (Zabriskie, 1964). Furthermore, phage curing techniques result in a return to the original phenotype. Johnson and Schlievert (1983) obtained a physical map of the lysogenic phage from T12gl and determined that the map is circularly permuted and 36 KB in size. Further analysis revealed that this phage carries the structural gene for SPEA, designated *speA*, which was subsequently cloned in *Escherichia coli*. *speA* is located adjacent to the phage attachment (*att*) site and was probably obtained by the phage upon abnormal excision from the bacterial genome (Johnson, et al., 1986). Interestingly, analysis of SPEA-positive strains by probing with T12 phage DNA-specific probes shows that they all contain phage DNA adjacent to *speA*. Even strains which do not produce phage when induced, presumably harbor defective phages since they also contain segments of phage T12 sequences.

A similar situation occurs with SPEC (Goshorn and Schlievert, 1989). The transformation of strains unable to elaborate SPEC to a SPEC-producing phenotype by lysogenic conversion has been demonstrated by two laboratories (Colon-Whitt, et al., 1979; Johnson, et al., 1980). Both groups were able to induce phage from SPEC-producing streptococcal strains and show that infection by this phage causes streptococci that previously did not express SPEC to produce the toxin. The high percentage of induced phage capable of causing this switch indicates that lysogenic conversion is responsible rather than specialized or generalized transduction. Unlike SPEA and SPEC, little work has been done concerning the relationship between the gene encoding SPEB, designated *speB*, and streptococcal bacteriophage. Although it was initially reported that SPEB production is influenced by bacteriophage in some streptococcal strains (Nida and Ferretti, 1982), it is unlikely that *speB* itself is carried by a bacteriophage since subsequent studies have shown that nearly all group A streptococci carry a single copy of this gene (Hauser, et al., 1989).

The genetic mobility of SEA and SEE appears to resemble the situation with SPEA and SPEC. Betley and Mekalanos confirmed earlier predictions that the SEA structural gene (*sea*) is carried by a lysogenic phage by directly demonstrating *sea* associated with bacteriophage (Betley, et al., 1984; Betley and Mekalanos, 1985). This gene, as well as the highly related *see* gene, both appear to be encoded by genes located near the *att* site on phage genomes. In a high percentage of cases, phages are not inducible despite the presence of phage sequences in the staphylococcal genome. This again suggests that the phages are defective (Couch, et al., 1988).

The role of transmissible plasmids, especially for the staphylococcal toxins, has received considerable attention, although it appears that other than for SED, plasmids are not common vehicles for harboring PT genes. Iandolo has reported that the *sed* gene is located on a stable 27.6KB plasmid that also encodes penicillinase and cadmium resistance (Bayles and Iandolo, 1989; Iandolo, 1989). Although both SEC and SEB have reportedly been associated with transmissible penicillin resistance plasmids (Altboum, et al., 1985), it is generally agreed that both of these toxins and TSST-1 are chromosomally encoded (Kreiswirth, et al., 1983; Bohach and Schlievert, 1987). This prediction does not preclude the possibility that they may be harbored on an uncharacterized variable genetic element as predicted by several investigations. In fact, in some instances one toxin genetic locus may serve as the insertion site for another. This is predicted based on the results of DeBoer and Chow (1993) who showed that although they appear to be mutually exclusive at the phenotypic level, both SEB and TSST-1 genes (*seb* and *tst*, respectively) apparently often coexist in *Staphylococcus aureus*. The lack of $SEB^+/TSST\text{-}1^+$ strains is apparently due to the insertion of *tst* into the *seb* locus, in some way interfering with *seb* expression. *seb* and *sec* apparently have distinct insertion sites since SEC can be coexpressed with either TSST-1 or SEB (Bohach, et al., 1989b).

ANTIGENIC EPITOPES

The mapping of both conserved and unique antigenic epitopes on PTs, in addition to identifying regions of the molecules that determine their toxicity, has potential significance from the standpoint of our basic understanding of the relatedness of the toxins. This line of investigation also has potential applications toward rational development of nontoxic vaccines which could be used to protect against toxigenic staphylococcal and streptococcal diseases. Individually, each PT has a certain degree of antigenic distinctiveness. It is possible to differentiate among toxins in the family using highly specific hyperimmune polyclonal antisera and monoclonal antibodies. Some degree of cross-reactivity can be demonstrated, the level of which generally correlates with shared primary sequences. This phenomenon is best demonstrated for the SEs in immunodiffusion assays. For example, the SEC subtypes and molecular variants exhibit significant cross-reactivity with antisera produced against other SEC subtypes. Generally, most of the SEC variants can be distinguished immunologically by the use of monoclonal antibodies or by immunodiffusion assays in which they produce lines of partial identity (Turner, et al., 1992). A similar situation occurs with the highly related SEA and SEE (Bergdoll, et al., 1971). Except for these two examples, and an occasional cross-reaction reported by some investigators between other PTs (Hynes, et al., 1987), immunodiffusion assays are not often sufficiently sensitive to routinely demonstrate cross-reactivity of antisera with heterologous PTs.

Antigenic relatedness can be more efficiently demonstrated using a variety of sensitive assays. For example, immunoblotting consistently results in cross-reactions between SEC1, SEB, and SPEA, three PTs with significant amino acid sequence homology (Bohach, et al., 1988). In fact, several cross-reactive epitopes on these highly related toxins have been partially localized to various tryptic fragments derived from the toxins (Bohach, et al., 1989a; Spero and Morlock, 1978; Spero, et al., 1973; Spero, et al., 1975; Spero, et al., 1976). Perhaps the most distantly related toxins that have been shown to cross-react immunologically are SEA and SED (Bergdoll, et al., 1971). However, despite significant advances toward predicting antigenic epitopes on SEs, SPEs, and TSST-1, the identification of a uniformly conserved epitope has not yet been successful.

GENERAL STRUCTURAL AND PHYSICOCHEMICAL CONSIDERATIONS

Based on what is known about their sequences, biochemistry, and functional aspects, all of the classical staphylococcal and streptococcal PTs are monomeric proteins. There is some conflicting data regarding the abilities of certain PTs to act intracellularly (Deresiewicz, et al., 1994). However, there is no evidence for division of PT molecules into functional A-B subunit arrangements. In this regard, PTs do not resemble most intracellular-acting bacterial toxins which often have distinct cell binding and enzymatically-active regions. All of the well characterized PTs have molecular masses ranging from approximately 22-28 kD. However, as is typical of exotoxins, these molecules are translated as larger precursors with a classical signal peptide that is cleaved during export.

Recently, the biochemical and biological characteristics of several novel or unique streptococcal PTs have been reported. In regard to molecular weight, with one possible exception, they are similar to the classical PTs. Based on predictions from sequence studies, the protein designated streptococcal superantigen (SSA) has a molecular weight of 26,892 and shares approximately 60% identity with SEB and SEC1 and 49% identity with SPEA (Reda, et al., 1994). Because of the highly significant sequence similarity with other PTs, and because this toxin was shown to be a superantigen, the SSA conformation is likely to be similar to other members of the PT family. All group A streptococci produce another recently described superantigenic PT, designated as streptococcal mitogenic factor or SPEF (Iwasaki, et al., 1993; Norrby-Teglund, 1994). The gene for this toxin codes for a mature protein of 228 residues (molecular weight 25,363), following removal of its 43 residue signal peptide. The SPEF primary sequence has little identity to the other PTs. Sequencing of the SPEB gene (*speB*), cloned from the chromosome of *Streptococcus pyogenes,* indicated that it encodes a 398 amino acid protein containing a 27 residue signal peptide (Hauser and Schlievert, 1990). The 371 residue mature protein with a molecular weight of 40,314 was readily proteolytically cleaved to yield a 253 residue peptide (molecular weight 27,580). A comparison of the predicted amino acid sequence of SPEB with the published sequence of streptococcal proteinase precursor (SPP), a cysteine proteinase elaborated by group A streptococci (Tai, et al., 1976), showed that the two proteins are closely related and probably represent variants of the same protein. Although initial studies did not detect protease activity associated with purified SPEB, subsequent data showing SPEB protease activity have been presented. Unlike the genes encoding all of the other pyrogenic exotoxins examined to date, SPEB and SPEF do not appear to be highly variable traits. This, along with their lack of sequence similarity with other PT genes, indicates that they may have evolved separately from the other toxins. At this time, one cannot make adequate predictions about their structural relatedness to other members of the toxin family.

One potential deviation from the typical PTs is the recently described SPE from Group B streptococcal isolates that have been implicated in several cases of group B streptococcal TSS (Schlievert, et al., 1993). The toxin isolated from these strains was shown to have a lower molecular weight of 14,000 compared to other PTs. However, since the gene for this toxin has not yet been cloned, the possibility that this protein represents a proteolytic cleavage fragment of a larger precursor still exists. Future structural studies of this and other novel uncharacterized PTs produced by streptococci other than Group A (Table 1) should determine their relatedness to the classical toxins in the family.

Although detailed studies have not been performed with all of the PTs, generally they are considered to be relatively stable molecules compared to many of the other bacterial exotoxins. Regarding stability, the best characterized PTs are the SEs, TSST-1, and SPEA, all of which are resistant to boiling. Although these toxins have susceptible bonds for cleavage by many common proteases, proteolysis is often not sufficient to cause a loss of biological activity unless the fragments are separated in the presence of denaturing agents. This is representative of the inherent molecular stability of the PTs which can be demonstrated by renaturation studies. For example, the SEs can be significantly denatured only under strong denaturing conditions (i.e. 6M urea) and spontaneously renature into a molecule that retains both emetic and T cell stimulating ability (Spero, et al., 1973; Spero, et al., 1976; Spero and Morlock, 1978). Differences in stability do exist among the toxins. For example, SEB and SEC1 are approximately 50-fold more stable under denaturing conditions than SEA

(Warren, 1977). There has been some speculation that the SE disulfide bond is responsible for the inherent molecular stability. However, Warren et al. (1974) showed that the closed disulfide bond is likely to contribute only a minor degree of conformational stabilization and that its disruption has only minimal effects on the overall stability and activity of the molecule. These results have been confirmed recently through the analysis of mutants unable to form the disulfide linkage (Harris, et al., 1993; Hovde, et al., 1994). Furthermore, TSST-1 is also very stable but has no cysteine residues and thus cannot form a disulfide bond (Blomster-Hautamaa, et al., 1986).

Biophysical studies have long suggested that the molecular topology of members of the PT family were similar. Although this would have been predicted for pairs of toxins (i.e. SEA/SEE and SEB/SEC) that share extensive sequence similarities, this prediction would not have been so obvious for less related toxins such as TSST-1. The initial significant progress came from predicting and comparing the secondary structures of several PTs by spectral and computational techniques. Although most of this work has been done with staphylococcal toxins, especially the SEs, one can make inferences for other PTs considering their functional and occasional sequence similarities. One strikingly consistent feature of results obtained from spectroscopic analyses of SEA, SEB, SEC, and SEE is the comparatively low α-helical content (<10%) compared to β-pleated sheet/β-turn structures (approximately 60-85%) (Middlebrook, et al., 1980; Munoz, et al., 1976; Singh and Betley, 1989; Singh, et al., 1988a). As expected, all the SEs studied displayed considerable similarities in overall secondary structural features, although sophisticated topographical analyses were able to define unique aspects for even highly related toxins. Using similar techniques with TSST-1, Singh et al. (1988b) showed that the conformational content of TSST-1 is similar to that of the enterotoxins.

Considerably less is known about the biophysical characteristics of the other PTs, especially some of the more recently characterized family members. It seems reasonable to suggest that one should be able to make logical assumptions about some PTs such as SSA, SPEA and SPEC which share stretches of sequences in common with the SEs. However, it is unclear whether less related PTs will show the structural conservation that TSST-1 does. Two PTs in this category are SPEB and mitogenic factor (SPEF). Like TSST-1, the sequences of both of these toxins have been determined and they contain no significant amino acid identity (less than 30%) with other members of the PT family. Although SPEs have been partially characterized from species of streptococci other than *Streptococcus pyogenes* (Groups C, F, and G), their cloning or sequencing have not yet been reported. Future work in this rapidly developing area will determine whether they conform to the general structural characteristics demonstrated for the classical PTs.

CRYSTALLOGRAPHIC ANALYSES OF PTs

Although they differ in some regards, the recently solved crystal structures of several PTs have been in general agreement with predictions based on prior spectral methods, confirming that the folding patterns within the PT family are similar. To date, the crystal structures of two SE, SEB and SEC3, have been solved (Swaminathan, et al., 1992, Chi, et al., Submitted). Considering their high level of sequence similarity and results of earlier structure predictions, it is not surprising that the crystal structures of both molecules are very similar. In fact, Swaminathan, et al. (1992) proposed that the folding of the α-carbon backbone demonstrated with the SEB structure could represent a general motif adopted by all the SEs. However, it is also possible that this prediction for structural conservation may be extended to other nonemetic PT superantigens. Recently, the structure of TSST-1 has also been reported from two independent crystallographic investigations (Acharya, et al., 1994;

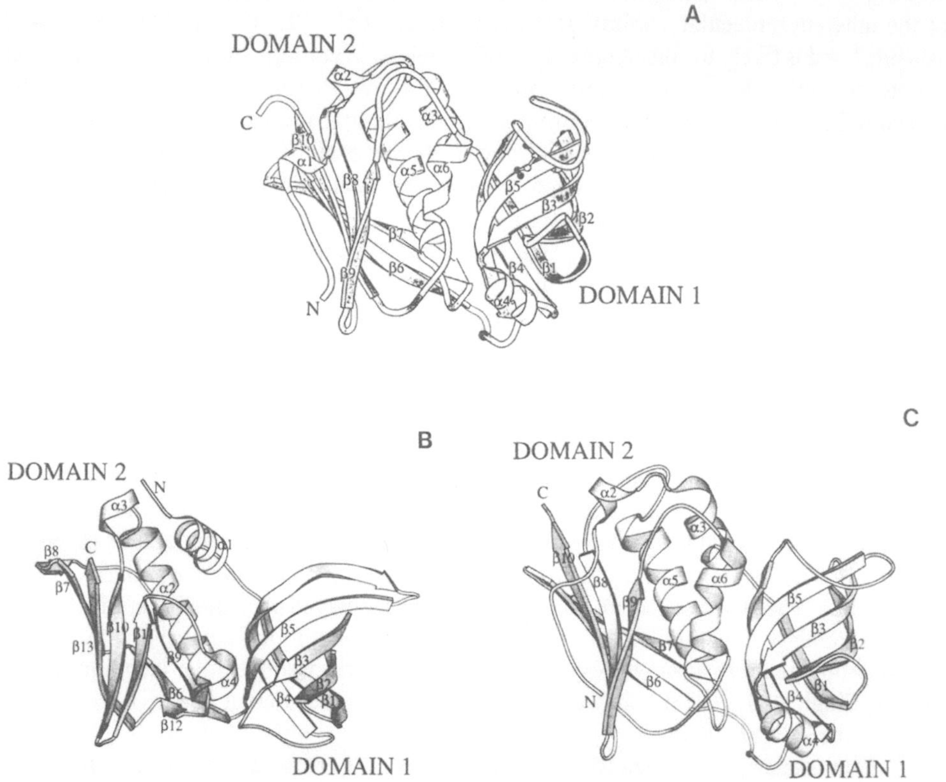

Figure 1. Schematic diagrams of the SEC3 (A), TSST-1 (B), and SPEA (C) structures illustrating major structural features. In the view shown, the SEC3 α5 groove is on the back side of the molecule. The structures of SEC3 and TSST-1 were adopted from crystallographic data described by Chi, et al., (submitted) and Prasad, et al., (1993). The SPEA structure was modeled for this report using the SEC3 atomic coordinates.

Prasad, et al., 1993). The TSST-1 molecule has a backbone fold similar to SEB and SEC3, despite the lack of significant shared sequences, providing additional evidence that a certain basic conformation is required for elaboration of shared PT biological activities.

The coordinates of the SEC3 structure were available to us, and will be used to describe general SE structure, although it is clear that both SEC3 and SEB are quite similar. The crystal structure of SEC3 has been refined to a resolution of 1.9Å (Chi, et al., submitted). Overall, the molecule has an ellipsoidal shape with maximal dimensions of 43Å × 38 Å × 32Å. The toxin molecule folds into two domains of unequal sizes and contains a mixture of α and β structures (Figure 1A). The smaller of the two domains (Domain 1) contains residues 35-120 near the N-terminus and forms a five stranded β-barrel, capped at one end by an α-helix. This domain has the same size and topology as the recently defined oligonucleotide/oligosaccharide binding (OB) fold (Murzin, 1993). Other proteins with this same folding feature include staphylococcal nuclease and the binding domains of several exotoxins produced by certain enteric bacterial pathogens. The internal portion of the β-barrel is rich in hydrophobic residues and the potential oligomer binding surface contains primarily hydrophilic residues. At one end of Domain 1, at the opposite end as the helix, is the disulfide loop characteristic of all SEs. Although its conformation is well defined in the SEC3 structure, values of the refined temperature factors for this region indicate it is quite flexible. A similar conclusion was drawn from the analysis of the SEB structure.

Both the N- and C-termini of SEC3 are located on the outer edge of the larger Domain 2 which encompasses residues 1-33 and 123-239. It is composed of a five-strand antiparallel β-sheet wall over which a group of α-helices is laid. The N-terminal 20 residues of SEC3 form a loosely attached structure which drapes over the edge of this β-sheet wall. Three of the helices in the SEC3 structure define the interface between the two domains. One helix, immediately downstream from the N-terminus, borders a shallow (α3) cavity implicated in T cell binding. The other two helices define long grooves on opposite sides of the molecule, one of which (α5 groove) has been implicated in superantigen activity of the SEC3 toxin.

The SEC3 and TSST-1 crystal structures are compared in Figure 1. The TSST-1 structure is also folded into an ellipsoidal shape with dimensions of 55Å × 44Å × 35Å and is somewhat simpler in construction than SEC3. It is characterized by a small N-terminal α-helix followed by a β-barrel analogous to the β-barrel which comprises Domain 1 of SEC3. Like both SEB and SEC3, striking structural features of the TSST-1 molecule are deep grooves formed by the diagonal α-helix and the β-sheet walls, particularly the groove on the back side of the molecule in the view shown in Figure 1B. Although overall this structure is similar to the SEs, several features are lacking in TSST-1. These include an extended N-terminus which folds back over the top of Domain 2 in the SE, an α-helix in the loop at the base of the β-barrel structure in Domain 1, and a second central α-helix which sits in the groove between both domains of the toxin. Also, conspicuously absent in the TSST-1 molecule is the long loop at the top of Domain 1 which is formed by the conserved disulfide linkage in the SEs.

Based on the high degree of sequence similarity that both SEB and SEC share with SPEA, it has been possible to produce a preliminary structure of this streptococcal toxin by homology modeling using the coordinates for SEC3 (Figure 1C). The major features of SPEA include a short N terminal α-helix, followed by a β-barrel structure (Domain 1), a central diagonal α-helix that is exposed to the surface on the back of the molecule, and finally ending in a wall of β-strands. In comparison to the two conserved cysteines of SEB and SEC3 which form the disulfide loop structure on the top of Domain 1, SPEA has 3 cysteines in the same general sequence location as the enterotoxins (Weeks and Feretti, 1986). Despite this similarity, to date there is no strong evidence to suggest there is a disulfide loop formed in SPEA.

Singh, et al. (1994) have pointed out discrepancies between the crystal structure data and results obtained from spectral analysis of toxins in solution. For example, the crystal structure of SEB reportedly contains 20% α-helical structures, which is nearly twice the content predicted by Singh et al. (1988a) and Singh and Betley (1989) based on the SEB circular dichroic and FT-IR spectral analysis. A similar amount of discrepancy was also observed with the β-sheet content predicted by both methods for analysis of SEB. Singh, et al., (1994) have suggested that crystal packing or pH conditions used for crystallization of PTs may result in artificially biased structural results. Given that the SEB and SEC3 structures have been solved from crystals with very different packing interactions and still show the same structures, the effect of crystal packing may be eliminated. Also, the SEC3 structure was solved with crystals grown under two different pH conditions, including physiological (pH 7.4), and neither form shows differences in structure or oligomeric association. Additional structural and biophysical analyses of other PTs should help to clarify this interesting question.

OTHER POTENTIALLY SIGNIFICANT PT STRUCTURAL FEATURES

One recent development in this field is the implication for zinc binding in the conformational integrity and biological activity of PTs. A requirement for zinc in superan-

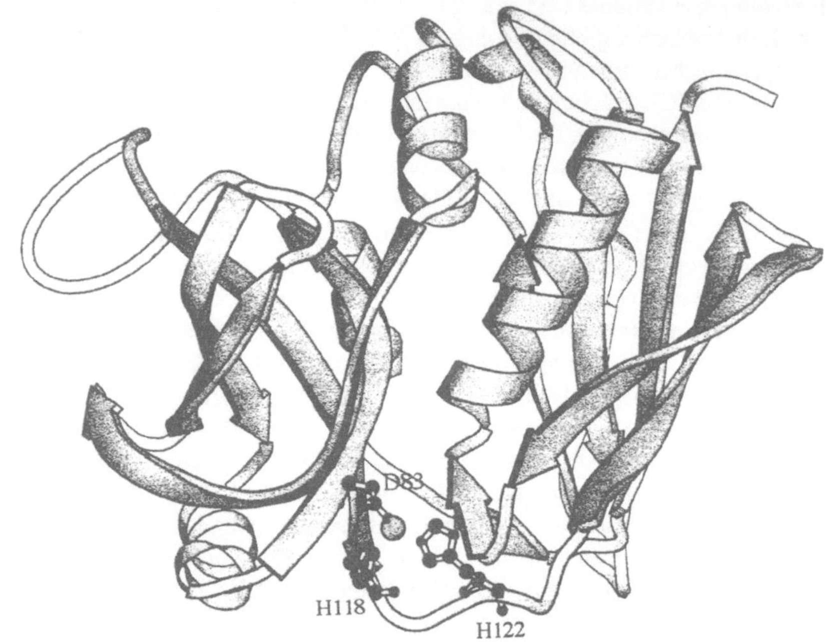

Figure 2. Schematic diagram of the SEC3 molecule rotated 180° compared to figure 1A showing the zinc site at the base of the α5 groove.

tigenicity was first proposed for SEA and SEE (Fraser, et al., 1992), although at that time no evidence was obtained to support the notion that other members of the PT had the same requirement. These observations were extended by refining the SEC3 crystal structure to 1.9Å (Chi, et al., Submitted). Interestingly, this work revealed the unexpected observation that the molecule contains a clearly defined zinc binding site at the base of the α5 groove situated between the two domains (Figure 2). The residues involved in zinc atom coordination are aspartate 83, histidine 118, and histidine 122. The zinc sits in the center of these residues approximately 0.75Å above the three liganding atoms and the fourth tetrahedral ligand site of the zinc is unoccupied.

Some of the features of the environment of the zinc atom are very similar to those found in zinc metalloenzymes, the best characterized of which is thermolysin. The most striking shared characteristic is the invariant zinc binding motif H-E-X-X-H (Hase and Finkelstein, 1993; Figure 3). This motif contains the two histidine coordination sites for the zinc in SEC3 and encompasses residues 118 through 122. Interestingly, a number of bacterial exotoxins such as the botulinum and tetanus neurotoxins (Binz, et al., 1990) and *Bacillus anthracis* lethal factor (Klimpel, et al., 1994) have reportedly identical zinc binding motifs. With each of these other toxins, evidence that they act as zinc metalloproteases has been presented. However, protease activity has not yet been defined for SEC3, despite the fact that the position of the zinc is in a region of the molecule suspected to be important in immune cell interactions and in emesis (see below).

The available evidence suggests that the PT may have evolved so that the requirement for zinc binding may not be uniform. Further complicating the situation is the potential that those PTs which bind zinc may have developed different mechanisms for zinc coordination. When Fraser, et al. (1992) first demonstrated zinc binding by SEA and SEE, they also implicated binding of zinc by these two toxins in the interaction with MHC Class II on

		Cystine loop or analogous region	Conserved Downstream Sequences												
					*	*	*		*		*			*	
SEC1	93	CYFSSKDNVGKVTGG---KTC	111	M	**Y**	**G**	**G**	I	**T**	K	**H**	E	G	**N**	H
SEC2	93	CYFSSKDNVGKVTGG---KTC	111	M	**Y**	**G**	**G**	I	**T**	K	**H**	E	G	**N**	H
SEC3	93	CYFSSKDNVGKVTGG---KTC	111	M	**Y**	**G**	**G**	I	**T**	K	**H**	E	G	**N**	H
SEB	93	CYFSKKTNDINSHQTDKRKTC	114	M	**Y**	**G**	**G**	V	**T**	E	**H**	N	G	**N**	Q
SSA	93	CYYSEGNSCKNAKKTC	109	M	**Y**	**G**	**G**	V	**T**	E	**H**	H	R	**N**	Q
SED	92	CYGGEIDRTAC	103	T	**Y**	**G**	**G**	V	**T**	P	**H**	E	G	**N**	K
SEA	96	CAGGTPNKTAC	107	M	**Y**	**G**	**G**	V	**T**	L	**H**	D	N	**N**	R
SEE	93	CAGGTPNKTAC	104	M	**Y**	**G**	**G**	V	**T**	L	**H**	D	N	**N**	R
SPEA	87	CYLCENAERSAC	99	I	**Y**	**G**	**G**	V	**T**	N	**H**	E	G	**N**	H
SPEC	74	GLFYILNSHTGE	86	Y	I	Y	**G**	G	I	T	P	A	Q	**N**	N
TSST-1	69	KRTKKSQHTSEGTYYH	85	Q	I	S	**G**	V	T	N	T	E	K	L	P

Figure 3. A comparison of cystine loop and downstream sequences for SEs and the analogous regions in nonenterotoxin PTs revealed by analysis of toxin sequences using computer alignment programs. The numbers shown provide reference points for the location of residues in their respective toxins. Highly conserved residues in the region downstream from the cysteine loop are indicated (*).

antigen presenting cells. The crystallographic demonstration of zinc binding by SEC3 and the residues involved in the coordination raises several interesting questions. First, although Fraser, et al. (1992) were unable to implicate zinc in binding of other superantigens to cell surface HLA-DR molecules, their results were inconclusive in regard to SEB and SEC1, due to low levels of binding by iodinated toxins in their assays. It is also interesting to note that neither SEA nor SEE contain the H-E-X-X-H motif in the interdomain groove of SEC3 (Figures 2 and 3) . The evidence suggests that SEA and SEE have an alternate zinc binding region located on the external face of the molecule on Domain 2. Mutagenesis of SEA has implicated histidine 187, histidine 225, and aspartic acid 227 as the zinc binding residues of the toxin (Fraser, et al., 1993). Although Swaminathan, et al. (1992) did not report zinc binding by in the crystal structure of SEB, it is possible that the resolution of their structure determination was insufficient to allow such a determination. Undoubtedly, crystallographic structure analysis of other SE, especially SEA, plus additional mutagenesis experiments with SEC3 and SEB, will clarify whether this group of toxins has multiple mechanisms for binding zinc, and whether zinc binding by PTs other than SEA and SEE is important for toxicity.

PT STRUCTURE-FUNCTION COMPARISONS (SUPERANTIGENICITY)

Considering the proposed mechanism of action for superantigens, it is logical to predict that receptors present on both antigen presenting cells and T cells are required. Thus far, most of the attention has focused on defining regions of PTs that interact with the TCR and MHC Class II. This does not preclude the possibility that other receptors are involved; in fact the current data suggests that this is likely.

The ability to interact with a defined TCR repertoire, determined by the Vβ region of the receptors, is a unique characteristic of each superantigen (Choi, et al., 1989; Marrack and Kappler, 1990; Norrby-Teglund, 1994). Considering that the panel of Vβ sequences recognized by each superantigen is different, it is likely that the divergent residues near the N-terminus of the SEs determine their unique Vβ interactions (Hoffmann, et al., 1994). In regard to TCR and MHC Class II binding by other SE, the most complete information is available for SEA and SEB. Consistent with the prediction that the nonconserved residues near the N-terminus determine Vβ interactions, it has been fairly well established that TCR

Table 3. Proposed models for binding to MHC Class II by four PTs

Property	SEA[a]	SEB[b]	SEC[c]	TSST-1[d]
MHC Chain Bound	β	α	?	α & slight β
PT domains implicated	2	1	1 & 2 (α5 groove)	1
Zinc binding demonstrated?	Yes	No[e]	Yes	No[f]
Location of Zn binding	H187 H225 & D227 (in Domain 2)	None[e]	D83 H118 & H122 between domains 1 & 2	None[e]
Zinc requirement for MHC binding demonstrated?	Yes	No[e]	No[e]	No

[a]Adopted from Fraser, et al. (1992) and Fraser and Hudson (1993) demonstrating the role of zinc in MHC binding and localization of residues involved by mutagenesis. The role of the MHC β-chain was reported by Herman, et al. (1991) and Karp and Long (1992).

[b]Based on crystal structure of SEB::HLA-DR1 complex reported by Jardetzky, et al. (1994).

[c]Reported by Hoffmann, et al. (1994) who used site-ditected mutagenesis and synthetic peptides to localize molecular regions involved in T cell and antigen presenting cell interactions. Zinc binding was determined by the 1.9Å SEC3 crystal structure (Chi, et al., submitted).

[d]Based on crystal structure of TSST-1::HLA-DR1 complex reported by Kim, et al. (1994).

[e]Not reported but inconclusive based on currently available data.

[f]Zinc binding is conditional. Recent studies have shown that one crystal form of TSST-1 incorporates zinc into intermolecular contacts (D. Ohlendorf, unpublished results)

binding by these two toxins occurs in a region analogous to the α3 cavity of SEC3 (see above and Figure 1A).

Attempts to define PT regions involved in MHC Class II binding have produced conflicting results. Part of the reason for these discrepancies is undoubtedly the different approaches used by investigators working with various PTs. For example, the use of mutagenesis and synthetic peptides has relied heavily on the loss or retention of activity by mutant toxins or fragments of toxins. Although these approaches can yield useful results, one must interpret them with caution in the absence of specific information about how they compare structurally to the native protein. Also, results of experiments which employ proteolytic fragments are open to the criticism that small amounts of residual uncleaved toxin could potentially be carried over into biological assays.

One further cause of conflicting conclusions is the possibility that, although MHC binding is a shared property of all PTs, the combined evidence suggests that each toxin may have evolved a different mechanism for interacting with the receptor. For example, work with the SECs suggests that MHC Class II binding is not entirely dependent upon residues within the N-terminal region of SEC1, but that the residues located in the α5 groove of this toxin may also be important (Figure 1; Hoffmann, et al., 1994). Peptide analyses have also indicated a potential role for residues in the α5 groove in T cell stimulation. Presumably these residues are required to interact with the antigen presenting cell. Other evidence for a potential role of this region in SEC binding to antigen presenting cells is that residues involved in zinc binding, identified in crystallographic experiments, are also located in the α5 groove at its base. The empty liganding site on the zinc atom could provide a binding site for a biologically important ligand, as it does in other structures with similar zinc sites. Site-directed mutagenesis of SPEA has identified a similar region in the sequence in which mutations result in less MHC binding (Hartwig and Fleischer, 1993).

Several investigators, supported by convincing data, have proposed models for SEA and SEB binding to MHC through residues located on opposite ends of the two toxins (Fraser et al., 1993; Jardetzky et al., 1994). Both of these models implicate residues outside the α5 groove implicated in other studies for SEC. Considering the reported differences in binding properties of these three toxins, it is possible that each has a unique mechanism for interacting with antigen presenting cells. For example, these three toxins have reportedly different properties in regard to a requirement for zinc in interaction with MHC Class II, location of zinc binding residues in the toxin molecules, and differences in regard to the chain of MHC Class II involved in the interaction. Table 3 summarizes the four published models for binding of three SEs and TSST-1 to MHC Class II. Mutagenesis results have suggested that residues of SEB outside its large interdomain groove (analogous to the SEC3 α5 groove) are involved in MHC II binding. Similarly, no contact points between SEB and HLA-DR were evident in this groove from the recently published SEB/HLA-DR complex structure (Jardetzky, et al., 1994). Instead, residues in Domain 1 (including the cystine loop and nearby residues) were involved in binding to the receptor α-chain. On the other hand, the model of SEA binding to MHC Class II implicates the zinc binding residues at the opposite end of the molecule in Domain 2 and its binding appears to involve the β-chain of the receptor (Fraser, et al., 1993).

A more complicated picture has been provided for TSST-1 binding in the recently published toxin-MHC complex crystal structure (Kim, et al., 1994). The mode of TSST-1 binding shares some similarities with that for SEB in regard to overall orientation of the toxin molecule. However, although the regions of MHC involved in the interaction overlap for the two toxins, they are not identical. For example, similar to the situation with SEB, residues in TSST-1 that are involved in the complex formation are located on the external surface of Domain 1 with the larger Domain 2 oriented up and away from the MHC molecule. Although the major receptor involvement is provided by residues in the α-chain, the toxin also makes contact with the β-chain and peptide bound in the groove. Furthermore, the orientation of TSST-1 in this complex structure precludes direct MHC:TCR interactions. Considering these features, Kim, et al. (1994) also have proposed a different mechanism for interaction of TSST-1 with the TCR.

RELATIONSHIP BETWEEN SE-INDUCED EMESIS AND SUPERANTIGENICITY

Emesis is a biological property that has been observed only for the SEs. Significant progress has been achieved in regard to understanding the unique emetic function of this group of PTs. Although patients with staphylococcal food poisoning may present with diarrhea, nausea, dizziness and prostration, the ability of SEs to induce an emetic response upon oral ingestion is considered to be the best correlation of their association in this intoxication. Until recently, our understanding of structural aspects of the SEs that determine their ability to induce an emetic response has progressed relatively slowly. The main reason for this has been the lack of an inexpensive and convenient animal model for staphylococcal food poisoning (Bergdoll, 1988). Humans and nonhuman primates are the only animals that vomit when the SEs are administered orally. Very early studies into the mode of SE action concluded that emesis occurs through stimulation of neural receptors in the abdomen by the toxins. The available evidence suggested that transmission through the vagus and sympathetic nerves stimulates the medullary emetic center (Bayliss, 1940; Clark, et al., 1962; Elwell, et al., 1975). Unfortunately, to date, the lack of a convenient animal model has prevented the characterization of receptors in the gut that interact with SEs. Although mast cells have been proposed (see above) to be involved in the response, it is still unclear whether this represents a direct action of the toxins or whether these cells are activated indirectly.

It is interesting that emesis can be induced in other vertebrate animals upon intravenous inoculation of SE (Clark, et al., 1962). Although this is a well documented response, this activity probably does not directly reflect the action of the toxins in the gastrointestinal tract. Other PTs which are not associated with food poisoning, such as TSST-1, cause an emetic response when they gain access to the systemic circulation (Bohach, et al., 1990). In fact, gastrointestinal abnormalities are commonly observed in patients with TSS. The most likely explanation for this phenomenon is that it is a consequence of the multiorgan effect of the toxins linked to their role in this disease.

Some investigators have suggested that emesis induced by SEs results from their function as superantigens, and that food poisoning symptoms are the result of superantigen induction of interleukin-2 (Johnson, et al., 1992). Despite this prediction, recent evidence from several laboratories has supported earlier work which indicated that T cell mitogenicity and emesis are determined by different functional regions of SE molecules. The strongest evidence to support this notion has come from investigation of several SEs including SEA, SEB, and SEC1 (Alber, et al., 1990; Harris, et al., 1993; Hovde, et al., 1994). The general conclusion from work with these toxins is that it is possible to dissociate the emetic function of the toxins from that of T cell stimulation.

Another approach that has been used in combination with mutagenesis to predict SE regions involved in emesis has been to identify structural features of enterotoxigenic PTs that are not shared by other members of the PT family. One shared feature of SEs, the intramolecular disulfide linkage, has received the most attention. Of the classical PTs, only the SEs and the streptococcal toxin SPEA have at least two cysteine residues. Accordingly, these toxins have the potential to form an intramolecular disulfide bond and presumably the characteristic SE cystine loop. Although all SEs form a loop between the two cysteines, it seems less likely that the amino acid residues in the loop are directly involved in the emetic response. Neither the number of residues in the loop nor the properties of the loop residues are highly conserved (Figure 3). Furthermore, several investigators have shown that disruption of the loop by proteolysis of residues between the cysteines does not dramatically alter emetic potency of the SE (Spero and Morlock, 1978; Spero, et al., 1973; Spero, et al., 1975; Spero, et al., 1976).

Hovde, et al. (1994) employed a mutagenesis strategy to investigate the requirement for the SE disulfide bond using SEC1 as their model. Their approach to this problem involved construction of mutants unable to form the linkage by substituting either alanine or serine for cysteine at positions 93 and/or 110. These investigators were able to draw several conclusions about the importance of the disulfide bond by testing the purified mutants for biological activity and stability. First, they showed that the disulfide bond is not absolutely required for emesis, since mutants in which serine substituted for cysteine were able to induce vomiting in a standard monkey feeding assay. Since the corresponding alanine mutants were nonemetic, they proposed that in the absence of a disulfide bond, residual native structure promoted by serine hydrogen bonds sufficiently holds critical residues in a proper orientation. This differential activity displayed by serine vs. alanine mutants was not due to differences in stability. In fact, some of the alanine mutants were more stable in gastric fluid. The second major conclusion from this report was that emesis and T cell stimulation were separable. There was no correlation between the loss of emetic potency, T cell stimulatory capability, or stability. Similar conclusions were obtained by several other investigators who have also dissociated emetic and superantigen functions in SEA and SEB (Alber, et al., 1990; Harris, et al., 1993).

Based on the requirement for structural stability provided by serine substitutions, one may propose that a region close to the disulfide linkage is needed for emesis. Several groups of residues could potentially be implicated. First, it is possible that the residues in the cystine loop themselves are involved in interaction with receptors in the gut. Although SE cystine loops are not highly conserved, residues of SEB have been shown to be important for the interaction of this toxin with MHC Class II. Another strong possibility is that the stretch of conserved residues (Figure 3) immediately downstream from the second SE cysteine (residues 111-123 in SEC1) is involved. These residues form a β-strand and loop in the SEB and SEC3 structures (Swaminathan, et al., 1992 and Figure 1) and their conformation could be affected by the presence of the disulfide linkage. However, this region is also somewhat conserved among nonemetic toxins. Therefore, one may assume that only toxins with critical residues in this region, plus a proper orientation of these residues maintained by the disulfide bond, would have the ability to induce emesis. Despite a lack of correlation between emesis and T cell proliferation, the regions implicated to be important for emesis in this model as well as the nearby zinc atom have been implicated in T cell stimulation. Taken together, these results support the prediction of Harris, et al. (1993) that both emesis and T cell stimulation are dependent upon distinct but overlapping regions on SE.

The fact that histidine residues 118 and 122, which make up the zinc binding motif of SEC3, are located within this conserved stretch of residues suggests that zinc binding may play a role in the emetic response of SEC. We have recently solved the crystal structure of the SEC3 C110S and C110A (emetic and nonemetic, respectively) mutants. The most obvious structural difference between these two mutants lies in the general accessibility of the zinc binding pocket in each of their structures (unpublished results).

The relatedness of several streptococcal toxins to the SEs is noteworthy and raises many interesting questions. First, although SPEA and the recently described SE-like streptococcal superantigen toxin produced by *Streptococcus pyogenes* have at least two cysteine residues and could potentially form a disulfide linkage (Reda, et al., 1994; Weeks and Ferretti, 1986), their emetic ability has not been thoroughly investigated. SPEA has not been implicated in food poisoning and it is unclear whether this toxin is produced in food. Our laboratories have demonstrated that the SPEA toxin is stable in monkey gastric fluid (manuscript in preparation), raising the possibility that since they are similar to SEs, especially SEB and SEC, some of the streptococcal toxins could also have emetic properties.

STRUCTURAL ASPECTS OF OTHER PT BIOLOGICAL PROPERTIES

The mechanisms of systemic pathogenesis of the PTs are undoubtedly multifactorial. In addition to superantigenicity, several potential toxicities could contribute to life-threatening symptoms observed in many patients. Although several of the activities listed in Table 2 have been proposed to contribute to PT-induced lethality, little structure-function information is available to adequately identify the responsible molecular regions.

One well documented shared property of the PTs that could play a significant role in lethality is their ability to enhance susceptibility of hosts to lethal endotoxin shock (Schlievert, 1982; Stiles, et al., 1993). The fact that PTs and endotoxin can synergistically induce lethal shock is evident from experiments in rabbits that consistently show that prior administration of PTs reduces the amount of endotoxin required for shock and lethality by as much as a factor of 10^5. Although the potential role for endotoxin in TSS remains controversial (Melish, et al., 1989), protection against lethal TSS can be demonstrated in both human patients and animal models by administering lipopolysaccharide (LPS)-specific antibodies (Fomsgaard, et al., 1987; Priest, et al., 1989). The susceptible cells involved in this phenomenon are not known with certainty, although cytotoxicity for both renal tubular cells and lymphocytes requires a synergistic action of both toxins (Leonard and Schlievert,

1992). A potential role for toxicity to other organs, such as the liver and heart, is certainly worthy of investigation considering the lack of present knowledge in this area.

Work with SPEA has revealed that at least some PTs are capable of binding specifically to LPS (Leonard and Schlievert, 1992). Presumably, cytotoxicity for susceptible cells requires formation of the PT-LPS complex. This interaction has been localized to the ketodeoxyoctanoate residues of LPS. Identification of the residues on the PTs which are involved in this interaction will require additional investigation. However, it is tempting to speculate that the typical binding site on the O/B binding domain (Murzin, 1993 and discussed above), common to PT structures, will contain the LPS contact points.

SUMMARY AND CONSIDERATIONS FOR THE FUTURE

The goal of this review was to summarize our current knowledge of the structure-function relationships of the PT family. This group of toxins has classically been implicated in numerous toxigenic staphylococcal and streptococcal diseases including toxic shock syndrome and scarlet fever. It is now clear that these diseases are a combined result of multiple biological effects linked to their superantigenic properties. However, other diseases, most notably staphylococcal food poisoning and others discussed above, may be due to additional biological properties unrelated to their T cell stimulation.

It is becoming increasingly clear that, to exert these biological properties, certain shared and unique structural features are required. Spectral and crystallographic data of SEC3 and SEB shows that the SE protein folding properties are rather unique compared to other types of bacterial toxins. T cell stimulation seems to be dependent upon a conserved conformation in which PTs are folded into two functional domains. This crucial folding presumably allows the PT to bridge both the antigen presenting cell and T cell, despite the fact that it does not require shared protein sequences.

Investigations into the functional regions involved in T cell stimulation have focused primarily on localization of residues that interact with MHC Class II on antigen presenting cells and the Vβ region of the TCR. Perhaps the most unexpected conclusion from these combined investigations is that, despite the fact that binding to both cell types is a shared property of PTs, each individual molecule may have evolved a somewhat different mechanism for interacting with their receptors. Work with four different PTs, (SEA, SEB, SEC1, and TSST-1) has produced several potential models of T cell stimulation; each of these is supported by convincing experimental data. This is especially true for interactions with antigen presenting cells, where there may be differences in binding to MHC Class II molecules. In this case, each toxin may bind to different locations and chains of the receptor, through different regions of the toxins themselves, and with a potentially differential involvement of zinc cations. However, our present knowledge is still compatible with a potential involvement of additional or alternate receptors on antigen presenting cells. In these cases, differences in binding to MHC Class II for individual toxins may be only minimal.

The study of biological properties other than superantigenicity has progressed more slowly. The only activity that has received significant attention in regard to structure-function analysis is the SE-induced emetic response. Recent sequence and structure information for the SEs has allowed for rational design and interpretation of mutagenesis experiments. The combined current evidence suggests that the emetic ability of SEs is distinct from the ability of the toxins to cause T cell stimulation. The conformation of the molecular region required for emesis is dependent on the disulfide bond, which is a feature of all known SEs. The amino acid residues most likely to be involved in invoking emesis are those immediately down-stream from the disulfide linkage, although the possibility for involvement of the cystine loop has not been ruled out.

The PT field has progressed rapidly during the last decade with the significant amount of new knowledge about the structural and biological features of the toxins. However, the demonstration of several unexpected molecular features has raised a number of important questions that could yield surprising new information about the mechanism of action of these proteins. First, the observation that SEC3 and possibly other related SEs share a zinc binding motif with a class of metalloenzymes could be very significant, considering that several bacterial exotoxins produced by pathogenic *Clostridium* and *Bacillus* species share this motif. Since proteolysis has been shown to be important for these other important exotoxins, it is conceivable that at least some PTs have similar requirements. Furthermore, considering the fact that SEs reportedly transmit stimulatory impulses through the vagus nerve, the possibility of a direct effect of nerve function by SE should be investigated. One other interesting feature in regard to zinc binding by SEs is that SEA and SEE bind zinc, albeit through a mechanism that is apparently different from SEC3. Although zinc binding by these two toxins is required for interacting with MHC Class II, they do not share the metalloprotease like motif, and mutagenesis results suggest that the cation is bound to a different place on the molecule. One other potentially important, but thus far uncharacterized molecular feature is the O/B binding domain demonstrated in the crystal structure of both SEB and SEC3 and the related structure of TSST-1. All previously characterized proteins, including several bacterial exotoxins, with this same folding motif interact with either oligosaccharides or oligonucleotides. They share this ability despite having no significant overall amino acid homology. It remains to be determined whether residues in this domain provide the toxins a mechanism for binding to LPS, an unidentified SE receptor, or are involved in regulation of gene expression.

ACKNOWLEDGMENTS

This manuscript was supported by grants from the Public Health Service; AI28401, AI22159, HL36611, and RR00166, and from Kimberly-Clark (Neenah, WI), Personal Products Company, (New Brunswick, NJ), Tambrands, (Palmer, MA), the United Dairymen of Idaho, and the Lucille P. Markey Foundation for the support and expansion of structural biology research (Purdue University). We also acknowledge support of the University of Washington Primate Field Station (Medical Lake, WA) and the Idaho Agriculture Experiment Station.

REFERENCES

Abe, J., Kotzin, B.L., Jujo, K., Melish, M.E., Glode, M.P., Kohsaka, T., and Leung, D.Y.M. 1992. Selective expansion of T cells expressing T-cell receptor variable regions Vβ 2 and Vβ 8 in Kawasaki Disease. *Proc. Natl. Acad. Sci. (USA)* 89:4066-4070.

Acharya, K.R., Passalacqua, E.F., Jones, E.Y., Harlos, K., Stuart, D.I., Brehm, R.D. and Tranter, H.S. 1994. Structural basis of superantigen action inferred from the crystal structure of toxic shock syndrome toxin-1. *Nature* 367:94-97.

Alber, G., Hammer, D. and Fleischer, B. 1990. Relationship between enterotoxic- and T lymphocyte-stimulating activity of staphylococcal enterotoxin B. *J. Immunol.* 144:4501-4506.

Altboum, Z., Hertman, I. and Sarid, S. 1985. Penicillinase plasmid linked genetic determinants for enterotoxin B and C1 production in *Staphylococcus aureus*. *Infect. Immun.* 47:514-521.

Arko, R.J., Rasheed, J.K., Broome, C.V., Chandler, F.W., and Paris, A.L. 1984. A rabbit model of toxic shock syndrome: clinicopathological features. *J. Infect.* 8:205-211.

Bayles, K.W. and Iandolo, J.J. 1989. Genetic and molecular analyses of the gene encoding staphylococcal enterotoxin D. *J. Bacteriol.* 171:4799-4806.

Bayliss, M. 1940. Studies on the mechanism of vomiting produced by staphylococcal enterotoxin. *J. Exp. Med.* 72:669-684.

Bergdoll, M.S. 1983. Enterotoxins. *In: Staphylococci and staphylococcal infections.* (Eds: Easmon, C.S.F., and Adlam, C.) Academic Press, Inc., London, 559-598.

Bergdoll, M.S. 1988. Monkey feeding assay for staphylococcal enterotoxins. *Methods. Enzymol.* 165:324-333.

Bergdoll, M.S., Borja, C.R., Robbins, R., and Weiss, K.F. 1971. Identification of enterotoxin E. *Infect. Immun.* 4:593-595.

Bergdoll, M.S., Robbins, R.N., Weiss, K., Borja, C.R., Huang, I.Y., and Chu, F.S. 1973. The staphylococcal enterotoxins: similarities. *Contrib. Microbiol. Immunol.* 1:390-396.

Bergdoll, M.S. and Schlievert, P.M. 1984. Toxic shock syndrome toxin. *Lancet.* 2:691.

Betley, M.J., Lofdall, S., Kreiswirth, B.N., Bergdoll, M.S., and Novick, R.P. 1984. Staphylococcal enterotoxin A gene is associated with a variable genetic element. *Proc. Natl. Acad. Sci. (USA).* 81:5179-5183.

Betley, M.J., and Mekalanos, J.J. 1985. Staphylococcal enterotoxin A is encoded by phage. *Science.* 229:185-187.

Betley, M.J., and Mekalanos, J.J. 1988. Nucleotide sequence of the type A staphylococcal enterotoxin gene. *J. Bacteriol.* 170:34-41.

Binz, T., Kurazono, H., Popoff, M.R., Eklund, M.W., Sakaguchi, G., Kozaki, S., Krieglstein, K., and Henschen, A. 1990. Nucleotide sequence of the gene encoding *Clostridium botulinum* neurotoxin type D. *Nucl. Acids Res.* 18:5556.

Blomster-Hautamaa, D.A., Kreiswirth, B.N., Kornblum, J.S., Novick, R.P., and Schlievert, P.M. 1986. The nucleotide and partial amino acid sequence of toxic shock syndrome toxin-1. *J. Biol. Chem.* 261:15783-15786.

Bohach, G.A., Fast, D.J., Nelson, R.D., and Schlievert, P.M. 1990. Staphylococcal and streptococcal pyrogenic toxins involved in toxic shock syndrome and related illnesses. *Crit. Rev. Microbiol.* 17:251-272.

Bohach, G.A., Handley, J.P., and Schlievert, P.M. 1989a. Biological and immunological properties of the carboxyl terminus of staphylococcal enterotoxin C1. *Infect. Immun.* 57:23-28.

Bohach, G.A., Hovde, C.J., Handley, J.P., and Schlievert, P.M. 1988. Cross-neutralization of staphylococcal and streptococcal pyrogenic toxins by monoclonal and polyclonal antibodies. *Infect. Immun.* 56:400-404.

Bohach, G.A., Kreiswirth, B.N., Novick, R.P., and Schlievert, P.M. 1989b. Analysis of toxic shock syndrome isolates producing staphylococcal enterotoxins B and C1 with use of southern hybridization and immunologic assays. *Rev. Infect. Dis.* 11 (Suppl. 1): S75-S82.

Bohach, G.A. and Schlievert, P.M. 1987. Nucleotide sequence of the staphylococcal enterotoxin C gene and relatedness to other pyrogenic toxins. *Mol. Gen. Genet.* 209:15-20.

Boyle, T., Lancaster, V., Hunt, R., Gemski, P., and Jett, M. 1994. Method for simultaneous isolation and quantitation of platelet activating factor and multiple arachidonate metabolites from small samples: analysis of effects of *Staphylococcus aureus* enterotoxin B in mice. *Anal. Biochem.* 216:373-382.

Chi Y.I., Bohach G.A., and Stauffacher C.V. A zinc binding site in a bacterial superantigen revealed by the 1.9Å X-ray structure of staphylococcal enterotoxin C3. Submitted for publication.

Chi, Y.I., Bohach, G.A., and Stauffacher, C.V. A refined crystal structure of staphylococcal enterotoxin C1 at 1.9C . Submitted for publication:.

Choi, Y., Kotzin, B., Herron, L., Callahan, J., Marrack, P. and Kappler, J. 1989. Interaction of *Staphylococcus aureus* toxin (superantigens) with human T cells. *Proc. Natl. Acad. Sci. (USA)* 86:8941-8945.

Clark, W.G., Vanderhooft, G.F., and Borison, H.L. 1962. Emetic effect of purified staphylococcal enterotoxin in cats. *Proc. Soc. Exp. Biol. Med.* 111:205-207.

Colon-Whitt A., Whitt R.S., Cole R.M. 1979. Production of an erythrogenic toxin (streptococcal pyrogenic exotoxin) by a nonlysogenized group A streptococcus. *In Pathogenic streptococci*, (Ed: Parker, M.T.) Reedbooks, England, 64-65.

Cone, L.A., Woodard, D.R., Schlievert, P.M., and Tomory, G.S. 1987. Clinical and bacteriologic observations of a toxic shock-like syndrome due to Streptococcus pyogenes. *N. Engl. J. Med.* 317:146-149.

Couch, J.L. and Betley, M.J. 1989. Nucleotide sequence of the Type C3 staphylococcal enterotoxin gene suggests that intergenic recombination causes antigenic variability. *J. Bacteriol.* 171:4507-4510.

Couch, J.L., Soltis, M.T., and Betley, M.J. 1988. Cloning and nucleotide sequence of the type E staphylococcal enterotoxin gene. *J. Bacteriol.* 170:2954-2960.

DeAzavedo, J.C., Foster, T.J., Hartigan, P.J., Arbuthnott, J.P., O'Reilly, M., Kreiswirth, B.N., and Novick, R.P. 1985. Expression of the cloned toxic shock syndrome toxin 1 gene (*tst*) in vivo with a rabbit uterine model. *Infect. Immun.* 50:304-309.

De Boer, M.L. and Chow, A.W. 1993. Toxic shock syndrome toxin-1 producing staphylococcus aureus isolates contain the staphylococcal enterotoxin B genetic element but do not express staphylococcal enterotoxin B. *J. Infect. Dis.* 170:818-827.

Demers, B., Simor, A.E., Vellend, H., Schlievert, P.M., Byrne, S., Jamieson, F., Walmsley, S., and Low, D.E. 1993. Severe invasive group A streptococcal infection in Ontario, Canada; 1987-1991. *Clin. Infect. Dis.* 16:792-800.

Deresiewicz, R.L., Flaxenburg, J.A., Chan, M., Finberg, R.W., and Kasper, D.L. 1994. Intracellular expression of toxic shock syndrome toxin 1 in *Saccharomyces cerevisiae*. *Infect. Immun.* 62:2202-2207.

Dolman, D.E. and Wilson, R.J. 1938. Experiments with staphylococcal enterotoxin. *J. Immunol.* 35:13-30.

Elwell, M.R., Liu, C.T., Spertzel, R.O., and Beisel, W.R. 1975. Mechanisms of oral staphylococcal enterotoxin B-induced emesis in the monkey. *Soc. Exp. Bio. Med.* 148:424-427.

Fast, D.J., Schlievert, P.M., and Nelson, R.D. 1988. Nonpurulent response to toxic shock syndrome toxin 1-producing *Staphylococcus aureus*. Relationship to toxin-stimulated production of tumor necrosis factor. *J. Immunol.* 140:949-953.

Fomsgaard, A., Neilsen, R., Froberg, K.D., Baek, L., and Deghn, H.K. 1987. Endotoxinaemia in toxic shock syndrome treated with anti-endotoxin antibodies. *Lancet I:514-515.*

Fraser, J.D. 1989. High affinity binding of staphylococcal enterotoxins A and B to HLA-DR. *Nature* (London) 339:221-223.

Fraser, J.D. and Hudson, K.R. 1993. Superantigens-remnants of a past process. *Res. Immunol.* 144:188-193.

Fraser, J.D., Lowe, S., Irwin, M.J., Gascoigne, N.R.J., and Hudson, K.R. 1993. Structural Model of Staphylococcal Enterotoxin A Interaction with MHC Class II Antigens. *In: Superantigens: A Pathogen's View of the Immune System.* (Eds: Huber, B.T. and Palmer, E.) Cold Spring Harbor Laboratory Press, Plainview, NY, 7-30.

Fraser, J.D., Urban, R.G., Strominger, J.L., and Robinson, H. 1992. Zinc regulates the function of two superantigens. *Proc. Natl. Acad. Sci. (USA)* 89:5507-5511.

Frobisher, M. and Brown, J.H. 1927. Transmissible toxigenicity of streptococci. *Bull. Johns Hopkins Hosp.* 41:167-173.

Goshorn, S.C. and Schlievert, P.M. 1988. Nucleotide sequence of streptococcal pyrogenic exotoxin type C. *Infect. Immun.* 56:2518-2520.

Goshorn, S.C. and Schlievert, P.M. 1989. Phage association of streptococcal pyrogenic exotoxin type C. *J. Bacteriol.* 171:3068-3073.

Harris, T.O., Grossman, D., Kappler, J.W., Marrack, P., Rich, R.R., and Betley, M.J. 1993. Lack of complete correlation between emetic and T-cell-stimulatory activities of staphylococcal enterotoxins. *Infect. Immun.* 61:3175-3183.

Hartwig, U.F. and Fleischer, B. 1993. Mutations affecting MHC Class II binding of the superantigen streptococcal erythrogenic toxin A. *International Immunol.* 5:869-875.

Hase, C.C. and Finkelstein, R.A. 1993. Bacterial extracellular zinc-containing metalloproteases. *Microbiol. Rev.* 57:823-837.

Hauser, A.R. and Schlievert, P.M. 1990. Nucleotide sequence of the streptococcal pyrogenic exotoxin type B gene and toxin relationship to streptococcal proteinase precursor. *J. Bacteriol.* 172: 4536-4542.

Hauser, A.R., Stevens, D.L., Kaplan, E.L., and Schlievert, P.M. 1989. Molecular analysis of pyrogenic exotoxins from *Streptococcus pyogenes* isolates associated with toxic shock-like syndrome. *J. Clin. Microbiol.* 29:1562-1567.

Hauser, A.R., Vath, G.M., Ohlendorf, D.H., and Schlievert, P.M. 1995. Structural studies of streptococcal pyrogenic exotoxin superantigens. *In: Bacterial Superantigens: Structure, Function and Therapeutic Potential.* (Eds: Thibodeau, J. and Sekaly, R.P.) Springer-Verlag, New York, 39-48.

Herman, A., Labrecque, N., Thibodeau, J., Marrack, P., Kappler, J.W., and Sekaly, R.-P. 1991. Identification of the staphylococcal enterotoxin A superantigen binding site in the β1 domain of the human histocompatibility antigen HLA-DR. *Proc. Natl. Acad. Sci.* (USA) 88:9954-9958.

Hoffmann, M.L., Jablonski, L.M., Crum, K.K., Hackett, S.P., Chi, Y.I., Stauffacher, C.V., Stevens, D.L., and Bohach, G.A. 1994. Predictions of T cell receptor and major histocompatibility complex binding sites on staphylococcal enterotoxin C1. *Infect. Immun.* 62:3396-3407.

Hoge, C.W., Schwartz B., Talkington D.F., Brieman, R.F., MacNeill, E.M., and Eglender, S.J. 1993. The changing epidemiology of invasive group A streptococcal infections and the emergence of streptococcal toxic shock-like syndrome. *JAMA* 269: 384-389.

Hovde, C.J., Hackett, S.P., and Bohach, G.A. 1990. Nucleotide sequence of the staphylococcal enterotoxin C3 gene: sequence comparison of all three type C staphylococcal enterotoxins. *Mol. Gen. Genet.* 220:329-333.

Hovde, C.J., Marr, J.C., Hoffmann, M.L., Hackett, S.P., Chi, Y.-I.,Crum, K.K., Stevens, D.L., Stauffacher, C.V., and Bohach, G.A. 1994. Investigation of the role of the disulfide bond in activity and structure of staphylococcal enterotoxin C1. *Molec. Microbiol.* 13:897-909.

Hynes, W.L., Weeks, C.R., Iandolo, J.J., and Feretti, J.J. 1987. Immunologic cross-reactivity of Type A streptococcal exotoxin (erythrogenic toxin) and staphylococcal enterotoxins B and C1. *Infect. Immun.* 55:837-838.

Iandolo, J.J. 1989. Genetic analysis of extracellular toxins of *Staphylococcus aureus*. *Annu. Rev. Microbiol.* 43:375-401.

Iwasaki, M., Igarashi, H., Hinuma, Y., and Yutsudo, T. 1993. Cloning, characterization and overexpression of a *Streptococcus pyogenes* gene encoding a new type of mitogenic factor. *FEBS Lett.* 331: 187-192.

Jardetzky, T.S., Brown, J.H., Gorga, J.C., Stern, L.J., Urban, R.G., Chi, Y.I., Stauffacher, C.V., Strominger, J.L., and Wiley, D.C. 1994. Three-dimensional structure of a human class II histocompatibility molecule complexed with superantigen. *Nature (London)* 368:711-718.

Johnson, H.M., Russell, J.K. and Pontzer, C.H. 1992. Superantigens in human disease. *Sci. Amer.* 266:92-101.

Johnson, L.P. and Schlievert, P.M. 1983. A physical map of the group A streptococcal pyrogenic exotoxin bacteriophage T12 genome. *Mol. Gen. Genet.* 189:251-255.

Johnson, L.P., Schlievert, P.M., and Watson, D.W. 1980. Transfer of group A streptococcal pyrogenic exotoxin production to non-toxigenic strains by lysogenic conversion. *Infect. Immun.* 28:254-257.

Johnson, L.P., Tomai, M.A. and Schlievert, P.M. 1986. Molecular analysis of bacteriophage involvement in group A streptococcal pyrogenic exotoxin A production. *J. Bacteriol.* 166:623-627.

Jones, C.L. and Khan, S.A. 1986. Nucleotide sequence of the enterotoxin B gene from *Staphylococcus aureus*. *J. Bacteriol.* 166:29-33.

Karp, D.R., and Long, E.O. 1992. Identification of HLA-DR1 β chain residues critical for binding staphylococcal enterotoxins A and E. *J. Exp. Med.* 175:415-424.

Kim, J. Urban, R.G., Strominger, J.L. and Wiley, D.C. 1994. Toxic shock syndrome toxin-1 complexed with a Class II major histocompatibility molecule HLA-DR. *Science* 266:1870-1874.

Klimpel, K.R., Arora, N. and Leppla, S.H. 1994. Anthrax toxin lethal factor contains a zinc metalloprotease consensus sequence which is required for lethal toxin activity. *Molec. Microbiol.* 13:1093-1100.

Kreiswirth, B.N., Lofdahl, S. Betley, M.J., O'Reilly, M., Schlievert, P.M., Bergdoll, M.S. and Novick, R.P. 1983. The toxic shock syndrome exotoxin structural gene is not detectably transmitted by a prophage. *Nature (London)* 305:707-712.

Lee, P.K., Deringer, J.R., Kreiswirth, B.N., Novick, R.P. and Schlievert, P.M. 1991. Fluid replacement protection of rabbits challenged subcutaneously with toxic shock syndrome toxins. *Infect. Immun.* 59:879-884.

Lee, P.K., Kreiswirth, B.N., Deringer, J.R., Projan, S.J., Eisner, W., Smith, B.L., Carlson, E., Novick, R.P., and Schlievert, P.M. 1992. Nucleotide sequences and biological properties of toxic shock syndrome toxin 1 from ovine- and bovine-associated *Staphylococcus aureus*. *J. Infect. Dis.* 165:1056-1063.

Leggiadro, R.J., Bugnitz, M.C., and Peck, B.A. 1993. Group A streptococcal bacteremia in a Mid-South children's hospital. *South Med. J.* 86:615-618.

Leonard, B.B. and Schlievert, P.M. 1992. Immune cell lethality induced by streptococcal pyrogenic exotoxin A. *Infect. Immun.* 60:3747-3755.

Marr, J.C., Lyon, J.D., Roberson, J.R., Lupher, M., Davis, W.C. and Bohach, G.A. 1993. Characterization of novel type C staphylococcal enterotoxins: Biological and evolutionary implications. *Infect. Immun.* 61:4254-4262.

Marrack, P. and Kappler, J. 1990. The staphylococcal enterotoxins and their relatives. *Science* 248:705-711.

Melish, M.E., Murata, S., Fukunaga, C., Frogner, K., Hirata, S. and Wong, C. 1989. Endotoxin is not an essential mediator in toxic shock syndrome. *Rev. Infect. Dis.* 11 (Suppl.): S219-S230.

Middlebrook, J.L., Spero, L. and Argos, P. 1980. The secondary structures of staphylococcal enterotoxins A, B, and C. *Biochem. Biophys. Acta.* 621:233-240.

Miethke, T., Wahl, C., Heeg, K., Echtenacher, B., Krammer, P.H. and Wagner, H. 1992. T-cell-mediated lethal shock triggered in mice by the superantigen staphylococcal enterotoxin B: critical role of tumor necrosis factor. *J. Exp. Med.* 175:91-98.

Mollick, J.A., Cook, R.G., and Rich, R.R. 1989. Class II MHC molecules are specific receptors for staphylococcal enterotoxin A. *Science* 244:817-820.

Mollick, J.A., Miller, G.G, Musser, J.M., Cook, R.G., Grossman, D., and Rich, R.R. 1993. A novel superantigen isolated from pathogenic strains of *Streptococcus pyogenes* with amino terminal homology to staphylococcal enterotoxins B and C. *J. Clin. Invest.* 92:710-719.

Munoz, P.A., Warren, J.R. and Noelken, M.E. 1976. Structure of aqueous staphylococcal enterotoxin B by spectropolarimetry and sequenced based conformational predictions. *Biochemistry* 15:4666-4671.

Murzin, A.G. 1993. OB(oligonucleotide/oligosaccharide binding)-fold: common structural and functional solution for non-homologous sequences. *EMBO* 12:861-867.

Musser, J.M., Hauser, A.R., Kim, M., Schlievert, P.M., Nelson, K. and Selander, R.K. 1991. *Streptococcus pyogenes* causing toxic shock-like syndrome and other invasive diseases: clonal diversity and pyrogenic exotoxin expression. *Proc. Natl. Acad. Sci. USA.* 88: 2668-2672.

Nelson, K., Schlievert, P.M., Selander, R.K., and Musser, J.M. 1991. Characterization and clonal distribution of four alleles of the *speA* gene encoding pyrogenic exotoxin A (scarlet fever toxin) in *Streptococcus pyogenes*. *J. Exp. Med.* 174:1271-1274.

Nida, S.K. and Ferretti, J.J. 1982. Phage influence on the synthesis of extracellular toxins in group A streptococci. *Infect. Immun.* 36:745-750.

Norrby-Teglund, A., Newton, D., Kotb, M., Holm, S.E., and Norgren, M. 1994. Superantigenic properties of the group A streptococcal exotoxin SPEF (MF). *Infect. Immun.* 62:5227-5233.

Prasad, G.S., Earhart, C.A., Murray, D.L., Novick, R.P., Schlievert, P.M. and Ohlendorf, D.H. 1993. Structure of toxic shock syndrome toxin 1. *Biochem* 32:13761-13766.

Priest, B.P., Schlievert, P.M. and Dunn, D.L. 1989. Treatment of toxic shock syndrome with endotoxin-neutralizing antibodies. *J. Surg. Res.* 46:527-531.

Reda, K.B., Kapur, V., Mollick, J.A., Lamphear, J.G., Musser, J.M., and Rich, R.R. 1994. Molecular characterization and phylogenetic distribution of the streptococcal superantigen gene (*ssa*) from Streptococcus pyogenes. *Infect. Immun.* 62: 1867-1874.

Scheuber, P.H., Golecki, J.R., Kickhofen, B., Scheel, D., Beck, G., and Hammer, D.K. 1987. Cysteinyl leukotrienes as mediators of staphylococcal enterotoxin B in the monkey. *Eur. J. Clin. Invest.* 17:455-459.

Schlievert, P.M. 1980. Purification and characterization of staphylococcal pyrogenic exotoxin type B. *Biochemistry* 9:6204-6208.

Schlievert, P.M. 1982. Enhancement of host susceptibility to lethal endotoxin shock by staphylococcal pyrogenic exotoxin type C. *Infect. Immun.* 36:123-128.

Schlievert, P.M. 1993. Role of superantigens in human disease. *J. Infect. Dis.* 167:997-1002.

Schlievert, P.M., Gocke, J.E. and Deringer, J.R. 1993. Group B streptococcal toxic shock-like syndrome: Report of a case and purification of an associated pyrogenic toxin. *Clin. Infect. Dis.* 17: 26-31.

Schlievert, P.M., Schoettle, K.J. and Watson, D.W. 1979. Purification and physicochemical and biological characterization of staphylococcal pyrogenic exotoxin. *Infect. Immun.* 23:609-617.

Schlievert, P.M. and Watson, D.W. 1978. Group A streptococcal pyrogenic exotoxin: pyrogenicity, alteration of blood brain barrier, and separation of sites for pyrogenicity and enhancement of lethal endotoxin shock. *Infect. Immun.* 21:753-759.

Singh, B.R. and Betley, M.J. 1989. Comparative structural analysis of staphylococcal enterotoxins A and E. *J. Biol. Chem.* 264:4404-4411.

Singh, B.R., Evenson, M.L. and Bergdoll, M.S. 1988a. Structural analysis of staphylococcal enterotoxins B and C1 using circular dichroism and fluorescence spectroscopy. *Biochemistry* 27:8735-8741.

Singh, B.R., Fu, F.-N. and Ledoux, D.N. 1994. Crystal and solution structures of superantigenic staphylococcal enterotoxins compared. *Nature Struct. Biol.* 1:358-360.

Singh, B.R., Kokan-Moore, N.P. and Bergdoll, M.S. 1988b. Molecular topography of toxic shock syndrome toxin as revealed by spectroscopic studies. *Biochemistry* 27:8730-8735.

Spero, L., Griffin, B.Y., Middlebrook, J.L. and Metzger, J.F. 1976. Effect of single and double peptide bond scission by trypsin on the structure and activity of staphylococcal enterotoxin C. *J. Biol. Chem.* 251:5580-5588.

Spero, L., Metzger, J.F., Warren, J.R. and Griffin, B.Y. 1975. Biological activity and complementation of the two peptides of staphylococcal enterotoxin B formed by limited tryptic hydrolysis. *J. Biol. Chem.* 250:5026-5032.

Spero, L. and Morlock, B.A. 1978. Biological activities of the peptides of staphylococcal enterotoxin C formed by limited tryptic hydrolysis. *J. Biol. Chem.* 253:8787-8791.

Spero, L., Warren, J.R. and Metzger, J.F. 1973. Effect of single peptide bond scission by trypsin on the structure and activity of staphylococcal enterotoxin B. *J. Biol. Chem.* 248:7289-7294.

Stevens, D.L., Tanner, M.H., Winship, J., Swarts, R., Ries, K.M., Schlievert, P.M., and Kaplan, E. 1989. Reappearance of scarlet fever toxin A among streptococci in the Rocky Mountain West: Association with severe streptococcal soft tissue infection, sepsis and the toxic shock-like syndrome. *N. Engl. J. Med.* 321:1-7.

Stiles, B.G., Bavari, S., Krakauer, T., and Ulrich, R.G. 1993. Toxicity of staphylococcal enterotoxins potentiated by lipopolysaccharide: major histocompatibility complex class II molecule dependency and cytokine release. *Infect. Immun.* 61:5333-5338.

Swaminathan, S., Furey, W., Pletcher, J. and Sax, M. 1992. Crystal structure of staphylococcal enterotoxin B, a superantigen. *Nature (London)* 359:801-806.

Tai, J.Y., Kortt, A.A., Liu, T.Y., and Elliott, S.D. 1976. Primary structure of streptococcal proteinase. III. Isolation of cyanogen bromide peptides: complete covalent structure of the polypeptide chain. *J. Biol. Chem.* 251:1955-1959.

Turner, T.N., Smith, C.L., and Bohach, G.A. 1992. Residues 20, 22, and 26 determine subtype specificities of staphylococcal enterotoxins C1 and C2. *Infect. Immun.* 60:694-697.

Warren, J.R. 1977. Comparative kinetic stabilities of staphylococcal enterotoxin types A, B, and C1. *J. Biol. Chem.* 252:6831-6834.

Warren, J.R., Spero, L. and Metzger, J.F. 1974. Stabilization of the native structure by the closed loop of staphylococcal enterotoxin B. *Biochem. Biophys. Acta* 359:351-363.

Weeks, C.R. and Ferretti, J.J. 1986. Nucleotide sequence of the type A streptococcal exotoxin (Erythrogenic toxin) gene from *Streptococcus pyogenes* bacteriophage T12. *Infect. Immun.* 52:144-150.

Wheeler, M.C., Roe, M.H., Kaplan, E.L., Schlievert, P.M., and Todd, J.K. 1991. Clinical, epidemiological, and microbiological correlates of an outbreak of group A streptococcal septicemia in children. *JAMA* 266:533-537.

White, J., Herman, A., Pullen, A.M., Kubo, R., Kappler, J.W. and Marrack, P. 1989. The Vβ-specific superantigen staphylococcal enterotoxin B: stimulation of mature T cells and clonal deletion in mice. *Cell* 56:27-35.

Zabriskie, J.B. 1964. The role of temperate bacteriophage in the production of erythrogenic toxin by group A streptococci. *J. Exp. Med.*119:761-780.

9

PRIMARY STRUCTURAL MOTIFS OF *Conus* PEPTIDES

Lourdes J. Cruz

Marine Science Institute
University of the Philippines
Diliman, Quezon City
1101 Philippines

INTRODUCTION

The venoms of predatory marine snails of the family Conidae are complex mixtures of proteins, peptides and smaller molecules; most biological activities are associated with peptides ranging in size from ~13 to 30 amino acid residues. Since the isolation of the first α-conotoxin from *Conus geographus* (Cruz et al., 1976), the structures and mode of action of many peptides from *Conus* venom have been elucidated (Myers et al., 1993). Among these are the μ- and δ-conotoxins that act on voltage-sensitive sodium channels (Cruz et al., 1985; Hillyard et al., 1989; Sato et al., 1983; Shon et al., 1994), the ω-conotoxins which inhibit voltage-sensitive calcium channels (Hillyard et al., 1992; Olivera et al., 1987; Olivera et al., 1985; Olivera et al., 1984; Ramilo et al., 1992), the α- and αA-conotoxins which inhibit nicotinic acetylcholine receptors (Gray et al., 1981; Hopkins et al., 1995; McIntosh et al., 1982; Myers, 1989; Ramilo et al., 1992; Zafaralla et al., 1988) and the conantokins that inhibit glutamate receptors of the N-methyl-D-aspartate (NMDA) subtype (Haack et al., 1990; McIntosh et al., 1984; Mena et al., 1990). It is now estimated that each *Conus* venom may contain 50 to 200 peptides. Since there are ~500 species, even a conservative projection of the total number of biologically-active peptides in the venom of this genus is a staggering 25,000. Indeed, the *Conus* system has continued to provide new peptide ligands with novel structures and pharmacological action.

Biological Action of *Conus* Venom

Initial interest on *Conus* venoms focused on the most dangerous species (*Conus geographus*), as well as other species that have inflicted serious envenomations in humans. *C. geographus*, the species responsible for most human deaths, has an estimated fatality rate of ~70% compared to the overall rate of <40% for all species (Yoshiba, 1984). Death is attributed mainly to paralysis of the diaphragm muscle with consequent respiratory failure (Cruz and White, 1995). Compared to neurotoxic snakebites, the onset of symptoms in *Conus*

Natural Toxins II, Edited by B. R. Singh and A. T. Tu
Plenum Press, New York, 1996

sting cases is much more rapid. In severe *C. geographus* envenomation, initial signs of paralysis are evident within thirty minutes, and death occurs within five hours after the sting. In snakebite victims, initial manifestations of paralysis take two to three hours, and death usually occurs in ~twenty hours. Among the symptoms generally observed after a cone sting are immediate numbness at the sting site which spreads to other parts of the body, paralysis of limbs and voluntary muscles, difficulty in speaking, blurring of vision, cyanosis, unconsciousness, respiratory failure and death.

The range of phylogenetic specificities in *Conus* venom was indicated early on by the experiments of Endean's group, and by Kohn and co-workers (Endean and Rudkin, 1963; Endean and Rudkin, 1965; Kohn, 1963; Kohn et al., 1960). In general, the venoms of fish-hunting species are more effective in vertebrates, and those from mollusc- and worm-hunters in invertebrates. Many examples of conotoxins that exhibit phylogenetic and subtype specificities are now known. For example, in mice, α-conotoxins GI from *C. geographus* and MI from *C. magus* are much more toxic than the α-conotoxins from *C. striatus* (Ramilo et al., 1992; Zafaralla et al., 1988), but they all have comparable activities in fish. Mice and rats also show significant differences in symptomatology when ω-conotoxin GVIA is injected intracranially. Mice do not die even at very high dose; they just continue shaking for long periods of time. On the other hand, rats shake at low dose, and will convulse and die at the higher dose. In a marine environment, the venom of fish-hunting species is tailored

Table 1. Conotoxin Cys/Loop Frameworks

Framework		Peptides Containing Framework
4-Cys/2-loop	CC---C---C	α-conotoxins
4-Cys/3-loop	C---C---C---C	*C. geographus* scratcher
6-Cys/3-loop	CC---C---C---CC	μ-conotoxins *C. textile* scratcher peptide Qc IIIA/B
	CCC---C---C---CC	α-conotoxin SII
6-Cys/4-loop	C---C---CC---C---C	ω-conotoxins δ-conotoxins conotoxin GS μO-Mr VIA/B *C. textile* convulsant peptide KK-1/KK-2 (*C. textile*)
	CC---C---C---C---C	αA-P IVA μ-Pn IVA/B

to paralyze its prey within a few seconds (Olivera et al., 1985; Olivera et al., 1990). Several strategies are believed to make this possible: the use of small and tightly-folded toxins that can be rapidly transported once injected into the prey, the synergistic effect of conotoxins that can simultaneously interfere with the function of several key macromolecules of the neuromuscular junction, and the presence of accessory components that may enhance transport and pharmacological activity. Recent work on the venom of other *Conus* species has indicated the use of similar strategies by mollusc- and worm-hunting species.

General Structural Features of *Conus* Peptides

Close to 90% of the naturally-occurring *Conus* peptides have four to six cysteine residues involved in disulfide bonds. Most of the peptides found so far belong to the three structural classes (see Table 1) corresponding to different arrangements of cysteine residues. The most abundant structural class appears to be the 6-Cys/4-loop cysteine framework typified by the ω-conotoxins (**C—C—CC—C—C**), followed by the 6-Cys/3-loop framework represented by μ-conotoxins (**CC—C—C—CC**), and the 4-Cys/2-loop framework of the α-conotoxins (**CC—C—C**). This structural motif, i.e., the presence of multiple disulfide bonds, is responsible for maintaining many *Conus* peptides in a fairly rigid and compact conformation needed for the rapid transport of the peptides to their macromolecular targets.

In contrast, two *Conus* peptides (conantokin-T and conantokin-G) have been found to contain no cysteine residues at all. Instead, these peptides contain four to five γ-carboxyglutamate (Gla) residues (Haack et al., 1990; McIntosh et al., 1984), which are believed to be important for maintenance of the peptides' α-helical conformation. The helical wheel projection of conantokins and the corresponding space-filling model show the orientation of all Gla residues on one face of the peptide (Gray et al., 1988); Raman spectroscopy and circular dichroism measurements indicate increased helicity of conantokins in the presence of Ca^{++} ions (Hernandez et al., 1990; Myers et al., 1990). Presumably, chelation of the carboxyl groups with Ca^{++} ions stabilize the α-helical conformation of peptides in this class. Recently, peptides combining the multiple disulfide bond motif and the multiple γ-carboxyglutamate motif have also been found in *Conus textile* (Ramilo et al., unpublished data). Since very little is yet known about the latter structural motif, this review will focus on a comparison of *Conus* peptides containing the multiple disulfide bond motif.

PEPTIDES WITH THE ω–TYPE 6-CYS/4-LOOP FRAMEWORK

Conotoxins with the ω-type 6-Cys/4-loop framework (**C—C—CC—C—C**) are the most abundant group of peptides isolated from *Conus* venoms so far. This structural class encompasses at least four known pharmacological classes: ω-conotoxins that block voltage-sensitive calcium channels, δ-conotoxins that retard sodium channel inactivation, the sodium channel blocker conotoxin-GS, and two peptides recently found in *C. marmoreus* that affect both sodium and calcium currents. In addition, three ω-type 6-Cys/4-loop peptides with still unknown targets have been found in *C. textile* (Cruz et al., 1992) and *C. quercinus* (Abogadie et al., 1990; Olivera et al., 1990).

ω-Conotoxins

Among *Conus* peptides, ω-conotoxins have received the most attention because they are widely used as pharmacological tools for differentiating subtypes of calcium channels. Since two reviews discussing ω-conotoxins have been published recently (Olivera et al.,

Table 2. Conotoxin Blockers of Calcium Channels

	Sequence
ω-Conotoxins	
GVIA	CKSOGSSCSOTSYNCCR-SCNOYTKRCY*
GVIB	CKSOGSSCSOTSYNCCR-SCNOYTKRCYG*
GVIC	CKSOGSSCSOTSYNCCR-SCNOYTKRC*
SVIA	CRSSGSOCGVTSI-CC-GRC--YRGKCT*
SVIB	CKLKGQSCRKTSYDCCSGSCGRS-GKC*
GVIIA	CKSOGTOCSRGMRDCCT-SCLLYSNKCRRY*
GVIIB	CKSOGTOCSRGMRDCCT-SCLSYSNKCRRY*
MVIIA	CKGKGAKCSRLMYDCCTGSCRS--GKC*
MVIIB	CKGKGASCHRTSYDCCTGSCNR--GKC*
MVIIC	CKGKGAPCRKTMYDCCSGSCGRR-GKC*
MVIID	CQGRGASCRKTMYNCCSGSC--NRGRC*

1991; Olivera et al., 1994), a relatively brief description of the structure of ω-conotoxins will be presented in this section.

The primary sequences of ω-conotoxins are diverse (see Table 2); the only absolutely conserved amino acid residues are the six cysteines and the glycine residue in the first intercysteine loop. Studies of structure-activity relationships are relatively complicated. Nevertheless, comparison of the effect of single Ala substitutions for various amino acids residues of ω-conotoxins GVIA and MVIIA indicate Tyr^{13} to be the most critical residue for toxin binding to brain membranes from chicks and rats (Becker et al., 1992; Kim et al., 1995; Sato et al., 1991). Replacement of this amino acid with Phe drastically reduced activity, indicating an important role for the hydroxyl group of tyrosine in the interaction between the toxins and the vertebrate N-type calcium channel (Kim et al., 1994). Curiously, Tyr^{13} is not conserved in all ω-conotoxins; it is replaced by Arg in GVIIA and GVIIB and by Ile in SVIA. In mice, SVIA is much less effective than MVIIA (Ramilo et al., 1992).

Acetylation of the amino group of the N-terminal Cys significantly affected activity; Lys^{2} of GVIA and MVIIA appears to be important also (Lampe et al., 1993). In GVIA, however, Ala substitutions in the other positions (including other basic groups) did not significantly affect activity (Sato et al., 1993). Optimal binding of MVIIA to rat brain membranes is affected by Ala replacement of Arg^{10} and Arg^{21} (Kim et al., 1995). Substitution of Leu^{11} by Ala reduced MVIIA binding to rat brain membrane 100-fold. However, this position is occupied by Thr in most ω-conotoxins and by Gly in GVIIA and GVIIB.

It is difficult to make generalizations with respect to the importance of specific amino acid residues in ω-conotoxins. The variety of natural ω-conotoxin sequences is a reflection of the diversity of groups that can fit the antagonist binding site (or the macrosite) of fish calcium channels. Phylogenetic and subtype specificities of ω-conotoxins in higher animals may reflect fine variations in the structure of calcium channels from different organisms and cell types.

δ-Conotoxins and Other Peptides with the ω-Type Cys/Loop Framework

δ-Conotoxins are the major venom components in two closely-related mollusc-hunting species, *C. textile* and *C. gloriamaris* (Shon et al., 1994) They induce retraction of the animal into the shell, followed by secretion of a green slime, and the undulation of the animal in and out of the shell. When tested in molluscan neurons, δ-conotoxins were found to broaden action potentials by delaying sodium current inactivation (Fainzilber et al., 1994). Structurally, δ-conotoxins belong to the same class as the calcium channel blocking ω-conotoxins; the Cys/loop framework and the Gly residue of the first intercysteine loop are conserved in the two groups of conotoxins. Beyond this, however, no other similarities exist. The basic residues of ω-conotoxins are distributed over the whole length of the peptide chain, whereas in δ-conotoxins the Arg and Lys residues are mainly located close to the N-terminal. Furthermore, δ-conotoxins are not as highly charged (+2 for GmVIA and -2 for TxVIA) as ω-conotoxins (net charge of +5 to +7), and they are much more hydrophobic.

Conotoxin-GS, a sodium channel blocker isolated from *C. geographus*, has the 6-Cys/4-loop framework (**C—C—CC—C—C**) of ω-conotoxins rather than the 6-Cys/3-loop arrangement (**CC—C—C—CC**) of the μ–conotoxins, the major Na channel blocker from *C. geographus* (Yanagawa et al., 1988). Conotoxin-GS also has the conserved Gly residue in the first loop typical of ω-conotoxins. In fact, Yanagawa and co-workers have pointed out the greater sequence similarity of conotoxin-GS to the calcium channel blocker ω-conotoxin MVIIA than to the μ-conotoxin GIII series. Although the affinity of conotoxin-GS to sodium channels is much less than that of μ-conotoxins, conotoxin-GS was also found to preferentially bind to Site I of the muscle rather than the neuronal subtype of sodium channels.

Recently, two new peptides, MrVIA and MrVIB, were isolated from *C. marmoreus* venom (Fainzilber et al., 1995; McIntosh et al., 1995). These conotoxins were reported to affect both sodium and calcium currents in the freshwater snail *Lymnaea stagnalis*, although the toxins block Na currents more potently than calcium currents (Fainzilber et al., 1995). Of the groups of peptides with the 6-Cys/4-loop framework, MrVIA and MrVIB are the only ones that do not have the "conserved" Gly in the first intercysteine loop. The peptides have an unusual concentration of basic residues in the first loop with mainly hydrophobic amino acid residues in the other loops (Fainzilber et al., 1995), a feature mentioned above for the δ-conotoxins.

Other peptides containing the ω-type 6-Cys/4-loop framework have been reported: conotoxin QcVIA from *C. quercinus*, a worm hunter (Abogadie et al., 1990), and three peptides from *C. textile* venom (Tx convulsant peptide, KK-1 and KK-2) (Cruz et al., 1992; Hillyard et al., 1989). The physiological targets of these toxins are unknown.

Table 3. Conotoxin Blockers of Sodium Channels

	Sequence
μ-Conotoxins	
GIIIA	RD**CC**TOOKK**C**KDRQ**C**KOQR**CC**A*
GIIIB	RD**CC**TOORK**C**KDRR**C**KOMK**CC**A*
GIIIC	RD**CC**TOOKK**C**KDRR**C**KOLK**CC**A*
PnIVA	**CC**KYGWT**C**LLG-**C**SP**C**G**C**
PnIVB	**CC**KYGWT**C**WLG-**C**SP**C**G**C**
μO-Conotoxins	
MrVIA	A**C**RKKWEY**C**IVPIIGFIY**CC**PGLI**C**GPFV**C**V
MrVIB	A**C**SKKWEY**C**IVPILGFVY**CC**PGLI**C**GPFV**C**V
Conotoxin GS	A**C**SGRGSR**C**OOQ**CC**MGLR**C**GRGNPQK**C**IGAHEDV

CONOTOXINS WITH THE μ-TYPE 6-CYS/3-LOOP FRAMEWORK

The 6-Cys/3-loop framework **(CC—C—C—CC)** has been found in several peptides including the seven μ-conotoxins isolated from fish-hunting species, a peptide from the mollusc-hunter *C. textile*, and two peptides from the worm-hunter *C. quercinus*. The best-studied group are the μ-conotoxins (see Table 3), which block the TTX-sensitive skeletal muscle subtype of voltage-sensitive sodium channels. The physiological targets of the other peptides belonging to this structural group have not yet been elucidated. However, the *C. quercinus* and *C. textile* peptides are structurally very interesting since they represent some of the most tightly-folded peptides known. For example, in the "scratcher" peptide of *C. textile,* six out of twelve amino acid residues are cysteines.

The μ-conotoxins of *C. geographus* are highly homologous peptides containing twenty-two amino acid residues of which six are cysteine and three are hydroxyprolines (Cruz et al., 1985; Sato et al., 1983). The presence of several Lys and Arg residues in μ-conotoxins results in a high positive charge of +6 for GIIIA and +7 for GIIIB and GIIIC. Structure-activity studies done by Sato et al. (Sato et al., 1991; Sato et al., 1990) and Becker et al. (Becker et al., 1992) on μ-conotoxins indicate that the most important group for toxin activity is Arg^{13}. The activity of GIIIA on isolated rat diaphragm was reduced ~6-fold when this group was replaced by Lys and ~200-fold when replaced by Ala (Becker et al., 1992). Substitution of Arg^{13} by Gln (R13Q replacement) reduced binding to electroplax membranes by at least 38,000-fold.

In rat skeletal muscle Na channels incorporated in lipid bilayers, different synthetic analogs of GIIIA were found to produce zero conductances of single channels (Becker et al., 1992) with the exception of $[Gln^{13}]$μ-conotoxin GIIIA. In this system, R13Q substitution produced an 80- to 100-fold decrease in "efficacy" but the "blocked" events still had conductances of about 20-40% of the open state. It was therefore suggested (Becker et al., 1992) that the guanidino group of Arg^{13} is important not only for tight binding but also for the complete occlusion of the channel pore. This analogy of μ-conotoxin block to the physical occlusion of the sodium channel pore by the protruding guanidino group of saxitoxin and tetrodotoxin has also been suggested by Lancelin et al. (Lancelin et al., 1991) on the basis of the loss of activity when Arg^{13} was replaced by Ala.

The other amino acids in the third intercysteine loop of GIIIA (—$Lys^{11}Asp^{12}Arg^{13}Gln^{14}$—) are relatively unimportant for activity. In fact, individual replacements of Asp^{12} and Gln^{14} with Ala increased the activity by about 3- and 2-fold, respectively, in isolated rat diaphragm and rat muscle Na channel inserted in lipid bilayers. Perhaps removal of the negative group at position 12 and substitution of a smaller group for the more bulky Gln allows a closer approach and a stronger interaction of Arg^{13} with the negative group of the channel pore. Replacement of K^{11} by Ala or Gln reduced the activity by about 2- or 7-fold, suggesting the importance of a positive group in position 11 (Becker et al., 1992; Sato et al., 1990).

After Arg^{13}, the most important amino acid residues appear to be those in the fourth intercysteine loop, which has the sequence —$Lys^{16}Hyp^{17}Gln^{18}Arg^{19}$—. Although an R19K replacement did not affect activity, the analogs with R19A and R19Q substitutions had 25-fold and 40-fold less activity respectively (Becker et al., 1992; Sato et al., 1990). The importance of positively charged residues in this loop is also indicated by a 6-fold or 25-fold reduction of activity when Lys^{16} is replaced by Ala or Gln. Hyp^{17} replacement by Ala reduced the activity 6-fold in rat diaphragm system; its substitution by Pro reduced the activity 25-fold. In view of this, Wakamatsu et al. (Wakamatsu et al., 1992) suggested the correspondence of the hydroxyl group of Hyp^{17} to the essential hydroxyl group of tetrodotoxin and saxitoxin.

In the other segments of μ-conotoxins, the amino acids most affected by Ala and Gln substitutions are the basic residues (Becker et al., 1992; Sato et al., 1990). In the N-terminal region, R1Q and R1A replacements reduced the activity ~7-fold and five-fold, respectively. Arg^1 replacement by Lys led to a 2-fold decrease in activity, indicating not only the requirement for a positive charge but possibly a special fit of the guanidino group to a microsite within the "macro" binding site. In the first intercysteine loop, K8A and K8Q replacements reduced activities 4-fold and 25-fold respectively. Substitution of Asp^2 by Asn reduced the activity by 3-fold, but the activity increased ~2-fold when the residue was replaced by Ala. Double substitution of Hyp^6 and Hyp^7 by Pro reduced the activity 2-fold in rat Na channels. In general, replacement of Arg and Lys by neutral amino acid residues decreased activity and substitution of Asp by Ala increased μ-conotoxin activity. Lys substitution for Gln^{18} increased activity almost 2-fold in eel electroplax membranes. (No corresponding data on rat muscle sodium channel was reported.)

The effect of several amino acid replacements on toxin binding to eel electroplax membranes (Becker et al., 1992) differs from the corresponding effects on rat muscle sodium channels. Thus, proline substitution for Hyp^{17} did not affect activity in eel electroplax, and double replacement of Hyp^6 and Hyp^7 by Pro even increased activity by ~27%. Asn replacements of Asp^2 and Asp^{12} also slightly increased activity. These same substitutions decreased activities in rat muscle sodium channels, suggesting important structural differences between the active site on sodium channels in rat skeletal membrane and in the eel electroplax.

CONOTOXINS WITH THE 4-CYS/2-LOOP FRAMEWORK

α-Conotoxins are inhibitors of the nicotinic acetylcholine receptors at post-synaptic membranes. The α-conotoxins compete with α-bungarotoxin and *d*-tubocurarine for the binding sites on the nicotinic acetylcholine receptor. Like d-tubocurarine, α-conotoxins can distinguish between the two nonequivalent agonist binding AChR sites residing at the α/γ and α/δ subunit interfaces (Groebe et al., 1995; Hann et al., 1994). Dimethyl-*d*-tubocurarine

Table 4. Conotoxin Inhibitors of Acetylcholine Receptors

	Sequence
α-Conotoxins	
GI	E**CC**NPA**C**GRHYS**C***
GIA	E**CC**NPA**C**GRHYS**C**GK*
GII	E**CC**HPA**C**GKHFS**C***
MI	GR**CC**HPA**C**GKNYS**C***
SI	I**CC**NPA**C**GPKYS**C***
SIA	Y**CC**HPA**C**GKNFD**C***
ImI	G**CC**SDPR**C**AWR**C***
PnIA	G**CC**SLPP**C**AANNPDY**C***
PnIB	G**CC**SLPP**C**ALSNPDY**C***
SII	G**CCC**NPA**C**GPNYG**CC**GTS**C**S^
αA-Conotoxins	
αA-PIVA	G**CC**GSYONAA**C**HO**C**S**C**KDROSY**C**GQ*

has an 89-fold higher preference for the α/γ site over the α/δ. In contrast, α-conotoxins MI, GI and SIA have much higher affinities (15,000-, 12,000- and 26,000-fold) for the α/δ site than for the α/γ. Specificity is reversed in the AChR of the *Torpedo* electroplax, where MI showed higher affinity for the α/γ site. α-Conotoxins SI and SII have a preference <400-fold for the α/δ versus the α/γ binding site. Consistent with the higher activity of MI, GI and SIA in mammalian systems, the affinities of these α–conotoxins for the mouse muscle-derived AChR are much higher than those of α-conotoxins SI and SII.

With the exception of α-conotoxin SII from *Conus striatus*, all α-conotoxins have the cysteine pattern, **CC—C—C** (Myers et al., 1993) (see Table 4). Peptides from the fish-hunting species, *C. geographus*, *C. striatus* and *C. magus,* have the consensus core sequence **CC**(N/H)PA**C**GXX(Y/F)X**C** and two disulfide bonds that connect Cys^2 to Cys^7 and Cys^3 to Cys^{13}. Comparison with the α-conotoxins recently isolated from *C. pennaceus*, a mollusc-hunter, and *C. imperialis*, a worm-hunter, indicates variations in the size of the intercysteine loops. The second loop has seven amino acid residues in α-conotoxins PnIA and PnIB from *C. pennaceus* (Fainzilber et al., 1994), and only three residues in α-conotoxin ImI from *C. imperialis* (McIntosh et al., 1994), compared to five residues in the other α-conotoxins.

The amino terminal position (residues before the first cysteine) of α-conotoxins may be occupied by a wide range of amino acids (Gly, Glu, Ile, Tyr and the dipeptide -Gly-Lys-). In fact, (des-Glu^1)α-conotoxin GI was found to have practically the same activity as the natural peptide (Gray et al., 1985). It is believed that the positive amino group at the terminus is important for binding to the receptor and a variation of side chains can be tolerated.

The common structural feature of the first intercysteine loop of α-conotoxins from different species is the proline residue, which is involved in a β-turn. For this loop, the consensus sequence of the α-conotoxins from the fish-hunters is —(N/H)**P**A-. The corresponding sequence in the peptide of the worm eater *C. imperialis* (ImI) is —SD**P**R— and in the mollusc-hunting species, *C. pennaceus* (PnIA and PnIB) the sequence is —SL**P**P— (Fainzilber et al., 1994; McIntosh et al., 1994). Replacement of this proline residue by Gly was found by Nishiuchi and Sakakibara (Hashimoto et al., 1989; Nishiuchi and Sakakibara, 1984) to completely abolish the activity of α-conotoxin MI on i.p. injection in mice.

In the second intercysteine loop, the shortest sequence (—AWR—) has been found in α-conotoxin ImI and the longest (—A(A/L)(N/S)NPDY—) in the mollusc-hunter, *C. pennaceus.* α-Conotoxins that are active on i.p. injection in mice (GI, GIA, GII, MI, and SIA) have the consensus sequence —G(R/K)X(Y/F)(S/K)— for this segment. Replacement of the Gly residue by Ala reduced the activity of MI to 3% (Hashimoto et al., 1989; Nishiuchi and Sakakibara, 1984). α-Conotoxins that are not active on the neuromuscular system of mice have either Ala instead of Gly or Pro instead of the Arg/Lys in the second loop. However, replacement of the Gly residue by D-Ala in (des-Glu^1)α-conotoxin GI affected the activity by only 40%. Substitution of L-Tyr by D-Tyr, as well as disruption of the disulfide loops by derivatization of Cys pairs with Acm completely abolished the activity of α-conotoxin MI (Hashimoto et al., 1989; Nishiuchi and Sakakibara, 1984). (des-Glu^1)α-conotoxin GI activity was similarly eliminated by substitution of Ala for the second Cys and deletion of the last Cys residue to remove the second disulfide bridge (Almquist et al., 1990). Activity was partially restored when D-Ala was substituted for Gly in the analog of (des-Glu^1)α-conotoxin GI containing one disrupted disulfide bond. Thus, although the characteristic disulfide framework is necessary for maintaining a fully-active conformation of the α-conotoxins, an alternative structure may present similar surfaces to the binding site of the receptor.

OTHER CYS/LOOP FRAMEWORKS OF CONOTOXINS

Among α-conotoxins, SII from *C. striatus* venom (Ramilo et al., 1992) is unique in having the unusual 6-Cys/4-loop framework (**CCC—C—C—C**) in place of the regular α–type pattern (**CC—C—C**). However, its main structure clearly fits the core consensus sequence (**CC**(N/H)PA**C**GXX(Y/F)X**C**) of the α-conotoxins from fish-hunting species discussed above.

A new 6-Cys/4-loop framework (**CC—C—C—C—C**) has been reported for three peptides: conotoxins μPnIVA and μPnIVB from *C. pennaceus* (Fainzilber et al., 1994) and αA-conotoxin PIVA of *Conus purpurascens* (Hopkins et al., 1995). μPnIVA and μPnIVB are sodium channel blockers selective for molluscan neurons: they have no effect on sodium currents in bovine chromaffin cells and rat brain synaptosomes (Fainzilber et al., 1994). αA-Conotoxin PIVA is an acetylcholine receptor inhibitor.

A 4-Cys/3-loop framework (**C—C—C—C**) was previously reported (Olivera et al., 1992) for a "scratcher" peptide from *C. geographus*. However, the physiological target of this peptide has not yet been elucidated.

GENERAL FEATURES OF THE MULTIPLE DISULFIDE BOND MOTIF

The multiple disulfide bond motif is a very effective strategy for maintaining *Conus* peptides in a compact and fairly rigid structure that can be rapidly transported from the sting site to the macromolecular target. Although many different arrangements of cysteine residues are found in *Conus* peptides, a majority belong to one of three types: the ω-type 6-Cys/4-loop arrangement (**C—C—CC—C—C**), the μ-type 6-Cys/3-loop framework (**CC—C—C—CC**) and the α-type 4-Cys/2-loop arrangement (**CC—C—C**). Several other minor arrangements and variations in the size of intercysteine loops contribute to the diversity of structures found in *Conus* venoms.

Peptides with the same Cys/loop framework may have a variety of physiological targets. Thus, the μ–type 6-Cys/3-Loop framework is shared by the μ-conotoxins of *C. geographus* and three other peptides with unknown physiological targets including the very tightly folded "scratcher" peptide of *C. textile* (Myers et al., 1993). *Conus* peptides that have the ω-type 6-Cys/4-loop framework include ω-conotoxins, δ-conotoxins, μO-conotoxins, conotoxin-GS and four peptides with unknown physiological action (conotoxin QcVIA from *C. quercinus* and three peptides from *C. textile*, namely Tx convulsant peptide, KK-1 and KK-2). The 6-Cys/4-loop framework, **CC—C—C—C—C** is found in αA-conotoxin PIVA of *C. purpurascens* (Hopkins et al., 1995) and conotoxins μPIVA and μPIVB of *C. pennaceus* (Fainzilber et al., 1994).

Peptides belonging to the same pharmacological class may have different structural frameworks. Indeed, acetylcholine receptor inhibitors may have the Cys/loop framework **CC—C—C** of α-conotoxins, **CC—C—C—C—C** of αA-conotoxins or **CC—C—C—C** of α-conotoxin SII. The known sodium channel blockers also have three types of structural framework: **CC—C—C—CC** (μ-conotoxins), **C—C—CC—C—C** (conotoxin-GS, μO-conotoxins MrVIA and MrVIB) or **CC—C—C—C—C** (μ-conotoxins PnIVA and PnIVB).

Typically, the 3-dimensional structures of conotoxins as determined by NMR have disulfide bridges at the core and peptide loops with residues protruding out of the globular structure. For α-conotoxins active on the vertebrate neuromuscular ACh receptor, it is believed that two positive centers on the toxin surface (the amino terminus and the GI Arg^9

guanidyl side chain) may be positioned ~11 Å apart to satisfy the structural requirements for curare-like acetylcholine antagonist activity (Kobayashi et al., 1989; Pardi et al., 1989). The 3-dimensional structure of μ-conotoxin GIIIA suggests the interaction of the sodium channels with one face which contains the residues Arg^{13}, Lys^{16}, Hyp^{17} and Arg^{19}, which are important for activity (Lancelin et al., 1991; Wakamatsu et al., 1992).

The model for ω-conotoxin GVIA (Davis et al., 1993; Kobayashi et al., 1988; Pallaghy et al., 1993; Skalicky et al., 1993) shows an asymmetric distribution of residues to form a positively-charged hydrophilic face (N-terminus, Lys^2, Arg^{17}, Lys^{24} and Arg^{25}), and a small hydrophobic face containing Tyr residues on the opposite side; the hydroxyl-containing amino acids (Thr, Ser, and Hyp) are distributed over the peptide surface. As discussed above, the interaction of ω-conotoxins GVIA and MVIIA with calcium channels involves mainly the critical Tyr^{13} residue as well as the positively-charged amino groups at the N-terminus and at Lys^2. Thus, opposite faces of ω-conotoxins GVIA and MVIIA may be involved in their interaction with the calcium channel.

A common structural motif consisting of a "cystine knot and a small triple-stranded β sheet" has been found in ω-conotoxins and some biologically diverse peptides (Pallaghy et al., 1994). These include a 29-residue cyclic polypeptide from a tropical plant that causes uterine contraction, a 29-residue trypsin inhibitor from pumpkin seed, other protease inhibitors and a spider toxin (Pallaghy et al., 1994). The common structure is believed to be one of the smallest stable globular domains commonly used in toxins and inhibitors that block the function of certain macromolecules. Alternative frameworks or scaffolds for efficiently presenting various functional groups to the ligand binding site of receptors and ion channels are provided by the various Cys/loop frameworks found in *Conus* peptides. The availability of many different scaffolds for developing new toxins may contribute to the success of *Conus* species in evolving numerous ligands for a variety of macromolecular targets.

ACKNOWLEDGMENTS

The author wishes to acknowledge the support of the Marine Science Institute and the Olivera Lab at the University of Utah, Department of Biology, Salt Lake City, Utah, 84112, USA.

REFERENCES

Abogadie, F. C., Ramilo, C. A., Corpuz, G. P. and Cruz, L. J. 1990. Biologically active peptides from *Conus quercinus*, a worm-hunting species. *Trans. Natl. Acad. Sci. & Tech. (Philippines)* 12:219-232.

Almquist, R. G., Kadambi, S. R., Yasuda, D. M., Weith, F. L., Polgar, W., Toll, L. R. and Uyeno, E. T. 1990. Development of antagonists of des-Glu^1-Conotoxin GI. In: *Peptides: Chemistry, Structure and Biology* Eds. Rivier, J. E. and Marshall, G. R.). Leiden, Escom.

Becker, S., Prusak-Sochaczewski, E., Zamponi, G., Beck-Sickinger, A. G., Gordon, R. D. and French, R. J. 1992. Action of derivatives of μ-conotoxin GIIIA on sodium channels. Single amino acid substitutions in the toxin separately affect association and dissociation rates. *Biochemistry* 31:8229-8238.

Cruz, L. J., Corpuz, G. and Olivera, B. M. 1976. A preliminary study of *Conus* venom protein. *Veliger* 18:302-308.

Cruz, L. J., Gray, W. R., Olivera, B. M., Zeikus, R. D., Kerr, L., Yoshikami, D. and Moczydlowski, E. 1985. *Conus geographus* toxins that discriminate between neuronal and muscle sodium channels. *J. Biol. Chem.* 260:9280-9288.

Cruz, L. J., Ramilo, C. A., Corpuz, G. P. and Olivera, B. M. 1992. *Conus* peptides: phylogenetic range of biological activity. *Biol. Bull.* 183:159-164.

Cruz, L. J. and White, J. 1995. Clinical toxicology of *Conus* snail stings. In: *CRC Handbook on Clinical Toxicology of Animal Venoms and Poisons* Eds. Meier, J. and White, J.). Boca Raton, FL, CRC Press. 117-127.

Davis, J. H., Bradley, E. K., Miljanich, G. P., Nadasdi, L., Ramachandran, J. and Basus, V. J. 1993. Solution structure of ω-conotoxin GVIA using 2-D NMR spectroscopy and relaxation matrix analysis. *Biochemistry* 32:7396-7405.

Endean, R. and Rudkin, C. 1963. Studies of the venoms of some Conidae. *Toxicon* 1:49-64.

Endean, R. and Rudkin, C. 1965. Further studies of the venoms of Conidae. *Toxicon* 2:225-249.

Fainzilber, M., Hasson, A., Oren, R., Burlingame, A. L., Gordon, D., Spira, M. E. and Zlotkin, E. 1994. New mollusc-specific α-conotoxins block *Aplysia* neuronal acetylcholine receptors. *Biochemistry* 33:9523-9529.

Fainzilber, M., Kofman, O., Zlotkin, E. and Gordon, D. 1994. A new neurotoxin receptor site on sodium channels is identified by a conotoxin that affects sodium channel inactivation in molluscs and acts as an antagonist in rat brain. *J. Biol. Chem.* 269:2574-2580.

Fainzilber, M., van der Schors, R., Lodder, J. C., Li, K. W., Geraerts, W. P. M. and Kits, K. S. 1995. New sodium channel-blocking conotoxins also affect calcium currents in *Lymnaea* neurons. *Biochemistry* 34:5364-5371.

Gray, W. R., Luque, A., Olivera, B. M., Barrett, J. and Cruz, L. J. 1981. Peptide toxins from *Conus geographus* venom. *J. Biol. Chem.* 256:4734-4740.

Gray, W. R., Middlemas, D. M., Zeikus, R., Olivera, B. M. and Cruz, L. J. 1985. Structure-activity relationships in α-conotoxins: A model. In: *Peptides: Structure and Function. The Ninth American Peptide Symposium* Eds. Deber, C. M., Hruby, V. J. and Kopple, K. D.). Rockford, IL. Pierce Chemical Pubs. 823-932.

Gray, W. R., Olivera, B. M., Cruz, L. J. and Rivier, J. 1988. A model for "sleeper peptide" (conotoxin GV) and other Gla-containing molecules. In: *Peptide Chemistry 1987* Eds. Shiba, T. and Sakakibara, S.). Osaka, Japan, Protein Research Foundation. 105-113.

Groebe, D. R., Dumm, J. M., Levitan, E. S. and Abramson, S. N. 1995. α-Conotoxins selectively inhibit one of the two acetylcholine binding sites of nicotinic receptors. *Mol. Pharmacol.* 48:105-111.

Haack, J. A., Rivier, J., Parks, T. N., Mena, E. E., Cruz, L. J. and Olivera, B. M. 1990. Conantokin T: a γ-carboxyglutamate-containing peptide with N-methyl-D-aspartate activity. *J. Biol. Chem.* 265:6025-6029.

Hann, R. M., Pagán, O. R. and Eterovic, V. A. 1994. The α-conotoxins GI and MI distinguish between the nicotinic acetylcholine receptor agonist sites while SI does not. *Biochemistry* 33:14058-14063.

Hashimoto, K., Uchida, S., Yoshida, H., Nishiuchi, Y., Sakakibara, S. and Yukari, K. 1989. *Eur. J. Pharmacol.* 118:351.

Hernandez, J.-F., Olivera, B. M., Cruz, L. J., Myers, R. A., Abbott, J. and Rivier, J. 1990. Synthesis, characterization and biological activity of conantokin-G analogs. In: *UCLA Symposia Conference on Biochemical and Biomedical Engineering Synthetic Peptides: Approaches to Biological Problems* Frisco, CO, February 22-March 4,

Hillyard, D. R., Monje, V. D., Mintz, I. M., Bean, B. P., Nadasdi, L., Ramachandran, J., Miljanich, G., Azimi-Zoonooz, A., McIntosh, J. M., Cruz, L. J., Imperial, J. S. and Olivera, B. M. 1992. A new *Conus* peptide ligand for mammalian presynaptic Ca^{2+} channels. *Neuron* 9:69-77.

Hillyard, D. R., Olivera, B. M., Woodward, S., Gray, W. R., Corpuz, G. P., Ramilo, C. A. and Cruz, L. J. 1989. A molluscivorous *Conus* toxin: conserved frameworks in conotoxins. *Biochemistry* 28:358-361.

Hopkins, C., Grilley, M., Miller, C., Shon, K.-J., Cruz, L. J., Gray, W. R., Dykert, J., Rivier, J., Yoshikami, D. and Olivera, B. M. 1995. A new family of *Conus* peptides targeted to the nicotinic acetylcholine receptor. J. Biol. Chem., in press.

Kim, J. I., Takahashi, M., Ohtake, A., Wakamiya, A. and Sato, K. 1995. Tyr^{13} is essential for the activity of ω-conotoxin MVIIA and GVIA, specific N-type calcium channel blockers. *Biochem. Biophys. Res. Comm.* 206:449-454.

Kim, J. I., Takahasi, M., Ogura, A., Kohno, T., Kudo, Y. and Sato, K. 1994. Hydroxyl group of Tyr^{13} is essential for the activity of ω-conotoxin GVIA, a peptide toxin for N-type calcium channel. *J. Biol. Chem.* 269:23876-23878.

Kobayashi, Y., Ohkubo, T., Kygoaku, Y., Nishiuchi, Y., Sakakibara, S., Braun, W. and Gö, N. 1989. Solution conformation of conotoxin GI determined by ^{1}H nuclear magnetic resonance spectroscopy and distance geometry calculations. *Biochemistry* 28:4853-4860.

Kobayashi, Y., Ohkubo, T., Nishimura, S., Kyoguku, Y., Shimada, K., Minobe, M., Nishiuchi, Y., Sakakibara, S. and Gö, N. 1988. Two-dimensional ^{1}H-NMR investigation of the tertiary structure of ω-conotoxin.

In: *Protein Chemistry 1987* Eds. Shiba, T. and Sakakibara, S.). Osaka, Japan, Protein Research Foundation. 85-88.

Kohn, A. J. 1963. Venomous marine snails of the genus *Conus*. In: *Venomous and Poisonous Animals and Noxious Plants of the Pacific Region* Eds. Keegan, H. C. and McFarlane, W. V.). Oxford, Pergamon Press. 83.

Kohn, A. J., Saunders, P. R. and Wiener, S. 1960. Preliminary studies on the venom of the marine snail *Conus*. *Ann. N.Y. Acad. Sci.* 90:706-725.

Lampe, R. A., Lo, M. M., Keith, R. A., Horn, M. B., McLane, M. W., Herman, J. L. and Spreen, R. C. 1993. Effects of site-specific acetylation on ω-conotoxin GVIA binding and function. *Biochemistry* 32:3255-3260.

Lancelin, J.-M., Kohda, D., Tate, S., Yanagawa, Y., Abe, T., Satake, M. and Inagaki, F. 1991. Tertiary structure of conotoxin GIIIA in aqueous solution. *Biochemistry* 30:6908-6916.

McIntosh, J. M., Hasson, A., Spira, M. E., Li, W., Marsh, M., Hillyard, D. R. and Olivera, B. M. 1995. A new family of conotoxins which block sodium channels. *J. Biol. Chem.* 270:16796-16802.

McIntosh, J. M., Olivera, B. M., Cruz, L. J. and Gray, W. R. 1984. γ-Carboxyglutamate in a neuroactive toxin. *J. Biol. Chem.* 259:14343-14346.

McIntosh, J. M., Yoshikami, D., Mahe, E., Nielsen, D. B., Rivier, J. E., Gray, W. R. and Olivera, B. M. 1994. A nicotinic acetylcholine receptor ligand of unique specificity, α-conotoxin ImI. *J. Biol. Chem.* 269:16733-16739.

McIntosh, M., Cruz, L. J., Hunkapiller, M. W., Gray, W. R. and Olivera, B. M. 1982. Isolation and structure of a peptide toxin from the marine snail *Conus magus*. *Arch. Biochem. Biophys.* 218:329-334.

Mena, E. E., Gullak, M. F., Pagnozzi, M. J., Richter, K. E., Rivier, J., Cruz, L. J. and Oivera, B. M. 1990. Conantokin-G: a novel peptide antagonist to the N-methyl-D-aspartate acid (NMDA) receptor. *Neurosci. Lett.* 118:241-244.

Myers, R. A. 1989. Toxins of *Conus* marine snails bind the nicotinic acetylcholine receptor at sites not identical to those of traditional antagonists. *J. Neurosci.* 15:678.

Myers, R. A., Cruz, L. J., Rivier, J. and Olivera, B. M. 1993. *Conus* peptides as chemical probes for receptors and ion channels. *Chem. Rev.* 93:1923-1936.

Myers, R. A., McIntosh, J. M., Imperial, J., Williams, R. W., Oas, T., Haack, J. A., Hernandez, J. F., Rivier, J., Cruz, L. J. and Olivera, B. M. 1990. Peptides from *Conus* venoms which affect Ca^{++} entry into neurons. *J. Toxin.-Toxin Rev.* 9:179-202.

Nishiuchi, Y. and Sakakibara, S. 1984. The structure-function relationship in conotoxins. In: *Peptide Chemistry 1983* (Ed. Munekata, E.). Osaka, Japan, Protein Research Foundation. 191.

Olivera, B. M., Cruz, L. J., de Santos, V., LeCheminant, G. W., Griffin, D., Zeikus, M., McIntosh, J., Galean, R., Varga, J., Gray, W. R. and Rivier, J. 1987. Neuronal calcium channel antagonists. Discrimination between calcium channel subtypes using ω-conotoxins from *Conus magus* venom. *Biochemistry* 26:2086-2090.

Olivera, B. M., Gray, W. R., Zeikus, R., McIntosh, J. M., Varga, J., Rivier, J., de Santos, V. and Cruz, L. J. 1985. Peptide neurotoxins from fish-hunting cone snails. *Science* 230:1338-1343.

Olivera, B. M., Imperial, J. S., Cruz, L. J., Bindokas, V. P., Venema, V. J. and Adams, M. E. 1991. Calcium channel-targeted polypeptide toxins. *Ann. N.Y. Acad. Sci.* 635:114-122.

Olivera, B. M., Johnson, D. S., Azimi-Zoonooz, A. and Cruz, L. J. 1992. Peptides in the venom of the geography cone, *Conus geographus*. In: *Toxins and Targets* Eds. Watters, D., Lavin, M., Maguire, D. and Pearn, J.). Victoria, Australia, Harwood Academic Publishers. 19-28.

Olivera, B. M., McIntosh, J. M., Cruz, L. J., Luque, F. A. and Gray, W. R. 1984. Purification and sequence of a presynaptic peptide toxin from *Conus geographus* venom. *Biochemistry* 23:5087-5090.

Olivera, B. M., Miljanich, G., Ramachandran, J. and Adams, M. E. 1994. Calcium channel diversity and neurotransmitter release: The ω-conotoxins and ω-agatoxins. *Ann. Rev. Biochem.* 63:823-867.

Olivera, B. M., Rivier, J., Clark, C., Ramilo, C. A., Corpuz, G. P., Abogadie, F. C., Mena, E. E., Woodward, S. R., Hillyard, D. R. and Cruz, L. J. 1990. Diversity of *Conus* neuropeptides. *Science* 249:257-263.

Pallaghy, P. K., Duggan, B. M., Pennington, M. W. and Norton, R. S. 1993. Three-dimensional structure in solution of the calcium channel blocker ω-conotoxin. *J. Mol. Biol.* 234:405-420.

Pallaghy, P. K., Nielsen, K. J., Craik, D. J. and Norton, R. S. 1994. A common structural motif incorporating a cystine knot and triple-stranded β-sheet in toxic and inhibitory polypeptides. *Protein Sci.* 3:1933-1839.

Pardi, A., Goldes, A., Florance, J. and Maniconte, D. 1989. Solution structures of α-conotoxins determined by two-dimensional NMR spectroscopy. *Biochemistry* 28:5494-5501.

Ramilo, C. A., Zafaralla, G. C., Nadasdi, L., Hammerland, L. G., Yoshikami, D., Gray, W. R., Ramachandran, J., Miljanich, G., Olivera, B. M. and Cruz, L. J. 1992. Novel α- and ω-conotoxins from *Conus striatus* venom. *Biochemistry* 31:9919-9926.

Sato, K., Ishida, Y., Wakamatsu, K., Kato, R., Honda, H., Ohizumi, Y., Nakamura, H., Ohya, M., Lancelin, J.-M., Kohda, D. and Inagaki, F. 1991. Active site μ-conotoxin GIIIA, a peptide blocker of muscle sodium channels. 266:16989-16991.

Sato, K., Ishida, Y., Wakamatsu, K., Ohizumi, Y., Kato, R., Honda, H. and Nakamura, H. 1990. Structure-activity relationship of geographutoxin (III). Search for an active site by the synthesis of a series of analogs substituted with a single alanine. *Peptide Chemistry 1990*, Osaka, Japan, Protein Research Foundation.

Sato, K., Park, N.-G., Kohno, T., Maeda, T., Kim, J. I., Kato, R. and Takahashi, M. 1993. Role of basic residues for the binding of ω-conotoxin GVIA to N-type calcium channels. *Biochem. Biophys. Res. Comm.* 194:1292-1296.

Sato, S., Nakamura, H., Ohizumi, Y., Kobayashi, J. and Hirata, Y. 1983. The amino acid sequences of homologous hydroxyproline containing myotoxins from the marine snail *Conus geographus* venom. *FEBS Lett.* 155:277-280.

Shon, K.-J., Hasson, A., Spira, M. E., Cruz, L. J., Gray, W. R. and Olivera, B. M. 1994. δ-Conotoxin GmVIA, a novel peptide from the venom of *Conus gloriamaris*. *Biochemistry* 33:11420-11425.

Skalicky, J. J., Metzler, W. J., Ciesla, D. J., Galdes, A. and Pardi, A. 1993. Solution structure of the calcium channel antagonist ω-conotoxin GVIA. *Protein Sci.* 2:1591-1603.

Wakamatsu, K., Kohda, D., Hatanaka, H., Lancelin, J.-M., Ishida, Y., Oya, M., Nakamura, H., Inagaki, F. and Sato, K. 1992. Structure-activity relationships of μ-conotoxin GIIIA: structure determination of active and inactive sodium channel blocker peptides by NMR and simulated annealing calculations. *Biochemistry* 31:12577-12584.

Yanagawa, Y., Abe, T., Satake, M., Odani, S., Suzuki, J. and Ishikawa, K. 1988. A novel sodium channel inhibitor from *Conus geographus*: Purification, structure, and pharmacological properties. *Biochemistry* 27:6256-6262.

Yoshiba, S. 1984. An estimation of the most dangerous species of cone shell *Conus geographus* venoms lethal dose in humans. *Jpn. J. Hyg.* 39:565.

Zafaralla, G. C., Ramilo, C., Gray, W. R., Karlstrom, R., Olivera, B. M. and Cruz, L. J. 1988. Phylogenetic specificity of cholinergic ligands: α-conotoxin SI. *Biochemistry* 27:7102-7105.

10

HYMENOPTERA VENOM PROTEINS

Donald R. Hoffman

Department of Pathology and Laboratory Medicine
East Carolina University School of Medicine
Greenville North Carolina 27858

INTRODUCTION

There has been much interest in studying and characterizing the venom proteins of social Hymenoptera, since stings from many of these insects can cause serious allergic reactions in man. Because most of the venom proteins are enzymes, specific toxins or other bioactive molecules, their study has led to a number of significant findings in other areas of biology and medicine, beside allergy and immunology. The primary structures of a large series of venom proteins have been determined by the techniques of molecular biology and protein sequencing, allowing the comparative study of families of related venom proteins. Although the primary purpose of the studies was to understand the immunological relationships among these proteins, these families of characterized proteins should provide powerful tools for other types of investigations.

SOCIAL HYMENOPTERA

The social Hymenoptera that commonly interact with man are members of the superfamilies, *Apoidea* and *Vespoidea*, bees and wasps (1). The *Vespoidea* include the social wasps and hornets, *Vespidae*, as well as ants, *Formicidae*. The bees that have been studied include honeybees, *Apis mellifera*, and bumble bees of the species, *Bombus pennsylvanicus*. Important wasps studied include yellowjackets of the genus, *Vespula*, hornets of the genera, *Dolichovespula* and *Vespa*, and paper wasps of the genus, *Polistes*. Yellowjackets can be divided into three species groups: the vulgaris group including *vulgaris*, *maculifrons*, *germanica*, *pensylvanica* and *flavopilosa*, the squamosa group including *squamosa* and *sulphurea*, and the rufa group including *vidua*, *acadica, atropilosa, austriaca,* and *consobrina* (2). The subclassification of *Polistes* is still somewhat complex (3). Ants studied include members of the subfamily, *Myrmicinae*, of the genera *Solenopsis*, fire ants (4), and *Pogonomyrmex*, harvester ants, as well as primitive Australian ants of the genus *Myrmecia* (5). Although other bees, wasps and ants occasionally sting people and may cause allergic reactions, their venoms have not been as thoroughly studied.

Natural Toxins II, Edited by B. R. Singh and A. T. Tu
Plenum Press, New York, 1996

TOXIC REACTIONS

Local toxic reactions normally appear at the site of a sting. In almost all cases these reactions are caused by the low molecular weight components of the venoms. Many of the features of a honeybee sting can be generated by the injection of the peptide melittin (6). The pain and dermal inflammation of wasp stings appears to be due to peptides and other small pharmacoactive molecules in the venom. Stings from imported fire ants of the genus, *Solenopsis*, are somewhat different, since about 90-95% of the volume is alkyl and alkenyl substituted piperidine alkaloids (7). These water insoluble organic molecules induce an immediate wheal and flare reaction, followed by the development of a sterile pustule in the next 6 to 24 hours (8). This reaction is pathognomic for fire ant stings. Most Hymenoptera venoms are not highly lethal, since cases have been reported of hundreds and even thousands of stings with only local reactions.

Systemic toxic reactions to Hymenoptera stings are relatively uncommon, although the more aggressive behavior of the Africanized honeybee has generated a large amount of sensational media attention. The intravenous LD_{50} of most Hymenoptera venoms is in the 1 to 10 mg/kg range (9, 10, 11). The weights of toxins delivered in stings ranges from about 50 micrograms for honeybees to 1 to 10 micrograms for yellowjackets and hornets to 10 to 100 nanograms for fire ants (12, 13). The great majority of cases of systemic toxic reactions to stings have been reported as descriptive anecdotal cases with no systematic study. The number of stings reported have ranged from 20 to over 1000 (14, 15). The culprit insects have included European honeybees (14, 16), yellowjackets and wasps (16, 17), Asian *Vespa* hornets (18, 19), fire ants (20, 21), and Africanized honeybees (15, 22). Toxic reactions have included intravascular hemolysis (16, 22, 23, 24, 25), rhabdomyolysis (16, 22), renal failure (16, 22, 26, 27, 28), neurologic reactions (20, 17, 29, 30, 31), coagulopathy (22, 32), cardiopathy (15, 22, 33), angioedema (16) and Henoch-Schonlein purpura (34). The most extensive and well-studied series of systemic toxic reactions are from stings of Africanized honeybees (15, 22, 26, 27, 28, 35, 36). In a carefully documented series from Brazil (22), circulating venom proteins were detectable in serum and urine more than 50 hours after the stings, and one victim had an estimated circulating unbound whole venom of 27 mg. Presentations included intravascular hemolysis, respiratory distress, hepatic dysfunction, rhabdomyolysis, hypertension, myocardial damage, shock, coma, acute renal failure and bleeding. Three of five patients died between 22 and 71 hours following the attacks, and autopsy findings included adult respiratory distress syndrome, hepatocellular necrosis, acute tubular necrosis, focal subendocardial necrosis and disseminated intravascular coagulation. In a series of twelve patients from Texas (15) clinical features included nausea, vomiting, weakness, fatigue, hypotension, pulmonary edema, tachycardia and three cases of unconsciousness. Two of the Texas patients with about 800 and 1000 stings suffered acute renal failure, three had evidence of hemolysis and three of rhabdomyolysis. In many of the descriptive cases it is very difficult to distinguish toxic reactions from allergic reactions or combinations of both. However, it appears that intravascular hemolysis, rhabdomyolysis and possibly liver damage are direct toxic effects. Renal failure may be a consequence of cell lysis and hepatic damage may also be secondary. Cardiac, pulmonary and neurological reactions may be primary or result from allergic and/or other immunological reactions.

ALLERGIC REACTIONS

The great majority of serious and fatal reactions to Hymenoptera venoms are due to allergic reactions to venom proteins (37). The area of allergy to Hymenoptera venoms has been recently reviewed (38, 39, 40, 41, 42, 43, 44) and is the subject of a current monograph (45), which should be consulted for information about the current standards and practices in the medical management of allergic individuals. Allergic reactions are the result of allergen combining with mast cell bound IgE antibodies, which causes a cascade of mediator release. These mediators including histamine, leukotrienes, platelet activating factor, enzymes, peptides and other substances generate the adverse effects. Many immediate local reactions are followed by a second phase which requires hours to develop, and which involves mediators from inflammatory cells including eosinophils and neutrophils. This is called the late phase allergic reaction. There are two categories of allergic reactions to stings, local reactions and systemic reactions. Local reactions are contiguous with the sting site and vary from a slightly enlarged and more persistent form of the normal sting reaction to large local reactions, which may involve an entire limb and require over 24 hours to develop. A large local reaction is usually hot to the touch, somewhat painful and hard, due to deposition of fibrin in the angioedematous area. Large local reactions typically take a few days to as long as two weeks to resolve.

Systemic allergic reactions include signs and symptoms that are not contiguous with the sting site. Among the more common manifestations are urticaria, pruritus and angioedema. Life-threatening reactions involve respiratory problems, angioedema of the throat or larynx, shock and cardiac symptoms. Other less commonly reported manifestations include gastrointestinal upset, central nervous system symptoms, and incontinence. Reactions almost always begin within seconds to minutes after stinging; in occasional cases a reaction does not appear until one to two hours later. Prompt and appropriate medical treatment can prevent many deaths and reverse most serious reactions (45). The use of epinephrine following a sting, either self-administered or adminstered by lay personnel, is recommended for individuals with histories of life-threatening allergic reactions (48). Medical management of the patient severely allergic to insect stings is best carried out by an allergy specialist. The evaluation consists of a clinical history and verification of IgE-mediated sensitivity by either intradermal skin testing with venoms (or whole body extract for imported fire ants) or detection of specific IgE antibodies in the clinical laboratory (39, 45, 46). Positive tests by either method are accepted as confirming the clinical history. In some countries patients are subjected to sting challenge tests to verify serious clinical reactivity; a number of studies have shown that only 25 to 65% of patients with definitive histories of life-threatening allergic reactions have a reaction to a deliberate challenge sting (39, 46, 47). After completion of the diagnostic work-up, the patient and physician decide whether immunotherapy is appropriate. Prophylactic immunotherapy consists of injection of increasing amounts of venom until a single dose tolerance of 100 micrograms of a single venom is reached (46). There are several protocols in use ranging from rush schedules, which take a few days to traditional schedules requiring several months (45, 49, 50). The efficacy of venom immunotherapy has been verified in numerous controlled sting challenge studies (38, 42, 46, 50). Among the incompletely resolved issues in venom immunotherapy are the selection of venoms for treatment and the duration of immunotherapy (45). Because of immunologic cross-reactivity and multiple exposure most patients test positively to multiple venoms (46, 51, 52). Some recommend treatment with all of the venoms that give positive reactions on testing, while others recommend using only the most probable (40, 41) or in some cases performing inhibition testing (53). Extensive longitudinal studies carried out at Johns Hopkins University with repeated sting challenges suggest that almost all allergic patients remain protected from reactions to stings for a period of at least two years after completing a course of four to five years of immunotherapy with maintenance injections every six weeks (54). However, many practioners still recommend an indefinite course of immunotherapy. Laboratory follow-up of venom immunotherapy is of little value, length of therapy is a better predictor of protection from sting reactions than serum IgE antibody, serum

IgG antibody or their ratio (50). Management of the fire ant venom allergic patient and the studies supporting efficacy of treatment with whole body extracts of verified venom allergen content are discussed in detail in a recent review (42).

VENOM PROTEINS

Phospholipase A_2

The major protein in bee venoms is a phospholipase of an apparent molecular weight of 16000 that specifically cleaves the fatty acid attached to the 2 position of the gycerol backbone of phospholipids (55, 56). A primary structure of *Apis mellifera* phospholipase was reported from protein sequence studies (56, 57, 58), but it contained a number of errors. The correct amino acid sequence was reported from translation of the DNA sequence cloned from a honeybee cDNA library (59). The structure of the N-linked carbohydrate has also been determined (60). Recently the structure of *Bombus pennsylvanicus* phospholipase has been reported from protein sequencing (61). The bumble bee venom protein has 136 amino acid residues with N-linked carbohydrate at position 16. The bumble bee and honeybee phospholipases are 53.7% identical, with a further 33% conservative substitutions. The first 64 residues show 72% identity, consistent with the high degree of immunological cross-reactivity between honeybee and bumble bee venoms (62). Bee venom phospholipases belong to the same family as phospholipase A_2 of many vertebrates found in venoms, cells and digestive system. A comparison of *Bombus pennsylvanicus* phospholipase with a gila monster venom phospholipase (63) is shown in Fig. 1. The molecules show 38% identity with all of the cysteines conserved. Twenty two of the first 38 residues are identical. Bee venom phospholipases also have significant similarity to the proposed common structure of phospholipase A_2's with four of the proposed 6 cysteine residues in the common region.

```
                   47.5% identity in 99 aa overlap

                        10        20        30        40        50        60
Bom p PLA      IIYPGTLWCGNGNIANGTNELGLWKETDACCRTHDMCPDIIEAHGSKHGLTNPADYTRLN
               .: :::::::.:: :.. ..::  :.:: :::.:: : . :.: . :::. :     : .
Gila PLA       FIMPGTLWCGAGNAASDYSQLGTEKDTDMCCRDHDHCENWISALEYKHGMRNYYPSTISH
                      10        20        30        40        50        60

                        70        80        90       100       110       120
Bom p PLA      CECDEEFRHCLHNSGDAVSAAFVGRTYFTILGTQCFRLDYPIVKCKVKSTILRECKEYEF
               :.::..:: :: .  :. .:..::.:::..: ..:: :.     :      : :: :
Gila PLA       CDCDNQFRSCLMKLKDG-TADYVGQTYFNVLKIPCFELEEGEG-CVDWNFWL-ECTESKI
                      70        80        90       100       110

                       130
Bom p PLA      DTNAPQKYQWFDVLSY
                  :
Gila PLA       MPVAKLVSAAPYQAQAET
                        130
```

Figure 1. Comparison of the sequences of venom phospholipases from bumble bee, *Bombus pennsylvanicus* (61), and gila monster, *Heloderma suspectum* (63). The double dots indicate identical residues and the single dots residues that differ by only a single base (conservative substitutions). Dashes indicate positions inserted to maximize alignment.

```
                          31.0% identity in 226 aa overlap

                      10        20        30        40        50
Ves m 1       GPKCPFNSDTVSIIIETRENRNRDLYTLQTLQNHPEFKKKT--ITRPVVFITHGFTSSAS
                  :........ .   .: :.. : . .  ...... ..  ..: . ::.:::.....
Mouse LPL     SKIFPWSPEDIDTRFLLYTNENPNNYQIISATDPATINASNFQLDRKTRFIIHGFIDKGE
                  60        70        80        90       100       110

              60        70        80        90       100       110
Ves m 1       EKNFINLAKALVDKDNYMVISIDWQTAACTNEYPGLKYAYYPTAASNTRLVGQYIATITQ
              :  .... : . . ..   :..::. ..  .::.  .:.        ::.::. :: ..:
Mouse LPL     EGWLLDMCKKMFQVEKVNCICVDWKRGS-RTEYTQASYN--------TRVVGAEIAFLVQ
                 120       130       140        150               160

             120       130       140       150       160       170
Ves m 1       KLVKDYKISMANIRLIGHSLGAHVSGFAGKRVQELKLGKYSEIIGLDPARPSFDSNHCSE
               : ..   : .:..:::::::.::.: ::.:... ..:.   :.::::: :.:.. . .
Mouse LPL     VLSTEMGYSPENVHLIGHSLGSHVAGEAGRRLEG-HVGR---ITGLDPAEPCFQGLPEEV
                170       180       190        200          210        220

             180       190              200       210       220       230
Ves m 1       RLCETDAEYVQIIHT-----SNYL--GTEKILGTVDFYMNNGKNNPGCGRFFSEVCSHTR
              ::  .:: .:..:::       ::  : .. .: .::. :.::. ::: .
Mouse LPL     RLDPSDAMFVDVIHTDSAPIIPYLGFGMSQKVGHLDFFPNGGKEIPGCQKNILSTIVDIN
                     230       240       250       260       270       280

                    240       250       260       270       280       290
Ves m 1       AVIYMAECIKHECCLIGIPRSKSSQPISRCTKQECVCVGLNAKKYPSRGSFYVPVESTAP

Mouse LPL     GIWEGTRNFAACNHLRSYKYYASSILNPDGFLGYPCSSYEKFQHNDCFPCPEQGCPKMGH
                     290       300       310       320       330       340

                    300
Ves m 1       FCNNKGKII

Mouse LPL     YADQFEGKT
                    350
```

Figure 2. Comparison of the sequences of *Vespula maculifrons* venom phospholipase (71) and mouse lipoprotein lipase (73). The double dots indicate identical residues and the single dots residues that differ by only a single base (conservative substitutions). Dashes indicate positions inserted to maximize alignment.

Phospholipase A_1B

The venoms of wasps of the family Vespidae contain a phospholipase of an apparent molecular weight of 34000 (64, 65, 66, 67). This phospholipase is not immunologically cross-reactive with bee phospholipases (68) and initially cleaves the fatty acid from the 1-position of phospholipids, followed by cleavage of the 2-fatty acid (lysophospholipase activity) (66). The amino acid sequence of one isozyme of venom phospholipase from *Dolichovespula maculata* was determined from the cDNA sequence obtained by anchor PCR cloning (69). This venom enzyme is a member of the lipoprotein lipase family, rather than the phospholipase A_2 family. The sequence of most of *Vespula vulgaris* phospholipase has been obtained by molecular biology and those of the second isozyme of *Dolichovespula maculata*, *Vespula maculifrons*, *Vespula squamosa* and *Vespa crabro* venom phospholipases by protein sequencing (70, 71, 72). A comparison of the sequences of *Vespula maculifrons* phospholipase (71) and mouse lipoprotein lipase (73) is shown in Fig. 2. An important feature of both wasp venom phospholipases and lipoprotein lipases is the presence of a labile aspartic acid-proline bond at about position 165 and 210 respectively. Cleavage at this site leads to the half size molecules seen in sodium dodecyl sulfate polyacrylamide gel electrophoresis (64, 65).

The two isozymes from *Dolichovespula maculata* venom have 66.7% sequence identity with almost all of the differences found in the first 131 residues. This is comparable

```
                   26.4% identity in 193 aa overlap

                        10        20        30        40        50        60
Sol i 1         GAASSNTYDIGQLMAKYTVDLMNKNLGNQQSCQDINASLPVVFITHGFTSSAQVSTFK
                                                       :::::::::::::  ..:
Dol m 1.01      FSNDTVKMIFLTRENRKHDFYTLDTMNRHNEFKKSIIKRPVVFITHGFTSSATEKNFVAM
                   10        20        30        40        50        60

                        70        80        90       100       110       120
Sol i 1                      IVDWSEAACTDGLPGVQ                            WK
                             .:::. :::::. ::..                            .:
Dol m 1.01      SEALMHTGDFLIIMVDWRMAACTDEYPGLKYMFYKAAVGNTRLVGNFIAMIAKKLVEQYK
                   70        80        90       100       110       120

                       130       140       150       160         170
Sol i 1         IPLNNIQYVGHSLGSHVSGFAAK           NDNTFEGEYHAF--PDCARNLFKGRQ
                .:..::. ::::::.:.::::.:           .  ... . ..:   ::. .. .. .
Dol m 1.01      VPMTNIRLVGHSLGAHISGFAGKRVQELKLGKFSEIIGLDPAGPSFKKNDCSERICETDA
                  130       140       150       160       170       180

               180       190       200       210        220       230
Sol i 1         QE                           SQPACSW-YDVPCSHSE  VY TGVPS
                .                            .::.:..  . .:::.    : :
Dol m 1.01      HYVQILHTSSNLGTERTLGTVDFYINNGSNQPGCRYIIGETCSHTRAVKYFTECIR
                  190       200       210       220       230       240
```

Figure 3. Alignment of proteolytically derived peptides from *Solenopsis invicta* venom phospholipase (76) with *Dolichovespula maculata* venom phospholipase isozyme 1 (69). The double dots indicate identical residues and the single dots residues that differ by only a single base (conservative substitutions). Dashes indicate positions inserted to maximize alignment.

to the differences found between the *Dolichovespula maculata* isozymes and *Vespula maculifrons* venom phospholipase, 68.3% and 58.7%, or *Vespa crabro* phospholipase, 75.7% and 65.7%. *Vespula maculifrons* phospholipase is 71.0% identical to that from *Vespa crabro*, 70.4% identical to that from *Vespula squamosa* and 95.3% identical for the first 257 residues of that from the sister species *Vespula vulgaris*.

The venoms of fire ants of the genus *Solenopsis* also contain a phospholipase, e.g. Sol i 1 (13, 74, 75). This phospholipase is immunologically cross-reactive with vespid venom phospholipases. Although the complete amino acid sequences of fire ant phospholipases have not been reported, the sequences of a number of proteolytic fragments are known and clearly establish that Sol i 1 and Sol r 1 are members of the same protein family as vespid venom phospholipase (76) (Fig. 3).

The sequences of two proteins from *Vespa orientalis* venom claimed to be lysophospholipase and phospholipase A_2 have been reported (77). Comparison of the sequence reported for the 152 amino acid lysophospholipase with that of *Vespa crabro* phospholipase A_1B indicates that it is the N-terminal 73 residues, residues 119-133, some of the C-terminal section fragments and the rest is not clearly related. Similar analysis of the reported structure of the 139 amino acid phospholipase A_2 shows that it consists of the N-terminal 49 residues, residues 119-133, and some fragments of the C-terminus of phospholipase A_1B; the rest is unrelated. It is probable that these reported sequences are fragments of the intact venom phospholipase, and not different proteins. The sequence data reported for these enzymes is not consistent with that found for other vespid venom phospholipases.

Hyaluronidase

Venoms of both bees and vespid wasps contain the enzyme hyaluronidase with an apparent molecular weight of about 42000 (55, 64, 65, 66, 67). The enzymes from bee and wasp are immunologically cross-reactive (65, 78, 79). The N-terminal regions of the molecules are quite similar among several species (79). The complete sequences of two venom hyaluronidases have been determined from cDNA clones, that of *Apis mellifera* (80) and *Dolichovespula maculata* (81). The honeybee venom enzyme has 350 amino acid residues and the hornet enzyme 331; and the sequences are 55.3% identical with many stretches of contiguous identical residues. Searches of the sequence databases showed that the venom hyaluronidases are related to a guinea pig sperm protein, PH-20 (80, 82, 83) involved in sperm-egg adhesion. Further studies have shown that similar proteins are found in all mammalian sperm and that they have hyaluronidase activity (83).

Antigen 5

In 1978 King et al. (64) described the isolation of a protein of about 23000 molecular weight from venoms of *Dolichovespula maculata*, *Dolichovespula arenaria*, and *Vespula vulgaris*, which he named antigen 5. These proteins are members of the same family as the presynaptic invertebrate neurotoxin isolated from venom of *Vespa mandarinia* (84). Antigen 5's have been isolated from three species of North American *Polistes* (66, 67), seven species of *Vespula* (64, 65, 67, 85) and *Vespa crabro* (86). In a classic paper, Fang et al. (87) reported on the cDNA cloning of three variants of *Dolichovespula maculata* antigen 5. The two major variants showed 23% difference between their sequences. Antigen 5's from twelve other species of vespid wasps have been sequenced by either cloning or protein sequencing (85, 88). The proteins from three North American species of *Polistes* were found to be 92.7% to 98.5% identical. The five species of yellow jackets of the *Vespula vulgaris* group had 93.1% to 98.0% sequence identity among themselves, but only 70.6% to 71.5% sequence identity with antigen 5 from *Vespula squamosa* and 73.0% to 74.5% with that from *Vespula vidua*; *Vespula squamosa* and *vidua* antigen 5's were 80.5% identical. Antigen 5's from hornets of the genus *Dolichovespula* showed 76.0% and 85.1% identity between species and 66.3% to 72.1% identity with antigen 5 from *Vespa crabro*. Antigen 5's from *Vespula* were about 63% to 73% identical to those from *Dolichovespula*, 65% to 72% identical to those from *Vespa* and 57% to 62% identical to those from *Polistes*. These results correlate well with the observed degrees of immunologic cross-reactivity (51, 89). The protein sequences of antigen 5 are related to two groups of known proteins (90): in plants a group of pathogenesis related proteins (91, 92) and in animals an androgen dependent sperm coating glycoprotein (93). Recombinant *Dolichovespula maculata* antigen 5's have been produced in a baculovirus expression system (94), the products reacted with polyclonal mouse antiserum to natural antigen 5 and were mildly toxic to fifth instar insect larvae. The recombinant Dol m 5.02 did not have the four cross-linking disulfide bonds formed, and did not induce antibody formation against the discontinuous epitopes in mice (95). The recombinant protein was able to induce antibodies against continuous B cell epitopes, which were of significantly lower affinity than the antibodies against discontinous epitopes induced by immunization with the native protein.

Proteins Sol i 3 (13, 75) and Sol r 3 (96, 97), isolated from *Solenopsis invicta* and *richteri* venoms were found to be members of the antigen 5 family. The two fire ant proteins are 95.8% identical, and are from 43.3% to 50.5% identical to antigen 5's from vespid wasps. The fire ant proteins have two regions of inserted residues, but all of the cysteine residues can be easily aligned as in shown in Fig. 4.

```
                       10         20         30         40         50
Ves f 5        1 NNYCKI-KCL KGG-VHTACK YG--SLKPNC GNKVVVSYGL TKQEKQDILK
Dol a 5        1 ******--*P **--T**L** **-T*M**** *G*I*K***V *ND**NE*V*
Vesp c 5       1 ******-**R S*--I**L** **-T*T**** *KN**KAS** ****NLE***
Sol r 3        1 T***NLQS*K RNNAI**M*Q *TSPTPG*M* LEYSN*--*F *DA**DA*VN

                       60         70         80         90        100
Ves f 5       51 EHNDFRQKIA RGLETRGNPG PQPPAKNMKN LVWNDELAYV AQVWANQCQY
Dol a 5       51 R**E****V* Q********* ********NL ********KI **T*****NF
Vesp c 5      51 Q**E****V* ********** ******S*NT ********QI ********N*
Sol r 3       51 K**EL**RV* S*K*M**TN* *****VK*P* *T*DP***TI **R*****TF

                      110        120        130        140        150
Ves f 5      101 GHDTCRDIAK YQVGQNVALT GSTAAKYDDP VKLVKMWEDE VKDYNPKKKF
Dol a 5      101 ***Q**NT** *P******IA ST*GNS*QTM SY*I****** ******H*DL
Vesp c 5     101 ***N**NS** *S****I*EG ST**DNFGSV SNM******* ****QYGS--
Sol r 3      101 E**A**NVER FA****I*A* S*SGKNKSTL SDMILL*YN* ***FDNRWIS

                      160        170        180        190        200
Ves f 5      151 ----SGNNFL KTGHYTQMVW ANTKEVGCGS IKFIQEKWHK HYLVCNYGPS
Dol a 5      151 ----MH***S *V******** GK***I**** V*Y*EN***T *********A
Vesp c 5     151 ----PK*KLN *V******** *K***I**** **Y*ENG**R *********A
Sol r 3      151 SFPSD**ILM HV*****I** *K**KI***R *M*KEDN*N* *********A

                      210
Ves f 5      201 GNFQNEELYQ TK
Dol a 5      201 **YM*QPV*E R*
Vesp c 5     201 **VG**PI*E R*
Sol r 3      201 **VLGAQI*E I*K
```

Figure 4. Comparison of the sequences of antigen 5 from *Vespula flavopilosa* (85), *Dolichovespula arenaria* (88), *Vespa crabro* (85) and *Solenopsis richteri* (97). The asterisks indicate residues identical to those in *Vespula flavopilosa* antigen 5. Dashes indicate positions inserted to maximize alignment.

Acid Phosphatase

Acid phosphatase is found in bee venoms as a dimer of a protein chain of about 49000 molecular weight (61, 98, 99, 100, 101, 102). Acid phosphatase activity cannot be consistently demonstrated in venoms from vespid wasps. The N-terminal sequence and the sequences of a number of proteolytic peptides from honeybee venom acid phosphatase have recently been reported (103). The venom acid phosphatase is more closely related to mammalian prostatic type acid phosphatases (104, 105, 106, 107, 108) than to lysosomal type acid phosphatases (109, 110), as is illustrated in Fig. 5.

Protease

Honeybee venom and vespid venoms do not contain significant amounts of protease; but venom from the bumble bee, *Bombus pennsylvanicus*, contains a large amount of a tryptic amidase (61, 62). This enzyme does not cleave ester substrates, but only amide substrates with a tryptic specificity; and it easily solubilizes casein. This protease is also allergenic (61). The sequence of 243 amino acids has recently been determined (62). The bumble bee venom enzyme is a serine protease that is related to a large number of other serine proteases. The closest match was with a horseshoe crab proclotting enzyme (111) with 41.1% identity. The next nearest matches were 37.7% and 37.0% with human (112) and mouse (113) acrosins, the sperm head proteases. The alignment of bumble bee venom protease with human acrosin is shown in Fig. 6. Other matches were found at 36.5% identity with human kallikrein (114)

```
                          10         20         30         40         50
Am Acid Phos    1                                          ELKQINV IFRHGDRIPD
Hum PAP         1 MRAAPLLLAR AASLALASCF CFFCWLDRSV LAKELKFVTL VFRHGDRSPI
Rat LAP         1   MAGRQSGW SQAALLQFLL GMCLMVMPPI QARSLRFVTL LYRHGDRSPV

                          60         70         80         90        100
Am Acid Phos   51 EKNEMYPKKL EEWTDK                                FVDESANNL
Hum PAP        51 DTFPTDPIKE SSWPQWIWPT HPAGMEQHYE LGEYIRKRYR KFLNESYKHE
Rat LAP        51 KAYPKDPYQE EKWPQGFGQL TKEGMLQHWE LGQALRQRYH GFLNASYHRQ

                         110        120        130        140        150
Am Acid Phos  101 SIEEIDFVK                         LQ QWNEDLNWQP IATK
Hum PAP       101 QVYIRSTDVD RTLMSAMTNL AALFPPEGVS IWNPILLWQP IPVHTVPLSE
Rat LAP       101 EVYVRSTDFD RTLMSAEANL AGLFPPTEVQ HFNPNISWQP IPVHTVPITE

                         160        170        180        190        200
Am Acid Phos  151                        GKYEFSKR                        YN
Hum PAP       151 DQLLYLPFRN CPRFQELESE TLKSEEFQKR LHPYKDFIAT LGKLSGLHGQ
Rat LAP       151 DRLLKFPLGP CPRYEQLQNE TRQTPEYQNM SIQNAQFLDM VANETGLMNL

                         210        220        230        240        250
Am Acid Phos  201 IFAGTWK
Hum PAP       201 DLFGIWSKVY DALYCESVHN FTLPSWATED TMTKLRELSE LSLLSLYGIH
Rat LAP       201 TLETIWNVYD TLFCEQTHGL LLPPWASPQT VQALSQLKDF SFLFLFGIHD

                         260        270        280        290        300
Am Acid Phos  251        LYG GPLLRDNYVG DER                    D ITTPKDYYYI
Hum PAP       251 KQKEKSRLQG GVLVNEILNH MKRATQIPSY KKLIMYSAHD TTVSGLQMAL
Rat LAP       251 QVQKARLQGG VLLAQILKNL TLMATTSQFP -KLLVYSAHD TTLVALQMAL

                         310        320        330        340        350
Am Acid Phos  301 YHTLVAENEY SSCIIMEYHN I-EGTHYVKI VYYLGIPSEA RELQLPGCEV
Hum PAP       301 DVYNGLLPPY ASCHLTELYF E-KGEYFVEM YYRNETQHEP YPLMLPGCSP
Rat LAP       301 NVYNGKQAPY ASCHIFELYQ EDNGNFSVEM YFRNDSKKAP WPLTLPGCPH

                         360        370        380        390        400
Am Acid Phos  351 LCPLEKYLQL IENVIPSNEE LICDKR
Hum PAP       351 SCPLERFAEL VGPVIPQDWS TECMTTNSHQ GTEDSTD
Rat LAP       351 RCPLQDFLRL TEPVIPKDWQ KECQLASDTA DTEVIVALAV CGSILFLLIV +
```

Figure 5. Comparison of sequences of peptides isolated from honeybee venom acid phosphatase (103) with human prostatic (107) and rat lysosomal (109) acid phosphatases. Residues identical to those in bee venom acid phosphatase are indicated in bold face type.

and 33.6% with human factor XI (115). Significant matches were seen over shorter stretches of sequence with various tryptic and chymotryptic enzymes. The catalytic residues typical of serine proteases were all present.

Solenopsis Proteins 2 and 4

Two proteins of apparent molecular weights of 30000 and 20000 were isolated from the venom of imported fire ants, *Solenopsis invicta* (13). The proteins were named Sol i 2 and Sol i 4. A protein similar to Sol i 2 was isolated from venom of the sister species, *Solenopsis richteri*, but no protein similar to Sol i 4 was found (96). Sera from patients allergic to the native fire ant species, *Solenopsis xyloni* (116) and *aurea* (117), show immunologic cross-reactivity with Sol i 2 and Sol r 2, suggesting that a similar protein is present in the venom of these two species. The amino acid sequence of Sol i 2 has been determined by both protein sequencing (75) and cDNA cloning (118). The molecule consists of a disulfide linked homodimer of 119 amino acids. The sequence of Sol r 2 is similar with 78.2% identity (97). The sequence of Sol i 4 (75) was 117 amino acids with one fewer cysteine than in Sol i 2. Sol i 4 was 34.7% identical to Sol i 2 and 33.5% identical to Sol r 2 (Fig. 7). Sol i 2, Sol r 2 and Sol i 4 were not significantly related to any other known

```
                            37.7% identity in 247 aa overlap

                         10        20        30        40        50
Bp Protease     VVGGKPAKLGAWPWMVALGFHNYRQPKKSPEWKCGGSLRISRHVLTAAHC-----AIHRS
                .::::.:. :::::::.: . .: .... ..  :::::  :: :::::::     ..:
Acrosin         IVGGKAAQHGAWPWMVSLQIFTY-NSHRYHT--CGGSLLNSRWVLTAAHCFVGKNNVHDW
                       50        60        70        80        90

                    60        70        80        90       100       110
Bp Protease     LYVVRIADLNLKRDDDGAHPIQ-MGIESKLIHPDYVYSEHHDDIAILKLEKDVSFSEYIR
                  :    ...  ...      :.: . .:. .:: .:   ... .:::.....   .: . .:
Acrosin         RLVFGAKEITYGNNKPVKAPVQERYVEKIIIHEKYNSATEGNDIALVEITPPISCGRFIG
               100       110       120       130       140       150

                    120       130       140       150        160       170
Bp Protease     PICLP-IEESLRNNNFIGYNPFVAGWGRLRYKGPLSDA-LMEVQVPVVRNKVCKRA--YS
                : ::: ....:....     . .::::: .  :.: ... :::..: ..  ..:...  :.
Acrosin         PGCLPHLKAGLPRGS---QSCWVAGWGYIEEKAPRPSSILMEARVDLIDLDLCNSTQWYN
               160       170          180       190       200       210

                        180       190        200       210       220
Bp Protease     D-VSDTVICAGYPKGRKDSCQGDSGGPLM--IPQESTYYEIGVVSYGHECALPKYPGVYT
                . :  : .::::: :. :.::::::::::    ..::.:  .:..:.: .:: .: ::.::
Acrosin         GRVQPTNVCAGYPVGKIDTCQGDSGGPLMCKDSKESAYVVVGITSWGVGCARAKRPGIYT
                 220       230       240       250       260       270

                230       240
Bp Protease     RVTSYLDSFILPALKK
                 . .::.
Acrosin         ATWPYLNWIASKIGSN
                 280       290
```

Figure 6. Comparison of the sequences on *Bombus pennsylvanicus* venom protease (62) and human acrosin (112). The double dots indicate identical residues and the single dots residues that differ by only a single base (conservative substitutions). Dashes indicate positions inserted to maximize alignment.

proteins in either the Protein Identification Resource (PIR) or SWISSPROT databases. The complete leader sequence and coding region for Sol i 2 has been incorporated into a baculovirus expression system; and a recombinant Sol i 2 has been produced that is immunoreactive with monoclonal antibodies specific for the native molecule and with an amino acid sequence that begins at the same point as in the native molecule.

```
                        33.9% identity in 118 aa overlap

                         10        20        30        40        50        60
Sol r 2         DIEAQRVLRKDIAECARTLPKCVNQPDDPLARVDVWHCAMSKRGVYDNPDPAVVKEKNSK
                ::.. ..... ...: ::.::  :.: .:: .:.: .::..:::..  .  .: ...  .
Sol i 4         DIKEISIMNRILEKCIRTVPKRENDPINPLKNVNVLYCAFTKRGIF--TPKGVNTKQYIN
                        10        20        30        40         50

                         70        80        90       100       110
Sol r 2         MCPKIITDPADVENCKKVVSRCVDRETQRPRSNRQKAINITGCILRAGVVEATVLARE
                 : :.:..:::.. :::..:.::..  .:: .   ... :. .:.:. :..: :: ...
Sol i 4         YCEKTIISPADIKLCKKIASKCVKKVYDRPGPVIERSKNLLSCVLKKGLLELTVYGKN
               60        70        80        90       100       110
```

Figure 7. Comparison of amino acid sequences of imported fire ant venom proteins, Sol r 2 (97) and Sol i 4 (75). The double dots indicate identical residues and the single dots residues that differ by only a single base (conservative substitutions). Dashes indicate positions inserted to maximize alignment.

```
                  42.9% identity in 35 aa overlap

                   10        20        30        40        50
Myr p 1    KDLADPESEAVGFADAFGEADAVGEADPNAGLGSVFGRLARILFRVIPKVAKKLGPK
                             ::.: .::::.::.:.:.  :.  :  .:. . .:
Melittin   VYISYIYAAPEPEPAPEPEAEADAEADPEAGIGAVLKVLTTGLPALISWIKRKRQQG
                subject  20        30        40        50        60        70
```

Figure 8. Comparison of the amino acid sequences of Myr p 1 (141) and prepromelittin (144). The double dots indicate identical residues and the single dots residues that differ by only a single base (conservative substitutions).

Melittin

Melittins are peptides of 26 amino acids found in *Apis* venoms (119, 120). Melittin comprises about 40% to 50% of the dry weight of *Apis mellifera* venom. At moderate and high concentrations, it exists primarily in the form of tetramers (120, 121), which can be immunogenic and allergenic (123). Tetrameric melittin is a potent lytic agent for cells and can also function as an ion channel (124, 125, 126). Melittin has profound effects on intracellular calcium and interacts with calmodulin (127, 128, 129). It is an amphiphilic and amphipathic molecule that has been found to be very useful in studies of cell lysis, membrane function, calcium regulation and as a model for proteins and peptides. Natural and synthetic melittin have been used in over a 1000 published studies in many areas of science.

Melittin also modulates phospholipase A_2 function. Other venoms also contain modulating peptides, which may also interact with ions, membranes and calmodulin, but none of the others are known to aggregate or be naturally antigenic. The peptides in *Bombus* venom are called bombolitins (122). Those in vespid venoms include crabrolin, mastoparans, vespulakinins, vespakinins and polistikinins (130, 131, 132 ,133, 134, 135, 136, 137, 138, 139, 140).

Myrmecia Venom Proteins

Ants of the Australian genus, *Myrmecia*, which are 10 to 40 mm long and are commonly called jack-jumpers, bullants, bulldog ants, inch ants, inchmen and sergeant ants, are a serious cause of venom allergy (5). The venom allergens are a series of proteins of molecular weights, 8500, 7500, 5500, 4000 and 2000 (141). An allergen cloned from *Myrmecia pilosula*, and named Myr p 1 estimated to be 86 amino acids long, appeared to be immunologically related to both the 7500 and 5500 molecular weight allergens (141, 142). A second allergen, Myr p 2, has also been cloned (143). A comparison of the sequence of Myr p 1 with all other known Hymenoptera venom proteins, showed that it was related to prepromelittin (Fig. 8). The cDNA for melittin corresponds to an amino acid chain 70 amino acids long (144), where residues 1 to 21 are the leader sequence and 44 to 69 are melittin (145). Residue 70 is converted to the C-terminal amide (146). Residues 1 to 43 are cleaved off in two stages to generate promelittin, and then melittin (147, 148, 149). Myr p 1 shows 42.9% identity and 34% conservative substitutions with residues 32 to 66 of the melittin precursor, which are in the promelittin and melittin regions. This amount of relatedness is suggestive that both proteins may originate from the same gene or gene family.

DISCUSSION

The study of venom proteins from social Hymenoptera has added much to scientific and medical knowledge, not only in the fields of toxinology and allergy, but also in fields as diverse as membrane biophysics and human fertility. Venom proteins and peptides have become important tools in many areas of bioscience, and they have been used in thousands of studies.

The structural studies of the venom proteins reviewed here have had profound effects in entomology as well. The observations that ant venoms contain antigen 5 and phospholipase A_1B molecules have strengthened the arguments for the contemporary taxonomy based upon cladistic analysis (150) that ants are a family of the superfamily *Vespoidea*, and do not belong in a superfamily of their own (75). The observations that many venom proteins are derived from male reproductive system genes challenges the traditional belief that the venom apparatus is a modified ovipositor (151, 152, 153, 154, 155). This belief is based upon the anatomic studies and functional observations of 19th century entomologists (156, 157) and has become traditional, despite the well-known fact that female social Hymenoptera have both a complete female reproductive system and a venom apparatus. The antigen 5 protein family is related to an androgen dependent epididymal protein (87, 90); venom hyaluronidases are very closely related to sperm hyaluronidase (80, 81, 82, 83); venom acid phosphatase is of the prostatic type (103); and bumble bee venom protease is related to acrosin (62). This protein structural evidence strongly suggests that the venom apparatus is derived from the male genital system, rather than the female. Male Hymenoptera are produced from unfertilized eggs and have a haploid chromosome complement, while both workers and queens are produced from fertilized eggs and have a diploid chromosome number (158, 159). Since there are no Y chromosomes, both queens and workers carry the genes for all male antigens.

REFERENCES

1. Krombein, K.V., Hurd, P.D., Smith, D.R., and Burks, B.D.,1979, *Catalog of hymenoptera in America North of Mexico*. Smithsonian Institution Press, Washington.
2. Akre, R.D., Greene, A., MacDonald, J.F., Landolt, P.J. and Davis, H.G.,1981, *The Yellowjackets of America North of Mexico*. USDA Handbook 552.
3. Richards, O.W., 1978, *The social wasps of the Americas*. British Museum (Natural History), London.
4. Rhoades, R.B., 1977, *Medical Aspects of the Imported Fire Ant*. University of Florida, Gainesville.
5. Ford, S.A., Baldo, B.A., Weiner, J., and Sutherland, S., 1991, Identification of jack-jumper ant (Myrmecia pilosula) venom allergens. *Clin. Exp. Allerg.* 21:167-171.
6. Prince, R.C., Gunson, D.E., and Scarpa, A., 1985, Sting like a bee - the ionophoric properties of melittin. *Trends Biochem. Sci.* 10:99.
7. Brand, J.M., Blum, M.S., Fales, H.M., and MacConnell, J.G., 1972, Fire ant venoms: Comparative analyses of alkaloidal components. *Toxicon* 10:259-271.
8. DeShazo, R.D., Griffing, C., Kwan, T.H., Banks, W.A., and Dvorak, H.F., 1984, Dermal hypersensitivity reactions to imported fire ants. *J. Allergy Clin. Immunol.* 74:841-847.
9. Schmidt, J.O., Blum, M.H., and Overal, W.L., 1980, Comparative lethality of venoms from stinging hymenoptera. *Toxicon* 18:469-474.
10. Schmidt, J.O., Yamane, S., Matsuura, M., and Starr, C.K., 1986, Hornet venoms- Lethalities and lethal capacities. *Toxicon* 24: 950-954.
11. Schumacher, M.J., Schmidt, J.O., and Egen, N.B., 1989, Lethality of killer bee stings. *Nature* 337:413.
12. Hoffman, D.R., and Jacobson, R.S., 1984, Allergens in hymenoptera venom XII. How much protein is in a sting? *Ann. Allergy* 52:276-278.
13. Hoffman, D.R., Dove, D.E., and Jacobson, R.S., 1988, Allergens in Hymenoptera venom XX. Isolation of four allergens from imported fire ant (*Solenopsis invicta*) venom. *J. Allergy Clin. Immunol.* 82:818-827.
14. Ring, J., Gottsmann, M., Przybilla, B., and Eisenmenger, W., 1986, Tod nach 1000 Bienenstichen. *Munch. med. Wschr.* 128:339.
15. McKenna, W.R., 1994, Characteristics of multiple-massive honeybee sting cases. *J. Allergy Clin. Immunol.* 93:224. (abstract)
16. Bousquet, J., Huchard, B., and Michel, F.B., 1984, Toxic reactions induced by hymenoptera venom. *Ann. Allergy* 52:371-374.

17. Riggs, J.E., Ketonen, L.M., Wymer, J.P., Barbano, R.L., Valanne, L.K., and Bodenstein, J.B., 1994, Acute and delayed cerebral infarction after wasp sting anaphylaxis. *Clin. Neuropathol.* 17:284-388.
18. Ghosh, K., Singh, S., Pereira, B.J., and Singhi, S.C., 1988, Acute systemic toxic reactions caused by hornet stings. *Indian Pediatr.* 25:796-798.
19. Severino, M.G., Manfredi, M., and Zerboni, R., 1991, Toxic reactions induced by Hymenoptera venom. *Ethol. Ecol.* 1991:49-50.
20. Fox, R.W., and Lockey, R.F., 1982, Neurologic sequelae following the imported fire ant sting. *J. Allergy Clin. Immunol.* 70:120-124.
21. Hardwick, W.E., Royall, J.A., Petitt, B.A., and Tilden, S.J., 1992, Near fatal fire ant envenomation of a newborn. *Pediatrics* 90:622-624.
22. Franca, F.O.S., Benvenutti, L.A., Fan, H.W., Dossantos, D.R., Hain, S.H., Picchi-martins, F.R., Cardoso, J.L.C., Kamiguti, A.S., Theakston, R.D., and Warrell, D.A., 1994, Severe and fatal mass attacks by killer bees (Africanized honeybees, *Apis mellifera scutellata*) in Brazil - Clinicopathological studies with measurements of serum venom concentrations. *Quart. J. Med.* 87:269-282.
23. Scragg, F.R., and Szent-Ivany, J.J.H., 1965, Fatalities caused by multiple hornet stings in the territory of Papua and New Guinea. *J. Med. Entomol.* 2:309-313.
24. Tan, K.H., Chew, I.S. and Chao, T.C., 1966, Fatal hemolysis from wasp and hornet stings. *Singapore med. J.* 7:122-6.
25. Korman, S.H., Jabbour, S., and Harari, M.D., 1990, Multiple hornet (*Vespa orientalis*) stings with fatal outcome in a child. *J. Paediat. C.* 26:283-285.
26. Mejia, G., Arbelaez, M., Henao, J.E., Sus, A.A., and Arango, J.L., 1986, Acute renal failure due to multiple stings by Africanized bees. *Ann. Int. Med.* 104:210-211.
27. Patrick, A., Roberts, L., Poonking, P., and Jeelal, V., 1987, Acute renal failure due to multiple stings by Africanized bees. Report of the 1st case in Trinidad. *West Ind. Med. J.* 36:43-44.
28. Munoz-Arizpe, R., Valencia-Espinoza, L., Velazquez-Jones, L., Abarco-Franco, C., Gamboa-Marrufo, J., and Valencia-Mayoral, P., 1992, Africanized bee stings and pathogenesis of acute renal failure. *Nephron* 61:478.
29. Catola, G., 1928, Crises vasomotrices cephaliques et menierformes par venin d'abeilles. *Rev. Neurol.* 35:260-262.
30. L'Epee, P., Lazarini, H.J., Bezian, J., and Doignon, J., 1971, Reflexions sur de multiples piqures d'abeilles et une sclerose en plaques. *Med. Liege* 4:235-237.
31. Candiotti, K.A., and Lamas, A.M., 1993, Adverse neurological reactions to the sting of the imported fire ant. *Int. Arch. Allergy Immunol.* 102:417-420.
32. Kini, P.G., Baliga, M., and Bhakaskara, N., 1994, Severe derangement of the coagulation profile following multiple bee stings in a 2 year old boy. *Ann. Trop. Para.* 14:153-155.
33. Wagdi, P., Mehan, V.K., Burgi, H., and Salzmann, C., 1994, Acute myocardial infarction after wasp stings in a patient with normal coronary arteries. *Amer. Heart J.* 128:820-823.
34. Burke, D.M., and Jellinek, H.L., 1954, Nearly fatal case of Schoenlein-Henoch syndrome following insect bite. *Am. J. Dis. Child.* 88:772-774.
35. Ariue, B.K., 1994, Multiple Africanized bee stings in a child. *Pediatrics* 94:115-117.
36. McKenna, W.R., 1992, Africanized honeybees: Proposed antivenom treatment and first case of massive sting in the U.S. *J. Allergy Clin. Immunol.* 89:294. (abstract)
37. Hoffman, D.R., Wood, C.L., and Hudson, P., 1983, Demonstration of IgE and IgG antibodies against venoms in the blood of victims of fatal sting anaphylaxis. *J. Allergy Clin. Immunol.* 71:193-196.
38. Mueller, U.R., 1990, *Insect sting allergy. Clinical picture, diagnosis and treatment.* Stuttgart, Gustav Fischer Verlag.
39. Yunginger, J.W., 1993, Insect Allergy in Middleton, E., Reed, C.E., Ellis, E.F., Adkinson, N.F., Yunginger, J.W., and Busse, W.W., eds., *Allergy Principles and Practice*, 4th Ed. St Louis, Mosby-Year Book, pp1511-1524.
40. Reisman, R.E., 1994, Insect stings. *N. E. J. Med.* 331:523-527.
41. Reisman, R.E., 1994, Venom hypersensitivity. *J. Allergy Clin. Immunol.* 94:651-658.
42. Hoffman, D.R., 1995, Fire ant allergy. *Allergy* 50: 535-544.
43. Lichtenstein, L.M., 1994, A reappraisal of sting challenges: To whom should we offer venom immunotherapy. *J. Allergy Clin. Immunol.* 94:137-138.
44. Essayan, D.M., Kagey-Sobotka, A., and Lichtenstein, L.M., 1994, Nearly fatal anaphylaxis following an insect sting. *Ann. Allergy* 73:297-300.
45. Levine, M.I., and Lockey, R.F., eds., 1995, *Monograph on Insect Allergy*. 3rd Ed., Milwaukee, American Academy of Allergy and Immunology.

46. Hunt, K.J., Valentine, M.D., Sobotka, A.K., Benton, A.W., Amodio, F.J., and Lichtenstein, L.M., 1978, A controlled trial on immunotherapy in insect hypersensitivity. *N. E. J. Med.* 299:157-161.
47. Van der Linden, P.-W., Kack, C.E., Struyvenburg, A., and van der Zwan, J.K., 1994, Insect sting challenge in 324 subjects with a previous anaphylactic reaction: Current criteria for insect venom hypersensitivity do not predict the occurrence and severity of anaphylaxis. *J. Allergy Clin. Immunol.* 94:151-159.
48. Frazier, C.A., 1985, In support of a model bill allowing the use of insect sting kits by trained laymen. *N. C. Med. J.* 46:371.
49. Bernstein, J.A., Kagen, S.L., Bernstein, D.I., Bernstein, I.L., 1994, Rapid venom immunotherapy is safe for routine use in the treatment of patients with Hymenoptera anaphylaxis. *Ann. Allergy* 73:423-428.
50. Hoffman, D.R., Gillman, S.A., Cummins, L.H., Kozak, P.P. and Oswald, A., 1981, Correlation of IgG and IgE antibody levels to honeybee venom allergens with protection to sting challenge. *Ann. Allergy* 46:17-23.
51. Hoffman, D.R., 1981, Allergens in hymenoptera venom VI. Cross reactivity of human IgE antibodies to the three vespid venoms and between vespid and paper wasp venoms. *Ann. Allergy* 46:304-309.
52. Kern, F., Sobotka, A.K., Valentine, M.D., Benton, A.W., and Lichtenstein, L.M., 1976, Allergy to insect sting III. Allergenic cross-reactivity among the vespid venoms. *J. Allergy Clin. Immunol.* 57:554-559.
53. Hamilton, R.G., Wisenauer, J.A., Golden, D.B.K., Valentine, M.D., and Adkinson, N.F., 1993, Selection of Hymenoptera venoms for immunotherapy on the basis of patient's IgE antibody cross-reactivity. *J. Allergy Clin. Immunol.* 92:651-659.
54. Golden, D.B.K., Addison, B.I., Gadde, J., Kagey-Sobotka, A., Valentine, M.D., and Lichtenstein, L.M., 1989, Prospective observations on stopping prolonged venom immunotherapy. *J. Allergy Clin. Immunol.* 84:162-167.
55. Habermann, E., 1972, Bee and wasp venoms: The biochemistry and pharmacology of their peptides and enzymes are reviewed. *Science* 177:314-322.
56. Shipolini, R.A., Callewaert, G.C., Cottrell, R.C. and Vernon, C.A., 1971, The primary sequence of phospholipase A from bee venom. *FEBS Lett.* 17:39-40.
57. Shipolini, R.A., Callewaert, G.L., Cottrell, R.C. and Vernon, C.A., 1974, The amino acid sequence and carbohydrate content of phospholipase A2 from bee venom. *Eur. J. Biochem.* 48:465-476.
58. Shipolini, R.A., Doonan, S. and Vernon, C.A., 1974, The disulfide bridges of phospholipase A2 from bee venom. *Eur. J. Biochem.* 48:477-483.
59. Kuchler, K., Gmachl, M., Sippl, M.J., and Kreil, G., 1989, Analysis of the C-DNA for phospholipase A2 from honeybee venom glands. The deduced amino acid sequence reveals homology to the corresponding vertebrate enzymes. *Eur. J. Biochem.* 184: 249-254.
60. Kubelka, V., Altmann, F., Staudacher, E., Tretter, V., Marz, L., Hard, K., Kammerling, J.P., and Vliegenter, J.F., 1993, Primary structures of the N-linked carbohydrate chains from honeybee venom phospholipase A2. *Eur. J. Biochem.* 213:1193-1204.
61. Hoffman, D.R., 1982, Allergenic cross-reactivity between honeybee and bumble bee venoms. *J. Allergy Clin. Immunol.* 69:139. (abstract)
62. Jacobson, R.S., and Hoffman, D.R., 1993, Characterization of bumble bee venom allergens. *J. Allergy Clin. Immunol.* 91:187. (abstract)
63. Vandermeers, A., Vandermeers-Piret, M.C., Vigneron, L., Rathe, J., Stievenart, M., and Christophe, J., 1991, Differences in primary structure among five phospholipases A(2) from *Heloderma suspectum*. *Eur. J. Biochem.* 196:537-544.
64. King, T.P., Sobotka, A.K., Alagon, A., Kochoumian, L. and Lichtenstein, L.M., 1978, Protein allergens of white-faced hornet, yellow hornet and yellow jacket venoms. *Biochemistry* 17:5165-5174.
65. Hoffman, D.R., and Wood, C.L., 1984, Allergens in hymenoptera venom XI. Isolation of protein allergens from *Vespula maculifrons* (yellow jacket) venom. *J. Allergy Clin. Immunol.* 74: 93-103.
66. King, T.P., Kochoumian, L. and Joslyn, A., 1984, Wasp venom proteins: Phospholipase A1 and B. *Arch. Biochem. Biophys.* 230:1-12.
67. Hoffman, D.R., 1985, Allergens in Hymenoptera venom XIII: Isolation and purification of protein components from three species of vespid venoms. *J. Allergy Clin. Immunol.* 75:599-605.
68. Hoffman, D.R., 1986, Allergens in Hymenoptera venom XVI: Studies of the structures and cross-reactivities of vespid venom phospholipases. *J. Allergy Clin. Immunol.* 78:337-343.
69. Soldatova, L., Kochoumian, L., and King, T.P., 1993, Sequence similarity of a hornet (*D. maculata*) venom allergen, Phospholipase A1 with mammalian lipases. *FEBS Lett.* 320:145-149.
70. Soldatova, L., Kochoumian, L., and King, T.P., 1993, Sequence similarity of vespid venom allergens, Phospholipase A1 and antigen 5 with mammalian proteins. *J. Allergy Clin. Immunol.* 91:283. (abstract)
71. Hoffman, D.R., 1994, Allergens in Hymenoptera venom XXVI: The complete amino acid sequences of two vespid venom phospholipases. *Int. Arch. Allergy Immunol.* 104:184-190.

72. Hoffman, D.R., 1994, The structure of vespid phospholipases and the basis for their immunologic crossreactivity. *J. Allergy Clin. Immunol.* 93:223. (abstract)
73. Grusby, M.J., Nabavi, N., Wong, H., Dick, R.F., Bluestone, J.A., Schotz, M.C., and Glimcher, L.H., 1990, Cloning of an interleukin-4 inducible gene from cytotoxic T lymphocytes and its identification as a lipase. *Cell* 60:451-459.
74. Hoffman, D.R., Dove, D.E., Moffitt, J.E., and Stafford, C.T., 1988, Allergens in Hymenoptera venom XXI. Cross-reactivity and multiple reactivity between fire ant venom and bee and wasp venoms. *J. Allergy Clin. Immunol.* 82:828-834.
75. Hoffman, D.R., 1993, Allergens in Hymenoptera venom XXIV: The amino acid sequences of imported fire ant venom allergens Sol i II, Sol i III and Sol i IV. *J. Allergy Clin. Immunol.* 91:71-78.
76. Hoffman, D.R., 1995, Imported fire ant venom allergen Sol i 1 is a phospholipase similar to vespid wasp venom phospholipases. *J. Allergy Clin. Immunol.* 95:372. (abstract)
77. Vesor, A.S., Korneev, Sh.I., Salikhov and Tuichibaev, M.U., 1989, [Amino acid sequence of orientotoxins I and II from the venom of the hornet *Vespa orientalis*]. *Bioorg. Khim.* 15, 127-9.
78. Wypych, J.I., Abeyounis, C.J., and Reisman, R.E., 1989, Analysis of differing patterns of cross-reactivity of honeybee and yellow jacket venom specific IgE. Use of purified venom fractions. *Int. Arch. Allerg. App. Immunol.* 89:60-66.
79. Jacobson, R.S., Hoffman, D.R., and Kemeny, D.M., 1992, The cross- reactivity between bee and vespid hyaluronidases has a structural basis. *J. Allergy Clin. Immunol* 89:292. (abstract)
80. Gmachl, M., and Kreil, G., 1993, Bee venom hyaluronidase is homologous to a membrane protein of mammalian sperm. *Proc. Nat. Acad. Sci. US* 90:3569-3573.
81. Lu, G., Kochoumian, L., and King, T.P., 1994, White face hornet venom allergen hyaluronidase: Cloning and its sequence similarity with other proteins. *J. Allergy Clin. Immunol.* 93:224. (abstract)
82. Lathrop, W.F., Carmichael, E.P., Myles, D.G., and Primakoff, P., 1990, cDNA cloning reveals the molecular structure of a sperm surface protein, PH-20, involved in sperm-egg adhesion and the wide distribution of its gene among mammals. *J. Cell Biol.* 111:2939-2949.
83. Gmachl, M., Sagan, S., Ketter, S., and Kreil, G., 1993, The human sperm protein PH-20 has hyaluronidase activity. *FEBS Lett* 336:545-548.
84. Abe, T., Kawai, N., and Niwa, A., 1982, Purification and properties of a presynaptically acting neurotoxin, Mandaratoxin, from hornet (*Vespa mandarinia*). *Biochemistry* 21:1693-1697.
85. Hoffman, D.R., 1993, Allergens in Hymenoptera venom XXV: The amino acid sequences of antigen 5 molecules and the structural basis of antigenic cross-reactivity. *J. Allergy Clin. Immunol.* 92:707-716.
86. Hoffman, D.R., Jacobson, R.S., and Zerboni, R., 1987, Allergens in Hymenoptera venom XIX. Allergy to *Vespa crabro*, the European hornet. *Int. Archs. Allergy Appl. Immunol.* 84:25-31.
87. Fang, K.S.Y., Vitale, M., Fehlner, P., and King, T.P., 1988, CDNA cloning and primary structure of a white face hornet venom allergen, Antigen 5. *Proc. Nat. Acad. Sci. US* 85:895-899.
88. Lu, G., Villalba, M., Coscia, M.R., Hoffman, D.R., and King, T.P., 1993, Sequence analysis and antigenic cross-reactivity of a venom allergen, antigen 5, from hornets, wasps and yellow jackets. *J. Immunol.* 150:2823-2830.
89. Hoffman, D.R., 1985, Allergens in hymenoptera venom XV: The immunologic basis of vespid venom cross-reactivity. *J. Allergy Clin. Immunol.* 75:611-613.
90. King, T.P., Moran, D., Wang, D.F., Kochoumian. L., and Chait, B.T., 1990, Structural studies of a hornet venom allergen antigen 5, Dol m V, and its sequence similarity with other proteins. *Protein Seq. Data Anal.* 3:263-266.
91. Lucas, J., Camacho-Henriquez, A., Lottspeich, F., Henschen, A., and Sanger, H.L., 1985, Amino acid sequence of the 'pathogenesis-related' leaf protein P14 from viroid infected tomato reveals a new type of structurally unfamiliar proteins. *EMBO J.* 4:2745-2749.
92. Cornelissen, B.J.C., Hooft van Huijsduijnen, R.A.M., Van Loon, L.C., and Bol, J.F., 1986, Molecular characterization of messenger RNAs for 'pathogenesis-related' proteins 1a, 1b and 1c induced by TMV infection of tobacco. *EMBO J.* 5:37-40.
93. Brooks, D.E., Means, A.R., Wright, E.J., Singh, S.P., and Tiver, K.K., 1986, Molecular cloning of the cDNA for androgen-dependent sperm-coating glycoproteins secreted by the rat epididymis. *Eur. J. Biochem.* 161:13-18.
94. Tomalski, M.D., King, T.P., and Miller, L.K., 1993, Expression of hornet genes encoding venom allergen antigen 5 in insects. *Arch. Insect Biol.* 22:303-313.
95. King, T.P., Kochoumian, L., and Lu, G., 1995, Murine T and B cell responses to natural and recombinant hornet venom allergen Dol m 5.02 and its recombinant fragments. *J. Immunol.* 154:577-584.

96. Hoffman, D.R., Smith, A.M., Schmidt, M., Moffitt, J.E., and Guralnick, M., 1990, Allergens in Hymenoptera venom XXII. Comparison of venoms from two species of imported fire ants, *Solenopsis invicta* and *richteri*. *J. Allergy Clin. Immunol.* 85:988-996.
97. Smith, A.M., and Hoffman, D.R., 1992, Further characterization of imported fire ant venom allergens. *J. Allergy Clin. Immunol.* 89:293. (abstract)
98. Hoffman, D.R., Shipman, W.H., and Babin, D., 1977, Allergens in Bee venom II. Two new high molecular weight allergenic specificities, *J. Allergy Clin. Immunol.* 59:147-153.
99. Hoffman, D.R., 1977, Allergens in bee venom III. Identification of allergen B as an acid phosphatase. *J. Allergy Clin. Immunol.* 59:364-366.
100. Marz, L., Kuhne, C., and Michl, H., 1983, The glycoprotein nature of phospholipase A2, hyaluronidase and acid phosphatase from honeybee venom. *Toxicon* 21:893-896.
101. Shkenderov, S.V., and Ivanova, I.V., 1987, Alpha glucosidase and acid phosphatase of bee venom - A new method of purification and partial characterization of the enzymes. *Dan. Bolg.* 40:63-6.
102. Barboni, E., Kemeny, D.M., Campos, S., and Vernon, C.A., 1987, The purification of acid phosphatase from honeybee venom (*Apis mellifica*). *Toxicon* 25:1097-1103.
103. Jacobson, R.S., and Hoffman, D.R., 1995, Honey-bee acid phosphatase is a member of the prostatic acid phosphatase family. *J. Allergy Clin. Immunol.* 95:372. (abstract)
104. Roiko, K., Jaenne, O.A., and Vihko, P., 1990, Primary structure of rat secretory acid phosphatase and comparison to other acid phosphatases. *Gene* 89:223-229.
105. Van Etten, R.L., Davidson, R., Stevis, P.E., MacArthur, H., and Moore D.L., 1991, Covalent structure, disulfide bonding, and identification of reactive surface and active site residues of human prostatic acid phosphatase. *J. Biol. Chem.* 266:2313-2319.
106. Sharief, F.S., Lee, H., Leuderman, M.M., Lundwall, A., Deaven, L.L., Lee, C., and Li, S.S.L., 1989, Human prostatic acid phosphatase: cDNA cloning, gene mapping and protein sequence homology with lysosomal acid phosphatase. *Biochem. Biophys. Res. Commun.* 160:79-86.
107. Vihko, P., Virkkunen, P., Henttu, P., Roiko, K., Solin, T., and Huhtala, M.L., 1988, Molecular cloning and sequence analysis of cDNA encoding human prostatic acid phosphatase. *FEBS Lett.* 236:275-281.
108. Tailor, P.G., Govindan, M.V., and Patel, P.C., 1990, Nucleotide sequence of human prostatic acid phosphatase determined from a full-length cDNA clone. *Nucleic Acids Res.* 18:4928.
109. Himeno, M., Fujita, H., Noguchi, Y., Kono, A., and Kato, K., 1989, Isolation and sequencing of a cDNA clone encoding acid phosphatase in rat liver lysosomes. *Biochem. Biophys. Res. Commun.* 162:1044-1053.
110. Geier, C., von Figura, K., and Pohlmann, R., 1991, Molecular cloning of the mouse lysosomal acid phosphatase. *Biol. Chem. Hoppe-Seyler* 372:301-304.
111. Muta, T., Hashimoto, R., Miyata, T., Nishimura, H., Toh, Y., and Iwanaga, S., 1990, Proclotting enzyme from horseshoe crab hemocytes. cDNA cloning, disulfide locations, and subcellular localization. *J. Biol. Chem.* 265:22426-22433.
112. Baba, T., Watanabe, K., Kashiwabara, S.I., and Arai, Y., 1989, Primary structure of human proacrosin deduced from its cDNA sequence. *FEBS Lett.* 244:296-300.
113. Watanabe, K., Baba, T., Kashiwabara, S., Okamoto, A., and Arai, Y., 1991 Structure and organization of the mouse acrosin gene. *J. Biochem.* 109:828-833.
114. Chung, D.W., Fujikawa, K., McMullen, B.A., and Davie, E.W., 1986, Human plasma prekallikrein, a zymogen to a serine protease that contains four tandem repeats. *Biochemistry* 25:2410-2417.
115. Fujikawa, K., Chung, D.W., Hendrickson, L.E., and Davie, E.W., 1986, Amino acid sequence of human factor XI, a blood coagulation factor with four tandem repeats that are highly homologous with plasma kallikrein. *Biochemistry* 25:2417-2424.
116. Weiss, S.J., Hoffman, D.R., and Stafford, C.T., 1990, Allergy to the native fire ant, *Solenopsis xyloni*. *J. Allergy Clin. Immunol.* 85:212. (abstract)
117. Ellis, M.H., Jacobson, R.S., and Hoffman, D.R., 1992, Allergy to *Solenopsis aurea*, an uncommon native fire ant. *J. Allergy Clin. Immunol.* 89:293. (abstract)
118. Schmidt, M., Walker, R.B., Hoffman, D.R., and McConnell, T., 1993, Nucleotide sequence of cDNA encoding the fire ant venom protein Sol i II. *FEBS Lett.* 319:138-140.
119. Habermann, E., and Jentsch, J., 1967, Sequenzanalyze des melittins aus den tryptischen and peptischen spaltsuchen. *Hoppe-Seyler Z. Physiol. Chem.* 348:37-50.
120. Habermann, E., and Reiz, K.G., 1965, Zur Biochemie der Bienengift peptide melittin und Apamin. *Biochem. Z.* 343:192-203.
121. Vogel, H., Jarig, R., Hoffman, V., and Stumpfl, J., 1983, The orientation of melittin in lipid membranes. *Biochim. Biophys. Acta* 733:201-209.

122. Argiolas, A., and Pisano, J.J., 1985, Bombolitins, a new class of mast cell degranulating peptides from the venom of the bumble bee *Megabombus pennsylvanicus*. *J. Biol. Chem.* 260:1437-1444.
123. Paull, B.R., Yunginger, J.W., and Gleich, G.J., 1977, Melittin: An allergen of honeybee venom. *J. Allergy Clin. Immunol.* 59:334-338.
124. Olah, G.A., and Huang, H.W., 1988, Conformation studies of model ion channels (alamethicin, melittin) in perfectly aligned hydrated lecithin multilayers. *Biophys. J.* 53:A314. (abstract)
125. Tosteson, M.T., Alvarez, O., Hubbell, W., Bieganski, R.M., Attenbacher, C., Caporale, L.H., Levy, J.J., Nutt, R.F., Rosenblatt, M., and Tosteson, D.C., 1990, Primary structure of peptides and ion channels - Role of amino acid side chains in voltage gating of melittin channels. *Biophys. J.* 58:1367-1375.
126. Alder, G.M., Arnold, W.M., Bashford, C.L., Drake, A.F., Pasternak, C.A., and Zimmerman, U., 1991, Divalent cation sensitive pores formed by natural and synthetic melittin and by Triton X-100. *Biochem. Biophys. Acta* 1061:111-120.
127. Raynor, R.L., Bin, Z., and Kuo, J.F., 1991, Membrane interactions of amphiphilic polypeptides mastoparan, melittin, polymyxin-B and cardiotoxin. Differential inhibition of protein kinase C, Ca2+ calmodulin dependent protein kinase II and synaptosomal membrane Na,K-ATPase and Na+ pump and differentiation of HL60 cells. *J. Biol. Chem.* 266:2753-2758.
128. Folleniuswund, A., Mely, A., and Gerard, D., 1987, Spectroscopic evidence of 2 melittin molecules bound to Ca2+- calmodulin. *Biochem. Int.* 15:823-33.
129. Kataoka, M., Head, J.F., Seaton, B.A., and Engelman, D.M., 1989, Melittin causes a large calcium dependent conformational change in calmodulin. *Proc. Nat. Acad. Sci. USA* 86:6944-6948.
130. Argiolas, A., and Pisano, J.J., 1984, Isolation and characterization of two new peptides mastoparan C and crabrolin from the venom of the European hornet, *Vespa crabro*. *J. Biol. Chem.* 259:10106-10111.
131. Hirai, Y., Kuwada, M., Yasuhara, T., Yoshida, H., and Nakajima, T., 1979, A new mast cell degranulating peptide homologous to mastoparan in the venom of the Japanese hornet (*Vespa xanthoptera*) *Chem. Pharm. Bull.* 27:1945-1946.
132. Hirai, Y., Yasuhara, T., Yoshida, H., Nakajima, T., Fujino, M., and Kitada, C., 1979, A new mast cell degranulating peptide mastoparan in the venom of *Vespula lewisii*. *Chem. Pharm. Bull.* 27:1942-1944.
133. Hirai, Y., Yasuhara, T., Yoshida, H., and Nakajima, T., 1981, A new mast cell degranulating peptide, *Polistes* mastoparan in the venom of *Polistes jadwigae*. *Biomed. Res.* 1:185-187.
134. Kurihara, H., Kitajima, K., Senda, T., Fujita, H., and Nakajima, T., 1986, Multigranular exocytosis induced by phospholipase A2 activators, melittin and mastoparan in rat anterior pituitary cells. *Cell and Tissue Research* 243:311-6.
135. Higashijima, T., Wakamatsu, K., Takemitsu, M., Fujino, M., Nakajima, T., and Miyazawa, T., 1983, Conformational change of mastoparan from wasp venom on binding with phospholipid membrane. *FEBS Lett.* 152:227-230.
136. Kishimura, H., Yasuhara, T., Yoshida, H. and Nakajima, T., 1976, Vespakinin-M, a novel bradykinin analogue containing hydroxyproline, in the venom of *Vespa mandarinia* Smith. *Chem. Pharm. Bull.* 24:2896-2897.
137. Yoshida, H., Geller, R.G., and Pisano, J.J., 1976, Vespulakinins: New carbohydrate containing bradykinin derivatives. *Biochemistry* 15:61-64.
138. Nakajima, T., Uzu, S., Wakamatsu, K., Saito, K., Miyazawa, T., Yasuhara, T., Tsukamoto, T., and Fujino. M., 1986, Amphiphilic peptides in wasp venom. *Biopolymers* 25:S115. (abstract)
139. Sanyal, G., Richard, L.M., Carraway, K.L., and Puett, D., 1988, Binding of amphiphilic peptides to a carboxy terminal tryptic fragment of calmodulin. *Biochemistry* 27:6229-6236.
140. Nakajima, T., 1993, Arthropod venoms affecting cell signaling system. *Regulatory Peptides* 46:89-92.
141. Donovan, G.R., Baldo, B.A., and Sutherland, S., 1993, Molecular cloning and characterization of a major allergen Myr p I from the venom of the Australian jumper ant, *Myrmecia pilosula*. *Biochem. Biophys. Acta* 1171:272-280.
142. Donovan, G.R., Street, M.D., Baldo, B.A., Alewood, D., Alewood, P., and Sutherland, S., 1994, Identification of an IgE binding determinant of the major allergen Myr p I from the venom of the Australian jumper ant *Myrmecia pilosula*. *Biochim. Biophys. Acta* 1204:48-52.
143. Donovan, G.R., Baldo, B.A., Street, M.D., and Sutherland, S.K., 1993, Molecular cloning and allergenic determinant tructure of major venom allergens Myr p I and Myr p II from the jumper ant *Myrmecia pilosula*. *J. Leuk. Biol.* suppl:95. (abstract)
144. Vlasak, R., Unger-Ullmann, C., Kreil, G., and Frischauf, A.M., 1983, Nucleotide sequence of cloned cDNA coding for honeybee prepromelittin. *Eur. J. Biochem.* 135:123-126.
145. Mollay, C., Vilas, U., and Kreil, G., 1982, Cleavage of honeybee prepromelittin by an endoprotease from rat liver microsomes, identification of intact signal peptide. *Proc. Nat. Acad. Sci. US* 79:2260-2264.

146. Suchanek, G., and Kreil, G., 1977, Translation of melittin messenger RNA in vitro yields a product terminating with glutaminyl-glycine rather than with glutaminamide. *Proc. Nat. Acad. Sci. US* 74:975-978.
147. Kreil, G., and Bachmayer, H., 1971, Biosynthesis of melittin, a toxic peptide from bee venom, detection of a possible precursor. *Eur. J. Biochem.* 20:344-350.
148. Kreil, G., 1973, Biosynthesis of melittin, a toxic peptide from bee venom, amino acid sequence of the precursor. *Eur. J. Biochem.* 33:558-566.
149. Kreil, G., Mollay, C., Kaschnitz, R., Haiml, L., and Vilas, U., 1980, Prepromelittin: Specific cleavage of the pre- and propeptide in vivo. *Ann. N. Y. Acad. Sci.* 343: 338-346.
150. Holldobler, B., and Wilson, E.O., 1990, *The Ants.* Cambridge Mass. Belknap Press of Harvard University Press.
151. Bischoff, H., 1927, *Biologie der Hymenopteran*, Berlin
152. Maschwitz, U.W.J., and Kloft, W., 1971, Morphology and function of the venom apparatus of insects-bees, wasps, ants and caterpillars. in Bucherl, W., and Buckley, E.E., *Venomous Animals and their Venoms* Vol III Academic Press New York, p.1-60.
153. Fenger, H., 1863, Anatomie and Physiologie des Giftapparatus bei den Hymenopteren. *Arch. Naturgesch.* 29:139-178.
154. Robertson, P.L., 1968, A morphological and functional study of the venom apparatus in representatives of some major groups of Hymenoptera. *Aust. J. Zool.* 16:133-166.
155. Van Marle, J., and Piek, T., 1986, Morphology of the venom apparatus. in Piek, T., (ed) *Venoms of the Hymenoptera.* Academic Press, Orlando, p.17-44.
156. Lacaze-Duthiers, H., 1849, Recherches sur l'armure genitale des insectes. *Ann. Nat. Sci. Zool.* [3] 12:353-374.
157. Zander, E., 1899, Beitrage zur Morphologie des Stachapparates der Hymenopteran. *Z. Wiss. Zool.* 66:289-333.
158. Siebold, C.T.E., 1871, *Beitrage zur Parthenogenesis der Arthropoden. I. Ueber die bei Polistes Wahrzunehmende Parthogenesis II. Pathenogenesis bei Vespa holsatica.* Leipzig, Wilhelm Engelmann.
159. Marchal, P., 1896, La reproduction et l'evolution des guepes sociales. *Arches. Zool. exp. gen.* 4:1-100.

11

STRUCTURE AND FUNCTIONS OF COAGULATION FACTOR IX/FACTOR X-BINDING PROTEIN ISOLATED FROM THE VENOM OF *Trimeresurus flavoviridis*

Takashi Morita, Hideko Atoda, and Fujio Sekiya

Department of Biochemistry
Meiji College of Pharmacy
1-22-1, Yato-cho, Tanashi, Tokyo 188, Japan

INTRODUCTION

Snake venoms contain various anticoagulants that affect blood coagulation system (1, 2). The anticoagulant is classified into two categories: anticoagulant enzymes and non-enzymatic anticoagulants. The examples of the former category are phospholipases, fibrinogenolytic enzymes, protein C activators, and proteolytic enzymes that convert coagulation factors into degraded inactive forms. An interesting example of the non-enzymatic anticoagulants is an inhibitor of the activation of prothrombin. The inhibitors in this category have been first found in the venom of *Agkistrodon acutus* (3) and *Trimeresurus gramineus* (4). The anticoagulant protein isolated from *A. acutus* inhibits the participation of factor Xa in the prothrombinase complex formation (5, 6). In order to understand the anticoagulant mechanism we have isolated and characterized another anticoagulant protein with *M*r 27,000 from the venom of the habu snake *Trimeresurus flavoviridis* (7). This anticoagulant protein forms a complex with either coagulation factor IX or factor X with a stoichiometry of 1 to 1 in a calcium-dependent fashion, and thereby blocks the amplification of the coagulation cascade (7). Thus, we named it IX/X-bp (factor IX/factor X-binding protein). This chapter describes the structural and functional properties of IX/X-bp from the venom of the habu snake (*T. flavoviridis*).

ISOLATION OF HABU IX/X-BP

Habu IX/X-bp as anticoagulant protein has been isolated originally from the crude venom of *T. flavoviridis* using an affinity chromatography column of insolubilized bovine factor X (7). We are now able to prepare this protein in a large scale by ion-exchange

Natural Toxins II, Edited by B. R. Singh and A. T. Tu
Plenum Press, New York, 1996

chromatographies of S-Sepharose Fast Flow and Q-Sepharose Fast Flow. Twenty mg of the protein are usually isolated from 500 mg of the crude venom.

STRUCTURAL PROPERTIES

Analysis of isolated IX/X-bp by sodium dodecyl sulfate-polyacrylamide gel electrophoresis has revealed a 27.0-kDa band under unreduced condition and this anticoagulant protein gives a 16.0-kDa band (designated A chain) and a 15.5-kDa band (designated B chain) under reduced condition. *S*-Pyridylethylated A and B chains have been separated by reversed-phase HPLC and their complete amino acid sequences are determined by sequencing the peptides obtained by digestion with lysyl endopeptidase, chymotrypsin, or *S. aureus* V8 protease and chemical cleavage with cyanogen bromide (8). The A chain has the amino-terminal sequence of Asp-Cys-Leu-Ser-Gly- and consists of 129 amino acid residues with a *M*r 14,830, and B chain has the amino-terminal sequence of Asp-Cys-Pro-Ser-Asp- and consists of 123 amino acid residues with a *M*r 14,400, as shown in Fig. 1. There is 47 % identity between A chain and B chain. The sequence of IX/X-bp shows 25-37% identity with that of the calcium-dependent (C-type) carbohydrate recognition domain (CRD)-like structure of acorn barnacle lectin, human and rat asialoglycoprotein receptors, proteoglycan core protein, pancreatic stone protein, and tetranectin (8). CRDs consist of approximately 130 amino acid residues, of which about 20 are invariant or well conserved. The amino acid sequence of IX/X-bp and the location of seven disulfide bridges in the molecule of IX/X-bp are shown in Fig. 2 (9). IX/X-bp is a heterogeneous two-chain protein linked by a single disulfide bond and each chain contains three intrachain disulfide bonds.

There are totally seven proteins containing CRDs of which the primary structure and the pattern of disulfide bridges have been determined to date. Fig. 3 shows the disulfide-bonding patterns of IX/X-bp and six other CRDs from, respectively, pancreatic stone protein (10), acorn barnacle lectin (11), echinoidin (12), galactose-specific lectin from *Crotalus atrox* (rattlesnake) venom (13), proteoglycan core protein (14), and tetranectin (15). All these domains share key structural features, including the presence of a small loop at the amino-terminal region and another small loop within a large loop in the carboxy-terminal region. Intrachain disulfide-bonding patterns of both the A and the B chains of IX/X-bp are similar to those of CRDs, indicating that IX/X-bp consists of two covalently linked, heterogeneous, CRD-like subunits. The chain structures of proteins that include CRD-like subunits can be classified into four types. First, pancreatic stone protein has a single CRD. Second, there are covalently linked two-chain proteins which consist of two homogeneous CRD-like subunits, such as acorn barnacle lectin, echinoidin, and the lectin from *C. atrox*. As shown in Fig. 3, the locations of interchain disulfide bridges in these proteins are different.

```
          1        10        20        30        40
A chain   DCLSGWSSYEGHCYKAFEKYKTWEDAERVCTEQAKGAHLVSIES
B chain   DCPSDWSSYEGHCYKPFSEPKNWADAENFCTQQHAGGHLVSFQS

              50        60        70        80
A chain   SGEADFVAQLVTQNMKRLDFYIWIGLRVQGKVKQCNSEWSDGSS
B chain   SEEADFVVKLAFQTFGHSIF--WMGLSNVWN--QCNWQWSNAAM

           90       100       110       120      129
A chain   VSYENWIEAESKTCLGLEKETDFRKWVNIYCGQQNPFVCE--A
B chain   LRYKAWAE-ES-YCVYF-KSTNN-KWRSRACRMMAQFVCEFQA
```

Figure 1. Sequence homology between the A chain and the B chain of habu IX/X-bp.

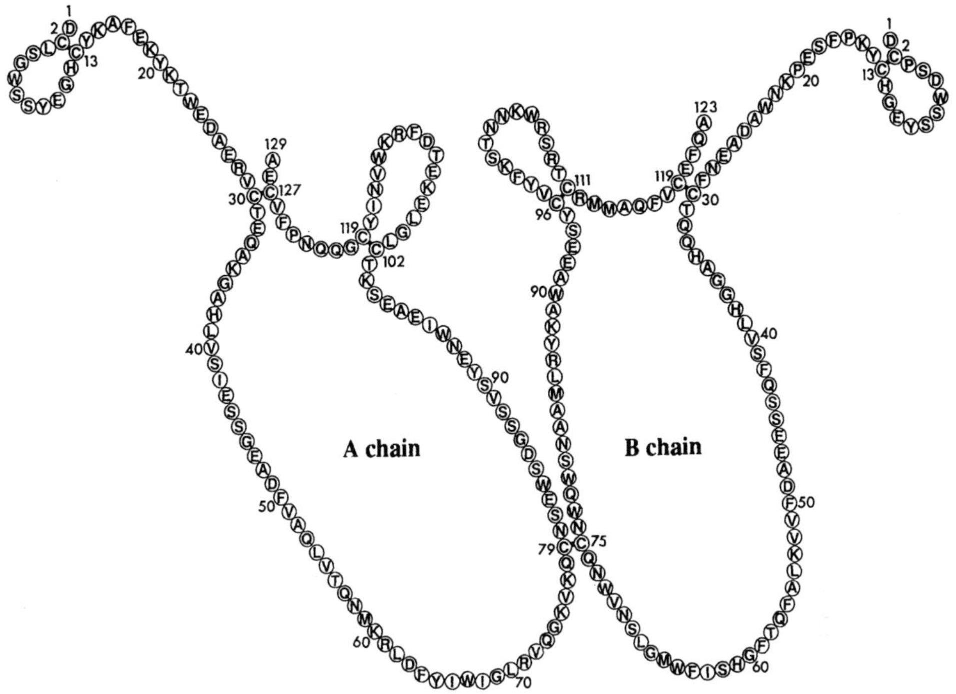

Figure 2. Schematic representation of habu IX/X-bp.

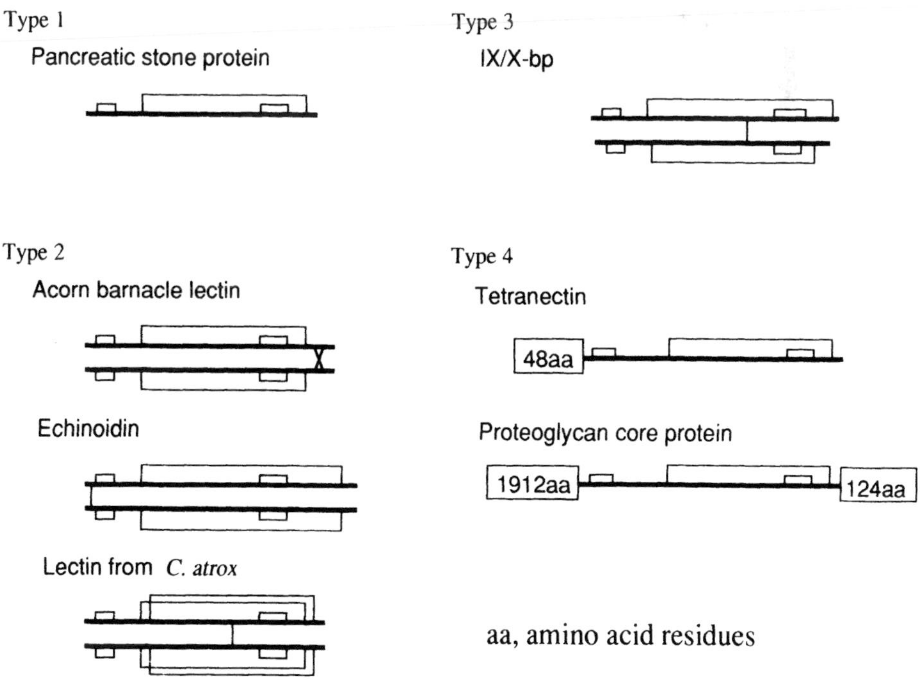

Figure 3. Disulfide-bonding patterns of CRD-like subunits in C-type lectin-like proteins.

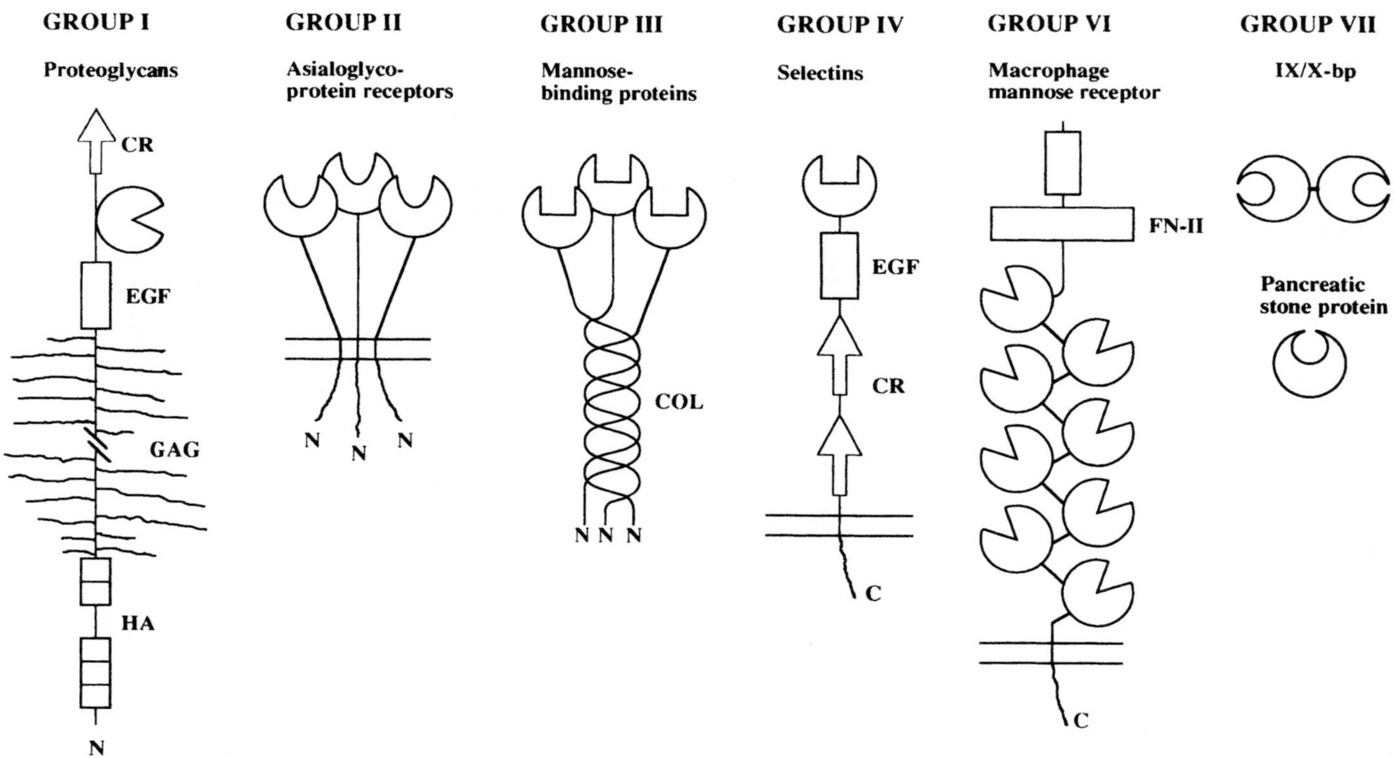

CR: complement-regulatory repeat,
EGF: epidermal growth factor-like domains,
GAG: glycosaminoglycan attachment sites,
HA: hyaluronic acid-binding domains,
COL: collagen-like sequences,
FN-II: fibronectin type-II repeats

Figure 4. Summary of the structures of several groups of C-type animal lectins. Figure 1 in reference 16 is modified.

IX/X-bp is a third type of protein and consists of two heterogeneous CRD-like subunits. There is also a fourth type of protein, exemplified by proteoglycan core protein and tetranectin, and these proteins contain distinct functional group(s) in addition to CRDs.

Drickamer summarizes the structures of several groups of C-type animal lectins as shown in Fig. 4 (16). The C-type lectins are the most diverse in overall organization and C-type CRDs are found in association with many other domains (16). Numerous proteins that contain the CRD-like structure have been isolated from mammalian tissues as well as snake venoms and various invertebrates. Most such mammalian proteins are integrated in membranes or are tightly associated with the extracellular matrix; they have relatively large structural elements in addition to CRDs; and some are composed of multiple subunits. For example, mammalian asialoglycoprotein receptor (hepatic lectin) in Group II is a heterohexameric protein consisting of two different but very similar polypeptides. Each of the polypeptides has a carboxy-terminal CRD region and a rather long non-CRD extension that includes trans-membrane and intracellular regions. Although physiological significance of these mammalian proteins has been recognized and many researchers have undertaken extensive investigations, the abundance of non-CRD regions and the low solubility of these proteins in aqueous solutions, as well as the difficulties encountered in preparing them in large quantities, have all hampered efforts to elucidate structure/function relationships of these proteins. However, IX/X-bp, a venom binding-protein in Group VII, consists simply of isolated free CRDs (16). IX/X-bp is a soluble protein without any posttranslationally modified amino acid residues (8). It can be easily obtained by conventional chromatographic procedures in large quantities. In addition, IX/X-bp from habu snake is able to crystallize by the hanging drop vapor diffusion method from 60% saturated ammonium sulfate (17). Because of these advantageous features of IX/X-bp, we believe that it should provide an excellent model system for gaining insights into the general nature of the binding of CRDs to their ligands.

FUNCTIONAL PROPERTIES

A simplified diagram of the cascade of mammalian blood coagulation is given in Fig. 5. A complex of coagulation initiator factor VIIa, tissue factor, phospholipids, and calcium ions, activates factor IX. Then, factor X and prothrombin are sequentially activated to factor Xa and alpha-thrombin, respectively. These coagulation factors such as factor VII, factor IX, factor X, and prothrombin are vitamin K-dependent coagulation factors, and contain gamma-

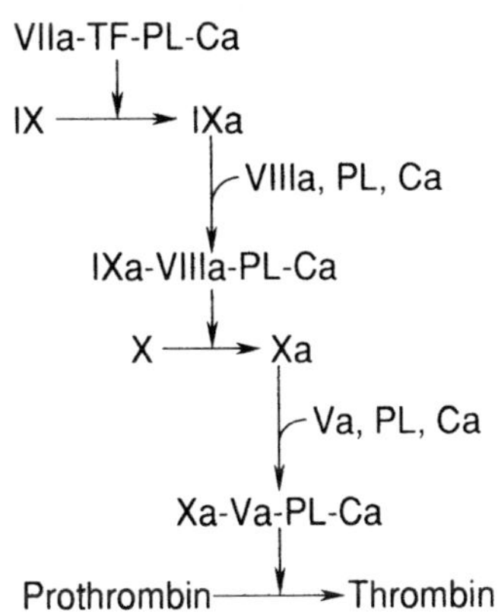

Figure 5. Cascade of mammalian blood coagulation. TF: tissue factor; PL: phospholipids.

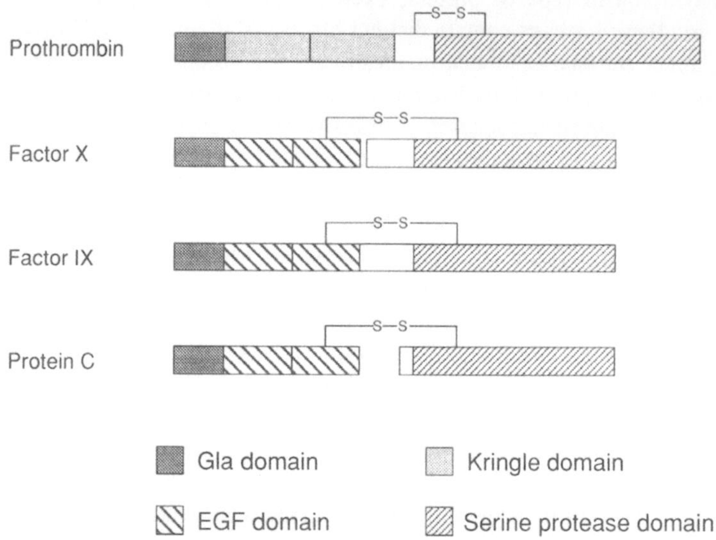

Figure 6. Chain structures of blood coagulation factors.

carboxyglutamic acid (Gla) residues. IX/X-bp binds to factor IX and factor X as well as factor IXa and factor Xa. The polypeptide chains of four typical vitamin K-dependent coagulation factors are shown in Fig. 6. These coagulation factors have a Gla domain in their amino-terminal part followed by a tandem repeat of Kringle domains or EGF domains, and a serine protease domain in their carboxy-terminal part. Gla domains of these factors are especially important structure for their function.

Affinities of Binding of Habu IX/X-Bp to Factor IX and Factor X

Binding of IX/X-bp to factors IX and X is dependent on the concentration of calcium ions. Half-maximal binding to factors IX and X of IX/X-bp occurs at 0.39 mM and 0.35 mM calcium ions, respectively. Maximal binding is observed at 1 mM calcium ions. The binding affinity of IX/X-bp for factor IX is compared with that of IX/X-bp for factor X in the presence of 1 mM calcium ions (18). IX/X-bp binds to factor IX at lower concentrations than to factor X. The apparent Kd values for binding of IX/X-bp to solid-phase bovine factors IX and X are calculated to be 0.36 nM and 1.13 nM, respectively (Table I). IX/X-bp binds to the activated forms of factors IX and X (bovine factor IXa and bovine factor Xa) in a similar manner. The binding affinities of IX/X-bp for human factors IX and X are also essentially the same. IX/X-bp does not bind to solid-phase bovine factor VII or to other vitamin K-dependent Gla-containing proteins, such as bovine prothrombin, protein C, protein Z, and human protein C at the concentrations tested (up to 670 nM).

Localization of the Binding Site of IX/X-Bp on Coagulation Factor IX and Factor X

To localize the binding site of IX/X-bp on factor IX, binding of IX/X-bp to Gla domain peptide of factor IX and Gla-domainless factor IXa has been examined (18). IX/X-bp does bind to the solid-phase Gla domain peptide of factor IX in the presence of 1 mM calcium ions, but it does not bind to Gla-domainless factor IXa. The binding affinity of IX/X-bp to the Gla domain

Table I. Apparent Kd values for the binding of IX/X-bp to the solid-phase coagulation factors

Proteins	Apparent Kd values[a] (nM)
factor IX	0.36 ± 0.14 (n= 5)
factor X	1.13 ± 0.37 (n= 6)
Gla domain peptide of factor IX	9.19 ± 1.90 (n= 7)

[a] Parameters are expressed as mean ± S. E.

peptide is greatly lower than that to intact factor IX (Table I). For the inhibition of IX/X-bp binding to factors IX and X by the isolated Gla domain peptide, higher concentrations of calcium are required. In the presence of 5 mM calcium, the IC_{50} of factor IX Gla domain peptide in the fluid-phase for the binding of IX/X-bp to solid-phase factor IX was 95 nM whereas that of intact factor IX was 1.4 nM. In the case of factor X, the IC_{50} of the Gla domain peptide of factor X for the binding of IX/X-bp to factor X was 400 nM whereas that of intact factor X was 2.2 nM. PCGFX is a recombinant human protein C whose Gla domain (residues 1-43) has been replaced by the corresponding portion of human factor X (19). IX/X-bp bound to PCGFX with the binding affinity about a tenth of that for intact human factor X. IX/X-bp bound to PCGFX with greatly higher affinity than to isolated Gla-domain peptide. From these observations, it appears that IX/X-bp is a protein that recognizes Gla domain region, the amino-terminal part in factors IX and X, in the presence of calcium ions.

IX/X-BPs AND THE STRUCTURALLY RELATED PROTEINS

We have isolated three IX/X-bps from the venoms of various crotalid snakes, namely *T. flavoviridis* (7), *Deinagkistrodon acutus* (20), and *Bothrops jararaca* (21). The amino-terminal amino acid sequences of these proteins reveal an apparent homology between the amino-terminal regions of these proteins (Fig. 7).

There are other structurally related but functionally distinct proteins isolated from the snake venoms (Fig. 7). The properties of IX/X-bps and the structurally related proteins

			1 10 20
habu IX/X-bp	A chain		D C L S G W S S Y E G H C Y K A F E K Y K
	B chain		D C P S D W S S Y E G H C Y K P F S E P K
acutus IX/X-bp	A chain		D C S S G W S S Y E G H C Y K V F K Q S K
	B chain		D C P S D W S S Y E G H C Y K P F N E P K
jararaca IX/X-bp	A chain		D C P S D W S P Y E G H C Y K H F I K W M
	B chain		D C P S D W S P Y E G H C Y R V F T E P Q
botrocetin	α chain		D C P S G W S S Y E G N C Y K F F Q Q K M
	β chain		D C P P D W S S Y E G H C Y R F F K E W M
bothrojaracin	15 kDa		D C P S D W S P Y G Q G C Y X F
	13 kDa		D C P X D W S X Y
RVV-X L chain (LCl)		V L	D C P S G W L S Y E Q H C Y K G F N D L K
alboaggregin-B	peak 2		D C P S D W S S F K Q Y C Y Q I I K Q L K
	peak 1		D C P S D W S S Y D L Y C Y R V F Q E K K
echicetin	β chain		N C L P D W S V Y E G Y C Y K V F K E R M
C. atrox lectin		N	N C P L D W L P N M G L C Y K I F N Q L K

Figure 7. Amino-terminal sequences of IX/X-bps and other related venom proteins.

Table II. Summary of IX/X-bps and structurally related proteins.
T., Trimeresurus, D., Deinagkistrodon, B., Bothrops, E., Echis, C., Crotalus

Proteins	Properties	Ligands	Origin
habu IX/X-bp	anticoagulant	factors IX/X	*T. flavoviridis*
acutus IX/X-bp	anticoagulant	factors IX/X	*D. acutus*
jararaca IX/X-bp	anticoagulant	factors IX/X	*B. jararaca*
bothrojaracin	hirudin-like anticoagulant	α-thrombin	*B. jararaca*
botrocetin	platelet agglutination inducer	von Willebrand factor	*B. jararaca*
alboaggregin-B	platelet agglutination inducer	platelet gbIb	*T. albolabris*
echicetin	platelet agglutination inhibitor	platelet gbIb	*E. carinatus*
lectin	C-type lectin	galactose	*C. atrox*

are summarized in Table II. Recently, it is reported that the sequences of the heterodimeric form of botrocetin exhibit an apparent homology to that of C-type lectins (22). Botrocetin is a von Willebrand factor-dependent platelet coagglutinin (agglutination inducer) which is isolated from the venom of *B. jararaca* (23). The hirudin-like anticoagulant protein bothrojaracin is also isolated from the venom of *B. jararaca* (24). In addition, the sequences have been reported of several proteins with C-type lectin-like structures that have been isolated from the venom of Crotalidae snakes, such as a light chain of coagulation factor X-activating enzyme (RVV-X) from the venom of *Vipera russelli* (Russell's viper) (25) and a lectin from the venom of *C. atrox* (13). *C. atrox* lectin is a C-type galactose-specific lectin composed of two covalently-linked identical CRDs (13).

The extent of the identity of the amino acid sequences between the alpha chain of botrocetin and the A chain of IX/X-bp is 61% and that between the beta chain of botrocetin and the B chain of IX/X-bp is 51% (22; Fig. 8). The extent of the identity between the A chain of IX/X-bp and the RVV-X light chain and between the A chain and *C. atrox* lectin is

```
                    1        10        20        30        40       50
IX/X-bp A chain     DCLSGWSSYEGHCYKAFEKYKTWEDAERVCTEQAKGAHLVSIES-SGEADF
IX/X-bp B chain     DCPSDWSSYEGHCYKPFSEPKNWADAENFCTQQHAGGHLVSFQS-SEEADF
Botrocetin β        DCPPDWSSYEGHCYRFFKEWMHWDDAEEFCTEQQTGAHLVSFQS-KEEADF
Botrocetin α        DCPSGWSSYEGNCYKFFQQKMNWADAERFCSEQAKGGHLVSIKIYSKEKDF
RVV-X L chain     VLDCPSGWLSYEQHCYKGFNDLKNWTDAEKFCTEQKKGSHLVSLHS-REEEEF
Echicetin β         NCLPDWSVYEGYCYKVFKERMNWADAEKFCMKQVKDGHLVSFRN-SKEVDF

                    51       60        70        80        90       100
IX/X-bp A chain     VAQLVTQNMKRLDFYIWIGLRVQGKVKQCNSEWSDGSSVSYENWIEAESK
IX/X-bp B chain     VVKLAFQTFGH--SIFWMGLSNVWN--QCNWQWSNAAMLRYKAWAE-ES-
Botrocetin β         VRSLTSEMLKG--DVVWIGLSDVWN--KCRFEWTDGMEFDYDDYYLIAFY
Botrocetin α        VGDLVTKNIQSSDLYAWIGLRVENKEKQCSSEWSDGSSVSYENVVERTVK
RVV-X L chain       VVNLISENLEY--PATWIGLGNMWK--DCRMEWSDRGNVKYKALAE-ESY
Echicetin β         MISLAFPMLKM--ELVWIGLSDYWR--DCYWEWSDGAQLDYKAWDN-ERH

                    101      110       120      129
IX/X-bp A chain     TCLGLEKETDFRKWVNIYCGQQNPFVCE--A
IX/X-bp B chain     YCVYF-K-STNNKWRSRACRMMAQFVCEFQA
Botrocetin β        ECVAS-K-PTNNKWWIIPCTRFKNFVCEFQA
Botrocetin α        KCFALEKDLGFVLWINLYCAQKNPFVCKSPPP
RVV-X L chain       -CLIM-I-THEKEWKSMTCNFIAPVVCKF
Echicetin β         -CFAA-K-TTDNQWMRRKCSGEFYFVCKCPA
```

Figure 8. Alignment of the complete amino acid sequence of habu IX/X-bp and those of CRD-like subunits of venom proteins.

41% and 30%, respectively. The B chain of IX/X-bp and the RVV-X light chain are 45% identical and *C. atrox* lectin and the B chain are 26% identical. A platelet agglutination inducer alboaggregin-B (26, 27) and a platelet agglutination inhibitor echicetin (28) are isolated from the venoms of *T. albolabris* and *Echis carinatus*, respectively. The primary structure of the alpha chain in echicetin molecule is not yet published, but the similarity with C-type lectin CRD in the beta chain of echicetin is indicated (29). It is interesting that these structurally related proteins, which probably share a common origin, have a variety of functions that differ from one another.

CONCLUSION

From all the results described above, it is apparent that IX/X-bp recognizes Gla domain regions of factors IX/X in a calcium-dependent fashion. The structural analyses of IX/X-bp reveal to exhibit similarity to that of carbohydrate-recognition domains (CRDs) of C-type lectins. It is noteworthy that CRD-like structures are being identified in more and more proteins and the functions of such structures are not restricted to the recognition of carbohydrates. These observations suggest that the CRD-like structure is a fundamental motif in proteins that has evolved from a single ancestral protein and now exhibits very divergent features. Moreover, increasing numbers of proteins that are structurally related to IX/X-bp and have interesting biological activities, which are totally different from that of IX/X-bp, have recently been isolated from snake venoms. Thus, evaluation of the properties of IX/X-bp should contribute to our understanding of the unique pharmacological actions of these venom proteins.

ACKNOWLEDGEMENTS

This work was supported in part by a Grant-in-Aid for Scientific Research from the Ministry of Education, Science and Culture of Japan.

REFERENCES

1. Tu, A. T., 1977, *Venom: Chemistry and Molecular Biology*, John Wiley and Sons, New York.
2. Pirkle, H., and Markland, F. S. Jr., 1987, *Hemostasis and Animal Venoms*, Marcel Dekker, New York.
3. Ouyang, C., and Teng, C.-M., 1972, Purification and properties of the anticoagulant principle of *Agkistrodon acutus* venom, *Biochim. Biophys. Acta* 278: 155-162.
4. Ouyang, C., and Yang, F.-Y., 1975, Purification and properties of the anticoagulant principle of *Trimeresurus gramineus* venom, *Biochim. Biophys. Acta* 386: 479-492.
5. Ouyang, C., and Teng, C.-M., 1973, The effect of the purified anticoagulant principle of *Agkistrodon acutus* venom on blood coagulation, *Toxicon* 11: 287-292.
6. Teng, C. -M., and Seegers, W. H., 1981, *Agkistrodon acutus* snake venom inhibits prothrombinase complex formation, *Thromb. Res.* 23: 255-263.
7. Atoda, H., and Morita, T., 1989, A novel blood coagulation factor IX/factor X-binding protein with anticoagulant activity from the venom of *Trimeresurus flavoviridis* (habu snake):Isolation and characterization, *J. Biochem.* (*Tokyo*) 106: 808-813.
8. Atoda, H., Hyuga, M., and Morita, T., 1991, The primary structure of coagulation factor IX/factor X-binding protein isolated from the venom of *Trimeresurus flavoviridis*. Homology with asialoglycoprotein receptors, proteoglycan core protein, tetranectin, and lymphocyte Fc epsilon receptor for immunoglobulin E, *J. Biol. Chem.* 266: 14903-14911.
9. Atoda, H., and Morita, T., 1993, Arrangement of the disulfide bridges in a blood coagulation factor IX/factor X-binding protein from the venom of *Trimeresurus flavoviridis*, *J. Biochem.* (*Tokyo*) 113: 159-163.

10. Rouimi, P., De Caro, J., Bonicel, J., Rovery, M., and De Caro, A., 1988, The disulfide bridges of the immunoreactive forms of human pancreatic stone protein isolated from pancreatic juice, *FEBS Lett.* 229: 171-174.
11. Muramoto, K., and Kamiya, H., 1990, The positions of the disulfide bonds and glycosylation site in a lectin of the acorn barnacle *Megabalanus rosa, Biochim. Biophys. Acta* 1039: 52-60.
12. Giga, Y., Ikai, A., and Takahashi, K., 1987, The complete amino acid sequence of echinoidin, a lectin from the coelomic fluid of the sea urchin *Anthocidaris crassispina, J. Biol. Chem.* 262: 6197-6203.
13. Hirabayashi, J., Kusunoki, T., and Kasai, K., 1991, Complete primary structure of a galactose-specific lectin from the venom of rattlesnake *Crotalus atrox. J. Biol. Chem.* 266: 2320-2326.
14. Sandy, J. D., Flannery, C. R., Boynton, R. E., and Neame, P. J., 1990, Isolation and characterization of disulfide-bonded peptides from the three globular domains of aggregating cartilage proteoglycan, *J. Biol. Chem.* 265: 21108-21113.
15. Fuhlendorff, J., Clemmensen, I., and Magnusson, S., 1987, Primary structure of tetranectin, a plasminogen kringle 4 binding plasma protein: Homology with asialoglycoprotein receptors and cartilage proteoglycan core protein, *Biochemistry* 26: 6757-6764.
16. Drickamer, K., 1993, Evolution of Ca^{2+}-dependent animal lectins. *Prog. Nucleic Acid Res. and Mol. Biol.*, 45: 207-232.
17. Mizuno, H., Atoda, H., and Morita, T., 1991, Crystallization and preliminary X-ray study of blood coagulation factor IX/factor X-binding protein with anticoagulant activity from habu snake venom, *J. Mol. Biol.* 220: 225-226.
18. Atoda, H., Yoshida, N., Ishikawa, M., and Morita, T., 1994, Binding properties of the coagulation factor IX/factor X-binding protein isolated from the venom of *Trimeresurus flavoviridis, Eur. J. Biochem.* 224: 703-708.
19. Ohsawa, K., Kimura, M., Kurosawa-Ohsawa, K., Takahashi, M., Koyama, M., Abiko, Y., Hirahara, K., Matsuishi, T., and Tanaka, S., 1988, Purification of sufficiently gamma-carboxylated recombinant protein C and its derivatives. Calcium-dependent affinity shift in immunoaffinity and ion exchange chromatography, *J. Chromatography* 597: 285-291.
20. Morita, T., and Atoda, H., 1993, The structural characterization of coagulation factor IX/factor X-binding protein isolated from the venom of *Trimeresurus flavoviridis, Currents Aspects of Blood Coagulation, Fibrinolysis, and Platelets* (Shen, M.-C., Teng, C.-M., and Takada, A. eds.) pp.35-40, Springer-Verlag, Tokyo.
21. Sekiya, F., Atoda, H., and Morita, T., 1993, Isolation and characterization of an anticoagulant protein homologous to botrocetin from the venom of *Bothrops jararaca, Biochemistry* 32: 6892-6897.
22. Usami, Y., Fujimura, Y., Suzuki, M., Ozeki, Y., Nishio, K., Fukui, H., and Titani, K., 1993, Primary structure of two-chain botrocetin, a von Willebrand factor modulator purified from the venom of *Bothrops jararaca, Proc. Natl. Acad. Sci. USA* 90: 928-932.
23. Fujimura, Y., Titani, K., Usami, Y., Suzuki, M., Oyama, T., Matsui, T., Fukui, H., Sugimoto, M., and Ruggeri, Z. M., 1991, Isolation and chemical characterization of two structurally and functionally distinct forms of botrocetin, the platelet coagglutinin isolated from the venom of *Bothrops jararaca, Biochemistry* 30: 1957-1964.
24. Zingali, R. B., Jandrot-Perrus, M., Guillin, M.-C., and Bon, C., 1993, Bothrojaracin, a new thrombin inhibitor isolated from *Bothrops jararaca* venom: Characterization and mechanism of thrombin inhibition, *Biochemistry* 32: 10794-10802.
25. Takeya, H., Nishida, S., Miyata, T., Kawada, S., Saisaka, Y., Morita, T., and Iwanaga, S., 1992, Coagulation factor X activating enzyme from Russell's viper venom (RVV-X). A novel metalloproteinase with disintegrin (platelet aggregation inhibitor)-like and C-type lectin-like domains, *J. Biol. Chem.* 267: 14109-14117.
26. Peng, M., Lu, W., and Kirby, E. P., 1991, Alboaggregin-B: A new platelet agonist that binds to platelet membrane glycoprotein Ib, *Biochemistry* 30: 11529-11536.
27. Yoshida, E., Fujimura, Y., Miura, S., Sugimoto, M., Fukui, H., Narita, N., Usami, Y., Suzuki, M., and Titani, K., 1993, Alboaggregin-B and botrocetin, two snake venom proteins with highly homologous amino acid sequences but totally distinct functions on von Willebrand factor binding to platelet, *Biochem. Biophys. Res. Commun.* 191: 1386-1392.
28. Peng, M., Lu, W., Beviglia, L., Niewiarowski, S., and Kirby, E. P., 1993, Echicetin: A snake venom protein that inhibits binding of von Willebrand factor and alboaggregins to platelet glycoprotein Ib, *Blood* 81: 2321-2328.
29. Peng, M., Holt, J. C., and Niewiarowski, S., 1994, Isolation, characterization and amino acid sequence of echicetin beta-subunit, a specific inhibitor of von Willebrand factor and thrombin interaction with glycoprotein Ib, *Biochem. Biophys. Res. Commun.* 205: 68-72.

12

STRUCTURE AND FUNCTION RELATIONSHIP OF CROTOXIN, A HETERODIMERIC NEUROTOXIC PHOSPHOLIPASE A_2 FROM THE VENOM OF A SOUTH-AMERICAN RATTLESNAKE

V. Choumet, C. Bouchier, E. Délot, G. Faure, B. Saliou, and C. Bon

Unité des Venins
Institut Pasteur
25, rue du Dr Roux
75724 Paris Cedex 15, France

INTRODUCTION

Snake neurotoxins that interfere with the release of transmitter at the neuromuscular junction have a clinical and a scientific interest and, therefore, their structure and their mode of action have been extensively studied. Phospholipase A_2 β-neurotoxins have been found in the venom of *Elapidae* and *Viperidae* snakes. They are potent neurotoxins which specifically inhibit the release of acetylcholine from peripheral cholinergic synapses. Although they all possess phospholipase A_2 activity, their quaternary structure is rather heterogeneous. Some of them consist of a single chain phospholipase A_2 polypeptide chain, as ammodytoxin A from the venom of *Vipera ammodytes ammodytes* (Lee *et al.*, 1984, Ritonja and Gubensek, 1985) or agkistrodotoxin from the venom of *Agkistrodon blomhoffii brevicaudus* (Chen *et al.*, 1981). Multichain β-neurotoxins, such as crotoxin, from the venom of *Crotalus durissus terrificus*, are made of the non covalent association of several subunits homologous to phospholipases A_2, at least one of which possessing enzymatic activity. Finally, β-bungarotoxins, which are only found in *Bungarus* venoms, covalently associate a phospholipase A_2 subunit with a non-enzymatic polypeptide homologous to Kunitz-type protease inhibitors (Kondo *et al.*, 1982a; 1982b). The mechanism of action of β-neurotoxins has been extensively studied *in vivo* and *in vitro* . This brief review focusses on crotoxin, the structure-function relationship of which was examined using the complementary approaches of biochemistry, pharmacology, immunochemistry and molecular biology.

Natural Toxins II, Edited by B. R. Singh and A. T. Tu
Plenum Press, New York, 1996

THE CROTOXIN COMPLEX, A STRIKING MODEL OF COMPLEMENTARITY BETWEEN TWO SUBUNITS

Crotoxin, the major toxic component from the venom of the South-American rattlesnake *Crotalus durissus terrificus*, is made up of two non-identical subunits: a basic and weakly toxic phospholipase A_2 subunit, CB, and an acidic, non-toxic and non-enzymatic subunit, CA (Rübsamen *et al.*, 1971; Hendon and Fraenkel-Conrat, 1971). Crotoxin is in fact a non-homogeneous protein. Several isoforms were isolated from the venom of a single snake and were shown to result from the association of several isoforms of its two subunits (Faure and Bon, 1987,1988). Four CA isoforms were purified and characterized. They are made of three disulfide-linked polypeptide chains (α, β and γ) that correspond to three different regions of a phospholipase A_2-like precursor (Faure *et al.*, 1991). CA isoforms were shown to be the consequence of different posttranslational events occurring on an unique precursor, pro-CA, the cDNA of which has been identified from a cDNA library prepared from one venom gland. On the other hand, two cDNAs encoding precursors of CB isoforms were isolated from this library, indicating that CB isoforms result from the expression of different mRNAs (Bouchier *et al.*, 1988; 1991).

The two subunits of crotoxin act in a synergistic manner. CB alone is weakly toxic and may block neuromuscular transmission although larger doses are required to cause the same effect. CB was shown to bind in a non-saturable manner on various membranes. CA behaves as a chaperon by preventing the non-saturable binding of CB, thus restricting it to saturable binding sites present on synaptic membranes (Bon *et al.*, 1979). On *Torpedo* electric organ, CA has no effect, whereas CB, either alone or complexed with CA, elicits a rapid and dose dependent acetylcholine release and a depolarization of the preparation (Délot and Bon, 1992). Subsequent acetylcholine release evoked by high levels of K^+ or by a Ca^{2+} ionophore was unaffected by CA but reduced after the action of crotoxin or CB.

The synergistic behavior of crotoxin subunits was further demonstrated by studying the differential stability of crotoxin isoforms differing in their toxicity and their enzymatic activity (Faure *et al.* 1993). More toxic isoforms are less enzymatically active, block neuromuscular transmission of chick *biventer cervicis* more efficiently, are more stable and dissociate more slowly than less toxic complexes. Therefore, less toxic crotoxin isoforms, because of their lower stability, are more likely to dissociate "en route" before reaching their target, CA being no more able to prevent CB non specific binding.

Upon interaction with membranes or phospholipid vesicles, the crotoxin complex dissociates: CB binds, while CA is released in solution (Bon *et al.*, 1979; Radvanyi *et al.*, 1989; Délot and Bon, 1993). The time scale of crotoxin dissociation in the presence of presynaptic membranes was studied by determining the kinetics of binding of radiolabelled crotoxin in the presence of CA or crotoxin. These studies suggest that CA is temporarily involved in the binding of CB to its acceptor. Crotoxin would bind to the acceptor, therefore forming a ternary complex, afterwhich a conformationl change could occur, inducing the dissociation of CA (Délot and Bon, 1993). High affinity specific binding sites for crotoxin have been demonstrated in brain tissue (Degn *et al.*, 1991).

Although negatively charged phospholipids could be an important component of the crotoxin target (Radvanyi *et al.*, 1989), the specific acceptor of crotoxin on synaptosomal membranes might involve other structural elements such as proteins (Hseu *et al.*, 1990; Tzeng *et al.*, 1989; Yen and Tzeng, 1991). Crotoxin was also shown to block potassium currents and might specifically interact with potassium channels (Rowan and Harvey, 1988). Recently, a potent inhibitor of crotoxin neurotoxicity (CICS) has been isolated from the blood of *Crotalus durissus terrificus* (Perales *et al.*, 1995). It is a glycoprotein which binds CB but not CA, and induces the dissociation of the crotoxin complex. The molecular mechanism of the interaction of crotoxin with CICS ressembles that of crotoxin with its acceptor on nerve terminal membranes. This suggests that CICS might be a false acceptor which would retain the toxin in the vascular system, therefore, preventing its action on the neuromuscular system.

IMMUNOCHEMICAL STUDIES OF CROTOXIN AND ITS SUBUNITS: MECHANISM OF NEUTRALIZATION OF CROTOXIN BY ANTIBODIES AND SEARCH FOR A "TOXIC SITE"

Rabbit polyclonal antibodies were prepared against the two crotoxin subunits (Choumet *et al.*, 1989). Analysis of the immunocrossreactions between CA and CB suggested that CA exposes determinants of low immunogenicity, which are also present on CB. However, the major antigenic determinants of CB are not present on CA and might be located on one of the three poplypeptides which are removed from the pro-CA precursor during posttranslational modifications. Anti-CB antibodies were shown to neutralize the lethal potency of crotoxin and to inhibit its enzymatic activity. The fact that anti-CB non precipitating Fab are as potent as anti-CB immunoglobulins suggests that crotoxin neutralization results from the binding of antibodies to the catalytic subunit, rather than from the formation of an immunoprecipitate.

Monoclonal antibodies were raised against both isolated subunits (Choumet *et al.*, 1992). Three epitopic regions were determined on CA and four on CB and were shown to be conformational. The observation that several epitopic regions of CB are involved in the neutralization of the lethal action of crotoxin suggests that different mechanisms can be implicated in this process: 1) an inhibition of the enzymatic activity of CB, which is necessary for the neurotoxic action of crotoxin; 2) a dissociation of the crotoxin complex, since some neutralizing monoclonal antibodies are directed against the zone of interaction between the two subunits; 3) an inhibition of crotoxin binding to its acceptor.

Immunocrossreactions between crotoxin and other β-neurotoxins were also studied. Anti-CB polyclonal antibodies were shown to crossreact with crotoxin-like toxins from the venom of several *Crotalus* species as well as with single chain phospholipase A_2 neurotoxins from *Viperidae* venom (agkistrodotoxin and ammodytoxin). There was no crossreaction with *Elapidae* β-neurotoxins. The immunochemical crossreaction between CB and agkistrodotoxin, which exhibit 80% of similarity in their polypeptidic sequences (Kondo *et al.*, 1989), was further analyzed using anti-agkistrodotoxin antibodies (Choumet *et al.*, 1991). The results indicated that the majority of agkistrodotoxin antigenic determinants are present on CB and that some of these common determinants are involved in the neutralization of the lethal potency and in the inhibition of the enzymatic activity of agkistrodotoxin and crotoxin. It is more surprising that antigenic determinants which are specific of each toxins are also involved in the neutralization of the toxicity of crotoxin or of agkistrodotoxin, showing that different mechanisms are involved in their neutralization.

Immunological similarities observed between CB and ammodytoxin were used to investigate the toxic site of crotoxin. In the case of ammodytoxin A, the search of a toxic site has taken advantage of the existence of two other isoenzymes, ammodytoxin B and C, which are 28 and 17 times less toxic (Ritonja *et al.*, 1986; Krizaj *et al.*, 1989). Ammodytoxin A, B and C, which are all composed of 122 amino acids, differ only in their C-terminal parts. Polyclonal antibodies prepared against three peptides covering the C-terminal part revealed that a stretch of amino acids (106 -113) could be involved in the interaction of ammodytoxin A with its acceptor (Curin-Serbec *et al.*, 1991). The interaction of these antipeptide antibodies with CB and crotoxin, as well as their ability to reduce the lethal potency but not the enzymatic activity, suggest that the C-terminal part of CB is involved in the toxicity of crotoxin, as in the case of ammodytoxin A. However, the residues involved in the toxicity of ammodytoxin A may differ to some extent from those of CB and crotoxin. On the other hand, the C-terminal part of CB could be masked in the crotoxin complex.

EVOLUTIONARY RELATIONSHIPS OF β-NEUROTOXINS: CROTOXIN CA CAN INTERFERE WITH SINGLE CHAIN β-NEUROTOXINS FROM *VIPERIDAE* VENOMS AND MODIFY THEIR PHYSIOLOGICAL PROPERTIES

Unlike *Elapidae* venoms which are known to contain single chain and multichain β-neurotoxins as well as β-bungarotoxins, *Viperidae* venoms contain only two kinds of β-neurotoxins: single chain β-neurotoxins such as ammodytoxin or agkistrodotoxin and dimeric β-neurotoxins which associate a chaperon subunit with a phospholipase A_2. CB and agkistrodotoxin which show the highest sequence similarity among phospholipase A_2 β-neurotoxins (80%) behave similarly on *Torpedo* synaptosomes (Délot and Bon, 1992). In good correlation, a strong immunocrossreaction was observed between CB and agkistrodotoxin (Choumet *et al.*, 1991). In the case of CB and ammodytoxin, however, the similarity is lower. CA was further reported to selectively interact with agkistrodotoxin, to modify its action on cholinergic nerve endings, and to increase its toxicity as it does for CB, whereas less marked effects were observed with ammodytoxin (Choumet *et al.*, 1993). These observations raise the question of the evolutionary relationship of single chain and dimeric β-neurotoxins in *Viperidae* venom. The various β-neurotoxins from *Viperidae* venoms could represent different stages of evolution of phospholipases A_2. Ammodytoxin A might derive from a crotoxin-like complex which lost its chaperon while evolving into a very potent neurotoxin on its own. Conversely, acidic non toxic phospholipases A_2 from *Agkistrodon* venoms, which show a 80% sequence similarity with the chaperon CA precursor, could be on the way to acquire CA-like chaperon properties.

CONCLUSION

The β-neurotoxins, such as crotoxin, are very interesting models to study how phopsholipases A_2, the primary function of which is to catalyze the hydrolysis of the 2-acyl ester bond of 1,2-diacyl-3-sn-phosphoglycerol lipids, could have acquired such a high specificity towards cholinergic structures. Furthermore, the molecular characterization of their binding sites as well as the elucidation of the mechanisms by which they block the release of acetylcholine will help in the understanding of the different stages involved in the release of the neurotransmitter at the neuromuscular junction.

REFERENCES

Bon, C., Changeux, J.-P., Jeng, T.W. and Fraenkel-Conrat, H. (1979) Postsynaptic effects of crotoxin and of its isolated subunits. *Eur. J. Biochem.* 99, 471-481.

Bouchier, C., Ducancel, F., Guignery-Frelat, G., Bon, C., Boulain, J.C., and Ménez, A. (1988) Cloning and sequencing of cDNAs encoding the two subunits of crotoxin. *Nucleic Acid Research*, 16, 9050.

Bouchier, C., Boulain, J.C., Bon, C., and Ménez, A. (1991) Analysis of cDNAs encoding the two subunits of crotoxin, a phospholipase A_2 neurotoxin from rattlesnake venom: the acidic non enzymatic subunit derives from a phospholipase A_2-like precursor. *Biochim. Biophys. Acta* 1088, 401-408.

Chen, J., Wu, X., Zhang, G., Jiang, M. and Hsu, K. (1981) Further purification and biochemical properties of a pre-synaptic neurotoxin from snake venom of *Agkistrodon halys* (Pallas). *Acta Biochim. Biophys. Sin.* 13, 205-212.

Choumet, V., Jiang, M.S., Radvanyi, F., Ownby, C., and Bon, C. (1989) Neutralization of lethal potency and inhibition of enzymatic activity of a phospholipase A_2 neurotoxin, crotoxin, by non-precipitating antibodies (Fab). *FEBS Lett.*, 244, 167-173.

Choumet, V., Jiang, M.S., Specker, I., and Bon, C. (1991) Immunochemical cross-reactivity of two phospholipase A_2 neurotoxins, agkistrodotoxin and crotoxin. *Toxicon,* 29, 441-451.

Choumet, V., Faure, G., Robbe-Vincent, A., Saliou, B., Mazié, J.C. and Bon, C. (1992) Immunochemical analysis of a snake venom phospholipase A_2 neurotoxin, crotoxin, with monoclonal antibodies. *Mol. Immunol.,*29, 871-882.

Choumet, V., Saliou, B., Fideler, L., Chen, Y.C., Gubensek, F., Bon, C., and Délot, E. (1993) Snake-venom phospholipase A_2 neurotoxins: potentiation of a single-chain neurotoxin by the chaperon subunit of a two-component neurotoxin. *Eur. J. Biochem.*, 211, 57-62.

Curin-Serbec, V., Novak, D., Babnik, J., Turk, D. and Gubensek, F. (1991) Immunological studies of the toxic site in ammodytoxin A. *FEBS Lett.* 280, 175-178.

Curin-Serbec, V., Délot, E., Faure, G., Saliou, B. Gubensek, F., Bon, C. and Choumet, V. (1993) Antipeptide antibodies directed to the C-terminal part of ammodytoxin A react with the PLA_2 subunit of crotoxin and neutralize its pharmacological activity. *Toxicon* 32, 1337-1348.

Degn, L. L., Seebart, C. S. and Kaiser, I. I. (1991) Specific binding of crotoxin to brain synaptosomes and synaptosomal membranes. *Toxicon* 29, 973-988.

Délot, E. and Bon, C. (1992) Differential effects of presynaptic phospholipase A_2 neurotoxins on *Torpedo* synaptosomes. *J. Neurochem.*, 58, 311-319.

Délot, E. and Bon, C. (1993) Model for the interaction of crotoxin, a phospholipase A_2 neurotoxin, with presynaptic membranes. *Biochemistry,* 32, 10708-10713.

Faure, G., and Bon, C. (1987) Several isoforms of crotoxin are present in individual venoms from the South American Rattlesnake, *Crotalus durissus terrificus*. *Toxicon,* 25, 229-234.

Faure, G. and Bon, C. (1988) Crotoxin, a phospholipase A_2 neurotoxin from the South American rattlesnake, *Crotalus durissus terrificus*: purification of several isoforms and comparison of their molecular structure and of their biological activities. *Biochemistry* 27, 730-738.

Faure, G., Guillaume, J.-L., Camoin, L., Saliou, B., and Bon, C. (1991) Multiplicity of acidic subunit isoforms of crotoxin, the phospholipase A_2 neurotoxin from *Crotalus durissus terrificus* venom, results from posttranslational modifications. *Biochemistry,*30, 8074-8083.

Faure, G., Harvey, A.L., Thomson, E., Saliou, B., Radvanyi, F. and Bon, C. (1993) Comparison of crotoxin isoforms reveals that stability of the complex plays a major role in its pharmacological action. *Eur. J. Biochim.* 214, 491-496.

Hendon, R. A. and Fraenkel-Conrat, H. (1971) Biological role of the two components of crotoxin. *Proc. Natl. Acad. Sci. USA* 68, 1560-1563.

Hseu, M.J., Guillory, R. J. and Tzeng, M.-C. (1990) Identification of a crotoxin-binding protein in membranes from guinea pig brain by photoaffinity labelling. *J. Bioenerg. Biomemb.* 22, 39-50.

Kondo, K., Toda, H., Narita, K. and Lee, C. Y. (1982a) Amino acid sequence of β2-bungarotoxin from *Bungarus multicinctus* venom: the amino acid substitution in the B chains. *J. Biochem.* 91, 1519-1530.

Kondo, K., Toda, H., Narita, K. and Lee, C. Y. (1982b) Amino acid sequences of three β-bungarotoxins (β3-, β4- and β5-bungarotoxins) from *Bungarus multicinctus* venom: amino acid substitutions in the A chains. *J. Biochem.* 1991, 1531-1548.

Kondo, K., Zhang, J.-K., Xu, K., and Kagamiyama, H. (1989) Amino acid sequence of a presynaptic neurotoxin, agkistrodotoxin, from the venom of *Agkistrodon halys* Pallas. J. Biochem. (Tokyo) 105, 196-203.

Krizaj, I., Turk, D., Ritonja, A. and Gubensek, F. (1989) Primary structure of ammodytoxin C further reveals the toxic site of ammodytoxin. *Biochim. Biophys. Acta* 999, 198-202.

Lee, C.Y., Tsai, M. C., Chen, M., Ritonja, A. and Gubensek, F. (1984) Mode of neuromuscular blocking action of toxic phospholipase A_2 from *Vipera ammodytes* venom. *Arch. Int. Pharmacodyn. Ther.* 268, 313-324.

Perales, J., Villela, C., Domont, G., Choumet, V., Saliou, B., Moussatché, H., Bon, C. and Faure, G. (1995) Molecular structure and mechanism of action of the crotoxin inhibitor from *Crotalus durissus terrificus* serum. *Eur. J., Biochem.* 227, 19-26.

Radvanyi, F., Saliou, B., Lembezat, M.-P., & Bon, C. (1989) Binding of crotoxin, a presynaptic phospholipase A_2 neurotoxin, to negatively charged phospholipid vesicles. *J. Neurochem.*, 53, 1252-1260.

Ritonja, A and Gubensek, F. (1985) Ammodytoxin A, a highly lethal phospholipase A_2 from *Vipera ammodytes ammodytes* venom. *Biochim. Biophys. acta* 828, 306-312.

Ritonja, A., Machleidt, W., Turk, V. and Gubensek, F. (1986) Amino acid sequence of ammodytoxin B partially reveals the location of the site of toxicity of ammodytoxins. *Biol. Chem. Hoppe-Seyler* 367, 919-923.

Rowan, E.G. and Harvey, A.L. (1988) Potassium channel blocking actions of β-bungarotoxin and related toxins on mouse and frog motor nerve terminals. *Br. J. Pharmacol.* 94, 839-847.

Rübsamen, K., Breithaupt, H. and Habermann, E. (1971) Biochemistry and pharmacology of the crotoxin complex. *Naunyn Schmiedebergs Arch. Pharmacol.* 270, 274-288.

Yen, C.-H. and Tzeng, M.-C. (1991) Identification of a new binding protein for crotoxin and other neurotoxic phospholipases A_2 on brain synaptic membranes. *Biochemistry* 30, 11473-11477.

13

ATROXASE–A FIBRINOLYTIC ENZYME ISOLATED FROM THE VENOM OF WESTERN DIAMONDBACK RATTLESNAKE

Isolation, Characterization and Cloning

Brenda J. Baker[1] and Anthony T. Tu[2]

[1] Department of Chemistry
University of Southern Colorado
Pueblo, Colorado 81001
[2] Department of Biochemistry and Molecular Biology
Colorado State University
Fort Collins, Colorado 80526

INTRODUCTION

Blood coagulation disorders are a worldwide medical problem. Among them, thrombosis, or the occlusion of a blood vessel, is a leading killer(1). Current thrombolytic agents used to reduce thrombosis are t-PA (tissue plasminogen activator), urokinase, and streptokinase. Although effective, these agents have undesirable side effects such as hemorrhage and reocclusion(2,3). Therefore, the search for safer thrombolytic agents from other sources continues(4,5,6).

Snake venom is a complex mixture of both toxic and nontoxic components with profound pharmacological properties which can be a source for potential pharmaceutical agents. Several proteases may be present in a single venom (7-11). Recently, numerous zinc metalloproteases containing a conserved zinc binding region have been isolated from the venom of various poisonous snakes(11-24). Most of these proteases hydrolyze the Aα and Bβ chain of fibrinogen and are also hemorrhagic *in vivo*. However, a few of these zinc metalloproteases are nonhemorrhagic while retaining fibrinogenolytic activity(12,23,24).

Atroxase is a nonhemorrhagic enzyme isolated from the venom of *C. atrox* which is capable of hydrolyzing fibrin and fibrinogen both *in vivo* and *in vitro*(25). In an earlier report, atroxase was found to recannalize the occluded blood vessels in a rat (26) (Figure 1). This study presents the nucleotide sequence of the cDNA encoding atroxase and the translation of the open reading frame for the primary structure of the protein.

Natural Toxins II, Edited by B. R. Singh and A. T. Tu
Plenum Press, New York, 1996

200 -GAGGATCAGCAGAACCTATCCFCAAAGATACATAGAGCTTGTCGTAGT
400 - ---

200 -TGCAGATCACAGAGTGTTCATGAAATACAACAGCGATTTAAATATTATA
400 - ---

200 - AGAAAACGGGTACATGAACTTGTCAACACTATAAATGGGTTTTACAGAT
400 - ---

200 - CTTTGAATATTGATGTCTCACTGACTGACCTAGAAATTTGGTCCGATCA
400 - --GATCA

200 - GGACTTCATCACT---------------------------------
400 - GGACTTCATCACTGTGCAGTCATCAGCAAAAAATACTTTGAACTCATTT

200 - ---
400 - GGAGAATGGAGAGAGGCAGATTTGCTGAGGCGCAAAAGTCATGATCA

200 - ---
400 - TGCTCAGTTACTCACGGCCATTAACTTCGAAGGAAAAATTATAGGAAG

200- ---
400 - AGCTTACACAAGCAGCATGTGCAACCCAAGGAAATCTGTAGGAATTGT

200 - ---
400 - TAAGGATCATAGTCCAATAAATCTCTTGGTGGGAGTTACCAATGGCCC

200 - ---
400 - ATGAGCTGGGTCATAATCTGGGCATGAACCATGATGGAGATAAGTGTC

200 - ---
400 -TTCGTGGTGCTTCCTTATGCATTATGCGTCCCGGGTTAACACCAGGAC

200 - ---
400 - GTTCCTATGAGTTCAGCGATGATAGTATGGGTTATTATCAGAGTTTTCT

200 - ---
400 - TAAGCAGTACAATCCC

Figure 1. Nucleotide sequence of overlapping PCR fragments.

MATERIALS AND METHODS

Isolation of Atroxase

Atroxase was isolated according to the method of Willis and Tu (25). Homogeneity was established by two independent methods: SDS-polyacrylamide gel electrophoresis, and reverse phase(RP) HPLC. *Crotalus atrox* venom was purchased in lyophilized form from Miami Serpentarium Laboratories, lot 16 AZ. RP HPLC was performed using Beckman System Gold Chromatography connected to binary pump model 26 and model 167 detector through a Vydac analytical reverse phase column (C_{18} with 300 Å pore) Samples were eluted with a linear gradient from 10% solvent A, (0.1% TFA in water) to 90% solvent B(0.1% TFA in acetonitrile) at a flow of 1 ml/min. Fractions were monitored at 280 and 220 nm.

Enzymatic and Biological Activity Assays

Atroxase was assayed for fibrinogenolytic, fibrinolytic, and hemorrhagic activities (25,27). Specific cleavage of fibrinogen was performed by incubating 500 µl of 2% human fibrinogen solution with 500 µl of protein (50 µg/ml) at 37°C in 5 mM imidazole-saline buffer (1:9), pH 7.4. At various time intervals, 0.1 ml of the above incubation mixtures were withdrawn and added to 0.1 ml of denaturing solution (8 M urea, 4% SDS, and 4% 2-mercaptoethanol). The samples were reduced and denatured overnight at 37°C before electrophoresis on a 11% SDS polyacrylamide gel.

Fibrin degradation was demonstrated on a SDS PAGE. Thrombin, 0.1 ml containing 10 units/ml, was added to 0.1 ml of a 1% fibrinogen solution in 5 mM imidazole-saline buffer (1:9), pH 7.4. The fibrin clot was allowed to form for 1 h at room temperature. After 1 h, 0.1 ml of protein (50 µg/ml) was added to the clot, and incubated at 37°C for various time intervals. The reaction was stopped by the addition of 0.3 ml of denaturing solution, incubated overnight, and run on an 11% SDS polyacrylamide gel.

Hemorrhagic activity was tested by subcutaneously injecting Swiss Webster mice with the test samples dissolved in 0.05 ml of 0.9% saline. The minimum hemorrhagic dose is defined as the least amount of protein that causes a hemorrhagic reaction (a releasing of blood into the surrounding tissue) of 5 mm in diameter 6 h after injection.

Construction of cDNA Library

The venom glands of one mature specimen of *C. atrox* were removed four days after venom extraction. The glands were immediately frozen in liquid nitrogen and stored at -70 C. mRNA was isolated using the Ribosep mRNA isolation kit (Collaborative Research, Inc. cat.# 30030).

The cDNA library was constructed from 5 µg of C. atrox venom gland mRNA using the Uni Zap-cDNA synthesis kit (Stratagene). Linkers containing a 3' EcoR 1 restriction site and a 5' Xho 1 site were added to the cDNAs and the DNA fragments were inserted into the XR site of λ-ZAP (Stratagene). The lambda library was packaged *in vitro* using Gigapack II Gold packaging extract (Stratagene) and plated on *E. coli* SURE (Stratagene). The primary library was amplified once, before in *vivo* excision to package the library in the plasmid vector pBluescript. XL Blue *E. coli* cells containing the recombinant plasmid library were stored as glycerol stock at -20 C.

Polymerase Chain Reaction

The polymerase chain reaction (PCR) was used to amplify gene fragments unique to atroxase using a combination of the synthetic degenerate oligonucleotide probes as upstream and downstream primers.

20 pmole of each upstream primer and downstream primer, 10 μl of 10X Hot Tub polymerase buffer (USB), 10 μl of 0.5 mM equal mixture of dideoxynucleotides, 0.1 μg of plasmid template and deionized water up to 99 μl were mixed in a 0.5 ml eppendorf tube. The mixture was overlayed with mineral oil. One μl (1 unit/reaction) of Hot Tub polymerase (USB) was added after the first denaturing step. DNA was recovered by phenol:chloroform extraction followed by ethanol precipitation. The entire PCR reaction was resuspended in 10 μl of TE and analyzed on 1% agarose gel.

Subcloning Into pBluescript

PCR products were isolated on a 1% agarose gel in TAE buffer. Wells were cut directly in front of desired DNA bands, TAE buffer was removed from the wells and replaced with TAE buffer containing 10% PEG. The DNA of interest migrated into the wells, was recovered by pipetting and was ethanol precipitated. The DNA was resuspended in TE and double digested with EcoR I and Xho I at 16 C overnight. The sticky ended PCR fragments were recovered on 1% agarose gel as previously described and ligated into precut pBluescript plasmid overnight at 16 C. Recombinant plasmid was electrotransformed into XL Blue strain *E. coli*. Inserts in pBluescript disrupt the LacZ gene fragment, preventing expression of β-galactosidase. Therefore, XL-Blue colonies which are transformed with recombinant pBluescript plasmid will appear white on LB/amp plates with X-gal and IPTG, while nonrecombinant transformed XL-Blue colonies will appear blue. Positive XL-Blue colonies were isolated from selection plates and recombinant plasmid was isolated using Qiagen plasmid prep columns (QIAGEN).

Isolation of cDNA Clone for Atroxase

The ^{32}P labeled PCR fragments were radiolabeled with 32 P and used to probe the cDNA library for cDNA atroxase. Positive colonies were isolated and sequenced as described above. Recombinant plasmids containing PCR fragment inserts and positive cDNA clones were sequenced using USB Sequenase Version 2.0 sequencing kit. Sequence manipulations, translations and comparisons were performed using PCGENE software.

RESULTS

Isolation of Atroxase

Atroxase was purified according to the previously published method. Homogeneity was established by gel electrophoresis and by HPLC. Biological activity assays confirmed the identity of atroxase. 25 μg of atroxase in 1 ml of a 1% solution of fibrinogen hydrolyzed the Aα chain of fibrinogen within 5 minutes followed by hydrolyzation of the Bβ chain. Atroxase had no effect of the γ-γ dimer. 5 μg of atroxase in 0.1 ml hydrolyzed the α polymer and unpolymerized α and β chains of a 1% fibrin clot within 30 minutes. Atroxase showed no hemorrhagic activity up to 50 μg of protein.

mRNA Isolation and cDNA Construction

Total RNA isolated from approximately 1 gram of venom gland tissue amounted to 175 μg. From this total RNA, 7 μg of poly (A)+ RNA was prepared by oligo d(T) cellulose chromatography. Double stranded cDNA was prepared from a 5 ug template of RNA and was ligated into the UNI-Zap vector purchased from Stratogene. The cDNA library was packaged and plated with XL-Blue *E. coli* cells with a resulting primary library titer of 2.6 x 10^5 recombinant/μl. pBluescript plasmids (Stratogene) are cloning vectors that have a multiple cloning region which interupts the LacZ gene, allowing for blue/white colony selection for positive recombinants. pBluescript phagemid can be circularized from the Uni-Zap phage vector to allow for a variety of cloning and sequencing procedures. Therefore, for ease of manipulation, the library was excised from its prophage form by *in vivo* excision into pBluescript phagemid form.

PCR to Obtain cDNA of Atroxase

Two major PCR products of approximately 200 and 400 nucleotides respectively were isolated on a PEG-agarose gel. These fragments were ligated into pBluescript and transformed into XL-Blue *E.coli*. Transformation of 200 and 400 PCR fragments was confirmed by double digestion with EcoR 1 and Xho I to release the inserted fragment. Electrophoresis on 1% agarose shows that both the 200 and 400 PCR fragment ligated into pBluescript plasmids. Recombinant plasmids containing each cDNA PCR fragment were isolated using Qiagen plasmid prep columns in preparation for nucleotide sequencing. Each PCR fragment with insert was sequenced 2-10 times to confirm base order. The nucleotide sequences have a region of overlap where the same oligonucleotide sense and antisense primers were used for the PCR reaction. The resulting cDNA was 598 nucleotide long (Figure 1). Translation of the two overlapping sequences results in an amino acid sequence of 199 amino acids (Figure 2).

Isolation of cDNA Atroxase. The PCR derived cDNA fragments were radiolabeled and used to probe the cDNA library for a full length cDNA encoding atroxase via colony

```
            10          20          30              40          50          60
GAG GAT CAG CAG AAC CTA TCC CAA AGA TAC ATA GAG CTT GTC GTA GTT GCA GAT CAC AGA
E   D   Q   Q   N   L   S   Q   R   Y   I   E   L   V   V   V   A   D   H   R

            70          80          90              100         110         120
GTG TTC ATG AAA TAC AAC AGC GAT TTA AAT ATT ATA AGA AAA CGG GTA CAT GAA CTT GTC
V   F   M   K   Y   N   S   D   L   N   I   I   R   K   R   V   H   E   L   V

            130         140         150             160         170         180
AAC ACT ATA AAT GGG TTT TAC AGA TCT TTG AAT ATT GAT GTC TCA CTG ACT GAC CTA GAA
N   T   I   N   G   F   Y   R   S   L   N   I   D   V   S   L   T   D   L   E

            190         200         210             220         230         240
ATT TGG TCC GAT CAG GAC TTC ATC ACT GTG GAC TCA TCA GCA AAA AAT ACT TTG AAC TCA
I   W   S   D   Q   D   F   I   T   V   Q   S   S   A   K   N   T   L   N   S

            250         260         270             280         290         300
TTT GGA GAA TGG AGA GAG GCA GAT TTG CTG AGG CGC AAA AGT CAT GAT CAT GCT CAG TTA
F   G   E   W   R   E   A   D   L   L   R   R   K   S   H   D   H   A   Q   L

            310         320         330             340         350         360
CTC ACG GCC ATT AAC TTC GAA GGA AAA ATT ATA GGA AGA GCT TAC ACA AGC AGC ATG TGC
L   T   A   I   N   F   E   G   K   I   I   G   R   A   Y   T   S   S   M   C

            370         380         390             400         410         420
AAC CCA AGG AAA TCT GTA GGA ATT GRR AAG GAT CAT AGT CCA ATA AAT CTC TTG GTG GGA
N   P   R   K   S   V   G   I   V   K   D   H   S   P   I   N   L   L   V   G

            430         440         450             460         470         480
GTT ACA ATG GCC CAT GAG CTG GGT CAT AAT CTG GGC ATG AAC CAT GAT GGA GAA AAG TGT
V   T   M   A   H   E   L   G   H   N   L   G   M   N   H   D   G   D   K   C

            490         500         510             520         530         540
CTT CGT GGT GCT TCC TTA TGC ATT ATG CGT CCC GGG TTA ACA CCA GGA CGT TCC TAT GAG
L   R   G   A   S   L   C   I   M   R   P   G   L   T   P   G   R   S   Y   E

            550         560         570             580         590
TTC AGC GAT GAT AGT ATG GGT TAT TAT CAG AGT TTT CTT AAG CAG TAC AAT CCC CAG
F   S   D   D   S   M   G   Y   Y   Q   S   F   L   K   Q   Y   N   P   Q
```

Figure 2. Translated amino acid sequence of atroxase.

```
        10        20        30        40        50        60        70        80        90       100
TTCGGCACGAGGCAGTTCAGCCAAAGTATGAAGACGCCATGCAATATGAATTGAAAGTGAATGGAGAGCCAGTGGTCCTTCACTTGGAAAAAAATAAAGA

       110       120       130       140       150       160       170       180       190       200
ACTTTTTTCAAAAGATTACAGTGAGACTCATTATTACCCCTGATGGCAGAAAAATTACAACAAACCCTTCGGTGAGGATCACTGCTATTATCGTGGACGC

       210       220       230       240       250       260       270       280       290       300
ATCGAGAATGATGCTGACTCAACTGCAAGCAATCAGTGCATGCAACGGTTTGAAAGGACATTTCAAGCTTCAAGGGGAGATGTACCTTATTGAACCCTTG

       310       320       330       340       350       360       370       380       390       400
GAGCTTTCCGACAGTGAAGCTCATGCAGTCTTCAAATATGAAAATGTAGAAAAAGGATGAGGCCCCCAAAATGTGTGGGGTAACCCAGAATTGGGAATCA

       410       420       430       440       450       460       470       480       490       500
TATGAGCCCATCAAAAAGGCCTCTGATTTAAATCTTAATGAGGACCAACAAAATTTATCCCAAAGATACATAGAGCTTGTCGTAGTTGCAGATCACAGAG
                                       E  D  Q  Q  N  L  S  Q  R  Y  I  E  L  V  V  V  A  D  H  R
                                       E  D  Q  Q  N  L  S  Q  R  Y  I  E  L  V  V  V  A  D  H  R

       510       520       530       540       550       560       570       580       590       600
TGTTCATGAAATACAACAGCGATTTAAATATTATAAGAAAACGGGTACATGAACTTGTCAACACTATAAATGGGTTTTACAGATCTTTGAATATTGATGT
V  F  M  K  Y  N  S  D  L  N  I  I  R  K  R  V  H  E  L  V  N  T  I  N  G  F  Y  R  S  L  N  I  D  V
V  F  M  K  Y  N  S  D  L  N  I  I  R  K  R  V  H  E  L  V  N  T  I  N  G  F  Y  R  S  L  N  I  D  V

       610       620       630       640       650       660       670       680       690       700
CTCACTGACTGACCTAGAAATTTGGTCCGACCAAGATTTCATCACTGTGCAGTCATCAGCAAAAAATACTTTGAACTCATTTGGAGAATGGAGAGAGGCA
 S  L  T  D  L  E  I  W  S  D  Q  D  F  I  T  V  Q  S  S  A  K  N  T  L  N  S  F  G  E  W  R  E  A
 S  L  T  D  L  E  I  W  S  D  Q  D  F  I  T  V  D  S  S  A  K  N  T  L  N  S  F  G  E  W  R  E  A

       710       720       730       740       750       760       770       780       790       800
GATTTGCTGAGGCGCAAAAGTCATGATCATGCTCAGTTACTCACGGCCATTAACTTCGAAGGAAAAATTATAGGAAGAGCTTACACAAGCAGCATGTGCA
 D  L  L  R  R  K  S  H  D  H  A  Q  L  L  T  A  I  N  F  E  G  K  I  I  G  R  A  Y  T  S  S  M  C
 D  L  L  R  R  K  S  H  D  H  A  Q  L  L  T  A  I  N  F  E  G  K  I  I  G  R  A  Y  T  S  S  M  C
```

```
       810       820       830       840       850       860       870       880       890       900
ACCCAAGGAAATCTGTAGGAATTGTTAAGGATCATAGTCCAATAAATCTCTTGGTGGGAGTTACAATGGCCCATGAGCTGGGTCATAATCTGGGCATGAA
N  P  R  K  S  V  G  I  V  K  D  H  S  P  I  N  L  L  V  G  V  T  M  A  H  E  L  G  H  N  L  G  M  N
N  P  R  K  S  V  G  I  V  K  D  H  S  P  I  N  L  L  V  G  V  T  M  A  H  E  L  G  H  N  L  G  M  N

       910       920       930       940       950       960       970       980       990      1000
CCATGATGGAGATAAGTGTCTTCGTGGTGCTTCCTTATGCATTATGCGTCCCGGGTTAACACCAGGACGTTCCTATGAGTTCAGCGATGATAGTATGGGT
 H  D  G  D  K  C  L  R  G  A  S  L  C  I  M  R  P  G  L  T  P  G  R  S  Y  E  F  S  D  D  S  M  G
 H  D  G  E  K  C  L  R  G  A  S  L  C  I  M  R  P  G  L  T  P  G  R  S  Y  E  F  S  D  D  S  M  G

      1010      1020      1030      1040      1050      1060      1070      1080      1090      1100
TATTATCAGAGTTTTCTTTCTTATCAGTATAAGCCACAATGCATTCTCAACCCTTGAGATAGTCCTGTTTCAACTCCAGTTTCTGGAATGAACTTTTGGA
 Y  Y  Q  S  F  L  S  Y  Q  Y  K  P  Q  C  I  L  N  P
 Y  Y  Q  S  F  L  K  -  Q  Y  N  P  Q

      1110      1120      1130      1140      1150      1160      1170      1180      1190      1200
GGCGGGGAAGAATGATACTGTGGCGCTCCTGCAGTCTGCAGCAACAGGCAGTGGCGATGTGACTACAGCCTAATAACAACCTCTGGCTTCTCTCAGATTT

      1210      1220      1230      1240      1250      1260      1270      1280      1290      1300
GATCTTGGAGTTCCTTCTTTCAGAAGGTTTGGCTTCCCTGTAGTCCAAAGAGACCCATCTGCCTGCATCCTACTAGTAAATCACTCTTAGCTTTCATATG

      1310      1320      1330      1340      1350      1360      1370      1380      1390      1400
GAATCTAAATTCTGCAATATTTCTTCTCCATATTTAACCTGTTTACCTTTTGCTGTAATCAAACCTTTTTCCCACCACAAAGCTCCATAGGCATATACAA

      1410      1420      1430      1440      1450      1460      1470      1480      1490      1500
CACCAAGGGCTTATTTGCTGTCAAGAAAAAAAAATGGCCATTTTACCGTTTGCCAAAGCACATTTAATGCAACAAGTTCTGCCTTTTGAGCTGGTGTATT

      1510      1520      1530      1540      1550      1560      1570      1580
CGAAGTGAATGCTTCCTCTCCCAAAATTTCATGCTGGCTTTCCAAGATGTAGCTGCTTCCATCGAATAAACTAACTATTGCATCAT
```

Figure 3. Nucleotide bases obtained from cDNA via colony hybridization. Translated region of atroxase is also included.

hybridization. The largest positive clone resulted in a nucleotide sequence of 1600 bases, and included regions of the start of the coding for the mature protien, the termination site and polyA site (Figure 3).

DISCUSSION

Atroxase has 205 amino acids obtained by translation of a truncated cDNA clone of atroxase isolated from a cDNA library constructed from activated rattlesnake venom glands.

Atroxase has sequence homology to two other known nonhemorrhagic fibrinogenolytic proteins isolated from snake venom; fibrolase from the venom of *A. c. contortrix (23)* and adamalysin, from the venom of *C. adamanteus* (12). The proteins are fairly conserved throughout, with the most conserved region being the putative zinc metal ion binding region at residues (142-163). A Met turn region just past the zinc metal binding region is conserved in atroxase and adamalysin.(12) Atroxase has an additional aspartic acid and glutamine residue on the N-terminal end.

These proteins have high sequence homology to each other and to known hemorrhagic toxins isolated from snake venoms. It is not clear what the structural difference is between the two types of proteases. When the oxidized insulin B or fibrinogen was used as a substrate, it is hydrolyzed by both hemorrhagic and nonhemorrhagic venom proteases (25, 28-30). However, these proteins have different proteolytic activities *in vivo*.

The highly conserved zinc binding region has been shown to be important in proteolytic activity(12). The zinc consensus regions can be regrouped into three general catagories from highest sequence similarity to lowest sequence similarity. The nonhemorrhagic proteases atroxase, fibrolase, and adamalysin fall into the same general category, indicating that the primary structure in this region may be related to nonhemorrhagic activity. Atroxase, fibrolase and adamalysin are identical in sequence in this region with a few exceptions. Fibrolase contains an aspartic acid instead of an asparagine (atroxase and adamalysin) after the conserved methionine, and four of the last five residues are not conserved for fibrolase compared to atroxase and adamalysin. The three histidines, which have been shown to be important in complexing the zinc ion, are conserved, as is the cysteine implicated in a disulfide bond.

However, three hemorrhagic toxins also fall into this category of sequence similarity to atroxase. So a direct comparison of the primary structure of the zinc bnding region does not give any clues as to why there is a difference in proteolytic activities between hemorrhagic and nonhemorrhagic zinc metalloproteinases.

In conclusion, the primary structure of atroxase has been elucidated and atroxase has been determined by direct sequence comparison to be a zinc metalloproteinase with high sequence similarity to two other known snake venom nonhemorrhagic fibrinogenolytic proteins. The reason for difference in hemorrhagic acitivity for these types of metalloproteins appears to be more complex than just a simple comparison of primary structure and is a subject for further study. Our goal is to express biologically active atroxase and to study the structure function relationship using site directed mutagenesis to determine the differences in biological activities of these homologous proteins.

ACKNOWLEDGEMENTS

This work is supported by NIH MERIT award 5RR 37 GM15591 (ATT) NIH grant TW 00211 (ATT) the American Heart Association grant (ATT) and Postdoctroral Fellowship (BJB).

REFERENCES

1. Granger, C., Califf, R., and Repol, E. (1992) *Drugs* **44**, 293-325.
2. Grahm, J., et al. (1993) *Coronary Art. Dis.* **4**, 371-377.
3. Sietz, R., et al. (1993) *Fibrinolysis 7*, 109-115.
4. Huang, D., Gai, L. Y., Wang, S. R., Li, T. D., Yang, T. S., Zhi, G., Du, L. S., and Li, L. H. (1992) *Acta Cardiol. 47*, 445-458.
5. Unemura, K., and Nakashima, M. (1993) *Jap. J. Psychopharmacol. 13*, 9-17.
6. Siigur, E., and Siigur, J. (1991) *Biochem. Biophys. Acta 1074*, 223-229.
7. Bjarnason, J. B. and Tu, A.T. (1978) *Biochemistry* **17**, 3395-3407.
8. Nikai, T., Mori, N., Kishida, M., Tsuboi, M., and Sugihara, H. (1985) *Am. J. Trop. Med. Hyg.* **34**, 1167-1172.
9. Ohsaka, A., Just, M., and Haberman, E. (1973) *Biochim. Biophys. Acta* **323**, 415-428.
10. Bjarnasson, J., and Fox, J. (1989) *J. Toxicol. Toxin Rev.* **7**, 121-209.
11. Kini, R., and Evans, C.(1992) *Toxicon* **30**, 265-293.
12. Gomis-Ruth F., Kress, L., and Bode, W.(1993) *EMBO* **12**, 4151-4157.
13. Takeya, H., Onikura, A., Nikai, T., Sugihara, H., and Iwanaga, S. (1990) *J. Biochem.* **108**, 711-719.
14. Shannon, J. D., Baramova, E. N., Bjarnason, J. B., and Fox, J. W. (1989) *J. Biol. Chem.* **264**, 11575-11583.
15. Hite, L. A., Shannon, J. D., Bjarnasson, J., and Fox, J. (1992) *Biochemistry* **31**, 6203-6211.
16. Sanchez, E. F., Diniz, C. R., and Richardson, M. (1991) *FEBS Lett.* **282**, 178-182.
17. Au, L., Huang, Y., Huang, T., Teh, G., Lin, H., and Choo, K. (1991) *Biochem. Biophys. Res. Comm.* **181**, 585-593.
18. Paine, M., Desmond, H., Deakssten, R., and Crampton, J. (1992) *J. Biol. Chem.* **267**, 22869-22876.
19. Takeya, H., Oda, K., Miyata, T., Omori-Satoh, T., and Iwanaga, S. (1990) *J. Biol. Chem.* **265**, 16068-16073.
20. Miyata, T., Takeya, H., Ozaki, Y., Arakawa, M., Tokunaga, F., Iwanaga, S., and Omori-Satoh, T. (1989) *J. Biochem.* **105**, 847- 853.
21. Neeper, M., and Jacobson, M. (1990) *Nucl. Acid Res.* **14**, 5843-5855.
22. Takeya, H., Arakawa, M., Miyata, T., Iwanaga, S., and Omori-Satoh, T. (1989) *J. Biochem. 106*, 151-157.
23. Randolph, A., Chamberlan, S., Chu, H., Retzus, A., Markland, F., and Masiarz, F. (1992) *Prot. Sci. 1*, 590-600.
24. Takeya, H., Nishika, S., Miyata, T., Kanada, S., Saisaka, Y., Morita, T., and Iwanaga, S. (1992) *J. Biol. Chem. 267*, 14109-14117.
25. Willis, T. W., and Tu, A. T. (1988) *Biochemistry 27*, 4769-4777.
26. Willis, T. W., Tu, A. T., and Miller, C. W. (1989) *Thrombosis Res. 53*, 19-29.
27. Kondo, H., Kondo, S., Ikezawa, H., Murata, R., and Ohsaka, A. (1960) *Japan J. Med. Sci. Biol. 13*, 43-51.
28. Tu, A. T., Nikai, T., and Baker, J. O. (1981) *Biochemistry 20*, 7004-7009.
29. Takahashi, T., and Ohsaka, T. (1970) *Biochim. Biophys. Acta 198*, 293-307.
30. Satake, M., Omori, T., Iwanaga, S., and Suzuki, T. (1963) *J. Biochem.* (Tokyo) *54*, 8-16.

14

ISOLATION OF A NOVEL LECTIN FROM THE GLOBIFEROUS PEDICELLARIAE OF THE SEA URCHIN *Toxopneustes pileolus*

H. Nakagawa,[1] T. Hashimoto,[1] H. Hayashi,[1] M. Shinohara,[2] K. Ohura,[2] E. Tachikawa,[3] and T. Kashimoto[3]

[1] Department of Life Science
University of Tokushima
Tokushima 770, Japan
[2] Department of Pharmacology
Osaka Dental University
Osaka 540, Japan
[3] Department of Pharmacology
School of Medicine
Iwate Medical University
Morioka 020, Japan

INTRODUCTION

The sea urchin, *Toxopneustes pileolus* (Lamarck), a member of the family *Toxopneustidae*, is found on the sea bed in the Indo-Pacific area. It possesses extremely well-developed globiferous pedicellariae which contain pharmacologically active substances and cause deleterious effects (Fujiwara, 1935; Okada., 1955a; Kimura et al., 1975; Nakagawa et al., 1982, 1991 and 1992; Takei et al., 1993). It is suggested that a sea urchin having large pedicellariae is more dangerous than one having numerous small pedicellariae (Halstead, 1987). Okada et al. (1955b) reported that there were sub-types of globiferous pedicellariae on the external surface of *T. pileolus*: ordinary pedicellaria, trumpet pedicellaria and giant pedicellaria. We have also observed the trumpet pedicellaria (globiferous pedicellaria) and giant pedicellaria (large globiferous pedicellaria). In the course of comparative study on venom proteins from the globiferous pedicellariae and large globiferous pedicellariae of *T. pileolus*, we found that the crude venom extracts from the pedicellariae of the two sub-types caused agglutination with mouse erythrocytes in a Ca^{2+}-independent manner.

Many lectins have been detected and isolated from various plants and animals, including invertebrates. Lectins are generally considered to nonenzymatic proteins which recognize specific carbohydrate structures and agglutinate a variety of animal cells. In recent years, several lectins have also been isolated from marine invertebrates (Muramoto and Kamiya, 1986; Giga et al., 1985; Suzuki et al., 1990; Hatakeyama et al., 1993; Kawagishi

Natural Toxins II, Edited by B. R. Singh and A. T. Tu
Plenum Press, New York, 1996

et al., 1994). Here we report on the isolation of a novel lectin from the giant pedicellariae (large globiferous pedicellariae) of *T. pileolus*. *Toxopneustes* lectin is the first lectin purified from a species of the family *Toxopneustidae*, which is galactose-specific and calcium-independent.

MATERIALS AND METHODS

Toxopneustes pileolus sea urchins were collected along the coast of Tokushima Prefecture, Shikoku Island, Japan, from 1991 through 1993 (Fig. 1). Horse blood sample was obtained from Nippon Bio-Test Laboratories (Tokyo, Japan). SephadexG-200 was from Pharmacia LKB. Diethylaminoethyl cellulose (DE52) was a product of Whatman Laboratory. All the other chemicals were reagent grade.

Isolation of Lectin

Thirty large globiferous pedicellariae (giant type pedicellariae) per sea urchin specimen were removed with fine forceps (Fig. 1). In the case of globiferous pedicellariae (normal type pedicellariae), one hundred pedicellariae were removed. They were extracted with 20 ml of distilled water at 4°C for 48 hr. The 20 ml aliquot of each extract was centrifuged at 12,000 g for 20 min and the supernatant was considered as the crude lectin extract. The crude lectin extract was applied to a Sephadex G-200 column (2.6 x 80 cm) equilibrated with 0.15 M NaCl solution and was eluted with the same solution at a flow rate of 15 ml/hr. Fractions of 10 ml each were collected and analyzed for absorption at 280 nm and agglutinating activity. For the second step of purification, the gel chromatographic fraction (the third protein peak) was dissolved in 0.016 M Tris-HCl buffer at pH 7.6 and placed on a DEAE-cellulose column (1.5 x 10 cm). The sample was eluted with a linear gradient of NaCl from 0 to 0.7 M in the same buffer. The 10 ml elution fractions were collected and analyzed for absorption at 280 nm and agglutinating activity. The major peak fractions were collected. Final purification was achieved by HPLC using a reverse-phase C_8 column. Two solvent, 0.1 % trifluoroacetic acid (TFA) and acetonitrile in 0.08 % TFA, were used. The fraction was monitored at 230 nm. The main peak was pooled and analyzed for agglutinating activity, and used as the purified lectin.

Assay of Agglutinating Activity

The agglutinating activity was assayed by using horse or mouse erythrocytes in microtiter plates. In some experiments, human erythrocytes was also used. 25 μl of 2.5 % (v/v) suspension of erythrocytes in 6.4 mM phosphate buffer saline, pH 7.2 (PBS) was added to 50 μl of serial 2-fold dilutions of the lectin fractions in PBS. The plates were incubated at room temperature for 1 hr. The results were expressed as the minimum concentration of lectin fractions (μg/ml). The agglutination inhibition was expressed as the minimum concentration of each sugar required for inhibition of the lectin fractions.

Contractile Activity

Longitudinal muscle was prepared from the isolated ileum of male guinea-pigs (Nakagawa et al., 1994). The muscle was suspended in an organ bath (10 ml capacity) containing Krebs-Ringer bicarbonate solution (NaCl 118.0, $NaHCO_3$ 25.0, KCl 4.7, KH_2PO_4 1.2, $MgSO_4$ 1.2, $CaCl_2$ 2.5, and glucose 5.5 mM) at 37°C, through which 95 %/5 % was bubbled continually. The muscle was loaded with 0.3 g wt, and the contraction of the muscle

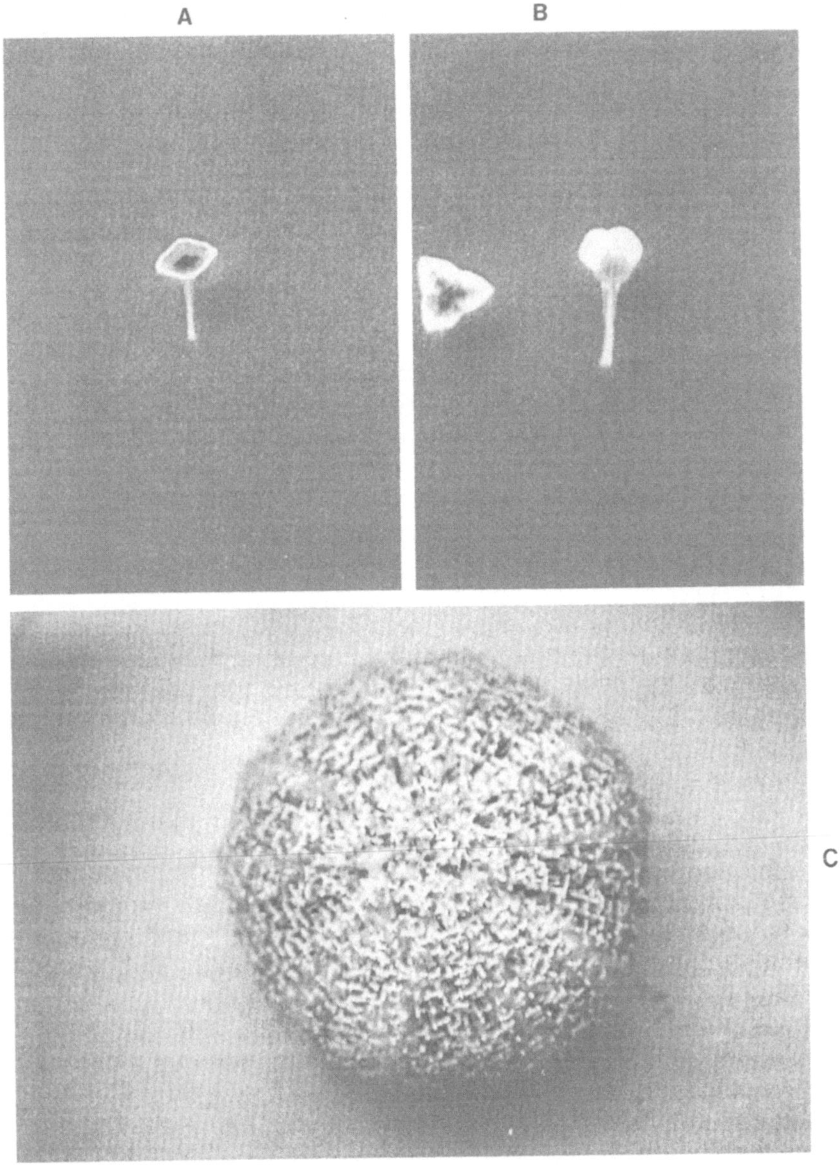

Figure 1. The globiferous pedicellaria (A, normal type) and large globiferous pedicellaria (B, giant type) of *T. pileolus* (C).

strips were recorded using an isometric transducer (Nihon Kohden, TB-651). Maximal response was induced by 40 mM KCl.

Gel Electrophoresis

Electrophoresis on sodium dodecyl sulfate-polyacrylamide gel electrophoresis (SDS-PAGE) was carried out by the method of Laemmli (1970) using a 15-25 % gradient gel. The protein samples were heated in the presence or absence of 2-mercaptoethanol for 4 min at 98°C. Gels were stained with Coomassie Brilliant Blue. The molecular weight standards (Amersham) used were ovalbumin (M_r 46kDa), carbonic anhydrase (30kDa), trypsin inhibitor (21.5kDa), lysozyme (14.3kDa), aprotinin (6.5kDa) and insulin (b) chain (3.4kDa).

Protein Measurement

Protein was measured according to the method of Bradford (1976) using bovine serum albumin as a standard.

Amino Acid Analysis

The analysis of the amino acid composition was carried out on a Model L-8500 Hitachi automatic analyzer by the standard HCl hydrolysis method. Three analyses with 22-, 48-, and 72-hr hydrolysates were used, and serine and threonine contents were obtained by extrapolating to zero hydrolysis times. The values of the other amino acid residues were the average of the values obtained at three hydrolysis times. Methionine was used as the minimum residue to calculate the number of amino acid residues.

Chemotaxis Assay

Polymorphonuclear leukocytes (neutrophils) were induced by intraperitoneal injection of 1 % glycogen solution into male guinea-pigs, and collected by centrifugation in PBS (Ohura et al., 1990). The washed cells were resuspended in Dulbecco's modified Eagle

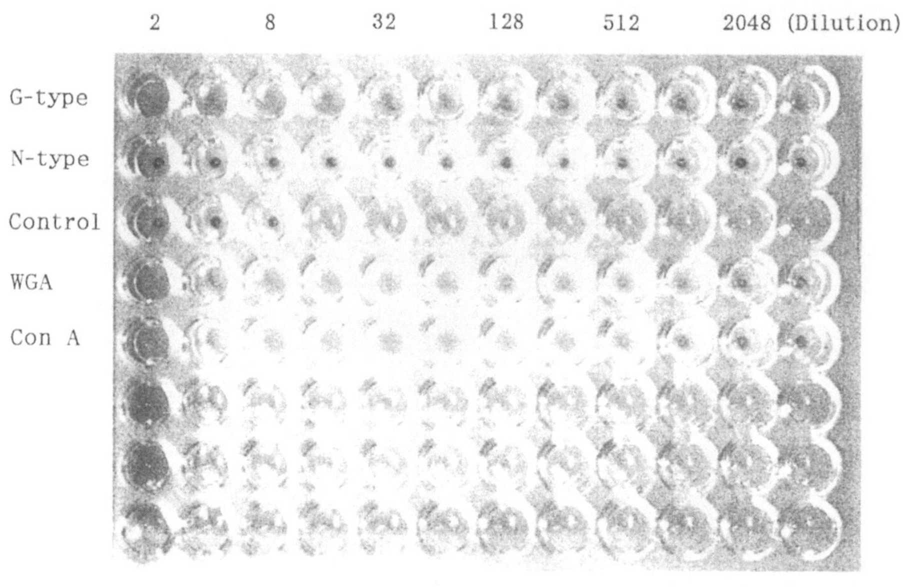

Figure 2. Agglutinating effect of increasing dilutions of the crude extracts from the pedicellariae of *T. pileolus* and plant lectins. The assay of agglutinating activity with mouse erythrocytes was carried out using a 2.5 % suspension in PBS. G-type, the crude extract (5 μg/ml) from the giant type pedicellariae; N-type, the crude extract (5 μg/ml) from the normal type pedicellariae; Control, PBS; WGA, wheat germ agglutinin (50 μg/ml); Con A, concanavalin A (100 μg/ml).

medium (DMEM) and adjusted to a neutrophil density of 2 x 10^6 cells/ml. Neutrophil migration was measured by the membrane filter method using 48-well microchemotaxis chambers (Neuroprove, MD, U.S.A.). A cell suspension was placed in the upper wells of the chamber separated by a polycarbonate membrane filter (3 μm pore size). The lower wells contained the chemotactic stimuli or DMEM as a negative control. After a 60-min incubation at 37°C in 5 % CO_2, the filters were removed, fixed, stained, mounted on slides and examined under a microscope. Neutrophil migration to the lower surface of the filter in 20 oil-immersion fields were quantitated.

RESULTS AND DISCUSSION

Biological Activities of the Crude Lectin Extract

The crude lectin extract from the giant type pedicellariae caused agglutination with mouse and horse erythrocytes. The agglutination with mouse erythrocytes caused by the giant type extract was significantly more intense at concentrations up to 0.33 μg/ml (Fig. 2),

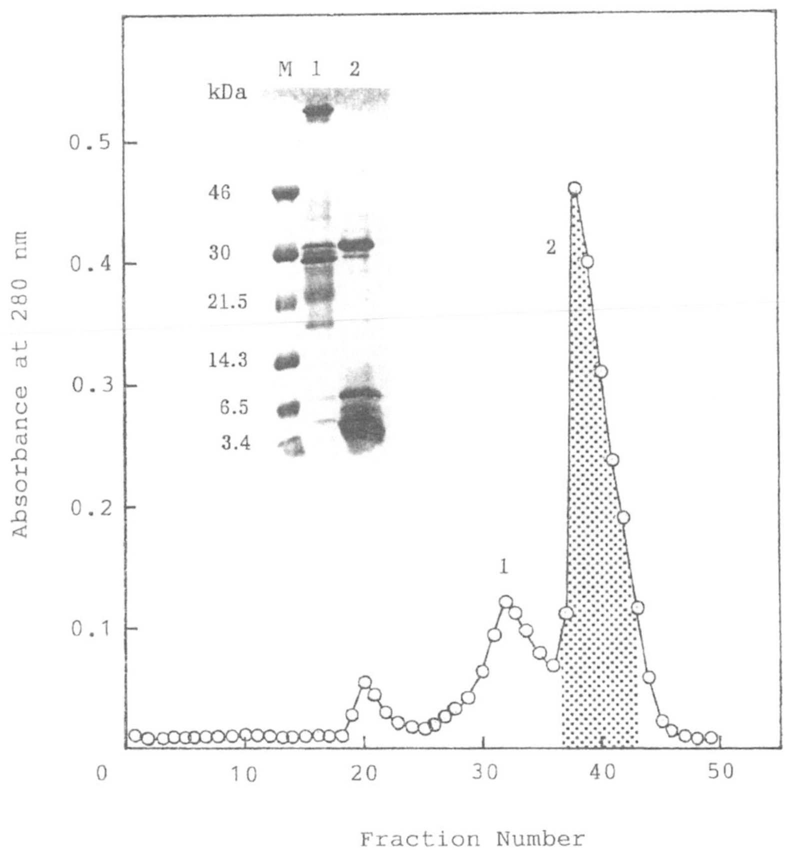

Figure 3. Gel chromatography on a Sephadex G-200 column of the crude lectin extract from the large globiferous pedicellariae (giant type pedicellariae) of *T. pileolus*. Agglutinating activity was localized in the dotted peak (the P-III fraction). Inset panel shows SDS-PAGE of the P-II fraction and P-III fraction. M, mol. wt markers; 1, the P-II fraction; 2, the P-III fraction.

while the normal type extract showed a weak agglutination at concentrations up to 250 μg/ml (data not shown). The agglutinating activity of the giant type extract was about 3 times higher than that of the plant lectin, wheat germ agglutinin (WGA) (Fig. 2).

Furthermore, the giant type extract produced the contraction of the longitudinal smooth muscle of the isolated guinea-pig ileum in a dose-dependent manner. The giant type extract at the concentration of 1 μg/ml produced 50 % contraction of the maximum contraction induced by 40 mM KCl in lower doses against the normal type extract (3 μg/ml) (data not shown).

Purification of *T. pileolus* Lectin

The giant type extract that was dialyzed against 0.15 M NaCl solution was applied to a Sephadex G-200 column equilibrated with the same solution. Fig. 3 shows an elution pattern with three peaks. The individual three peaks were assayed for the agglutinating activity with mouse and horse erythrocytes. The agglutinating activity of third peak (the P-III fraction) was higher than that of second peak (the P-II fraction), while that of first peak (the P-I fraction) was almost insensitive.

From the analysis on SDS-PAGE, the P-III fraction appeared to have several proteins with molecular weights of 5 kDa to 32 kDa. For further fractionation, the P-III fraction was applied on a DEAE-cellulose column equilibrated with 0.016 M Tris-HCl buffer at pH 7.6,

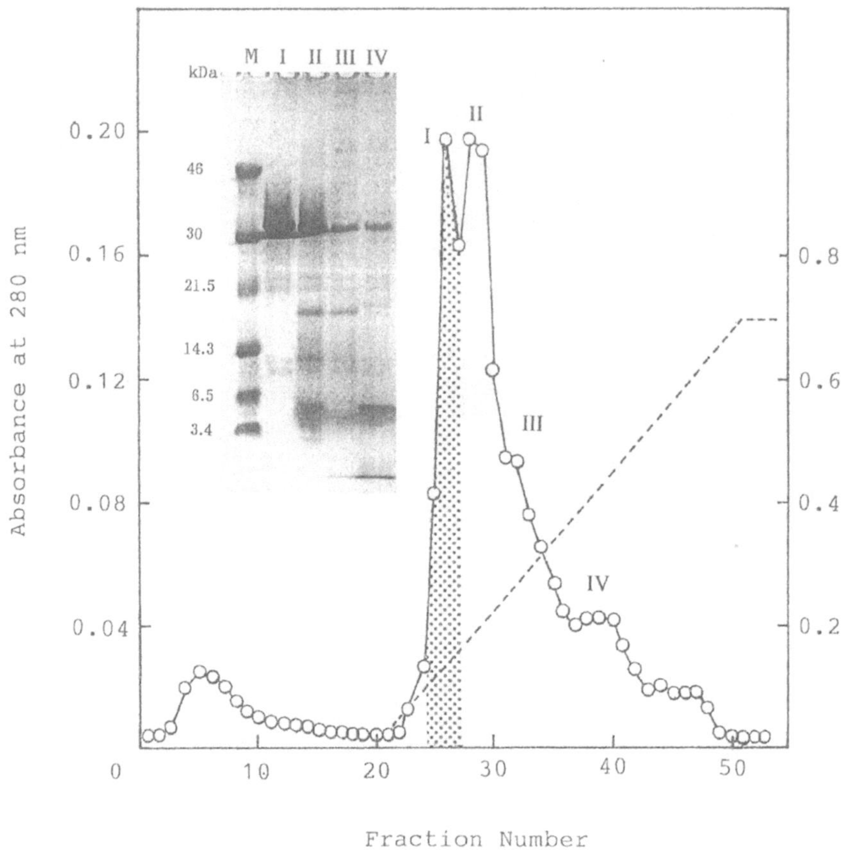

Figure 4. Anion-exchange chromatography on a DEAE-cellulose column of the P-III fraction from Sephadex G-200 gel chromatography and SDS-PAGE of the DEAE fractions (I-IV). The dotted peak (the DEAE-I fraction) was pooled. M, mol. wt markers.

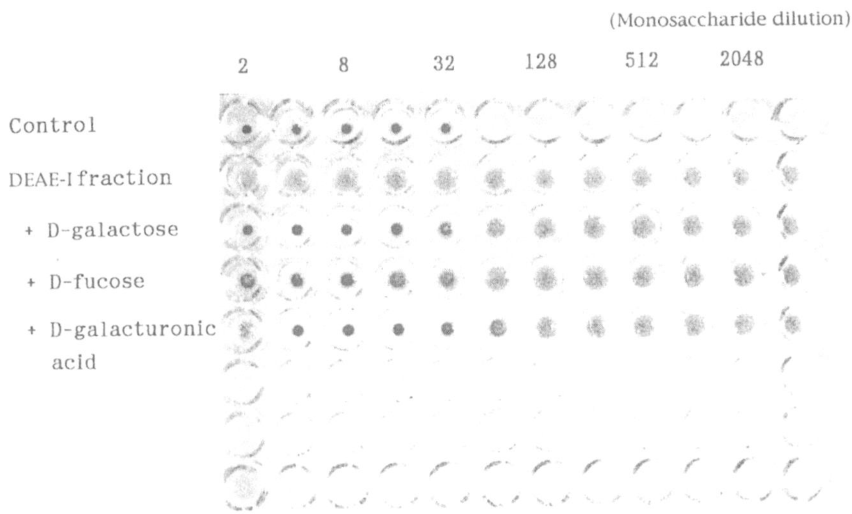

Figure 5. Inhibition of agglutination with mouse erythrocytes by monosaccharides. DEAE-I fraction, 1 μg/ml; monosaccharides, 100 mM.

and eluted with a linear gradient of NaCl from 0 to 0.7 M in the same buffer. Fig. 4 shows an elution pattern obtained from DEAE-cellulose column chromatography. Approximately four fractions (I-IV) were observed. The DEAE-I fraction that was eluted with 0.1 M NaCl had the highest activity of agglutination with mouse erythrocytes among the four DEAE fractions. The minimum concentration of the agglutination by the DEAE-I fraction was 0.1 μg/ml. This fraction also agglutinated horse and human erythrocytes. On SDS-PAGE, the DEAE-I fraction containing 32 kDa lectin was considered to be relatively pure in comparison with the other fractions (Fig. 4). The effects of monosaccharides on agglutination by the DEAE-I fraction were examined. As shown in Fig. 5, the agglutination with mouse erythrocytes by the DEAE-I fraction (1 μg/ml) were effectively inhibited by D-galactose (3.13 mM), D-galacturonic acid (3.13 mM) and D-fucose (12.5 mM) but not by N-acetyl-D-galactosamine, suggesting that the C-2 animo group is unfavorable for binding to the lectin.

The 32 kDa lectin, contained within the DEAE-I fraction, eluted as a single symmetrical peak from RP-HPLC and was found to be electrophoretically homogeneous by SDS-PAGE under reduced conditions (Fig. 6). Under nonreduced conditions, a single band corresponding to a molecular mass of 32 kDa was somewhat diffuse (data not shown), suggesting that the lectin might have some intramolecular S-S linkage(s). From these observations, the 32 kDa lectin appears to be a monomer. This 32 kDa lectin, sea urchin lectin-I (SUL-I) agglutinated mouse and horse erythrocytes.

Amino Acid Composition of Sea Urchin Lectin-I (SUL-I)

Sea urchin lectin (SUL-I) was hydrolyzed for 22-, 48-, and 72-hr. Serine and threonine was obtained by extrapolating to zero hydrolysis time. SUL-I consists of 294 amino acid residues with the molecular weight of 31,500 (Table 1). This was in good agreement with the value of 32,000 Da measured by the SDS-PAGE analysis. The amino acid composition indicates that SUL-I is particularly rich in serine and has 38 serine residues. Serine comprises about 13 % of the total residues. It is also rich in glycine residue, which gives the value of 23. Thus, there are relatively high amounts of neutral amino acids.

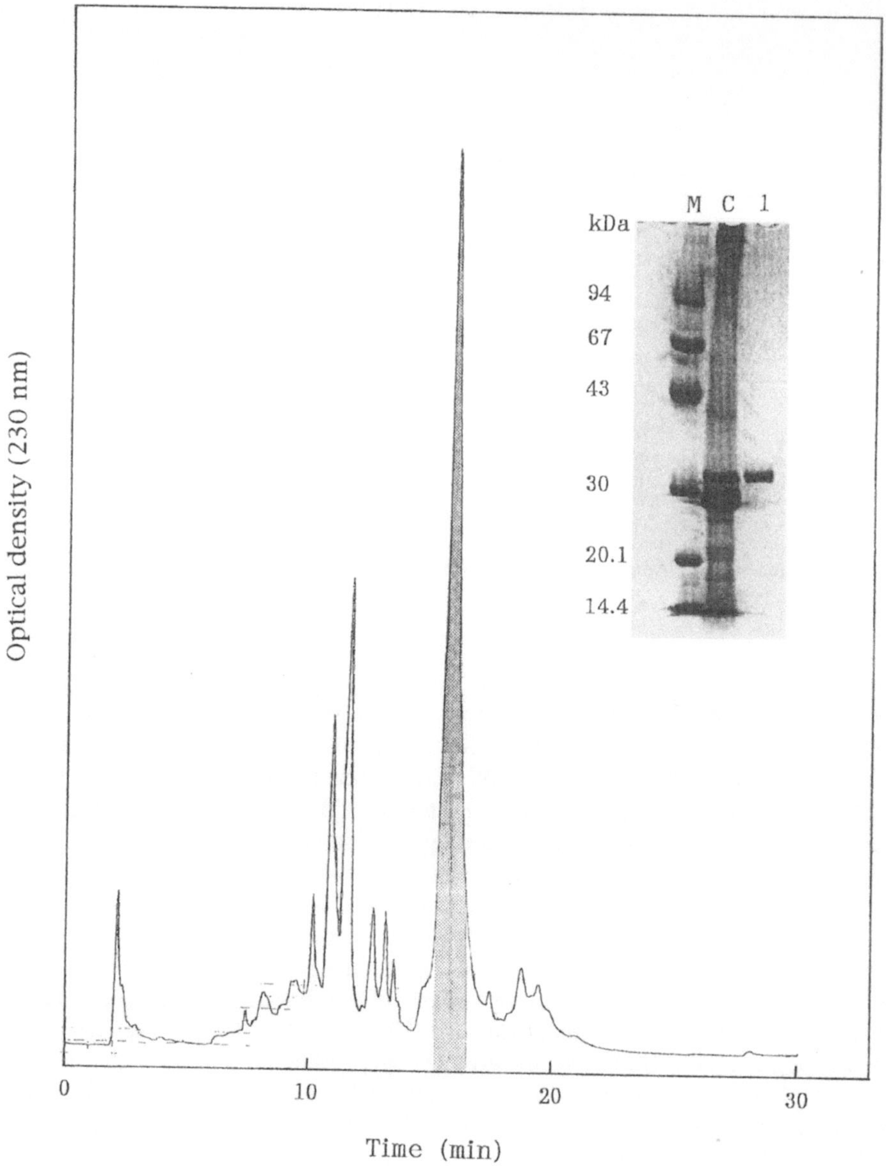

Figure 6. HPLC separation of the DEAE-I fraction. The sample was resolved on a reverse-phase HPLC column (C_8, 4.6 x 150 mm), monitored at 230 nm, using a linear gradient of acetonitrile (0-80 %) in 0.08 % TFA at 1 ml/min. Inset, SDS-PAGE of the crude extract (C) and the purified lectin (1). M, mol. wt markers.

Chemotactic Activity of SUL

We found that the P-III fraction from the giant type pedicellariae had chemotactic activity of guinea-pig neutrophils in a dose-dependent manner (25-200 μg/ml) (Fig.7A). Therefore, the chemotactic response of SUL-I was also examined. As can be seen in Fig. 7B, SUL-I induced chemotactic response in lower dose ranges (0.625-10 μg/ml) than those in the P-III fraction. Thus, the chemotactic activity by SUL-I was about 20 times higher than that by

Table 1. Amino acid composition of SUL-I from *T. pileolus*

Amino acid	Time of HCl hydrosis 24h	48h	72h	Corrected value	Nearest integer
Asx	5.98	5.94	5.97	5.97	33
Thr	2.08	1.98	1.89	2.16	12
Ser	6.35	5.62	5.07	7.00	38
Glx	4.27	4.26	4.27	4.26	23
Pro	2.54	2.63	2.52	2.56	14
Gly	4.13	4.14	4.15	4.14	23
Ala	3.86	3.86	3.87	3.86	21
Val	2.65	2.69	2.72	2.72	15
Met	0.37	0.36	0.37	0.37	2
Ile	2.55	2.74	2.81	2.81	15
Leu	2.62	2.62	2.62	2.62	14
Tyr	2.91	2.91	2.92	2.92	16
Phe	0.74	0.74	0.74	0.74	4
His	0.67	0.72	0.72	0.70	4
Lys	1.45	1.48	1.49	1.47	8
Arg	3.09	3.07	3.08	3.08	17
Cys	6.27	6.34	6.39	6.33	35
NH4	5.64	6.47	7.24	6.45	
			Total	residues	294
			Molecular	weight	31,500

the P-III fraction in comparison with the maximal responses. Chemotaxis by neutrophils plays an important role in the defense reaction to infection or injury of the organisms. Therefore, it is expected that SUL-I as a chemoattractant can be used as a valuable tool for biomedical research.

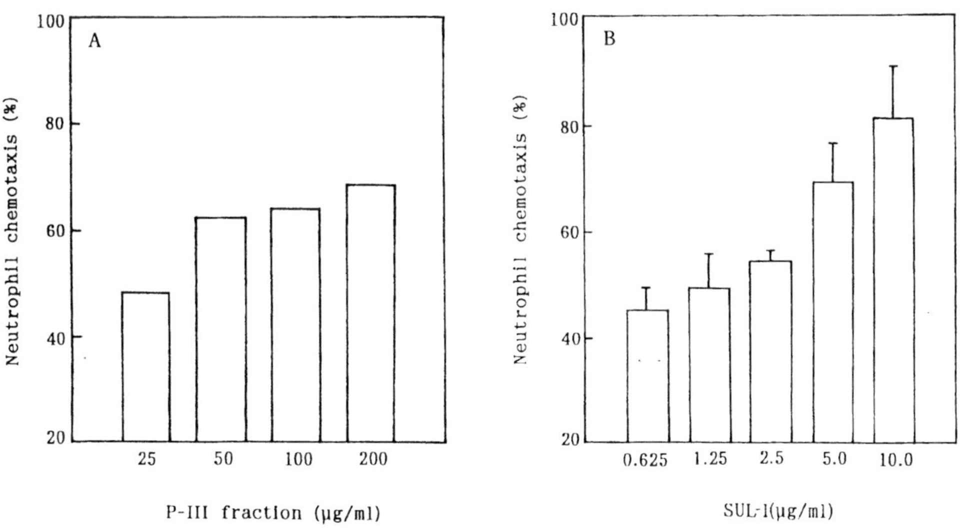

Figure 7. Effects of the P-III fraction (A) and sea urchin lectin (SUL-I) (B) on neutrophil chemotaxis. Neutrophils (2 x 10^6 cells/ml) were incubated at 37°C for 60 min with or without the P-III fraction or SUL-I, and migrated cells were counted. The chemotactic response to FMLP (10^{-7} M) was expressed as 100 %. A, data represent the mean of two experiments of triplicate determinations; B, data represent the mean ± SD of 5 to 9 experiments of triplicate determinations.

Drickamer (1988) have classified as C-type and S-type animal lectins, which are characterized by the requirement of Ca^{2+} and reducing reagents, and by their primary structures. Recently, some lectins from marine invertebrates belonging to Echinodermata, have been isolated and classified as C-type animal lectins (Giga et al., 1987; Hatakeyama et al., 1993, 1994). In the present study, we have isolated a calcium-independent lectin, SUL-I (32 kDa lectin) from Echinoderm, sea urchin *T. pileolus* by gel chromatography, ion-exchange chromatography and HPLC. In addition, SUL-I was also purified by affinity chromatography using an immobilized D-galactose or immobilized lactose (data not shown). The mammalian/vertebrate S-type lectins (14-16 kDa) are present both intra- and extracellularly, and show galactose/lactose specificity (Drickamer, 1988). The above observations suggest that SUL-I, a galactoside-binding lectin appears to be a S-type lectin, which is active in the absence of Ca^{2+}. However, it is uncertain if SUL-I shares structural homology with other S-type lectins in respect of the amino acid sequences. At present, we are attempting the amino acid sequencing of SUL-I. On the other hand, it has been reported that echinoidin from the coelomic fluid of the sea urchin *Anthocidaris crassispina* is a C-type lectin (Giga et al., 1987). More recently, we have also isolated a 32 kDa lectin from the coelomic fluid of *T. pileolus* as well as that from the giant type pedicellariae by affinity chromatography (unpublished data). This is surprising because the two lectins may be similar to each other.

Although the physiological roles of the 32 kDa lectin from the giant type pedicellariae of *T. pileolus* are not clear, it is suggested that the primary role of pedicellarial lectin may be defense and offense against a foreign body. In addition, the lectin from the coelomic fluid may be related to defense mechanisms of invertebrates, like vertebrate immune recognition.

Further structural studies would advance the understanding of biological functions of invertebrate lectins.

SUMMARY

Sea urchin lectin-I (SUL-I), a 32 kDa lectin was purified from the large globiferous pedicellariae of the sea urchin, *Toxopneustes pileolus* by using gel permeation chromatography, ion-exchange chromatography and reverse-phase HPLC. SDS-PAGE showed that SUL-I is a monomeric protein with a molecular mass of 32 kDa. Amino acid analysis indicates SUL-I to contain 294 residues. SUL-I was shown to have chemotactic properties for guinea-pig neutrophils at concentrations of 0.625 μg/ml. These data suggest that a 32 kDa lectin from *T. pileolus* may be related to defensive role.

ACKNOWLEDGMENTS

We are grateful to Prof. F. Tomiyoshi and Dr. M. Owhashi for the helpful discussion. We also thank Mr. T. Nagata and Mr. T. Nada for the collections of sea urchins, and Mr. S. Sakai for his assistance. This work was supported in part by a Grant-in-Aid from the University of Tokushima.

REFERENCES

Bradford, M. M., 1976, A rapid and sensitive method for the quantitation of microgram quantities of protein utilizing the principle of protein-dye binding. *Analyt. Biochem.* 72:248-254.

Drickamer, K., 1988, Two distinct classes of carbohydrate-recognition domains in animal lectins. *J. Biol. Chem.* 263: 9557-9560.

Fujiwara, T., 1935, On the poisonous pedicellaria of *Toxopneustes pileolus* (Lamarck). *Annot. Zool. Jpn.* 15: 62-69.

Giga, Y., Sutoh, K., and Ikai, A., 1985, A new multimeric hemagglutinin from the coelomic fluid of the sea urchin *Anthocidaris crassispina. Biochemistry.* 24: 4461-4467.

Giga, Y., Ikai, A., and Takahashi, K., 1987, The complete amino acid sequence of echinoidin, a lectin from the coelomic fluid of the sea urchin *Anthocidais crassispina. J. Biol. Chem.* 262: 6197-6203.

Halstead, B. W., 1988, *Poisonous and Venomous Marine Animals of the World.* pp.196-212. Darwin Press, Princeton.

Hatakeyama, T., Himeshima,T., Komatsu, A., and Yamasaki, N., 1993, Purification and characterization of two lectins from the sea cucumber *Strichopus japonicus. Biosci. Biotech. Biochem.* 57: 1736-1739.

Hatakeyama, T., Kohzaki, H., Nagatomo, H., and Yamasaki, N., 1994, Purification and characterization of four Ca^{2+}-dependent lectins from the marine invertebrate, *Cucumaria echinata. J. Biochem.* 116: 209-214.

Kawagishi, H., Yamawaki, M., Isobe, S., Usui, T., Kimura, A., and Chiba, S., 1994, Two lectins from the marine sponge *Halichondria akadai* : An N-acetyl-sugar-specific lectin (HOL-I) and an N-acetyllactosamine-specific lectin (HOL-II). *J. Biol. Chem.* 269: 1375-1379.

Kimura, A., Hayashi, H., and Kuramoto, M., 1975, Studies of urci-toxins : Separation, purification and pharmacological actions of toxinic substances. *Japan J. Pharmacol.* 25: 109-120.

Laemmli, U, K., 1970, Cleavage of structural proteins during the assembly of the head of bacteriophage T4. *Nature.* 227: 680-685.

Muramoto, K., and Kamiya, H., 1986, The amino-acid sequence of a lectin of the acorn barnacle *Megabalanus rosa. Biochim. Biophys. Acta.* 874: 285-295.

Nakagawa, H., Kimura, A., Takei, M., and Endo, K., 1982, Histamine release from rat mast cells induced by an extract from the sea urchin *Toxopneustes pileolus. Toxicon.* 20: 1095-1097.

Nakagawa, H., Tu, A. T., and Kimura, A., 1991, Purification and characterization of Contractin A from the pedicellarial venom of sea urchin, *Toxopneustes pileolus. Arch. Biochem. Biophys.* 284: 279-284.

Nakagawa, H., Yanagihara, N., Izumi, F., Wada, A., and Kimura, A., 1992, Inhibition of nicotinic acetylcholine receptor-mediated secretion and synthesis of catecholamines by sea urchin toxin in cultured bovine adrenal medullary cells. *Biochem. Pharmacol.* 44: 1779-1785.

Nakagawa, H., Tomihara, Y., Araki, Y., and Hayashi, H., 1994, Preliminary studies on venom proteins in the pedicellariae of the toxopneustid sea urchins, *Toxopneustes pileolus and Tripneustes gratilla. J. Natural Toxins.* 3: 25-34.

Ohura, K., Shinohara, M., Ogata, K., Nishiyama, A., and Mori, M., 1990, Leucocyte function in rats with naturally occurring gingivitis. *Archs oral Biol.* 35: Suppl. 185s-187s.

Okada, K., 1955a, Biological studies on the practical utilities of poisonous marine invertebrates. I. A preliminary note on the toxical substance detected in the trumpet sea urchin, *Toxopneustes pileolus. Rec. Oceanogr. Wks.* 2: 49-52.

Okada, K., Hashimoto, T., and Miyauchi, Y., 1955b, A preliminary report on the poisonous effect of *Toxopneustes* toxin upon the heart of oyster. *Bull. Marine Biol. Asamushi.* 7: 133-140.

Suzuki, T., Takagi, T., Furukohri, T., Kawamura, K., and Nakauchi, M., 1990, A calcium-dependent galactose-binding lectin from the Tunicate *Polyandrocarpa misakiensis*: Isolation, characterization, and amino acid sequence. *J. Biol, Chem.* 265: 1274-1281.

Takei, M., Nakagawa, H., and Endo, K., 1993, Mast cell activation by pedicellarial toxin of sea urchin, *Toxopneustes pileolus. FEBS Lett.* 328: 59-62.

INDIAN CATFISH (*Plotosus canius*, HAMILTON) VENOM

Occurrence of Lethal Protein Toxin (Toxin-PC)

B. Auddy and A. Gomes

Department of Physiology
University College of Science and Technology
92, A.P.C. Road, Calcutta, India 700009

INTRODUCTION

India possessed long coastal areas where venomous catfish are abundant. But almost none of these venomous fish species has been touched for detail investigation regarding their venoms and the active components present in the venom. *Plotosus Canius* is one of such major venomous catfish species that is present in the estuaries of eastern India. Sting of these fish produces intense pain, swelling and gangrene of the affected areas. On experimental animals the venom produced hypertension and respiratory failure, cardiac arrest, neuromuscular blockage and the LD_{50} was found to be 3.9 mg/kg (ip) in male albino mice (Auddy *et al*, 1993). An attempt was made to isolate and purify the lethal factor from the crude venom and its probable involvement in pathophysiology.

MATERIALS AND METHODS

Live fish of both sex averaging of 150 gm were obtained commercially during June to November. The pectoral and dorsal spines along with the covering membrane were cut and stored at -20°C. The pooled spines were ground using a heavy mortar and pestle, homogenized with 0.1M phosphate buffer pH 7.2 and centrifuged at 10,000 g for 20 mins at 4°C. The supernatant was used as crude venom. The venom concentration has been expressed in terms of protein, estimated after Lowry *et al* (1951). Crude venom was subjected to precipitation by addition of crystalline ammonium sulphate to 50% saturation (8 hr at 4°C). Ion Exchange Chromatography was carried out with a DEAE-cellulose column (15 ID x 60 mm). The column was eluted with a stepwise gradient of the 0.01M Phosphate buffer containing NaCl from 0.025 to 0.8 (M) followed by a pH gradient of bicarbonate buffer pH 8.0 and acetate buffer pH 5.0. The homogeneity of the protein toxin was checked by Polyacrylamide gel electrophoresis as described by Davis (1964). The gels were stained with

Natural Toxins II, Edited by B. R. Singh and A. T. Tu
Plenum Press, New York, 1996

periodic acid and Schiff's reagent for glycoprotein staining. To determine the molecular weight of the DEAE purified toxin, sodium dodecyl sulphate (SDS) polyacrylamide gel electrophoresis was carried out as described by Weber and Osborn (1969). The reference proteins used were bovine serum albumin (68,000), RNAase (13,700), β-lactoglobulin (18,400), pepsin (35,000) and ovalbumin (45,000). Isolated guinea pig heart was prepared after Langendorff. Contractins were recorded on a smoked drum with a Starling heart lever. Effect of agonist (Acetylcholine 10 μg and noradrenaline 5μg) and antagonist (Atropine 100 μg/ml and propanolol 5 μg/ml) were studied before and after Toxin-PC exposure. Isolated Chick biventer cervicis and Rat phrenic nerve diaphragm were prepared after Ginsborg and Warriner (1960) and Bulbring (1946) and stimulated with a square-wave electronic stimulator (8V, 0.5 m sec duration, 0.2 Hz). Effect of High Ca^{++} ion, high K^+ ion, 4-aminopyridine and neostigmine were studed. The LD_{50} was determined in male albino mice (18-20 gm), toxin being injected intravenously through the tail vein. Toxin-PC was mixed with Freund's complete adjuvant (1:1 v/v) and injected (30 μg/injection) subcutaneously in male rabbits in the first, second and third weeks. after one week rest, booster injections of Toxin-PC without adjuvant were administered intravenously for another three weeks (one injection per week). With the serum Immunogel diffusion and immunogel electrophoresis were carried out after Ouchterlony (1948) and Graber and William (1953) respectively. The mixture of antiserum and toxin (1:1 v/v) was incubated at 37° for one hour and then injecting the mixture intravenously into male albino mice (20 gm) for the lethality antagonism study.

All results are expressed as mean ± SE. The significance of the difference between means was determined by student's t-test. P values <0.05 were considered significant.

RESULTS

Isolation and purification of Toxin-PC - Protein precipitated with 50% ammonium sulphate saturation possessed lethality (MLD 46 μg/20 gm mice, iv. bolus). The fold of purification achieved by the process was 1.95 (Table 1). The same precipitate of the venom was loaded on DEAE cellulose column. Seven distinct protein peaks were eluted. Peak 7 eluted with 0.2 M acetate buffer pH 5.0 and possessed lethality. The fold of purification was found to be 15 with a yield of 8.8% (Table 1), and it was provisionally designated as "Toxin-PC" (P = Plotosus, C = Canius).

Toxin-PC showed a sharp single PAS-negative band in polyacrylamide gel electrophoresis and was found homogeneous on SDS-PAGE the molecular weight of Toxin-PC was found to be near 15 kDa. Toxin-PC was the first lethal protein toxin isolated from Indian catfish species. The minimum lethal dose (MLD) of Toxin-PC was 6 μg/20 gm mice and LD_{50} was 225 μg/kg, i.v., bolus, in male albino mice. Several other lethal protein toxins has been isolated from venomous fish species like Stingray, Oriental catfish, Weeverfish, Stonefish and few other species. A comparison of lethal protein toxins from venomous fish were given in Table 2.

Toxin-PC (15 μg) produced immediate reversible cardiac arrest on isolated guinea pig heart and prior administration of atropine and propanolal failed to antagonise Toxin-PC induced cardiac arrest. On isolated rat phrenic nerve diaphragm preparation, Toxin-PC (15 μg/ml) produced 100% blockage of electrically induced twitch response within 12 ± 3.1 min, n = 6. On isolated chick biventer cervicis (CBC) preparation Toxin-PC (5 μg/ml) produced 100% inhibition of twitch response within 8±1.6 min, n = 6. On CBC preparation, Toxin-PC did not alter the acetylcholine and carbachol induced contractile response. Toxin-PC induced neuromuscular blockage was unaffected in presence of neostigmine. Action of Toxin-PC was potentiated by doubling the K^+ ion concentration (9.4 mM) of the medium as judged by reduced blockage time of CBC. A voltage dependent K^+ channel blocker 4-amino pyridine

Table 1. Purification of *P. canius* Venom by DEAE - Chromotography

Seperation Steps	Total Protin (mg)	MLD (μg/20 gm, i.v.)	Total No. of MLD	Fold of Purification	Yield (%)
Crude Venom	15	90	167	—	100
50% $(NH_4)_2SO_4$ppt	9	46	196	1.95	60
DEAE-purified 'Toxin-PC'	1.32	6	220	15	8.8

failed to antagonise Toxin-PC action. Toxin-PC induced tension was significantly altered by the Ca^{2+} ion concentration of the suspensioon medium. Doubling and trippling the Ca^{2+} ion concentration (5.2 and 7.8 mM) increased the tension by 100% and 300% respectively. Antiserum raised against Toxin-PC showed low titre antibody and failed to develop precipitation bands in immunogel diffusion and immuno electrophoresis. The antiserum also failed to provide significant protection against Toxin-PC induced lethality in male albino mice as judged by neutralization studies.

DISCUSSION

Plotosus Canius is one of the most deadly venomous Indian Catfish. Its sting produced several pathophysiological alteration and there is no specific treatment due to inadequate information on the venom toxic component. The present investigation isolated a lethal toxic principle and try to explore the underlying mechanism of lethality.

A two-step purification procedure was used for the isolation of lethal protein toxin, Toxin-PC, including 50% ammonium sulphate precipitation followed by DEAE-ion exchange chromatography. Toxin-PC was homogenous in polyacrylamide gel electrophoresis with a SDS-molecular weight of near 15 kDa and devoid of glycoprotein moiety. As described in the results section, lethal Toxin-PC was highly active on isolated heart and nerve

Table 2. A comparative representation of some of the lethal protein toxins isolated from fish venom

Species	Toxic Component	Major Characteristic	Lethality
Stingray (*U.halleri*)	Lethal protein (Russel, 1965)	MW 100 kd (apx)	LD_{50} = 2.9 mg/kg, i.v. in mice
Stonefish (*S. horrida*)	Stonustoxin (Poh et al 1991)	MW 150 kd exist as heterodimer	LD_{50} = 17ng/gm i.v. in mice
(*S. trachynis*)	Cytolytic toxin (Kreger 1991)	MW 158	Lethal and haemolytic
Weeverfish (*T. Vipera*)	Trachinine (Perrier et al. 1988)	MW 324 kd, composed of four identical subunit	LD_{100} = 100 mg/kg i.v. in mice
(*T. draco*)	Dracotoxin (Chhatwal & Dreyer 1992)	MW 105 Kd consist of a single poly-peptide chain	Lethal and haemolytic
Oriental Catfish (*P. lineatus*)	Lethal protein (*Shiomi et al 1986*)	MW 12 kd	Lethal having edema forming activity
Indian Catfish (*P.Canius*)	Toxin-PC (Present study)	MW 15 Kd	LD_{50} = 225 μg/kg i.v. in mice

muscle preparation. Probably these two factors synergistically contributed to the lethality. On isolated guinea pig heart, Toxin-PC produced blockade without interacting directly with the cholinergic and ß-adrenergic receptors. On isolated nerve-muscle preparation Toxin-PC was found to be highly active. On chick biventer cervicis and rat phrenic nerve diaphragm (RPND) Toxin-PC produced complete irreversible blockade of twitch response. The sensitivity of Toxin-PC to CBC was greater than that of RPND. There were some reports that showed the presence of neuromuscular blocking component in the fish venom. The venom of two stonefish varities (*S. horrida* and *S. trachynis*) possessed neuromuscular blocking activity. 'Stonustoxin' isolated from *S. horrida* was neurotoxic between 8 - 50 μg/ml (Low et al 1994), a dose which was much higher from its lethal dose. Lionfish (*P. Volitans*) also contained a toxin that effects neuromuscular transmission (Cohen & Olek, 1989). These reports indicate fish venom may contain specific neurotoxin though several other species, viz, Indian Catfish (*H. fossilis*) did not possess any neuromuscular blocking activity (Datta, 1980). Toxin-PC induced neuromuscular blockage was probably one of the major reasons of its lethality. Among the fish venom toxins, only Toxin-PC can be campared with the venom of the elapidae group of snakes due to its neurotoxicity as well as lethality. Further study is necessary to establish the actual mechanism of Toxin-PC action.

Lethality contributing factors of Toxin-PC were probably the cardiotoxic or neuromuscular blocking activity or a combination of these two factors. Respiratiory distress followed by respiratory failure induced by the *P. Canius* venom (Auddy *et al.* 1993) indicated respiratory muscle paralysis induced by neuromuscular blocking by the toxin. The cardiotoxic action of Toxin-PC may also contribute lethality in experimental animals.

Toxin-PC isolated from *Plotosus canius* venom showed potent cardiotoxicity and neuromuscular blocking activity. The nature of the venom action somewhat resembles that the venom of the elapidae group of snakes. The actual mechanism of Toxin-PC action was found to be very complex and we failed to antagonise toxin action by chemical antagonists. Immunological antagonism was not possible in this investigation probably due to poor antigenic property/low degree of purity of Toxin-PC.

ACKNOWLEDGEMENT

This investigation was supported partly by an award of a Senior Research Fellowship to one of the authors (AUDDY B) by the University of Calcutta (Sanction No. UGC/942/Jr Fellow (Sc)/91/92).

REFERENCES

Auddy B, Alam M I, Gomes A (1993) : Pharmacological actions of the venom of the Indian Catfish (*Plotosus canius*, Hamilton). Ind. J Med Res 99 : 47-51.-

Bulbring E (1946) : Observation of the isolated phrenic nerve diaphragm preparation of the rat. Br J Pharmacol 1 : 38-42.

Burn J H (1952) : In Practical Pharmacology (Blackwell Scientific Publications, Oxford) p 22.

Chhatwal I and Dreyer F (1992) : Isolation and characterization of dracotoxin from the venom of the greater weeverfish *Trachinus draco*. Toxicon, 30; 87-93.

Davis B J (1964) : Disc electrophoresis II. Method and application to human serum Proteins. Ann N Y Acad Sci 121 : 404-427.

Datta A (1980) : Studies on Fish venom Ph. D. thesis, University of Calcutta.

Ginsborg B L, Warriner J (1960) : The isolated chick biventer cervicis nerve muscle preparation. Br J Pharmacol 15 : 410-411.

Greber P, Williams L A (1953) : Method for combined investigations of electrophoretic and immunochemical properties of a protein. Biochem Biophys Acta 10 : 193-197.

Kreger, A S (1991) : Detention of a cytolytic toxin in the venom of the Stonefish (*S trachynis*). Toxicon, 29, 733-743.

Litchfield J T, Wilcoxon F (1949) : A simplified method of evaluation dose effects experiments. J Pharmacol Exp ther 96 : 99-102.

Lowry O H, Rosebrough N J, Farr A L, Randal R J (1951) : Protein measurement with folin phenol reagent. J Biol Chem 193 : 265-275.

Ouchterlony O (1948) : In vitro methods for testing the toxin producing capacity of diptheria bacteria. Acta pathol Microbiol Scand 25 : 186-190.

Perriere, C., Perriere, F. G. and Petek, F. (1988) : Purification of a lethal fraction from the venom of the weeverfish, *Trachinis Vipera*, Toxicon, 26, 1222-1227.

Poh, C. H., Yuen, R., Khoo, H. E., Chung, M., Gwee, M.C.E. and Gopalakrishnakone, P. (1991). Purification and partial characterization of Stonustoxin (lethal factor) from *Synaceja horrida* Venom. Comp. Biochem. Physiol. 99B, 793-798.

Russell, F. E. (1965) : Marine toxins and venoms poisonous marine animals. Advance Marine Biol., 3, 255-384.

Shiomi, K., Takamiya, M., Yamanaka, H., Kikuchi, T. and Konno, K. (1986). Haemolytic, lethal and edema forming activities of the skin secretion from the oriental catfish (*Plotonus lineatus*). Toxicon, 24, 1015-1018.

Weber K, Osborn M (1969) : The reliability of molecular weight determination by dodecyl sulphate polyacrylamide gel electrophoresis. J Biol Chem 224 : 4406-4412.

16

NEUROTOXIN FROM BLACK WIDOW SPIDER VENOM

Structure and Function

E. V. Grishin

Shemyakin and Ovchinnikov Institute of Bioorganic Chemistry
ul. Miklukho-Maklaya, 16/10, 117871 Moscow, Russia

INTRODUCTION

As known some natural neurotoxins can effectively influence the secretion process of a neurotransmitter from the nerve ending. Among such neurotoxins particular interest is presented by the toxic components of the black widow spider *Latrodectus mactans tredecimguttatus*. *Latrodectus* toxins affect on all synapse types of various classes of animals, causing a massive release of a transmitter from the nerve terminal (Longenecker et al., 1970; Frontali, 1972; Grasso et al., 1979). α-Latrotoxin (α-LTX) is major toxic component of the black widow spider venom selectively toxic to vertebrates. The toxin can induce the transmitter release from nerve terminals and PC cells both in calcium and calcium-free medium (Grasso, 1988). The cDNA encoding the putative α-LTX precursor contains the 4203 base-pair open reading frame corresponding to the 156,855 Da protein composed of 1401 amino acid residues (Kiyatkin et al., 1990). But the molecular mass of the toxin deduced from the cDNA considerably differs from that determined for α-LTX by means of SDS gel electrophoresis (Ushkarev & Grishin, 1986) assuming possible processing in the C-terminal part of the polypeptide chain during its maturation. The size of the α-LTX molecule as well as its multifunctional properties might testify to the existence of several functional domains in the toxin structure. The paper summarizes the experimental results, which allow us to define some functional domains of the α-LTX molecule.

STRUCTURAL ANALYSIS

Ankyrin-Like Repeats

Detailed analysis of amino acid sequences of the α-LTX reveals the central region being almost entirely composed of a series imperfect repeats. These 33 amino acid repeats are found in membrane-binding domains of human ankyrins (Lux et al., 1990). Toxin repeats

Natural Toxins II, Edited by B. R. Singh and A. T. Tu
Plenum Press, New York, 1996

can be viewed as 33-34 amino acid ankyrin-like repeats containing N-terminal conserved and C-terminal variable parts. The ankyrin-like repeat motif is observed in several other proteins involved in cell differentiation, cell cycle control and transcription. Identification of ankyrin-like repeats in the neurotoxin molecule suggests a structural basis of the neurotoxin-membrane interaction. Ankyrins constitute a family of proteins coordinating interactions between various integral membrane proteins and cytoskeletal elements (Lux et al., 1990). Recent studies indicate that the repetitive motif domain of the ankyrin molecule is responsible for high affinity binding to membrane proteins. Thus one proposes that the structural domain of α-LTX containing ankyrin-like repeats could also take part in binding to presynaptic receptor components. Structural analysis of α-latroinsectotoxin (α-LIT) reveals that the central part of the toxin molecule is also composed of ankyrin-like repeats (Kiyatkin et al., 1993). Both neurotoxins can be hypothetically divided into three structural domains: an N-terminal domain (Mr about 51-52 kD) including 464-469 amino acid residues essentially free of internal repeats; the ~80 kD domain extending from residues 465-470 to about 1180, comprised almost entirely of imperfect ankyrin-like repeats and apparently assembled as an integrated unit; the C-terminal domain (about 200 amino acids) with Mr about 22 kD evidently released at toxin maturation.

Possible Binding Site

The comparative amino acid sequences provide some reasonable ground for a structural basis of the selectivity of α-LIT and α-LTX for insect and vertebrate receptors. Despite a high structural similarity, strong divergence is observed in analogous regions of the neurotoxins extending from residues 926-935 to 1027-1030, containing both an unusual interval between two ankyrin-like repeats and a clustering of cysteine residues. This region is supposed to be differentially cross-linked within itself in α-LIT domains. Just this molecule region is apparently responsible for the specificity of the toxin binding to its presynaptic acceptor and serves as a potential binding domain. Noteworthy, the attempts to obtain monoclonal antibodies against this region were not a success. The binding region of the α-LTX appears to be buried deep in the toxin molecule, thus being inaccessible to antibodies. This region with coordinates 935- 1030 is situated in the body of ankyrin-like repeats which are proved to form a globular structure in other proteins. So, ankyrin domain is expected to have a globular shape in the toxin molecule as well.

Possible Transmembrane Fragments

The α-LTX ability to interact with the lipid bilayer membrane forming cation channels means the existence of some hydrophobic fragments in its molecule. According to the hydropathy profile analysis a number of hydrophobic regions can be identified of insufficient length to constitute a conventional membrane-spanning α-helix, but might provide membrane interaction. At the same time at least two possible transmembrane fragments might be located in the N-terminal part of the toxin (Fig. 1).

One may conclude that channel forming properties of the toxin are realized due to the existence of only these fragments. If so, two fragments with coordinates 34-60 and 228-253 may be inserted in the lipid or cell membrane. In this case the fragment of the toxin molecule composed of 166 amino acid residues with coordinates 61-227 should penetrate into the nerve cell ending. The given fragment might participate in the calcium independent stimulation of the transmitter release. One toxin molecule apparently cannot form ion channels. The electron microscopy of single molecules indicates the existence of tetra- or even octamer α-LTX complex in solution (Lunev et al., 1991). Therefore one assumes that

```
        10        20        30        40        50
EGEDLTLEEKAEICSELELQQKYVDIASNIIGDLSSLPIVGKIAGTIAAA
        60        70        80        90       100
AMTATHVASGRLDIEQTLLGCSDLPFDQIKEVLENRFNEIDRKLDSHSAA
       110       120       130       140       150
LEEITKLVEKSISVVEKTRKQMNKRFDEVMKSIQDAKVSPIISKINNFAR
       160       170       180       190       200
YFDTEKERIRGLKLNDYILKLEEPNGILLHFKESRTPTDDSLQAPLFSII
       210       220       230       240       250
EEGYAVPKSIDDELAFKVLYALLYGTQTYVSVMFFLLEQYSFLANHYYEK
       260       270       280       290       300
GYLEKYDEYFNSLNNVFLDFKSSLVGTGTSNNEGLLDRVLQVLMTVKNSE
```

Figure 1. N-Terminal sequence of α-LTX. Possible transmembrane toxin fragments are underlined.

the only toxin tetramer is one functional unit of the channel formation, as a result several segments membrane channel is formed.

LOW MOLECULAR WEIGHT PROTEIN

Highly purified α-LTX preparations contain two components: protein of 1401 (α-LTX) and 70 (LMWP) amino acid residues (Kiyatkin et al., 1992). As shown α-LTX and LMWP were present in the toxin preparation in practically equal quantity forming a rather stable complex. The LMWP structure can be partially aligned with the sequence of erabutoxin A from the snake venom, but a much higher analogy exists with the primary structure of crustacean hyperglycemic hormones, suggesting a possible structural relationship between these proteins (Gasparini et al., 1994). LMWP possesses no toxicity to mammals and insect. It is also inactive in the electrophysiological experiment on the frog muscle. To investigate the functional role of LMWP in the toxin complex specific antibodies against its 14 amino acid residue C-terminal peptide were prepared. These antibodies inhibit α-LTX induced calcium influx and enhance an induced GABA release on synaptosomes (Grishin et al., 1993). Therefore, in spite of the absence of the observed toxicity LMWP is supposed to contribute to the functional action of the α-LTX probably owing to stabilization of the toxin complex. In any case it might be postulated that there should be a special structural region cooperating with LMWP in the α-LTX molecule.

FUNCTIONAL STUDIES

Functional Expression of α-LTX Receptor

On *Xenopus laevis* oocytes injected with rat brain poly (A $)^{+}$-RNA, α-LTX induced a slow increasing transmembrane inward current usually unobserved on uninjected oocytes (Filippov et al., 1994). The most prominent toxin-induced inward current arose from the use of mRNA fraction larger than 6kB. High molecular mass toxin-binding components of the α–LTX receptor were related to the toxin-induced ion channel. The main question to be answered here is: what the nature of the channel forming molecule is. To clarify this point further patch clamp experiments on oocytes were carried out. Very long single openings were observed in approximately 30% of both cell-attached and inside-out patches on oocytes injected with rat brain mRNA fraction of 7-8 kB when α-LTX was in a pipette. The channel conductance determined from the slope of the current-voltage relation was 7 pS. Single channel openings assembled into groups of bursts. The reversal potential of single channel currents was near to 0 and did not change significantly when calcium ions were substituted for Na^+ ions. This indicates that the channel produced by α-LTX does not discriminate between Ca^{2+} and Na^+ and can conduct both the cations. Preliminary results showed that channel openings were completely blocked by 2 mM Cd^{2+}. α–LTX produced on injected oocytes the channel even at 0.1 nM, the concentration comparable to that effective on mammalian synapses. Monoclonal antibodies against the α-LTX receptor protein did not influence induced channel properties. On the contrary antitoxin antibodies possessed visible effects on channel openings. The binding of α-LTX to the receptor seemed to increase greatly the probability of the α-LTX insertion into the cell membrane. One might conclude that after the α-LTX binding to the receptor the toxin inserts its transmembrane fragments into the presynaptic membrane forming ion channels exactly in the very active zones of synaptic transmission, where docking and fusion of synaptic vesicles occur. As a result a Ca^{2+} influx through this permanently open cation channel triggers a vesicle exocytosis and a massive neurotransmitter release.

Immunochemical Studies

α-LTX induces the calcium influx and secretagogue action on rat brain synaptosomes, which can be determined by ^{45}Ca uptake and [^{14}C]GABA release (Pashkov et al., 1993). A panel of monoclonal antibodies (mAbs) against α-LTX was obtained. The study with mAbs provided evidence that the toxin binding site was not connected with its action on calcium uptake and GABA release. None of the tested mAbs had the ability to inhibit α-LTX binding to the synaptosomal membrane, but at least five mAbs could modify the main toxin effects on synaptosomes and artificial lipid membrane. The addition of mAbs to the synaptosomal preparation exerted different effect on channel-forming and secretagogue function of the α-LTX. Three out of five investigated mAbs, A4, A6 and A24, completely neutralized the toxin ability to increase the calcium permeability of the synaptosomal membrane, while A15 and A19 did not influence this property of the α–LTX. At the same time, A6 and A24 completely, and A4 only partially, inhibited the capacity of α-LTX for initiating the release of GABA from synaptosomes. The calculated rate-constants for the GABA release in the absence of α-LTX or in the presence of unmodified and A4-modified α-LTX indicated that toxin molecule treatment by A4 decreased this α-LTX effect practically by half. In this case the process of the GABA release occurred at low synaptosomal calcium, since α-LTX-A4 complex blocked toxin-induced calcium influx into synaptosomes. So, the above experiments on rat brain synaptosomes showed that monoclonal antibodies can uncouple the toxin effects on toxin-induced calcium influx and stimulated the neurotransmitter release. Our experiments on incorporation of α-LTX into a bilayer lipid membrane displayed that the toxin after treatment with Abs never lost completely its capacity for increasing the BLM conductivity. There were observed slight differences related only to the frequency of channel incorporation. So, immunochemical experiments clearly demonstrated the existence in the α-LTX molecule of distinguishable functional domains responsible for the calcium influx (ionophoric action) and GABA release effect (secretogenic function), respectively.

FUNCTIONAL DOMAINS

α-LTX molecule might be divided into several functional domains. The first domain includes two potential membrane regions (34-60 and 228-253) together with intracellular region (61-227) of the toxin (Fig. 2).

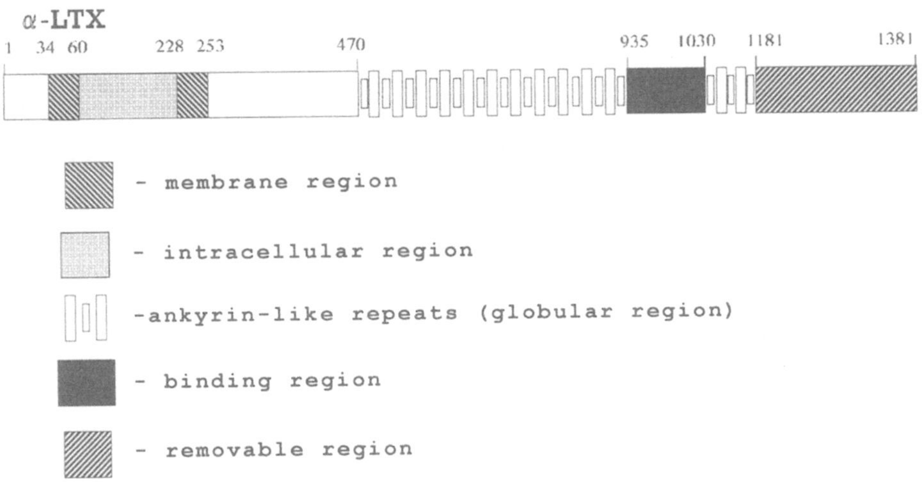

Figure 2. Schematic presentation of the α-LTX molecule.

The second domain (470-1180) consisting of ankyrin-like repeats contains the fragment (935-1030) presumably involved in binding to the presynaptic receptor. The C-terminal domain (1181-1381) has to be removed at the toxin maturation. Ionophoric action of the α-LTX and its secretogenic function are supposed to be linked with the first toxin domain. The toxin fusogenic effect is apparently connected with ankirin fragment situated in the second domain. Practically nothing is known on localization of the toxin fragment linked with interaction with LMWP.

REFERENCES

Frontali, N., 1972. Catecholamine-depleting effect of black widow spider venom on iris nerve fibers, Brain Res. 37:146-148.

Filippov, A.K., Tertishnikova, S.M., Alekseev, A.E., Tsurupa,G.P., Pashkov, V.N., and Grishin, E.V., 1994. Mechanism of α-latrotoxin action as revealed by patch-clamp experiments on *Xenopus* oocytes injected with rat brain messenger RNA, Neurosci. 61:179-189.

Gasparini, S., Kiyatkin, N., Drevet, P., Boulain, J.-C., Tacnet, F., Ripoche, P., Forest, E., Grishin, E., and Menez, A., 1994. The low molecular weight protein which co-purifies with α-latrotoxin is structurally related to crustacean hyperglycemic hormones, J.Biol.Chem. 269:19803-19809.

Grasso, A., and Senni, M.-I., 1979. A toxin purified from the venom of the black widow spider affects the uptake and release of radioactive gamma-amino butyrate and N-epinephrine from rat brain synaptosomes, Eur.J.Biochem. 102:337-344.

Grishin, E.V., Himmelreich, N.H., Pluzhnikov, K.A., Pozdnyakova, N.G., Storchak, L.G., Volkova, T.M., and Woll, P.G., 1993. Modulation of functional activities of the neurotoxin from black widow spider venom, FEBS Lett. 336:205-207.

Kiyatkin, N.I., Dulubova, I.E., Chekhovskaya, I.A., and Grishin,E.V., 1990. Cloning and structure of cDNA encoding α-latrotoxin from black widow spider venom, FEBS Lett.270:127-131.

Kiyatkin, N., .Dulubova, I., Chekhovskaya, I.,.Lipkin, A., and Grishin, E., 1992. Structure of the low molecular weight protein copurified with α–latrotoxin, Toxicon 30:771-774.

Kiyatkin, N., Dulubova, I., and Grishin, E., 1993. Cloning and structural analysis of α-latroinsectotoxin cDNA. Abundance of ankiryn-like repeats, Eur.J.Biochem. 213:121-127.

Longenecker, H.E., Hurlbut, W.P., Mauro, A., and Clark, A.W., 1970. Effects of black widow spider venom on the frog neuromuscular junction, Nature 225:701-703.

Lunev, A.V., Demin, V.V., Zaitsev, O.I., Spadar, S.I., and Grishin, E.V., 1991. Electron microscopy of α-latrotoxin from the venom of black widow spider *Latrodectus mactans tredecimguttatus*, Bioorg.Khim. 17:1021-1026.

Lux, S.E., John, K.M., and Bennet, V., 1990. Analysis of cDNA for human erythrocyte ankyrin indicates a repeated structure with homology to tissue-differentiation and cell-cycle control proteins, Nature 344:36-42.

Ushkaryov, Yu.A., and Grishin, E.V., 1986. Black widow spider neurotoxin and its interaction with rat brain receptors, Bioorg.Khim. 12:71-80.

17

STRUCTURAL AND FUNCTIONAL STUDIES OF LATRODECTIN FROM THE VENOM OF BLACK WIDOW SPIDER (*Latrodectus tredecimguttatus*)

A. Grasso and M. Pescatori

Institute of Cell Biology, CNR
Viale Marx 43, 00137 Rome, Italy

INTRODUCTION

Spider venoms are a rich source of potential probes for ion channels and receptors in nerve cells (Kawai and Nakajima,1993). From a biochemical point of view, the study of the molecular composition of spider venoms, is not easy because a large number of venom glands are necessary from animals which are generally small, and frequently, difficult to collect in large number. To avoid these limitations, we have been recently induced to use a molecular biology approach, to study spider venoms. In fact, cloning and expression of cDNAs offers an alternative to secure sufficient material to the purpose of detecting molecules having pharmacological activities. In the venom of the Mediterranean black widow spider (*Latrodectus tredecimguttatus),* we developed the initial idea to sequence and clone the most interesting neurotoxin, named alpha-latrotoxin. The potentialities of this project were based on a limited amino acid sequence available for the toxin (Pescatori et al., 1995). In fact, alpha-latrotoxin was considered to be a single-polypeptide toxin that exerts its neurotoxic action, by dramatically affecting synaptic vesicles exocytosis at the nerve endings (Grasso, 1988). The finding that this toxin could stimulate neurosecretion has been of great interest in neurobiology and the possibility of developing its use, as a pharmacological tool for the study of neurotransmitter release at various nerve terminals, has been greatly developed, afterwards. At nerve endings, the toxin stimulates and supports secretion even in the absence of external calcium, thus offering a great variation, not even considered before, of experimental conditions to test. Furthermore, alpha-latrotoxin binds with a high-affinity to receptor which is localised exclusively at the presynaptic plasma membrane (Petrenko et al., 1990). The molecular organisation of presynaptic structures has been derived by the development of functional analysis on toxin-receptor interactions (Petrenko, 1993; O'Connor et al., 1993). The receptor was characterised and defined as a member of a new class of extracellular matrix components known as neurexins (Geppert et al., 1992). It was suggested to bind to synaptotagmin (Petrenko et al., 1991), a protein specific of synaptic vesicles and thus implicated in synaptic vesicles exocy-

Natural Toxins II, Edited by B. R. Singh and A. T. Tu
Plenum Press, New York, 1996

tosis. It is now apparent that, with the definition of the toxin binding site, not only a series of biochemical correlations have been developed, as briefly summarised above but also our general understanding of the functional organisation of the nerve terminal has improved (Ushkariov et al., 1992; O'Connor et al., 1993). These comments mainly serve as an introduction for alpha-latrotoxin, the putative amino acid sequence of which has been now described (Kiyatkin et al., 1990), but they also serve as the basis for the development of new ideas on the study of the components of spider venom by a molecular biology approach. In fact, as a direct consequence of the study of the venom gland of spiders on the composition of nucleic acids of Latrodectus gland, a more complete picture of the protein components present in the venom fluid and venom gland extract has emerged. As mentioned earlier (Pescatori and Grasso, 1994), we have cloned by a RACE (rapid amplification of cDNA ends) methodology (Frohman et al., 1987), a cDNA encoding a small protein tightly associated (as indicated by its co-purification) with the high molecular mass component of the toxin. The low molecular mass protein was named Latrodectin (Pescatori et al., 1995) that corresponds to LMWP (low molecular weight protein) cloned in the Asiatic species of *Latrodectus*, (Kyiatkin et al., 1991). Being highly and specifically expressed only in the venom gland of the spider, it is considered to take part in toxic action of the venom (Pescatori and Grasso, 1994). Grishin et al. (1993) suggested that LMWP probably forms a functional part of alpha-latrotoxin and it could constitute the beta-subunit of the toxin. In view of the growing interest in the alpha-latrotoxin and its receptor, we deemed it essential to characterise Latrodectin further and to investigate the nature of its putative functional and structural interaction with latrotoxin. The present communication tends to answer some of the emerging questions on this matter by describing briefly some of the properties of Latrodectin as well as the immunochemical purification procedure developed for recombinant Latrodectin expressed in insect cells infected with baculovirus. This system of expression was chosen to secure sufficient, correctly folded, polypeptide to be utilised for studies directed towards a demonstration of a functional involvement of Latrodectin in the mechanism of action of alpha-latrotoxin.

EXPERIMENTAL PROCEDURES

Baculovirus Expression of Latrodectin Gene Product

Spodoptera frugiperda (SF9) cells were grown in TNM-FH medium (Gibco Lab.) with 10% inactivated calf serum. *Autographa californica* multiple nuclear polyhedrosis virus strain E2 (AcMNPV) was obtained from M.D. Summers (Texas A & M University). Recombinant baculoviruses that are capable of expressing Latrodectin (BV-XTx) were isolated by plaque purification following co-transfection of AcMNPV DNA and the baculovirus transfer vector pVL1392-XTx into SF9 cells, using described procedures (Summers and Smith, 1987). pVL1392-XTx transfer vector was constructed subcloning Latrodectin coding sequence into Eco-RI Bam-HI sites of pVL1392 plasmid (Pharmingen) according to standard methods (Sambrook et al., 1989).

Synthesis of Peptide, Antibody Production and Purification

A synthetic peptide corresponding to the predicted C-terminal tridecapeptide of Latrodectin was synthesised using a Fmoc solid phase synthesis strategy and Wang resin (Evans, 1993). The peptide was purified using Sephadex G-10. To facilitate the coupling of this peptide to a carrier protein, a cysteine residue which does not belong to the predicted sequence was included at the N-terminal. The complete sequence of the synthetic peptide thus being C- VYEEKDTPPVQE. For immunisation, the peptide was coupled to activated

Keyhole limpet hemocyanin (Sigma, USA) using maleimido-benzoyl-N-hydroxysuccinimide ester essentially as described by Harlow and Lane (1988). Antipeptide antiserum were raised in rabbits using a schedule of boosting at three weeks intervals followed by bleeding ten days after each boost. Antipeptide antibodies were affinity purified with immobilised synthetic peptide on epoxy-activated Sepharose 6B (Pharmacia).

Immunoaffinity Purification of Latrodectin

Anti-Latrodectin purified IgG were coupled and crosslinked to Protein A Sepharose with dimethylpimelimidate. These immunoaffinity columns were used to purify native and recombinant Latrodectin. Antigens (SF9 cell lysates infected with Baculovirus or black widow spider gland extract) were bound to the columns and eluted, after extensive washing, by a low pH step (0.1M glycine, pH 2.5) followed by a high pH step (0.1 M phosphate, pH 11.0). The presence of Latrodectin in the fractions was determined by Western blot analysis of the elution profile.

Western Blotting and Immunochemical Detection

Proteins were separated by electrophoresis on SDS-polyacrylamide gels mostly according to Schagger and von Jagow (1988), transferred to nitrocellulose (0.2 μm, Pharmacia), and probed with affinity purified anti-Latrodectin antiserum prepared as described above (1:2000 dilution). When needed, mouse monoclonal or polyclonal antibodies prepared in rabbit against alpha–latrotoxin, were used. Proteins were visualised by staining with goat anti-rabbit or rabbit anti-mouse secondary antisera conjugated to alkaline phosphatase.

RESULTS

Description and Properties of Latrodectin

The amino acid sequence of Latrodectin, deriving from *L. tredecimguttatus* or LMWP deriving from *L. mactans,* is schematically represented in Fig. 1a. As reported by Gasparini

a ISPAD IG[*C*]TD ISQAD FDEKN NN[*C*]IK

[*C*]GEDG FGEEM VNR[*C*]R DK[*C*]FT DNFYQ

S[*C*]VDL LN**KVY** **EEKDT PPVQE**

b

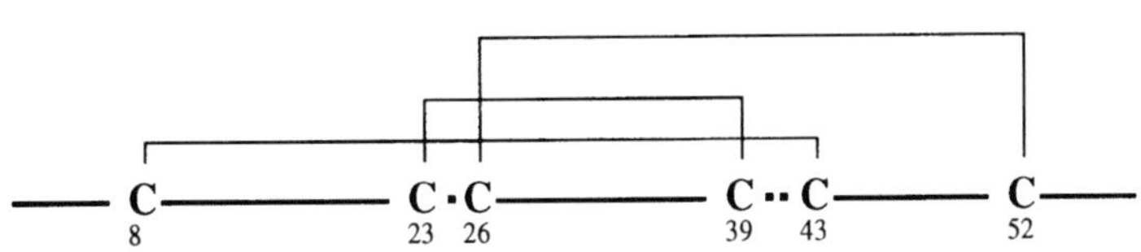

Figure 1. a) Amino acid sequence of mature Latrodectin, according to Pescatori et al., (1995). Cysteine residues are in italics and in brackets; in bold face is indicated the sequence of the synthetic peptide used as an immunogen for Latrodectin. b) The disulfide bridges assignment according to Gasparini et al. (1994), is given for a recombinant LMWP.

et al. (1994), initially no homology was found for LMWP to any other protein known. More recently a structural homology of LMWP has been suggested, with a crustacean hyperglycaemic hormone. The mature protein is characterised by being essentially hydrophilic, having an acidic isoelectric point, a molecular mass of about 8 000 Da, by the absence of histidine and tryptophan, and by the presence of six cysteine residues in its sequence (Kiyatkin et al., 1992; Gasparini et al., 1994; Pescatori et al., 1995). The formation of three disulfide bridges has been studied (Gasparini et al., 1994) and assigned as being of the following type: I-V; II-IV; III-VI (Fig. 1b).

Northern analysis for Latrodectin using recombinant cDNA indicated that the molecule is specific for the venom gland and that is present in *Latrodectus* species geographically remote from *L.tredecimguttatus* (Pescatori and Grasso, 1994; Pescatori et al., 1995).

Antisera Production and Immunoblot Analysis of Native Latrodectin

In an attempt to understand the role of Latrodectin in venom action, we produced antisera against the synthetic peptide mentioned earlier in rabbits (Fig. 1a). Antibodies were used as tools first to localise Latrodectin molecule in the venom or in various steps of the purification procedure for alpha-latrotoxin, and second to study the interference of Latrodectin on toxin action. The anti-Latrodectin antibodies produced were highly specific for the peptide and were displaced by an excess of immunogen. A Western blot of gland and cephalothoracic extracts confirmed that Latrodectin is present exclusively in the gland extract. Gel filtration of gland extracts showed that Latrodectin appears in two sets of fractions, a high molecular mass fraction and a low molecular mass fraction corresponding to free Latrodectin. This would suggest that Latrodectin tends to associate with other components of the venom. However, the high molecular mass component is not detectable in the venom fluid. Fig. 2, compares the Western blot analysis of the protein present in venom gland extracts and in the venom fluid obtained by stimulating the living spiders to secrete their venom directly in a capillary. Latrodectin is present in the venom fluid exclusively as a low molecular mass component, whereas a cross-reactive band of high molecular mass is visible in the venom gland extract. It is interesting to mention here that the native and recombinant Latrodectin are not recognised in Western blots unless the gel at the end of the electrophoretic run, or nitrocellulose after blotting, are reduced by treatment with 0.01 M dithioerythritol or 2–mercaptoethanol for 15-20 min. This suggests that, after electrophoresis, the C terminal portion of Latrodectin is not exposed unless some of the intrachain disulfide bonds are reduced.

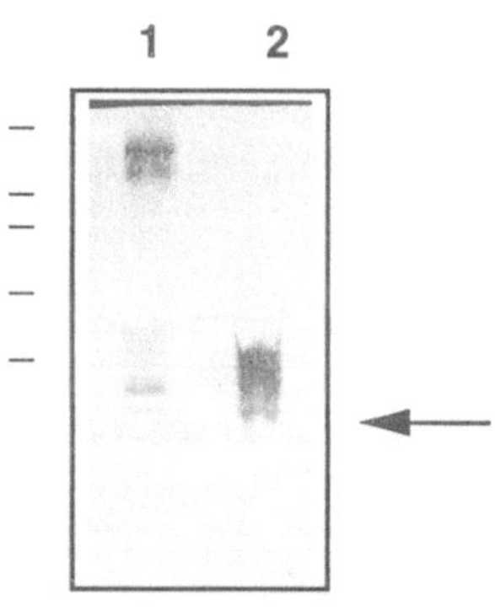

Figure 2. Comparison of Latrodectin immunoreactivity in venom glands extract and venom fluid. Comparable amount of proteins (about 10 μg) obtained from the venom gland extract of frozen spiders (lane 1) or from living spiders stimulated to secrete the venom in a capillary (lane 2) were analysed. The mobility of prestained standard proteins (Kaleidoscope prestained standard-BioRad) is reported on the left.

Latrodectin Immunoreactivity in the Cell System Infected with Baculovirus

The alternative to purifying Latrodectin from venom glands or venom fluid was to synthesise the recombinant polypeptide using a baculovirus expression system (Summers

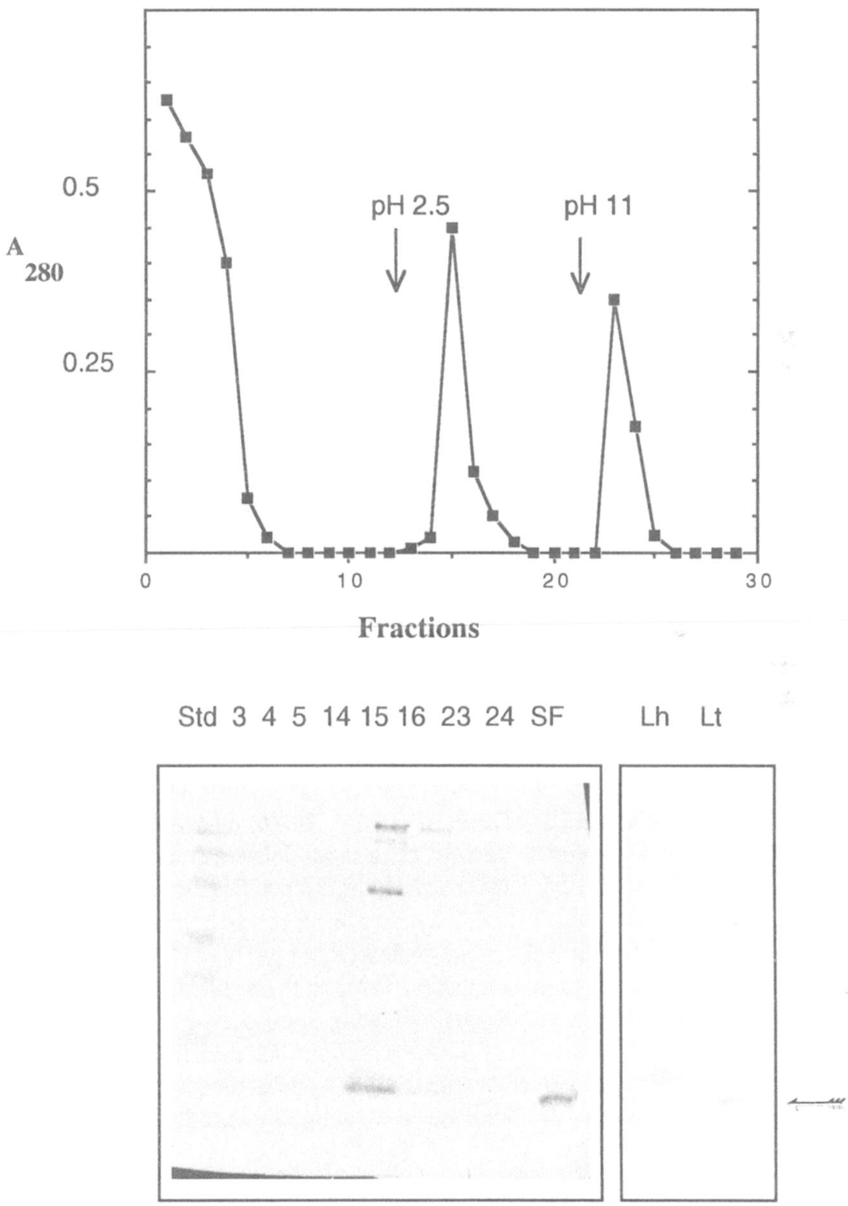

Figure 3. Immunochemical detection of Latrodectin in various fractions of the immunoaffinity column. Top) elution profile of the immuno-affinity column; Bottom) Western blot analysis of the indicated fractions and comparison with the pattern shown by extracts of *L. hesperus* (Lh); *L.tredecimguttatus* (Lt); standard proteins (Std); total SF9 cell extract (SF).

and Smith, 1987). The recombinant protein expressed by insect cells infected with a recombinant *Autographa californica* virus, was characterised by Western blot. The insect cell lysates infected with Latrodectin-baculovirus not only contain proteins capable of being recognised by antibodies against the C-terminus of Latrodectin, but after immunoaffinity purification, the expressed proteins had electrophoretic mobility in the same range as that of native proteins after electrophoresis of the gland extract. Fig. 3 shows that the yield for Latrodectin is greatly increased after immunoaffinity chromatography of cell lysates. A comparison with positive bands deriving from venom gland extracts of two Latrodectus species is also made.

DISCUSSION

The RACE method [Frohman et al., 1988] is useful for cloning cDNA for which limited sequence data is available. In this method, specificity of the PCR product is enhanced by using two nested oligonucleotides. In the case of protein sequence data, it is often difficult to derive two nested oligonucleotides due to the limitations of protein sequence length and the degeneracy of the amino acid code. Therefore, we have used a modification of the RACE method to clone the cDNA for the small component of *Latrodectus* venom, which employs overlapping and degenerate oligonucleotides. On the basis of the obtained sequence (Pescatori et al., 1995), the antiserum produced against the C-terminal portion of the molecule was used in immunoblot analysis to asses the specificity of the protein to the venom gland, and to detect its expression in insect cell cultures infected with baculovirus carrying the cDNA of the protein. In an attempt to understand the function of the protein in the toxic action of the venom, the antibodies were used to study whether they modify some of the biological responses attributed to alpha-latrotoxin. Although it is thought that this specific relationship between the cloned protein and alpha-latrotoxin (Pescatori and Grasso, 1994; Kiyatkin et al., 1992) exists, pilot experiments using recombinant Latrodectin or its antibodies have been unsuccessful, so far. Furthermore, preliminary studies of Gasparini et al. (1994) regarding the toxicity as well as channel forming ability of a recombinant analog of LMWP, in which the unique Met was changed in Leu and obtained as a fusion product of *E. coli*, did not show any modification of these two well known properties of alpha-latrotoxin.

On the other side, comparison of protein families having disulfide bridges arrangement similar to those described for Latrodectin [I-V; II-IV; III-VI] indicates that to this group toxins of sea anemone and rattlesnake venoms or protease inhibitors have been ascribed (Warne and Laskowski, 1990). We think that Latrodectin, so specific to the venom gland and so massively expressed in the venom fluid, may have functions in a way similar to those shown by sea anemone toxins in affecting ion channels (Harvey et al., 1993). We should add that, whereas this suggestion is mainly conjectural, we trust the significance of the positive response in Western and Northern blot analysis showing that Latrodectin is unique to the venom gland as an indication of specificity to venom action. The availability of the protein expressed by the insect cell system, together with the future production of specific antibodies against the entire molecule, should facilitate the verification of this suggestion.

REFERENCES

Evans, C.J., 1993, Peptide Synthesis, in: *Selected Methods for Antibodies and Nucleic Acid Probes* (Hockfield S., Carlson S., Evans C., Levitt P., Pintar J. and Silberstein L. Eds) Cold Spring Harbor Laboratory Press, Cold Spring Harbor, N.Y., pp. 503-546.

Frohman, M.A., Dush, M.K. and Martin, G.R., 1988, Rapid production of full-length c-DNA from rare transcripts: amplification using a single gene-specific oligonucleotide primer, *Proc. Natl. Acad. Sci. USA 85*, 8998-9002.

Gasparini, S., Kiyatkin, N., Drevet, P., Boulain, J.C., Tacnet, F., Ripoche, P., Forest, E., Grishin, E. and Menez, A., 1994, The low molecular weight protein which co-purifies with α-latrotoxin is structurally related to crustacean hyperglicemic hormones, *J. Biol. Chem. 269,* 19803-19809.

Geppert, M., Ushkaryov, Y.A., Hata, Y., Davletov, B., Petrenko, A.G., and Sudhof, T.C., 1992, Neurexins, *Cold Spring Harb. Symp. Quant. Biol.* 57, 483-490.

Grasso, A., 1976. Preparation and properties of a neurotoxin purified from the venom of the black widow spider (*Latrodectus mactans tredecimguttatus*), *Biochim. Biophys. Acta* 439, 409-412.

Grasso, A., Alemà, S., Rufini, S., and Senni, M.I., 1980, black widow spider toxin-induced calcium fluxes and transmitter release in a neurosecretory cell line, *Nature* 283, 774-776.

Grasso, A., 1988, α-Latrotoxin as a tool for studying ion channels and transmitter release process, in: *Neurotoxins in Neurochemistry* (Dolly, J.O. Ed.) Ellis Horwood Limited, Chichester, pp. 67-7

Grishin, E.V., Himmelreich, N.H., Pluzhnikov, K.A., Pozdnyakova, N.G., Storchak, L.G., Volkova, T.M., and Woll, P.G., 1993, Modulation of functional activities of the neurotoxin from black widow spider venom, *FEBS Lett. 336,* 205-207.

Harlow, E., and Lane, D., 1988, *Antibodies: A Laboratory Manual,* Cold Spring Harbor Press, Cold Spring Harbor, N.Y.

Harvey, A.L., Anderson, A.J., and Rowan, E.G., 1993, Toxins affecting ion channels in, *Natural and Synthetic Neurotoxins* (Harvey, A. Ed), Academic Press, London, New York, pp. 319-345.

Kawai, N., and Nakajima, T., 1993, Neurotoxins from spider venoms, in: *Natural and Synthetic Neurotoxins* (Harvey, A. Ed), Academic Press, London, New York, pp. 319-345.

Kiyatkin, N.I., Dulubova, I.E., Chekhovskaya I.A., and Grishin, E.V., 1990. Cloning and structure of cDNA encoding α-latrotoxin from black widow spider venom, *FEBS Lett. 270,* 127-131.

Kiyatkin, N., Dulubova, I., Chekhovskaya, I., Lipkin, A. and Grishin, E., 1992, Structure of the low molecular weight protein copurified with α-latrotoxin. *Toxicon 30,* 771-774.

O'Connor, V.M., Shamotienko, O., Grishin, E., and Betz, H., 1993, On the structure of the synaptosecretosome. *FEBS Lett.* 326, 255-260.

Pescatori, M., and Grasso, A. 1994, A tissue specific protein of the venom gland of black widow spider affects α-latrotoxin action. *Ann. N.Y. Acad. Sci. 710,* 38-47.

Pescatori, M., Bradbury, A., Bouet, F., Gargano, N., Mastrogiacomo, M., and Grasso, A., 1995, The cloning of a cDNA encoding a protein (Latrodectin) which co-purifies with the α-latrotoxin from the black widow spider *Latrodectus tredecimguttatus* (Theridiidae). *Eur.J.Biochem.* 230, 322-328.

Petrenko, A.G., Kovalenko, V.A., Shamotienko, O.G., Surkova, I.N., Tarasyuk, T.A., Ushkaryov, Y.A., and Grishin, E.V., 1990, Isolation and properties of the α-latrotoxin receptor, *EMBO J. 9,* 2023-2027.

Petrenko, A.G., Perin, M.S, Davletov, B.A., Ushkaryov, Y.A., Geppert, M., and Sudhof, T.C., 1991, Binding of synaptotagmin to the α–latrotoxin receptor implicates both in synaptic vesicles exocytosis, *Nature 353,* 65-68.

Petrenko, A.G. 1993, α-Latrotoxin receptor. Implications in nerve terminal function. *FEBS Lett. 325,* 81-85.

Sambrook, J., Fritsch, E.F., and Maniatis,T., 1989, *Molecular Cloning. A Laboratory Manual.* Cold Spring Harbor Laboratory Press. Cold Spring Harbor, N.Y.

Schagger, H., and von Jagow, G., 1987, Tricine-sodium dodecyl sulfate-polyacrylamide gel electrophoresis for the separation of proteins in the range from 1 to 100 kDa. *Anal. Biochem. 166,* 368-379.

Summers, M.D., and Smith, G.E., 1987, A manual of methods for baculovirus vectors and insect cell culture procedures, *Texas Agricultural Experiment Station Bulletin.* Vol. 1555.

Ushkaryov, Y.A., Petrenko, A.G., Geppert, M., and Sudhof, T., 1992, Neurexins: synaptic cell surface proteins related to the α–latrotoxin receptor and laminin, *Science 257,* 50-56.

Warne, N.W., and Laskowski, M., 1990, All fifteen possible arrangements of three disulfide bridges in proteins are known. *Biochem. Biophys. Res. Com. 172: 1364-1370.*

18

EFFECTS OF TOXIC SHOCK SYNDROME TOXIN-1 AND A SITE-DIRECTED MUTANT, H135A, IN MICE

B. G. Stiles,[1] T. Krakauer,[1] and P. F. Bonventre[2]

[1] U. S. Army Medical Research Institute of Infectious Diseases
Frederick, Maryland 21702-5011
[2] University of Cincinnati Medical Center
Cincinnati, Ohio 45267-0524

INTRODUCTION

Staphylococcus aureus is a medically important, gram-positive bacterium commonly found in soil, water, and on human skin. The organism can produce a variety of protein toxins that are intimately linked to many cases of toxic shock syndrome and food poisoning. Staphylococcal food poisoning is often attributed to contaminated meats and dairy products stored at improper temperatures. During the growth of *S. aureus* in food, the bacterium can make staphylococcal enterotoxins (SE) which are ~30 kD proteins that are heat stable and protease resistant. Symptoms of staphylococcal food poisoning occur within 6 h after ingesting contaminated food and include severe cramping, nausea, vomiting, and diarrhea for an additional 8 h.

Another toxin designated as toxic shock syndrome toxin-1 (TSST-1), but originally named enterotoxin F (1) and pyrogenic exotoxin C (21), has also been purified and characterized from the supernatants of *S. aureus* cultures isolated from toxic shock patients. Toxic shock syndrome was first described by Todd *et al.* (26) and consists of hypotension, fever, and multisystem dysfunction (4). Although most commonly associated with women and menstruation, toxic shock syndrome attributed to TSST-1 has been reported in men, usually as a result of a skin wound subsequently contaminated by *S. aureus*. TSST-1 is a 24 kD protein that has "superantigenic" characteristics, like the SE and pyrogenic exotoxins of *Streptococcus pyogenes*, which include binding to major histocompatibility complex class II (MHC II) and a subsequent proliferative effect upon specific T-cell populations. TSST-1 is serologically distinct from the seven SE serotypes. However, there are common epitopes among the SE (13, 14), and even between the SE and streptococcal pyrogenic exotoxins (3, 11). Interestingly, TSST-1 and SEB share very little amino acid sequence homology yet have remarkably similar crystal structures (20, 25).

In vivo studies with bacterial superantigens have been difficult, as many non-primate animals, such as mice, require large doses of toxin for an effect (15). Mice are relatively

Natural Toxins II, Edited by B. R. Singh and A. T. Tu
Plenum Press, New York, 1996

insensitive to SE and TSST-1, perhaps because their MHC II molecules have lower binding affinities than human MHC II. Endogenous viral and/or bacterial superantigens may also desensitize mice towards a subsequent challenge with another bacterial superantigen (8, 10).

The toxic effects of SE or TSST-1 are potentiated in mice with D-galactosamine (17), actinomycin-D (28), or lipopolysaccharide (LPS) (22, 24). LPS, also known as endotoxin, is an integral component of the outer membrane from gram-negative bacteria and normal flora found in the intestinal tract may naturally potentiate the toxicity of TSST-1 and SE (23). The biological activity of TSST-1 is intimately linked to cytokines such as tumor necrosis factor (TNF), interleukin 1, interleukin 2, and interferon gamma (IFNγ). Toxic shock syndrome is likely due to abnormally elevated levels of various cytokines (18). This paper describes the use of an LPS-potentiated mouse model for studying TSST-1 and a site-directed mutant, H135A.

MATERIALS AND METHODS

TSST-1 and H135A

Protein concentrations of purified TSST-1 (Toxin Technologies, FL) and H135A were determined by A_{277} using an extinction coefficient ($E^{1\%}{}_{277}$) of 9.7. The mutation of TSST-1 at histidine 135 was achieved by site-directed mutagenesis as described previously (5). Aliquots of TSST-1 and H135A were diluted in phosphate buffered saline (PBS), pH 7.4 and stored at -50°C.

Murine Bioassay

The toxicity of TSST-1 and H135A was tested in 20 g BALB/c mice (n=15 per dose) injected intraperitoneally (ip) with either protein diluted in PBS. Four hours later, mice were injected ip with 75 µg of *E. coli* 055:B5 LPS (Difco Laboratories, MI) and lethality was observed over 72 h. The serum cytokine levels of tumor necrosis factor α and β (TNF), as well as interferon gamma (IFNγ), were measured over 14 h (2 h intervals) in mice injected with either 15 µg TSST-1, 15 µg TSST-1 plus 75 µg LPS, 15 µg H135A plus LPS, or 75 µg LPS only. Serum levels of TNF were determined by an ELISA (Genzyme, MA) with triplicate samples ± the standard deviation (SD). Data for IFNγ were collected from an L929 cell lysis assay with duplicate samples ± 10%.

The vaccine study consisted of mice immunized ip, every two weeks for a total of three injections, with TSST-1 or H135A (8 µg/injection) mixed in RIBI adjuvant (RIBI ImmunoChem Research, MT). Control mice were injected with RIBI adjuvant alone. Each mouse was challenged 7 days after the final boost with a 15 LD_{50} dose of TSST-1 plus LPS.

Proliferation assays were done with naive BALB/c lymphocytes (6 x 10^6 cells/ml) incubated with a 100 ng/ml concentration of TSST-1 plus a 1:32 dilution of pooled sera from mice immunized with either TSST-1, H135A, or adjuvant alone. Cells were incubated with ^{3}H-thymidine, harvested, and radioactivity measured in a liquid scintillation counter.

ELISA

The serological similarity of H135A and TSST-1 was determined by serially diluting each protein in carbonate buffer, pH 9.6 and adsorbing them onto Immulon II microtiter plates (Dynatech Laboratories, VA). After incubating overnight at 4°C, the wells were aspirated and blocked with PBS-1% gelatin. Rabbit anti-TSST-1 serum (5) was diluted 1:200 in PBS containing 0.1% Tween-20 plus 0.1% gelatin (PBSTG) and added to wells for 1 h at

37°C. Wells were washed with PBS containing 0.1% Tween-20 (PBST) and subsequently incubated with anti-rabbit-alkaline phosphatase conjugate for 1 h at 37°C. Para-nitrophenyl phosphate substrate was added to wells after a final PBST wash, and the absorbance read at 405 nm after 30 min at room temperature. Data are presented as the mean reading of triplicate samples ± SD.

The anti-TSST-1 titers of immunized mice were determined by an ELISA for each injection group with pooled sera taken 5 days after the final boost. Pooled sera were diluted serially in PBSTG and added to ELISA wells previously coated with a 1 μg TSST-1/ml of carbonate buffer solution. Data are presented as the mean reading of triplicate samples ± SD.

RESULTS AND DISCUSSION

Biological Activity of TSST-1 and H135A in BALB/c Mice

The lethal effects of TSST-1 and H135A, potentiated by LPS (75 μg/mouse), was tested in BALB/c mice. TSST-1 doses of 10, 1.0, 0.1, or 0.01 μg/mouse (n=15 per dose) were 73%, 53%, 33%, and 0% lethal, respectively. The calculated LD_{50} of TSST-1 was 47.2 μg/kg (0.94 μg/mouse). In contrast, a H135A dose of 10 μg per mouse (500 μg/kg) was not lethal (n=15). Controls injected with either TSST-1 or H135A alone (10 μg/mouse) did not show any adverse reactions, but animals given only LPS (75 μg/mouse) had ruffled fur and were lethargic. None of the control mice died.

Another method of determining the biological activity of H135A in mice was to measure serum levels of cytokines, like TNF and IFNγ, after an injection of H135A plus LPS. Cytokines are important markers for toxic shock in humans and various animals, including mice (12, 16), following a bacterial superantigen stimulus. The maximum serum levels of TNF and IFNγ were 10- and 50-fold higher, respectively, in LPS-potentiated mice injected with 15 μg of TSST-1 versus H135A. The highest serum concentrations of TNF and IFNγ were detected at 2 h and 4 h, respectively, after LPS potentiation. Controls injected with TSST-1 alone had background serum levels of IFNγ and TNF equivalent to those found in naive mice. Mice given only LPS also had normal serum levels of IFNγ. However, there was a slight increase in TNF concentrations following an injection of LPS alone that was equivalent to 10% of the TNF levels found in mice given TSST-1 plus LPS. Mice injected with H135A plus LPS did not have elevated serum concentrations of IFNγ, and the TNF levels were identical to those found in mice injected with only LPS.

Serological Similarity of H135A and TSST-1

H135A was nontoxic as determined by a lack of mouse lethality or elevated serum levels of TNF or IFNγ. Therefore, we wanted to determine if the mutant was serologically similar to native TSST-1 and thus provide a relevant antigen for developing neutralizing antibodies against TSST-1. An ELISA with polyclonal anti-TSST-1 resulted in nearly identical dose response curves for H135A and TSST-1 (Fig. 1).

These data suggest that TSST-1 and H135A share some common epitopes, and a mutation at residue 135 did not greatly alter the conformation of TSST-1. The three dimensional shape of an immunogen is important as most epitopes are conformationally dependent (27). Historically, peptide vaccines have not been very efficacious. This may reflect the inability of most peptides to truly mimic an epitope conformation naturally found on a whole molecule antigen.

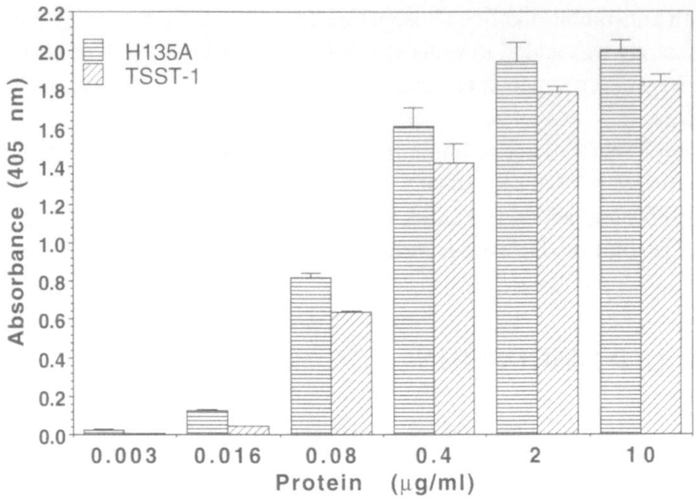

Figure 1. Serological similarity of H135A and TSST-1 by ELISA.

Vaccine Potential of H135A

The vaccine study consisted of mice (n=12 per group) immunized with H135A plus RIBI adjuvant, TSST-1 plus RIBI adjuvant, or RIBI adjuvant alone. Interestingly, four mice among the TSST-1 plus adjuvant group died during the immunization protocol, which suggests some synergistic effect between the toxin and adjuvant. Monophosphoryl lipid A, a less toxic version of diphosphoryl lipid A naturally found in bacterial LPS, is present in RIBI adjuvant and likely potentiates TSST-1 activity in immunized mice. None of the mice injected with H135A plus RIBI adjuvant died during the vaccination period, which further suggests that H135A is nontoxic. Future vaccine studies with bacterial superantigens will be done with other adjuvants, like alum, that should not potentiate the biological activity of these toxins. Mice immunized with either H135A or TSST-1 developed antibodies that recognized TSST-1 (Fig. 2).

When each mouse was challenged with 15 LD_{50} of TSST-1 plus LPS, the survival rate of mice immunized with TSST-1 (n=8) or H135A (n=12) was 75% and 67%, respectively, which contrasts with the 8% survival rate of RIBI adjuvant controls (n=12). Additionally, pooled sera (1:32 final dilution) from mice immunized with TSST-1 or H135A prevented 80% and 96% of the T-cell proliferation due to TSST-1 (100 ng/ml), respectively, versus cells co-incubated with TSST-1 and control sera from mice immunized with RIBI adjuvant alone.

There is a strong correlation between a lack of antibodies against TSST-1 and the incidence of toxic shock syndrome (7, 9). Data presented in this paper, and that previously reported (5, 6), suggest that the H135A mutant of TSST-1 is nontoxic and a likely vaccine candidate against TSST-1. Another group (19) has more recently described the importance of residues 132 and 140 in the biological activity of TSST-1. In contrast with H135A, mutation of TSST-1 at other amino acid residues either do not have a profound effect upon biological activity, as determined by mitogenicity and a rabbit infection model (2, 5), or the biological effects are significantly reduced with an unfortunate loss in serological similarity to native toxin. A protein that structurally mimics TSST-1, yet is devoid of toxicity, might

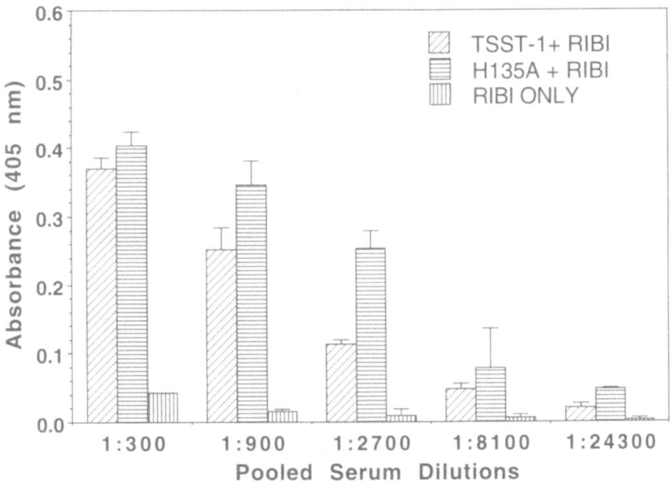

Figure 2. Comparison of anti-TSST-1 titer in mice immunized with either H135A, TSST-1, or RIBI adjuvant.

provide a long-term cure (i.e. vaccine) for those who suffer from recurrent bouts of toxic shock syndrome due to TSST-1.

The LPS-potentiated mouse model for SE and TSST-1 has enabled our laboratory to examine the biological effects of toxic bacterial superantigens and their mutants *in vivo*. Further work will provide more information regarding mice as a viable model for studying this intriguing, yet complex, group of bacterial toxins.

REFERENCES

1. Bergdoll, M., Crass, B., Reiser, R., Robbins, R., and Davis, J. (1981) Lancet 1, 1017-1021.
2. Blanco, L., Choi, E., Connolly, K., Thompson, M., and Bonventre, P. (1990) Infect. Immun. 58, 3020-3028.
3. Bohach, G., Hovde, C., Handley, J., and Schlievert, P. (1988) Infect. Immun. 56, 400-404.
4. Bohach, G., Fast, D., Nelson, R., and Schlievert, P. (1990) Crit. Rev. Microbiol. 17, 251-272.
5. Bonventre, P., Heeg, H, Cullen, C., and Lian, C-J. (1993) Infect. Immun. 61, 793-799.
6. Bonventre, P., Heeg, H., Edwards, C., and Cullen, C. (1995) Infect. Immun. 63, 509-515.
7. Bonventre, P., Linnemann, C., Weckbach, L. S., Staneck, J. L., Buncher, C. R., Vigdorth, E., Ritz, H., Archer, D., and Smith B., (1984) J. Infect. Dis. 150, 662-666.
8. Cole, B., Kartchner, D., and Wells, D. (1990) J. Immunol. 144, 425-431.
9. Crass, B., and Bergdoll, M. (1986) J. Infect. Dis. 153, 918-926.
10. Frankel, W., Rudy, C., Coffin, J., and Huber, B. (1991) Nature, 349, 526-528.
11. Hynes, W., Weeks, C., Iandolo, J., and Ferretti, J. (1987) Infect. Immun. 55, 837-838.
12. Jupin, C., Anderson, S., Damais, C., Alouf, J., and Parant, M. (1988) J. Exp. Med. 167, 752-761.
13. Lee, A., Robbins, R., and Bergdoll, M. (1978) Infect. Immun. 21, 387-391.
14. Lee, A., Robbins, R., Reiser, R., and Bergdoll, M. (1980) Infect. Immun. 27, 431-434.
15. Marrack, P., Blackman, M., Kushnir E., and Kappler, J. (1990) J. Exp. Med. 171, 455-464.
16. Miethke, T., Duschek, K., Wahl, C., Heeg, K., and Wagner, H. (1993) Eur. J. Immunol. 23, 1494-1500.
17. Miethke, T., Wahl, C., Heeg, K., Echtenacher, B., Krammer, P., and Wagner, H. (1992) J. Exp. Med. 175, 91-98.
18. Miethke, T., Wahl, C., Regele, D., Gaus, H., Heeg, K., and Wagner, H. (1993) Immunobiology 189, 270-284.
19. Murray, D., Prasad, G., Earhart, C., Leonard, B., Kreiswirth, B., Novick, R., Ohlendorf, D., and Schlievert, P. (1994) J. Immunol. 152, 87-95.

20. Prasad, G., Earhart, C., Murray, D., Novick, R., Schlievert, P., and Ohlendorf, D. (1993) Biochemistry 32, 13761-13766.
21. Schlievert, P., Shands, K., Dan, B., Schmid, G., and Nishimura, R. (1981) J. Infect. Dis. 143, 509-516.
22. Stiles, B., Bavari, S., Krakauer, T., and Ulrich, R. (1993) Infect. Immun. 61, 5333-5338.
23. Stone, R., and Schlievert, P. (1987) J. Infect. Dis. 155, 682-689.
24. Sugiyama, H., McKissic, E. M., Bergdoll, M., and Heller, B. (1964) J. Infect. Dis. 114, 111-118.
25. Swaminathan, S., Furey, W., Pletcher, J., and Sax, M. (1992) Nature 359, 801-806.
26. Todd, J., Fishaut, M., Kapral, F., and Welch, T. (1978) Lancet 2, 1116-1118.
27. Van Regenmortel, M. H. V. (1992) Structure of Antigens, Vol. I. CRC Press.
28. Yok-Jen, J., Qiao, Y., Komisar, J., Baze, W., Hsu, I-C., and Tseng, J. (1994) Infect. Immun. 62, 4626-4631.

19

THE RELATIONSHIP BETWEEN HISTIDINE RESIDUES AND VARIOUS BIOLOGICAL ACTIVITIES OF *Clostridium perfringens* ALPHA TOXIN

Masahiro Nagahama, Sadayuki Ochi, Keiko Kobayashi, and Jun Sakurai

Department of Microbiology
Faculty of Pharmaceutical Sciences
Tokushima Bunri University, Yamashiro-cho, Tokushima 770, Japan

INTRODUCTION

Clostridium perfringens alpha toxin is known to possess various biological activities (hemolysis, lethality, necrosis) and, in addition, phospholipase C (PLC) activity (4). It has been reported that these biological activities of the toxin are due to the PLC activity (4). We have reported that the toxin-induced contraction of isolated rat ileum or aorta and the toxin-induced hemolysis of rabbit erythrocytes are closely related to phospholipid metabolism stimulated by the toxin (1, 6, 7, 8). However, little is known about the relationship between these biological activities and PLC activity. The gene encoding alpha toxin has been isolated from chromosome of the organism and characterized (9). The genes encoding PLC's of *C. bifermentans* and *Bacillus cereus* also have been isolated. From these findings, the deduced amino acid sequences of alpha toxin and these enzymes were found to have significant homology up to approximately 250 residues from N-terminus. Hough *et al.* have reported the three-dimensional structure of the *B. cereus* PLC, which binds three Zn^{2+}(2). Little and Otnass reported that the *B. cereus* PLC has two tightly bound zinc ions and one loosely bound zinc ion (3). It is reported that these ions are coordinated to fixed histidine residues in the enzyme (2). Alpha toxin is known to be one of zinc-enzymes. Vallee and Auld (10) reported the possibility that the ligand spacers and vicinal amino acids of the *B. cereus* PLC resemble the likely location of the metal binding sites in the toxin (2). Recently, it has been reported that histidine residues in zinc-proteins such proteases and botulinum toxin are ligands to Zn^{2+}. Therefore, to clarify the mode-of-action of alpha toxin, we replaced all of histidine residues in the toxin to glycine by site-directed mutagenesis, and then investigated the relationship between histidine residues and Zn^{2+} contents or the biological and PLC activities.

Natural Toxins II, Edited by B. R. Singh and A. T. Tu
Plenum Press, New York, 1996

MATERIALS AND METHODS

An 1.3kb *Ssp I/Hind III* fragment of alpha toxin gene was isolated by the method of Titball *et al.* (9). Mutagenesis was carried out by the unique restriction enzyme site elimination technique using the Transformer mutagenesis kit and various synthetic oligonucleotides for construction of the mutants. The 1.3kb fragments of wild type and variant toxins were transformed into *B. subtilis* strain ISW1214, as described previously (5). Transformants carrying wild type or variant toxin genes were cultured in LB broth, and wild type and variant alpha toxins were purified form the culture, as described previously (5). Hemolytic, lethal and PLC activities were determined as described previously (5). Erythrocyte membranes in Tris-buffered saline were incubated with wild type or variant toxin in the presence or absence of 3 mM $CaCl_2$ at 37°C for 30 min. After incubation, the membranes were collected by centrifugation. After SDS-PAGE of the pellet, the bands were transferred to polyvinylidene fluoride membrane. The membrane was incubated with polyclonal anti-alpha toxin antiserum (mouse) to alpha toxin and then with anti-mouse IgG-peroxidase conjugate. Wild type or variant toxins were dot-blotted onto nitrocellulose paper strips. The strips were rinsed in 25 mM Tris-HCl buffer (pH 7.5) containing 100 mM sodium chloride for 5 min and then incubated for 60 min at room temperature in the presence of 50 nM $^{65}ZnCl_2$ in the same buffer. The amount of bound $[^{65}Zn]^{2+}$ was determined using Fuji BAS 2000 system.

RESULTS AND DISCUSSION

Alpha toxin reversibly inactivated by EDTA, as well as the untreated toxin, contained 2 moles of zinc metals per mole of the toxin, suggesting that alpha toxin purified from culture supernatant of *C. perfringens* contains two tightly bound zinc ions, although, as judged from structural studies of the *B. cereus* PLC (2), the toxin seems to possess three metals. EDTA-inactivated alpha toxin is known to be activated by divalent cations such as Co^{2+} and Zn^{2+}. It is likely that alpha toxin binds a divalent metal cation removed by EDTA. Therefore, the third metal binding site may bind Zn^{2+} or other divalent cations.

The change of H-68, -126, -136 or -148 residue to glycine caused drastic decrease or complete loss of the biological activities, indicating that these residues are important for the activities. However, the replacement of H-46, -207, -212 and -241 residues resulted in no effect on the biological activities, showing that these four histidine residues are not essential for the activities. The characterization of each variant toxin is shown in Table 1. The mutation at H-68, -126 or -136 resulted in a loss or drastic decrease in hemolysis and binding to membranes in the presence of Ca^{2+}, Co^{2+}, and Mn^{2+} dose-dependently stimulated the hemolytic activity and the binding of the variant toxin at H-68, -126 or -136 to membranes in the presence of Ca^{2+}, but did not in the absence of Ca^{2+}, suggesting that hemolytic activity of these variant toxins is dependent on the binding to membranes. Accordingly, replacement of H-68, -126 or -136 residue appears to result in reduction in binding to erythrocyte membranes rather than damage of the catalytic site in alpha toxin. H-68, -126 and -136 residues in wild type toxin seem to be directly or indirectly required for binding of the toxin to membranes. Exposure of wild type toxin with $[^{65}Zn]^{2+}$ caused $[^{65}Zn]^{2+}$ binding to the toxin. The binding of $[^{65}Zn]^{2+}$ to the toxin was competitively inhibited by cold Zn^{2+}, Co^{2+} or Mn^{2+} ion, suggesting that Zn^{2+} specifically binds to the toxin under the condition and that the Co^{2+} or Mn^{2+}-binding site is similar to the zinc-binding site. The evidence supports that wild type toxin has a divalent cation such as Co^{2+}, Mn^{2+} and Zn^{2+} in the third metal binding site. From these results, binding of $[^{65}Zn]^{2+}$ to wild type toxin appears to be due to the reversible exchange of a divalent metal cation in the binding site. The variant toxin at H-68, -126 or

Table 1. Biological activities of wild type and variant alpha toxins

Toxin	Activity (%) [a]						
	Hemolytic activity	Phospholipase C activity	SMase[b] activity	Lethal activity	Binding to erythrocytes[c]	Activation by metal ions[d]	^{65}Zn binding[e]
Wild type	100	100	100	100	+	+	+
H46G	99	86	89	100			
H68G	ND	ND	ND	ND	-	+	-
H126G	0.8	1.4	0.4	1.6	-	+	-
H136G	0.7	1.6	0.5	1.6	-	+	-
H148G	ND	ND	ND	ND	+	-	+
H207G	100	95	90	100			
H212G	106	91	95	100			
H241G	99	93	105	100			

[a]Activity (%) was expressed as a percentage of each activity in wild type toxin. Results for hemolytic, phospholipase C and sphingomyelinase activities were all means of five or six determinations. Lethal activity was expressed as a percentage of the activity (the lethal dose for 50 % of mice) of wild type toxin.
[b]SMase ; sphingomyelinase.
[c]Binding to sheep erythrocyte membranes in the presence of 3 mM $CaCl_2$; binding (+), no binding (-).
[d]Activation of hemolytic activity by cobalt or mangenese ion in the presence of 3mM $CaCl_2$; activation (+), no activation (-).
[e]$[^{65}Zn]^{2+}$ binding to wild type and variant toxins ; binding (+), no binding (-).
ND ; not detected.

-136 had 2 zinc metals per mole of the protein. Therefore, the replacement of H-68, -126 or -136 residue seems to result in no loss of two tightly bound zinc metals. Moreover, binding of the labeled metal to H68G, H126G and H136G variants showed about 90 % decrease of wild type toxin, indicating that the loosely bound metal in alpha toxin is coordinated by H-68, -126 and -136 at least. On the basis of the effect of Co^{2+} and Mn^{2+} on the activity and binding of these variant toxins, it is suggested that H-68, -126 and -136 residues bind the exchangeable and labile metal ion which is required for the binding. Furthermore, the metal binding site in the variant toxins bound Co^{2+} or Mn^{2+} with higher affinity than Zn^{2+}. It therefore appears that the site is not only specific for Zn^{2+}.

Alpha toxin is able to bind to sheep erythrocytes in the presence of Ca^{2+}, but is unable to do in the absence of Ca^{2+}. The toxin causes hemolysis in the presence of Ca^{2+}, but does not in the absence of Ca^{2+}. It therefore is likely that the toxin specifically binds to erythrocyte membranes in the presence of Ca^{2+}. However, Ca^{2+} alone did not stimulate binding and hemolytic activity of variant toxins at H-68, -126 and -136 or EDTA-treated alpha toxin. The effect of Co^{2+} and Mn^{2+} in the presence of Ca^{2+} shows that divalent cations such as Co^{2+}, Mn^{2+} and Zn^{2+} in the third site are essential for the binding of toxin and that these divalent cations are not replaced by Ca^{2+} in the site. Accordingly, alpha toxin binding to membranes seems to be produced by both Ca^{2+} and a divalent cation (Co^{2+}, Mn^{2+} and Zn^{2+}) in the third site.

The variant toxin at H-148 contained one zinc metal, suggesting that H-148 in alpha toxin tightly binds one zinc ion out of two tightly bound zinc metals. The variant toxin was found to bind to erythrocytes in the presence of Ca^{2+}, but indicated no activities of alpha toxin. In addition, the variant toxin inhibited competitively binding of wild type toxin to membranes and hemolysis induced by alpha toxin. Incubation of the variant toxin with $[^{65}Zn]^{2+}$ caused $[^{65}Zn]^{2+}$ binding, indicating that the third site is not damaged in the variant toxin. Therefore, the zinc ions which is tightly bound to H-148 seems to be important to the active site of alpha toxin, but not to be required for binding to erythrocytes.

As mentioned above, the feature of H136G variant was coincident with that of H68G and H126G variants, indicating that the role of H-136 resembles that of H-68 and -126.

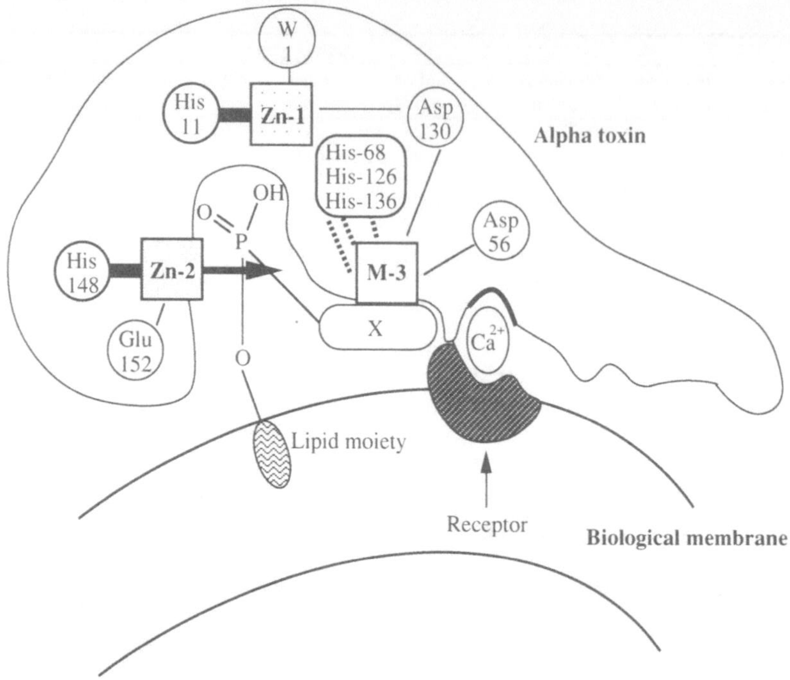

Metal ions in alpha toxin

Metal	Role of divalent cations	Binding power
Zn -1	Maintenance of structure	Strong
Zn -2	Catalytic site	Strong
M-3	Binding to phospholipids	Weak

X : Polar head group (ex. choline or ethanolamine)

Figure 1. Mode of action of alpha toxin on biological membranes.

H-136 is reported to correspond to H-128 in the *B. cereus* PLC, which binds to Zn^{2+} in the second site (2, 10). We are unable to explain the discrepancy between H-136 in the toxin and H-128 in the *B. cereus* enzyme. H11G was not detected in culture supernatant or in cells of *B. subtilis* transformant carrying the variant toxin gene. However, these variant toxins except H11G were purified from culture supernatant of *B. subtilis* transformants. We have reported that the N-terminal region of alpha toxin may be buried in the protein (5), as reported in the *B. cereus* PLC (2). Accordingly, it is possible that replacement of H-11 in alpha toxin results in a less compact structure of the molecule due to lack of zinc binding so that H11G is destroyed by attack of proteases in B. subtilis. H-11 in alpha toxin may be regarded as essential and integral parts of the toxin. From the present work and crystallography of *B. cereus* PLC (2), the role of histidine residues in the toxin is showed in Figure 1.

As judged from the relationship between Zn^{2+} and histidine residues in *B. cereus* PLC, it appears that the mutations at H-68, -126,- 136 and -148 result in damage in PLC activity of the toxin. Changes in PLC activity of all variant toxins resemble those of other biological activities, suggesting that the PLC activity is essential for various activities of the toxin. On the basis of the findings from chemical modification and neutralization of anti-alpha toxin against the toxin, and comparison of the toxin with *B. cereus* PLC, however, the PLC activity seems to be not all of biological activities of toxin.

REFERENCES

1. Fujii, Y., and Sakurai, J. 1989. Br. J. Pharmacol. **97**: 119-124.
2. Hough, E., Hansen, L.K., Birknes, B., Jynge, K., Hansen, S., Hordvik, A., Little, C., Dodson, E., and Derewenda, Z. 1989. Nature **338**: 357-360.
3. Little, C., and Otnass, A. B. 1975. Biochim. Biophys. Acta **391**: 326-333.
4. McDonel, J. L. 1986. In Dorner, F., and Drews, J. (eds.), p 477-517 Pharmacology of bacterial toxins, Pergamon Press, Oxford.
5. Nagahama, M., Iida, H., Nishioka, E., Okamoto, K., and Sakurai, J. 1994. FEMS Microbiol. Lett. **120**: 297-302.
6. Sakurai, J., Fujii, Y., and Shirotani, M. 1990. Toxicon **28**: 411-418.
7. Sakurai, J., Ochi, S., and Tanaka, H. 1993. Infect. Immun. **61**: 3711-3718.
8. Sakurai, J., Ochi, S., and Tanaka, H. 1994. Infect. Immun. **62**: 717-721.
9. Titball, R.W., Hunter, S.E.C., Martin, K.L., Morris, B.C., Shuttleworth, A.D., Rubidge, T., Anderson, D.W., and Kelly, D.C. 1989. Infect. Immun. **57**: 367-376.
10. Vallee, B.L., and Auld, D.S. 1993. Biochemistry. **32**: 6493-6500.

20

MECHANISM OF ACTION OF *Clostridium perfringens* ENTEROTOXIN

N. Sugimoto, Y. Horiguchi, and M. Matsuda

Department of Bacterial Toxinology
Research Institute for Microbial Diseases
Osaka University
3-1 Yamadaoka Suita
Osaka 565, Japan

INTRODUCTION

An enterotoxin produced by *Clostridium perfringens* during sporulation is a simple protein with a molecular weight of about 35,000 and is a causative agent for the food poisoning by the organism (Todd, 1987). Production of the enterotoxin by the organism and the methods for purification were studied intensively in 1960s and 1970s (Duncan and Strong, 1968; Stark and Duncan, 1972; Sakaguchi et al., 1973; Granum and Whitaker, 1980a; Sugimoto et al., 1985a). These studies disclosed some unique characteristics of the enterotoxin protein, e.g. low solubility and anomalous aggregation in the presence of SDS (Enders and Duncan, 1976; Granum and Whitaker, 1980a,b). Recently, the primary structure of the enterotoxin was clarified (Richardson and Granum, 1985) and the gene encoding the enterotoxin was also cloned (Hanna et al., 1989; Iwanejko et al., 1989; Jongsten et al., 1989). Thus, it is expected that protein chemistry of the enterotoxin will clarify the physicochemical background of unique characteristics of the enterotoxin. On the other hand, studies on the biological action of the enterotoxin also have made a great progress in the last two decades. The purified enterotoxin was shown to cause diarrhea in experimental animals and human volunteers (Hauschild et al., 1971; Skjelkvåle and Uemura, 1977). It also causes fluid accumulation in ligated ileal loops of rabbits and mice when administered into the ileal lumen (Nillo, 1971; McDonel and Duncan, 1977). Besides the action on the gastrointestinal tract, the enterotoxin was reported to kill mice rapidly by intravenous or intraperitoneal injection (Genigeorgis et al., 1973; Nillo, 1975). It was also shown that the enterotoxin caused morphological alterations and detachment of some mammalian cells in culture (Keusch and Donta, 1975). In this article, we overview the action of the enterotoxin at animal, organ, cellular and molecular levels.

Natural Toxins II, Edited by B. R. Singh and A. T. Tu
Plenum Press, New York, 1996

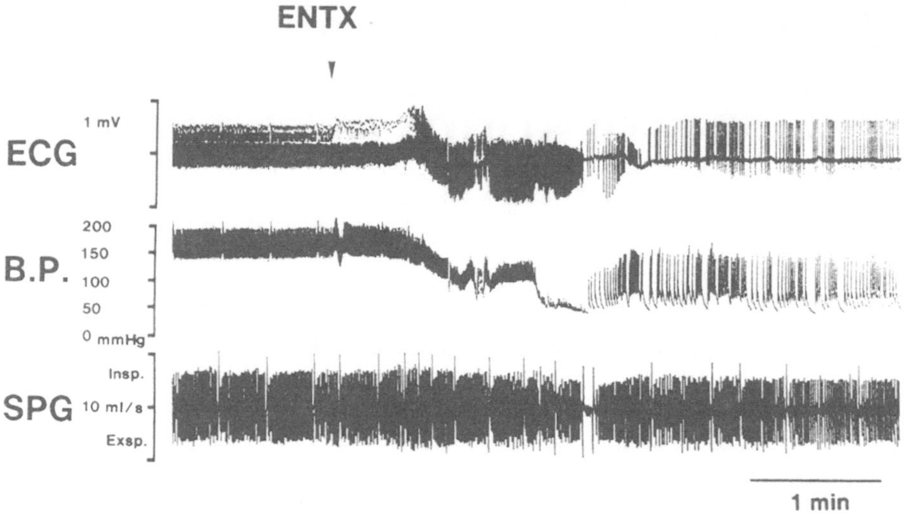

Figure 1. Effects of i.v. injection of the enterotoxin on ECG, arterial blood pressure (B.P.) and respiration pattern (SPG).

MECHANISM OF THE LETHAL ACTION OF THE ENTEROTOXIN

In 1970s, groups of Nillo and McDonel studied the mechanism of food poisoning by the enterotoxin using ligated rabbit ileal loops. They showed that the enterotoxin causes severe disturbance of metabolism and transportation of water and electrolytes in the intestine (McDonel, 1974; McDonel and Asano, 1975; McDonel and Duncan, 1975a), and that the intestine treated with the enterotoxin showed serious histological damage including bleb formation of epithelial cells (Nillo, 1971; McDonel and Duncan, 1975b; McDonel et al., 1978). From these experimental results, it is clarified that direct cytotoxic action of the enterotoxin on intestinal epithelial cells is the cause of diarrhea which is the most striking symptom of the food poisoning caused by *Clostridium perfringens*. On the other hand, lethal intoxication of animals by the enterotoxin was little studied. The minimal lethal dose of the toxin to a mouse by i.v. injection is about 2 μg. Death of animals usually occurs within 30 min after the injection and animals that survive for 30 min rarely die later (Hauschild and Hilsheimer, 1971; Sakaguchi et al., 1973). To clarify this characteristic *in vivo* toxicity of the enterotoxin, we studied the cause of acute death induced by i.v. injection of the enterotoxin in rats (Sugimoto et al., 1991). Figure 1 shows the changes of ECG, arterial blood pressure, and respiration pattern of a rat before and after the injection of a lethal dose of the enterotoxin. Just after the injection of the enterotoxin, ECG showed striking changes. Then, progressive hypotension and apnea were observed.

The early changes in ECG pattern were prolongation of P-R interval, low P wave, tall T wave, and biphasic QRS-T complex. These changes in ECG suggested hyperpotassemia, and actually measurement of serum components of the intoxicated rats demonstrated the elevation of potassium level while serum sodium, calcium and hemoglobin levels remained normal. Moreover, injection of KCl mimicked the changes in ECG elicited by the injection of the enterotoxin. Thus, it is clarified that the disturbance of circulatory function elicited by i.v. injection is caused by hyperpotassemia.

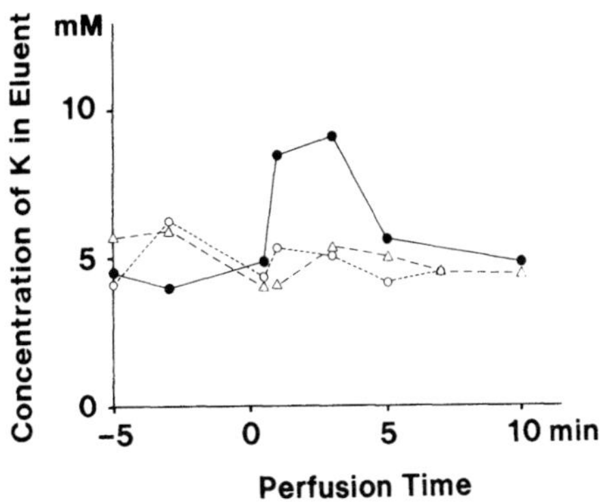

Figure 2. Effects of the enterotoxin on perfusates from isolated organs. Symbols used are: O, lower extremities; ●, liver; Δ, lung.

The question arose next was which organ or tissue is responsible for potassium liberation into the blood in intoxicated animals. To answer the question, we perfused the enterotoxin into isolated organs or tissue. As shown in Fig. 2, concentration of potassium in perfusates from isolated lungs and lower extremities showed no obvious change after the perfusion of the enterotoxin. However, the concentration of potassium in perfusate from isolated liver showed a remarkable increase after the perfusion of the enterotoxin. The leakage of potassium from the liver occurred within a minute and finished 5 min after the perfusion of the toxin (Fig. 3). The amount of the potassium liberated from the liver was estimated to be about 133 μmol, which is enough to raise plasma level of potassium to more than lethal level. In addition, activities of transaminases in the perfusate from the enterotoxin-treated liver increased following to the elevation of potassium concentration (Fig. 3). The results indicated that potassium is liberated from hepatocytes, and that the enterotoxin makes a small 'pore'

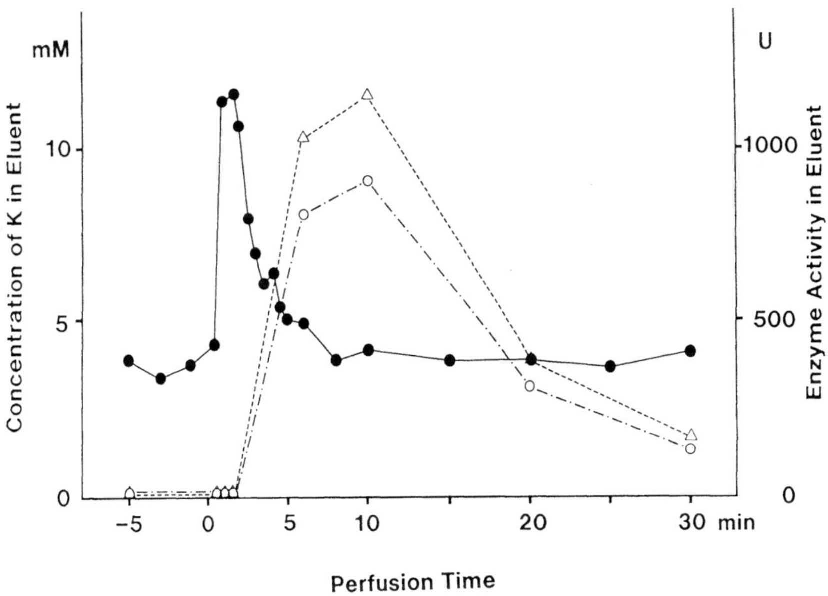

Figure 3. Leakage of potassium and transaminases by the enterotoxin from the isolated rat liver. Symbols used are: ●, potassium; O, aspartate aminotransferase; Δ, alanine aminotransferase.

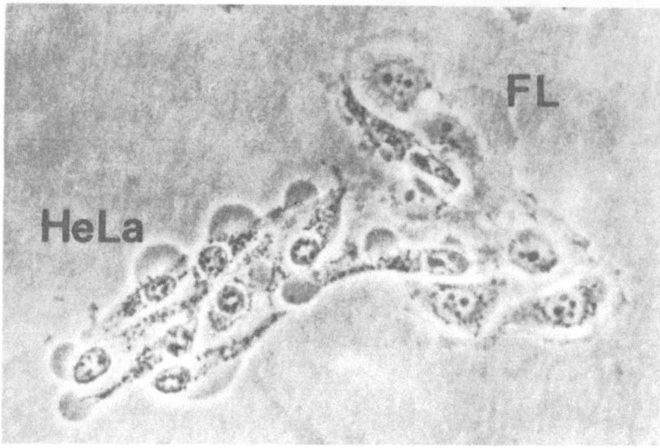

Figure 4. Phase contrast micrograph of HeLa and FL cells treated with the enterotoxin.

which allows the leakage of potassium from cytoplasm of hepatocytes to blood stream and then the 'pore' increases gradually its size enough to liberate transaminases. Thus, it is concluded that hyperpotassemia elicited by the cytotoxic action of the enterotoxin on hepatocytes caused cardiac failure leading to the death of the intoxicated animals.

MECHANISM OF CYTOTOXIC ACTION OF THE ENTEROTOXIN

Keusch and Donta reported in 1975 that the enterotoxin killed HeLa cells. This opened the way for the study on the action of the enterotoxin at cellular level. In 1978, Giger and Pariza reported disturbance of amino acid transport in hepatocyte by the enterotoxin. Matsuda and Sugimoto (1979) and McClane and McDonel (1979) reported that morphological alterations induced by the enterotoxin on HeLa and Vero cells looked like those of epithelial cells in sections of ileal loops treated with the enterotoxin.

Figure 4 shows the morphological alterations of HeLa cells induced by the treatment with the enterotoxin. Typical bleb formation is observed in HeLa cells but not in FL cells. This indicates the selectivity of cytotoxic action of the enterotoxin. It was reported that a variety of established cell lines could be clearly classified into the sensitive and the resistant cells to the enterotoxin (Giugliano et al., 1983; Horiguchi et al., 1985). The sensitivity of the cells is attributable to a specific receptor for the enterotoxin on their cell membrane (Horiguchi et al., 1985). It was reported that the carboxyl-terminal part of the enterotoxin molecule carries a binding domain to the receptor (Horiguchi et al., 1987; Hanna et al., 1991). Sugii and Horiguchi (1988) and Wnek and McClane (1983, 1986, 1989) reported their attempts for the extraction and identification of the enterotoxin-receptor from cell membrane. The substances to which the enterotoxin binds have a protein moiety with molecular weights ranging from 45-50 kDa to 60 kDa. However, characteristics of these substances have not been clarified in detail. Further studies will clarify the concrete feature of the receptor.

It was shown that the morphological alterations elicited by the toxin is dependent on calcium in extracellular fluid, and that the cytotoxic action of the enterotoxin is divided into two steps; calcium-independent early step and calcium-dependent late step (Matsuda and Sugimoto, 1979). Calcium dependency of the enterotoxin action was later confirmed by

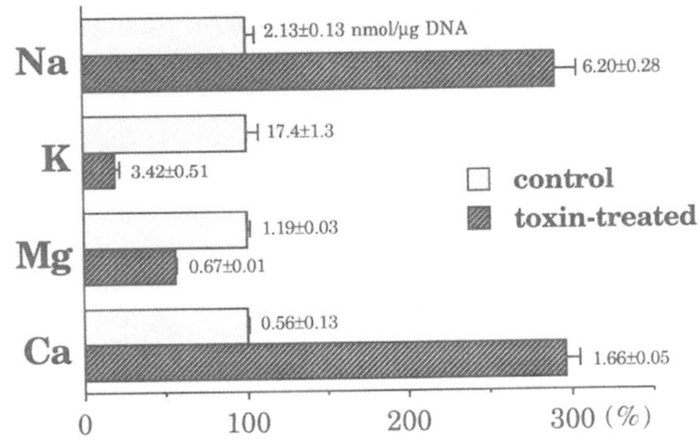

Figure 5. Changes of cation contents in HeLa cells after the treatment with the enterotoxin.

Granum (1985), Horiguchi et al. (1986) and McClane (1988). Similar morphological changes were induced in HeLa cells by calcium-ionophore A23187. Therefore, it was expected that the treatment with the enterotoxin induced an increase in intracellular calcium in the cells. This was demonstrated by direct measurement of calcium in the cells treated with the enterotoxin (Sugimoto et al., 1985b). As shown in Fig. 5, cellular calcium actually increased after the treatment with the enterotoxin. In addition, cellular sodium, potassium and magnesium contents were all influenced by the treatment with the enterotoxin (Fig. 5).

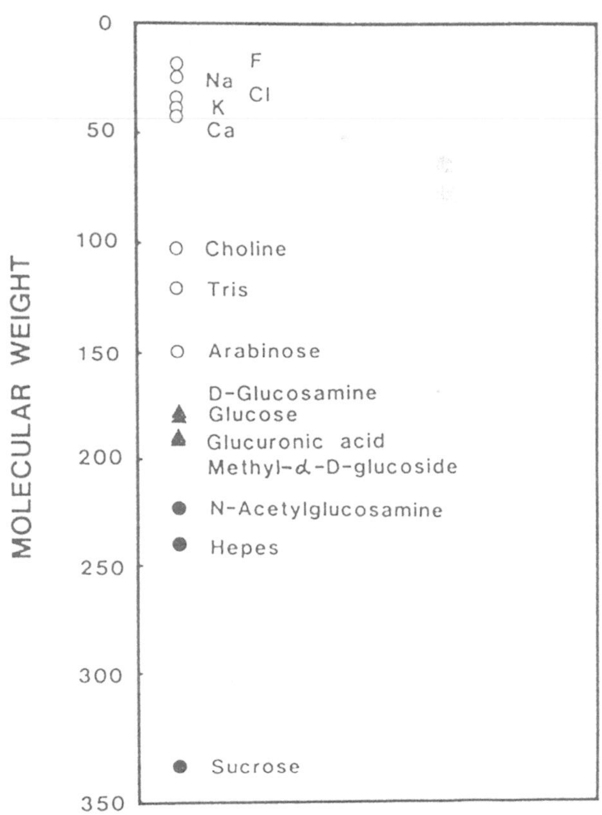

Figure 6. Effects of substances with different molecular weight in extracellular fluid on the enterotoxin-induced bleb-balloon formation in HeLa cells.

Cellular magnesium and potassium, of which concentrations are normally higher inside than outside, decreased after the treatment with the enterotoxin. On the other hand, calcium and sodium, of which concentrations are normally lower inside than outside, increased. This suggests that extracellular calcium moves passively into the cells according to the concentration gradient after the treatment of the cells with the enterotoxin. These changes in electrolytes permeability of the cell membrane by the enterotoxin were shown to occur without extracellular calcium (Matsuda et al., 1986; Horiguchi et al., 1986). Thus, calcium-independent early step of the cytotoxic action of the enterotoxin was disclosed to be the changes in electrolyte permeability of the plasma membrane of the cell.

The bleb-balloon formation elicited in the cells treated with the enterotoxin was inhibited by increasing osmolarity of extracellular fluid with sucrose (Matsuda and Sugimoto, 1979; McClane, 1984). Figure 6 summarizes the inhibitory effects of substances with different molecular weights in extracellular fluid on the morphological alterations of HeLa cells induced by the treatment with the enterotoxin. Substances with molecular weights higher than about 200 inhibited the morphological alteration, but those with molecular weights less than 200 did not. From the results obtained in these studies, the mechanism of the cytotoxic action of the enterotoxin is disclosed as follows: 1) the enterotoxin forms a kind of 'pore' in the membrane of the susceptible cells, which allows substances with molecular weight less than 200 to pass freely through the membrane, 2) through the 'pore', calcium in extracellular fluid fluxes into the cells according to its concentration gradient between intra- and extra-cellular fluid, 3) rapid and massive increase in intracellular calcium concentration disturbs the cellular function to maintain the structure, 4) this disturbance of the cellular function, in combination with water influx, results in morphological alteration of the cells.

MOLECULAR BASIS OF THE ACTION OF THE ENTEROTOXIN

What is the molecular feature of the 'pore' which is formed in the plasma membrane by the action of the enterotoxin? Three possibilities for the component(s) of the 'pore' are;

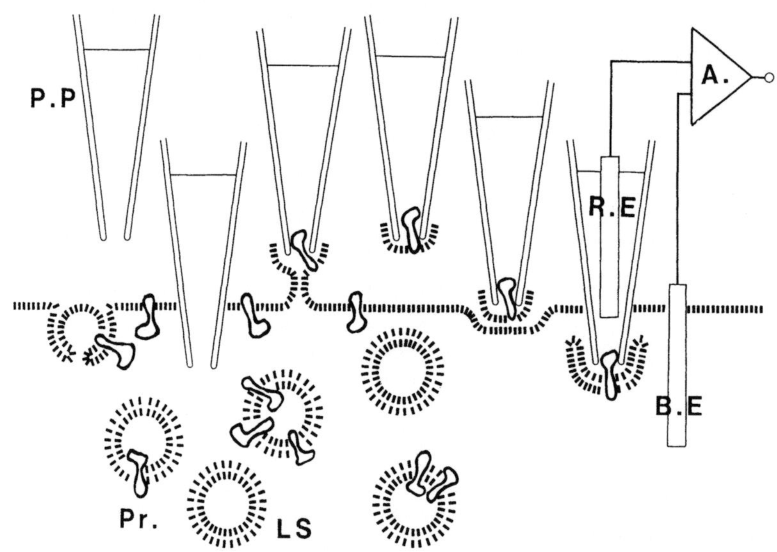

Figure 7. Schematic diagram of an experimental system using artificial lipid bilayer. P.P, patch pipette; Pr., protein; LS, liposome; R.E, recording electrode; B.E, bath electrode; A. amplifier.

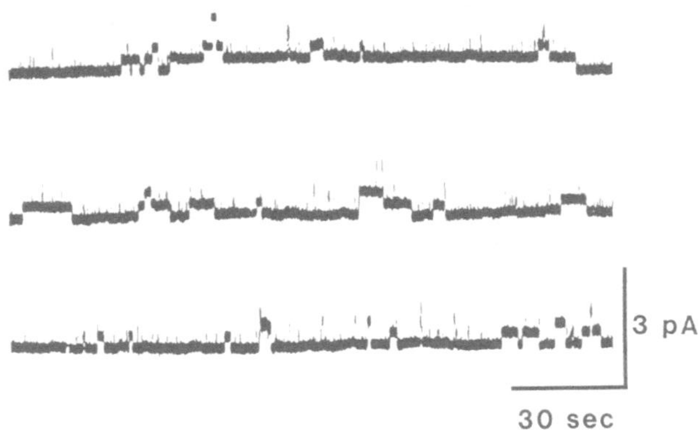

Figure 8. A current record of asolectin bilayer which incorporated the enterotoxin protein. Holding potential; +11.5 mV.

1) cellular membrane protein, 2) complex of membrane protein and the enterotoxin protein, 3) toxin molecule alone. For the examination of the latest possibility, artificial lipid bilayer system were used (Sugimoto et al., 1988).

Figure. 7 shows a scheme of the experiment. The purified enterotoxin protein was integrated in liposomes of asolectin by sonication. By using these liposomes, artificial lipid bilayer was formed at the tip of a glass micropipette. Electrolytes passing through the bilayer was monitored as an electric current by a voltage-clamp technique. Figure 10 shows a typical example of membrane current recorded from the lipid bilayer. The result shows that the enterotoxin protein forms ion-permeable channels which have constant electric conductance and randomly repeat opening and closing. The conductance of the enterotoxin-channel ranged 40-450 pS and the opening time of the channel sometimes lasts for more than several seconds. Current-voltage relationship recorded from the enterotoxin-channel showed no rectification, indicating that ions can pass both direction through the channel.

To characterize the enterotoxin-channel formed in the asolectin bilayer, ion selectivity of the channel was examined by making a concentration gradient between inside and outside of the patch pipette. Figure 9 shows current voltage relationships recorded from an enterotoxin-integrated asolectin membrane before, during and after the change of concentration of electrolytes in the fluid outside of the pipette. Dilution of electrolytes concentration to 1/10 in the fluid outside of the pipette shifted the line of current-voltage relationship toward right by +6.6 mV. As shown in Fig. 10 schematically, when the electrode inside the pipette is charged positively, positively charged sodium ion is forced to move outside by both diffusion and electrostatic expulsion while negatively charged chloride ion is retained in balance between force toward outside by diffusion and electrostatic attraction. The result showing that the current through the channel became zero at +6.6 mV indicates that the enterotoxin-channel has a tendency of passing negatively charged ions more easily. However, the equilibrium potential of chloride estimated from Nernst's equation under the experimental condition is +60 mV which is much larger than that obtained from the result. This indicates that the enterotoxin- channel has little ion-selectivity. These characteristics of the enterotoxin-channel are consistent with those of 'pore' observed in the enterotoxin-treated cells. Therefore it is likely that apparent 'pore' in the membrane of the enterotoxin-treated cells is the sum of ion-permeable channels formed in the plasma membrane by the enterotoxin proteins.

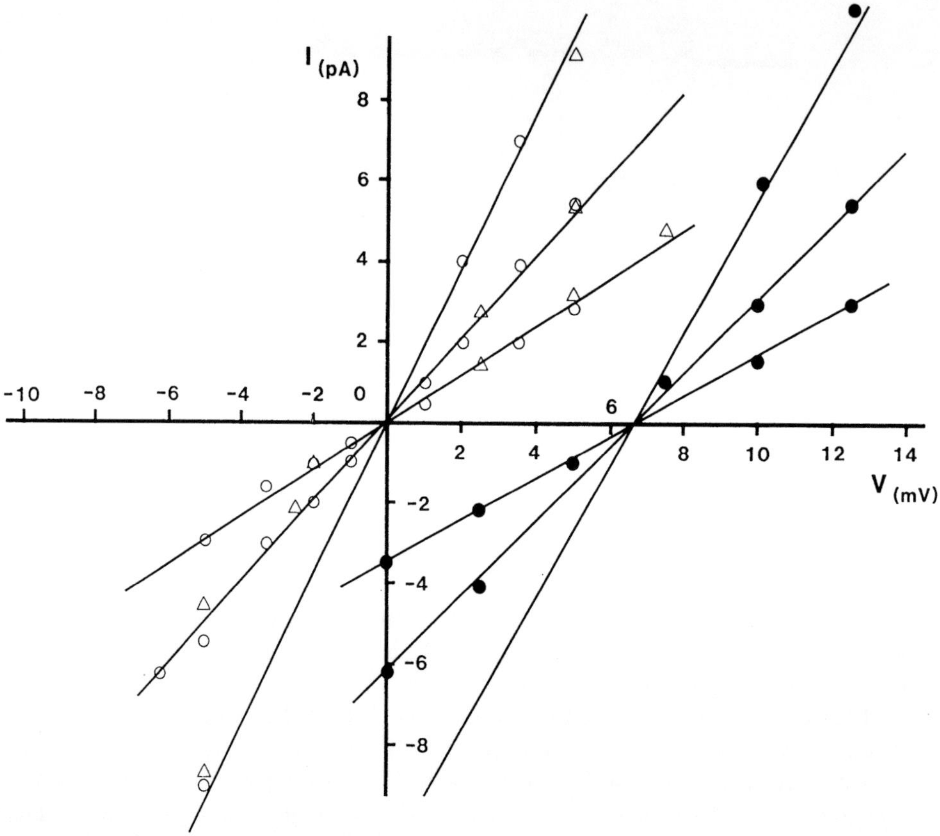

Figure 9. Current-voltage relationships of the enterotoxin-channel before, during and after the dilution of electrolytes in the fluid outside of the pipette. Concentration of sodium chloride inside the pipette was 130 mM. The concentration of sodium chloride outside the pipette was first 130 mM(○), then diluted to 13 mM(●) and returned to 130 mM(△).

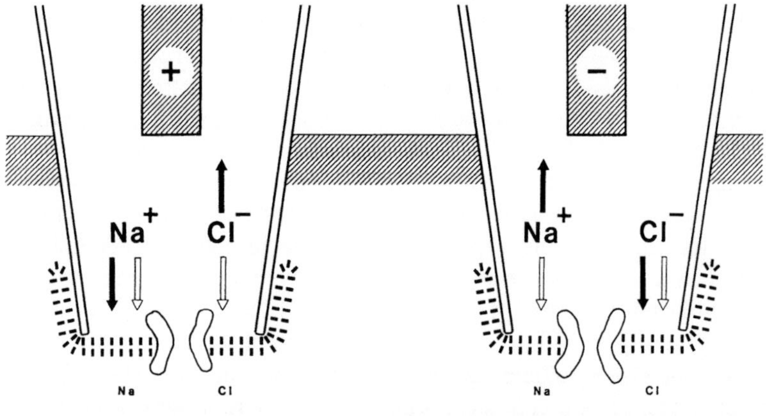

Figure 10. Scheme of driving forces acting on sodium and chloride ions by diffusion and electrostatic pressure. White arrows indicate the direction of driving force by diffusion and black arrows indicate the direction of electrostatic pressure.

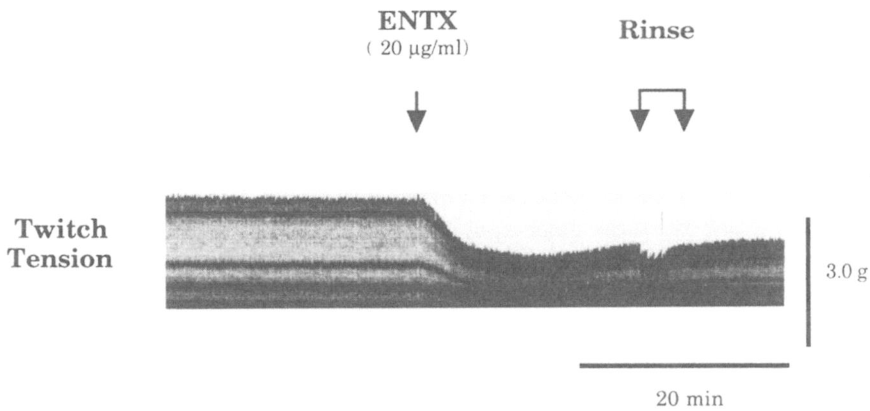

Figure 11. Effects of the enterotoxin on twitch tension of an isolated phrenic nerve- semidiaphragm preparation. The preparation was equilibrated to 0.5 mM calcium in bathing solution and phrenic nerve was stimulated by square pulses(1 V, 10 msec) at a frequency of 0.25 Hz. The enterotoxin(20 μg/ml) was added to the bathing solution at the time indicated by an arrow and ENTX.

A NOVEL ACTION OF THE ENTEROTOXIN ON NEUROMUSCULAR TRANSMISSION

We happened to find that the enterotoxin inhibited twitch tension of diaphragm elicited by the nerve stimulation under low calcium conditions (Sugimoto et al., 1992). In Krebs solution containing 0.5 mM calcium, the enterotoxin reduced within 10 min the amplitude of the twitch tension to about 34% of that recorded before the treatment. The degree of reduction of the twitch tension by the enterotoxin was inversely related to the concentration of calcium in bathing solution and dependent on temperature and concentra-

Table 1. Effects of the enterotoxin on end-plate potentials of phrenic nerve- semidiaphragm preparations in Krebs solution containing 0.5 mM calcium

	Before toxin-treatment	After toxin-treatment*
Resting potential (mV)	66±2	68±4
Miniature end-plate potential		
Amplitude (mV)	0.51±0.02	0.48±0.02
Frequency (s^{-1})	0.91±0.07	0.72±0.07†

Values are mean±S.E. of five end-plates.

*Preparations were treated with 2.2 μg/ml of the enterotoxin for 10 min by superfusion, then rinsed with toxin-free Krebs solution containing 0.5 mM calcium.

†P<0.005. Significance of difference from before toxin-treatment was calculated by two-tailed Student's *t*-test for paired variates.

tion of the toxin. The tension of muscular twitch elicited by direct stimulation to the muscle was not affected by the enterotoxin. The effects of the enterotoxin were antagonized by 2 μM physostigmine or 0.1 mM guanidine. Unlike the effects of curare, pretreatment of the preparation with enterotoxin did not antagonize the neuromuscular block by decamethonium. These pharmacophysiological characteristics suggest that the enterotoxin inhibits neuromuscular transmission prejunctionally.

To confirm the prejunctional action of the enterotoxin, end-plate potentials before and after the treatment of the diaphragm preparation with the enterotoxin. The resting membrane potentials of the muscle fibers were not affected by the enterotoxin measurement. However, the enterotoxin reduced frequency, but not mean amplitude or amplitude distribution, of miniature end-plate potentials (Table 1). The results indicate that the release of transmitter was inhibited by the treatment with the enterotoxin, but that the response of acetylcholine-receptor in the postjunctional membrane to the transmitter was unaffected.

Immunohistochemical studies showed that most of the synaptophysin-positive sites were also positive with anti-enterotoxin monoclonal antibody (Senda et al., 1995). With higher magnification, the configurations of the enterotoxin-positive sites were coincide with those of synaptophysin-positive sites (Figure 12). These results indicate that the enterotoxin binds to prejunctional nerve terminal and inhibits prejunctionally neuromuscular transmission. Pathophysiological role of this finding in the intoxication is, at the moment, not known. But the finding seems to be important for the studies on the distribution and physiological role of the enterotoxin-receptor.

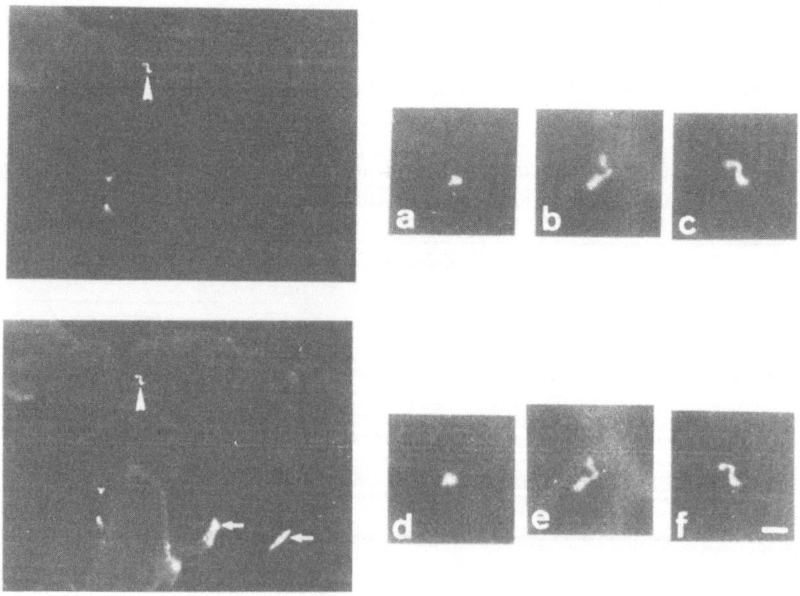

Figure 12. Immunohistochemistry of enterotoxin binding sites with synaptophysin- positive sites. A phrenic nerve-diaphragm preparation showing inhibition of neuromuscular transmission after the treatment with the enterotoxin was stained by a double immunofluorescent method. a, b, and c show the fluorescence of FITC conjugated anti-enterotoxin antibody. d, e, and f show the fluorescence of rhodamine conjugated anti-synaptophysin antibody corresponding to the sites of a, b, and c, respectively. Note the coincidence of configurations of anti-enterotoxin fluorescence and those of anti-synaptophysin.

CONCLUSION

The enterotoxin of *Clostridium perfringens* is a channel-forming toxin. The fact that the characteristics of the enterotoxin-channel formed in the artificial lipid bilayer are consistent with those of 'pore' formed in the cell membrane indicates that the enterotoxin is likely to form similar ion-permeable channels in the plasma membrane of the cells. At the step of formation of ion-permeable channels, the specific enterotoxin-receptor on the cell membrane probably plays a key role. Through the ion-permeable channels, extracellular calcium and water move into the cells and induce morphological alterations in the cells. When the enterotoxin is injected intravenously, it acts on hepatocytes and induces potassium leakage from the cells. Potassium thus leaked into the blood stream causes hyperpotassemia in the animal and cardiac dysfunction elicited by hyperpotassemia results in death of the animal. An outline of the action of the enterotoxin from animal to molecular levels is now clarified as formation of ion-permeable channels in the lipid membrane. One of the major problems remained to be solved is the identification of the enterotoxin-receptor. The facts that the prejunctional nerve terminals have enterotoxin-binding sites and that the enterotoxin inhibits neuromuscular transmission suggest that the enterotoxin-receptor plays an important physiological role in cellular function. It is expected that whole affairs of the action of the enterotoxin will soon be brought to light and the toxin will be used as a useful tool for the understanding of cellular functions.

REFERENCES

Duncan, C. L., and Strong, D. H., 1968, Improved medium for sporulation of *Clostridium perfringens*, *Appl. Microbiol.* 16:82-89.

Enders, Jr. G. L., and Duncan, C. L., 1976, Anomalous aggregation of *Clostridium perfringens* enterotoxin under dissociating conditions, *Can. J. Microbiol.* 22:1410-1414.

Genigeorgis, C., Sakaguchi, G., and Riemann, H., 1973, Assay methods for *Clostridium perfringens* type A enterotoxin, *Appl. Microbiol.* 26:111- 115.

Giger, O., and Pariza, M. W., 1978, Depression of amino acid transport in cultured rat hepatocytes by purified enterotoxin from *Clostridium perfringens*, *Biochem. Biophys. Res. Commun.* 82:378-383.

Giugliano, L. G., Stringer, M. F., and Drasar, B. S., 1983, Detection of *Clostridium perfringens* enterotoxin by tissue culture and double-gel diffusion methods, *J. Med. Microbiol.* 16:233-237.

Granum, P. E., 1985, The effect of Ca++ and Mg++ on the action of *Clostridium perfringens* enterotoxin on Vero cells, *Acta Path. Microbiol. Scand.* Sect. B. 93:41-48.

Granum, P. E., and Whitaker, J. R., 1980a, Improved method for purification of enterotoxin from *Clostridium perfringens* type A, *Appl. Environ. Microbiol.* 39:1120-1122.

Granum, P. E., and Whitaker, J. R., 1980b, Perturbation of the structure of *Clostridium perfringens* enterotoxin by sodium dodecyl sulfate, guanidine hydrochloride, pH and temperature, *J. Food Biochem.* 4:219-234.

Hanna, P. C., Mietzner, T. A., Schoolnik, G. K., and McClane, B. A., 1991, Localization of the receptor-binding region of *Clostridium perfringens* enterotoxin utilizing cloned toxin fragments and synthetic peptides. *J. Biol. Chem.* 266:11037-11043.

Hanna, P. C., Wnek, A. P., and McClane, B. A., 1989, Molecular cloning of the 3' half of the *Clostridium perfringens* enterotoxin gene and demonstration that this region encodes receptor-binding activity, *J. Bacteriol.* 171:6815-6820.

Hauschild, A. H. W., and Hilsheimer, R., 1971, Purification and characteristics of enterotoxin of *Clostridium perfringens* type A, *Can. J. Microbiol.* 17:1425-1433.

Hauschild, A. H. W., Walcroft, M. T., and Campbell, W., 1971, Emesis and diarrhea induced by enterotoxin of *Clostridium perfringens* type A in monkeys, *Can. J. Microbiol.* 17:1141-1143.

Horiguchi, Y., Akai, T., Sakaguchi, G., 1987, Isolation and function of a *Clostridium perfringens* enterotoxin fragment, *Infect. Immun.* 55:2912-2915.

Horiguchi, Y., Uemura, T., Kozaki, S., and Sakaguchi, G., 1985, The relationship between cytotoxic effect and binding to mammalian cultured cells of *Clostridium perfringens* enterotoxin, *FEMS Microbiol Lett.* 28:131-135.

Horiguchi, Y., Uemura, T., Kozaki, S., and Sakaguchi, G., 1986, Effects of Ca2+ and other cations on the action of *Clostridium perfringens* enterotoxin, *Biochim. Biophys. Acta.* 889:65-71.

Iwanejko, L. A., Routledge, M. N., and Stewart, G. S. A. B., 1989, Cloning in Escherichia coli of the enterotoxin gene from *Clostridium perfringens* type A, *J. Gen. Microbiol.* 135:903-909.

Jongsten, M. D., Werners, K., and Notermans, D., 1989, Cloning and sequencing of the *Clostridium perfringens* enterotoxin gene, *Antonie Leeuwenhoek.* 56:181-190.

Keusch, G. T., and Donta, S. T., 1975, Classification of enterotoxins on the basis of activity in cell culture, *J. Infect. Dis.* 131:58-63.

Matsuda, M., Ozutsumi, K., Iwahashi, H., and Sugimoto, N., 1986, Primary action of *Clostridium perfringens* type A enterotoxin on HeLa and Vero cells in the absence of extracellular calcium: rapid and characteristic change in membrane permeability, *Biochem. Biophys. Res. Commun.* 141:704-710.

Matsuda, M., and Sugimoto, N., 1979, Calcium-independent and dependent steps in action of *Clostridium perfringens* type A enterotoxin on HeLa and Vero cells, *Biochem. Biophys. Res. Commun.* 91:629-636.

McClane, B. A., 1984, Osmotic stabilizers differentially inhibit permeability alterations induced in Vero cells by *Clostridium perfringens* enterotoxin, *Biochim. Biophys. Acta.* 777:99-106.

McClane, B. A., and McDonel, J. L., 1979, The effects of *Clostridium perfringens* enterotoxin on morphology, viability and macromolecular synthesis in Vero cells, *J. Cell Physiol.* 99:191-200.

McClane, B. A., Wnek, A. P., Hulkower, K. I., and Hanna, P. C., 1988, Divalent cations involvement in action of *Clostridium perfringens* type A enterotoxin, *J. Biol. Chem.* 263:2423-2425.

McDonel J. L., 1974, In vivo effects of *Clostridium perfringens* enteropathogenic factors on the rat ileum, *Infect. Immun.* 10:1156-1162.

McDonel, J. L., and Asano, T., 1975, Analysis of unidirectional fluxes of sodium during diarrhea induced by *Clostridium perfringens* enterotoxin in the rat terminal ileum, *Infect. Immun.* 11:526-529.

McDonel, J. L., Chang, L. W., Poundsm, J. G., and Duncan, C. L., 1978, The effects of *Clostridium perfringens* enterotoxin on rat and rabbit ileum. *Lab. Invest.* 39:210-218.

McDonel, J. L., and Duncan, C. L., 1975a, Effects of *Clostridium perfringens* enterotoxin on metabolic indexes of everted rat ileal sacs, *Infect. Immun.* 12: 274-280.

McDonel, J. L., and Duncan, C. L., 1975b, Histopathological effect of *Clostridium perfringens* enterotoxin in the rabbit ileum, *Infect. Immun.* 12:1214- 1218.

McDonel, J. L., and Duncan C. L., 1977, Regional localization of activity of *Clostridium perfringens* type A enterotoxin in the rabbit ileum, jejunum and duodenum, *J. Infect. Dis.* 136:661-666.

Nillo, L., 1971, Mechanism of action of the enteropathogenic factor of *Clostridium perfringens* type A, *Infect. Immun.* 3:100-106.

Nillo, L., 1975, Measurement of biological activities of purified and crude enterotoxin of *Clostridium perfringens*, *Infect. Immun.* 12:440-442.

Richardson, M., and Granum, P. E., 1985, The amino acid sequence of the enterotoxin from *Clostridium perfringens* type A, *FEBS Lett.* 182:479-485.

Sakaguchi, G., Uemura, T., and Riemann, H. P., 1973, Simplified method for purification of *Clostridium perfringens* type A enterotoxin, *Appl. Microbiol.* 26:762-767.

Senda, T., Sugimoto, N., Horiguchi, Y., and Matsuda, M., 1995, The enterotoxin of *Clostridium perfringens* type A binds to the presynaptic nerve endings in neuromuscular junctions of mouse phrenic nerve-diaphragm, *Toxicon*, in press.

Skjelkvålc, R., and Uemura, T., 1977, Experimental diarrhea in human volunteers following oral administration of *Clostridium perfringens* enterotoxin, *J. Appl. Bacteriol.* 43:281-286.

Stark R. L., and Duncan, C. L., 1972, Purification and biochemical properties of *Clostridium perfringens* type A enterotoxin, *Infect. Immun.* 6:662-673.

Sugii, S., and Horiguchi, Y., 1988, Identification and isolation of the binding substance for Clostridium pefringens enterotoxin on Vero cells, *FEMS Microbiol. Lett.* 52:85-90.

Sugimoto, N., Chen, Y-M., Lee, S-Y., Matsuda, M., and Lee, C-Y., 1991, Pathdynamics of intoxication in rats and mice by enterotoxin of *Clostridium perfringens* type A, *Toxicon.* 29:751-759.

Sugimoto, N., Miyamoto, A., Horiguchi, Y., Okabe, T., and Matsuda, M., 1992, Inhibition of neuromuscular transmission in isolated mouse phrenic nerve-diaphragm by the enterotoxin of *Clostridium perfringens* type A, *Toxicon.* 30:825-834.

Sugimoto, N., Ozutsumi, K., and Matsuda, M., 1985a, Purification by high performance liquid chromatography of *Clostridium perfringens* type A enterotoxin prepared from high toxin producers selected by a toxin-antitoxin halo, *Eur. J. Epidemiol.* 1:131-138.

Sugimoto, N., Ozutsumi, K., and Matsuda, M., 1985b, Morphological alterations and changes in cellular cations induced by *Clostridium perfringens* type A enterotoxin in tissue culture cells, *Eur. J. Epidemiol.* 1:264-273.

Sugimoto, N., Takagi, M., Ozutsumi, K., Harada, S., and Matsuda, M., 1988, Enterotoxin of *Clostridium perfringens* type A forms ion-permeable channels in a lipid bilayer membrane, *Biochem. Biophys. Res. Commun.* 156:551- 556.

Todd, E. C. D., 1987, Foodborn disease in six countries - a comparison, *J. Food Protect.* 41:559-565.

Wnek, A. P., and McClane, B. A., 1983, Identification of 50,000 Mr protein from rabbit brush border membranes that binds *Clostridium perfringens* enterotoxin, *Biochem. Biophys. Res. Commun.* 112:1099-1105.

Wnek, A. P., and McClane, B. A., 1986, Comparison of receptors for *Clostridium perfringens* type A and cholera enterotoxins in isolated rabbit intestinal brush border membranes, *Microb. Pathogenesis.* 1:89-100.

Wnek, A. P., and McClane, B. A., 1989, Preliminary evidence that *Clostridium perfringens* type A enterotoxin is present in a 160,000 Mr complex in mammalian membrane, *Infect. Immun.* 57:574-581.

21

BINDING PROTEINS ON SYNAPTIC MEMBRANES FOR CERTAIN PHOSPHOLIPASES A_2 WITH PRESYNAPTIC TOXICITY

Mu-Chin Tzeng,[1,2*] Chon-Ho Yen,[1] and Ming-Daw Tsai[3]

[1] Institute of Biological Chemistry
Academia Sinica
[2] Institute of Biochemical Sciences
National Taiwan University
P.O. Box 23-106, Taipei, Taiwan, Republic of China
[3] Department of Chemistry
The Ohio State University
120 W. 18th Ave.
Columbus, Ohio 43210

Many steps in the process of neurotransmitter release are vulnerable to various neurotoxins. Some of these presynaptic toxins exhibit phospholipase A_2 (PLA_2) activity. These neurotoxic PLA_2s (or PLA_2 neurotoxins) are members of a group of extracellular (or secreted) PLA_2 proteins found in most if not all animals. Besides phospholipid metabolism, these PLA_2s exhibit a variety of biological effects, including host defense, neurotoxicity (presynaptic and/or postsynaptic), myotoxicity, and alteration of coagulation, which may or may not be related to hydrolysis of phospholipids. Despite large differences in biological actions, the PLA_2 chains of these proteins show high degrees of homology in the primary, secondary and possibly tertiary structures. A small number of these proteins, mostly isolated from the venoms of a number of snakes, act primarily at the presynaptic level to cause synaptic blockade by inhibiting the release of neurotransmitters, though most of them also produce postsynaptic toxicity and other effects. These presynaptic PLA_2 toxins may be classified into three classes. The toxins differ in their subunit structures, but in every case, at least one subunit is an active PLA_2 with M.W. of 12,000 to 16,000. Each toxin in the first class is a single-chained protein. In the second class, a toxin may comprise 2 to 4 homologous subunits associated noncovalently. The toxins in the third class are composed of a phospholi-

* To whom correspondence may be addressed

Natural Toxins II, Edited by B. R. Singh and A. T. Tu
Plenum Press, New York, 1996

pase chain in disulfide linkage to a non-PLA_2 chain. Different facets of these toxins have been studied by many investigators (see Ref. 1-9 for recent reviews).

Most of the earlier studies on these toxins were concerned about the structural, pharmacological and electrophysiological aspects. At the neuromuscular junction, these presynaptic neurotoxins have been shown to affect neuromuscular transmission. The time courses vary with the toxins and the experimental conditions. An initial decrease in neurotransmission followed by a transient increase is most common. And in the final stage, when sufficient amounts were used, all of these toxins cause neuromuscular blockade.

A number of studies indicate that the PLA_2 activity is required for the blocking effect of these toxins under normal condition. On the other hand, many PLA_2s bear little if any neurotoxicity. There are wide differences in toxic potencies between the PLA_2s, but the differences in enzymatic activities are small. Furthermore, the relative toxic potencies of the PLA_2 toxins are not proportional to their PLA_2 activities. In order to explain this phenomenon, it has been postulated that the distinction lies in disparities in the strength of binding to the presynaptic membranes, strong for the presynaptic toxins against poor for the non-neurotoxic PLA_2s. Indirect evidence for the existence of high-affinity binding sites in the target membranes has been provided by the observations that neuromuscular blockade still occurs when the nerve-muscle preparation has been exposed to the toxins for only a brief period followed by thorough washing. Since the PLA_2 activity is greatly attenuated by lowering the temperature to near 0°C or when Ca^{2+} is replaced by Sr^{2+}, whereas toxin binding persists, it is even possible to temporally separate the binding step from the subsequent event of synaptic blockade. As all known toxic PLA_2s are from snake venoms, most of the many pharmacological studies of these toxins were done on the peripheral tissues with the consideration that in snakebites the toxins do not reach the brain because of the blood-brain barrier. However, when the toxins were injected directly into the brain or tested on its isolated preparations, the neurotoxic PLA_2s showed high degrees of toxicity.

Direct evidence for binding to presynaptic membranes has been obtained for several toxins in recent years (10-22). Membrane preparations from the brain have been used in most of the endeavors. One study has been conducted with nerve ending preparation from the electric organ. The scarcity of the mammalian peripheral nerve endings has impeded their use for studies in this respect.

Specific binding of high affinity to synaptosomes or to synaptic membrane fragments prepared from the brain has been reported (in chronological order) for β-bungarotoxin, crotoxin, Mojave toxin, taipoxin, PLA_2 from *Naja nigricollis*, PLA_2 from *Naja naja atra*, *Pseudocerastes* neurotoxin, OS_2 (a PLA_2 from *Oxyuranus scutellatus*), Pa-11 (a PLA_2 from *Pseudechis australis*), ammodytoxin C, and Pseudexin B (10-21). Among the above PLA_2s, it is not known whether OS_2 is a presynaptic toxin, and the PLA_2s from *Naja nigricollis* and from *Naja naja atra* showed little effect on the peripheral nervous system. Our own studies in this respect (10-13) will be described below. In our experiments, crotoxin, Mojave toxin, taipoxin, β-bungarotoxin, Pa-11 and pseudexin B were purified from the crude venoms and then radioactively labeled with Na[^{125}I]. The neurotoxicity of each of the iodinated toxins was found to be fully retained when assayed on nerve-muscle preparations from the chick. Each ^{125}I-toxin was incubated with synaptosome or synaptic membrane (23,24) fractions from the cerebral cortex of guinea pig in the absence or in the presence of large excess (≥100-fold) of the unlabeled toxin for a suitable time. The membrane bound radioactivity was separated from the unbound by centrifugation. For each radioactive toxin, specific binding, i.e., that displaceable by the unlabelled toxin, to synaptosomes or synaptic membranes was observed. The specific binding is saturable for each toxin, being consistent with the presence of a limited number of binding sites. The degree of nonspecific binding is low (generally <10%) in each case. In sharp contrast, when pancreatic PLA_2, which is non-toxic, was tested, specific binding was undetectable. The affinities of the toxins to the membrane

Table I. A list of the parameters reported for the binding of PLA_2 neurotoxins

Toxin	Membrane preparation	K_D (nM)	B_{max} (pmol/mg protein)	References
β-bungarotoxin	rat synaptic membrane	high affinity, 0.26	0.16	14, 18
		low affinity, 6.1	2.6	
	chick synaptic membrane	0.47	~0.05	15
	guinea pig synaptosome	0.7	0.1	
crotoxin	guinea pig synaptosome	< 10	~10	10, 11
	guinea pig synaptosome	high affinity, 4	2	20
		low affinity, 87	4	
	Torpedo presynaptic membrane	700	240	22
Mojave toxin	guinea pig synaptosome	< 8	~10	10
OS_2	rat synaptic membrane	high affinity, 0.0015	1	19
		low affinity, 0.045	3	
Pa-11	guinea pig synaptic membrane	high affinity, 0.25	6.9	12
		low affinity, 4.6	23	
ammodytoxin C	bovine synaptic membrane	6.0	5.7	21
pseudexin B	guinea pig synaptosome	high affinity, 1.2	2	13
		low affinity, 10	77	
taipoxin	guinea pig synaptosome	< 12	~60	

preparations and the density of the binding sites are shown in Table I, which also includes data obtained for other toxins by other investigators. We want to stress that we have chosen the condition that minimizes the PLA_2 activity by using a solution containing 10 mM Sr^{2+}, 0.5 mM EGTA and no Ca^{2+} for the binding assay in our studies to avoid unwanted complications which may arise from hydrolysis of membrane phospholipids (see Ref. 1 for more information). Ca^{2+} is known to be required for the enzymatic activity of the secreted PLA_2s, while Sr^{2+} is inhibitory (see Ref. 1-4 for reviews).

Crotoxin is produced by the South American rattlesnake *Crotalus durissus terrificus* and is the first neurotoxin isolated and crystalized. This toxin is composed of two differnt subunits, an acidic subunit A of 9,000 Da and a basic subunit B of 14,400 Da. Subunit B is an active PLA_2 and is weakly toxic on its own. Subunit A can potentiate the toxicity of subunit B, but otherwise no biological activity by itself is known (Ref. 25 and 26 for recent reviews). Since there are two subunits in the crotoxin molecule, it is of interest to know whether the whole complex or only one subunit is involved in the binding. For this purpose, we subjected both the pellets and the supernatants after centrifugation to sodium dodecyl sulfate-polyacrylamide gel electropheresis (SDS-PAGE). From the autoradiogram of the gel, it was found that only the subunit B bound to the synaptosomal membrane, while the subunit A remained in the supernatant (10, 11). Previously this phenomenon was also demonstrated with the postsynaptic membrane from electric fish using a different technique by Bon *et al.* (27). Since it is difficult to accurately determine the concentration of the unassociated subunit B in the reaction mixture, only IC_{50} value of ~10 nM for unlabeled crotoxin to compete for the binding of labeled toxin was given by us. The dissociation constant (K_D) would be considerably lower. The density of binding sites (B_{max}) was ~10 pmol/mg of total protein (10, 11).

Since subunit A did not bind to the membranes in the end of the equilibrium binding assay, obviously the dissociation of the crotoxin complex at some point is an important step in the action of the toxin. Hendon and Tu (28) converted crotoxin complex into a non-dissociable conjugate by cross-linking the two subunits with a bifunctional cross-linker, dimethyl suberimidate. This conversion abolished the lethality of crotoxin but did not significantly reduce its PLA_2 activity. When we used the non-dissociable crotoxin in binding assay, we found that it did not bind to the membranes. On the other hand, subunit A was capable of inhibiting the binding of subunit B (20). This may not be surprising, for excess subunit A will reduce the effective concentration of free subunit B by mass action. We may deduce from these observations that subunit A competes with membrane binding sites for combining with subunit B, and thus prevents subunit B from binding to low-affinity sites at various places. Accordingly, the presence of subunit A allows subunit B to bind only to the high-affinity site. This scheme is compatible with the proposal that subunit A acts as a chaperone, as put forward by Folman, Habermann and others (29,30), although more than one mechanism is possible with that proposition. A different view has been expressed by other investigators (22).

We then examined the effects of other PLA_2s on the binding of crotoxin. β–Bungarotoxin had no effect at all (11). Pancreatic PLA_2 was also without effect. On the other hand, in contrary to our preliminary report (11), taipoxin inhibited up to 40% of the binding of ^{125}I-crotoxin (1). Caudoxin, notexin, and pseudexin A behaved like taipoxin in this respect. Unlabeled crotoxin inhibited to completion, as would be expected.

Taipoxin, found in the venom of the Australian taipan *Oxyuranus scutellatus*, is made of three subunits, α, β, and γ of 13,750, 13,500 and 18,350 Da, respectively, each showing sequence homologous to other PLA_2s. The α–subunit is weakly toxic itself. The β and γ–subunits are non-toxic, but can greatly enhance the toxicity of the α–subunit (29, 31). And similar to the case of crotoxin, only the α–subunit was found to bind to the synaptosome (1). By analogy with crotoxin, it may be proposed that dissociation of the taipoxin complex occurs before binding of the α–subunit. The β– and γ–subunits may well act like the subunit A of crotoxin. The IC_{50} for the unlabeled taipoxin to inhibit the binding of the radioactive toxin was ~12 nM (11), and the binding sites of taipoxin appeared to be about 60 pmol/mg protein of synaptosomes. The binding of taipoxin was not antagonized by β–bungarotoxin. Unexpectedly, high concerntration of crotoxin appeared to increase the binding to some extent. The nonreciprocal antagonism between crotoxin and taipoxin binding would indicate that some binding site (or sites) is common for the two toxins, while others are not shared.

As the binding of crotoxin is strongly protease sensitive, the involvement of proteins in the plasma membrane of the synaptosome is indicated. We went on to identify the binding protein(s) for crotoxin (11, 32) first by using the photoaffinity labeling technique in collaboration with Dr. R. J. Guillory of the University of Hawaii. We were able to attach a photoactivatable group to crotoxin by coupling iodinated crotoxin with N-hydroxysuccinimidyl-4-azidobenzoate (HSAB) without impairing the neurotoxicity. This crotoxin derivative was allowed to bind to the synaptosomal membrane and then irradiated with UV light. When the membrane pellet was analyzed by SDS-PAGE, a major radioactive conjugate of 100 kDa was found in the autoradiogram. The formation of this conjugate was prevented by unlabeled crotoxin, indicating the binding is specific. Prior treatment of the membrane by trypsin, papain, or protease V8 from *Staphylococcus aureus* also abolished the conjugation. Therefore this putative receptor is a protein. The M.W. of this protein was estimated to be about 85 k by subtracting the mass of subunit B, which is 15 k, from that of the conjugate. The formation of this conjugate was also inhibited by serveral other neurotoxic PLA_2s, such as Mojave toxin, taipoxin, caudoxin, but not by β–bungarotoxin and pancreatic PLA_2. The IC_{50} of unlabeled crotoxin to abolish this conjugation was ~10 nM (32). When the synaptosome was subfractionated, this polypeptide was chiefly found in the synaptic membrane fraction, and not present in the intra-synaptosomal mitochondrial fraction. We have also tested membranes from non-neural tissues, such as liver, kidney, erythrocytes, and did not find the 85-k polypeptide in these membranes.

When purely chemical cross-linking reagent disuccinimidyl suberate was used, a different conjugate of ~60 kDa was found. As prior treatment of the synaptosome with proteinase K, trypsin, subtilisin, α-chymotrypsin, elastase or heat destroyed this conjugation capability of the membrane, the membrane component in the conjugate was a single-chained protein of ~45 kDa. Similar results were obtained when dithio*bis*(succinimidylpropionate), or ethylene glycol*bis*(succinimidylsuccinate) was used as the cross-linker. Of the subfractions of synaptosomes, this binding protein was mostly found in the synaptic membrane fraction, and not present in the mitochondrial fraction. Plasma membranes from several non-neural tissues also did not contain this binding protein. The IC_{50} of unlabeled crotoxin to inhibit the formation of this adduct was ~10 nM. Mojave toxin, taipoxin, Pa-11, and pseudexin A were also highly inhibitory to in this conjugation, notexin and caudoxin were less effective, while β-bungarotoxin and pancreatic PLA_2 were totally ineffective. The results may signify that the 45-k polypeptide preferentially present in neuronal membranes is another putative receptor responsible for the binding of crotoxin, but it is also possible that this polypeptide is another subunit of the receptor (33).

Besides membrane proteins, crotoxin appears to bind also to lipid components of the membrane. Radvanyi *et al.* (34) have shown that incubating crotoxin with negatively charged vesicles of phospholipids resulted in the binding of subunit B with K_D ranging from <1 to 100 μM, depending on the lipid composition, whereas binding was not observed with uncharged vesicles. Therefore, it is possible that some negatively charged lipids may be components of the binding sites for crotoxin in its target membranes, but the relative contribution remains to be determined.

When similar cross-linking experiments were performed with ^{125}I-taipoxin, a 60-k radioactive conjugate was also formed using disuccinimidyl suberate as the cross-linker. The IC_{50} for the unlabeled taipoxin to compete for the formation of this conjugate was somewhere between 1 and 5 nM. This adduct was not observed when the synaptosomes were treated priorly with trypsin, proteinase K or chymotrypsin. From these experiments, we deduced that a 45-k polypeptide in the synaptic membrane is a taipoxin-binding protein or a subunit of it. Again, this polypeptide was not present in the nonneural membranes from erythrocytes, kidney and liver. Although the M.W. of this polypeptide is indistinguishable from that of one of the crotoxin-binding proteins, crotoxin was not very efficacious in blocking the formation of the taipoxin-conjugate. Hence the two 60-k polypeptides are not identical, and crotoxin is a poor ligand for this taipoxin-binding protein. In contrast, taipoxin appears to be a good ligand for the two crotoxin-binding proteins, as taipoxin can inhibit the conjugation of crotoxin to its two binding proteins with high potency. Caudoxin, notexin, Pa-11 behaved like crotoxin, whereas β–bungarotoxin and pancreatic PLA_2 showed no effects at all (35).

We have also undertaken purification of the binding proteins for crotoxin. The synaptic membrane fraction from guinea pig brain was extracted with 4% Triton X-100. Crotoxin-binding activities were solubilized in active form as assayed by a gel filtration procedure and by affinity labeling experiment. The detergent extract was then subjected to gel filtration with Sepharose 6B followed by affinity chromatography with crotoxin linked to Sepharose 4B. After elution from the affinity column, only three bands with molecular weights of ~85 k, 65 k and 50 k were present when analyzed by SDS-PAGE under reducing condition. For the purpose of purifying the binding proteins in large scale, porcine brain was used instead. Only two bands of ~65 k and 50 k were observed when the material purified by affinity chromatography was analyzed by SDS-PAGE. A partial amino acid sequence of KPTEKKDRVHHEPQLL has been obtained for the 50-k band.

We are also interested in learning the determinants in the crotoxin molecule involved in binding to synaptic membranes. An illuminating result was obtained by modifying the toxin with *p*-nitrobenzenesulfonyl fluoride, which is known to react quite selectively with Tyr residues. When crotoxin complex was allowed to react with the reagent, one *p*-nitrobenzenesulfonyl group was readily incorporated into subunit B, while subunit A was modified insignificantly. After separating the modified subunit from the unmodified, we cleaved each protein with trypsin, then separated the peptide fragments with HPLC. When the elution patterns of the modified subunit B and the unmodified were compared, a new peak in the elution profile of the modified subunit B was identified. The amino acid sequence of this peptide was found to correspond to residues 70-77 (SGYITCGK) of subunit B, thus the residue modified was Tyr72. This modification had little effect on the PLA_2 activity, the affinity for binding to synaptosomes, or the neurotoxicity.

We then subjected the subunit B to further modification and separated the modified proteins into fractions modified to different extents by reversed-phase HPLC. When two *p*-nitrobenzenesulfonyl groups were incorporated, there were drastic reductions in the binding affinity and in neurotoxicity. When three residues were modified, these two activities were virtually completely abolished. In sharp contrast, the PLA_2 activity was largely retained. These results reinforced the importance of binding and confirmed that the PLA_2 activity alone is not sufficient for neurotoxicity of crotoxin. The second and the third modified residues were found to reside in the same peptide fragment (NAIP-FYAFYGCYCGWGGR) corresponding to residues 16-33, which contains three Tyr residues at positions 21, 24, and 27. The modifications occurred mostly at residues 21 and 24. Since Tyr24 and Tyr27 (25 and 28, respectively, in alignment number according to Ref. 36, 37) are conserved for all of the secreted PLA_2s sequenced to date, toxic or not, and Tyr21 (22 in alignment number) is found in the neurotoxic PLA_2s but not in the great majority of the nontoxic ones, Tyr21 would be more important for binding.

In the bovine pancreatic PLA_2, as well as many other nontoxic PLA_2s, the residue at the position corresponding to Tyr21 of subunit B is Phe (and the number is 22 for the bovine enzyme). We found that when the bovine pancreatic PLA_2 mutant F22Y, in which the Phe22 of the wild type was replaced by Tyr (38) by site-directed mutagenesis utilizing the phosphorothioate method (39), was present during the binding period, the subsequent formation of the 60-k radioactive conjugate was suppressed with an IC_{50} of 1 μM. In sharp contrast, the wild-type PLA_2 purified from the bovine pancreas gave no effect even at concentration as high as 50 μM. The wild-type pancreatic enzyme produced by cloning techniques and another mutant F22A, which has Ala at residue 22, were also without effect at the highest concentration used. The one- and two-dimensional NMR spectra of F22Y and the wild type are almost identical except for the obvious changes arising from the new phenolic OH group and a 0.19 ppm change in one of the three chemical shifts of Phe106, which is in close proximity to residue 22, forming the second half of the Phe22-Phe106 aromatic sandwich (40). The enzymatic activities of the two mutants are also comparable with that of the wild-type enzyme (38). Besides, we have performed the competitive binding assay under Ca^{2+}-free condition. Hence it is unlikely that the inhibitory effect of the F22Y mutant is due to the hydrolysis of membrane phospholipids. We thus conclude that the F22Y mutant blocked the formation of the radioactive conjugate by competing the binding of ^{125}I-crotoxin to the binding protein (40). However, because the binding affinity of the F22Y mutant was not high relative to that of crotoxin ($K_D < 10$ nM), there must be other residues also involved in binding and thereby in neurotoxicity. We are currently trying to find them out.

Judging from the affinity of F22Y, we would not expect it to be neurotoxic. All mice injected with the F22Y mutant, either intraperitoneally or intracisternally, lived and behaved normally even at a dose of 18 mg of protein/kg of body weight. We have iodinated the F22Y mutant and attempted to demonstrate its binding to the synaptic membrane directly. Specific binding was not evident, apparently because the affinity is too low.

Obviously, nerve cells produce these toxin-binding proteins not for the purpose of self-destruction. In order to learn the normal function of these binding proteins, we have looked into the effects on the synaptosomes of the two toxins not related to their PLA_2 actions. We found that in a Ca^{2+}-free, Sr^{2+}-containing Tyrode solution, taipoxin caused a marked inhibition of the synaptosomal Na^+/Cl^--dependent uptake of tritiated norepinephrine and glycine, with IC_{50}s of ~20 and 30 nM, respectively, to be compared with low potencies (IC_{50} > 0.2 μM) in blocking the uptake of dopamine, serotonin, lysine, and phenylalanine, and no effect on choline uptake into synaptosomes under the same condition. Suppression of the uptake of tritiated norepinephrine was also observed with crotoxin, although with low potency (IC_{50} = 1 μM). Again, choline uptake was not affected.

Taipoxin was also capable of antagonizing the binding of desipramine, which is a specific blocker of norepinephrine transporter, to its high-affinity site(s) with IC_{50} of ~15 nM. Taken together, these results strongly indicate that the Na^+/Cl^--dependent norepinephrine and glycine transporters act as two high-affinity binding proteins for taipoxin. The genes for norepinephrine, glycine, dopamine, and serotonin transporters have been cloned just recently (Ref. 41-45 for reviews). They are homologous to each other. But purification of them from the nervous tissue has been difficult. We hope that the PLA_2 neurotoxins can also contribute to the study of these transporters. It is probable that the structural similarity of these transporters is one of the reasons for the multiplicity in the binding proteins for these PLA_2 toxins. As to the receptors for the two toxins at the neuromuscular junction, we consider it possible that they are members (as yet unidentified ?) of the Na^+/Cl^--dependent transporters or molecules related to these transporters in structure, for it is becoming canonical that many different proteins may be grouped into family or superfamily evolutionarily related according to their structures.

ACKNOWLEDGMENTS

This work was supported in part by grants from the National Science Council, R.O.C. to M.-C. Tzeng (NSC81-0211-B001-07, 82-0211-B001-041, and 83-0203-B001-093), and from the National Institutes of Health, U.S.A. to M.-D. Tsai (GM41788). We thank J. C. Chuang, C. M. Dupureur, R. J. Guillory, M.-J. Hseu, Y. Y. Ko, C.-C. Tseng, and J. H. Yang for their contributions and S. H. Rao for typing the manuscript.

REFERENCES

1. Tzeng, M.-C. (1993) J. Toxicol.-Toxin Reviews **12**, 1-62.
2. Harris, J. B. (1991) in *Snake Toxins* (A. L. Harvey, Ed.), pp. 91-129, Pergamon Press, New York.
3. Hawgood, B., and Bon, C. (1991) in *Handbook of Natural Toxins, Vol. 5, Reptile Venoms and Toxins* (A. T. Tu, Ed.), pp. 3-52, Marcel Dekker, New York.
4. Davidson, F. F., and Dennis, E. A. (1991) in *Handbook of Natural Toxins, Vol. 5, Reptile Venoms and Toxins* (A. T. Tu, Ed.), pp. 107. Marcel Dekker, New York.
5. Harvey, A. L. (1990) Int. Rev. Neurobiol. **32**, 201-239.
6. Rosenberg, P. (1990) in *Handbook of Toxinology* (W. T. Shier and D. Mebs, Eds.), pp. 67-277, Marcel Dekker, New York.
7. Dennis, E. A. (1994) J. Biol. Chem. **269**, 13057-13060.
8. Mayer, R. J., and Marshall, L. A. (1993) FASEB J. **7**, 339-348.
9. Kudo, I., Murakami, M., Hara, S. and Inoue, K. (1993) Biochim. Biophys. Acta **1170**, 217-231.
10. Hseu, M. J., Yang, J. H., Guillory, R. J., and Tzeng, M.-C. (1985) 13th Intl. Congr. Biochem. p.143, Amsterdam.
11. Tzeng, M.-C., Hseu, M. J., Yang, J. H., and Guillory, R. J. (1986) J. Protein Chem. **5**, 221-228.
12. Chuang, J. C. (1992) Master Thesis, National Taiwan University
13. Ko, Y. Y. (1994) Master Thesis, National Taiwan University
14. Othman, I. B., Spokes, J. W., and Dolly, J. O. (1982) Eur. J. Biochem. **128**, 267-276.
15. Rehm, H., and Betz, H. (1982) J. Biol. Chem. **257**, 10015-10022.
16. Rapuano, B. E., Yang, C.-C., and Rosenberg, P. (1986) Biochim. Biophys. Acta **856**, 457-470.
17. Shabo-Shina, R., and Bdolah, A. (1987) Toxicon **25**, 253-266.
18. Breeze, A. L., and Dolly, J. O. (1989) Eur. J. Biochem. **178**, 771-778.

19. Lambeau, G., Barhanin, J., Schweitz, H., Qar, J., and Lazdunski, M. (1989) J. Biol. Chem. **264**, 11503-11510.
20. Degn, L. L., Seebart, C. S., and Kaiser, I. I. (1991) Toxicon **29**, 973-988.
21. Krizaj, I., and Gubensek, F. (1994) Biochemistry **33**, 13938-13945.
22. Delot, E., and Bon, C. (1993) Biochemistry **32**, 10708-10713.
23. Whittaker, V. P. (1959) Biochem. J. **72**, 694-706.
24. De Robertis, E., Rodriguez de Lores Arnaiz, G., Salganicoff, L., Pellegrino de Iraldi, A., and Zieher, L. M. (1963) J. Neurochem. **10**, 225-235.
25. Bieber, A. L., Mills, J. P., Jr., Ziolkowski, C., and Harris, J. (1990) Toxin Rev. **9**, 285.
26. Bon, C., Bouchier, C. *et al.* (1989) Acta Physiol. Pharmacol. Latino-Amer. **39,** 439.
27. Bon, C., Changeux, J. P., Jeng, T. W., and Fraenkel-Conrat, H. (1979) Eur. J. Biochem. **99**, 471-481.
28. Hendon, R. A., and Tu, A. T. (1979) Biochim. Biophys. Acta **578**, 243-252.
29. Fohlman, J., Eaker, D., Karlsson, E., and Thesleff, S. (1976) Eur. J. Biochem. **68**, 457-469.
30. Habermann, E., and Breithanpt, H. (1978) Toxicon **16,** 19-30.
31. Fohlman, J., Lind, P., and Eaker, D. (1977) FEBS Lett. **84**, 367-371.
32. Hseu, M. J., Guillory, R. J., and Tzeng, M.-C. (1990) J. Bioenerg. Biomembr. **22**, 39-50.
33. Yen, C.-H., and Tzeng, M.-C. (1991) Biochemistry **30**, 11473-11477.
34. Radvanyi, F., Saliou, B., Lembezat, M. P., and Bon, C. (1989) J. Neurochem. **53,** 1252-1260.
35. Tzeng, M.-C., Hseu, M. J., and Yen, C.-H. (1989) Biochem. Biophys. Res. Commun. **165**, 689-694.
36. Mebs, D., and Klaus, I. (1991) in *Snake Toxins* (A. L. Harvey, Ed.), pp. 425-447, Pergamon Press, New York.
37. Heinrikson, R. L. (1991) in *Methods in Enzymology,* vol. 197 (E. A. Dennis, Ed.), pp. 201-214, Academic Press, New York.
38. Dupureur, C. M., Yu, B. Z., Mamone, J. A., Jain, M. K., and Tsai, M.-D. (1992) Biochemistry **31**, 10576-10583.
39. Sayers, J. R., Krekel, C., and Eckstein, F. (1992) BioTechniques **13**, 592-596.
40. Tzeng, M.-C., Yen, C.-H., Hseu, M.-J., Dupureur, C. M., and Tsai, M. D. (1995) J. Biol. Chem. **270,** 2120-2123.
41. Schloss, P., Puschel, A. W., and Betz, H. (1994) Curr. Opin. Cell Biol. **6**, 595-599.
42. Amara, S. G., and Kuhar, M. J. (1993) Annu. Rev. Neurosci. **16**, 73-93.
43. Amara, S. G., and Arriza, L. (1993) Curr. Opin. Neurobiol. **3**, 337-344.
44. Uhl, G. R. (1992) Trends Neurosci. **15**, 265-268.
45. Wright, E. M., Hager, K. M., and Turk, E. (1992) Curr. Opin. Cell Biol. **4**, 696-702.
46. Attwell , D., and Bouvier, M. (1992) Curr. Biol. **2**,541-543.

22

PYRULARIA THIONIN*

Physical Properties, Binding to Phospholipid Bilayers and Cellular Responses

Leo P. Vernon

Chemistry Department
Brigham Young University
Provo, Utah 84602

INTRODUCTION

There is an increasing interest in the interaction of small basic peptides with cellular and synthetic membranes, both in terms of the initial binding process and subsequent alterations seen in the phospholipid bilayer order and organization. In terms of its general structure and mode of action with cellular membranes, pyrularia thionin (PT) can be directly compared to both melittin and cardiotoxin (CTX), two other toxic peptides which have been extensively studied. It is generally accepted that all three peptides show electrostatic binding to membranes through strongly charged regions of the peptide, followed by insertion of a hydrophobic portion of the molecule into the membrane bilayer, causing alterations of phospholipid order and structure. The properties of melittin (1,2) and CTX (3,4) have been reviewed.

PT, isolated in our laboratory in 1985, is a strongly basic peptide containing 47 amino acids, including 5 Lys and 4 Arg as well as 8 Cys which participate in 4 disulfide bonds (5). These properties place PT in the family of peptides known as thionins, which are closely related to the viscotoxins. All peptides in these two groups are strongly basic and evoke several related responses with cells, which derive primarily from their binding to and alterations of cellular membranes in terms of structure and properties. Experiments to date show that the biological responses of PT are very similar to those elicited by CTXs isolated from cobra venoms (6-8), and the similar biological responses will be considered briefly in terms of the general structural properties of these two toxic peptides which come from widely divergent sources. The thionins occur in seeds, leaves and roots of plants and recent reviews have covered those derived from leaves (9) and from seeds (10).

---

* Abbreviations: CTX, cardiotoxin; PA, phosphatidic acid; PC, phosphatidylcholine; PS, phosphatidylserine; PLA_2, phospholipase A_2; PT, pyrularia thionin.

Natural Toxins II, Edited by B. R. Singh and A. T. Tu
Plenum Press, New York, 1996

PT is a member of the same order (*Santalales*) as mistletoe, but of a different family. It contains 47 amino acids and has a high degree of sequence homology with the thionins from wheat and barley endosperm as well as *Crambe* and viscotoxin (5). Similar homology is also shown with endosperm thionins from oats and leaf thionins from barley (10). This distribution of similar peptides throughout the plant world indicates an important function for these peptides. Although the function(s) are not definitely known, some plausible suggestions have been given.

A strikingly similar yet different peptide has been isolated from seeds of the crucifer *Crambe abyssinica*, and is called crambin (11). This peptide contains only three disulfide bonds, whose positions correspond to those of viscotoxin (12). It is very hydrophobic, is not positively charged as are the other thionins and consequently is not toxic. Because of its hydrophobic nature it is easy to crystallize, and the three dimensional structure has been determined by X-ray and neutron diffraction (13). Its three dimensional structure is very similar to that determined for the related thionin α 1-purothionin, as discussed below (14,15).

Because of their wide distribution in the plant world and the high degree of structure conservation, one would expect the thionins would play an important role in the economy of plant cells. The thionins isolated from seeds have been proposed to play a role as either a storage protein or as a toxin which protects the seed from insect, bacterial or fungal infestation. A role as a storage protein could better be performed by proteins which are not toxic, and this does not seem to be a major function of the thionins. It could be the role, however, for crambin which is non-toxic.

The toxicity of the thionins is well documented for small animals, insects, bacteria, fungi and mammalian cells. It has been proposed that thionins could protect the starchy endosperm against the action of bacteria and fungi (16). A role as an inhibitor of DNA synthesis has been proposed for thionins, since purothionin inhibits ribonucleotide reductase when reduced thioredoxin served as the hydrogen donor (17). Other studies show that thionins affect cellular membrane permeability and inhibit growth as well as DNA, RNA and protein synthesis in mammalian cells (18). They are also hemolytic for human and animal erythrocytes (6).

A good case can be made for a protective role for leaf thionins. The thionins of barley leaf cells are toxic for pathogenic fungi and their synthesis is triggered by the same infecting fungi (19,20). The toxicity of barley leaf thionins was tested against two fungi, *Thieraliopsis paradoxa*, a pathogen of sugar cane, and *Drechslera seres*, a pathogen of barley. Upon inoculation with spores of the fungus *Erisiphe graminis f.sp hordei*, which causes powdery mildew, barley plants showed a rapid increase in the transcript level of the leaf specific thionins. These leaf thionins are also found predominantly in the outer cell wall of the epidermal cell layer of etiolated barley leaves. A more detailed discussion of the response of barley plants to infection with powdery mildew as it relates to the generation and levels of plant thionin mRNA transcripts is contained in the review by Bohlmann and Apel (10.)

Structure and Relationship to Other Thionins

PT has structural and biological response similarities to both the *Gramineae* thionins and the viscotoxins. It has 8 Cys residues and 4 disulfide bonds which are characteristic of the *Gramineae* thionins. It has a similar sequence, RNCYN, for the highly conserved amino acids 10-14, yet it is not from a monocotyledonous plant. Both PT and the viscotoxins are from dicotyledonous plants of the same order, *Santalales*, yet differ in the number of disulfide bonds (3 for the viscotoxins) and the sequence for amino acids 10-14 (RNIYN for the viscotoxins). The amino acid homology between PT and the viscotoxins is greater than for the cereal thionins (5).

All of the thionins which have been examined to date show a similar three dimensional structure, and except for crambin show a high degree of sequence homology. The homology is highest for the Cys residues and the amino terminus portion of the peptides. The 6 Cys of the viscotoxins, phoratoxins and crambin coincide with 6 of the 8 Cys residues found in the *Gramineae* thionins and PT. The amino acid sequences of three representative thionins, α 1-purothionin, pyrularia thionin and viscotoxin A3 are given below:

α 1-purothionin:	KSCCRSTLGRNCYNLCRARGA-QK-LCAGVCRCKISSGLSCPKGFP-K
Pyr. thionin:	KSCCRNTWARNCYNVCRLPGTISREICAKKCDCKIISGTTCPSDYP-K
viscotoxin A3:	KSCCPNTTGRNIYNACRLTGA-PRPTCAKLSGCKIISASTCPS-YPDK

Additional thionin sequences are given in (5) and more completely in the review by Bohlmann and Apel (10).

The three dimensional structure of the thionins is dictated by the location of the Cys residues and the resulting disulfide bonds, which are highly conserved. Because of the ease of crystallization, the structure of crambin has been most intensively studied, and its crystal structure solved to very high resolution by X-ray and neutron diffraction (21). Its structure has also been determined by NMR spectroscopy (22-25). Other thionins which have been shown by NMR spectroscopy to have similar structures include the purothionins (24,25), hordothionins (24), viscotoxins (25,26) and phoratoxin (26,27). All these studies reveal a similar three dimensional structure for the thionins, which is a very compact L-shaped molecule in which the long arm is formed from two α-helical regions and the short arm by two antiparallel beta sheets. The crystal structure of α 1-purothionin was determined at a resolution of 2.5 Å and compared to the models predicted by NMR, CD and Raman spectroscopy (28). Fig. 1 is taken from the report of Teeter et al (28) and shows the structure of α 1-purothionin in ribbon form and as a space filling model. The conformations of the three common disulfide bonds between purothionin and crambin are similar. The first helix contains amino acids 7-18, the second helix amino acids 22-29 and the anti-parallel beta-sheet contains the segments 1-4 and 31-34. The surface of the molecule has amphiphathic character, with a cluster of charged and polar groups well segregated from the hydrophobic surface. The face of the inner bend between the helical stem and the beta-arm consists mostly of charged and polar residues and is almost entirely hydrophilic. A hydrophobic region is located on the surface of the long helical stem, including four fully exposed Leu, Ile and Val. This explains the fundamental physical and biological properties of thionin molecules, which include heat stability due to disulfide bonds, and their strong binding to cellular membranes by means of a positively charged hydrophilic domain and their interaction with phospholipid hydrocarbon tails through the hydrophobic domain. This could explain the lytic activity of α 1-purothionin. The amphipathic nature of the thionins is illustrated by the fact that they can be extracted from plant tissue with either aqueous salt solutions or in the form of proteolipids by organic solvents. Although the three dimensional structure of PT has not yet been determined, we assume that because of the similar location of the disulfide bonds and amino acid homology it will be similar to the thionins whose structures are already known.

Cellular Responses to Pyrularia Thionin

Toxicity. PT shows toxicity toward all mammalian cell lines tested to date (29). The ID_{50} values obtained for the human cell lines HeLa and MRC-5 were 17 and 0.46 μg/ml respectively. The murine cell lines of B16 melanoma, P388 and L1210 exhibited values of 3.0, 0.62 and 3.9 μg/ml of PT respectively, and the monkey Vero line had a value of 9.3 μg/ml. Viscotoxin is more toxic than PT, with ID_{50} values of 0.2 - 1.7 μg/ml for human tumor

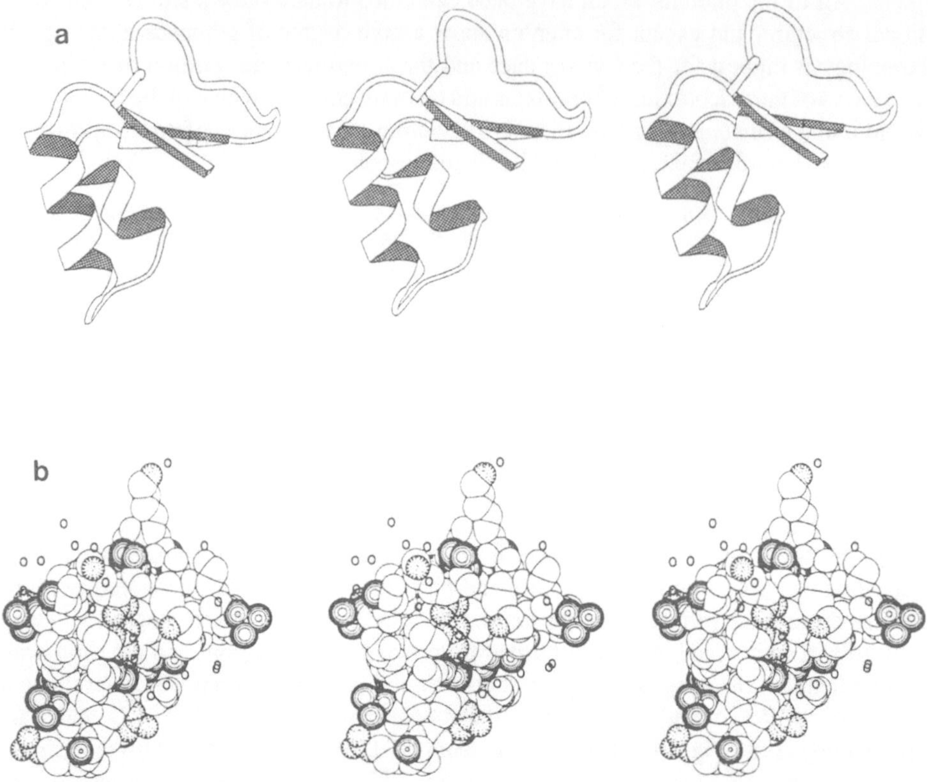

Figure 1. A: Stereo cartoon of α 1-purothionin. B: Stereo van der Waals' surface diagram of the refined 2.5 A α 1-purothionin structure, containing the 369 non-hydrogen atoms and 39 water molecules (small circles). Taken from reference 28 and reprinted by permission of John Wiley and Sons, Inc. The left pair are for divergent viewing and the right pair are for convergent viewing.

cells KB and HeLa (30), and Nakanishi et al (31) reported that 6 ng/ml of purothionin killed transformed mouse fibroblasts, but 10 ng/ml did not affect contact-inhibited cells. The LD_{50} values for mice for PT and viscotoxin have been reported as 1.5 and 0.5 mg/kg, respectively (29,32). Thus, where comparisons can be made, the toxicity of PT with cultured cells and mice is about that of viscotoxin and purothionin. PT does not show any toxicity toward bacteria (29).

Hemolytic Activity. A ready and reproducible test for biological activity of PT is the hemolysis of erythrocytes (6,33). As is observed with snake venoms and CTX isolated from snake venoms (3,4), PT acts synergistically with added phospholipase A_2 (PLA_2) in catalyzing hemolysis. Adding PT to erythrocytes does elicit slight PLA_2 activity (33), but since the PT used in the hemolysis assay does not contain any contaminating PLA_2 (as is the case with most cardiotoxins) (3,4), the hemolytic activity observed in the absence of added PLA_2 must be due to PT. Kinetic analysis shows that the hemolysis reaction follows typical Michaelis-Menten kinetics with a K_m value of 1.6 μM. Binding studies with ^{125}I-PT show by Scatchard analysis a K_d value of 2.1 μM.

The hemolytic activity of PT has been compared to that of several snake venoms and a CTX prepared from *Naja naja kaouthia*, which was devoid of contaminating PLA_2 (6,33). The CTX was slightly less active than PT in the hemolysis assay, and inhibition studies performed with iodinated PT (which was inactive) showed a competitive inhibition, indicating that PT and CTX bound to the same site on the membrane to initiate hemolysis. The Km observed for CTX was 0.69 μM. All these data indicate a specific and common binding site on the erythrocyte membrane for both PT and CTX.

As discussed below, both PT and CTX most likely bind to the phospholipid bilayer to initiate hemolysis, and if this is the case the hemolysis activity and peptide binding should be sensitive to the electrostatic environment of the membrane bilayer. This was studied by adding Ca^{2+} and phosphate ions and by removing membrane-bound Ca^{2+} by the addition of EGTA (7). The basal rates of hemolysis induced by added PT and CTX in physiological saline solution increased by the addition of phosphate in the 5-10 mM range, and also by removing membrane-bound Ca^{2+} by incubation of the cells with 10 mM EGTA. These treatments do not change the apparent K_m values, but increase the Vmax, indicating that more sites are made available with both treatments. On the other hand, added Ca^{2+} in the 5-10 mM range competitively inhibited the reactions by preventing PT and CTX binding to the membrane.

Binding Sites. A companion study utilizing the same treatments involved the binding of ^{125}I-PT and ^{125}I-CTX to human erythrocytes (8). Similar results were obtained, with added phosphate or removal of membrane-bound Ca^{2+} stimulating the binding, while added Ca^{2+} at 10 mM competitively inhibited the binding of both iodinated peptides. Competition studies showed that both toxins bound to the same site on the membrane, and the Scatchard analysis showed the number of sites/cell ranged from 0.7 to 1.7 x 10^5 for PT, and from 0.82 to 1.6 x 10^5 for CTX. Sheep erythrocytes, which are only slightly hemolyzed by PT and CTX (33), showed fewer binding sites/cell for both PT and CTX (8). Significantly more binding sites were observed with murine P388 cells, which had 7.0 and 9.5 x 10^6 sites/cell for PT and CTX respectively.

Activation of Calcium Channel. The addition of PT to P388 cells causes an influx of Ca^{2+}, as measured by changes in Fura-2 fluorescence (34). A later study showed that PT addition also caused the release of some Fura-2 from the cell, and it combines with Ca^{2+} in the suspending medium (35), thus complicating this assay for Ca^{2+} influx. By adding Mn^{2+} to the medium, fluorescence from external Ca^{2+} bound to Fura-2 was prevented, and although the observed Ca^{2+} influx was less than reported previously, there was a net Ca^{2+} influx into the cells upon the addition of PT. Experiments performed with rat anterior pituitary cells also showed that the Ca^{2+} channel inhibitor methoxyverapamil inhibits the PT-induced release of growth hormone from these cells, thus giving indirect evidence for induction by PT of a Ca^{2+} channel in these cells (36).

Activation of Cellular Phospholipase A_2. The early data obtained with erythrocytes indicated an activation of an endogenous PLA_2 by added PT (33). This is not in itself surprising since extensive work has been done on the activation of endogenous cellular PLA_2 by peptide toxins, including melittin and CTX (37,38). Studies of this type with either melittin or CTX isolated from bee or snake venoms are now considered with some caution, however, since samples of both toxins are contaminated by PLA_2 present in the original venom (4,39). A claim for PLA_2 activation requires proof of noncontamination with venom PLA_2 or inactivation of such contaminating PLA_2 by treatment with *p*-bromophenacyl bromide (39). PT is well suited for studies of this nature since it has no contaminating PLA_2. This was shown by the lack of pH change using egg yolk phosphatidylcholine (PC) as substrate (33). Angerhofer et al also reported no detectable PLA_2 activity using radiolabeled PC in an assay that would have detected 20 picomoles of PC hydrolyzed per minute (40).

PT induces a PLA_2 activity in NIH 3T3 fibroblast cells, measured by the release of radiolabeled arachidonic acid (40). Arachidonic acid release by the cells was inhibited by *p*-bromophenacyl bromide, and this coupled with the lack of contaminating PLA_2 shows an activation of an endogenous PLA_2 by membrane-adsorbed PT. There is a time lag of about 20 minute for release of arachidonic acid and cell killing, a lag in the dose response up to 10 of PT, and no response to added Ca^{2+} was noted. A similar activation was observed for purothionin, but not for phoratoxin or crambin. Similar experiments performed in our laboratory (35) gave essentially the same results, except for the lack of Ca^{2+} sensitivity reported by Angerhofer et al (40).

PT also stimulates a PLA_2 in rat anterior pituitary cells (36). Thionin increases the amount of arachidonic acid released from loaded cells as well as increasing the levels of membrane lysophospholipids. PT also stimulates the release of prolactin and growth hormone, but whereas these activities are inhibited by Ca^{2+} channel blockers, the release of arachidonic acid is not. This indicates two roles for PT, one an activation of a Ca^{2+} channel leading to release of prolactin and growth hormone and the other an effect on the membrane itself, leading to PLA_2 activation. Perturbation of membrane phospholipids as a means of inducing PLA_2 activity will be discussed below.

Adenylate Cyclase. Another secondary cellular effect of PT is the enhancement of adenylate cyclase activity in S49 lymphoma cell membranes (41). The addition of PT to isolated cell membranes resulted in an increase in the forskolin-induced cAMP production in both wild type and cyc^- variant cell membranes. The stimulatory G protein (G_s) was not involved in the action of PT on adenylate cyclase, but PT did appear to reduce the effect of G_i on the enzyme. CTX also enhanced the activity of the enzyme, and both toxins appear to exert their effect by acting on the membrane, followed by a secondary activation of adenylate cyclase in the membrane.

Direct Effects on Biological Membranes

Binding Site for Pt. Although it is natural to think of an integral membrane protein as the receptor for PT and other thionins, there is no evidence to date supporting this presumption. A search has been made for a protein receptor with erythrocyte and P388 membranes, including direct binding of PT and cross linking with glutaraldehyde (35). In contrast, there is abundant evidence that PT and other thionins react readily with synthetic phospholipid bilayers as well as cellular membrane phospholipids, which leads to the thesis that thionins bind to a specific phospholipid domain on the membrane surface.

Binding of thionins to phospholipids is in agreement with the fact that thionins can be extracted from seeds by either an aqueous salt or dilute acid solution or by extraction with organic solvents like petroleum ether or chloroform/methanol. The original procedure of Balls et al (42) involved extraction with petroleum ether to yield a "proteolipid" containing both protein and lecithin. This procedure with some modifications has been used by others. The stability of PT to heat and acid allows dilute acid to be used as an extracting medium, and PT is readily extracted from ground seeds with either dilute salt or acid (5), with the acid extraction giving the greatest initial purification. Extraction with organic solvents has not been tried, but should be effective.

There is evidence for interaction of PT with membrane phospholipids. When murine P388 cells are treated with PT, the cell membranes become ruffled and small blebs form on the external surface. These blebs increase in size with increasing time, sometimes becoming as large as the cell itself (35) A similar response was observed with CTX treatment. Blebbing is readily observed with hepatocytes under stress conditions which prevent the synthesis of the phospholipids required to maintain membrane integrity (43). Activation of cellular PLA_2 activity can produce the same result (43).

The number of PT binding sites on various cell types is not consistent with a protein being the binding site. The observed number of sites with human erythrocytes is in the range of 1.0 to 1.5 x 10^5/cell (8), but this increases by a factor of about 10 with erythrocyte ghosts (unpublished data). A major change that takes place during ghost formation is the redistribution of phosphatidylserine (PS) from its preferential location on the interior leaflet to give a symmetrical distribution. There are about 1 x 10^6 band 3 molecules per erythrocyte, and theoretically this could bind PT with the native erythrocytes, but there are not enough to bind the number of PT molecules bound to ghosts. A reasonable assumption is that the PT binds to PS which has redistributed to the external face of the ghost membrane. The large number of binding sites for P388 cells, 7 and 9.5 x 10^6 for PT and CTX is also not consistent with a protein as the binding site for PT or CTX with this cell.

Membrane Permeability. All evidence indicates that PT does not bind to a protein receptor site, but binds in a specific manner to a phospholipid area or domain in the phospholipid bilayer. This is supported by data indicating that the cereal thionins as well as PT have an immediate effect on the permeability of cellular membranes. The purothionins, hordothionins and viscotoxins all elicit ^{86}Rb leakage from BHK cells connected to an inhibition of protein synthesis (31). Other low molecular weight compounds such as uridine and its nucleotide derivative also leaked from the cells, with maximal effect taking place after 2-3 hours. The translation inhibitor hygromicin B was also shown to enter the cells after thionin treatment. Another study with purothionin showed that it caused adrenal medullary cells to release protein and to take up ^{14}C-inulin (44). Similar effects on membrane permeability have been observed with PT.

P388 cells loaded with ^{51}Cr release it into the medium upon addition of PT (35). The release is immediate and from the dose response curve a half response concentration of 2.5 μM was calculated, which would correspond to the K_m value. Treatment of P388 cells with PT causes an immediate influx of Ca^{2+} via a channel which earlier experiments showed is blocked by Ni^{2+} (34). Evidence obtained with rat anterior pituitary cells also indicates the opening of a Ca^{2+} channel after PT treatment (36).

Membrane Depolarization. Another indication of a rapid and direct interaction of PT with cellular membranes is the depolarization of P388 cellular membranes and sartorius muscle caused by PT action (34). Depolarization of muscle cells is also a primary effect of CTX (3,4). A comparison of the ability of CTX and PT to induce depolarization of myotubules has been made, and it shows similar responses of the two peptides 45).

Viscosity Changes and Phase Separation in Erythrocyte Membranes

Experiments with P388 cells (described above) showed that addition of PT caused significant changes in the phospholipid membrane, observed by the extensive formation of blebs (35). Other experiments which show a direct effect on the phospholipid bilayer of cells involved adding PT to erythrocyte ghosts previously loaded with the fluorescent phospholipid derivative NBD-PC. Addition of PT caused an increase in the fluorescence polarization of the fluorophore, indicating an increase in phospholipid order in the membrane (46). This activity was abolished by iodination of PT, which treatment also abolishes the ability of PT to hemolyze erythrocytes, indicating that the property of PT which caused the increase in membrane order was also involved in the hemolysis reaction, which is the ability of PT to partially insert into the phospholipid bilayer. CTX was more active than PT in this reaction, which agrees with the ability of CTX to readily interact with phospholipid bilayers, as discussed below.

The technique of fluorescence digital imaging microscopy, using fluorescent derivatives of PS (NBS-PS) and PC (NBD-PC) was used to show a profound change in phospholipid distribution in erythrocyte ghosts upon addition of PT (47). The addition of the peptide to erythrocyte membranes, as well as to phospholipid vesicles prepared from erythrocyte phospholipids, caused an enhancement of phospholipid domains made visible with the fluorescent phospholipid derivatives. There was formation and enrichment of a NBD-PS domain accompanied by a corresponding formation of a NBD-PC domain. Double labeling experiments with a Texas Red derivative of PT showed that the peptide binds to the domain enriched in NBD-PS. When inactivated PT was used (Trp 8 modified by treatment with N-bromosuccinimide), domain formation was not observed, although the peptide bound to the membrane with the same affinity. The separation of the membrane phospholipids into separate and discrete domains required an unmodified Trp 8, which is involved in insertion of the hydrophobic portion of the peptide into the phospholipid bilayer, as discussed below.

PT Reaction with Synthetic Phospholipid Bilayers

The fact that thionins, although readily soluble in water, can be extracted from ground seeds with petroleum ether and other organic solvents to form proteolipids shows that the thionin peptides are amphipathic and do combine with membrane phospholipids. There is some information concerning the direct interaction of thionins with phospholipids. Binding studies have shown that ^{125}I-PT binds readily to synthetic vesicles composed of PS, but not with PC (unpublished data). The experiments cited above (47) describe the effect of PT on phospholipid vesicles formed from extracted erythrocyte phospholipids to induce domain formation.

The interaction of PT with phospholipid bilayers containing acidic phospholipids has been studied by measuring phosphorescence quenching, ESR spectroscopy with spin labels, and ^{1}H- and ^{31}P-NMR at different phospholipid compositions (48). With PC bilayers containing 20 mole % cardiolipin or phosphatidylinositol, PT induces an increase in membrane viscosity, and at higher PT concentrations nonbilayer structures are formed. PC bilayers containing 20 mole % PS respond differently in that the nonbilayer structures form at lower PT concentrations, and after an initial small rise the membrane viscosity decreases. PT also induces membrane fusion, indicative of formation of a H_{II} structure in the membrane as a consequence of PT-induced phospholipid perturbation.

A study has been made on the effects of PT and CTX on PLA_2 hydrolytic activity of two PLA_2 isolated from *Crotalus molossus molossus* venom (49). Unilamellar liposomes of various compositions were used in the study, which showed that PC liposomes were hydrolyzed faster in the absence of added PT or CTX. Addition of PT or CTX inhibited the activity on PC liposomes. The activities of the enzymes were profoundly enhanced on thionin-pretreated liposomes containing PS and on CTX-pretreated liposomes containing cardiolipin or phosphatidic acid. These data show that the two basic peptides modify phospholipid structure to affect the activity of added PLA_2.

Recent experiments have been performed on the fluorescence properties of Trp8 of PT in the absence and presence of dipalmitoyl phosphatidylglycerol (DPPG) synthetic vesicles (50). Addition of PT to the vesicles caused a significant increase in fluorescence intensity and a 10 nm shift of the fluorescence maximum, as shown in Fig. 2. This shows that the single Trp molecule on PT does insert into the hydrophobic region of phospholipid bilayers, and supports the concept that it is this insertion of Trp8 on the hydrophobic face of PT that causes phospholipid perturbation in the bilayer, leading to the many cellular responses observed by treating cells with this peptide.

There is much more information available about the interaction of CTX with phospholipid bilayers. The similarity of PT and CTX in terms of their having a common

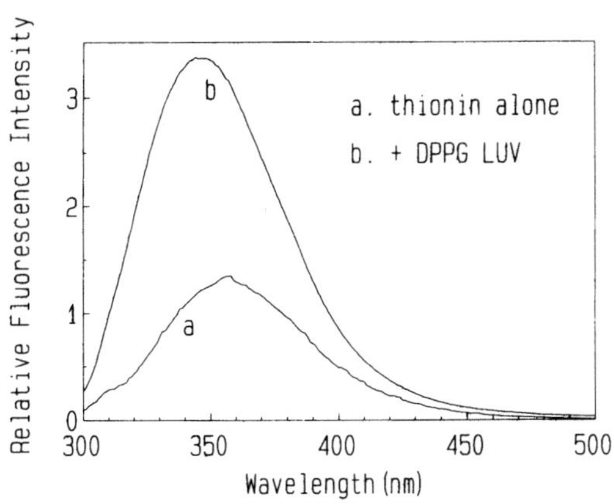

Figure 2. Emission spectrum of thionin. Tryptophan emission spectra of 6 μM thionin were obtained ± 30 μM DPPG at 42°C (ex=280 nm) and pH 8.0.

binding site on erythrocyte membranes and the fact that both PT and CTX elicit hemolysis, an increase in membrane order, depolarization of cellular membranes and activation of PLA_2, as discussed above, is strong evidence for very similar mechanisms of action of both PT and CTX with cellular membranes. Although the CTXs do not have much amino acid homology with PT and other thionins (3,4), they are similar in the sense that both groups are strongly basic, are very stable due to the presence of 4 disulfide bonds, show similar biological responses and a Trp or Tyr in the same region of the molecules has been implicated in the cellular responses of both toxins.

CTX II from *Naja mossambica mossambica* binds to phospholipid vesicles containing the acidic phospholipids PS, phosphatidic acid (PA) or phosphatidylinositol, leading to changes in fluorescence of the Trp residue contained in CTX. This indicates an insertion of the Trp into the hydrocarbon phase of the membrane. A phase separation in PC/PS and PC/PA vesicles is also observed (51,52). The same CTX increases probe motion of diphenylhexatriene in the gel phase and increases it in the fluid phase of membranes containing acidic phospholipids (53,54). A primary result of membrane-CTX interaction is most likely the insertion of a Trp-containing portion of the first loop into the bilayer, since significant changes in Trp fluorescence result from CTX binding (55-57).

Experiments primarily from the laboratory of Aripov on a CTX isolated from Asian cobra venom also show a direct CTX interaction with phospholipid bilayers. Using spin label probes,the CTX (designated cytotoxin V_c5) was shown to partially insert into phospholipid bilayers to cause an increase in the order parameter of PA liposomes (58). The CTX also interacts with PC/PS liposomes to cause aggregation, increased membrane permeability and enhanced fusion (59). Insertion of CTX into a PC bilayer containing 10% PA was also shown by x-ray small angle scattering (60).

Mechanism of PT Action on Cells

There are several responses observed when cells are treated with PT, including the early responses of an increased order parameter of the phospholipids, an increase in membrane permeability, depolarization of the membrane and the opening of Ca^{2+} channels. Longer term responses include the formation of membrane blebs and finally activation of endogenous PLA_2 and adenylate cyclase. For reasons discussed above we feel the initial binding of PT to membranes does not involve a protein receptor, but does involve a membrane site with a specific phospholipid environment. This site would most likely include an acidic phospholipid such as PS. The binding by itself would not cause the observed effects, but insertion of a portion of the peptide into the bilayer would initiate the observed responses. The three dimensional structure of α 1-purothionin, as determined by Teeter et al (28) strongly supports this concept. They propose that the monomer of the thionin, which has exposed hydrophobic regions, associates with the membrane in the plane of the bilayer. The amphipathic helices have hydrophobic groups directed towards one side of the molecule, with hydrophilic groups on the other side. This would allow the hydrophobic groups of the amino acid segment 8-15, including Leu 8 and Leu 15, to insert into the bilayer and bind to the hydrophobic tail groups of the phospholipids. The polar side of the helices would then position on the surface of the bilayer where they could interact with the charged head groups of the phospholipids as well as water in the aqueous phase. The binding of PT to the membrane is not by itself sufficient to elicit the biological responses, since we observe no difference in the binding properties of moderately iodinated PT (which is still biologically active) and extensively iodinated PT which has lost its biological activity (unpublished data).

The biological activity of CTX involves the first loop of the highly constrained peptide, and all evidence indicates that the Trp (or a Tyr) on this first loop is inserted into the bilayer to perturb the phospholipid order (55-57).

Similar physical data are not available for PT, but there is evidence for involvement of a corresponding Trp in PT activity. The highly conserved region of thionins involves amino acids 9-14. The preceding amino acid in PT is Trp. For the cereal thionins this amino acid is the highly nonpolar Leu, and for the viscotoxins the amino acid in this position is Thr. Extensive iodination of PT inactivates PT for all its biological activities (34), and recent experiments on the sequence of Tyr iodination show that limited iodination causes diiodoTyr formation from Tyr 45 with no loss of activity, so Tyr 45 is not required for activity (61). The reaction can also be followed by NMR spectroscopy, since resonance bands corresponding to Tyr 45, Tyr 13 and Trp 8 are identified and can be used to follow modification of these three amino acid residues. More extensive iodination causes iodination of Tyr 13 (most likely to the monoiodoTyr derivative), also results in Trp 8 destruction and inactivates the peptide for all biological activities. The essential role of Trp in PT biological activity was shown by reacting the peptide with N-bromosuccinimide, which selectively oxidized Trp. The loss of activity correlated with destruction of Trp 8, showing the essential nature of this amino acid (61). Modification of Tyr 13 of wheat purothionin has been accomplished by nitration with tetranitromethane at neutral pH and by iodination (62). These modifications reduced but did not eliminate the toxicity of purothionin. Although not directly supporting our data with PT (61), the data on purothionin show that the peptide segment from amino acids 9-14 is required for activity. The experiments of Wada et al (62) included modification of amino groups with acetic or succinic anhydride, and this results in a loss of activity. This is in agreement with our concept that the basic amino acids are involved in the binding process, and a large non-polar amino acid residue like Trp or Tyr then interacts with and perturbs the phospholipid bilayer to cause the observed cellular responses.

The available data for both PT and CTX indicate that the first loop of CTX, containing Trp of Tyr, and the corresponding region of PT, containing Trp 8, are the active sites of the two peptides which show very similar responses with cells. These regions of both peptides have neighboring positively charged residues Lys and Arg which bind to the phospholipid bilayer in a specific fashion. The result of the binding is the insertion of a portion of the peptide chain containing Trp into the bilayer, causing an alteration in the phospholipid order. This initial perturbation is translated into an increased permeability of the bilayer, depolarization of the membrane, changes in the order parameter and opening of a Ca^{2+} channel, as discussed above.

The longer term responses include formation of membrane blebs and a delayed activation of an endogenous PLA_2 and an adenylate cyclase. Any of these effects could be harmful to the cell, but it is probably the activation of the PLA_2 which results in toxicity to the cell. Measuring cell vitality by means of trypan blue exclusion shows that the toxicity of PT closely follows the release of radiolabeled arachidonic acid, which is a measure of PLA_2 activation (35). The activation of PLA_2 can also be explained in terms of perturbation of the phospholipid bilayer resulting from the binding of the toxic peptide to the membrane bilayer. The time lag for PLA_2 activation clearly separates it from the initial rapid responses described above. The nonlinear dose response for PT-induced PLA_2 activity shows that low concentrations of the basic peptide are not effective. This would agree with the results of a study on the action of isolated PLA_2 on synthetic bilayers, which showed a time lag before significant activity was observed (63). The lag was interpreted in terms of the time required to accumulate sufficient products of the enzymatic reaction to significantly perturb the bilayer and allow the enzyme ready access to the phospholipid bilayer. This need to build up reaction products which significantly perturb the phosholipid bilayer could also explain the lag observed for PLA_2 activation with 3T3 cells. A discussion of PT-induced activation of PLA_2 is beyond the scope of this review, but has been considered in another recent review (64).

LITERATURE CITED

1. Dempsey, C. E., 1990, The actions of melittin on membranes, *Biochim. Biophys Acta*, 1031:143-161.
2. Fletcher, J. E., 1993, Possible mechanisms of action of cobra snake venom cardiotoxins and bee venom melittin, *Toxicon*,31:669-695.
3. Harvey, A. L., 1985, Cardiotoxins from cobra venoms, *J. Toxicol. Toxin Rev.*, 4:41-69.
4. Harvey, A. L., 1991, Cardiotoxins from cobra venoms, In Handbook of Natural Toxins, vol. 5, Reptile and Amphibian Venoms, pp. 85-106. Tu, A. T. (ed) Marcel-Dekker, New York.
5. Vernon, L. P., Evett, G. E., Zeikus, R. D., and Gray, W. R., 1985, A toxin thionin from *Pyrularia pubera*: Purification, properties, and amino acid sequence, *Arch. Biochem. Biophys.*, 238:18.
6. Osorio e Castro, V. R. and Vernon, L. P., 1989, Hemolytic activity of thionin from *Pyrularia* pubera nuts and snake venom toxins of Naja Naja species: pyrularia thionin and snake venom cardiotoxin compete for the same membrane site, *Toxicon*, 27:511.
7. Vernon, L. P. and Rogers, A., 1992, Effect of calcium and phosphate ions on hemolysis induced by pyrularia thionin and *Naja Naja Kaouthia* cardiotoxins, *Toxicon*, 30:701.
8. Vernon, L. P. and Rogers, A., 1992, Binding properties of *pyrularia* thionin and *Naja Naja Kaouthia* cardiotoxins to human and animal erythrocytes and P388 cells, *Toxicon*, 30:711.
9. Garcia-Olmedo, F., Rodriguez-Palenzuela, P., Hernandez-Lucas, C., Ponz, F., and Marana, C., 1989, The Thionins: A protein family that includes purothionins, viscotoxins and crambins, *Oxford Surv. Plant Mol. Cell Biol.*, 6:31.
10. Bohlmann, H. and Apel K., 1991, Thionins, *Ann. Rev. Plant. Physiol.* 42:227.
11. Van Etten, C. H., Nielsen, H. C. and Peters, J. E., 1965, A crystalline polypeptide from the seed of *crambe abyssinica, Photochemistry*, 4:467.
12. Teeter, M. M., Mazer, J. A. and L'Italien, J. J., 1981, Primary structure of the hydrophobic plant protein crambin, *Biochemistry*, 20:5437.
13. Hendrickson, W. A. and Teeter, M. M., 1981, Structure of the hydrophobic protein crambin determined directly from the anamalous scattering of sulphur, *Nature*, 290:107.
14. Clore, G. M., Nilges, M., Sukumaran, D. K., Brunger, A. T., Karplus, M. and Gronenborn, A. M., 1986, The three-dimensional structure of α 1-purothionin in solution: combined use of nuclear magnetic resonance, distance geometry and restrained molecular dynamics, *EMBO J.*, 5:2729.
15. Clore, G. M., Sukamaran, D. K., Gronenborn, A. M., Teeter, M. M., Whitlow, M. and Jones, B. L., 1987, Nuclear magnetic resonance study of the solution structure of α 1-purothionin. Sequential resonance assignment, secondary structure and low resolution tertiary structure, *J. Mol. Biol.*, 193:571.
16. Fernandez de Caleya, R., Gonzalez-Pascual, B., Garcia-Olmedo, F. and Carbonero, P., 1972, Susceptibility of phytopathogenic bacteria to wheat purothionins in vitro, *Appl. Microbiol.*, 23:998.

17. Johnson, T. C., Wada, K., Buchanan, B. B. and Holmgren, A., 1987, Reduction of purothionin by the wheat seed thioredoxin system, *Plant Physiol.*, 85:446.
18. Carrasco, L., Vazquez, D., Hernandez-Lucas, C., Carbonero, P. and Garcia-Olmedo, F., 1981, Plant peptides that modify membrane permeability in cultured mammalian cells, *Eur. J. Biochem.*, 116:185.
19. Bohlmann, H., Clausen, S., Behnke, S., Giese, H. and Hiller, C., 1988, Leaf-specific thionins of barley - a novel class of cell wall proteins toxic to plant-pathogenic fungi and possibly involved in the defense mechanism of plants, *EMBO J.*, 7:1559.
20. Ebrahim-Nesbat, F., Behnke, S., Kleinhofs, A. and Apel, K., 1988, Cultivar-related differences in the distribution of cell-wall-bound thionins in compatible and incompatible interactions between barley and powdery mildew, *Planta*, 179:203.
21. Brunger, A. T., Campbell, R. L., Clore, G. M., Gronenborn, A. M., and Karplus, M., 1987, Solution of a protein crystal structure with a model obtained from NMR interproton distance restraints, *Science*, 235:1049.
22. Clore, G. M., Brunger, A .T., Karplus, M., and Gronenborn, A. M., 1986, Applications of molecular dynamics with interproton distance restraints to three-dimensional protein structure determination. A model study of crambin, *J. Mol. Biol.*, 191:523.
23. Williams, R. W. and Teeter, M. M., 1984, Raman spectroscopy of homologous plant toxins: Crambin and α 1- and β-Purothionin secondary structures, disulfide conformation, and tyrosine environment, *Biochemistry*, 23:6796.
24. Lecomte, J. T. J., Jones, B. L. and Llinas, M., 1982, Proton magnetic resonance studies of barley and wheat thionins: structural homology with crambin, *Biochemistry*, 21:4843.
25. Whitlow, M. and Teeter, M. M., 1985, Energy minimization for tertiary structure predictions of homologous proteins: α_1-purothionin and viscotoxin A3 models from crambin, *J. Biochem.*, Struct. Dynam., 2:831.
26. Lecomte, J. T. J., Kaplan, D., Llinas, M., Thunberg, E. and Samuelsson, G., 1987, Proton magnetic resonance characterization of phoratoxins and homologous proteins related to crambin, *Biochemistry*, 26:1732.
27. Clore, G. M., Sukumaran, D. K., Nilges, M. and Gronenborn, A. M., 1987, Three-dimensional structure of phoratoxin in solution: combined use of nuclear magnetic resonance, distance geometry, and restrained molecular dynamics, *Biochemistry*, 26:1732.
28. Teeter, M. M., Ma, X. -Q., Rao, U. and Whitlow, M., 1990, Crystal sturcture of a protein toxin α_1-purothionin at 2.5A and a comparison with predicted models, *Proteins*, 8:118.
29. Evett, G. E., Donaldson, D. M. and Vernon, L. P., 1986, Biological properties of pyrularia thionin prepared from nuts of *Pyrularia pubera*, *Toxicon*, 24:622.
30. Konopa, J., Woynarowski, J. M., and Lewandowska-Gunieniak, M., 1980, Isolation of viscotoxins: cytotoxic basic polypeptides from *viscum album* L., *Hoppe Seyler's Z. Physiol. Chem.*, 131:1525.
31. Nakanishi, T., Yoshizumi, H., Tahara, S., Hakura, A., and Toyoshima, K., 1979, Cytotoxicity of purothionin A on various animal cells, *Jpn. J. Cancer Res.*, 70:323.
32. Samuelsson, G., 1974, Mistletoe Toxins, *Syst. Zool.*, 22:566.
33. Osorio e Castro, V. R., Van Kuiken, B. A., and Vernon, L. P., 1989, Action of a thionin isolated from nuts of *pyrularia pubera* on human erythrocytes, *Toxicon*, 27:501.
34. Evans, J., Wang, Y., Shaw, K. -P and Vernon, L. P., 1989, Cellular responses to pyrularia thionin are mediated by Ca^{2+} influx and phospholipase A2 activation and are inhibited by thionin tyrosine iodination, *Proc. Natl. Acad. Sci., USA*, 86:5849.
35. Evans, J. G. and Vernon, L. P, 1993, Cellular membrane responses and phospholipase A_2 activation induced by *pyrularia* thionin, *J. of Natural Toxins*, 2:143.
36. Judd, A. M., Vernon, L. P and MacLeod, R. M., 1992, *Pyrularia* thionin increases arachidonate liberation and prolactin and growth hormone release from anterior pituitary cells, *Toxicon*, 30:1563.
37. Shier, W. T., Activation of self-destruction as a mechanism of action for cytolytic toxins, In: Natural Toxins, D. Eaker and T. Wadstrom, eds. Permagon Press, Oxford and New York, pp. 193-200.
38. Shier, W. T., 1983, Toxins as research tools: potentials and pitfalls, *J. Toxicol. -Toxin Reviews*, 2:111.
39. Fletcher, J. E., Jiang, M. -S. and Gong, Q. -H., 1991, Snake venom cardiotoxins, and bee venom mellitin activate phsopholipase C activity in primary cultures of skeletal muscle, *Biochem. Cell Biol.*, 69:274.
40. Angerhofer, C. K., Shier, W. T., and Vernon, L. P., 1990, Phospholipase activation in cytotoxic mechanism of thionin purified from nuts of *Pyrularia pubera, Toxicon*, 28:547.
41. Huang, W., Vernon, L. P and Bell, J. D., 1994, Enhancement of adenylate cyclase activity in S49 lymphoma cell membranes by the toxin thionin from *Pyrularia pubera*, *Toxicon*, 32:789-797.
42. Balls, A. K., Hale, W. S. and Harris, T. H., 1942, A crystalline protein obtained from lipoprotein of wheat flour, *Cereal Chem.*, 19:279.

43. Florine-Casteel, K. F., Lemasters, J. J. and Herman B., 1991, Lipid order in hepatocyte plasma membrane blebs during ATP depletion measured by digitized video fluorescence polarization microscopy, *FASEB J.*, 5:2078.
44. Kashimoto T., Sakakibara, R., Huynh, Q. K., Wada, H. and Yoshizumi, H., 1979, The effect of purothionin on bovine adrenal medullary cells, *Res. Commun. Chem. Pathol. Pharmacol.*, 26:221.
45. Chen, X. -H and Harvey, A. L, 1993, Effects of different antagonists on depolarization of cultured chick myotubes by cobra venom cardiotoxins and *pyrularia* thionin from the plant *Pyrularia pubera, Toxicon,* 31:1229-1236.
46. Osorio e Castro, V. R., Ashwood, E. R., Wood, S. G. and Vernon, L. P., 1990, Hemolysis of erythrocytes and fluorescence polarization changes elicited by peptide toxins, aliphatic alcohols, related glycols and benzylidene derivatives, *Biochim. Biophys. Acta,* 1029:252.
47. Wang, F., Naisbitt, G. H., Vernon, L. P and Glaser, M., 1993, Pyrularia thionin binding to and the role of tryptophan-8 in the enhancement of phosphatidylserine domains in erythrocyte membranes, *Biochemistry*, 32:12283-12289.
48. Gasanov, S. E., Vernon, L. P. and Aripov, T. F., 1992, Modification of phospholipid membrane structure by the plant toxic peptide pyrularia thionin, *Arch. Biochem. Biophys.*, 30:367.
49. Gasanov, S. E., Rael, E. D., Martiniz, M., Baeza, G. and Vernon, L. P, 1994, Modulation of phospholipase A_2 activity by membrane-active peptides on liposomes of different phospholipid composition, *Gen. Physiol. Biophys.* 13:275-286.
50. Bell, J. Personal Communication.
51. Dufourcq, J. and Faucon, J. -F., 1978, Specific binding of a cardiotoxin from *Naja mossambica mossambica* to charged phospholipids detected by intrinsic fluorescence, *Biochemistry*, 17:1170.
52. Vincent, J. -P., Balerna, M. and Lazdunski, M., 1978, Properties of association of cardiotoxin with lipid vesicles and natural membranes, *FEBS Lett.*, 85:103.
53. Faucon, J. F., Bernard, E., Dufourcq, J., Pexolet, M. and Bougis, P., 1981, Perturbation of charged phospholipid bilayers induced by melittin and cardiotoxins. A fluorescence, differential scanning calorimetry and raman spectroscopy study, *Biochmie,* 63:857.
54. Faucon, J. F., Dufourcq, J., Bernard, E., Duchesneau, L. and Pexolet, M., 1983, Abolition of the thermotropic transition of charged phospholipids induced by a cardiotoxin from *Naja mossambica mossambica* as detected by fluorescence polarization, differential scanning calorimetry, and raman spectroscopy, *Biochemistry*, 22:2179.
55. Bougis, P, Tessier, M., Van Rietschoten, J., Rochat, H., Faucon, J.F. and Dufourcq, J., 1983, Are interactions with phospholipids responsible for pharmacological activities of cardiotoxins?, *Mol. Cell Biochem.*, 55:49.
56. Dufourcq, J., Faucon, J.F., Bernard, E. and Pezolet, M., 1982, Structure-function relationships for cardiotoxins interacting with phospholipids, *Toxicon*, 20:165.
57. Gatineau, E., Toma, F., Montenay-Garestier, T., Takechi, M., Fromageot, P. and Menez, A., 1987, Role of tyrosine and tryptophan residues in the structure-activity relationships of a cardiotoxin from *Naja nigrocollis* venom, *Biochemistry*, 26:8046.
58. Aripov, T. F., Salakhutdinov, B. A., Salkihova, Z. T., Sadykov., A. S. and Tashmukhamedov, B. A., 1984, Structural changes of liposome phospholipid packing induced by cytotoxin of the central asia cobra venom, *Gen. Physl. Biophys.*, 3:489.
59. Aripov, T. F., Gasanov, S. E., Salakhuldinov, B. A., Rozenshtein, I. A. and Kamaev, F. G., 1989, Central Asian cobra venom cytotoxins-induced aggregation, permeability and fusion of liposomes, *Gen. Physl. Biophys.*, 8:459.
60. Oimatov, M., Lyoy, Yu. M., Aripov, T. F. and Feigin, L. A., 1986, X-ray small-angle scattering structural studies of model membrane complexes with cytotoxin, *Stud. Biophys.* 112:237.
61. Fracki, W. S., Li, D., Owen, N., Perry, C., Naisbitt, G. H. and Vernon, L. P., 1992, Role of Tyr and Trp in membrane responses of pyrularia thionin determined by optical and NMR spectra following Tyr iodination and Trp modification, *Toxicon*, 30:1427.
62. Wada, K., Ozaki, Y., Matsubara, H. and Yoshizumi, H., 1982, Studies on purothionin by chemical modifications, *J. Biochem.*, 91:257.
63. Bell, J. D. and Biltonen, R. L., 1989, The temporal sequence of events in the activation of phospholipase A_2 by lipid vesicles, *J. Biol. Chem.*, 264:12194.
64. Vernon, L. P and Bell, J. D., 1992, Membrane structure, toxins and phospholipase A_2 activity, *Pharm. Therap.*, 54:269.

23

THE CHEMISTRY AND BIOLOGICAL ACTIVITIES OF THE NATURAL PRODUCTS AAL-TOXIN AND THE FUMONISINS

H. K. Abbas,[1] S. O. Duke,[1] W. T. Shier,[2] R. T. Riley,[3] and G. A. Kraus[4]

[1] USDA-ARS, SWSL
Stoneville, Mississippi 38776
[2] College of Pharmacy
University of Minnesota
Minneapolis, Minnesota 55455
[3] USDA-ARS, TMRU
Athens, Georgia 30613
[4] Department of Chemistry
Iowa State University
Ames, Iowa

INTRODUCTION

The fumonisins and AAL-toxin are structurally-related natural products produced by fungi (22, 23, 24). Fumonisin B_1 [FB_1] has been recognized since 1988 as a mammalian toxin (36, 37, 44, 52). More recently, FB_1 has been shown to be a phytotoxin as well (1, 2, 3). AAL-toxin has been known to produce symptoms of stem canker disease on susceptible (*asc/asc*) tomatoes (9, 48, 56, 58). Our research has focused on evaluating the potential of the class of natural products defined by AAL-toxin and the fumonisims for the development of improved weed control agents. This work has involved the development of improved methods for the isolation, purification and characterization of these toxins. Useful methods have been developed for assessing phytotoxicity of these toxins with various weed species, and for estimating mammalian toxicity. A series of analogs has been obtained, either derivatizing the parent toxins or by *de novo* synthesis, and evaluated for phytotoxicity and mammalian cytotoxicity. Characterization of the toxins has included studies on their mode of action in plant tissue, their ability to act as sphingosine analogs and their stereochemistry.

1. Isolation, Purification and Characterization

The genus *Alternaria* is widely distributed in nature, and a number of *Alternaria* spp are pathogenic to plants (66). *A. alternata* f. sp. *lycopersici* was isolated from susceptible (*asc/asc*) tomatoes with stem canker disease (33, 40, 48). The active phytotoxin responsible

Natural Toxins II, Edited by B. R. Singh and A. T. Tu
Plenum Press, New York, 1996

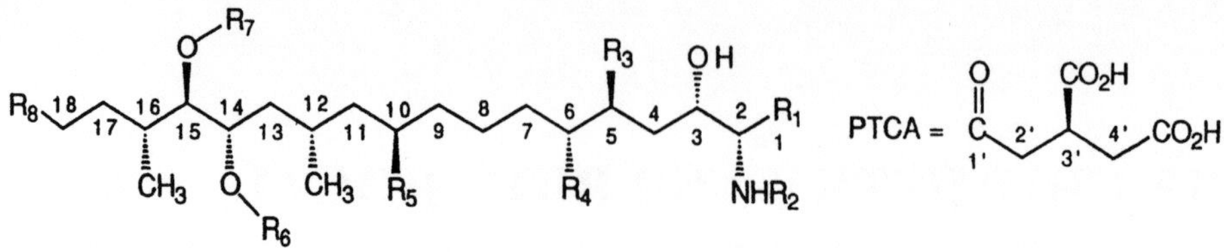

Toxin[a]	R_1	R_2	R_3	R_4	R_5	R_6	R_7	R_8
AAL-TA_1	H	H	OH	OH	H	PTCA	H	H
AAL-TA_2	H	H	OH	OH	H	H	PTCA	H
AAL-TB_1	H	H	OH	H	H	PTCA	H	H
AAL-TB_2	H	H	OH	H	H	H	PTCA	H
AAL-TC_1	H	H	H	H	H	PTCA	H	H
AAL-TC_2	H	H	H	H	H	H	PTCA	H
AAL-TD_1	H	$COCH_3$	OH	H	H	PTCA	H	H
AAL-TD_2	H	$COCH_3$	OH	H	H	H	PTCA	H
AAL-TE_1	H	$COCH_3$	H	H	H	PTCA	H	H
AAL-TE_2	H	$COCH_3$	H	H	H	H	PTCA	C_2H_5
FA_1	CH_3	$COCH_3$	OH	H	OH	PTCA	PTCA	C_2H_5
FA_2	CH_3	$COCH_3$	OH	H	H	PTCA	PTCA	C_2H_5
FB_1	CH_3	H	OH	H	OH	PTCA	PTCA	C_2H_5
FB_2	CH_3	H	OH	H	H	PTCA	PTCA	C_2H_5
FB_3	CH_3	H	H	H	OH	PTCA	PTCA	C_2H_5
FB_4	CH_3	H	H	H	H	PTCA	PTCA	C_2H_5
FC_1	H	H	OH	H	OH	PTCA	PTCA	C_2H_5
HFB_1 (AP_1)	CH_3	H	OH	H	OH	H	H	C_2H_5
HFB_2 (AP_2)	CH_3	H	OH	H	H	H	H	C_2H_5

[a] The indicated absolute configurations have been established only for AAL-TA_1, AAL-TA_2, FB_1, HFB_1 and HFB_2.

Figure 1. Chemical structures of various analogs of AAL-toxins isolated from *A. alternata* f. sp. *lycopersici*, various analogs of fumonisins A and B isolated from *F. moniliforme*, the hydrolyzed products (AP_1 and AP_2) and the absolute configurations of AAL-TA_1, AAL-TA_2, FB_1, HFB_1 and HFB_2.

for stem canker was found to be AAL-toxin (9,22, 23, 33). AAL-toxin was originally isolated from liquid fermentation cultures of *A. alternata* f. sp. *lycopersici* by thin-layer chromatography (22, 23). Two compounds were isolated, TA and TB, each having two structural isomers (Fig. 1). Since then TC, TD and TE, each with two isomers, have been discovered (28).

The AAL-toxins are hydroxylated, long-chain alkylamines with one esterified propane-1,2,3-tricarboxylic acid moiety. The difference between TA and TB is that TB lacks a hydroxyl group at R_4 (Fig. 1). TA and TB are equally active against susceptible tomatoes (56, 58). However, removal of the hydroxyl group at R_3, which produces TC, renders the compound much less active (300 ng vs. 10 ng per plant to cause visible symptoms) (28). TD and TE are acetylated forms of TB and TC, respectively. They represent up to 40% of the AAL-toxins produced by fermentation in liquid medium. TD and TE are essentially inactive in phytotoxicity testing and are probably metabolic products from detoxification of the active metabolites (28).

We have developed a fermentation process for *A. alternata* and other fungi on solid media, particularly rice (5, 8). The procedure is briefly described as follows. Rice was moistened with water and shaken and autoclaved twice. The flasks were inoculated with *A. alternata* NRRL 18822 grown on potato dextrose agar. Flasks were incubated for 4 wks in 12 h light/12 h dark at 28 C.

AAL-toxin was extracted from rice by initially pre-extracting with chloroform overnight and discarding the lipid extract. The lipid-free residue contained the AAL-toxin which was extracted three times with a water:methanol (60:40 v/v) mixture. Evaporation of the methanol leaves AAL-toxin in water, which is applied to an Amberlite XAD-2 column. Elution with methanol yields crude AAL-toxin which can be purified by flash column chromatography on reverse-phase C_{18} packing eluted with 60% aqueous methanol. This produces AAL-toxin isomer TA as a white solid at ≥ 95% purity determined by TLC. The advantage of this method is isolation of a pure compound, which can be used for standardized phytotoxicity testing. All of our subsequent studies have been done using purified AAL-toxin isomer TA.

The fumonisins are related structurally to AAL-toxin (Fig. 1), but are produced by *Fusarium moniliforme*, *F. proliferatum* and other *Fusaria* of the Liseola section (57, 76). Recently, it has been reported by Chen et al., (31) that small amounts of FB_1 were produced by *Alternaria alternata* f. sp. *lycopersici* as well, although this finding has been disputed by others. Recently, Abbas and Riley (14) reported that *A. alternata* produces small amounts of FB_1, FB_2 and FB_3 as well as AAL-toxin.

FB_1 was isolated from a phytotoxic extract of *F. moniliforme* NRRL 18738 and shown to be the compound responsible for phytotoxic symptoms on jimsonweed (1, 3 , 4, 7). FB_1 was confirmed to be the same toxin that caused mammalian toxicity including equine leukoencephalomalacia (52) and porcine pulmonary edema (34).

The structure of FB_1 was determined in 1988 by Bezeidenhout et al., (24) and Gelderblom et al., (36) (Fig. 1). FB_2 and FB_3 lack hydroxyl groups at R_5 and R_3 respectively, and FB_4 lacks hydroxyl groups at both sites (Fig. 1). FA_1 and FA_2 are the N-acetylated forms of FB_1 and FB_2, respectively, and are essentially inactive (Fig. 1). The hydrolysis products, the aminopentols [AP_1 (HFB_1) and AP_2 (HFB_2)], are weakly phytotoxic (6, 75).

FB_1 is also produced by fermentation of *F. moniliforme* on solid media, especially rice (1, 4). The isolate was grown on autoclaved, converted long-grain enriched rice by the following procedure. Fungus-infested rice was transferred to a screen-bottomed tray and allowed to air-dry at room temperature for 72 to 96 h in a ventilated hood. One kg of ground infested rice was soaked in aqueous methanol (40%) at a ratio 1:5 culture to extracting solvent (w/v) overnight at room temperature. Following the soaking, cultures were blended for 5 min at high speed, centrifuged at 10,000 xg for 10 min, and filtered through a double layer of cheese cloth. The filtrates were concentrated in a rotary evaporator at 40° C until all traces of methanol were removed. The water layer was applied to an XAD-2 column for clean-up, followed by a silica gel column. Fumonisin was purified as described in detail by Vesonder et al., (79). Branham and Plattner (26) recently reported that a new fumonisin C_1 has been obtained from a isolate of *F. moniliforme.* This compound lacks the *N*-terminal methyl group (R_1) of FB_1. Nothing is yet known about its toxicity (Fig. 1).

2. Biological Effects

AAL-toxin was first described as the active component of *A. alternata* culture filtrates that cause stem canker disease in *asc/asc* tomatoes (*Lycopersicon esculentum* Mill.) (39). At that time, because heterozygous (*Asc/asc*) and dominant (*Asc/Asc*) genotypes were resistant to AAL-toxin, it was felt that it was a host-specific toxin (33).

When FB_1 was found to be structurally related to AAL-toxin, it was decided to test both toxins for phytotoxicity with a variety of plant species. AAL-toxin was phytotoxic to both weed and crop species, including black nightshade (*Solanum nigrum* L.), redroot pigweed (*Amaranthus retroflexus* L.), northern jointvetch [(*Aeschynomene virginica* (L.) B.S.P.)], duckweed (*Lemna* spp.), prickly sida (*Sida spinosa* L.), jimsonweed (*Datura stramonium* L.), cocklebur (*Helianthus annuus* L.), hemp sesbania (*Sesbania exaltata* [(Raf.)

Table 1. Cytotoxicity of Fumonisin and Analogs with Cultured Mammalian Cell Lines

Fumonisin Analog	Molecular Weight	IC_{50} (μM) MDCK	H4TG	NIH3T3
FB_1	721.84	10	10	150
FB_2	705.81	20	2	—#
FB_3	705.81	50	5	—#
AP_1	405.62	100	—#	100
AP_2	389.59	30	40	35
FA_1	747.89	—*	—*	—*
FA_2	763.86	—*	—*	—*
$NAcAP_1$	447.66	300	150	150
Ac_6AP_1	657.84	400	400	300
1	705.81	100	75	50
2	617.83	300	400	200
3	677.67	400	300	200
4	691.79	100	100	15
5	375.57	100	150	25
6	509.01	300	300	300
7	357.60	25	15	15
8	705.81	15	15	15
9	733.79	200	200	150
Octadecylamine	269.52	—	400	400

*No detectable cytotoxicity at 132 μM
No detectable cytotoxicity at 70 μM.
(Modified table from ref nos. 7 and 71).

Rydb. ex A.W. Hill]), and sicklepod [*Senna obtusifolia* (L.) Irwin and Barneby]. Symptoms included chlorosis, necrosis, stunting and mortality (5, 12).

Symptoms caused by FB_1 on susceptible plants also included chlorosis, necrosis, stunting and mortality. The symptoms were similar to AAL-toxin with redroot pigweed, sunflower (*Helianthus annuus* L.), sicklepod, hemp sesbania, northern jointvetch, soybean [*Glycine max* (L.) Merr.], prickly sida, venice mallow (*Hibiscus trionium* L.), spurred anoda [*Anoda cristata* (L.) Schlecht.], susceptible tomatoes, jimsonweed and black nightshade being susceptible to ≤ 50 μg ml^{-1} (2).

Duckweed (*Lemna pausicostata* L.) has proved to be a sensitive bioassay for AAL-toxin, FB_1 and related compounds (75). Duckweed is a small aquatic plant that can be easily grown in the laboratory (75). Phytotoxic effects can be easily quantified by measuring chlorophyll loss and cellular leakage (74). AAL-toxin was about 10-fold more active than FB_1 in the duckweed bioassay, causing maximal effect at a 0.1 μM concentration. FB_1, FB_2 and FB_3 caused effects identical to those of AAL-toxin in duckweed at 1 μM. The hydrolysis products, AP_1 and AP_2 were much less active, and FA_1 and FA_2 were completely inactive in the duckweed bioassay (75).

FB_1 has been identified as the cause of leukoencephalomalacia, a degenerative brain disorder in horses (52), and pulmonary edema in swine (34). It was first isolated using a short-term assay for tumor promoters (36) and there is continuing concern that it may be a carcinogen (37, 65). Because FB_1 is a mammalian toxin, it is not suitable for use as a herbicide. Therefore, studies on structure-activity relationships for mammalian toxicity of fumonisins have been carried out in mammalian cell cultures in an attempt to find an analog without mammalian toxicity that retains herbicidal activity. NIH3T3 mouse fibroblasts are most resistant to all compounds studied with FB_1, FB_2, and FB_3 causing no significant cytotoxicity at 70 μM (Table 1). However, H4TG rat hepatoma cells were susceptible at low concentrations (~ 4 μM) of the fumonisins and MDCK dog kidney cells had somewhat higher IC_{50}'s (20 - 56 μM) for the same toxins (7, 71). The most active compound against mammalian cells in our study was FB_2. AAL-toxin was approximately 5-fold less toxic than FB_1 on most cell lines tested. AP_1 and AP_2 showed moderate toxicity against mammalian cell lines (7, 11). No compounds tested had low mammalian toxicity and high phytotoxicity.

3. Mode of Action

Abundant evidence now exists that FB_1, AAL-toxin and their analogs are potential inhibitors of sphingolipid biosynthesis (10, 63, 80, 81, 82). Sphingolipids are essential constituents of cell membranes in both animals and plants (54, 55, 69). Most of the research on the mode of action of this class of natural products has been carried out with FB_1 in mammalian systems. FB_1 has been shown to inhibit the enzyme sphinganine (sphingosine) N-acyl-transferase (ceramide synthase) (81, 82). Elevated levels of free sphinganine and the sphinganine: sphingosine ratio are observed in animal cells treated with FB_1 (64), and in the serum and tissue of animals exposed to dietary fumonisins (63, 80). It has been hypothesized that disruption of sphingolipid metabolism is a contributing factor in the animal diseases associated with consumption of fumonisin contaminated feeds and foods.

Sphingosine and sphingosine 1-phosphate can induce DNA synthesis in growth-arrested Swiss 3T3 cells. FB_1 incubated with the cells elevated sphingosine and induced an increase in [^{3}H] thymidine incorporation into DNA (59). Further studies showed that sphinganine was required for stimulation of DNA synthesis. Studies with LLC-PK renal cells showed a close correlation between fumonisin-induced cytotoxicity and inhibition of *de novo* sphingolipid biosyntheses (82). However, studies with primary rat hepatocytes have not shown any relationship between cytotoxicity and elevation of free sphingoid bases (38).

Our work has focused on confirmation of a similar mode of action in plant tissues (10). Following treatment with FB_1 or AAL-toxin, the free sphingoid bases, phytosphingosine and sphinganine, were found to be markedly elevated relative to controls in the duckweed bioassay (Table 2). Extracts of treated tomato plants and tobacco callus cultures showed the same results. FB_1 and AAL-toxin caused 18- and 45-fold increases in phytosphingosine and 76- and 129-fold increases in sphinganine, respectively, after 24 h (10).

Phytosphingosine and sphingosine also have phytotoxic effects on duckweed at higher concentrations (50 μM) causing increased conductivity and chlorophyll loss (75). This suggests that the build-up of free sphingoid bases is toxic to cells (65, 75). The

Table 2. Concentration (pmol/mg dry weight) of free phytosphingosine and free sphinganine in duckweed treated with 1 mM fumonisin B_1 or AAL-toxin; and susceptible tomato leaf discs (LA12) with 1 mM AAL-toxin for 24 h

Treatment	Phytosphingosine	Sphinganine
Duckweed		
Control	17	6
Fumonisin B_1	309	454
AAL-toxin	770	776
Tomato leaf discs		
Control	7	4
AAL-toxin	279	1749

Values are means from three to six independent experiments.
(Modified table from Ref. no. 10).

mechanisms of cellular toxicity of free sphingoid is unknown, but they do not appear to be acting as a cationic detergent (75).

The fumonisins have been associated with human cancer (32, 77), although the mechanism of carcinogenicity is unknown. FB_1 was not found to be mutagenic in the *Salmonella* mutagenicity assay and was not genotoxic in DNA repair assays (65). The fumonisins appear to act mainly as tumor promoters.

It is unknown how the elevated levels of free sphingoid bases caused by FB_1 and AAL-toxin in plants might mediate the physiological changes observed. It is possible that changes in membrane permeability are involved. The structural analogy with sphingosine

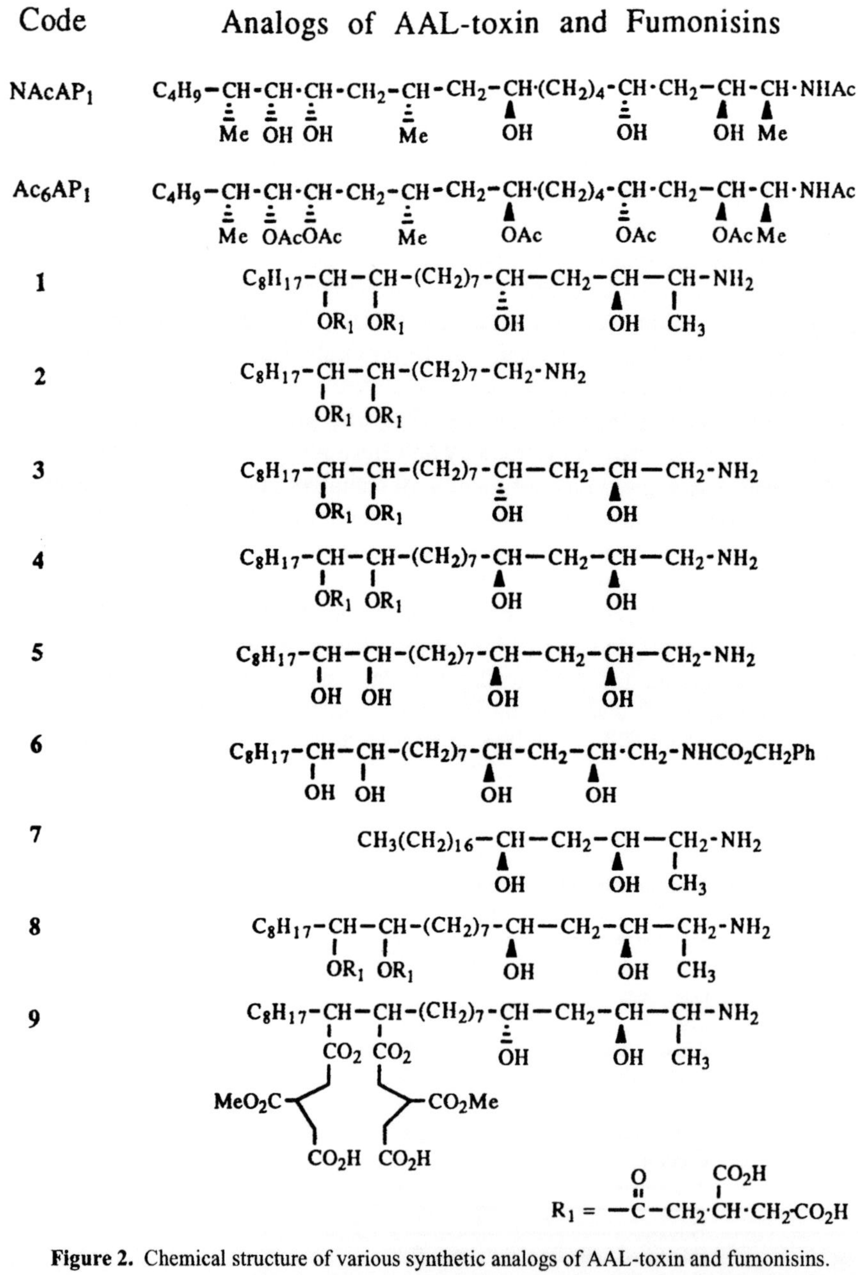

Figure 2. Chemical structure of various synthetic analogs of AAL-toxin and fumonisins.

also suggests an alternate mechanism of action for FB_1 and AAL-toxins. FB_1, AAL-toxin and notably the hydrolyzed toxins are structural analogs of all sphingolipids, but they are particularly closely analogous to sphingosine and sphinganine, a biosynthetic intermediate in the *de novo* synthesis of sphingolipids. Both sphingosine and sphingolipids play key roles in cells. Sphingosine is believed to be an important intracellular regulatory molecule (41) which has been hypothesized to be produced in cells by effector-stimulated degradation of sphingolipids. Sphingosine and FB_1 have been shown to inhibit the activity of protein kinases, notably protein kinase C (42, 46), a key regulatory enzyme which is activated by the well-studied tumor promoter, phorbol myristate acetate (29). The full extent of sphingosine's ability to regulate other protein kinases, such as MAP kinases that are thought to be important in the regulation of cell proliferation and the mechanism of tumor promotion, remains to be determined (54). Sphingolipids are essential structural components of cell membranes; for example, selective destruction of sphingolipids by sphingomyelinases such as staphylococcal ß-toxin is a very effective method for destroying membrane integrity (35). Therefore, disruption of sphingolipid metabolism could alter many biochemical processes essential to the normal function and viability of cells.

4. Analogs

A basic requirement for a safe herbicide is that it exhibit strong phytotoxicity and weak or no mammalian toxicity. The demonstration that FB_1 is a tumor promoter eliminates it from direct consideration as a herbicide. However, the finding that FB_1 is phytotoxic by a different mechanism than any commercially used herbicide makes it worthwhile to search for analogs with low mammalian toxicity and high phytotoxicity.

The fumonisin and AAL-toxin analogs were obtained from a variety of sources. Aminopentols, AP_1 and AP_2 were either the gifts of W.C.A. Gelderblom (7) prepared by hydrolysis of purified FB_1 and FB_2, respectively, or prepared in our laboratory in larger quantities by hydrolysis of crude fumonisins with subsequent purification of the aminopentols (70). The latter approach is somewhat simpler because it avoids the difficulties involved in purifying zwitterions. An extensive series of analogs (Fig. 2) was prepared from oleic acid (17, 49, 50, 74). These analogs were tested on duckweed and susceptible tomato leaf discs for phytotoxicity and on three mammalian cell lines for mammalian toxicity (11, 13).

The most notable thing about these results is that a wide range of structures retains biological activity in both the plant and mammalian systems. Of the compounds tested, analog 9 has the best combination of high phytotoxicity and low mammalian toxicity. The basis for the selectivity may simply reflect higher levels of non-specific carboxylesterase activity in plant tissues. Other analogs of the fumonisins and AAL-toxin will continue to be investigated for desirable characteristics.

5. Biosynthesis and Metabolism of AAL-Toxin and Fumonisins

The substantial structural analogy between fumonisins, AAL-toxin, sphingosine, and related sphingolipids was recognized early in the studies on the toxins, and it has shaped people's ideas about mode of action and probable biosynthetic and metabolic pathways.

Sphingolipids are important constituents of cell membranes in animals, plants, fungi and some bacteria (78). Some important long-chain sphingoid bases are shown in Figure 3.

The biosynthesis of sphingolipids has been studied in mammalian systems (47) and in yeast (73). They are constructed by condensation of the co-enzyme A derivative of a long chain fatty acid, usually palmitate (C-16) with serine with concomitant loss of the carboxyl group of serine. The ketone group is then reduced to give sphinganine, a close structural

AAL-TOXIN

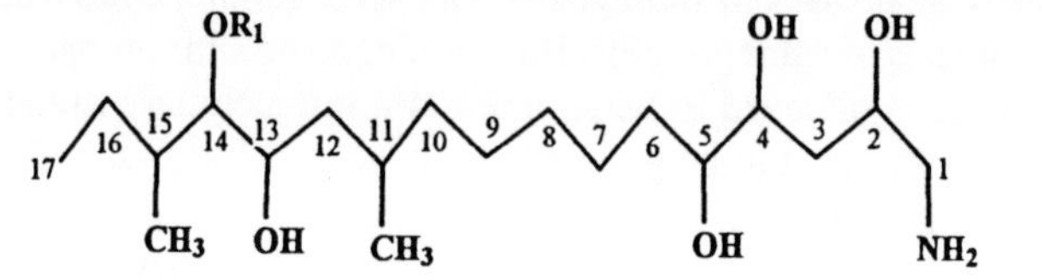

FUMONISIN B_1

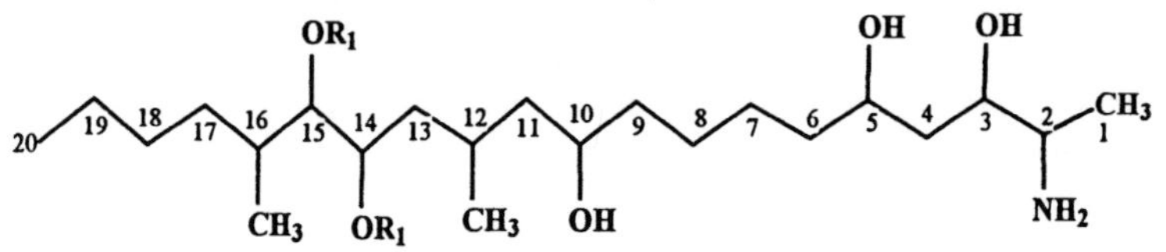

$R_1 = CO\text{-}CH_2\text{-}CH(CO_2H)CH_2CO_2H$

SPHINGOLIPIDS

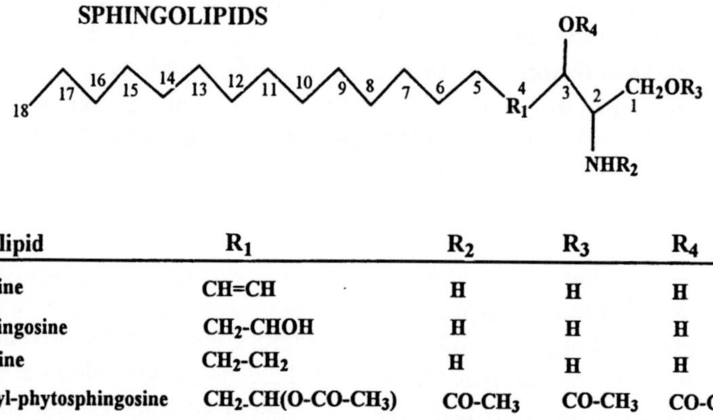

Sphingolipid	R_1	R_2	R_3	R_4
Sphingosine	CH=CH	H	H	H
Phytosphingosine	CH_2-CHOH	H	H	H
Sphinganine	CH_2-CH_2	H	H	H
Tetraacetyl-phytosphingosine	CH_2.CH(O-CO-CH_3)	CO-CH_3	CO-CH_3	CO-CH_3
N-Lignoceroyl-D,L-sphinganine	CH_2-CH_2	CO-$(CH_2)_{22}$-CH_3	H	H

Figure 3. Chemical structure of various sphingolipids.

analog of the fumonisins. It is subsequently acylated on the amino group, and a double bond introduced at the carbon adjacent to the hydroxyl group.

All evidence so far obtained indicates that fumonisin is biosynthesized by a pathway analogous to sphingosine rather than the acetogenin pathway that is used in the biosynthesis of aflatoxins and many antibiotics from closely-related *Streptomycetes* and fungi. We have observed (15) that propionate, which is predicted to be a good precursor by the acetogenin pathway, is not an effective precursor of FB_1, whereas methionine, which is predicted to be a good precursor in the sphingosine-type pathway, has been observed to be an excellent precursor of FB_1 in our studies (15) and the work of others (16, 20, 21, 61). Similarly, we (15) and others (27) have observed that alanine is a good precursor of fumonisins. It is predicated to be the precursor of C(1) and C(2) as well as entering the side chains of fumonisins and AAL-toxins via the citric acid cycle (Fig. 4).

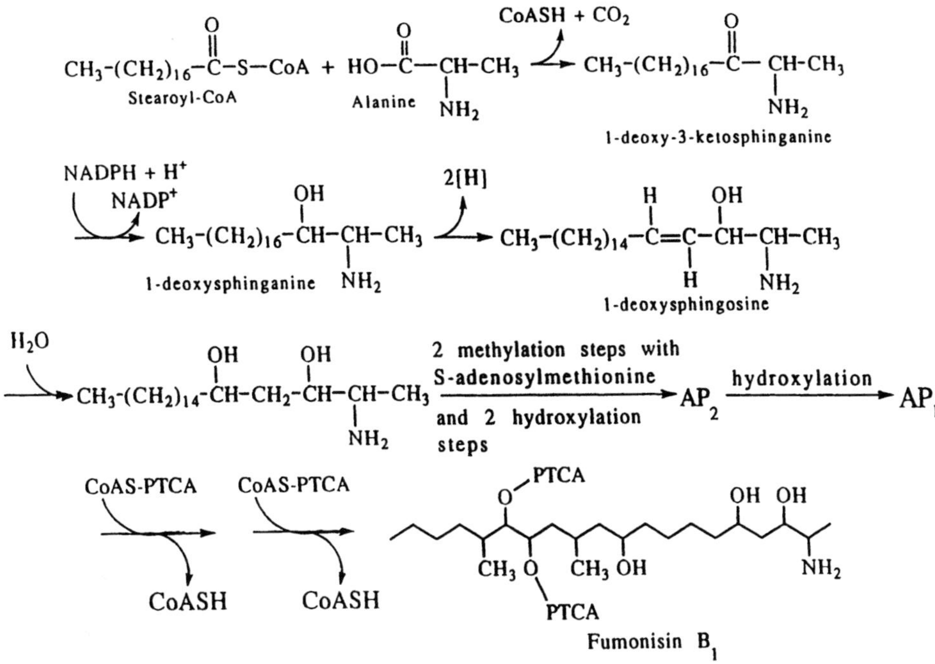

Figure 4. Proposed biosynthetic pathway for fumonisins.

Structural analogy to sphingosine has also shaped thinking about the metabolism of fumonisins. The major metabolic pathway for sphingosine in mammalian tissues occurs by phosphorylation on the hydroxyl of C(1) followed by a retro-aldol reaction which cleaves off a small fragment containing the amino group (53). Deviations from the sphingosine structure in both fumonisins and AAL-toxins block this metabolic pathway because the toxins lack the hydroxyl on C(1), and the result is substantially enhanced metabolic stability relative to sphingosine. Indeed, metabolic stability is the only thing that has been observed experimentally in studies of fumonisin metabolism. Radio-labelled fumonisin B_1, whether it is administered orally or by injection, is almost entirely eliminated unmetabolized in the feces of rats (51, 67) and primates (68), suggesting that it is being excreted in the bile. Cultured rat hepatocytes did not yield readily detected metabolites with radiolabelled FB_1 either (30). It must be concluded that either FB_1 exerts its numerous biological activities in an unmetabolized form, or there are highly active metabolite(s) formed in amounts too small to be detected by the methods used (Fig. 5).

Structural analogy with sphingosine prompts a predicted metabolic pathway for FB_1. In Fig. 5 it is predicted that FB_1 may be (i) acylated by ceramide synthetase with or without removal of the side chains by carboxylesterase action; and (ii) de-aminated by either monoamine oxidases or mixed function oxidases with or without esterase action to yield metabolites that would be expected to be inactive, since a free amino group appears to be required for activity. However, none of the metabolites predicted in Fig. 5 are likely to be highly active.

6. Absolute Stereochemistry of AAL-Toxin and Fumonisins

The 2D structure of AAL-toxin was determined by Bottini et al., (22, 23) in 1981 and of fumonisin B_1 and B_2 by Bezuidenhout et al., (24) in 1988, but the 3D structure has only

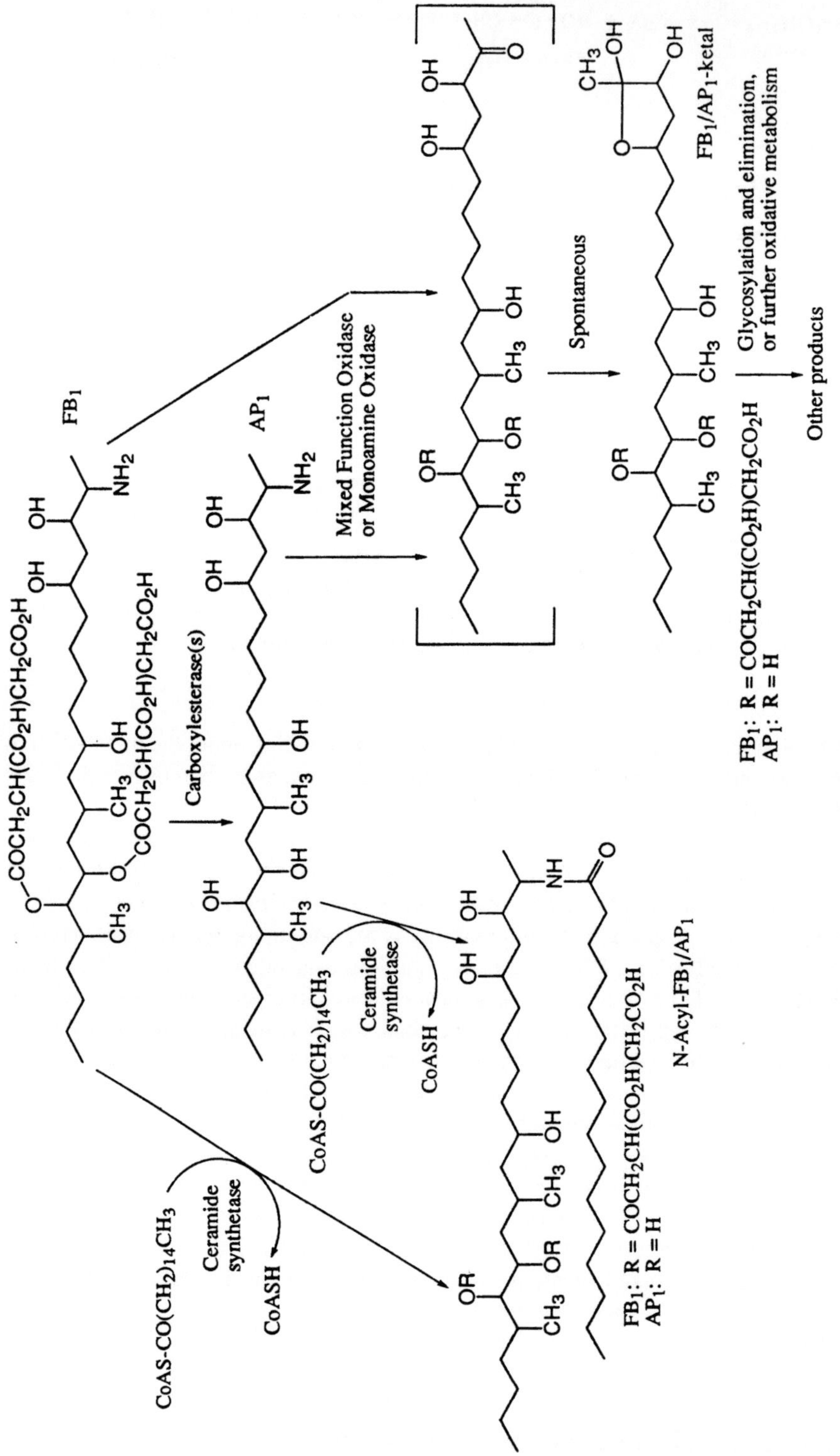

Figure 5. Predicted metabolic pathway of fumonisins.

recently been determined through the contributions of several research groups. The stereochemistry was determined in order to complete the structure determination, and because it can provide information about the biosynthesis of the toxins. The presence of a single configuration at a stereogenic carbon implies the presence of an enzyme to catalyze its introduction. The study of FB_1 backbone stereochemistry has provided a particularly useful piece of information about its biosynthesis in the case of the C(2) configuration. The configuration at C(2) was found to be the opposite to that of the C(2) in sphingolipids in four independent studies using three different methods (i) synthesis (43); (ii) NMR analysis of a ring derivative (18, 19, 62); and (iii) Mosher amide analysis (45). This observation makes it unlikely that the enzyme which couples stearoyl-CoA to alanine in FB_1 biosynthesis or palmitoyl-CoA to glycine in AAL-toxin biosynthesis is closely related to the corresponding sphingolipid biosynthesis enzyme which couples palmitoyl-CoA to serine.

The various research groups that have contributed to the understanding of fumonisin and AAL-toxin stereochemistry have used a variety of approaches. The approach used by Boyle et al., (25) and Oikawa et al., (58) has been to synthesize model compounds representing fragments of the toxins in all possible configurations and assign the correct one by detailed comparison of the NMR spectra. Pock et al., (62) and ApSimon et al., (18, 19), as well as ourselves (45) have made extensive use of NMR analysis to determine relative configurations, particularly using derivatives which lock two or more stereogenic centers into rings. Our group has also made extensive use of two other techniques, Mosher ester analysis (45) and chiral column chromatographic analysis of suitably derivatized fragments to provide information on the absolute configurations. The latter technique has recently been used to complete the determination of the absolute configuration of the last remaining unknown stereogenic centers in FB_1 and AAL-toxin (72). Because propane-1,2,3-tricarboxylic acid is a symmetrical molecule it was necessary to use a reagent that would selectively reduce either the ester or the free carboxylic acids while the side chains were still attached to the backbone. Selective reduction of the free carboxylic acids to alcohols with diborane was chosen. The alternative approach, selective reduction of the ester linkage with sodium borohydride, was not chosen because the alkaline conditions used with this reagent might cause racemization of the stereogenic centers. Several conversion steps were required after release of the side chains in order to convert them into a form which would separate on the chiral column. As shown in Fig. 6, the side chains were converted to 3-methylvaleric acid methyl ester, the racemic form of which could be resolved on a chiraldex GT-A chiral gas chromatography column. Identification of the absolute configuration was made by comparison of retention times with authentic (S)-3-methylvaleric acid methyl ester prepared by deamination of L-isoleucine, which has a known absolute configuration from x-ray crystallography studies. These results combined with previously reported studies on the FB_1 backbone in our group (44) and on the AAL-toxin backbone by Boyle et al., (25) completes the determination of the 3D structures of both toxins (Fig. 6).

7. Conclusions

We have reviewed a variety of aspects of the chemistry and biological activities of AAL-toxin and the fumonisins. Both AAL-toxin and the fumonisins are toxic to a wide range of weed and crop species. Investigation of the mode of action of this class of toxins in plant tissues demonstrates that an early effect is disruption of sphingolipid metabolism. FB_1 and AAL-toxin cause buildup of phytosphingosine and sphinganine, most probably due to inhibition of ceramide synthesis. This mechanism appears to be the same as that described in animals for FB_1. Thirteen analogs of AAL-toxin and FB_1 were obtained and tested for phytotoxicity and mammalian cell culture toxicity. Analog 9, a diester derivative, caused phytotoxic effects in leaf disc and duckweed bioassays, while showing less toxicity to

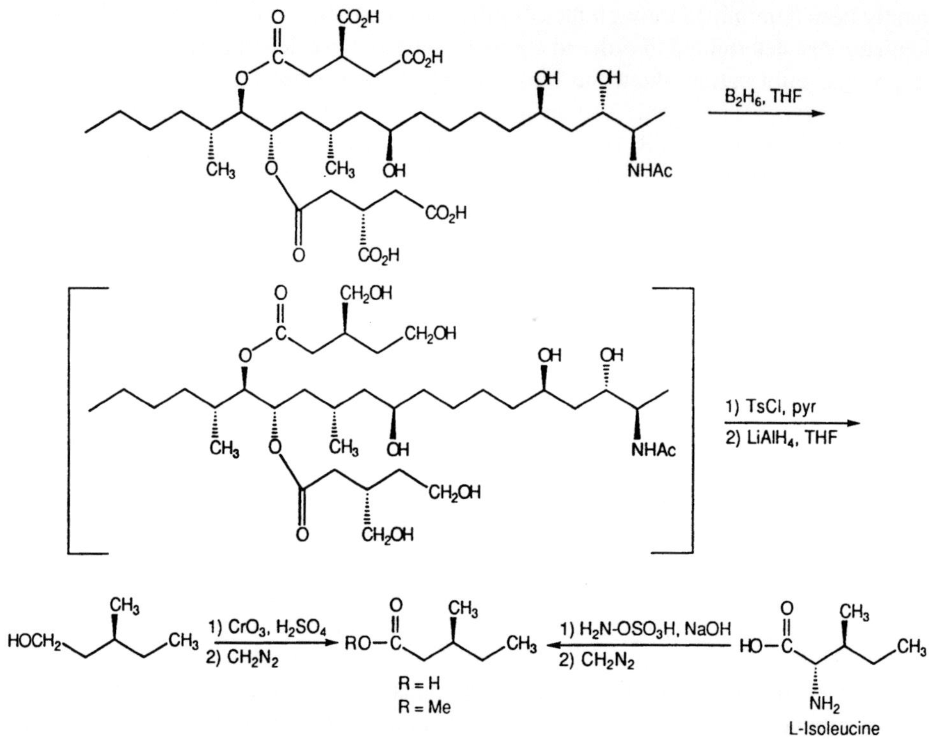

Figure 6. Determination of the absolute configuration of the side chains of fumonisin B_1. AAL-toxin was studied in the same manner.

mammalian cells than FB_1. AAL-toxin and the fumonisins are themselves sphingosine analogs. As such, this has helped us understand better the biosynthesis and metabolism of these toxins. Also, the role of sphingolipids in membrane integrity suggests the molecular basis of phytotoxicity. Finally, we can now state with confidence the absolute configuration of AAL-toxin and the fumonisins. This structure also suggests differences in the biosynthetic pathways of the toxins and sphingolipids. All these studies have contributed to our knowledge of AAL-toxins and the fumonisins especially with regard to their phytotoxicity. It is hoped that these studies will eventually lead to possible use of analogs of these compounds in weed control.

REFERENCES

1. Abbas, H.K., Boyette, C.D., Hoagland, R.E., and Vesonder, R.F. 1991. Bioherbicidal potential of *Fusarium moniliforme* and its phytotoxin, fumonisin. Weed Sci. 39:673-677.
2. Abbas, H.K. and Boyette, C.D. 1992. Phytotoxicity of fumonisin B_1 on weed and crop species. Weed Technol. 6:548-552.
3. Abbas, H.K., Paul, R.N., Boyette, C.D., Duke, S.O., and Vesonder, R.F. 1992. Physiological and ultrastructural effects of fumonisin on jimsonweed leaves. Can. J. Bot. 70:1824-1833.
4. Abbas, H.K., Vesonder, R.F., Boyette, C.D., Hoagland, R.F., and Krick, T. 1992. Production of fumonisins by *Fusarium moniliforme* cultures isolated from jimsonweed in Mississippi. J. Phytopathology 136:199-203.
5. Abbas, H.K., Boyette, C.D., and Vesonder, R.F. 1993. Biological control of weeds using AAL-toxin, United States Patent Number 5,256,628; Date of patent: October 26, 1993, pp. 1-10.

6. Abbas, H.K., Duke, S.O., and Tanaka, T. 1993. Phytotoxicity of fumonisins and related compounds. J. Toxicol. - Toxin Reviews 12:225-251.
7. Abbas, H.K., Gelderblom, W.C.A., Cawood, M.E., and Shier, W.T. 1993. Biological activities of various fumonisins in jimsonweed and mammalian cell cultures. Toxicon 31:345-353.
8. Abbas, H.K., and Vesonder, R.F. 1993. Isolation and purification of AAL-toxin from *Alternaria alternata* grown on rice. Toxicon 31:355-358.
9. Abbas, H.K., Tanaka, T., and Duke, S.O. 1995. Pathogenicity of *Alternaria alternata* and *Fusarium moniliforme* and phytotoxicity of AAL-toxin and fumonisin B_1 on tomato cultivars. J. Phytopathol. 143:329-334.
10. Abbas, H.K., Tanaka, T., Duke, S.O., Porter, J.K., Wray, E.M., Hodges, L., Sessions, A.E., Wang, E., Merrill, A.H., Jr., and Riley, R.T. 1994. Fumonisin and AAL-toxin-induced disruption of sphingolipid metabolism with accumulation of free sphingolipid bases. Plant Physiology 106:1085-1093.
11. Abbas, H.K., Tanaka, T., Duke, S.O., Kraus, G., Applegate, J.M., Su, Q., and Shier, W.T. 1994. Biological activities of synthetic analogs of fumonisin in plants and mammalian cell cultures. Phytopathology 84:1130.
12. Abbas, H.K., Tanaka, T., Duke, S.O., and Boyette, C.D. 1995. Susceptibility of various crop and weed species to AAL-toxin, a natural herbicide. Weed Technol. 9:125-130.
13. Abbas, H.K., Tanaka, T., and Shier, W.T. 1995. Biological activities of synthetic analogs of *Alternaria alternata* AAL-toxin and fumonisin in plants and mammalian cell cultures. Phytochemistry (In press).
14. Abbas, H.K. and Riley, R.T. 1995. The presence of the fumonisins and AAL-toxin in *Alternaria alternata* and *Fusarium moniliforme*. Toxicon (In press).
15. Abbas, H.K. and Shier, W.T. 1992. Evaluation of biosynthetic precursors for the production of radiolabeled fumonisin B_1 by *Fusarium moniliforme* on rice medium. Abstracts of the 106th Association of Official Analytical Chemists Meeting; Cincinnati, OH; AOAC International: Arlington, VA, pp. 236.
16. Alberts, J.F., Gelderblom, W.C.A., Vleggar, R., Marasas, W.F.O., and Rheeder, J.P. 1993. Production of [^{14}C] fumonisin B_1 by *Fusarium moniliforme* MRC 826 in corn cultures. Appl. Environ. Microbiol. 59:2673-2677.
17. Applegate, J.M. 1992. The synthesis of fumonisin B_1 analogs and silicon crosslinking agents. Ph.D. Thesis, Iowa State University, Ames, IA.
18. ApSimon, J.W., Blackwell, B.A., Edwards, O.E., and Fruchier, A. 1994. Relative configuration of the C-1 to C-5 fragment of fumonisin B_1. Tetrahedron Lett. 35:7703-7706.
19. ApSimon, J.W., Blackwell, B.A., Edwards, O.E., Fruchier, A., Miller, J.D., Savard, M., and Young, J.C. 1994. The chemistry of fumonisins and related compounds. Fumonisins from *Fusarium moniliforme*: chemistry, structure and biosynthesis. Pure & Appl. Chem. 66:2315-2318.
20. Blackwell, B.A., Miller, J.D., and Savard, M.E. 1994. Production of carbon $^{-14}$ labeled fumonisin in liquid culture. JAOAC Int. 77:506-511.
21. Blackwell, B.A., Edwards, O.E., ApSimon, J.W., and Fruchier, A. 1995. Relative configuration of the C-10 to C-16 fragment of fumonisin B_1. Tetrahedron Lett. 36:1973-1976.
22. Bottini, A.T. and Gilchrist, D.G. 1981. Phytotoxins I. A. 1-aminodimethylheptadecapentol from *Alternaria alternata* f. sp. *lycopersici*. Tetrahedron Lett. 22:2719-2722.
23. Bottini, A.T., Bowen, J.R., and Gilchrist, D.G. 1981. Phytotoxins II. Characterization of a phytotoxic fraction from *Alternaria alternata* f. sp. *lycopersici*. Tetrahedron Lett 22:2723-2726.
24. Bezuidenhout, S.C., Gelderblom, W.C.A., Gorst-Allman, C.P., Horak, R.M., Marasas, W.F.O., Spiteller, G. and Vleggaar, R. 1988. Structure elucidation of the fumonisins, mycotoxins from *Fusarium moniliforme*. J. Chem. Soc. Chem. Commun. 1984:743-745.
25. Boyle, C.D., Harmange, J.C., and Kishi, Y. 1994. Novel structure elucidation of AAL-toxin T_A backbone. J. Amer. Chem. Soc. 116:4995-4996.
26. Branham, B.E. and Plattner, R.D. 1993. Isolation and characterization of a new fumonisin from liquid cultures of *Fusarium moniliforme*. J. Natural Products 56:1630-1633.
27. Branham, B.E., and Plattner, R.D. 1993. Alanine is a precursor in the biosynthesis of fumonisin B_1 by *Fusarium moniliforme*. Mycopathologia 124:99-104.
28. Caldas, E.D., Jones, A.D., Ward, B., Winter, C., and Gilchrist, D.G., Jr. 1994. Structural characterization of three new AAL-toxins produced by *Alternaria alternata* f. sp. *lycopersici*. J. Agric. Food Chem. 42:327-333.
29. Castagna, M., Takai, Y., Kaibuchi, K., Sano, K., Kikkawa, U., and Nishizuka, Y. 1982. Direct activation of calcium-activated, phospholipid-dependent protein kinase by tumor-promoting phorbol esters. J. Biol. Chem 257:7847-7851.
30. Cawood, M.E., Gelderblom, W.C.A., Alberts, J.F., and Snyman, S.D. 1994. Interaction of C^{-14} or ^{14}C labelled fumonisin B mycotoxin with primary rat hepatocytes cultures. Food Chem. Toxic. 32:627-632.

31. Chen, J., Mirocha, C.J., Xie, W., Hogge, L., and Olson, D. 1992. Production of the mycotoxin fumonisin B_1 by *Fusarium alternata* f. sp. *lycopersici*. Appl. Environ. Microbiol. 58:3928-3931.
32. Chu, F.S., and Li, G.Y. 1994. Simultaneous occurrence of fumonisin B_1 and other mycotoxins in moldy corn collected from the people's Republic of China in regions with high incidences of esophageal cancer. Appl. Environ. Microbiol. 60:847-852.
33. Clouse, S.D. and Gilchrist, D.C. 1987. Interaction of the *asc* locus in F8 paired lines of tomato with *Alternaria alternata* f. sp. *lycopersici* and AAL-toxin. Phytopathology 77:80-82.
34. Colvin, B.M. and Harrison, L.R. 1992. Fumonisin-induced pulmonary edema and hydrothorax in swine. Mycopathologia 117:79-82.
35. Doery, H.M., Magnuson, B.J., Cheyne, I.M., and Galasekharam, J. 1963. A phospholipase in staphylococcal toxin which hydrolyses sphingomyelin. Nature 198:1091-1092.
36. Gelderblom, W.C.A., Jaskiewicz, K., Marasas, W.F.O., Theil, P.G., Horak, R.M., Vleggaar, R., and Kreik, N.P. 1988. Fumonisins - novel mycotoxins with cancer-promoting activity produced by *Fusarium moniliforme*. Appl. Environ. Microbiol. 54:1806-1811.
37. Gelderblom, W.C.A., Semple, E., Marasas, W.F.O., and Farber, E. 1992. The cancer-initiating potential of the fumonisin B mycotoxins. Carcinogenesis 13:433-437.
38. Gelderblom, W.C.A., Snyman, S. D., Westhuizen, L. van der, and Marasas, W. F. O. 1995. Mitoinhibitory effect of fumonisin B1 on rat hepatocytes in primary culture. Carcinogenesis 16: 625-631.
39. Gilchrist, D.G. and Grogan, R.G. 1976. Production and nature of a host-specific toxin from *Alternaria alternata* f. sp. *lycopersici*. Phytopathology 66:165-171.
40. Grogan, R.G., Kimble, K.A., and Misaghi, I. 1975. A stem canker disease on tomato caused by *Alternaria alternata* f. sp. *lycopersici*. Phytopathology 65:880-886.
41. Hannun, Y.A. and Bell, R.M. 1989. Function of sphingolipids and sphingolipid breakdown products in cellular regulation. Science 243:500-507.
42. Hannun, Y.A., Loomis, C.R., Merrill, A.H., and Bell, R.M. 1986. Sphingosine inhibition of protein kinase C activity and of phorbol dibutyrate binding *in vitro* and in human platelets. J. Biol. Chem. 261:12604-09.
43. Harmange, J.C., Boyle, C.D., and Kishi, Y. 1994. Relative and absolute stereochemistry of the fumonisin B_2 backbone. Tetrahedron Lett. 35:6819-6822.
44. Harrison, L.R., Colvin, B.M., Greene, J.T., Newman, L.E., and Cole, J.R. 1990. Pulmonary edema and hydrothorax in swine produced by fumonisin B_1, a toxic metabolite of *Fusarium moniliforme*. J. Vet. Diagn. Invest. 2:217-221.
45. Hoye, T.R., Jimenez, J.I., and Shier, W.T. 1994. Relative and absolute configuration of fumonisin B_1 backbone. J. Amer. Chem. Soc. 116:9409-9410.
46. Huang, C., Dickman, M., Henderson G., and Jones C. 1995. Repression of protein kinase C and stimulation of cyclic AMP response elements by fumonisin, a fungal encoded toxin which is a carcinogen. Cancer Res. 55:'655-1659.
47. Kanfer, J.N. 1970. Sphingolipid biosynthesis by a rat brain particulate fraction. Chem. Phys. Lipids. 5:159-177.
48. Kohmoto, K., Verma, V.S., Nishimura, S., Tagami, M. and Scheffer, R.P. 1982. New outbreak of *Alternaria* stem canker of tomato in Japan and production of host-selective toxins by the causal fungus. J. Fac. Agric., Tottori Univ. 17:1-8.
49. Kraus, G.A., Applegate, J.M., and Reynolds, D. 1992. Synthesis of analogs of fumonisin B_1. J. Agric. Food Chem. 40:2331-2332.
50. Kraus, G.A., Applegate, J.M., Su, Q., and Reynolds, D. 1995. Defining the functionality necessary for fumonisin toxicity, pp. 89-95 In: Natural Protectants Against Natural Toxicants, Eds. W. R. Bidlack and S. T. Omaye, Technomic Publishing Company, Inc., Lancaster, PA.
51. Lebepe-Mazur, S.M. 1993. Production of fumonisin B_1 by *Fusarium proliferatum* in liquid culture medium, and the toxicity and metabolism of fumonisin B_1 in rats. Ph.D. Thesis, Iowa State University, Ames, IA.
52. Marasas, W.F.O., Kellerman, T.S., Gelderblom, W.C.A., Coetzer, J.A.W., Thiel, P.G., and Van Der Lugt, J.J. 1988. Leukoencephalomalacia in a horse induced by fumonisin B_1 isolated from *Fusarium moniliforme*. Onderstepoort, J. Vet. Res. 55:197-203.
53. Merrill, A.H., and Jones, D.D. 1990. An update of the enzymology and regulation of sphingomyelin metabolism. Biochim. Biophys. Acta 1044:1-12.
54. Merrill, A.H., Jr. 1991. Cell regulation by sphingosine and more complex sphingolipids. J. Bioenerget Biomerb 23:83-104.
55. Merrill, A.H., Jr., Hanun, Y.A., and Bell, R.M. 1993. Sphingolipids and their metabolites in cell regulation. In R.M. Bell, Y.A. Hanun, and A.H. Merrill, Jr. (eds.), Advances in Lipid Research: Sphingolipids and their Metabolites, Vol. 25, 1-24, San Diego, CA, Academic Press, pp. 1-24.

56. Mirocha, C.J., Gilchrist, G.D., Shier, W.T., Abbas, H.K., Wen, Y., and Vesonder, R.F. 1992. AAL-toxins, fumonisins (biology and chemistry) and host-specificity concepts. Mycopathologia 117:47-56.
57. Nelson, P.E., Plattner, R.D., Shackleford, D.D., and Desjardins, A.E. 1992. Fumonisin B_1 production by *Fusarium* species other than *F. moniliforme* in section Liseola and by some related species. Appl. Environ. Microbiol. 58:984-989.
58. Nishimura, S. and Kohmoto, K. 1983. Host-specific toxins and chemical structures from *Alternaria* species. Annu. Rev. Phytopathol. 21:87-116.
59. Norred, W.P. 1993. Fumonisins - mycotoxins produced by *Fusarium moniliforme*. J. Toxicol. Environ. Health 38:309-328.
60. Oikawa, H., Matsuda, I., Ichihara, A., and Kohmoto, K. 1994. Absolute configuration of C(1)-C(5) fragments of AAL-toxin: conformationally rigid acyclic aminotriol moiety. Tetrahedron Lett. 35:1223-1226.
61. Plattner, R.D. and Branham, B.E. 1994. Labeled fumonisins: production and use of fumonisin B_1 containing stable isotypes. JAOAC Int. 77:525-532.
62. Pock, G.K., Powell, R.G., Plattner, R.D., and Weisleder, D. 1994. Relative stereochemistry of fumonisin B_1 at C-2 and C-3. Tetrahedron Lett. 35:7707-7710.
63. Riley, R.T., An, N.H., Showker, J.L., Yoo, H.-S., Norred, W.P. Chamberlain, W.J., Wang, E., Merrill, A.H., Jr., Motelin, G., Beasley, V.R., and Haschek, W.M. 1993. Alteration of tissue and serum sphinganine to sphingosine ratio: an early biomarker of exposure to fumonisin-containing feeds in pigs. Toxicol. Appl. Pharmacol. 118:105-112.
64. Riley, T.R., Wang, E., and Merrill, A.H. 1994. Liquid chromatography of sphinganine and sphingosine: use of the sphinganine to sphingosine ratio as a biomaker for consumption of fumonisins. J. AOAC. Internat. 77:533-540.
65. Riley, R.T., Voss, K.A., Koo H.S., Gelderblom, W.C.A., and Merrill, A.H., Jr. 1994. Mechanism of fumonisin toxicity and carcinogenicity. J. Food Protect 57:638-645.
66. Rotem, J. 1994. The genus *Alternaria*: biology, epidemiology, and pathogenicity. APS Press, The American Phytopathology Society, St. Paul, Minnesota, pp. 326.
67. Shephard, G.S., Thiel, P.G., Sydenham, E.W., Alberts, J.F., and Gelderblom, W.C.A. 1992. Fate of a single dose of the ^{14}C-labeled mycotoxin, fumonisin B_1 in rats. Toxicon 30:768-770.
68. Shephard, G.S., Thiel, P.G., Sydenham, E.W., Alberts, J.F., and Cawood, M.E. 1994. Distribution and excretion of a single dose of the mycotoxin fumonisin B_1 in a non-human primate. Toxicon 32:735-741.
69. Shier, W.T. 1992. Sphingosine analogues: An emerging new class of toxins that includes the fumonisins. J. Toxicol.-Toxin Rev. 11:241-257.
70. Shier, W.T. and Abbas, H.K. 1992. A simple procedure for the preparation of aminopentols (fumonisin hydrolysis products AP_1 and AP_2) from *Fusarium moniliforme* on solid media. Abstracts of the 106th Association of Official Analytical Chemists Meeting; Cincinnati, OH; AOAC International:Arlington, VA, pp. 237.
71. Shier, W.T., Abbas, H.K., and Mirocha, C.J. 1991. Toxicity of the mycotoxins fumonisins B_1 and B_2 and *Alternaria alternata* f. sp. *lycopersici* toxin in cultured mammalian cells. Mycopathologia 116:97-104.
72. Shier, W.T., Abbas, H.K., and Badria, F.A. 1995. Complete structures of the sphingosine analog mycotoxins fumonisin B_1 and AAL-toxin T_A: absolute configuration of the side chains. Tetrahedron Lett., 36:1571-1574.
73. Snell, E.E., Dimari, S.J., and Brady, R.N. 1970. Biosynthesis of sphingosine and dihydrosphingosine by cell-free systems from *Hansenula ciferri*. Chem Phys. Lipids 5:116-138.
74. Su, Q. 1994. 1. An approach to the total synthesis of leucothol A; 2. Synthesis of the analogs of fumonisin B_1. M.S. Thesis, Iowa State University, Ames, IA.
75. Tanaka, T., Abbas, H.K., and Duke, S.O. 1993. Structure-dependent phytotoxicity of fumonisins, and related compounds in a duckweed bioassay. Phytochemistry 33:779-785.
76. Thiel, P.G., Marasas, W.F.O., Sydenham, E.W., Shephard, G.S., Gelderblom, W.C.A., and Nieuwenhuis, J.J. 1991. Survey of fumonisin production by *Fusarium* species. Appl. Environ. Microbiol. 57:1089-1093.
77. Thiel, P.G., Marasas, W.F.O., Sydenham, E.W., Shephard, G.S., and Gelderblom, W.C.A. 1992. The implications of naturally occurring levels of fumonisins for human and animal health. Mycopathologia 117:3-9.
78. Vance, D.E., and Vance, J.E. 1985. Biochemistry of Lipids and Membranes, Benjamin/Cummings (Eds.), Menlo Park, CA. pp. 361-403.
79. Vesonder, R.F., Peterson, R.E., and Weisleder, D. 1990. Fumonisin B_1: Isolation from corn culture, and purification by high performance liquid chromatography. Mycotoxin Res. 6:85-88.

80. Wang, E., Norred, W.P., Bacon, C.W., Riley, R.T., and Merrill, A.H., Jr. 1992. Increases in serum sphingosine and sphinganine and decreases in complex sphingolipids in ponies given feed containing fumonisins, mycotoxins produced by *Fusarium moniliforme*. J. Nutr. 122:1706-1716.
81. Wang, E., Norred, W.P., Bacon, C.W., Riley, R.T., and Merrill, Jr., A.H. 1991. Inhibition of sphingolipid biosynthesis by fumonisins. J. Biol. Chem. 266:14486-14490.
82. Yoo, H., Norred, W.P., Wang, E., Merrill, A.H., Jr., and Riley, R.T. 1992. Fumonisin inhibition of *de novo* sphingolipid biosynthesis and cytotoxicity are correlated in LLC-PK1 cells. Toxicol. Appl. Pharmacol. 114:9-15.

24

NEW ASPECTS OF AMANITIN AND PHALLOIDIN POISONING

Heinz Faulstich and Theodor Wieland

Max-Planck-Institute for Medical Research
Heidelberg, Germany

AMATOXINS

The green death cap Amanita phalloides produces two families of toxic peptides, the amatoxins and the phallotoxins. Amatoxins are by far the more dangerous toxins and account for all fatalities occurring in human mushroom poisoning.

The molecular mechanism of amanitin toxicity has been studied in detail. The toxins form a tight complex (K_D ca. 1 nM) with eukaryotic RNA polymerases II and thereby inhibit transcription (1). It appears that in mechanical terms the action of amanitin is similar to that of a brakeshoe: According to Vaisius and Wieland, amatoxins block the movement of the enzyme on single-stranded DNA and thus inhibit the elongation step (2). The complex is not covalent and is in dynamic equilibrium with free toxin and free enzyme. There are reasons to believe that, after dissociation of amanitin from the complex, transcription by RNA polymerases II continues.

The formulae of the nine amatoxins found in A. phall. were elucidated by Wieland and coworkers more than 30 years ago. The structure of the main component, α-amanitin, is shown in figure 1.

From experiments with natural analogs as well as with chemically modified amatoxins, evidence has accumulated that some structural features of the peptide - among them the OH in hydroxy-proline in position 2 and the γ-OH in dihydroxyleucine in position 3 - are absolutely necessary for binding the toxin to RNA-polymerase II.

In a recent, detailed study on structure-activity relationships in amatoxins we investigated 24 structural analogs of α-amanitin for their binding properties to RNA polymerase II (3). We wished to know the amount of binding energy contributed to complex formation by particular structural features of the peptide. We further asked whether other proteins that can bind amanitin with similar affinity, such as for example antibodies, prefer the same structural features of amanitin for binding as mammalian RNA polymerase II. Monoclonal antibodies against amatoxins were raised in mice; three of these having K_D values comparable to that of RNA polymerase II, were selected for the binding studies. The antigen used in all cases was α-amanitin linked to a protein via the amide in position 1.

Natural Toxins II, Edited by B. R. Singh and A. T. Tu
Plenum Press, New York, 1996

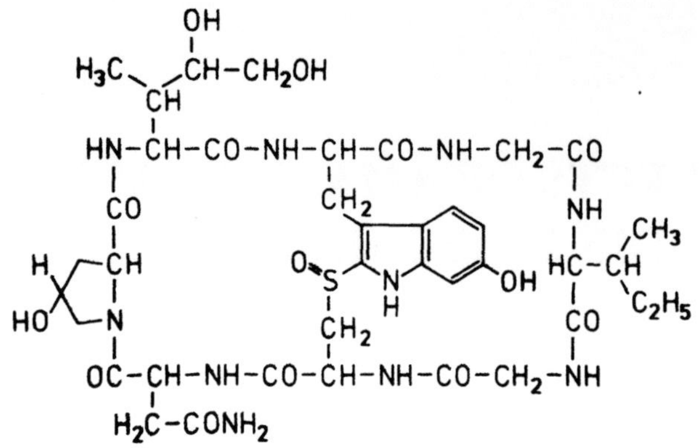

Figure 1. Formula of α-amanitin.

For each of the 24 amanitin-derivatives we determined the dissociation constant of their complexes with the 4 proteins. K_D values with the monoclonal antibodies were calculated from the concentration values of the amatoxins derivatives required to displace 50 % of a radioactively labeled amatoxin with known K_D from its binding site in the antibody (4). For RNA polymerase II K_i values for the isolated calf thymus enzyme were measured expecting that, given the molecular mechanism of amanitin action, K_i values would be identical to the K_D values. From the changes in K_D (ΔK_D) associated with distinct changes in structure, as for example the replacement of an OH for H, we calculated $\Delta\Delta G^o$ values, i.e. the loss of binding energy attributable to the replacement of a given structural detail with another one.

We found that the most important determinant for binding was isoleucine 6. Substitution for isoleucine led to loss of large amounts of binding energy ($\Delta\Delta G^o$ from -12 tro -17 kJ/mol for all 4 proteins). Apparently, the hydrophobic side-chain of this amino acid which, according to the X-ray structure of Kostansek et al (5), (fig. 2), is on top of a β-turn, finds a hydrophobic pocket in all 4 amatoxin-binding proteins.

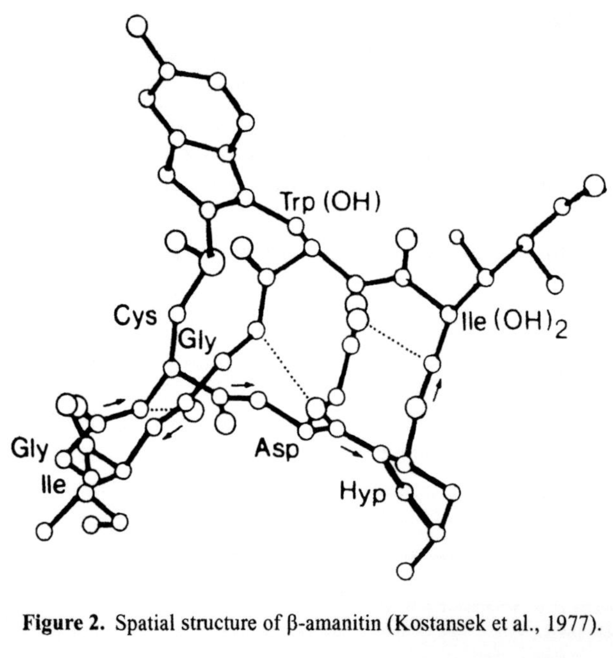

Figure 2. Spatial structure of β-amanitin (Kostansek et al., 1977).

Of similar importance for binding was glycine 5; when it was replaced with alanine, a considerable loss of binding energy (9-15 kJ/mol) occurred, except for one antibody, in which the loss of binding energy was smaller (3 kJ/mol). Glycine 5 is located adjacent to isoleucine 6 and is also part of the β-turn. Apparently an amino acid without a side chain is required in this position, either to maintain the correct structure of the β-turn, or to enable an approach between protein and peptide close enough to allow the formation of a strong hydrogen bond, possibly with the peptide backbone. Another structure that proved to be important for binding was the OH group of hydroxyproline. While in the complexes with RNA polymerase II and one of the antibodies the contribution of this OH to binding was large ($\Delta\Delta G^o$ = -9 to -15 kJ/mol), it was smaller ($\Delta\Delta G^o$ = -4 tro -5 kJ/mol) for the two other antibodies. There is no doubt that this OH group functions as part of a hydrogen bond with the binding proteins. As the binding capacity of this OH was lost not only on replacement with H but also after acetylation, we assume that the OH group is the donor part of this hydrogen bond. Of comparable importance, but exclusively for RNA polymerase II, is the OH group in dihydroxyleucine in position 3 ($\Delta\Delta G^o$ = -9 kJ/mol). In contrast to this, only one of the 3 antibodies had a (much weaker) interaction with this OH ($\Delta\Delta G^o$ = -5 kJ/mol), while the two others did not recognize it at all.

Interestingly, all three antibodies recognized the hydroxytryptophan residue of α-amanitin, while the RNA polymerase II did not. Particularly the indole NH (with $\Delta\Delta G^o$ values up to -12 kJ/mol) and the OH in 6'position (with $\Delta\Delta G^o$ values up to -8 kJ/mol) seem to be involved in hydrogen bonds with the antibodies. The fact that RNA polymerase II is unable to recognize the aromatic part of amanitin is in line with previous experiments, which showed that various alkylation reactions at the 6'hydroxy group, or azo-coupling in 7'position, did not seriously impair the inhibitory capacity of the amatoxin derivatives. We conclude that in the complex with the enzyme the hydroxytryptophan moiety is located juxtaposed to that part of amanitin that binds to RNA polymerases II.

The attribution of $\Delta\Delta G^o$ values to distinct changes in side chains of functional groups in amanitin provided reasonable results only if the chemical modifications had no effect on the overall conformation of the peptide. This condition was fulfilled in the majority of the amanitin derivatives investigated here, as proven by e.g. CD measurements. The conserved conformation in most amatoxin derivatives is a consequence of the bicyclic nature of these peptides, further stabilized by the presence of several intramolecular hydrogen bonds. Only in two chemically modified amatoxins investigated in this study was the overall shape of the cyclopeptide found to have changed as a consequence of the derivatization. One of them resulted from the reductive cleavage of the sulfur atom, which led to a monocyclic compound (dethioamanitin) which was no longer recognized by any of the 4 binding proteins. The second example of a changed conformation was seen when the side chain in position 3 was degraded to an aldehyde. In this case, a proposed intramolecular reaction of the aldehyde group with the backbone (6) probably distorts the typical amanitin conformation. In both cases the affinity to the amatoxin binding proteins was reduced by factors 10^3-10^4, i.e. was virtually lost.

From the chemical point of view not every side chain in α-amanitin is accessible to modifications. As a consequence, the number of contact sites between the peptide and the binding proteins that can be examined in such an assay is limited. Nonetheless, we were able to identify the main features of the amatoxin-binding sites in 4 amatoxin-binding proteins as composed of a hydrophobic pocket, flanked by at least two strong hydrogen bonds, in which amanitin is the donor part. One of these hydrogen bonds involves the OH of hydroxyproline, which is necessary for binding of all four proteins; the other hydrogen bond either includes the NH of the indole nucleus, as in the interaction with the antibodies, or is formed from the γ-positioned OH in dihydroxyleucine, as in the interaction with RNA polymerase II.

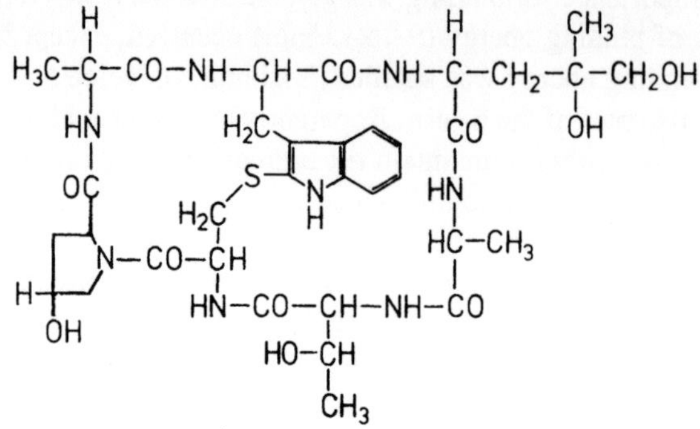

Figure 3. Formula of phalloidin.

For all contact sites that could be identified, the method used in this study allowed calculation of the strength of the interactions expressed in $\Delta\Delta G^o$ values. In this respect, the method is complementary to X-ray analysis, which gives the exact distance, but no values for binding energies. Work is in progress to characterize such amanitin protein complexes also by X-ray analysis. From one of the monoclonal antibodies the Fab fragment was prepared and crystallized in the presence and the absence of amanitin. The structure of the complex waits to be elucidated.

PHALLOTOXINS

The formulae of the nine members of the phallotoxin family were likewise elucidated by Wieland and coworkers (7). Fig. 3 shows the stucture of the main toxin, phalloidin.

Phalloidin binds to polymeric, or filamentous, actin and thereby enhances the stability of the filaments to such an extent that the rate of depolymerization is close to zero. All toxic effects caused by phalloidin are understood as consequence of this stabilization effect.

In contrast to amatoxins, phalloidin poisoning in laboratory animals is seen only after parenteral application. From this it was concluded that the absorption rate of phallotoxins after oral administration must be rather low. Besides this explanation however, one has to consider that the target protein of phallotoxins, actin in non-muscle cells and particularly in hepatocytes, is present in a ca. 10^3 times higher concentration (5×10^{-5} M) than RNA-polymerases II (5×10^{-8}M). Correspondingly, the intracellular threshold for toxic events caused by phallotoxins is distinctly higher than that for amatoxins and may not be reached after enteral administrations. This may explain why phallotoxins fail to contribute to human Amanita poisoning.

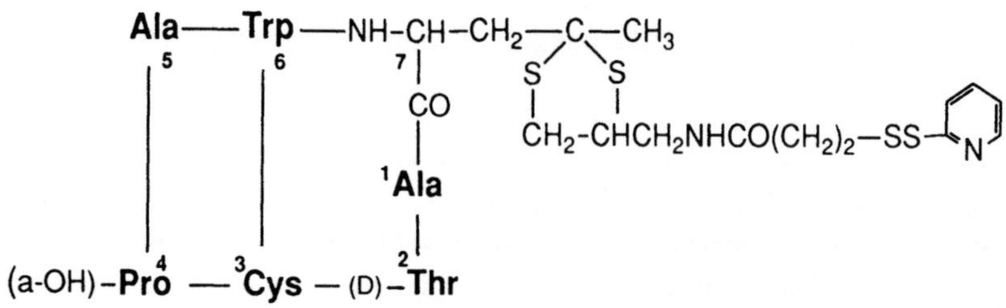

Figure 4. Structure of a thiol-capturing phalloidin derivative.

The influence of chemical modification on biological activity was examined for the phallotoxins also. We found that the amino acid which in these peptides appears to be juxtaposed to the binding site with actin is dihydroxyleucine, since intense derivatization of this side chain was found to be associated with only little effect on the actin-binding capacity. Therefore, numerous reporter groups such as radioactive or fluorescent labels were introduced into this position. Among them, the phalloidin derivatives with red or green fluorescence are nowadays widely used for visualizing microfilaments in the cytoskeleton of all kinds of cells (8).

In the experiments to be discussed here we introduced into this side-chain a thiol-capturing moiety as shown in fig. 4, which was reacted with cysteine 374 of monomeric actin. When this actin conjugate was subjected to polymerization conditions we observed that the phalloidin convalently attached to actin in this position had almost the same stabilization effect as free phalloidin (9). The parameter for filament stability that we followed in these experiments was the so-called steady state ATPase activity, which grossly reflects the number of breaks in the filaments repaired at the expense of ATP. In contrast to phalloidin linked to cysteine 374, phalloidin linked to cysteine 10, which is located in the same subdomain of actin as cysteine 374 but on the opposite side, showed no stabilization effect at all. Moreover, phalloidin fixed to this position badly disturbed the formation of actin filaments. Our finding that the binding site of phalloidin must be located close to the C-terminus of actin confirms previous results by Wieland and coworkers, who found that a photoaffinity-labeled phallotoxin reacted with methionine 355, an amino acid likewise positioned in close vicinity to the C-terminus.

As well as identifying the binding site of phalloidin in the actin filaments we made some progress in understanding phalloidin toxicity in vivo. Unlike the in vitro system normally used for studying phalloidin activity, which is based on isolated muscle actin (α-actin), hepatocytes contain β-actin together with a small actin-binding protein of 5 kDa, thymosin β4. While there is almost no difference in functionality of α- and β-actin, the in vitro system differs from the in vivo situation in that the thymosin β4, which in hepatocytes is present in concentrations similar to that of actin, keeps actin in the monomeric state. We simulated the in vivo situation using an in vitro system containing muscle actin monomers stabilized by 1.5 equivalents of thymosin β4. When subjected to polymerization conditions, the actin thymosin β4 complex was unable to polymerize, but started polymerization immediately when substoichiometric amounts of phalloidin were added (10). The most likely explanation for this is that phalloidin stabilizes not only actin polymers but also actin oligomers, for example trimeric actin aggregates, which are regarded as nuclei or seeds for polymerization. In chemical terms, the cell behave, with respect to actin polymerization, like an oversaturated solution, in which the solute cannot crystallize by itself but will do so immediately after crystal seeds are added.

If this model holds true and stable actin seeds really represent the limiting factor for actin polymerization in vivo, cells would have the possibility to regulate the de novo synthesis of microfilaments at any place in the cells and any time by making available, in a regulated way, stabilizers of actin nuclei at the place where microfilaments are intended to grow. Phalloidin would then represent the unregulated, toxic counterpart of a regulated, physiological event. One of the compounds, which under physiological conditions may exert such nuclei-stabilizing properties and thus regulate actin polymerization, is myosin (10).

In isolated hepatocytes the first toxic lesion is a change in cell shape. About 10 minutes after administration of phalloidin to the medium hepatocytes develop numerous blebs on their surface (11). We compared the kinetics of bleb formation in vivo with the kinetics of actin polymerization from the thymosin β4 complex in vitro, and found that the two processes develop at a comparable rate. Since polymerization occurs at a rate somewhat faster than the development of surface blebbing in vivo, the morphological lesions in the hepatocytes could well represent one of the consequences of a complete phalloidin-induced polymerization of actin from its monomer-stabilizing complex. This novel toxic mechanism of phalloidin is thought to account for the fast-developing lesions in hepatocytes such as surface blebbing. At a later stage, when larger amounts of the toxin have penetrated the cell and each of the monomers in the filaments is complexed by phalloidin, the total block of the depolymerization reaction, as already mentioned, may occur. We thus have to distinguish between two toxic mechanisms of phalloidin, both based on the same physical event,the stabilization of actin aggregates. Early effects of phalloidin, starting ca. 10 min after administration of the toxin, can best be explained by the activity of phalloidin-stabilized nuclei that induce chaotic polymerization, while the second mechanism, which comes into play after ca. 1 h, may explain toxic events like the inhibition of cell locomotion or the inhibition of cell growth, lesions, that are understood as based on the lack of actin monomers. A third possible mechanism of phalloidin toxicity that so far has not been excluded is that microfilaments complexed with phalloidin lose their affinity for a so far-unidentified F-actin-binding protein that contributes to the architecture of e.g. the cortical web. Loss of such a contact may likewise result in the rapid breakdown of complex cytoskeletal structures, and, in consequence, to the development of surface blebbing.

REFERENCES

1. Cochet-Meilhac, M., and Chambon, P., 1974, Animal DNA-dependent RNA polymerases. 11. Mechanism of the inhibiton of RNA polymerases B by amatoxins. *Biochim. Biophys. Acta* 353:160-184
2. Vaisius, A.C., and Wieland, Th., 1982, Formation of a single phosphodiester bond by RNA polymerase B from calf thymus is not inhibited by α-amanitin. *Biochemistry* 21:3097-3101
3. Baumann, K.H., Zanotti,G., and Faulstich, H., 1994, A β-turn in α-amanitin is the most important structural feature for binding to RNA polymerase II and three monoclonal antibodies. *Protein Science* 3:750-756
4. Baumann, K., Münter K., and Faulstich H., 1993, Identification of structural features involved in binding of α-amanitin to a monoclonal antibody. *Biochemistry* 32:4043-4050
5. Kostansek, E.C., Lipscomb, W.N., Yocum, R.R., and Thiessen, W.E., 1977. The crystal structure of the mushroom toxin β-amanitin. *J.Am.Chem.Soc.* 99:1273-1274
6. Mullersman, J.E., Bonetti, S.J., and Preston, J.F., 1991, Periodate oxidation products derived from methylated α-amanitin: evidence for distinct aldehydic and non-aldehydic forms. *Int.J.Peptide Protein Res.* 38:409-416
7. Wieland, Th., Peptides of poisonous Amanita mushrooms. Springer New York (1986)
8. Faulstich, H., Zobeley, S., Rinnerthaler, G., and Small, J-.V., 1988, Fluorescent phallotoxins as probes for filamentous actin. *J. Muscle Res.Cell Motil.* 9:370-383
9. Faulstich, H., Zobeley, S., Heintz, D., and Drewes, G., 1993, Probing the phalloidin binding site of actin. *FEBS Lett.* 318:218-222
10. Reichert, A., Heintz, D., Voelter, W., Mihelic, M., and Faulstich, H. 1994, Polymerization of actin from thymosin β4 complex initiated by the addition of actin nuclei, nuclei-stabilizeing agents and myosin S1. *FEBS Lett.* 347:247-250.
11. Weiss, G., Sterz, I.,Frimmer, M., and Kroker, K., 1973, Electron microscopy of isolated rat hepatocytes before and after treatment with phalloidin. *Beitr. Pathol.* 150:345-356.

ACTIONS OF BANANA TREE EXTRACT ON SMOOTH AND CARDIAC MUSCLES AND IN THE ANESTHETIZED RAT

Yadhu N. Singh,[1,2] Zhong-Ping Feng,[2] and William F. Dryden[2]

[1] College of Pharmacy
South Dakota State University
Brookings, South Dakota 57007
[2] Department of Pharmacology
University of Alberta
Edmonton, Alberta T6G 2H7

INTRODUCTION

Pygmy tribesmen of Central and Southern Africa have been reported to use the juice of the banana plant stem for hunting animals and in tribal warfare (Heymer, 1974). Wooden arrows or darts may be given a poisonous tip by driving them into the plant stem for 1-2 days. An animal hit by one of these weapons becomes paralyzed but its flesh is safe to eat. An extract prepared from the juice augments, then blocks contractions of both directly or indirectly stimulated skeletal muscles (Singh and Dryden, 1985, 1990). In higher concentrations, the extract induces a sustained muscle contracture which is slowly reversed by removal of the extract. It also increases the rate of spontaneous transmitter release, as measured by the frequency of miniature endplate potentials (Singh and Dryden, 1985). The effect of the extract has now been tested on smooth and cardiac muscle preparations and in the anesthetized rat to assess whether the presumed incapacitation of the animal by the extract-tipped arrows and darts of the Pygmy tribesmen was due to paralysis of limb skeletal muscles and the diaphragm or by a collapse of its cardiovascular activity.

MATERIALS AND METHODS

Preparation of the Banana Stem Extract

Pieces of the banana plant stem were mechanically squeezed to extract a colorless fluid which was allowed to stand overnight at room temperature (20-23°C) for the resin and debris to separate out and settle. The aqueous layer was removed, centrifuged at 4000g for 10-15 min, and the residue discarded. The supernatant was rotary evaporated at 55-60°C to

Natural Toxins II, Edited by B. R. Singh and A. T. Tu
Plenum Press, New York, 1996

form a viscous liquid, which on freeze-drying yielded a brown solid. The solid was further purified by gel chromatography using Sephadex G-50 and freeze-dried to provide a white, water soluble powder, which was used in all subsequent experiments.

Guinea Pig Ileum Preparation

The guinea pig ileum longitudinal muscle strip was prepared according to the method of Paton and Vizi (1969). A short segment of the ileum was obtained from guinea pigs of either sex (450-575g) which had been starved overnight, stretched over a 1.0 ml pipette and the longitudinal muscle layer was separated from the underlying circular muscle by gently stroking it away from its mesenteric attachment along the entire length of tissue with a piece of cotton wool soaked in Krebs-Henseleit solution. The ends of the muscle were tied and suspended in a water-jacketed tissue bath in Krebs-Henseleit solution, maintained at 37°C and gassed with oxygen containing 5% CO_2. The preparation was stimulated by two parallel platinum electrodes using square wave pulses of 1 msec duration, a frequency of 0.1 Hz and supramaximal voltage. Isometric tension was recorded after the tissues had been placed under a resting tension of 0.5 g. The preparation was stimulated for a 60 min equilibration period during which it was washed periodically by overflow with Krebs-Henseleit solution. A dose-response curve was prepared in the absence of muscle stimulation by adding banana tree extract (0.03-1.3 mg/ml) directly to the tissue bath. Contact period was 60 sec. The effect of hemicholinium-3 (HC-3) on contractile response of the preparation to the banana tree extract was also investigated. Stimulation frequency was increased to 0.5 Hz and HC-3 (10^{-4}M) was left in contact for 80-120 min, after which the response to the extract was again assessed. Muscle contraction was monitored with Grass force displacement transducers, model FT03 and Grass Polygraph, model 7.

Isolated Cardiac Muscles

Sprague-Dawley rats of either sex (295-375 g) were sacrificed by light ether anesthesia, followed by a blow to the head. The hearts were rapidly removed and placed in ice-cold Krebs-Henseleit solution bubbled with oxygen containing 5% CO_2, pH 7.4. The left and right atria and left ventricular papillary muscles were isolated and mounted on punctate platinum electrodes, then placed in 4 ml tissue baths containing Krebs-Henseleit solution, maintained at 37± 0.5°C. The papillary and left atrial muscles were electrically stimulated at 1 Hz by square wave pulses of 5 msec duration and supramaximal voltage. The right atria were used if they maintained a spontaneous rhythm for at least 30 min. After this equilibration period, dose-response curves for the extract (10^{-4}-1.3 mg/ml) were prepared. The developed tension was recorded isometrically by a Devices Mx4 pen recorder using force-displacement dynamometers (Searle Bioscience).

Isolated Perfused Hearts (Langendorff Preparation)

Hearts, isolated from Sprague-Dawley rats of either sex (280-368 g) with about 1 cm of the aorta attached, were gently squeezed to remove excess blood and then mounted on a Langendorff apparatus (Edinburgh University staff, 1971). Krebs-Henseleit solution at 37°C was perfused through the heart at a constant pressure and maintained at a constant temperature by circulating water (37°C) in the outer jacket. After an equilibration period of 30-50 min, responses to the extract were obtained by adding individual bolus doses (0.1-10 mg) to the perfusion medium. The developed tension was monitored isometrically by a Devices Mx4 pen recorder for 30-40 min after each dose, by which time the hearts had fully recovered from the effects of the extract.

Anesthetized Rat Preparation

The protocol for this part of the study followed an established procedure (Mishra and Ramzan, 1992) and was approved by the Animal Care and Use Committee of South Dakota State University. Sprague-Dawley rats (292-428 g) were temporarily anesthetized with ether inhalation and a catheter was placed in the right external jugular vein for drug administration and maintaining anesthesia throughout the experiment with urethane (1.25 g/kg, i.v.). A carotid artery was cannulated for continuous recording of arterial blood pressure using a Spectramed pressure transducer, model P23XL. The rats were then tracheostomized and mechanically ventilated with room air. The distal tendon of the left tibialis anterior muscle was dissected, separated from its insertion and attached to a Grass FT03 transducer for isometric recording of muscle contraction with a Grass 7 polygraph. The corresponding sciatic nerve was exposed, placed on a bipolar platinum electrode, tied with a heavy ligature and cut proximal to the ligature. The tension on the muscle was adjusted to evoke maximal contraction when the nerve was stimulated with supramaximal pulses of 0.2 msec duration and 0.1 Hz. Electrocardiogram (EKG) was recorded using two subcutaneous needle electrodes inserted on either side of the thorax and monitored by a Grass EKG Preamplifier (Model 7P4). The preparations were allowed to stabilize for at least 30 min after which baseline measurements were performed. Banana stem juice extract (32-58 mg) was then infused over a period of 2-5 min. Rectal temperatures were maintained at 35-38°C with a heated surgical table and a heat lamp.

Drugs and Chemicals

Drugs used were: acetylcholine chloride, atropine sulfate, hemicholinium-3 (all Sigma). All other chemicals were of reagent grade. Krebs-Henseleit solution had the following composition (g/1): NaCl 6.92, KCl 0.35, $MgSO_4.7H_2O$ 0.29, $CaCl_2$ 0.28, $NaHCO_3$ 2.12, KH_2PO_4 0.16, and dextrose 1.8, equilibrated with 95% oxygen, 5% CO_2.

Statistical Analysis

Data are presented as means ± S.E.M. of 4-7 experiments. Statistical analysis of the data was made using Students *t*-test, $P<0.05$ being regarded as significant.

RESULTS

Guinea Pig Ileum Muscle

The extract produced a contraction of the preparation which was rapid both in onset and inactivation (see inset, Fig. 1). The dose-response relationship is shown in Fig. 1. In the presence of atropine (10^{-6}M), the effect of 0.3 mg/ml of the extract was depressed by 40.7±6.3% of control (n=4) and for 1 mg/ml the depression was 19.9±7.2% of control (n=4). On the other hand, atropine in the same concentration (10^{-6}M) abolished the responses to equieffective doses of acetylcholine (3-10 μM). Furthermore, pretreatment with HC-3 (10^{-4}M) depressed the stimulatory actions of the extract, with the effects at the two highest doses being significantly different from control ($P<0.05$) (Fig. 1).

Isolated Cardiac Muscles

The effects of the banana stem extract on developed tension in the isolated preparations are shown in Fig. 2. The rate of spontaneously beating right atria was not significantly

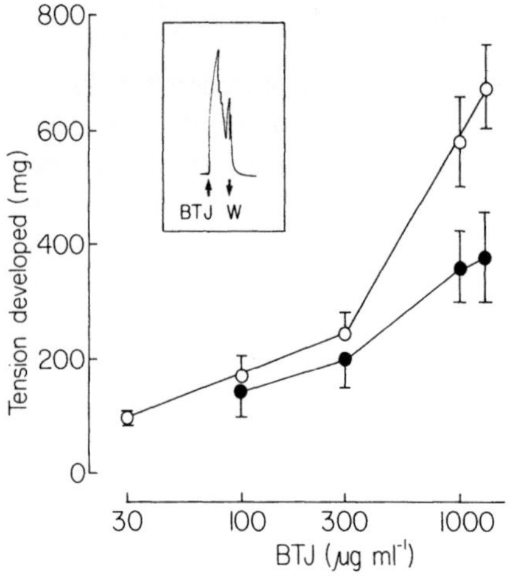

Figure 1. Stimulatory effect of the extract (BTJ) on guinea pig ileum longitudinal muscle strip before (O) and after treatment with hemicholinium-3 (10^{-4}M, ●). The inset shows the contractile effect of the extract (300 μg/ml) on an untreated muscle.

(P>0.05) affected by the extract (10^{-4}-1.3 mg/ml), even in the highest concentrations tested, while contractility of the left atria and ventricular papillaries was gradually depressed and eventually almost completely blocked. In 1 out of 5 left atria and 2 out of 7 papillary muscles, the extract transiently developed augmented tension before a decline in tension and eventual blockade.

Langendorff Preparation

In some experiments, a large but transient depression in the contractility of the preparation was noted in the first 60-90 sec after the addition of the extract. This probably was due to a bolus effect. Responses monitored at 2, 5, 10, and >10 min after the administration of each dose (Fig. 3) showed a dose- and time-dependent depression of developed tension, reaching a plateau at t>10 min (usually 15-25 min) before start of recovery which normally was complete at t=40 min. No initial potentiation of contractility was noted in any of the 5 hearts used.

Anesthetized Rats

The typical effects of infusing the extract (42 mg, 129 mg/kg) in an anesthetized rat (325 g) on contractility of tibialis anterior muscle (TIB), electrocardiogram (EKG), and

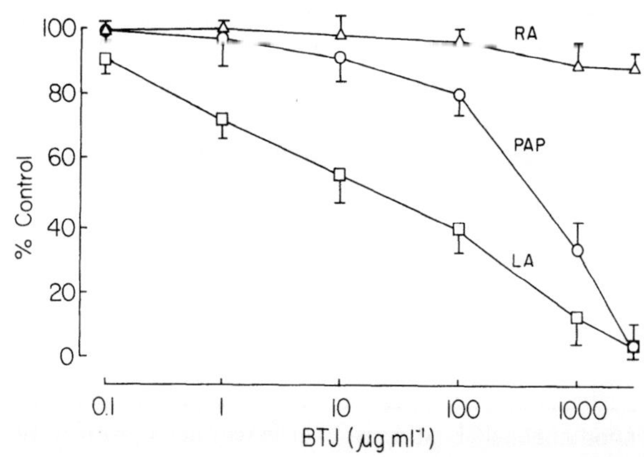

Figure 2. Inhibitory effect of the extract (BTJ) on spontaneous rate in the right atrium (RA) and on muscle contraction in papillary (PAP) and left atrial muscles (LA).

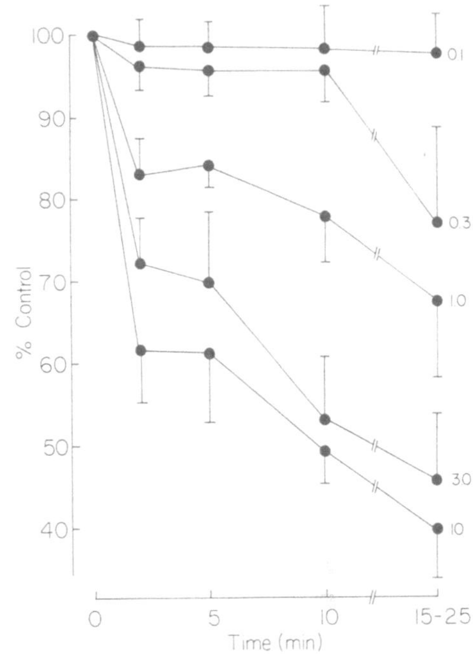

Figure 3. Time dependence of the action of bolus injections of the extract (0.1, 0.3, 1.0, 3.0, and 10 mg) on rat heart mounted in a Langendorff apparatus. The largest depression of contractility was reached 15-25 min after the application, after which recovery occurred.

blood pressure (BP) are shown in Fig. 4. For the EKG, the regular cardiac rhythm evolved into a tachycardia, then arrhythmia, followed by a loss of the characteristic configuration of the EKG, and a reduction and eventual loss of the signal. There was a simultaneous decrease in both systolic and diastolic blood pressures accompanied by an increase in the pulse pressure. The complete loss of pulse pressure coincided with the disappearance of the EKG. On the other hand, muscle contractility was not significantly ($P>0.05$) affected during this period or for at least 15 min after all cardiovascular function had ceased. Similar effects were obtained with the extract (89-146 mg/kg) in 4 other rats. In some animals, and especially in lower doses, the arrhythmia alternated with short periods of normal rhythm, and in one case the animal fully recovered about 40 min after treatment with the extract (87 mg/kg).

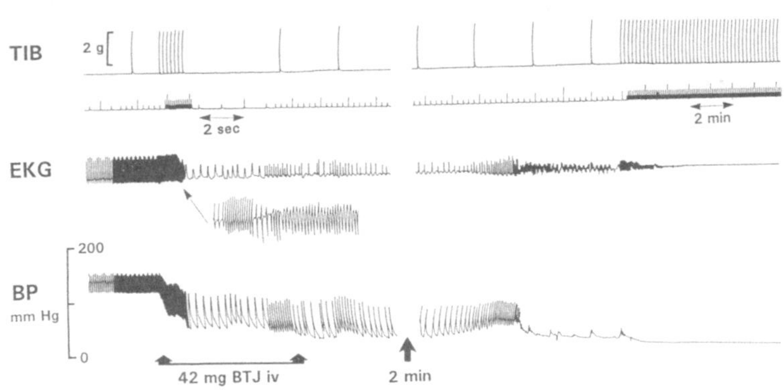

Figure 4. Effect of the extract (BTJ, 42 mg i.v.) on tibialis anterior muscle (TIB), electrocardiogram (EKG), and arterial blood pressure (BP) in a 325 g anesthetized rat. The records were made at different chart speeds, as indicated by the time base. The inset just below the EKG trace shows details of the onset of the arrhythmia during the early stages of extract infusion.

DISCUSSION

The banana plant is the source of several biologically active substances. Dopamine, norepinephrine, serotonin, and an antiulcerogenic factor have been detected in the pulp and skin of the fruit (Waalkes *et al.*, 1958; West, 1958; Best *et al.*, 1984). Acetic acid, gallic acid, tannin, sodium chloride, and salts of aluminum, calcium, and potassium have been associated with the juice of the plant stem (Herbert, 1896; Wehmer, 1929; Watt and Breyer-Brandwijk, 1962). Recently, the presence of dopamine, norepinephrine, and monopotassium oxalate (Benitez *et al.*, 1991) and potassium nitrate and sodium nitrate (Singh *et al.*, 1993) has been reported in extracts of the stem juice.

In the guinea pig ileum preparation, the extract produced a stimulatory contractile effect, which was significantly reduced ($P<0.05$) in the two highest concentrations by pretreatment with HC-3. These observations are consistent with the presence of K^+ in the extract. After HC-3 pretreatment, the extract had only a direct contractile effect on the muscle, whereas in the intact muscle, K^+ would additionally stimulate the release of acetylcholine which would account for the larger response. Furthermore, the depression of the contraction by atropine (10^{-6}M) would also be due to blockade of the activity of the release acetylcholine on muscarinic receptors on the muscle (Yamamura and Snyder, 1974). The effect of Mg^{2+} probably was masked by the stimulatory activity of K^+.

In the isolated cardiac muscles and the Langendorff heart, there was no initial enhancement of muscle contractility, in contrast to the report of Pereira *et al.* (1963) and observations in skeletal muscles (Singh and Dryden, 1985, 1990), only a dose-dependent blockade. Furthermore, there was no significant ($P>0.05$) change in the chronotropic response of the spontaneous right atria. The above observations are somewhat surprising, given the presence of K^+ in the extract; however, the concomitant presence of Mg^{2+} might have affected the activity of the extract.

According to Heymer (1974), Pygmy tribesmen in Africa have used the banana stem juice extract as an arrow or dart poison. In previous reports, it was assumed that the extract incapacitated animals by paralysis of the limb skeletal muscles and the diaphragm. The results of the *in vivo* experiments indicate that despite the rapid induction of cardiac arrhythmias, the marked concomitant fall in blood pressure, and the subsequent cessation of cardiovascular activity with high doses of the extract , nerve-evoked contractions of the tibialis muscle of the anesthetized rat were not significantly ($P>0.05$) affected. Thus, if Heymer's report is to be accepted, the primary cause for the incapacitating action of the banana stem juice extract, is unlikely to be due to skeletal muscle paralysis but rather to a deterioration of cardiovascular function caused by a severe cardiac arrhythmia, profound hypotension, and subsequent reduction in tissue perfusion with blood, resulting in a general collapse of vital physiological functions. Whether a sufficient amount of the extract to induce these effects can be injected into the animals by the hunting equipment of the Pygmy tribesmen is yet to be determined.

ACKNOWLEDGEMENTS

This work was supported in part by a University of Alberta Central Research Fund grant and the South Dakota State University College of Pharmacy.

REFERENCES

Benitez, M.A., Navarro, E., Feria, M., Trujillo, J. and Boada, J. Pharmacological study of the muscle paralyzing activity of the juice of the banana trunk. *Toxicon* **29**, 511-515 (1991).

Best, R., Lewis, D.A. and Nasser, N. The anti-ulcerogenic activity of the unripe plantain banana (*Musa* species). *Brit. J. Pharmacol.* **82,** 107-116 (1984).

Edinburgh University staff. *Pharmacological Experiments on Isolated Preparations.* pp. 116-119. Livingstone, Edinburgh (1971).

Hebert, A. (1896) quoted in Watt and Breyer-Brandwijk, 1962, p. 783.

Heymer, A. The hunting and food-gathering pygmies. *Image Roche* **62,** 17-26 (1974).

Mishra, Y. and Ramzan, I. Influence of cimetidine on gallamine-reduced neuromuscular paralysis in rats. *Clin. exp. Pharmacol. Physiol.* **19,** 803-807 (1992).

Paton, W.D.M. and Vizi, E.S. The inhibitory action of noradrenaline and adrenaline on acetylcholine output by guinea pig ileum longitudinal muscle strip. *Brit. J. Pharmacol.* **35,** 10-28 (1969).

Pereira, J.R., Bustos, R.E. and Zyngier, Z. Some pharmacological properties of the juice of the banana plant. *Arch int. Pharmacodyn.* **144,** 144-150 (1963).

Singh, Y.N. and Dryden, W.F. Muscle paralyzing effect of the juice from the trunk of the banana tree. *Toxicon* **23,** 973-981 (1985).

Singh, Y.N. and Dryden, W.F. The augmenting action of banana tree juice on skeletal muscle contraction. *Toxicon* **28,** 1229-1236 (1990).

Singh, Y.N., Inman, W.D., Johnson, A., and Linnell, E.J. Studies on the muscle-paralyzing components of the juice of the banana plant. *Arch. int. Pharmacodyn.* **324,** 105-113 (1993).

Waalkes, T.P., Sjoerdsma, A., Creveling, C.R., Weissbach, H. and Udenfriend, S. Serotonin, norepinephrine, and related compounds in bananas. *Science* **127,** 648-650 (1958).

Watt, J.M. and Breyer-Brandwijk, M.G. *The Medicinal and Poisonous Plants of Southern and Eastern Africa,* pp 783-784. E. & S. Livingston, Edinburgh, 2nd Ed. (1962),

Wehmer, C. *Die Pflanzenstoffe,* pp. 173-175. Fischer, Jena, 2nd Ed. (1929).

West, G.B. Tryptamines in edible fruits. *J. Pharm. Pharmacol.* **10,** 589-590 (1958).

Yamamura, H.I. and Snyder, S.H. Muscarinic cholinergic receptor binding in the longitudinal muscle of the guinea pig ileum. *Mol. Pharmacol.* **10,** 861-867 (1974).

THE EARLY EXPRESSION OF MYOTOXICITY AND LOCALIZATION OF THE BINDING SITES OF NOTEXIN IN THE SOLEUS MUSCLE OF THE RAT

Notexin and Muscle

R. W. Dixon and J. B. Harris

School of Neurosciences, University of Newcastle upon Tyne
and Muscular Dystrophy Research Laboratories
Newcastle General Hospital
Newcastle upon Tyne NE4 6BE, United Kingdom

INTRODUCTION

Notexin is a major toxic component of the venom of the Australian tiger snake, *Notechis scutatus scutatus*. It is a peptide of 119 residues with 7 disulphide bridges and a relative molecular mass of 13.6 KDa. Notexin is one of a group of neurotoxic myotoxins that exhibit phospholipase A_2 activity and are homologous with mammalian pancreatic phospholipases (Harris, 1991).

The pathological responses of skeletal muscle to notexin and related toxins have been well documented (Harris and Cullen, 1990) but little is known of the precise mechanism of action of the toxins. Based on a great deal of descriptive data, Harris (1984) suggested that the hydrolysis and fragmentation of the the plasma membrane was the primary event, leading to the loss of ionic gradients, influx of Ca^{2+}, hypercontraction and degeneration of major muscle fibre proteins. It was conceded, however, that the physical disintegration of the membrane was not a necessary prerequisite for the onset of degeneration, and that the hydrolysis of membrane lipids could lead to a "leaky" membrane without evidence of structural damage.

Another possible mechanism of action would be that the toxins were internalised and then released into the cytosol where they could disrupt the function of subcellular components such as mitochondria and sarcoplasmic reticulum (Harris, 1984).

In favour of the generally accepted view that the toxins are active on cell plasma membranes is the recognition of a number of putative high affinity binding proteins for the neurotoxic PLA_2. The proteins have a relative molecular mass of 45kDa, 50KDa, 65KDa and 85KDa respectively and have so far been identified only in synaptosomal preparations (Tzeng, 1993). No comparable studies have yet been made using skeletal muscle.

Natural Toxins II, Edited by B. R. Singh and A. T. Tu
Plenum Press, New York, 1996

In this study we have used a combination of scanning electron microscopy, transmission electron microscopy, immunoelectron microscopy and immunocytochemistry to determine the earliest pathological changes in skeletal muscle following the inoculation of notexin and to study the distribution of toxin binding sites.

MATERIALS AND METHODS

Adult female wistar rats (90-100g body weight) were obtained from a registered supplier. 2 μg of NTX (0.2ml of 10μg/ml in 0.9% w/v NaCl) was introduced subcutaneously into the dorso-lateral aspect of one hind limb of a rat so that the soleus muscle was exposed to toxin; the contralateral limb was used as a control.

Light Microscopy

Toxin binding sites were identified using indirect immunogold labelling. The belly of the muscle was removed and cut longitudinally to form blocks 3-4mm in length. These were affixed to strips of filter paper and frozen in melting iso-pentane cooled to -150°C in liquid nitrogen. Sections of tissue were cut 6μm thick using a cryostat (-25°C) and allowed to dry onto subbed slides for 15 mins. To prevent non-specific binding of the antibodies used for labelling the binding sites, the sections were incubated with normal goat serum (100μl of 10% in 0.9% w/v NaCl). The sections were rinsed with 0.1M phosphate buffered saline at pH 7.4 (P.B.S.) (3 x 5 mins) and then incubated with a rabbit-anti-notexin polyclonal Fab_2 fragment (50μl of 50μgml^{-1} in 0.9% w/v NaCl), at room temperature for 30 minutes. The sections were then washed (P.B.S.; 3 x 5mins) before the secondary antibody, a goat anti-rabbit IgG conjugated with 1nm colloidal gold (suspended in P.B.S. containing 1% bovine serum albumin (B.S.A.) and 0.1% sodium azide), was introduced for 30 minutes.

The sections were again washed 3 times in P.B.S. before being fixed in glutaraldehyde (2% w/v in P.B.S.; for 10 minutes). After washing 3 times in distilled water, the gold particles were enhanced with silver using a proprietary kit (British Biocell International SEKL15) for approximately 10 minutes. Once the enhancement was complete, the sections were dehydrated, cleared and mounted in the usual way and the results photographed.

Transmission Electron Microscopy

The material to be studied using transmission electron microscopy was labelled using a pre-embedding immuno-incubation. The soleus muscles were removed, pinned out at c. 1.2x resting length and fixed (2% paraformaldehyde, 0.0001% glutaraldehyde in 0.1M sodium cacodylate buffer, pH 7.2). After 1 hour, the tissue was cut into blocks of 1mm x 1mm x 2mm and then rinsed in P.B.S. (3 x 5 mins). The blocks were next treated with sodium borohydride (0.05% $NaBH_4$, 0.1% Glycine in P.B.S., 10 mins). To prevent non-specific binding the blocks were incubated in bovine serum albumin (0.5% B.S.A., 0.1% gelatine in P.B.S.; 1 hour). The blocks were next incubated with a rabbit-anti-notexin polyclonal Fab_2 fragment (50μgml^{-1} in 0.9% w/v NaCl) at 4°C overnight. The blocks were rinsed (0.5% B.S.A. + 0.1% gelatine in P.B.S.; 2 hours) and then incubated with the secondary antibody, a goat-anti-rabbit IgG conjugated with 1nm colloidal gold (suspended in P.B.S. containing 1% B.S.A. + 0.1% sodium azide) for 5 hours at 37°C. The blocks were again rinsed (0.5% B.S.A. + 0.1% gelatine in P.B.S.; 2 hours) before being fixed in glutaraldehyde (2.5% in P.B.S.; 10 mins). The blocks were next rinsed in distilled water (3 x 5 mins) before silver enhancement for 2 mins and a further washing in distilled water (3 x 5 mins).

The blocks were dehydrated in a graded series of ethanol and then acetone before embedding in araldite. 50nm thick sections were cut and mounted on copper grids and counterstained with 20% uranyl acetate in methanol for 5 mins and modified Reynold's lead citrate (Reynolds, 1963), for 5 mins.

The sections were viewed with a Jeol JEM-1200 EX using an accelerating voltage of 80kV.

Scanning Electron Microscopy

The soleus muscles were removed and pinned out as for the T.E.M. studies. They were fixed whole in Karnovsky's fixative for 1 hr, cut into 1mm x 1mm x 2mm blocks and fixed for a further hour (Karnovsky, 1965). After rinsing in P.B.S. (3 x 5 mins) the blocks were post-fixed for 2 hrs in OsO_4 (1% in P.B.S.). After rinsing in distilled water (3 x 5 mins) the connective tissue binding the muscle fibres together, including the collagen fibres and the basal lamina (Desaki and Uehara, 1981), were removed by treatment with hydrochloric acid (8N at 60°C) for 20 mins. The dissociated fibres were dehydrated in a graded series of ethanol before being left in amyl acetate overnight. After drying by the critical point method and sputter coating with gold, the fibres were examined using a Jeol JEM-1200 EX fitted with an EM-ASID10 scanning image device, using an accelerating voltage of 40kV.

RESULTS

Light Microscopy

Transverse sections of soleus muscle, prepared 3 hrs after the inoculation of notexin, exhibited dense dark staining of the periphery of the muscle fibres. This staining was absent in muscle sections exposed to only 1° and 2° antibodies and silver (fig. 1). The dark deposit clearly marked the distribution of notexin and localised binding to the sarcolemma.

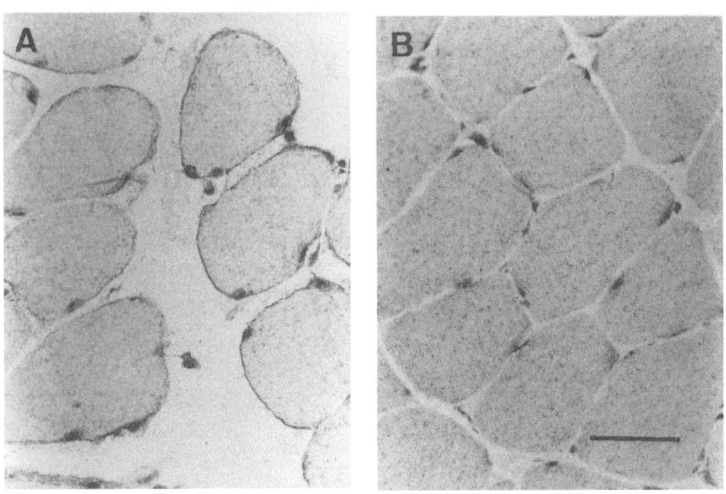

Figure 1. A. Light micrograph of a transverse section through a rat soleus muscle that was exposed to notexin then labelled indirectly with gold decorated 2° antibodies. B. Control muscle labelled using 1° and 2° antibodies only. Scale bar = 20μm.

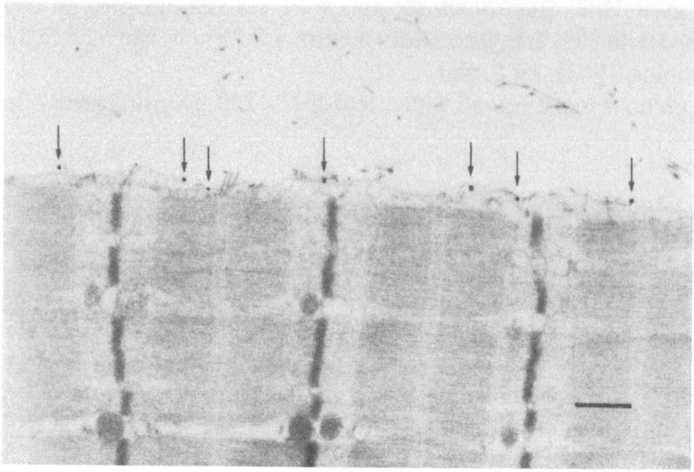

Figure 2. Transmission electron micrograph of a longitudinal section through a rat soleus muscle exposed to toxin *in vivo* and then labelled indirectly. The dark spots (arrows) mark silver enhanced gold particles decorating the 2° antibody. Scale bar = 500nm.

Transmission Electron Microscopy

An electron micrograph of a longitudinal section of a soleus muscle prepared 3 hrs after the inoculation of notexin and prior to significant evidence of degeneration is shown in figure 2. The preparation was made following exposure to a 1° antibody reactive against notexin and a 2° antibody decorated with gold. The silver-enhanced gold particles seen as discrete black spots located along the plasma membrane of the muscle fibre, confirming the observations made using light microscopy that binding is to the plasma membrane. We could find no evidence of internalisation in early stages of exposure. At late stages (6-24hr) there was considerable artefactual staining of necrotic material caused by the attraction of the gold particles for accumulations of basic materials in the degenerating fibres (not shown).

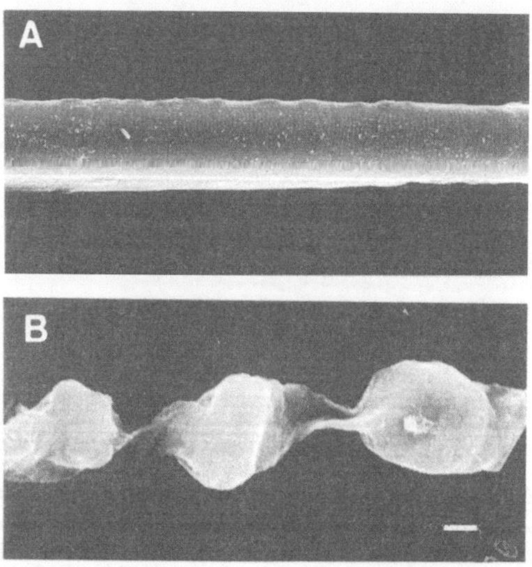

Figure 3. Scanning electronmicrograph of a rat soleus muscle fibre A = control, B = following 3 hrs *in vivo* exposure to notexin. Scale bar = 10μm.

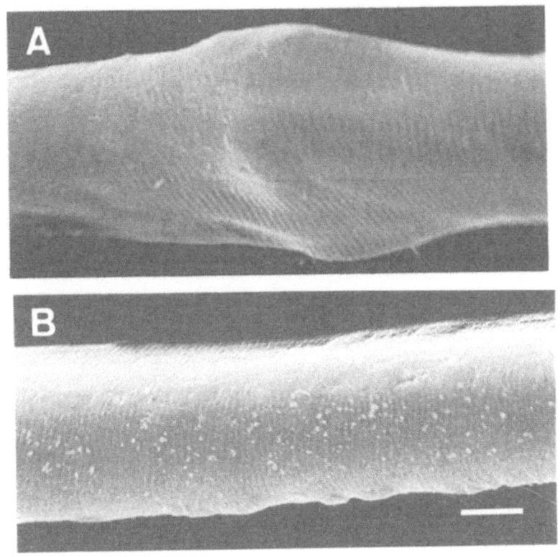

Figure 4. A. Scanning electronmicrograph of an area of hypercontraction of a rat soleus muscle fibre at high magnification following 3 hrs *in vivo* exposure to notexin. B. Control fibre. Scale bar = 10μm.

Scanning Electron Microscopy

Scanning electron microscopy was used to determine whether early stages of pathology were associated with gross damage to the plasma membrane. Normal fibres were regular in outline, with a smooth external membrane and a regular sarcomeric repeat (fig 3). 3 hours after exposure to notexin *in vivo* the pathological response of the fibre was characterised by areas of hypercontraction and areas where the myofibrils have been torn apart (fig 3). An area of hypercontraction is shown at high magnification in figure 4. The hypercontraction was characterised by a shortening of the sarcomeric length in the area of interest. Despite the deformation and local swelling caused by the toxin, there was no evidence of damage to the plasma membrane.

At more advanced stages of damage large lesions appeared in the plasma membrane (fig. 5) and the myofibrils could be seen exposed inside the remnants of the fibre.

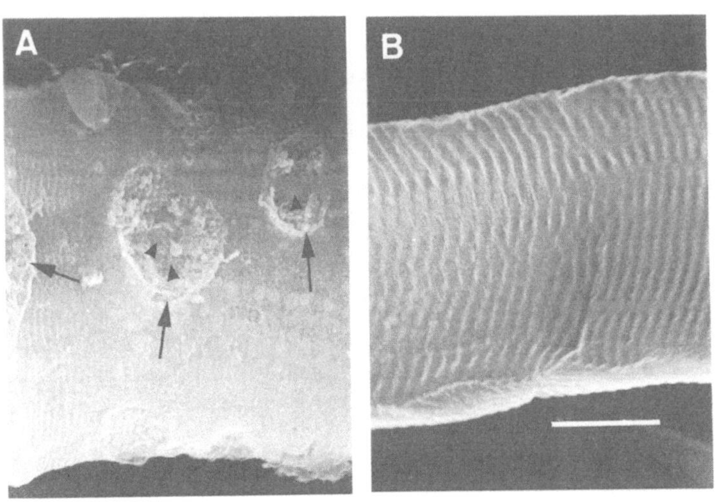

Figure 5. A. Scanning electronmicrograph of a lesion (arrow) in a rat soleus muscle fibre following 3 hrs *in vivo* exposure to notexin. Note the intact myofibrils (arrow heads). B. Control fibre. Scale bar = 10μm.

DISCUSSION

Notexin, and similar toxins are homologous with pancreatic phospholipases (Halpert and Eaker, 1975) and this PLA_2 activity is generally believed to be central to the action of the toxin (see Rosenberg (1986) for a dissenting view). The cell membrane is thought to be the primary target of the toxin (Harris and MacDonell, 1981; Brenes *et al,* 1987).

Our data are entirely consistent with this view. Muscle fibres exposed for brief periods to notexin bound the toxin to the plasma membrane. This was demonstrated using both light and electron microscopy. We found no evidence at all of internalisation of the toxin using the immunological technioques at our disposal. We can conclude, therefore, that although subcellular fractions such as mitochondria and S.R. are exquisitely sensitive to PLA_2-active toxins (Stringer *et al,* 1971; Ownby *et al,* 1982; Harris and Maltin, 1982; Gopalakrishnakone and Hawgood, 1984; Ng and Howard, 1980), they are not the primary targets and probably not involved in the onset of pathology.

The suggestion that the primary effect of the toxin is to hydrolyse and rupture the plasma membrane is clearly incorrect. The hypercontraction and disorganisation of the sarcomere preceded the appearance of major lesions in the plasma membrane. We conclude that the toxins bind to selective sites on the plasma membrane and probably initiate a localised hydrolysis of membrane lipids. We now suggest that the formation of lysophosphatides and free fatty acids with detergent-like properties results in increasing membrane fluidity and instability. This leads directly and indirectly to the movement of ions along a concentration gradient - which would include the influx of Na^+ and Ca^{2+}, depolarisation and hypercontraction and the onset of muscle necrosis.

ACKNOWLEDGEMENTS

Rupert Dixon is supported through a Wellcome Trust Toxicology Studentship.

REFERENCES

Brenes, F., Gutierrez, J.M. and Lomonte, B. Immunohistochemical demonstration of the binding of *Bothrops asper* myotoxin to skeletal muscle sarcolemma. *Toxicon* 25:574-577, 1987.

Desaki, J. and Uehara, Y. The overall morphology of neuromuscular junctions as revealed by scanning electron microscopy. *J.Neurocytol.* 10:101-110, 1981.

Gopalakrishnakone, P. and Hawgood, B.J. Morphological changes induced by crotoxin in murine nerve and neuromuscular junction. *Toxicon* 22:791-804, 1984.

Halpert, J. and Eaker, D. Amino acid sequence of a presynaptic neurotoxin from the venom of *Notechis scutatus scutatus* (Australian tiger snake). *J.Biol.Chem.* 250:6990-6997, 1975.

Harris, J.B. Polypeptides from snake venoms which act on nerve and muscle. In: *Progress in Medicinal Chemistry*, edited by Ellis, G.P. and West, G.B. Amsterdam: Elsevier, 1984, p. 63-110.

Harris, J.B. Phospholipases in snake venoms and their effects on nerve and muscle. In: *Snake Toxins*, edited by Harvey, A.L. New York: Pergamon Press, 1991, p. 91-124.

Harris, J.B. and Cullen, M.J. Muscle necrosis caused by snake venoms and toxins. *Electron Microsc.Rev.* 3:183-211, 1990.

Harris, J.B. and MacDonell, C.A. Phospholipase A_2 activity of notexin and its role in muscle damage. *Toxicon* 19:419-430, 1981.

Harris, J.B. and Maltin, C.A. Myotoxic activity of the crude venom and the principal neurotoxin, taipoxin, of the Australian taipan, *Oxyuranus scutellatus*. *Br.J.Pharmac.* 76:61-75, 1982.

Karnovsky, M.J. A formaldehyde-glutaraldehyde fixative of high osmolarity for use in electron microscopy. *J.Cell Biol.* 27:137A, 1965.

Ng, R.H. and Howard, B.D. Mitochondria and sarcoplasmic reticulum as model targets for neurotoxic and myotoxic phospholipase A_2. *Proc.Natn.Acad.Sci.USA* 77:1346-1350, 1980.

Ownby, C.L., Gutierrez, J.M., Colberg, T.R. and Odell, G.V. Quantitation of myonecrosis induced by myotoxin α from Prairie rattlesnake (*Crotalus viridis viridis*) venom. *Toxicon* 20:877-885, 1982.

Reynolds, E.S. The use of lead citrate at high pH as an electron opaque stain in electron microscopy. *J.Cell Biol.* 17:208, 1963.

Rosenberg, P. The relationship between enzymatic activity and pharmacological properties of phospholipases in natural poisons. In: *Natural Toxins: animal, plant and microbial*, edited by Harris, J.B. Oxford: Oxford University Press, 1986, p. 129-174.

Stringer, J.M., Kainer, R.A. and Tu, A.T. Ultrastructural studies of myonecrosis induced by Cobra venom in mice. *Toxicol.Appl.Pharmacol.* 18:442-450, 1971.

Tzeng, M.C. Interaction of presynaptically toxic phospholipases A_2 with membrane receptors and other binding sites. *J.Toxicol.-Toxin Reviews* 12:1-62, 1993.

27

FUMONISIN B1–IMMUNOLOGICAL EFFECTS*

The Influence of FB1 on the Early Stage of Immune Response

E. A. Martinova

Institute of Nutrition RAMS
2/14 Ustinsky proezd
109240 Moscow
Russian Federation

1. INTRODUCTION

General Considerations

Fumonisin B1 (FB1) ($C_{34}H_{59}NO_{15}$, Mol.wt.721) is classified as a *Fusarium mycotoxin* produced by some related species including *F.moniliforme, F.proliferatum, F.anthophilum, F.dlamini, F.napiforme, and F.nygamai*. These strains have been isolated in Australia, Austria, Brazil, Canada, China, Indonesia, Italy, Nepal, New Caledonia, Peru, the Philippines, Poland, South Africa, Thailand, and the USA [37,38,61,62,63]. Exposure occurs mainly through dietary consumption of contaminated corn.

The toxicity of the FB1 for animals [19,40,43,49,64] is related to their cancer promoting potential in a short-term cancer initiation/ promotion bioassay. The FB1 can stimulate the outgrowth of initiated cells by selective inhibition of normal cell growth after the cell proliferation stimulus [19,20,21,23].

The correlative analysis between exposure to *Fusarium* toxins and oesophageal cancer rates in human presumes to be carcinogenicity of toxins derived from *F.moniliforme*, and indirect evidence for the same effect of the FB1 [29,30,48,60,69]. The FB1 is neither

* ABBREVIATIONS: Ab - antibody; Ag - antigen; APC - antigen presenting cell; CD - clasters of differentiation; CTL - cytotoxic lymphocyte; FB1 - fumonisin B1; Ig - immunoglobulin; IL - interleukin; i.p. - intraperitoneal injection; i.v. - intravenous injection; LPS - lipopolysaccharide; MLS - mixed lymphocyte culture; MHC - major histocompatibility complex; PFC - plague forming cell; PHA - phytohemagglutinin; PTK - phosphotyrosine kinase; PTP - phosphotyrosine phosphatase; PLC - phospholipase C; PLD - phospholipase D; SRBC - sheep red blood cells.

Natural Toxins II, Edited by B. R. Singh and A. T. Tu
Plenum Press, New York, 1996

genotoxic nor mutagenic, and does not appear to be metabolised to any significant extent *in vivo* [40,54].

In vivo, the proliferating foci have been observed in both the liver and kidney of rats exposed to the FB1 [21,49]. *In vitro*, mitogenic effect of the FB1 has been shown only on Swiss 3T3 fibroblasts [51]. For other cell lines and normal tissue, the FB1 did not appear to be mitogenic [23]. The FB1 effectively delayed hepatocyte cell proliferation after partial hepatectomy [21].

Known Immunological Effects of FB1 and Some Sphingolipids

An investigation of the FB1 immunological properties could discover the mechanisms of its toxicity and carcinogenicity. There are a little reports concerning this question. Significant cytotoxicity and depression in the phagocytic potential of chicken peritoneal macrophages occurred after 4h treatment with 0.5-10μg FB1/ml. However, exposure to the FB1 alone, as well as after stimulation with LPS, induced the secretion of cytolytic factors by chicken macrophage cell line MQ-NCSU [46].

Male White Leghorn chickens have been shown to decrease agglutinin response to sheep red blood cells (SRBC) (T-dependent Ag) and Brucella abortus (T-independent Ag) after supplementation of *F.moniliforme* culture material for 6 weeks [31].

Calves fed the FB1-contaminated corn for one month have been found to reduce the lymphocyte blasttransformation at the end of feeding of the maximal dose of the FB1 [43].

Turkey lymphocytes exposed *in vitro* to the FB1 for 48 and 72h revealed dose-dependent inhibition of cell proliferation. The 50% inhibitory dose was 0.4-5.0μg FB1/ml [14].

The molecular mechanism of action of the FB1 was presumed to be via disruption of sphingolipid metabolism [34]. The FB1 inhibited the activity of sphingosine N-acyltransferase (ceramide synthase), accumulated mainly sphinganine, and sphingosine to a smaller extent, decreased the formation of more complex sphingolipids, and so on [34]. Immunological action of the FB1 may be either via direct interaction between FB1 and immune cell surface receptors or via biological activity of accumulated sphingolipids, in particular, by inhibition of protein kinase C (PKC) [23]. Sphinganine was found to block the cell progression through the G1/S-phase and to appear the cytostatic effect on cell lines [25]. Sphingosine inhibited IL-8-induced *in vitro* human lymphocyte migration [3], as well as both the Ia-induced adhesion and adhesion via LFA-1-dependent pathway [18,56].

Sphingosine, ceramide, and sphingomyelin suppressed both the IL-2-dependent proliferation of cell lines HT-2 and CTLL, and the cell proliferation in MLC [10,15,27]. Sphingosine-1-phosphate inhibited integrin-dependent cell motility, but not integrin-dependent adhesion. Sphinganine-1-phosphate was less effective in this experiment [50].

2. ABSORBTION AND DISTRIBUTION OF FB1

The exposure of mammals to the FB1 can lead to distinct results depending upon the time and the level of contact of the FB1 with immune system. Both the confinement of the FB1 by immune organs and the short-time contact with immune cells can modulate the immune status and the immune response to foreign antigens. It is known that there are short-, and long-living lymphocytes [16]. B cells are renewed through continuous production and export from the bone marrow. The peripheral T cell pool is permeable, allowing the incorporation and expansion of recent thymic migrants [16]. In this connection, the circulation of the FB1 in the organism during 3-4 days after single exposure causes the contacts the FB1 with new generations of lymphocytes.

Table 1. Incorporation of [^{14}C]FB1 in the mouse organs after a single i.p. administration of 1 μg[^{14}C]FB1/g bw

Organ	Time after i.p. administration, hours		
	2 h	24 h	96 h
	% of administrated dose		
Thymus	0.010 ± 0.003	0.010 ± 0.002	0.005 ± 0.001
Spleen	0.010 ± 0.002	0.005 ± 0.002	0.003 ± 0.001
Plasma	0.800 ± 0.029	0.007 ± 0.003	0.005 ± 0.002
Intestine	0.800 ± 0.032	0.030 ± 0.015	0.040 ± 0.018
Kidney	0.080 ± 0.005	0.060 ± 0.003	0.050 ± 0.003
Liver	1.500 ± 0.005	1.200 ± 0.006	0.800 ± 0.007

^{14}C-labelled FB1 was purchased from Dr. B. Miller (Canada). Tissue solubilizer was from "Serva", and liquid scintillation cocktail was from "Packard". The organs from BALB/c mice after [^{14}C]FB1 i.p. injection were weighed, and the samples were analysed for radioactivity by liquid scintillation counting in the "Beta-2" (Russia) under the low-background regime; "dpm" was determined using internal standardisation. Residual radioactivity per organ was measured as percentage of administered dose. The results are given as ±S.E. from 3 experiments.

The FB1 undergoes mainly billiary excretion [41,53,54]. The remainder of radio-labelled FB1 was distributed in tissues, with the liver, kidney, and blood having the highest percentages [41]. The FB1 *dosed single* by i.v. was rapidly eliminated from plasma with a mean half-life during the elimination phase of 40 min. Residual radioactivity was recovered in low levels from skeletal muscle -1%, liver -0.4%, brain-0.2%, and kidney, heart, plasma, red blood cells, and bile - each 0.1%, while the contents of the intestines accounted for a further 12% of the radioactive dose [54].

We studied the distribution of [^{14}C]FB1 in the BALB/c mice over the 2h, 24h, and 96h period after a single i.p. administration of FB1.

The thymus and spleen appeared the lowest levels of residual radioactivity. The plasma can be inspected for place of short-term contact between the FB1 and immune cells. The significant radioactivity was recovered in plasma only over a 2h period. The intestinal epithelium acts as a place of extrathymic differentiation of T-lymphocytes in a parallel with a spleen [35,45]. The residual radioactivity was recovered in intestine over 96h period.

Although both the thymus and spleen were not shown to be the target for FB1 action, such transitory accumulation of the FB1 in these organs was enough for immune disruptions (see Fig.1.2).

The measurement of binding of the FB1 by immune cells is important for explanation of its effects on the immune system. Hypothetically, the FB1 can exhibit its action by transitory binding with cell membrane. Sphingolipid bases can be produced in cells in response to external stimuli. The levels of endogenous sphingosine and sphinganine can be increased up to 9.5-fold and 4.5-fold, respectively, within 60 min [28].

In our early experiments, the i.p. administration of 1 μg FB1/g bw caused a 2-fold activation of sphingomyelinase and a 2.5-fold decrease in sphingomyeline concentration in the thymus. Sphingomyelinase activity in the spleen has been shown to be increased insignificantly. A doubling of the ceramide concentration has been found in both the thymus and spleen after 2h [32]. In human T cells, the accumulation of cellular ceramide decreased the capacity of T lymphocytes to proliferate [4].

Table 2. Percentage of [^{14}C]FB1 binding by separated mouse thymus and spleen cells

t,°C	Time, h	Thymus cells		Spleen cells	
		[^{14}C]FB1 concentration			
		100 μg/ml	20 μg/ml	100 μg/ml	20 μg/ml
37°C	0.5*	0.120 ± 0.005	0.360 ± 0.005	0.250 ± 0.015	0.430 ± 0.001
	1*	0.160 ± 0.005	0.520 ± 0.020	0.280 ± 0.018	0.500 ± 0.010
	2*	0.360 ± 0.005	0.500 ± 0.020	0.600 ± 0.060	0.870 ± 0.050
37°C	0.5	nd	nd	nd	nd
	1	nd	nd	nd	nd
	2	tr	tr	tr	tr
4°C	0.5*	tr	-	tr	-
	1*	tr	-	0.020 ± 0.002	-
	2*	0.150 ± 0.020	-	0.250 ± 0.003	-

* - Albumin (albumin : FB1 = 1 M : 1 M)
tr - trace values
nd - nondetectable values

[^{14}C]FB1 was purchased from Dr. B. Miller (Canada). The spleen and thymus cells from BALB/c mice were cleared from erythrocytes by hypotonic lysis, washed 3 times, diluted in PBS, and 10x10^6 cells were placed into each culture tube. [^{14}C]FB1 was added either alone in final concentration 20 or 100 μg/ml/10x10^6 cells, or mixed with BSA (Sigma) mole/mole at same concentrations. Cells were incubated for 2h, 37°C or 4°C, and then washed 5 times in the cold PBS. The last supernatants were measured for control radioactivity, after that cells were lysed by 0.5 ml H_2Odest. Each sample was analysed for radioactivity by liquid scintillation counting in the "Beta-2" (Russia) under the low-background regime; "dpm" was determined using internal standardisation. The results are given as ±S.E. from 3 experiments.

The binding of FB1 by cells requires the availability of albumin which can be used as both a carrier and a ligand for double-receptor binding by cell membrane. Our data calls into questions the availability of specific receptors for FB1, however the distinct percentage of FB1 binding between spleen and thymus cells indicates the essential role of cell surface antigen composition for a similar interaction.

3. FB1 IS NOT MITOGENIC FOR MOUSE SPLEEN CELLS

Mitogenic action of FB1 has been shown only on Swiss 3T3 fibroblasts [51]. This line was also stimulated for DNA synthesis by sphingosine, sphingosine-1-phosphate, and ceramide [24,34,51,58]. The indicated sphingolipids have been found to be not mitogenic for other cell lines [22,23]. Increasing cAMP concentration in Swiss 3T3 cells caused both the MAP kinase activity and cell differentiation. The change from G1 to S phase of cell cycle arisen from cAMP augmentation in this cell line [11,33,65].

In normal murine cells, MHC class II molecules are associated with cAMP-generating signal transduction pathway [67]. These signals are involved in the regulation of cell proliferation. The binding of growth factor receptors leads to increasing free α-subunits of GTP-binding proteins with followed by activation of adenyl-cyclase and increasing cAMP concentration [9,11]. cAMP activates the PKA that phosphorylates Raf-1 and inhibits

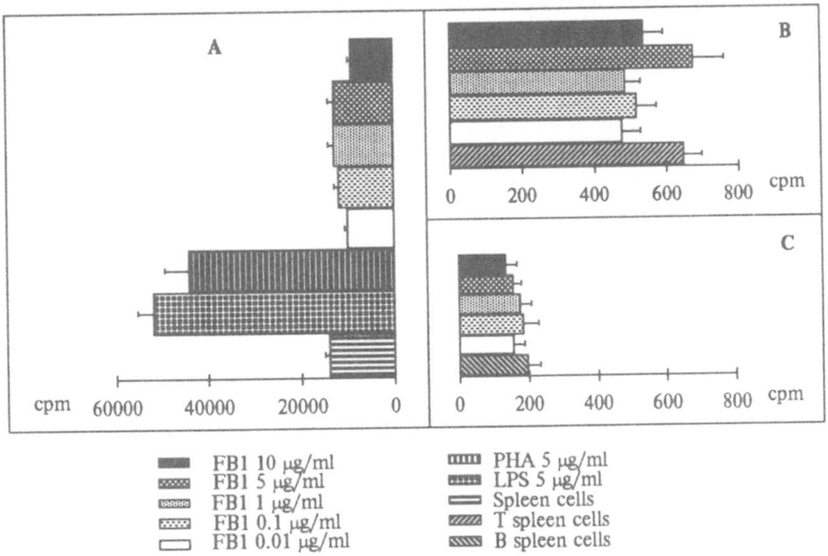

Figure 1. Effect of FB1 on [^{3}H]-thymidine incorporation into mouse spleen cells in vitro. For this experiment FB1 was purchased from Dr. Merrill, A.H., Jr, (Emory University, Atlanta, GA). For separation T and B spleen cells, we used anti-Thy1.2 Abs diluted 1:100 (obtained from Dr. Chervonsky, A.V., Cancer Research Center, Moscow, Russia) or anti-Ig serum (rabbit anti-mouse) diluted 1:10 (Caltag Laboratories) with Guinea pig complement (Sigma) in final concentration 1:10. Whole splenocyte suspension, T, and B cells were washed 3 times and plated at $2x10^5$/w and incubated either with different doses of FB1 or with known mitogens (PHA and LPS) at optimal concentrations. Control cells were untreated. After 48h incubation, 50 μCi of [^{3}H]thymidine (Amersham) was added to each well for 16h. Then the cells were transferred to filter paper, each filter disk was added to scintillation vials and counted. The results (cpm) are given as ±S.E. from 5 experiments.

Ras-pathway signal transduction [11,33,67,68]. Selective activation of cAMP-dependent protein kinase type I (cAKI), but not type II, is sufficient to mediate an inhibition of T cell replication induced through the TCR/CD3 complex [55]. It may be essential for a reverse reaction preventing excessive cell activation.

In the next part of our experiments *in vitro*, we added FB1 to the cells exposed to PHA or LPS. In this case, the stimulation of DNA synthesis was shown to decrease in comparison with PHA or LPS alone, but the stimulation was not completely blocked. The spleen cells exposed to forskoline have been found to revise the mitogenic action of FB1 (data is submitted for publication).

Opposite effects of the FB1 to DNA synthesis in normal murine splenocytes and Swiss 3T3 fibroblasts, as well as reduction of mitogenic action of PHA and LPS by FB1, led us to hypothesise that one of the mechanisms of FB1 effects on cell could involve an increase in cAMP concentration that leads to reduction of growth signals through Ras-pathway signal transduction.

4. FB1 MODULATES THE HUMORAL IMMUNE RESPONSE TO T-DEPENDENT ANTIGENS

The recognition of T-dependent antigen (in particular, SRBC) causes the clonal expansion of B cells into plasma cells resulting in the production of specific antibodies (Abs).

Table 3. Modulation of humoral immune response by a single i.p. administration of FB1

FB1 μg/g bw	PFC* per 10^6 Spleen cells		PFC* per 10^6 Thymic cells	
	BALB /c	SJL	BALB /c	SJL
FB1 0.05	130 ± 20	130 ± 25	80 ± 17	65 ± 12
FB1 0.25	85 ± 10	90 ± 10	80 ± 10	50 ± 8
FB1 1.0	76 ± 8	62 ± 5	20 ± 7	15 ± 7
FB1 2.5	32 ± 7	45 ± 8	6 ± 3	7 ± 4
Control: SRBC** 2x10^7 per mouse	120 ± 15	115 ± 13	15 ± 5	15 ± 5

* PFC - plaque forming cells
** SRBC - sheep red blood cells

Male BALB/c and SJL mice (8-9 weeks of age) were immunised i.p. by 2x10^7 SRBC. FB1 (Sigma) were injected in 0.5ml PBS simultaneously with Ag. Control (PBS alone), SRBC-, and SRBC plus FB1-treated mice were killed by cervical translocation on 4th day. The spleen or thymus cells cleared from erythrocytes by hypotonic lysis were added to Petri dishes (1x10^6 cells mixed with 50x10^6 SRBC per dish) covered by 0.4% agarose (Serva) and incubated for 1.5h at 37°C and 5%CO_2. Then Guinea pig complement in final concentration 1:10 was added to each dish for 1h. Plague forming cells (PFC) were counted in each dish and finally calculated as the number per 1x10^6 spleen cells. We treated three mice per each dose. The data received from 4 experiments.

The chemical or biological agent changing the level of antibody-secreting cells can be viewed as a immunomodulator. In our experiments, potential immune activity of FB1 was tested by single i.p. administration concurrent with suboptimal dose of SRBC.

The FB1 at the concentration 0.05 μg/g bw did not reduce the number of PFC. This dose has got light stimulant action, statistically insignificant. Doses of FB1 from 0.25 to 2.5 μg/g bw inhibited PFC in dose-dependent manner.

Under the same conditions, the sphingolipids, namely, sphingosine, phyto-sphinganine, and ceramide inhibited the humoral immune response. Both the sphingosine and phyto-sphinganine, at concentrations from 0.25 to 2.5 μg/g bw, have been found to reduce the number of PFC to 20.0±5.0% from control in a dose-independent manner. The ceramide at the concentration from 0.025 to 0.25 μg/g bw inhibited the PFC production in dose-dependent manner. Higher concentration of ceramide (0.5-10.0 μg/g bw) decline the PFC to 15.0±4.0% from control in dose-independent manner (data is submitted for publication).

FB1 has been found to affect the B cells in the thymus. The commonly accepted role of the small number of B cells in the thymus is the role of antigen presenting cells (APC) in the negative selection of the thymocytes, and they are not designed to secrete the antibodies to foreign antigens [17,52,59]. Change-over from APC to PFC could be essential for the role of thymic B cells in the clonal deletion of the $CD4^+CD8^+$ thymocytes.

5. FB1 CHANGES THE T CELL ANTIGEN EXPRESSION IN THE EARLY STAGE OF IMMUNE RESPONSE

Discovery of FB1 as stimulator of PFC in the thymus is closely related to the questions concerning both the apoptosis and modulation of T cell surface antigens.

Using the standard program for "MOP-Videoplan" (Reichert) we counted the number of thymus cells undergoing an apoptosis. The FB1 (i.p., 1 μg/g bw) increased the number of

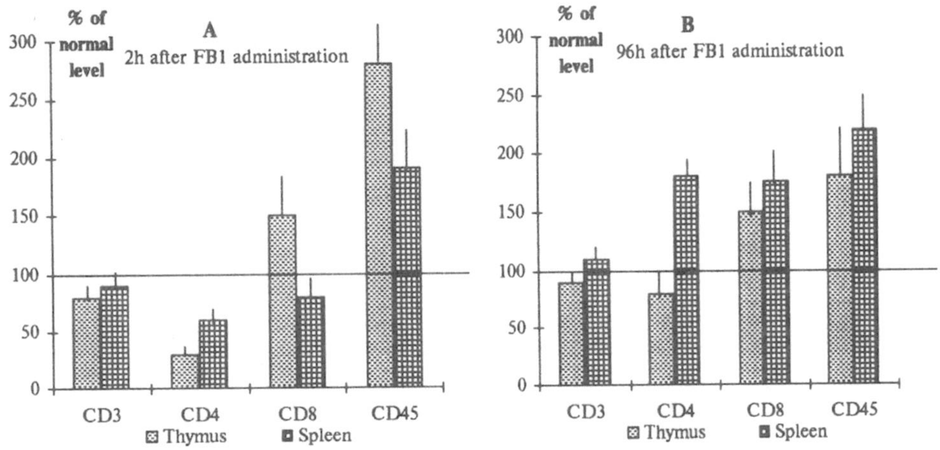

Figure 2. Modification of T cell surface antigen expression after a single i.p. administration of 1 μg FB1/g bw. Fluorochrome-conjugated mAb reactive with CD4(Leu-3), CD8(Leu-2), CD3(Leu-4), and CD45RA (Leu-18)Ag were from (Becton Dickinson), affinity-purified goat Ab against mouse Ig were from (Caltag Laboratories). Samples were analysed by flow cytometer FACScan (Becton Dickenson). FB1 (Sigma) in 0.25ml PBS was injected i.p. into BALB/c mice (8-9 weeks of age). Mice were killed by cervical translocation. Thymus and spleen cells cleared from erythrocytes were washed 3 times and incubated with fluorochrome-labeled antibodies in common use. The cell population was gated by size 5,000 events were measured. The fluorescence was analysed with excitation wavelength of 488 nm and an emission wavelength of 530 nm for FITC and 575 nm for PE. Data was expressed as percentage positive cells per gated event, and present as percentage of same CD-antigen expression on the cells obtained from mice without treatment. All results are given as the mean ± S.E.

apoptotic cells (AC) in the thymus subcortical area from 2.82±0.02 AC per 100 thymus cells, up to 3.80±0.03 AC after 6h, and up to 4.26±0.03 AC after 12h. This effect bore on mainly thymocytes incorporated into the thymic nurse cells.

The performance of T-dependent immune response leads to the functional maturation of all cells involved in this process. Each phase of maturation defines the basic characteristics of specific immune response. We tested the FB1 effects on T cell surface antigens both in the early stage and peak of humoral immune response to T-dependent Ag, as well as FB1 action alone.

The FB1 has been shown to increase CD8 and to decrease CD4 on the thymic cells in the early stage of immune response. CD4 and CD8 are glycoproteins expressed on both the thymocytes and mature T cells, and act as co-receptors for T cell receptor complex (TCR/CD3) [35]. CD4 and CD8 bind to monomorphic MHC class II and class I determinants, respectively [8,13]. Ag-induced binding of TCR/CD3 and CD4 or CD8 leads to activation of PTK $p56^{lck}$ (that closely linked with CD4/8), stimulates the tyrosine phosphorylation, and activates the PLD and PKC [7,8,26,36,57]. TCR/CD3 activation causes distinct results in the thymus and spleen. In the immature thymocytes, this activation leads to the apoptosis or selection. In the mature T cells, this activation causes either the proliferation or the anergy that depends on co-stimulative signal [66].

CD45 are the major membrane-associated phosphotyrosine phosphatases (PTP). There are 8 distinct CD45 isoforms that selectively expressed on the different cell types and subsets in the different stages of maturation. Association CD45 with CD2 or TCR/CD3 complex prevents these receptors from generating intracellular signals by dephosphorylating

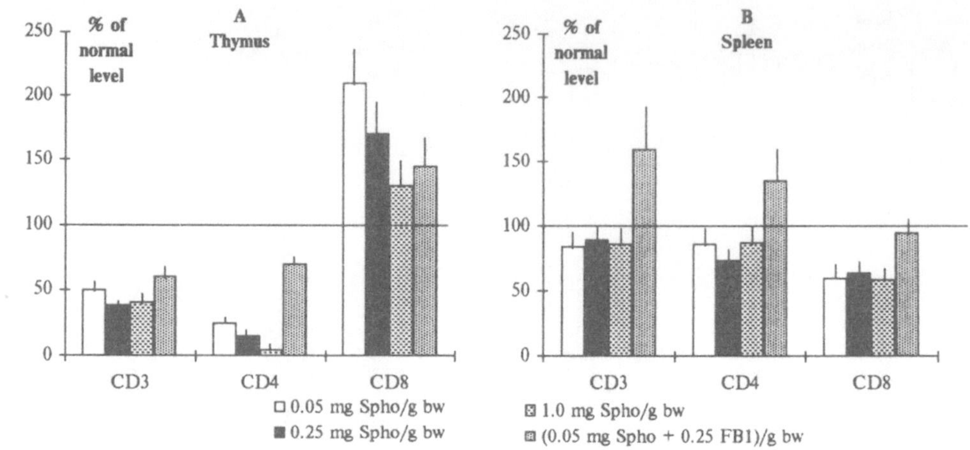

Figure 3. Modification of T cell surface antigen expression after a single i.p. administration of sphingosine (Spho). BALB/c mice (8-9 weeks of age) were i.p. injected either by D-sphingosine (Sigma) alone or simultaneously with FB1 in 0.25ml PBS. This method is identical to described for Fig.2.

one or more phosphoproteins involved in the signalling process [1,2,66,70]. MHC class I molecules do not depend on the membrane-bound CD45 [13].

The maturation of lymphocytes leads to loss of CD45RA expression and the augmentation of CD45RO expression followed by an increase of the proliferate capacity [42]. Increasing of CD45RO expression on immature $CD4^{+}CD8^{+}$ thymocytes leads to both the activation of the $p56^{lck}$ tyrosine kinase and an apoptosis, and loss of cortical thymocytes [42,44]. Binding of TCR and CD4/8 co-receptors induces the signal transduction through CD45 PTP.

In our experiments *in vivo*, FB1 caused the CD45 and CD8 increasing in the thymus that was correlated with augmentation of thymocyte apoptosis. The mature spleen T cells have been shown to be less susceptible to FB1 exposure, that could be explained by either T cell surface antigen composition (namely CD45 isoform linked by CD4/8 Ag) or the availability of the other receptors (for example, CD27, CD40, CD70 which are essential for T cell-dependent expansion of humoral immune response [5,12,39,47]). The CD Ag expression was distinguished by time after FB1 exposure, as well as by the concurrent present of Ag. FB1 changes the T cell surface Ag composition that can lead to disruption of immune response.

In our experiments *in vitro,* the addition of bacterial sphingomyelinase (EC 33.1.4.12. (Sigma), $10U/ml/10x10^6$ cells) to the spleen and thymus cells has been shown to modulate cell antigen expression: more than 2-fold increase in CD8 and CD4 in the thymus, 2 fold decrease in CD8 in the spleen. In same time, CD3 expression on both spleen and thymus cells was augmented (data is not shown).

Among the products generated by the catabolism of sphingomyelin, sphingosine displayed exactly the same effects as sphingomyelinase: they decrease both the cytosolic $[Ca^{2+}]$ and Ca^{2+} influx induced by the CD3 mAb [6].

CD3 and CD4 antigens on the spleen cells decreased moderately in dose-independent manner, while CD3 and especially CD4 on the thymic cells decreased more significantly. CD8 antigen on the thymic cells have been found to augment. Simultaneous i.p. injection of sphingosine and FB1 revised the CD-antigen expression on both spleen and thymus cells. This data calls into questions the strong correlation between immunological action of FB1 and disruption of sphingolipid metabolism. In any event, FB1 can directly modulate the T cell surface antigens in the early stage of cell recognition.

ACKNOWLEDGMENTS

The author is grateful Dr. Merrill, A. H., Jr.(Emory University School of Medicine, Atlanta, GA) for his irreplaceable help in the pursuance of research. We thank Dr. Zabotina, T.N. (Cancer Research Center, Moscow, Russia) for the help in the execution of flow cytometer analyses, and Dr. Zorin, S.N. (Institute of Nutrition, Moscow, Russia) for technician support in the measurements of radioactivity.

REFERENCES

1. Alexander, D., Shiroo, M., Robinson, A., Biffen, M., and Shivnan, E., 1992, The role of CD45 in T-cell activation - resolving the paradoxes?, *Immunol.Today*. 13(12):477-481.
2. Altin, J.G., Pagler, E. B., and Parish,C.R., 1994, Evidence for cell surface association of CD2 and LFA-1 (CD11a/CD18) on T lympocytes, *Eur.J.immunol.* 24(2):450-457.
3. Bacon, K.B., and Camp, R.D., 1990, Interleukin (IL)-8-induced *in vitro* human lymphocyte migration is inhibited by Cholera and Pertussis toxins and inhibitors of protein kinase C, *Biochem.Biophys.Res.* 169(3):1099-1104.
4. Borchardt, R.A., Lee, W.T., Kalen, A., Buckley, R.H., Peters, C., Schift,S., and Bell,R.M., 1994, Growth-dependent regulation of cellular ceramides in human T cells, *Biochim.Biophys.Acta.* 1212(3):327-336.
5. Bowman, M.R., Crimmins, M.A.V., Yetz-Aldape, J., Kris, R., Kelleher, K., and Herrmann, S., 1994, The cloning of CD70 and its identification as the ligand for CD27, *J.Immunol.* 152(4):1756-1761.
6. Breittmayer, J.P., Bernard, A., and Aussel, C., 1994, Regulation by sphingomyelinase and sphingosine of Ca2+ signals elicited by CD3 monoclonal antibody, thapsigargin, or ionomycin in the Jurkat T cell line, *J.Biol.Chem.* 269(7):5054-5058.
7. Cambier, J.C., 1992, Signal transduction by T- and B-cell antigen receptors: converging structures and concepts, *Current Opinion Immunol.* 4:257-264.
8. Cambier, J.C., and Jensen, W.A., 1994, The hetero-oligomeric antigen receptor complex and its coupling to cytoplasmic effectors, *Current Opinion Gen.Develop.* 4(1):55-63.
9. Chen, J., and Iyengar, R., 1994, Suppression of Ras-induced transformation of NIH 3T3 cells by activated G-α-s, *Science*. 263:1278-1281.
10. Chu, J.W., and Sharom, F.J. 1991, Effect of micellar and bilayer gangliosides on proliferation of interleukin-2-dependent lymphocytes, *Cell.Immunol.* 132(2):319-338.
11. Cook, S.J., and McCormick, F.,1994, Inhibition by cAMP of Ras-dependent activation of Raf, *Science*. 262:1069-1072.
12. Covey, L.R., Cleary, A.M., Yellin, M.J., Ware, R., Sullivan,G., Belko, J., Parker, M., Rothman,P., Chess, L., and Lederman, S., 1994, *Mol.Immunol.* 31(6):471-484.
13. Dasgupta, J.D., Granja, C.B., Yunis, E.J., and Relias,V., 1994, MHC class I antigens regulate CD3-induced tyrosine phosphorylation of proteins in T cells, *Int.Immunol.* 6(3):481-489.
14. Dombrink-Kurtzman, M.A., Bennett, G.A., and Richard, J.L., 1994, An optimized MTT bioassay for determination of cytotoxicity of fumonisins in Turkey lymphocytes, *J.ADAC Int.* 77(2):512-516.
15. Felding-Habermann,B., Igarashi,Y., Fenderson, B.A., Park, L.S., Radin, N.S., Inokuchi, J., Strassmann, G., Handa, K., and Hakomori, S., 1990, A ceramide analogue inhibits T cell proliferative response through inhibition of glycosphingolipid synthesis and enhancement of N,N-dimethylsphingosine synthesis, *Biochemistry*. 29(26):6314-6322.
16. Freitas, A.A., and Rocha, B.B., 1993, Lymphocyte lifespans: homeostasis, selection and competition, *Immunol.Today*. 14(1):25-29.
17. Fukuba,Y., Inaba, M., Taketani, S., Hitoshi, Y., Adachi, Y., Tokunaga, R., Inaba, K., Takatsu, K., and Ikehara, S., 1994, Functional analysis of thymic B cells, *Immunobiol.* 190(1-2):150-163.
18. Fuleihan, R., Spertini, F., Geha, R.S., and Chatila, T., 1992, Role of protein kinase activation in the induction of B cell adhesion by MHC class II ligands, *J.Immunol.* 149(6):1853-1858.

19. Gelderblom, W.C.A., Kriek, N.P.J., Marasas, W.F.O., and Thiel, P.G., 1991, Toxicity and carcinogenicity of the *Fusarium moniliforme* metabolite, fumonisin B1, in rats, *Carcinogenesis*. 12:1247-1251.
20. Gelderblom, W.C.A., Marasas, W.F.O., Vleggaar, R., Thiel, P.G., and Cawood, M.E., 1992, Fumonisins: isolation,chemical characterization and biological effects, *Mycopathologia*. 117:11-16.
21. Gelderblom, W.C.A., Cawood, M.E., Snyman, S.D., and Marasas, W.F.O., 1994, Fumonisin B1 dosimetry in relation to cancer initiation in rat liver, *Carcinogenesis*. 15(2):209-214.
22. Ghosh, T.K., Bian, J., and Gill, D.L., 1994, Sphingosine-1-phosphate generated in the endoplasmic reticulum membrane activates release of stored calcium, *J.Biol.Chem*. 269(36):22628-22635.
23. Hannum,Y.A., and Linardic, C.M., 1993, Sphingolipid breakdown products: anti-proliferative and tumor-suppressor lipids, *Biochim.Biophys.Acta*. 1154(3-4):223-236.
24. Hauser, J.M.L., Buehrer, B.M., and Bell, R.M., 1994, Role of ceramide in mitogenesis induced by exogenous sphingoid bases, *J.Biol.Chem*. 269(9):6803-6809.
25. Hui, E.K. and Yung, B.Y., 1993, Cell cycle phase-dependent effect of retinoic acid on the induction of granulocytic differentiation in HL-60 promyelocytic leukemia cells. Evidence for sphinganine potentiation of retinoic acid-induced differentiation, *FEBS Lett*. 318(2):193-199.
26. Iwashima, M., Irving, B.A., van Oers, N.S.C., Chan, A.C., and Weiss, A., 1994, Sequential interactions of the TCR with two distinct cytoplasmic tyrosine kinases, *Science*. 263:1136-1139.
27. Jordan, M.L., and Wright, J. 1991, Evidence that protein kinase C regulates allosensitized T lymphocyte function, *J.Surg.Res*. 50(6):569-573.
28. Lavie,Y., Blusztain, J.K., Liscovitch, M., 1994, Formation of endogenous free sphingoid bases in cells induced by changing medium conditions, *Biochim.Biophys.Acta*. 1220(3):323-328.
29. Marasas, W.F.O., Wehner, F.C., van Rensburg, S.J., and van Schalkwyk, D.J., 1981, Mycoflora of corn produced in human esophageal cancer areas in Transkei, southern Africa, *Phytopathology*. 71:792-796.
30. Marasas, W.F.O., Jaskiewicz, K., Venter, F.S., and van Schalkwyk, D.J., 1988, *Fusaruim moniliforme* contamination of maize in oesophageal cancer areas in Transkei, *S.Afr.med.J*. 74:110-114.
31. Marijanovic, D.R., Holt, P., Norred, W.P., Bacon, C.W., Voss, K.A., Stancel, P.C., and Ragland, W.L., 1991, Immunosuppressive effects of *Fusarium moniliforme* corn cultures in chickens, *Poultr.Sci*. 70:1895-1901.
32. Martinova, E.A., Soloviev, A., Khrenov, A., Zabotina, T., and Alesenko, A.V., 1995, Fumonisin B1 modulates sphingomyelin cycle product levels and the expression of CD3 receptors in immunocompetent organs, *Biochemistry (Moscow)*. 60(4): 461-465.
33. Marx, J., 1993, Two major signal pathways linked, *Science*. 262:988-990.
34. Merrill, A.H., Jr., Wang, E., Gilchrist, D.G., and Riley, R.T., 1993, Fumonisins and other inhibitors of *de novo* sphigolipid biosynthesis, *Adv.Lipid Res*. 26:215-234.
35. Nakayama, K.i., Nakayama, K., Nagishi, I., Kuida, K., Louie, M.C., Kanagawa, O., Nakauchi, H., and Loh, D.Y., 1994, Requirement for CD8-β chain in positive selection of CD8-lineage T cells, *Science*. 263:1131-1133.
36. Nel, A.E., Taylor, L.K., Kumar, G.P., Gupta, S., Wang, S.C.-T., Williams, K ., Liao, O., Swanson, K., and Landreth, G.E., 1994, Activation of a novel serine/threonine kinase that phosphorylates C-*fos* upon stimulation of T and B lymphocytes via antigen and cytokine receptors, *J.Immunol*. 152(9):4347-4357.
37. Nelson, P.E., Plattner, R.D., Shackelford, D.D., and Desjardins, A.E., 1991, Production of fumonisins by *Fusarium moniliforme* strains from various substrates and geographic areas, *Appl.environ.Microbiol*. 57:2410-2412.
38. Nelson, P.E., Plattner, R.D., Shackelford, D.D., and Desjardins, A.E., 1992, Fumonisin B1 production by *Fusarium* species other than *F.moniliforme* in section *Liseola* and by some related species, *Appl.environ.Microbiol*. 58:984-989.
39. Noelle, R., and Snow, E.C., 1992, *Current Opin.Immunol*. 4:333-337.
40. Norred, W.P., 1993, Fumonisins - mycotoxins produced by *Fusarium moniliforme*, *J.Toxicol.environ.Health*. 38:309-328.
41. Norred, W.P., Plattner, R.D., Chamberlain, W.J., 1993, Distribution and excretion of [14C]Fumonisin B1 in male Sprague-Dawley rats, *Nat.Toxins*. 1(6):341-346.
42. Ong, C.J., Chui, D., Teh, H.-S., and Marth, J.D., 1994, Thymic CD45 tyrosine phosphatase regulates apoptosis and MHC-restricted negative selection, *J.Immunol*. 152(8):3793-3805.
43. Osweiler, G.D., Kehrli, M.E., Stabel, J.R., Thurston, J.R., Ross, P.F., and Wilson, T.M., 1993, Effects of fumonisin-contaminated corn screening on growth and health of feeder calves, *J.Anim.Sci*. 71(2):459-466.
44. Paramithiotis, E., Tkalec, L., and Ratcliffe, M.J.H., 1991, High levels of CD45 are coordinately expressed with CD4 and CD8 on avian thymocytes, *J.Immunol*. 147(11):3710-3717.

45. Penninger, J., Kishihara, K., Molina, T., Wallace, V.A., Timms, E., Hedrick, S.M., and Mak, T.W. 1993, Requirement for tyrosine kinase p56lck for thymic development of transgenic γδ T cells, *Science.* 260:358-361.
46. Qureshi, M.A., and Hagler, W.M., 1992, Effects of fumonisin B1 exposure on chicken macrophage function *in vitro*, *Poultr.Sci.* 71:104-112.
47. Renard, N., Duvert. V., Blanchard, D., Banchereau, J., and Saeland, S., 1994, Activated CD4+ T cells induce CD40-dependent proliferation of human B cell precursors, *J.Immunol.* 152(4):1693-1701.
48. Rheeder, J.P., Marasas, W.F.O., Thiel, P.G., Sydenham, E.W., Shephard, G.S., and van Schalkwyk, D.J., 1992, *Fusarium moniliforme* and fumonisins in corn in relation to human esophageal cancer in Transkei, *Phytopathology.* 82:353-357.
49. Riley, R.T., Norred, W.P., and Bacon, C.W., 1993, Fungal toxins in foods: recent concerns, *Annu.Rew.Nutr.* 13:167-189.
50. Sadahira,Y., Zheng, M., Ruan, F., Hakomori, S., and Igarashi, Y., 1994, Sphingosine-1-phosphate inhibits extracellular matrix protein-induced haptotactic motility but not adhesion of B16 mouse melanoma cells, *FEBS Lett.* 340(1-2):99-103.
51. Schroeder, J.J., Crane, H.M., Xia, J., Liotta, D.C., and Merrill, A.H.,Jr., 1994, Disruption of sphingolipid metabolism and stimulation of DNA synthesis by fumonisin B1. A molecular mechanism for carcinogenesis assosiated with *Fusarium moniliforme*, *J.Biol.Chem.* 269(5):3475-3481.
52. Sebzda, E., Wallace, V.A., Mayer, J., Yeung, R.S.M., Mak, T.W., and Ohashi, P.S., 1994, Positive and negative thymocyte selection induced by different concentrations of a single peptide, *Science*, 263:1615-1618.
53. Shephard, D.S., Thiel, P.G., Sydenham, E.W., and Alberts, J.F., 1994, Biliary excretion of the mycotoxin fumonisin B1 in rats, *Food Chem.Toxicol.* 32(5):489-491.
54. Shephard, D.S., Thiel, P.G., Sydenham, E.W., Alberts, J.F., and Cawood, M.E., 1994, Distribution and excretion of a single dose of the mycotoxin fumonisin B1 in a non-human primate, *Toxicon.* 32(6):735-741.
55. Skalhegg, B.S., Tasken, K., Hansson, V., Huitfeldt, H.S., Jahnsen, T., and Lea, T., 1994, Location of cAMP-dependent protein kinase type I with the TCR-CD3 complex, *Science.* 263:84-87.
56. Spertini,F., Chatila, T., and Geha, R.S., 1992, Engagement of MHC class I molecules induces cell adhesion via both LFA-1-dependent and LFA-1-independent pathways, *J.Immunol.* 148(7):2045-2049.
57. Stewart, S.J., Cunningham, G.R., and House, F.S., 1993, Activation of phospholipase D following perturbation of the human T lymphocyte antigen receptor/CD3 complex is dependent upon protein kinase C, *Cell Signal.* 5(3):315-323.
58. Su, Y., Rosenthal, D., Smulson, M., and Spiegel, S., 1994, Sphingosine 1-phosphate, a novel signaling molecule, stimulates DNA binding activity of AP-1 in quiescent Swiss 3T3 fibroblasts, *J.Biol.Chem.* 269(23):16512-16517.
59. Swat, W., von Boehmer, H., and Kisielow, P., 1994, Central tolerance: clonal deletion or clonal arrest?, *Eur.J.Immunol.* 24(2):485-487.
60. Sydenham, E.W., Thiel, P.G., Marasas, W.F.O., Shephard, G.S., van Schalkwyk, D.J., and Koch, K.R., 1990, Natural occurrence of some *Fusarium* mycotoxins in corn from low and high esophageal cancer prevalence areas of the Transkei, southern Africa, *J.agric.Food Chem.* 38:1900-1903.
61. Sydenham, E.W., Shephard, G.S., Thiel, P.G., Marasas, W.F.O., and Stockenstrom, S., 1991, Fumonisin contamination of commercial corn-based human foodstuffs, *J.agric.Food Chem.* 39:2014-2018.
62. Sydenham, E.W., Marasas, W.F.O., Shephard, G.S., Thiel, P.G., and Hirooka, E.Y., 1992, Fumonisin concentrations in Brazilian feeds associated with field outbreaks of confirmed and suspected animal mycotoxicoses, *J.agric.Food Chem.* 40:994-997.
63. Thiel, P.G., Marasas, W.F.O., Sydenham, E.W., Shephard, G.S., Gelderblom, W.C.A., and Nieuwenhuis,J.J., 1991, Survey of fumonisin production by *Fusarium* species, *Appl.environ.Microbiol.* 57:1089-1093.
64. Voss, K.A., Chamberlain, W.J., Bacon, C.W., and Norred, W.P., 1993, A preliminary investigation on renal and hepatic toxicity in rats fed purified fumonisin B1, *Nat.Toxins.* 1(4):222-228.
65. Wang, D.-j., Huang, N.-n., Heller, E.J., and Heppel, L.A., 1994, A novel synergistic stimulation of Swiss 3T3 cells by extracellular ATP and mitogens with opposite effects on cAMP levels, *J.Biol.Chem.* 269(24):16648-16653.
66. Weiss, A., and Littman, D.R., 1994, Signal transduction by lymphocyte antigen receptors, *Cell.* 76(2):263-274.
67. Woodrow, M.A., Rayter, S., Downward, J., and Cantrell, D.A., 1993, p27ras function is important for T cell antigen receptor and protein kinase C regulation of Nuclear factor of activated T cells, *J.Immunol.* 150(9):3853-3861.

68. Wu, J., Dent, P., Jelikek, T., Wolfman, A., Weber, M.J., and Sturgill, T.W., 1993, Inhibition of the EGF-activated MAP kinase signaling pathway by Adenosine 3'5'-monophosphate, *Science.* 262:1065-1069.
69. Yoshizawa, T., Yamashita, A., and Luo, Y., 1994, Fumonisin occurrence in corn from high- and low-risk areas for human esophageal cancer in China, *Appl.Environ.Microbiol.* 60(5):1626-1629.
70. Zapata, J.M., Pulido, R., Acevedo, A., Sanchez-Madrid, F., and de Landazuri, M.O., 1994, Human CD45RC specificity. A novel marker for T cells at different maturation and activation stages, *J.Immunol.* 152(8):3852-3861.

BIOCHEMICAL STUDIES ON THE EFFECT OF *Plotosus lineatus* CRUDE VENOM (*in Vivo*) AND ITS EFFECT ON EAC-CELLS (*in Vitro*)

Fawzia A. Fahim,[1] Essam A. Mady,[1] Samira M. Ahmed,[2] and M. A. Zaki[2]

[1] Department of Biochemistry
Faculty of Science
Ain Shams University, Cairo, Egypt
[2] National Institute of Oceanography and Fisheries, Egypt

INTRODUCTION

The oriental catfish, *Plotosus lineatus*, is one of the most dangerous venomous fishes known which occasionally causes death (Halstead, 1978). This fish is responsible for approximately many hundreds of envenomation to fishermen during handling in Suez Gulf and Red Sea. The envenomation, from *P. lineatus*, resulted in severe pain, numbness, fever, weakness, nausea, local paralysis and dizziness. In extreme cases there may be a massive edema involving the entire limb and gangrene of the area around wounds (Shiomi *et al.*, 1986, 1987 and 1988 and Zerman, 1989).

Al-Hassan *et al.* (1985a & b) reported that the Arabian Gulf catfish, *Arius thallassinus*, possesses a proteinaceous toxin in the skin secretion, together with a venom gland. Shiomi *et al.* (1986) also reported that the skin secretion of *P. lineatus* contains at least one hemolysin, two lethal factors and two edema-forming factors but little is known about the chemical properties of the venom.

The interest for snake venoms in relation to malignant tumors goes back as far as 1911 (Grunbaum and Grunbaum). The anti-tumor effect of a proposed natural product is mostly determined firstly by screening test. Although it was suggested that no single tumor system could be expected to select all useful agents, but the Ehrlisch ascites tumors are preferred due to their enhanced sensitivity and the variability of their growth which is independent of the anti-tumor agents (Sellei *et al.*, 1970 and Goldin *et al.*, 1974). Such studies were thereafter dropped for a long while, until Fahim *et al.* (1988) reported 56% inhibition in tumor growth and 90% elongation of life in rats bearing EAC-cells using fraction III of *N. nigricollis* venom.

This work aimed to study some biochemical and haematological effects of two sublethal doses of the mixed crude venom of *P. lineatus* (LD_{10} and 0.1 LD_{10}) using rats as model animal, as this fish inhibits Suez Canal waters and no studies were reported on its

Natural Toxins II, Edited by B. R. Singh and A. T. Tu
Plenum Press, New York, 1996

effects, and also aimed to elucidate the effect of this mixed venom on the inhibition of EAC-cells *in vitro*.

MATERIALS AND METHODS

Catfish

Specimens of *P. lineatus* were collected along the coast of the Suez city during the Winter 1992/93. The specimens were then transported alive to our laboratory where they were maintained in an aerated aquaria till extraction of the venom (within 6 hr. post-catch).

Extraction of the Venom

The crude venom was extracted from the skin secretion and venom glands according to the method of Shiomi *et al.* (1986). Protein content of the crude mixed venom was determined (Weichelbaum, 1946) and was found to be 21.5±1 mg protein/ml.

Toxicity Studies

Fifty adult male Swiss albino rats of mean weight 100±10 g were used for the determination of the median lethal dose (LD_{50}) (Reed and Meuench, 1938). Results are shown in Table (1).

Animal Studies

A total number of 80 adult male Swiss albino rats mean weight 100±10 g were used in this study. The animals were maintained on standard pellets purchased from the "Nile Co. for Oils and Soap" (Cairo), and supplemented with fresh lettuce and carrots, water was supplied *ad libitum*. Rats were assorted 4/cage and divided at random into 3 main groups as follows:

GrI NC. Normal control (16 rats) injected intraperitoneally (i.p.) with 0.4 ml saline/100 g rat.

GrII LD_{10}. Included 32 rats, each received a single i.p. LD_{10} (0.6 mg/100 g rat) injection of the crude venom in 0.4 ml saline.

GrIII 0.1 LD_{10}. Included 32 rats, each received a single i.p. 0.1 LD_{10} (0.06 mg/100 g rat) injection of the crude venom in 0.4 ml saline.

Eight rats from each envenomated group (GrII and GrIII) were sacrificed by decapitation after 15, 30, 60 and 360 min. post-injection and only 4 rats from the control non-treated animals.

Collection of Blood Serum and Liver Tissue Preparation

Blood was collected at the end of the experimental treatment period of each group (15, 30, 60 and 360 min) and the individual serum samples, then stored at -20°C until assayed.

Liver organs were dissected, washed in ice-cold isotonic saline, blotted dry then cut into 2 weighted portions. One portion was dropped into a test tube containing 30% KOH for glycogen determination. The other portion was homogenized in ice-cold isotonic saline with a glass homogenizer for 2 min. The liver homogenate was then centrifuged at 3000 r.p.m. for 10 min. to give a final concentration of 5% (w/v) for the determination of enzymes activities.

Table 1. Determination of LD_{100} and LD_{50} of the crude venom of the oriental catfish, *Plotosus lineatus*

Dose mg/100 g Rat	No of Animals	Survivals (S)	Death (D)	Total (S)	Total (D)	% Mortility
2.637	10	6	4	13	4	23.5
3.164	10	4	6	7	10	58.8
3.797	10	2	8	3	18	85.7
4.556	10	1	9	1	27	96.4
5.468	10	0	10	0	37	100

Reed & Meuench. (1938).

Increasing Factor: 1.2
LD_{100} = 5.468 mg/100 g rat
LD_{50} = 3.010 mg/100 g rat
LD_{10} = 0.600 mg/100 g rat
0.1 LD_{10} = 0.060 mg/100 g rat

Biological Assays

Blood clotting time and prothrombin time were estimated immediately according to the methods of Martin (1983) and Quick (1940), respectively. Results are shown in Table(2).

Serum samples for all groups were subjected to the determination of glucose (Trinder, 1969), total lipids (Knight *et al.*, 1972), total bilirubin (Jenderassik and Grof, 1983), AST and ALT activities (Reitman and Frankle, 1957), GGT activity (Szasz, 1969) and ALP activity (Belfield and Goldberg, 1971). Liver glycogen content was assessed in the liver digest according to the method of Seifter *et al.* (1950).

EAC-Cells *in Vitro* Study

EAC-cells were kindly supplied from the National Cancer Institute, Cairo University. The cells were grown on the basal nutritional medium RPMI-1640 containing 1% gentamicin, 0.4% L-glutamine and 10% fetal bovine serum. Growth was continued till the cell density reached $1X10^6$ cells/cm^2 in sterile tissue culture flasks (Falcon).

The cells were thereafter incubated with 5 increasing doses of the crude toxin as follows: 0.01, .05, 0.1, 0.5 and 1 mg in CO_2 incubator (5%) at 37°C. after 24 h and 1 week, the cells were stained with trypan blue and their viability was examined microscopically.

Statistical Analysis

The statistical significance of the experimental results was determined by the Student's t-test (Fisher and Yates, 1943). For all analysis, $p<0.05$ was accepted as a significant probability level.

RESULTS

Table (1) shows the lethal dose (LD_{100}), the median lethal dose (LD_{50}) and both the sublethal doses (LD_{10}) & 0.1 LD_{10}) of the crude mixed venom as 5.47, 3.01, 0.6 and 0.06 mg/100 g rat, respectively.

Table (2) illustrates very highly significant prolongation in blood clotting time and prothrombin time 60 min. post-injection (p.i.)that amounted to 74% and 87% ($p<0.001$) in animals of GrII and 47% and 59% ($p<0.001$) in animals of GrIII, respectively.

Table (3) shows very highly significant increase in serum glucose, total lipids and total bilirubin in animals of the two envenomated groups (II & III) 60 min. p.i. Serum total protein showed variable changes, as it shows very highly significant increase (49%, $p<0.001$) in GrII 60 min p.i. met with a slight decrease in GrIII 15 min. p.i. Urea increased significantly in GrII (27%, $p<0.001$) while it shows non significant change in GrIII. Potassium shows highly significant increase 15, 30, 60 and 360 min. p.i. in the two envenomated groups II and III.

Table (4) shows a very significant increase in the activity of serum LDH, AST, ALT, GGT, ALP and pseudo choline esterase activities in animals of GrII and to a lesser extent in GrIII.

Table (5) illustrates a significant reduction in the liver glycogen content of animals of the two envenomated groups II & III (75% and 76%, $p<0.001$). Hepatic LDH activity shows highly significant increase (80% and 21%, $p<0.001$ respectively). Liver GGT activities shows the highest significant increase 30 and 60 min. p.i. in animals of both envenomated groups ($p<0.001$) followed by gradual decrease till the end of the experimental period (6 hr). Liver AST and ALT activities show very highly significant decrease (60% and 26%, respectively, $p<0.001$) 360 min. p.i. in animals of GrII. Group III shows an increase in AST activity (30%, 30 min. p.i.) and ALT non significant changes. Hepatic ALP shows very highly significant reduction (56%, 60 min p.i. in animals of GrII) and non significant change in animals of GrIII.

Table (6) shows the effect of the crude toxin on the cell viability of EAC-cells *in vitro*, where the 5 tested doses exerted a marked effect on the inhibition of cell growth. The most effective dose (1 mg) caused 41% inhibition in the total count of the cells. The same effect is reflected on the viability of the incubated cells as it resulted in about 50% reduction in cell viability.

DISCUSSION

The lethal dose (LD_{100}), the median lethal dose (LD_{50}) and the sublethal doses (LD_{10}) & 0.1 LD_{10}) of the crude mixed venom of the oriental catfish *P. lineatus* are 5.47, 3.01, 0.6 and 0.06 mg/100 g rat, respectively (Table 1). Shiomi *et al.* (1986) recorded a LD_{50} of 2.90 mg/100 g mouse of the crude venom of the same fish. However our results differ from those recorded by Datta *et al.* (1982) who reported that the LD_{50} of the Indian catfish, *Heteropneustes fossilis*, is 0.501 mg/100 g mouse, which may be due to difference in the species of the catfish, route of administration, extraction procedure or animal species.

The total protein content of the crude venom of the *P. lineatus* was 21.5±1 mg protein.ml which is in line with that of Shiomi *et al.* (1986) who reported a mean of 18.7 mg protein/ml of the crude venom of the same fish and with that of Al-Hassan *et al.* (1982) who reported a range of 20-30 mg protein/ml of the crude venom of the *A. thallassinus*.

Seven main fractions of the crude venom of *P. lineatus* were collected on Sephadex G_{200} using sodium phosphate buffer pH7.4, column 40X2 cm, flow rate 10 ml/h (Fig. 1). This result differs from that observed by Shiomi *et al.* (1986) who recorded 5 fractions only using Sephadex G_{75} and column dimensions of 50X3 cm. This difference may be due to the difference in the experimental conditions.

The blood clotting time and plasma prothrombin time prolongation in GrII and GrIII 60 min. p.i. (Table 2) may be attributed to the hemolysin factors present in the crude venom (Shiomi *et al.*, 1986; Al-Lahham *et al.*, 1987; Al-Hassan *et al.*, 1987b and Khoo *et al.*, 1992).

Table 2. Blood clotting time and prothrombin time of rats injected with LD_{10} & 0.1 LD_{10} of the crude venom of *Plotosus lineatus* compared with GrI N.C.

Parameter	Group I	Group II (LD_{10})				Group III (0.1 LD_{10})			
	Normal	Time after (i.p) injection				Time after (i.p) injection			
	Control	15 min.	30 min.	60 min.	360 min.	15 min.	30 min.	60 min.	360 min.
Blood clotting time (sec) Mean ± SD	49±4.3	75.1±3.2	80.0±3.3	85.4±6.2	54.4±1.6	56.1±5.1	64.1±5.2	71.8±6.3	52.0±1.6
%Change	-	53	63	74	11	15	31	47	6
P<	-	0.001	0.001	0.001	0.01	0.01	0.001	0.001	N.S.
Prothrombin time (sec) Mean ± SD	14.5±2.3	21.1±1.8	24.0±2.2	27.1±3.4	16.6±1.3	18.9±1.4	20.1±1.2	23.0±1.0	15.1±1.5
% Change	-	46	66	87	15	30	40	59	4
P<	-	0.001	0.001	0.001	0.05	0.01	0.001	0.001	N.S.

Table 3. Serum glucose, total lipid, total protein, urea, potassium and total bilirubin in rats of all groups

Parameter	Group I	Group II (LD_{10})				Group III (0.1 LD_{10})			
	Normal Control	Time after (i.p) injection				Time after (i.p) injection			
		15 min.	30 min.	60 min.	360 min.	15 min.	30 min.	60 min.	360 min.
Glucose(mg/dl) Mean ± SD % Change *P*<	96.1±7.7 - -	146.4±3.0 52 0.001	163.8±4.7 70 0.001	200.9±6.0 110 0.001	89.5±4.7 -7 N.S.	106.3±4.3 11 0.01	110.5±7.4 15 0.01	118.4±6.9 23 0.001	90.4±6.0 -5 N.S.
Total lipid (mg/dl) Mean ± SD % Change *P*<	309±12.8 - -	322.0±9.1 4 0.05	397±27.1 28 0.001	458±10.5 48 0.001	342±14.6 10 0.001	315±21.9 2 N.S.	326±19.1 6 N.S.	353±32.2 14 0.01	313±7.2 1 N.S.
Total protein (g/dl) Mean ± SD % Change *P*<	7.6±0.4 - -	7.9±0.7 4 N.S.	8.9±0.5 17 0.001	11.3±1.0 49 0.001	7.8±0.7 3 N.S.	6.9±0.6 -9 0.02	7.0±0.7 -8 N.S.	8.2±0.7 8 N.S.	7.4±0.5 -3 N.S.
Urea (mg/dl) Mean ± SD % Change *P*<	31.0±3.1 - -	32.6±3.4 3 N.S.	34.1±2.9 10 N.S.	39.4±4.0 27 0.001	28.1±3.3 -9 N.S.	30.8±3.9 -1 N.S.	30.3±4.5 -2 N.S.	28.6±2.9 -8 N.S.	31.3±4.3 0.03 N.S.
Potassium (mEq/l) Mean ± SD % Change *P*<	5.6±0.3 - -	10.1±0.2 80 0.001	9..7±0.5 73 0.001	8.9±0.9 59 0.001	9.0±0.4 61 0.001	6.2±0.7 11 0.05	7.0±1..0 25 0.01	7.9±0.5 41 0.001	6.8±0.8 21 0.01
Total bilirubin Mean ± SD % Change *P*<	0.331±0.01 - -	0.446±0.03 35 0.001	0.569±0.01 72 0.001	0.647±0.01 96 0.001	0.360±0.02 9 0.02	0.401±0.02 22 0.001	0.463±0.03 40 0.001	0.541±0.02 63 0.001	0.335±0.0 07 1 N.S.

Table 4. Serum LDH, AST, ALT, GGT, ALP and Ch.E activities in rats of all groups

Parameter	Group I	Group II (LD_{10})				Group III (0.1 LD_{10})			
	Normal	Time after (i.p) injection				Time after (i.p) injection			
	Control	15 min.	30 min.	60 min.	360 min.	15 min.	30 min.	60 min.	360 min.
LDH (U/ml) Mean ± SD % Change *P*<	1044±58.6 - -	1262±93.1 21 0.001	1616±173 55 0.001	2338±141 125 0.001	9724±141 832 0.001	1056±72.9 2 N.S.	1272±74.3 22 0.001	1503±107 44 0.001	1256±76.3 20 0.001
AST (U/ml) Mean ± SD % Change *P*<	187.4±29 - -	282.8±18 51 0.001	290.4±14 55 0.001	307.8±16 64 0.001	571.5±25 205 0.001	201.0±21 7 N.S.	218.9±17 17 0.05	229.5±15 22 0.01	194.8±20 4 N.S.
ALT (U/ml) Mean ± SD % Change *P*<	27.6±1.9 - -	29.8±3.2 8 N.S.	32.1±1.7 16 0.001	32.3±2.2 17 0.001	37.1±4.4 37 0.001	27.2±2.0 -1 N.S.	27.1±3.0 -2 N.S.	30.2±2.7 9 N.S.	27.6±2.6 0 N.S.
GGT (U/l) Mean ± SD % Change *P*<	4.2±0.8 - -	4.3±0.6 2 N.S.	4.4±0.5 5 N.S.	4.5±0.6 7 N.S.	5.7±0.5 36 0.001	4.5±0.5 7 N.S.	4.8±0.4 14 N.S.	4.8±0.6 14 N.S.	4.2±1.0 0 N.S.
ALP (U/l) Mean ± SD % Change *P*<	97.7±4.9 - -	101.9±5.0 4 N.S.	105.6±7.0 8 0.05	119.0±7.7 22 0.001	124.6±7.5 28 0.001	100.1±5.8 3 N.S.	108.1±4.6 11 0.001	121.7±6.0 25 0.001	103.1±5.8 6 N.S.
Ch.E (U/l) Mean ± SD % Change *P*<	71.2±6.2 - -	244.0±24.8 243 0.001	319.0±15.2 348 0.001	347.0±24.1 388 0.001	504.0±32.1 608 0.001	82.5±7.6 16 0.01	98.8±11.2 39 0.001	87.1±4.8 22 0.001	72.8±4.3 3 N.S.

Table 5. Liver glycogen, LDH, AST, ALT, GGT and ALP activities in rats of all groups

Parameter	Group I	Group II (LD_{10})				Group III (0.1 LD_{10})			
	Normal	Time after (i.p) injection				Time after (i.p) injection			
	Control	15 min.	30 min.	60 min.	360 min.	15 min.	30 min.	60 min.	360 min.
Glycogen (g/100g) Mean ± SD % Change *P*<	3.1±0.6 - -	2.7±0.8 -12 N.S.	1.8±0.5 -40 0.001	1.2±0.2 -61 0.001	0.8±0.1 -75 0.001	2.8±0.7 -10 N.S.	2.6±0.4 -16 N.S.	1.8±0.4 -42 0.001	0.7±0.2 -76 0.001
LDH (U/g) Mean ± SD % Change *P*<	451.3±37.4 - -	532.0±42.2 18 0.01	553.0±36.4 23 0.001	621.0±31.7 38 0.001	813.0±34.8 80 0.001	460.0±39.5 2 N.S.	509.0±33.7 13 0.01	545.0±34.9 21 0.001	494.0±44.8 10 N.S.
AST (U/g) Mean ± SD % Change *P*<	78.6±5.3 - -	103.1±11 31 0.001	93.1±4.7 19 0.001	82.2±9.8 5 N.S.	31.1±3.8 -60 0.001	73.6±7.1 -6 N.S.	75.1±7.0 -4 N.S.	101.8±8.3 30 0.001	90.4±11.9 15 0.05
ALT (U/g) Mean ± SD % Change *P*<	22.1±1.6 - -	36.3±3.9 64 0.001	24.2±4.4 10 N.S.	17.2±2.0 -22 0.001	16.4±3.1 -26 0.001	24.9±3.6 13 N.S.	21.4±2.1 -3 N.S.	22.0±2.1 -1 N.S.	22.0±2.4 -1 N.S.
GGT (U/g) Mean ± SD % Change *P*<	24.0±2.4 - -	25.6±0.1 7 N.S.	96.1±6.3 302 0.001	68.3±5.0 186 0.001	42.4±1.0 77 0.001	25.3±2.0 5 N.S.	46.9±3.2 95 0.001	36.6±3.7 53 0.001	33.6±3.7 40 0.001
ALP (U/g) Mean ± SD % Change *P*<	1.35±0.4 - -	1.48±0.3 10 N.S.	1.56±0.1 16 N.S.	0.59±0.1 -56 0.001	1.15±0.2 -15 N.S.	1.51±0.3 12 N.S.	1.55±0.3 15 N.S.	1.0±0.1 -26 N.S.	1.33±0.2 -2 N.S.

Table 6. Effect of the crude toxin of *Plotosus lineatus* on EAC-cells *in vitro*

A- Total count

Dose (mg)	24 hr			1 week		
	Control	After	% Change	Control	After	% Change
	$8X10^6$ cells/ml			$60X10^6$		
1		$6.0X10^6$	25%		$35.4X10^6$	41%
0.5		$6.6X10^6$	18%		$37.2X10^6$	38%
0.1		$6.9X10^6$	14%		$43.2X10^6$	22%
0.05		$7.2X10^6$	10%		$52.2X10^6$	13%
0.01		$7.4X10^6$	8%		$54.6X10^6$	9%

B- Viability

	Control		24 h		Control		1 week	
Dose	T	D	T	D	T	D	T	D
	8*	0.6*			60*	5*		
1			6*	1.3*			35.4*	17.7*
0.5			6.6*	1.22*			35.7*	14.9*
0.1			6.9*	1.03*			43.2*	15.6*
0.05			7.2*	0.94*			52.2*	13.1*
0.01			7.4*	0.66*			54.2*	7.64*

T total
D dead
* 10^6/ml

The hyperglycemia produced 60 min. p.i. in animals of GrII and GrIII (Table 3) followed by the significant reduction in their liver glycogen content 360 min. p.i. (Table 5), is in accord with Ramadan *et al.* (1980) results who found that an i.p. lethal dose of crude extract toxin from the Red Sea pufferfish, *Athron hispidus*, caused liver and muscle glycogenolysis, severe hyperglycemia, increased pyruvic acid and LDH activity in rabbits, indicating increased glycolysis and inhibition in the citric acid cycle.

In our opinion a combined factor is the increased stress on body mechanisms which enhanced catecholamine and Ca^{++} liberation leading eventually to increased glycogenolysis.

In this study serum total lipids were increased significantly in animals of GrII and GrIII 60 min. p.i. (Table 3) which may be partly due to enhanced c-AMP production due to stress and partly to the lipolytic activity of the venom (Ramadan *et al.*, 1980 and Ali *et al.* 1989).

The i.p. injection of LD_{10} of the crude venom in rats caused a highly significant elevation in total serum protein 30 and 60 min. p.i. (Table 3) which may be mainly attributed to gradual and steady increase in gamma globulin fraction to meet with the introduced toxic component(s). The slight decrease which occurred within the first 15 min. p.i. of the mild dose (0.1 LD_{10}) of the crude venom, may be due to an early reaction mechanism of the body leading to minor destruction effect or limited synthesis affected by the venom.

Nair (1981) observed a significant increase in serum protein in rats injected with the scorpion venom, *Heterometrus scaber*, and related this increase to the possible increase in antibody formation. The highly significant increase in the serum urea level (GrII 60 min. p.i., Table 3) may be due to increased free amino acid turnover or protein catabolism which goes with Varley *et al.* (1980) observations who stated that increased protein breakdown, which occurs in fevers, cardiac failure and other toxic conditions may cause a moderate increase in serum urea level.

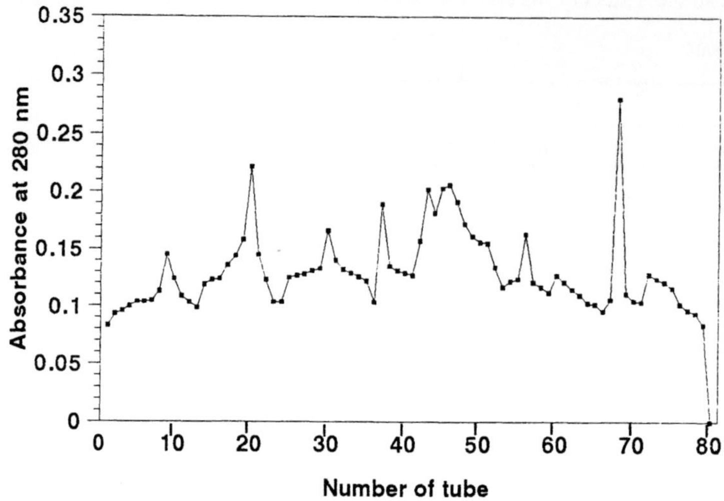

Figure 1. Fractionation of the crude venom of *Plotosus lineatus* on Sephadex G_{200} using sodium phosphate, pH 7.4, column 40X2 cm, flow rate/hr. Fractions anlaysed 4 proteins by measurements of absorbance at 280 nm.

A sudden and drastic increase in serum potassium level reaching 10.1 and 8.9 m Eq/l, 15 and 60 min. p.i. in animals of GrII (80% and 59%, $p<0.001$ respectively) and 7.9 m Eq/l in animals of GrIII, 60 min. p.i. The noticed hyperkalemia may be due to increased cell membrane permeability and hemolysis caused by the crude venom affection.

Varley *et al.* (1980) stated that hyperkalemia leads to arrhythmia including ventricular fibrillation, bradycardia and eventually cardiac arrest in diastole when the serum value reaches 8 m Eq/l.

Our results coincide with those of other authors who studied the produced hyperkalemia due to jellyfish and fish poison injection (Larsen and Price, 1978 and Mansour *et al.*, 1980) who attributed this result to cell membrane destruction.

An elevation in serum bilirubin in animals of GrII and GrIII, 60 min. p.i. (Table 3) was mainly due to the contained hemolytic factor (Shiomi *et al.,* 1986) which thus induces hemolytic disorders affecting the parenchymal cells leading to hyperbilirubinemia.

A very highly significant increase of 823% and 205% ($p<0.001$, Table 4) in the activities of serum LDH and AST respectively in animals of GrII 360 min. p.i. and only of 44% and 24% in animals of GrIII 60 min. p.i. respectively was noticed showing also the affection of the heart and that the higher dose was more effective and persisting than the lower one.

Our results are in line with those of Al-Hassan *et al.* (1987a) and Ali *et al.* (1989) who recorded a very highly significant increase in serum LDH and AST activities in rabbits injected with 4 mg/kg of the crude venom of *A. thalassinus* which was attributed to the toxic effect of the venom on heart and liver cells. A very highly significant increase in hepatic LDH activity being higher and persisting in rats of Gr II was noticed (Table 5), which may be due to the toxic effect of the venom on liver cells (Ali *et al.*, 1989).

A moderate increase was observed in the activities of serum ALT and ALP respectively in animals of GrII, 360 min. p.i. and a mild increase in serum ALP activity in GrIII animals 60 min. p.i. (Table 4).

Ali *et al.* (1989) recorded similar results induced in rabbits due to injection of lethal doses of the crude toxic secretion of *A. bilineatus, A. thalassinus* and *A. tenuispinis* and attributed this result to liver affection, these results were confirmed by histopathological studies on rabbits liver (Alnaqeeb *et al.*, 1989).

Hepatic AST and ALT activities showed very highly significant decrease (60% and 26%, respectively) 360 min. p.i. in animals of GrII while GrIII showed different changes, as liver AST activity increased by 30%, 30 min. p.i. and ALT showed non significant changes.

The significant reduction in activity of hepatic transaminases 360 min. p.i. may reflect the highest recorded increase in the same enzymes in serum exactly after the same time. This result confirmed the destructive effect of the crude venom on hepatocytes, as well as the marked increase in serum and liver GGT activity being higher in rats of GrII confirms the affection of the liver. A sudden and very highly significant increase in serum pseudo cholinesterase activity Ch.E (243%) was recorded 15 min. p.i. reaching its maximum activity (608%) 360 min. p.i. in animals of Gr II while a significant increase was recorded 30 min. p.i. in animals of GrIII which is mainly due to the neurotoxic effect of the venom, which is in accordance with explanation of Birkhed (1972), Shiomi *et al.* (1986) and Al-Hassan (1987a), results who observed neurotoxic factors in the crude venom of *I. catus; P. plotosus* and *A. thalassinus*, respectively. This also agrees with our observations that followed the injections of the crude venom in rats such as agitation, jerking motions of the body, erection of the hair, the hind limb paralysis and convulsions. Fahim and Zahran (1987a) observed an increase in serum pseudo Ch.E activity due to injection of LD_{50} of the toxic fraction (F. III) *N. nigricollis* in rats which was attributed to the venom's hemolytic effect. The cytotoxic effect of the crude toxin on EAC-cells *in vitro* was evaluated by the effect on the cell viability using 1 mg of the crude toxin. A similar but less significant reduction in the total number of EAC-cells was also obtained (41%, Table 6). These two experiments represent the preliminary trials elucidating the anti-tumor effect of the crude venom and the results obtained hereby are encouraging and could be expanded to explore the mode of action of the venom.

In conclusion, the crude venom of the oriental catfish, *Plotosus lineatus*, directly affects heart, nervous system, liver and also red blood cells, considering the higher dose (LD_{10}) is more effective and persisting than the lower dose (0.1 LD_{10}). This study confirms the hemolytic and neurotoxic effects of the crude venom.

REFERENCES

Al-Hassan, J. M.; Thomoson, M. and Criddle, R. S. (1982). Composition of the proteinaceous gel secretion from the skin of the Arabian Gulf catfish (*Arius. thalassinus).* Mar. Biol., 70: 27.

Al-Hassan, J. M.; Ali, M.; Thomson, M.; Fatima, T. and Gubler, C. J (1985a). Toxic effect of the soluble skin secretion from the Arabian Gulf catfish (*Arius. thalassinus)* on plasma and liver enzyme levels. Toxicon., 23: 532.

Al-Hassan, J. M.; Ali, M.; Thomson, M.; Fatima, T.; Gubler, C. J Criddle, R. S. and Summers, B. (1985b). Catfish epidermal secretions in response to threat or injury: A novel defense response. Mar. Biol., 88: 117.

Al-Hassan, J. M.; Ali, M.; Thomson, M.; Fatima, T.; Gubler, C. J and Criddle, R. S. (1987a). Prostaglandin associated mortality following intravenous injection of catfish epidermal secretions in rabbits. Prostaglandins, Leukotrienes. Med., 28: 95.

Al-Hassan, J. M.; Thomson, M.; Ali, M. and Criddle, R. S. (1987b). Toxic and pharmacologically active secretions from the Arabian Gulf catfish (*Arius. thalassinus,* Ruppell). J. Toxicol.-Toxin Rev., 6:1.

Ali, M.; Thomson, M.; Al-Hassan, J. M.; Al-Saleh, J.;Fayad, S.; Assad, H. and Criddle, R. S. (1989). Comparative biochemical and pharmacological properties of epidermal secretion from *Ariid* catfish of the Arabian Gulf. Comp. Biochem. Physiol., 92B: 205.

Al-Lahham; Al-Hassan, J. M.; Thomson, M.; and Criddle, R. S. (1987). A hemolytic protein secreted from epidermal cells of the Arabian Gulf catfish *Arius. thalassinus* (Ruppell). Comp. Biochem. Physiol., 87B: 321.

Al-Naqeeb, M. A.; Al-Hassan, J. M.; Ali, M.; Thomson, M. and Criddle, R. S. (1989). Histopathological observations on organs from rabbits injected with the skin toxin of the Arabian Gulf catfish *Arius. thalassinus*, Valenciennes). Toxicon., 27: 789.

Belfield, A. and Golderg, D. M. (1971). Revised assay for serum alkaline phosphatase activity using 4-amino antipyrine. Enzyme., 12: 561.

Birkhead, W. E. (1972). Toxicity of stings of *Ariid* and *Ictalurid* catfish. Copeia. 1972: 790.

Cabaud, P; Wroblewski, F. and Ruggiero, V. (1958). Calorimetric measurement of lactate dehydrogenase activity of body fluids. Am. J. Clin. Pathol., 30: 234.

Datta, A; Gomes, A.; Sarangi, B.; Kar, P. K. and Lahiri, S. C. (1982). Parmacodynamic actions of crude venom of the Indian catfish (*Hetropneustes fossilis*). Indian J. Med. Res., 76: 829.

Ellman, G. L.; Courtney, K. D.; Anders, V. J. and Featger stone, R. M. (1961). A new and rapid calorimetric determination of acetyl cholinesterase activity. Biochem. Pharmacol., 7: 88.

Fahim, F. A. and Zahran, F. I. (1987). Effect of the most lethal fraction (F. III) of *Naja nigricollis* venom: part I: on liver function, pp173. 12th international Congress for Statistic, Computer Science, Social and demographic research center. Ain Shams University, Cairo, Egypt.

Fahim, F. A.; Zahran, F. I. and Mady, E. A. (1988). Effect of *N nigricollis* venom and its fraction on EAC-cells in mice. The first Inter. Conf. (Feb. 15, 1988). The Egypt. Soc. of Tumor Markers Oncology.

Fisher, G. H. and Yates, F. (1943). Tests of significance based on the normal distraction. In: Statistical tables for biological, agricultural and medical research. Oliver and Boyd. Edinberg and London.

Goldin, A.; Carter, S. and Mantel, N. (1974). In: Antineoplastic and Immunosuppressive Agents. Eds. Sartorelli, A and John, D. G., Springer Verlag. Berlin Heidelberg. New York. Part 1. 12.

Grunbaum, H. G. and Grunbaum, A. S. (1911). Lancet 1:879.

Hald, P. M. (1951). Determination of sodium and potassium ions with flame photometric method. In: Medical research. Vol. IV.: 79.

Halstead, B.W. (1978). Venomous catfish. In: Venomous and poisonous marine animals of world. Vol. 3, Government printing office, Washington, DC.

Jendrassik, L. and Grof, P. (1983). A calorimetric method for bilirubin determination. Bioch. J., 297: 81.

Khoo, H. E.; Yuen, R.; Poh, C. H. and Tann, C. H. (1992). Biological activities of Synanceja horrida (stonefish) venom. Nat. Toxins., 1: 54.

Knight, J. A.; Anderson, S. and Rauk, J. M. (1972). Chemical basis of the Sulfo-Phosphovanillin reaction for estimating total serum lipids. Clin. Chem., 18: 199.

Larsen, J. B. and Price, W. J. Jr. (1978). Some physiological effects of fractionated jelly fish toxins. Toxicon (suppl)., 1: 517.

Mansour, M. A.; Abdel-Hamid, M. E.; Ramadan, M. A. and Al-Nagy, S. A. (1980). Physicochemical studies on the protein and mineral metabolism in rabbits poisoned by *Arthron hispidus* tetrodotoxin. Ann. Zool., 16: 139.

Marsh, W. H.; Fringerhut, B. and Miller, H. (1965). Determination of serum urea by diaoxal method. In: Practical clinical biochemistry, 5th (edn) by Varley London.

Martin, D. W. (1983). Blood plasma and clotting. In: Harper's review of Biochemistry. pp.635, 20th ed. Medical publication, Los Anglos, California.

Nair, R. B. (1981). Effect of sublethal dose of toxic protein isolated from venom of the scorpion *Heterometrus scaber*. Indian J. Exp. Biol., 19: 103.

Quick, A. J. (1940). Thromboplastin reagent for the determination of prothrombin time. Science. 92: 113.

Ramadan, M. A.; Tash, F. M.; Al-Nagdy, S. A.; Mahdy, Z. and Mansour, M. A. (1980). Effect of the toxin extracted from the Red Sea Fish *Arthron hispidus* on the metabolism of carbohydrates and lipid metabolism. Zool. Soc. Egypt. Bull., 22: 22.

Reed, L. J. and Meuench, H. (1938). A simple method for estimated fifty percent end points. Amer. J. Hygi., 37: 493.

Reitman, S. and Frankle, S. (1957). A colorimeteric method for the determination of glutamic oxaloacetic & glutamic pyruvate transaminase. Am. J. Clin. Pathol., 28: 56.

Seifter, S.; Dayton, S.; Novic, B. and Muntmyler, E. (1950). Estimation of glycogen with the anthrone reagent. Arch. Biochem., 25: 291.

Sellei, C. C.; Eckhardt, S. and Mette, L. (1970). In: Chemotherapy of neoplastic disease, Academy, Kado, Budapest.

Shiomi, K.; Takamiya, M.; Yamanaka, H.; Kikuch, T. and Konno, K. (1986). Short communications: Hemolytic, lethal and edema forming activities of the skin secretion from the oriental catfish (*Plotosus lineatus*). Toxicon. 24: 1015.

Shiomi, K.; Takamiya, M., Yamanaka, H. and kikuch, T. (1987). Purification of lethal factor in the skin secretion from the oriental catfish (*Plotosus lineatus*). Toxicon 53: 1275.

Shiomi, K.; Takamiya, M., Yamanaka, H. ; Kuch, T. and Suzuki, Y. (1988). Toxins in the skin secretion of the oriental catfish (*Plotosus lineatus*): Immunological properties and immunocytochemical identification of producing cells. Toxicon 26: 353.

Szasz, G. (1969). A kinetic photometric method for serum GGT. Clin. Chem., 15: 124.

Trinder, P. (1969). Determination of glucose in blood using glucose oxidase with an alternative oxygen receptor. Ann. Clin. Biochem. 6: 24.

Varley, H.; Gowenlock, A. H. and Bell, M. (1980). Practical Clinical Biochemistry. Vol. I. General topics and commoner test. 5th William Heinemann medical books LTD. London.

Weichelbaum, T. E. (1946). An accurate and rapid method for the determination of proteins in small amounts of blood and plasma. Am. J. Clin. Path. Tech. Sec., 10: 4.

Zerman, M. G. (1989). Catfish Stings, A report of 3 cases. Ann. Emerg Med, 18: 211.

INTERACTION OF LIPOPOLYSACCHARIDE WITH THE ANTIMICROBIAL PEPTIDE "CECROPIN A"

T. J. Jacks,[1] A. J. De Lucca,[1] and K. A. Brogden[2]

[1] USDA, ARS
New Orleans, Louisiana 70124
[2] USDA, ARS
Ames, Iowa 50010

INTRODUCTION

In our studies of preventing the contamination of plant commodities by the carcinogenic fungal toxin, aflatoxin, we found that growth of the toxigenic fungus, *Aspergillus flavus*, was strongly inhibited by cecropin A (CA). Cecropins, originally found in insects, are antimicrobial polypeptides inert to eukaryotes but biocidal to prokaryotes (Boman and Hultmark, 1987). Examing the antimicrobial activity of CA further, we found that the activity was inhibited by a bacterial membrane component, lipopolysaccharide (LPS). In this paper we describe the physicochemical interaction of LPS with CA and the concomitant inhibition of the microbicdal activity of CA.

MATERIALS AND METHODS

Materials

CA from pig intestine and Lipid A (LA) from *Escherichia coli* (LA-E) and *Salmonella minnesota* (LA-S) were obtained from Sigma Chemical Co. (St. Louis, MO). LPS (MW, 10,000 g mole^{-1}) was prepared from *Pantoea agglomerans* (ATCC 27996) as described earlier (De Lucca et al., 1992). All other chemicals were reagent grade. Water and laboratory-ware were pyrogen-free.

Dye Binding

Solutions containing 25 μM LPS and 0-25 μM CA were incubated for 2 hrs and then treated with dimethyl methylene blue (DMB) according to Keller and Nowotny (1986). Absorbances of DMB bound to LPS were measured at 535 nm and were amended for CA.

Natural Toxins II, Edited by B. R. Singh and A. T. Tu
Plenum Press, New York, 1996

Circular Dichroism (CD)

CD of CA with and without LPS was measured with a JASCO Model J 500A spectropolarimeter. Amounts of α-helical conformational modes in CA were determined from ellipticity values at 222°nm (Chen and Yang, 1971), adjusted for the presence of LPS. The mean residue weight of CA was 108.2 g mole^{-1}.

Solubility

Four μmoles of either LA-E or LA-S was mixed with 0.22 μmoles of CA in 10 ml and centrifuged at 7000 G for 15 min. The amount of soluble CA remaining in each supernatant was estimated from its polypeptide content quantified by the method of Waddell (1956).

Bacterial Viability

Germinating spores of *P. agglomerans* were incubated in 0-1.25 μM CA for 2 hr and then inoculated on trypticase agar plates to enumerate colony-forming units. Effects of LPS on CA-induced lethality were determined by adding 0.04-0.4 moles of LPS per mole of CA before bioassays.

RESULTS AND DISCUSSION

Competition from CA for Dye-Binding to LPS

LPS is an acidic polysaccharide that binds basic thiazine dyes such as DMB (Keller and Nowotny, 1986). To determine whether LPS also binds CA, we measured the competition between CA and DMB for binding to LPS as a function of CA concentration. Binding of DMB to LPS was halved by CA at a mole ratio of about two LPS per CA. These results show that CA bound to LPS and indicate anionic binding sites were involved.

Effect of LPS on the Conformation of CA

For additional assessment of binding, we determined by CD whether exposure of CA to LPS produced changes in the secondary structure of CA. The CD spectrum of CA was consistent with an absence of helical conformational modes (Fig. 1). In the presence of 3.5 moles of LPS per mole of CA, however, CA assumed a conformation with 44.8% helix (Θ_{222} = -1.59 X 10^4 deg cm^2 dmol^{-1}). A conformational change of this magnitude shows unequivocally that CA interacted with LPS. The interaction produced helical coiling in CA similar in extent to the 50% helix generated by dissolution in trifluoroethanol (Merrifield et al., 1982). Virtually identical spectra were obtained with other sources of LPS (data not shown).

Effect of LPS Subunit (LA) on CA Solubility

The role of LA in the binding of CA to LPS was estimated from losses of CA from solution after incubation with an 18-fold mole excess of LA. Only 25% and 29% of the original amount of CA remained in solution after exposure to LA-S and LA-E, respectively, indicating that CA bound to the LA moiety of LPS.

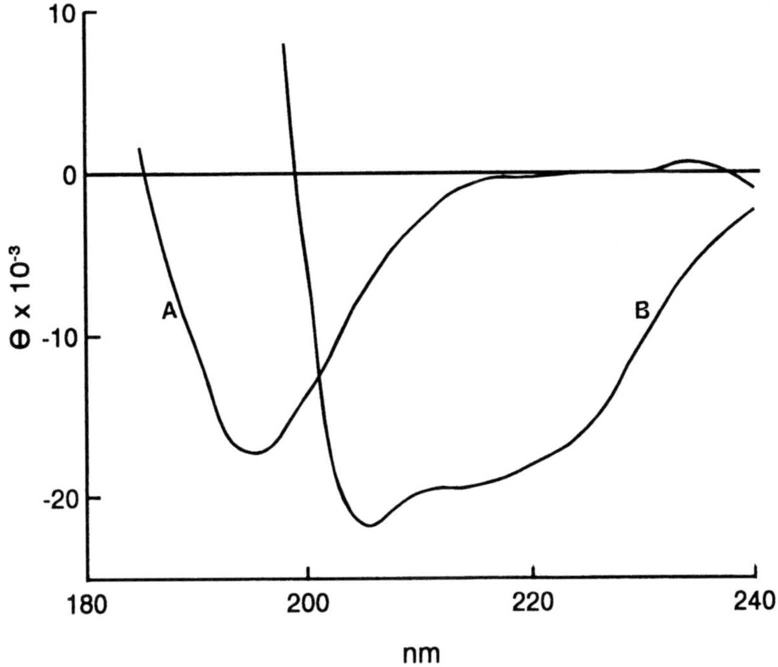

Figure 1. Far-ultraviolet CD spectra of CA and CA-LPS complex. Values on the ordinate refer to deg cm2/dmol. (A) CA; (B) CA-LPS.

Antibacterial Activity of CA Inhibited by LPS

To determine the effect of LPS on the antibacterial activity of CA, the effects of CA and CA-LPS mixtures on PA viability were measured. Total lethality occurred with 0.8 μM CA and 50% killing was obtained with about 0.16 μM CA (Fig. 2). LPS, however, inhibited the biocidal activity of CA. When CA was preincubated with LPS at a mole ratio of five CA

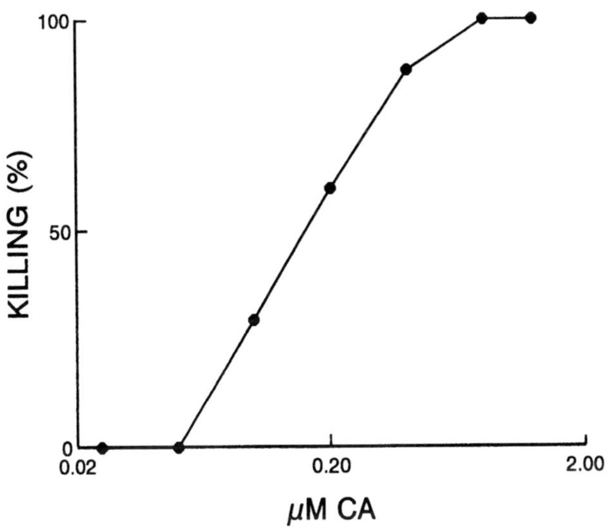

Figure 2. Killing *P. agglomerans* with CA.

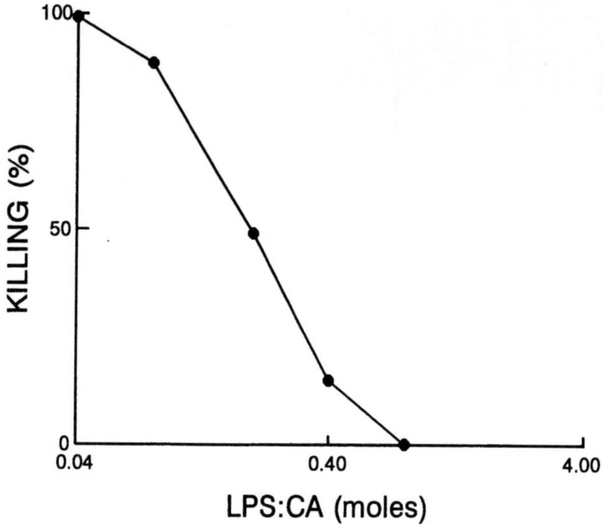

Figure 3. Inhibition by LPS of killing *P. agglomerans* with CA. Abscissa represents mole ratios of LPS:CA.

per LPS before bacterial exposure, 50% of the germinating spores normally killed by CA survived (Fig. 3), showing that LPS disarmed CA.

Our results show that LPS bound CA and concomitantly neutralized its antibacterial activity. LPS binds other peptides that inhibit its endotoxigenicity (Rana and Blazsk, 1991; Warren et al., 1985; Warren et al., 1992); whether this holds true for CA binding is being examined.

REFERENCES

Boman, H. G., and Hultmark, D., 1987, Cell-free immunity in insects, *Ann. Rev. Microbiol.* 31:103-126.

Chen, Y.-H., and Yang, J. T., 1971, A new approach to the calculation of secondary structures of globular proteins by optical rotary dispersion and circular dichroism, *Biochem. Biophys. Res. Commun.* 44:1285-1291.

De Lucca, A. J., Brogden, K. A., and French, A. D., 1992, Agglutination of lung surfactant with glucan, *Brit. J. Ind. Med.* 49:755-760.

Keller, T., and Nowotny, A., 1986, Metachromatic assay for the quantitative determination of bacterial endotoxins, *Anal. Biochem.* 156:189-193.

Merrifield, R. B., Vizioli, L. D., and Boman, H. G., 1982, Synthesis of the antibacterial peptide cecropin A (1-33), *Biochemistry* 21:5020-5031.

Rana, F.R., and Blazsk, J., 1991, Interactions between the antimicrobial peptide, magainin 2, and *Salmonella typhimurium* lipopolysaccharides, *FEBS* 293:11-15.

Waddell, W. R., 1956, A simple ultraviolet spectrophotometric method for the determination of protein, *J. Lab. Clin. Med.* 48:311-314.

Warren, H. S., Glennon, M. L., Wainwright, N., Amato, S. F., Black, K. M., Kirsch, S. J., Riveau, G. R., Whyte, R. I., Zapol, W. M., and Novitsky, T. J., 1992, Binding and neutralization of endotoxin by *Limulus* antilipopolysaccharide factor, *Infect. Immun.* 60:2506-2513.

Warren, H. S., Kania, S. A., and Siber, S., 1985, Binding and neutralization of bacterial lipopolysaccharide by colistin nonapeptide, *Antimicrob. Agents Chemother.* 28:107-112.

30

STUDY ON THE ACTION MECHANISM OF HEMORRHAGIN I FROM *Agkistrodon acutus* VENOM

Xun Xu,[1,2] Yuzhen Wang,[1] C. Wei,[1] and Xueliang Zhu[1]

[1] Department of Biology
University of Science and Technology of China
Hefei, 230027, Peoples Republic of China
[2] Department of Marine Biology
Third Institute of Oceanography
Xiamen, 361005, Peoples Republic of China

INTRODUCTION

Up to now, studies on the receptor of neural toxins and their functional mechanism were extensively reported (Tu, 1977a). Receptors of neural toxins have been proved to be present in the animal's cells. As to hemorrhagins from snake venoms, it is already known from previous observations (Ownby et al., 1974; Ohsaka et al., 1976) that after the hemorrhagins have entered the animal's body, they break the wall of the blood vessel or increase the gap between endothelial cells of capillaries and cause red cells flowing out. But, a few reports have been published about the hemorrhagic mechanism in detail or whether there are some receptors or binding sites existing in the animal's tissue.

Xu et al. (1981) have purified and characterized three hemorrhagic principles from the *Agkistrodon acutus* venom. Hemorrhagin I (AaHI) is the major hemorrhagic factor and can cause rupture of vascular system and extensive hemorrhage. In this paper, the hemorrhagic mechanism of AaHI is reported at the molecular level, and a new interpretation of the hemorrhagic mechanism is presented.

MATERIALS AND METHODS

AaHI was purified from *A. acutus* venom by the procedure described previously by Xu et al. (1981). Sephadex G-75 was purchased from Pharmacia, Sweden. SDS, acrylamide, plasmin, fibrinogen, and standard proteins were purchased from Sigma. Other reagents and chemicals were made in China and analytically pure.

Natural Toxins II, Edited by B. R. Singh and A. T. Tu
Plenum Press, New York, 1996

Preparation of HRP (Horseradish Peroxidase) - AaHI

HRP-AaHI was prepared with periodic acid oxidization (Nakane et al., 1974). 5 mg HRP and 5 mg AaHI were used. After labelling, HRP-AaHI was separated from HRP and AaHI with Sephadex G-75. The elution solution was 0.02 M PBS and the flow rate was 8 ml/hr. 2.5 ml fractions were collected in each tube.

Preparation of Samples for Electronmicroscope Observation

While the rabbit was killed, the leg muscles were dissected immediately. A small piece of muscle (about 2X5X7 cm) was dipped into hexane precooled in liquid nitrogen. Ultramicrotome sections were cut to a thickness of 25 μm and spread in distilled water briefly and then put on a wet slide. After that, these sections were incubated with HRP-AaHI for one hr at room temperature and washed thoroughly with 0.02 M phosphate buffer (pH 7.2). Afterwards, regular procedure for preparing electronmicroscopic sample was applied, and the samples were observed with a HI-TACHI 800 electron microscope.

Assay for Cell Monolayer Disrupting Effect

The culture of Hela cells was carried out with Eagle's minimal essential medium, and the fibroblast cell line derived from human kidney was cultured with RPMI medium 1640. Both were supplemented with calf serum to a concentration of 10% and with penicillin and streptomycin sulfate to concentrations of 100 units per ml. Before cells were cultured, several slides were immersed in the cell culture bottle. While covered with cultured cells, they were taken out and put in the bottles containing Hank's solution (3 ml) and AaHI (30-80 μg/ml). These bottles were incubated at 37°C and the slides were examined microscopically at various time intervals.

Assay for Cytotoxicity

Erythrocytes. The blood of a rabbit was collected in citrate and centrifuged to let the erythrocytes separate from the plasma; the packed cells were washed with Hank's solution for three times. The erythrocytes were diluted by Hank's solution and produced an absorbance of approximately 1.0 at 540 nm when complete hemolysis had been reached. AaHI (80 μg/ml) and erythrocytes were incubated for 2 hrs at 37°C; the cytotoxicity of the AaHI on erythrocytes was determined by measuring the increase in absorbance at 540 nm.

Lymphocytes and Tissue Culture Cells. The lymphocytes were squeezed out from the rabbit's spleen and mixed with Hank's solution. The Hela cells and the fibroblast cell line from human kidney mentioned above were treated with trypsin solution. All the cells were washed with Hank's solution for three times.

The cytocidal activity was examined by the dye exclusion method of McLimans et al. (1975). A cell suspension was prepared immediately before the start of each experiment. The suspension was adjusted to contain $2X10^5$ cells per ml. 0.5 ml of AaHI (80 μg/ml) were added to 4.5 ml of cell suspension. At 30-minute intervals thereafter 0.5 ml of the cell suspension was taken out from the tubes and mixed with 0.1 ml trypan blue (0.4%). Five minutes later, the viable and nonviable cells were counted under a microscope.

SDS-Polyacrylamide Gel Electrophoresis (PAGE)

Fibrinogen solution (20 mg/ml) and AaHI (0.1 mg/ml 0.03 M pH 7.4 ammonium acetate buffer) were incubated at room temperature. At various time intervals (from 0-36 hrs), 0.1 ml of reaction solution was taken out for SDS-PAGE. The sample buffer is 4% SDS/10% β-mercaptoethanol/0.05 M EDTA. The electrophoresis procedure was carried out according to the conventional method except that a 5% gel was used instead of a 10% gel.

Hydrolysis Specificity of AaHI

The oxidized β chain of insulin was used as a substrate for studying the hydrolysis specificity of AaHI. 5 mg of the oxidized insulin β chain and 0.2 mg of AaHI were dissolved in 0.5 ml $Na_2B_4O_7$-HCl buffer, pH 8.0, and incubated at 35°C for 3 hrs, boiling for 15 min, centrifuging at 6000 rpm for 10 min. The supernatant was stored at -20°C. Identification of peptide fragment and the hydrolysis sites were made by the following methods: Two-dimensional electrophoresis chromatography on minicrystal cellulose (Heiland et al., 1976); DNS-Cl method for determination of N-terminal amino acids (Tamura et al., 1973); separation of peptide fragments by HPLC; an LKB ultropac TSK-545 DEAE column was used and amino acid compositions were analyzed by ar LKB 4400 amino acid analyzer.

RESULTS

Electromicroscopic Observation of the Specific Attachment of HRP-AaHI

HRP-AaHI was separated from free HRP and AaHI chromatographically. Tubes wit high concentration of HRP-AaHI were collected. As shown in Figure 1, the specifi attachment of HRP-AaHI to the wall of capillary was observed.

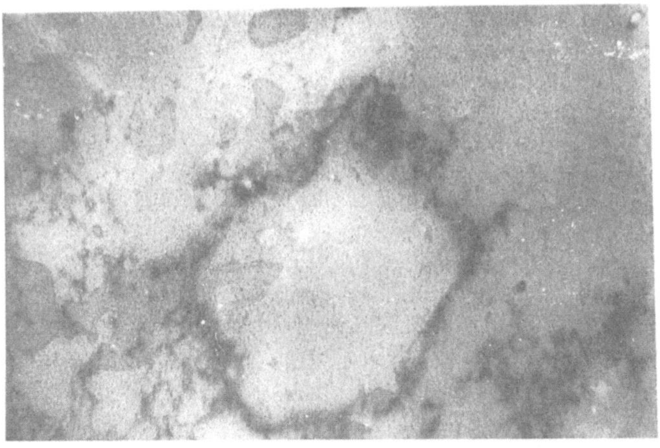

Figure 1. Electronmicroscopic view of HRP-AaHI deposited on capillaries - Specific attachment of H AaHI to the endothelial cells is shown X70000.

Cytotoxicity and the Effect of Cell Monolayer

AaHI showed no cytocidal activity and cannot disrupt the cell monolayer at the concentration of 80 μg/ml within 2 hrs.

Fibrinolytic Activity

According to the results of SDS-PAGE (Figure 2), α chain and most of the β chain were degraded within 10 minutes, after 2 hrs, γ chain was degraded too.

Hydrolysis Specificity of AaHI

The enzymolysis site of oxidized β chain of insulin could be shown as follows:

```
• F-V-N-Q-H-L-C-G-S-H-L-V-E-A-L-Y-L-V-C-G-E-R-G-F-F-Y-T-P-K-A.
          |          |     |  |          | |
```

This result indicated that AaHI attacked the amino end of Leu and probably Gly as well as Phe. Meanwhile, the sites of hydrolysis may also depend on the amino acid sequences near the cleavage site. This specificity of AaHI is similar to that of proteases from Crotalidae, such as protease C, α-protease, H2-protease, etc. (Tu, 1977b).

DISCUSSION

The most serious symptom of *A. acutus* bite is the endless bleeding at the wound and internal organs as well. Since 1980, several components related to its toxicity have been investigated in our laboratory; they are: AaH I, II, III; all can cause hemorrhage and lethality (Xu et al., 1981); depressor component, inducing rapid fall of blood pressure (Wang et al., 1988); hyaluronidase, referred to as the "spreading factor" (Xu et al., 1982); thrombin-like enzyme, the defibrination factor inducing nonclotting of blood. Among these components,

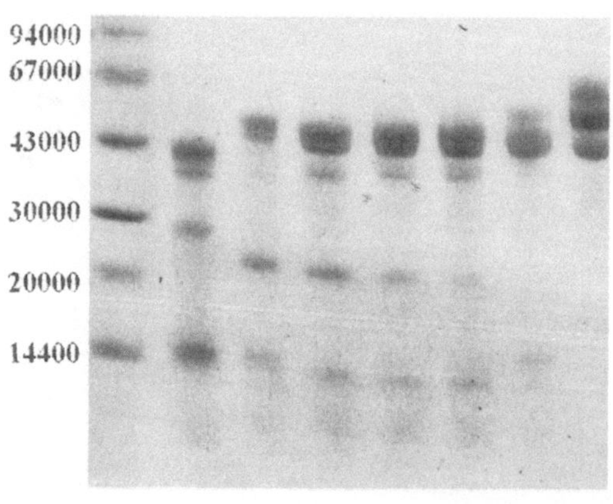

Figure 2. SDS-polyacrylamide gel electrophoresis of fibrinogen after digestion by AaHI. 10 μl reaction solution (200 μg of protein) mixed with 10 μl of 0.5 M EDTA/4% SDS/10% β-mercaptoethanol were applied to the gel. The electrophoresis was performed on 5% gel in 0.2% SDS-0.1 M phosphate buffer (7.0) for various times. The positions of the α, β, and γ chain of fibrinogen are indicated. From right to left: 0 time control, 10 min, 40 min, 60 min, 2 hrs, 6 hrs, 36 hrs, and standard proteins control.

AaHI is the major factor responsible for hemorrhage. This paper describes some of the results about the action mechanism of AaHI obtained in our lab in recent years. Specific attachment of HRP-AaHI to the endothelial wall of capillary was observed, which indicated that AaHI acted specially on the endothelial wall of capillary. Former work on the distribution of ^{125}I-labelled hemorrhagic toxin from *A. acutus* (^{125}I-AaHI) in rabbits and its pharmacokinetic characters also supported that there must have been some corresponding binding sites for the AaHI in the tissue (Huang et al., 1988). According to Copley's concept (Copley, 1983), capillary vessel consists of endothelial cell, basement membrane, and endo-endothelial fibrin lining (EEFL). Endothelial cell is surrounded by cement fibrin, and Ca^{2+}-fibrin is the essential compound of basement membrane, EEFL as well as cement fibrin. Calcium is an inhibitor of plasmin, so Ca^{2+}-fibrin of EEFL possesses the function of protecting the endothelial vessel from the degradation of plasmin (Haverkate et al., 1977). No hemorrhage occurred even though there was a high activity of plasmin in the blood.

Results of this paper show that AaHI could not disrupt the monolayer of cultured cells and had no cytoxic activity *in vitro*. It indicates the hemorrhage induced by AaHI is not due to actions in the cells. Also, AaHI is an endopeptidase and possesses fibrinogenolytic activity and fibrinolytic activity (data not shown), which is similar to the plasmin; however, calcium may play an important role in maintaining a proper structure for AaHI (Zhang et al., 1984). Calcium is essential for the hemorrhage and enzymolytic activities of AaHI. Moreover, EDTA could inhibit the activity of AaHI (Xu et al., 1981). Based on these facts, we suggest that the hemorrhagic mechanism of AaHI is: after entering into the blood, AaHI binds specifically to the endothelial wall of capillary and its activity is activated by Ca^{2+} in EEFL; consequently, EEFL and cement fibrin of endothelial cell as well as basement membrane are destroyed and hemorrhage takes place. Further work is needed to isolate the receptor or binding site from the animal cells and study on its properties and structures. Studies on the inhibitors of AaHI are also useful for the treatment of snakebite patients.

ACKNOWLEDGEMENT

We appreciate Bangxu Luo for critical reading of the manuscript.

REFERENCES

Copley, A. L., 1983, The endo-endothelial fibrin lining. A historical account. Thromb. Res. Supplement V.1.

Haverkate, F., et al., 1977, Effect of calcium in the plasmin degradation of fibrinogen and fragment, D. Thromb. Res. 10:803.

Heiland, I., et al., 1976, Primary structure of protein L_{10} from the large subunit of *E. coli* ribosome. Hoppe-Seyler's Z. Physiol. Chem. 357:1751.

Huang, Y., Jiang, B., Wang, Y. Z., and Xu, X., 1988, Studies on the distribution of ^{125}I-labelled hemorrhagic toxin (^{125}I-AaHI) from *A. acutus* in rabbits and its pharmacokinetic characters. J. Isotope 1(1):44.

Kondo, H., et al., 1960, Studies on the quantitative method for determination of hemorrhagic activity of Habu snake venom. Jap. J. Med. Sci. Biol. 13:43.

McLimans, W. F., et al., 1975, The submerged culture of mammalian cells: The spinner culture. J. Immunol. 79:482.

Nakane, O. K., and Karwaoi, A., 1974, Peroxidase-labelled antibody—A new method of conjugation. J. Histochem. 22(12):1084.

Ownby, C. L., Kainer, R. A., and Tu, A. T., 1974, Pathogenesis of hemorrhage induced by rattlesnake venom. Am. J. Pathol. 75:401.

Ohsaka, A., 1976, Biochemical aspect of increased vascular permeability for proteins and erythrocytes with special reference to the mechanism of hemorrhage. J. Japan. Biochem. Soc. 48:308.

Tamura, Z., Nakajima, T., and Nakayama, T., 1973, Identification of peptides with 5-dimethylaminoaphthalenesulfonyl chloride. Anal. Biochem. 52:595.

Tu, A. T., 1977a, Binding of neurotoxins to acetylcholine receptors. In Venoms: Chemistry and Molecular Biology. John Wiley & Sons, New York, p. 240.

Tu, A. T., 1977b, Proteolytic enzymes. In Venoms: Chemistry and Molecular Biology. John Wiley & Sons, New York, p. 104.

Wang, Y. Z., Xu, Y. Z., Luo, D., and Xu, X., 1988, Purification and characterization of depressor component of *A. acutus* venom. Acta Pharmacol. Sin. 9(4):334.

Xu, X., et al., 1981, Purification and characterization of the hemorrhagic components from *Agkistrodon acutus* venom. Toxicon 19(5):633.

Xu, X., et al., 1982, Purification and partial characterization of hyaluronidase from *A. acutus* venom. Toxicon 20(6):973.

Zhang, J. J., Chen, Z. X., He, Y. Z., and Xu, X., 1984, Effect of calcium on proteolytic activity and conformation of hemorrhagic toxin I from five pace snake (*A. acutus*) venom. Toxicon 22(6):931.

K_{252a} AND STAUROSPORINE MICROBIAL ALKALOID TOXINS AS PROTOTYPE OF NEUROTROPIC DRUGS

Philip Lazarovici,*[1] David Rasouly,[1] Lilach Friedman,[1] Rinat Tabekman,[1] Haim Ovadia,[2] and Yuzuru Matsuda[3]

[1] Department of Pharmacology and Experimental Therapeutics
School of Pharmacy, Faculty of Medicine
The Hebrew University of Jerusalem
POB 12065, Jerusalem, 91120, Israel
[2] Department of Neurology
Hadassah Medical Center
POB 12000, Jerusalem, 91120, Isarael
[3] Tokyo Research Labs, Kyowa Hakko Kogyo Co. Ltd.
3-6-6 Asahimachi, Machida-Shi, Tokyo, 194, Japan

1. INTRODUCTION

Protein kinases are enzymes that transfer a phosphate group from ATP to an acceptor amino acid in a substrate protein (Edelman et al., 1987). Protein kinases which transfer the phosphate to alcohol groups as acceptor are called serine/threonine kinases and those to a phenolyc group as acceptor are called protein tyrosine kinases. While certain tyrosine kinases are domains of growth factor receptors (Hunter, 1991), serine/threonine kinases are classified by their second messenger activators: cAMP dependent (PKA), cGMP dependent (PKG), calcium-phospholipid dependent (PKC), calcium-calmodulin dependent (CaMK), etc (Nairn et al., 1985). Protein phosphorylation of different cellular substrates by protein kinases is an important messenger switch in signal transduction for many plasma membrane receptors leading to a defined biological response (Kikkawa and Nishizuka, 1986). Therefore, in order to elucidate and manipulate receptor signal transduction pathways, potent and selective inhibitors would be of great value.

K_{252a} family of alkaloid toxins isolated from microbes represent the second generation of natural protein-kinase inhibitors (Rasouly et al., 1995). During the last decade they have been intensively used in studies aimed at elucidating the cellular role of protein kinases in signal transduction pathways and different biological functions (Parker et al., 1992).

* To whom correspondence should be adressed.

Natural Toxins II, Edited by B. R. Singh and A. T. Tu
Plenum Press, New York, 1996

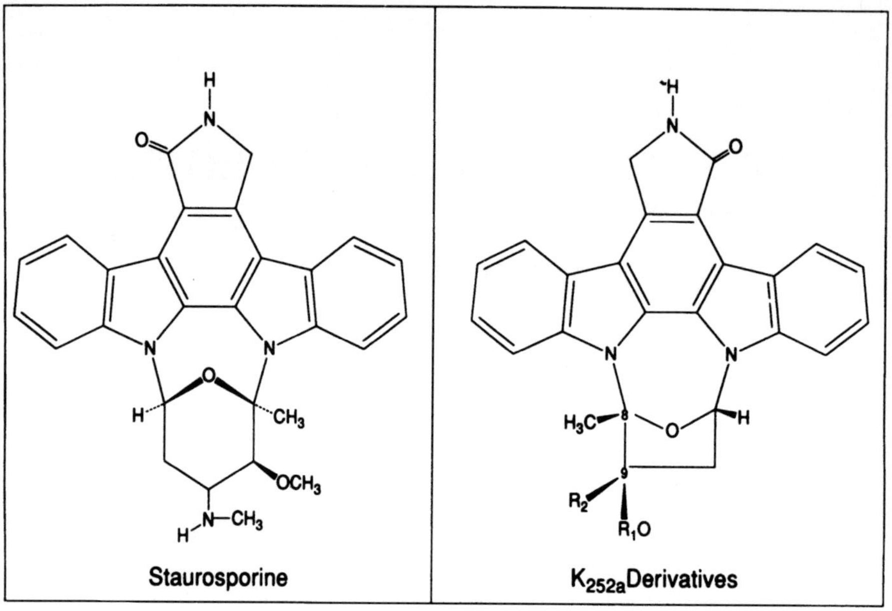

Figure 1. Chemical structures of Staurosporine and K_{252a} derivatives. $\mathbf{K_{252a}}$: R_2 - CH_3; R_1 - H; $\mathbf{K_{252b}}$: R_2 - H; R_1 - H; $\mathbf{KT_{5720}}$: R_2 - nHexane; R_1 - H; $\mathbf{KT_{5822}}$: R_2 - CH_3; R_1 - CH_3.

2. K_{252} FAMILY OF ALKALOID TOXINS AS PROTEIN KINASES INHIBITORS

K_{252a} derivatives and Staurosporine (figure 1) were isolated from the culture broath of *Nocardiopsis sp.* and *Streptomyces staurospores* microbes, respectively (Matsuda, 1993).

Using the chromatin-associated protein histone H_1 as a substrate and partially purified protein kinases from different biological sources, the inhibitory potency of these alkaloids was measured *in vitro* (table 1).

K_{252a} is a potent (1-25 nM), nonselective protein-kinases inhibitor. K_{252b} , the 9-carboxylic acid derivative of K_{252a} (figure 1), is less potent (10-150 nM). KT_{5720}, the n-hexyl ester derivative of K_{252a} (figure 1) appears to be a potent (60 nM) and more selective

Table 1. *In vitro* inhibitory effects of K_{252} alkaloid toxins on the most common serine/threonine protein kinases[(1)]

	Alkaloid toxin[(3)]				
Protein kinase[(2)]	K_{252a}	K_{252b}	KT_{5720}	KT_{5822}	Staurosporine
PKA	10-20	90	60	>40	7-20
PKG	15-20	100	>2000	2-4	8-10
PKC	25	20	>2000	80	0.7-10
MLCK	20	150	>2000	>10000	1.3
CaMK	2	12	nt	nt	10-100

(1) Inhibitory activity is expressed by Ki (nM) inhibition constants.
(2) PKA - cAMP dependent; PKG - cGMP dependent; PKC - calcium and phospholipid dependent; MLCK - myosin light chain kinase; CaMK - calcium-calmodulin type 2 kinase.
(3) Taken from Rasouly et al., 1995.

inhibitor of PKA than PKC or PKG. KT_{5822}, the 9-methoxy derivative of K_{252a} (figure 1) displays the highest relative selectivity towards PKG relative to the other derivatives (table 1). Staurosporine is the most potent (2-10 nM) inhibitor of PKC, yet it shows little selectivity towards inhibition of this particular kinase (table 1). It is accepted that staurosporine and K_{252a} derivatives compete with ATP binding site on the catalytic domains of the different protein kinases (Tamaoki, 1991).

3. K_{252a} AS MODIFIER OF NERVE GROWTH FACTOR (NGF) ACTION

Nerve growth factor (NGF) is the first neurotrophin found to induce profound biological effects such as the survival and maintenance of sympathetic and sensory neurons in the peripheral nervous system as well as in groups of cholinergic neurons within the central nervous system (Levi-Montalcini, 1987). The first specific and potent NGF inhibitor identified by neurochemical studies was K_{252a} (Koizumi et al., 1988). It was established that K_{252a} is not a competitive antagonist of NGF but efficiently inhibited the increase in **c-fos** oncogene transcription, the increase in intracellular calcium and the stimulation of the phosphorylation cascade produced by NGF in PC12 cells (Lazarovici et al., 1989). Recently, the "trk" proto-oncogene product $gp140^{trk}$ was found as the high affinity NGF receptor which transduced its signals by activating a cascade of protein phosphorylations initiated by the tyrosine kinase activity of trk (Kaplan et al., 1991). Studies of the action of K_{252a} on "trk" revealed that K_{252a} is a potent inhibitor of "trk" tyrosine kinase, resulting in inhibition of the cellular effects of NGF (Berg et al., 1992).

4. STAUROSPORINE AS A PROTOTYPE OF A NEUROTROPIC DRUG

In human, the nervous system stops dividing by the second postnatal year. Consquently, mechanically or chemically injured neurons would not be replaced by new neuronal tissue, and the ability of the injured neuronal circuit to regenerate would be totally dependent on the sprouting and formation of neurites from neighboring neurons. A neurotropic drug that could inhance the neurite-regenerating capacity of the neurons would result in faster restructuring of the injured neuronal circuit and recovery of the damaged neuronal physiological function.

Staurosporine possesses dual, dose-dependent effects (Rasouly et al., 1992). In the range of 100-500 nM staurosporine inhibited NGF action by reducing "trk" tyrosine kinase activity and caused cytotoxicity (figure 2). At 10-100 nM, staurosporine induced neurite outgrowth (figure 2) similarly to NGF action, both in PC12 cells and other neuronal cell cultures.

The dose-dependent dual effects of staurosporine characterize it as a functional partial agonist of NGF (Rasouly et al., 1992). Staurosporine does not activate NGF receptor-trk tyrosine kinase activity (figure 3A) (Rasouly and Lazarovici, 1994) or downstream trk cellular substrates such as erk's or SNT proteins (Rasouly et al., 1996). However, staurosporine was found to induce the tyrosine-phosphorylation of a 145 kD protein (figure 3B), related to the neurotropic effects, with an unknown physiological action (Rasouly and Lazarovici, 1994).

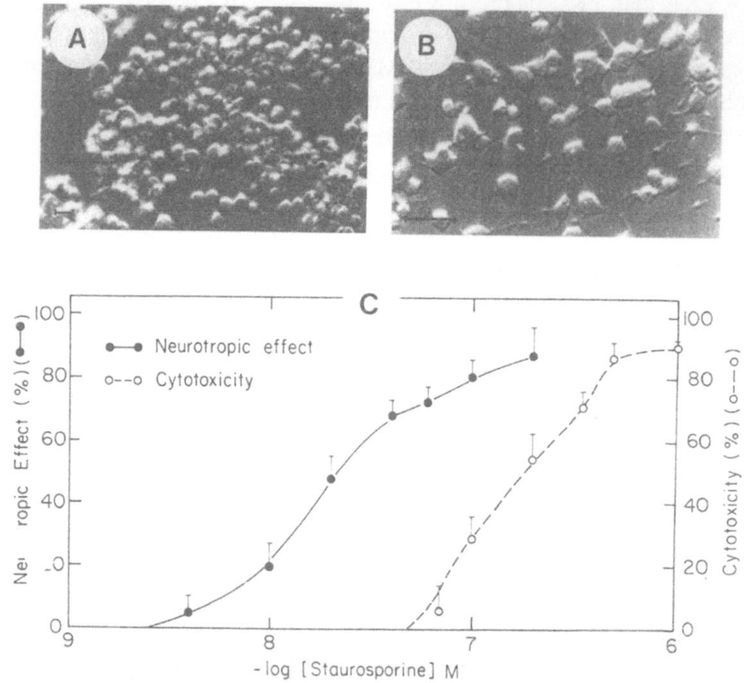

Figure 2. Dose-response neurotropic and cytotoxicity effects of staurosporine on PC12 cells. Light micrographs of untreated (A) and treated cells (B) for 24 hours with staurosporine (25 nM). (C) Neurotropic and cytotoxicity effects of staurosporine, evaluated 24 hours upon toxin addition.

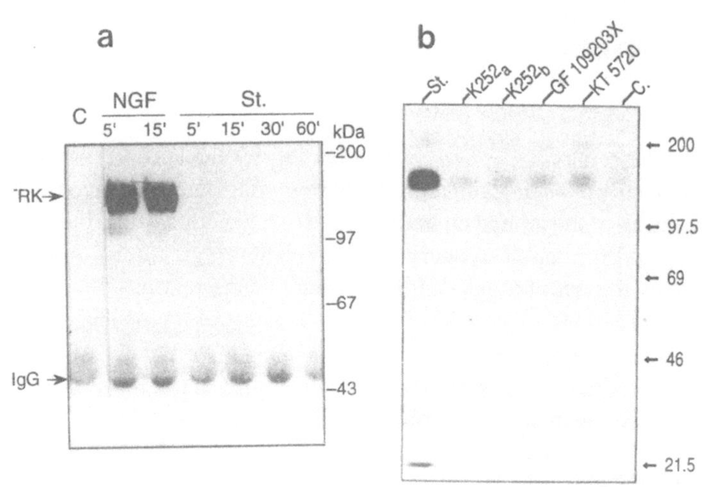

Figure 3. NGF induced "trk" (A) and staurosporine induced gp[145] tyrosine phosphorylation in PC12 cells. (A) PC12 cultures were treated for 5, 15, 30 or 60 minutes with either NGF (50 ng/ml) or staurosporine (50 nM), lysed, immunoprecipitated with a pan-anti trk antibody, electrophoresed on SDS-PAGE, western blotted with a monoclonal anti P-Tyr antibody and developed with the ECL-Amersham reagent. TRK and IgG arrows indicate the position of NGF receptor and IgG used for immunoprecipitation. (B) Phosphotyrosine blotts of PC12 cultures treated for 6 hours with either staurosporine (50 nM; St.), K_{252a} (200 nM), K_{252b} (200 nM), a bisindolemaleimide-staurosporine derivative (100 nM; GF, 109203X), KT_{5720} (500 nM) and control, DMSO 0.1% treated cultures. Numbers and arrows - the position of molecular weight markers.

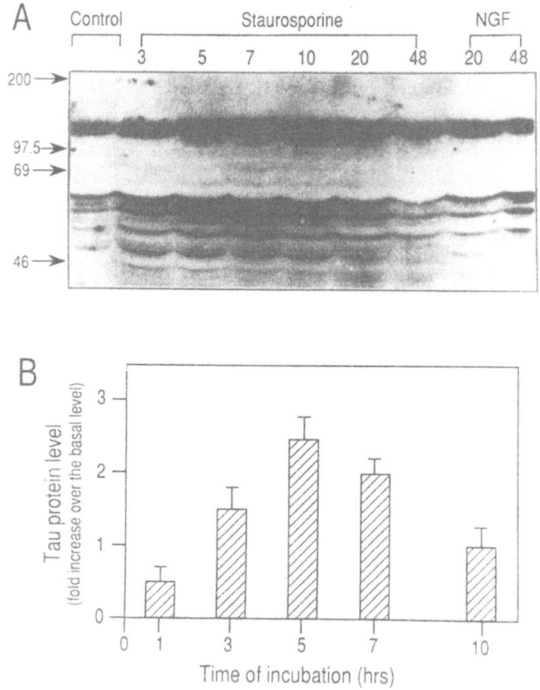

Figure 4. Kinetics of the staurosporine-induced increase in tau protein levels.One day before the experiment, PC12 cells were plated on 20-cm dishes coated with collagen/polylysine and treated with staurosporine (50 nM) or NGF (50 ng/ml) as indicated. After treatment, cells were harvested, lysed and subjected to electrophoresis on 10% SDS-PAGE, electrotransferred to nitrocellulose filters, and immunostained with anti-tau antibody and ECL detection system. (A) A representative Western blot autoradiograph of PC12 cell cultures treated with staurosporine or NGF. *Numbers* on top of the autoradiograph indicate hours of treatment. *Arrows* indicate mobility of rainbow-colored molecular weight markers. (B) Quantitation of tau protein levels by laser densitometry in four different kinetic experiments. Levels of 55- and 110-kD tau proteins were analyzed separately. Values (mean ± standard error) represent tau proteins level in the treated cells relative to their level in unstimulated control cells.

Our studies have indicated that staurosporine increases the levels of the cytoskeleton-associated tau proteins required to stabilize the neurite outgrowths (Rasouly et al., 1994). In contrast to NGF which increases tau protein levels upon more than 48 hours (figure 4), staurosporine increased tau protein levels during the first 10 hours of incubation (figure 4), in good correlation with its neurotropic effects (Rasouly et al., 1994). Staurosporine treatment also resulated in an increase in amount of the high-molecular weight (100-110 kDa) tau proteins supposed to be involved in neurite regeneration following mechanical crush injury (Couchie et al., 1992).

Treatment of PC12 cells with staurosporine for short-time intervals induces a rapid elevation in tau protein immunoreactivity, but with no change in tau mRNA levels. Moreover, neither actinomycin D nor cyclohexamide, which inhibit RNA and protein synthesis, respectively, affected the staurosporine-induced neurite outgrowth or increase in tau proteins (Sadot et al., 1995). On the other hand, analysis with tau-1 mAb after alkaline phosphatase treatment, as well as analysis with α-tau 333 mAb, revealed similar tau immunoreactivity in protein extracts prepared from control and staurosporine-treated cells (Sadot et al., 1995). Thus, the effects observed following staurosporine treatment can be explained mainly by the change in the phosphorylation state of the protein during the short term of the staurosporine-induced effects. These results suggest that the high molecular weight (HMW, 110 kDa) and the low molecular weight (LMW, 55 kDa) tau levels can be regulated independently, and that whereas the HMW tau isoforms are more sensitive to treatment with staurosporine, the low molecular weight tau isoforms are more sensitive to NGF (Sadot et al., 1995). The above data may suggest that staurosporine, being a protein kinase inhibitor, disrupts the delicate balance between kinases and phosphatases acting in PC12 cells, resulting in reduction of the extent of tau phosphorylation and leading to rapid neurite outgrowth.

5. STAUROSPORINE INHIBITED PHORBOL ESTER EFFECTS ON ALZHEIMER'S PRECURSOR PROTEIN LEVELS IN PC12 CELLS

The amyloid precursor proteins (APP) are transmembrane glycoproteins with apparent molecular weights of 90-100 kD on SDS-PAGE (Dyrks et al., 1988). Their physiological role is yet unknown and they are mainly expressed in neurons. During the progression of theAlzheimer's disease the APP is abnormally cleaved by yet unknown proteases, resulting in extracellular formation of the toxic β-amyloid and its deposition in the form of neurotoxic plaques (Ashall and Goate, 1994). Pheochromocytoma PC12 sympathetic cultures synthesise and process APP to β-amyloid, therefore providing a convenient neuronal culture model to study APP protein function and to develop or search new drugs to reduce β-amyloid formation. In untreated PC12 cultures, the level of APP is relatively of low abundance and the presence of β-amyloid in culture's medium is very poor (Caporaso et al., 1992). However, upon treatment of PC12 cells with phorbol esters (protein kinase C activators) such as PMA

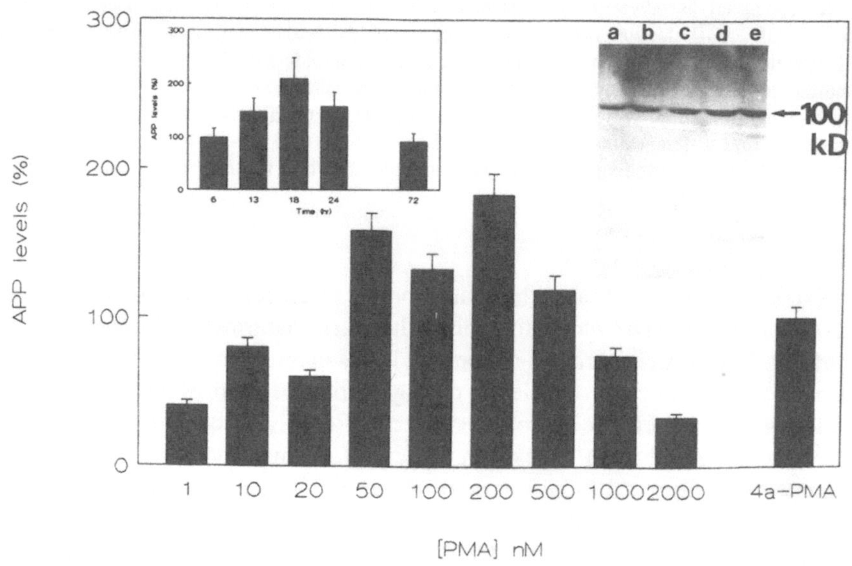

Figure 5. Kinetics and dose-response of PMA induced increase in APP levels in PC12 cells. PC12 confluent cultures grown in 10 cm petri dishes coated with poly L-lysine and collagen have been exposed for 18 hours at 37°C to the indicated concentrations of PMA, lysed, submitted for SDS-PAGE and western blotting using a monoclonal antibody directed to the amino acids sequence 511-608 of APP_{695}. The level of APP protein in each sample was estimated by laser densitometry scanning and is represented as percentages of untreated cultures. *Left insert* represents a kinetic experiment performed under similar conditions using 200 nM PMA. *Right insert* represents a western experiment photograph of a PMA dose-response experiment. *Arrows* - position of the ~100 kD APP protein. a - 1 nM; b - 20 nM; c - 50 nM; d - 100 nM; e - 200 nM.

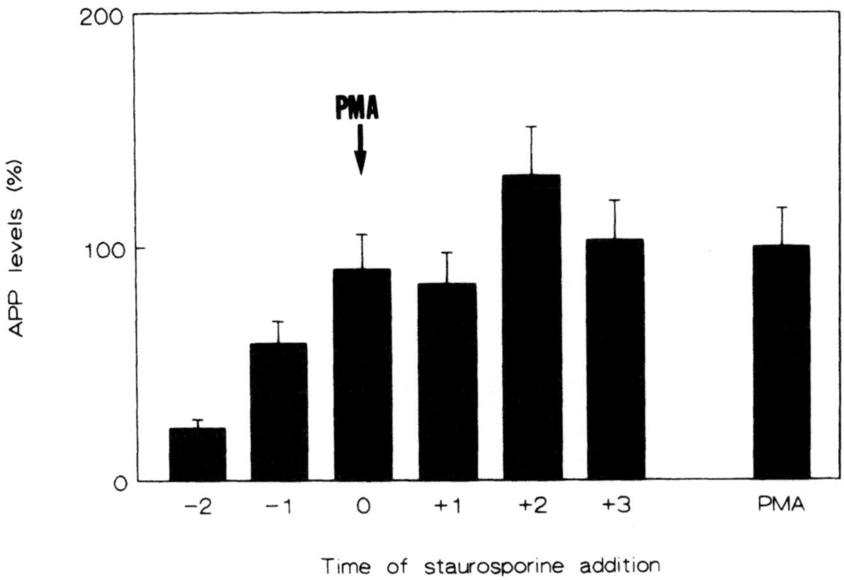

Figure 6. Kinetics of staurosporine effects on PMA induced increase in APP levels in PC12 cells. PC12 cultures have been treated at the indicated time periods with 100 nM staurosporine: before (-), concomitant (o) or after (+) addition of 50 nM PMA (arrow).

(figure 5), APP levels increased by 35% with slow kinetics, reaching maximal levels at 18 hours of PMA treatment (figure 5, left insert) at concentrations between 50-200 nM (figure 5 and right insert). It has been previously shown that this increase in APP level is the result of an increase in APP gene transcription (Slack et al., 1993), followed by APP synthesis (Buxbaum et al., 1992) and increase processing and secretion of APP peptides (Caporaso et al., 1992).

This PMA-induced increase in APP levels was blocked by staurosporine (100 nM) only when applied before PMA treatment (figure 6). These data support the concept that the activity of protein kinase C isoforms regulate APP level/processing in PC12 cells. Further characterization of the effects of staurosporine and other K_{252a} derivatives on the signals regulating APP metabolism in PC12 cells should contribute to the understanding of the physiological function of this molecule and to the development of more specific alkaloid compounds to control APP synthesis and derived amyloid fragments production.

6. K_{252a} DELAYS THE EXPERIMENTAL AUTOIMMUNE ENCEPHALOMYELITIS (EAE) SYMPTOMS IN SJL/J MICE

Multiple sclerosis is a central nervous system (CNS) idiotypic disease resulting from consistent damage to myelin sheets (Quarles et al., 1989). The common animal model to investigate the disease is the experimental autoimmune allergic encephalomyelitis in mice and rats (McFarlin et al., 1974). In the present study we injected SJL/J mice with 0.1 ml of the emulsion composed of the synthetic peptide sequence of the myelin sheet - proteolipid protein, PLP 139-151 (HCLGKWLGHPDKF, 10 mg/ml), disolved in PBS together with complete Freund adjuvant enriched in *Mycobacterium tuberculosis* (6 mg/ml). In parallel and 48 hours later the mice were i.v. injected with inactivated *B. pertusis* (2.7 X 10^9 bacteria) (Jansson et al., 1991). The neurological symptoms appear 9-11 days after immunization.

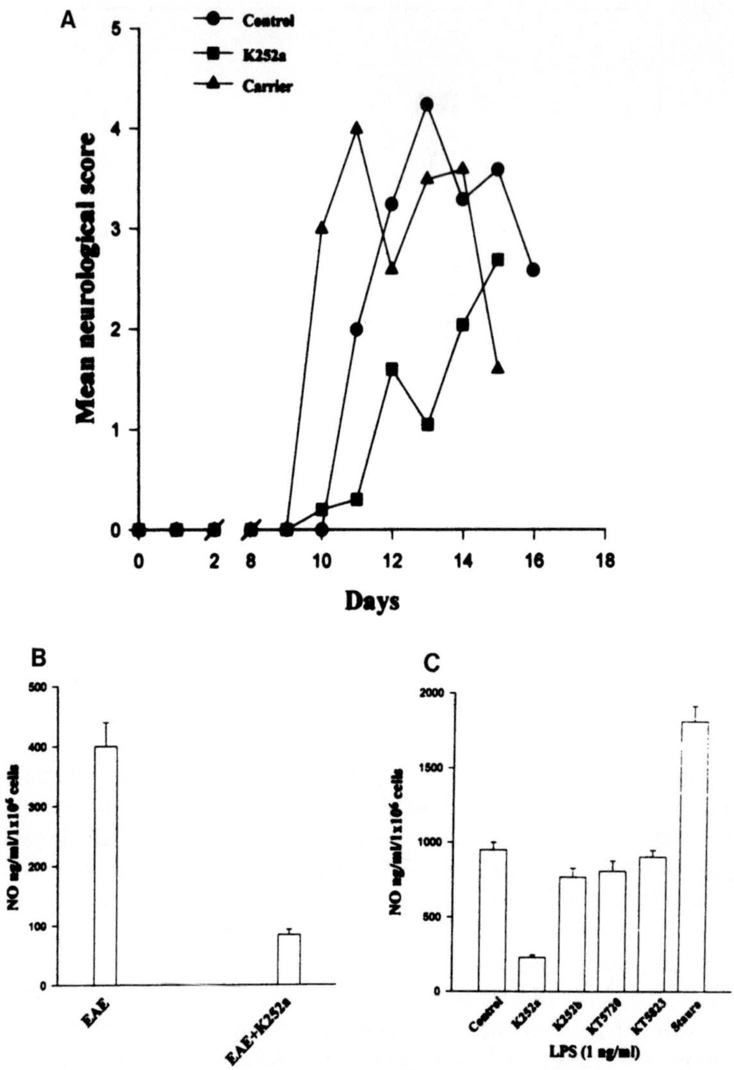

Figure 7. The effect of K_{252a} on the therapeutics and macrophage activation in the experimental autoimmune allergic encephalomyelitis. (A) EAE was induced by a common protocol, neurological examinations were daily performed and the mean neurological score was measured. The control animal groups received either carrier or immunized without any other treatment. (B) Peritoneal macrophages collected from the animal groups as presented in figure 7A, were isolated in F12 medium, divided in samples of 10^6 cells/ml and incubated with LPS (bacterial lipopolysacharides, 10 ng/ml) for 18 hours at 37°C. Thereafter NO release in the culture medium was measured as described by Denis, 1991. (C) NR-8383 rat alveolar macrophages were incubated for 18 hours at 37°C (10^6 cells/ml) with 1 ng/ml LPS in the presence of 1 μM mentioned alkaloids. NO release to the culture medium was measured as previously described.

They were evaluated based on an established neurological score (McFarlin et al., 1974) and expressed by floppy tail (score 1), tail paralysis (score 2), tail paralysis and hind leg paralysis (score 3), hind leg paralysis and mild four limbs weakness (score 4), quadriplegia (score 5) and death (score 6). K_{252} family of microbial toxins have been examined for a potential therapeutic benefit in this EAE animal model. The alkaloids have been applied per os 6-7 times, in a specific formulation, from the start of the immunization paradigm. When comparing K_{252a}, K_{252b}, staurosporine, KT_{5823} and KT_{5720} (figure 7A and data not shown) it has been found that K_{252a} (1 mg/kg body weight) was the only compound that caused a marked reduction in the incidence and severity of EAE symptoms and caused a delay of several days in their appeareance (figure 7A).

Since the macrophages have a key role in the inflammation process induced in EAE model (Carrieri, 1994) and their activation is supposed to be in part regulated by phosphorylation / dephosphorylation, we saught to investigate the effect of K_{252a} on macrophages activation in the sick or K_{252a} treated mice. As evident from (figure 7B), the macrophages isolated from K_{252a} treated mice are less activated as measured by their ability to synthesize nitric oxide (NO), an important factor in the inflammatory process (Moncada et al., 1991). To prove a direct effect of K_{252a} on macrophages nitric oxide synthetase (NOS) activity, cells of a rat alveolar macrophage line (NR-8383) have been exposed for 18 hours at 37°C to 1 μM of K_{252a} and derived alkaloids in the presence of the NOS activator (LPS, 1 ng/ml). Figure 7C indicates K_{252a} inhibited by 75% LPS-induced NO release. This result confirms our working hypothesis that K_{252a} beneficial effects on the EAE model are probably due to an inhibitory effect on macrophage activation. K_{252a} provides a pharmacological tool to investigate protein kinases involvement in the process of macrophages activation and their role in the pathophysiology of CNS diseases.

7. SUMMARY

K_{252} family of alkaloid toxins - kinase inhibitors are the most widely used compounds in biological research on the role of protein kinases in cellular transduction systems, biological functions and pathophysiology of neurological disorders. The wide research interest in these toxins is due to their potency in inhibiting cellular protein kinases. However, lack of kinase specificity is one of their major drawbacks. Synthesis of new K_{252} derivatives can be expected to open up a new generation of kinase inhibitors.

Staurosporine might be considered as a prototype neurotropic drug in view of its ability to induce neurite outgrowth and to increase tau protein levels. Because it mimics some of the neuroprotective effects of NGF and might blocks certain signals required to enhance cellular levels and/or beta amyloid processing, staurosporine might play a beneficial role in the treatment of Alzheimer's disease. The ability of staurosporine to promote neuronal regeneration and brain cholinergic neurons survival has been also demonstrated in animal studies (Nabeshima et al., 1991).

The beneficial effects of K_{252a} on the experimental autoimmune encephalomyelitis (EAE) disease mice model and it's ability to supress macrophage activation suggest an important role of protein kinases inhibitors as immunosupressive agents.

These results may also point to the potential clinical relevance of K_{252} microbial toxins as prototypes for the development of new drugs for the management of neuronal diseases.

REFERENCES

Albina, J.E., Abate, J.A., and Henry, W.L.Jr (1991). Nitric oxide production is required for murine resident peritoneal macrophages to supress mitogen stimulated T-cell proliferation. *The American Association of Immunologists*, *147*, 144-148.

Ashall, F. and Goate, A.M. (1994). Role of the β-amyloid precursor protein in Alzheimer's disease. *Trends in Biochem. Sci.*, *19*, 42-45.

Berg, M.M., Sternberg, D.W., Parada, L.F. and Chao, M.V. (1992). K_{252a} inhibits nerve growth factor induced trk proto-oncogene tyrosine phosphorylation and kinase activity. *J. Biol. Chem.*, *267*, 13-16.

Buxbaum, J.D., Oishi, M., Chen, H.I., Pinkas-Kramarski, R., Jaffe, E.A., Gandy, S.E. and Greengard, P. (1992). Cholinergic agonists and interleukin 1 regulate processing and secretion of Alzheimer β/A_4 amyloid protein precursor. *Proc. Natl. Acad. Sci. USA*, *89*, 10075-10078.

Caporaso, G.L., Gandy, S.E., Buxbaum, J.D., Ramabhadran, T.V. and Greengard, P. (1992). Protein phosphorylation regulates secretion of Alzheimer β/A_4 amyloid precursor protein. *Proc. Natl. Acad. Sci. USA*, *89*, 3055-3059.

Carrieri, P.B. (1994). The role of cytokines in the pathogenesis of multiple sclerosis. *Int. Multiple Sclerosis*, *1*, 53-59.

Couchie, D., Mavilia, C., Georgieff, I.S., Lien, R.K.H., Shelanski, M.L. and Nunez, J. (1992). Primary structure of high molecular weight tau present in the peripheral nervous system. *Proc. Natl. Acad. Sci. USA*, *89*, 4378-4381.

Denis, M. (1991). Interferon-gamma treated murine macrophages inhibit growth of tubercle bacilli via the generation of reactive nitrogen intermediates. *Cellular Immunology*, *132*, 150-157.

Dyrks, T., Weidemann, A., Multhaup, G., Salbaum, J.M., Lemaire, H.G., Kang, J., Muller-Hill, B., Masters, C.L. and Beyreuther, K. (1988). Identification, transmembrane orientation and biogenesis of the amyloid A4 precursor of Alzheimer's disease. *EMBO J.*, *7*, 949-957.

Edelman, A.M., Blumenthal, D.K. and Krebs, E.G. (1987). Protein serine/threonine kinases. *Ann. Rev. Biochem.*, *56*, 567-613.

Hunter, T. (1991). Protein kinases classification. *Methods Enzymol.*, *200*, 3-38.

Jansson, L., Olsson, T., Hojeberg, B. and Holmdahl, R. (1991). Chronic experimental autoimmune encephalomyelitis induced by the 89-101 myelin basic protein peptide in B10RIII (H-2^r) mice. *Eur. J, Immunol.*, *21*, 693-699.

Kaplan, D.R., Hempstead, B.L., Martin-Zanca, D., Chao, M.V. and Parada, L.F. (1991). The trk proto-oncogene product: a signal transduction receptor for nerve growth factor. *Science*, *252*, 554-557.

Kikkawa, U. and Nishizuka, Y. (1986). The role of protein kinase C in transmembrane signaling. *Ann. Rev. Cell. Biol.*, *2*, 149-178.

Koizumi, S., Contreras, M,L., Matsuda, Y., Hama, T., Lazarovici, P. and Guroff, G. (1988). K_{252a}: a specific inhibitor of the action of nerve growth factor on PC12 cells. *J. Neurosci.*, *8*, 715-721.

Lazarovici, P., Levi, B.Z., Lelkes, P.I., Kuizumi, S., Fujita, K., Matsuda, Y., Ozato, K. and Guroff, G. (1989). K_{252a} inhibits the increase in *c-fos* transcription and the increase in intracellular calcium produced by nerve growth factor in PC12 cells. *J. Neurosci. Res.*, *23*, 1-8.

Levi-Montalcini, R. (1987). The nerve growth factor: thirty-five years later. *Science*, *237*, 1154-1164.

Matsuda, Y. (1993). Screening of bioactive substances of microbial origin based on enzyme and receptor binding assays. *Actinomycetology*, *7*, 110-118.

McFarlin, D.E., Blank, S.E. and Kibler, R.F. (1974). Recurrent experimental allergic encephalomyelitis in the Lewis rats. *J. Immunol.*, *113*, 712-715.

Moncada, S., Palmer, R.M.J. and Higgs, E.A. (1991). Nitric oxide: physiology, pathophysiology and pharmacology. *Pharmacol. Rev.*, *43*, 109-136.

Nabeshima, T., Ogawa, S.I., Nishimura, H., Fuji, K., Kameyama, T. and Sasaki, Y. (1991). Staurosporine facilitates recovery from the basal forebrain-lesion-induced impairment of learning and deficit of cholinergic neurons in rats. *J. Pharmacol. Exp. Therap.*, *257*, 562-566.

Nairn, A.C., Hemmings, H.S. and Greengard, P. (1985). Protein kinases in the brain. *Ann. Rev. Biochem.*, *54*, 931-976.

Parker, P.J., Cook, P.P., Olivier, A.R., Pears, C, Ways, D.K. and Webster, C. (1992). Second messenger systems as pharmacological targets. *Biochem. Soc. Trans.*, *20*, 415-418.

Quarles, R.H., Morell, P. and McFarlin, D.E. (1989). Diseases involving myelin. In: *Basic Neurochemistry* (Siegel, G., Agranoff, G., Albeis, R.W. and Molinoff, P. eds.), 697-713, Raven press, New York.

Rasouly, D., Rahamim, E., Lester, D., Matsuda, Y. and Lazarovici, P. (1992). Staurosporine-induced neurite outgrowth in PC12 cells is independent of protein kinase C inhibition. *Mol. Pharmacol.*, *42*, 35-43.

Rasouly, D., Rahamim, E., Ringel, I., Ginzburg, I., Muarakata, C., Matsuda, Y. and Lazarovici, P. (1994). Neurites induced by staurosporine in PC12 cells are resistant to colchicine and express high levels of tau proteins. *Mol. Pharmacol.*, *45*, 29-35.

Rasouly, D. and Lazarovici, P. (1994). Staurosporine, a kinase inhibitor with neurotropic effects, induces the tyrosine phosphorylation of a 145 kD protein, but does not activate trk in PC12 cells. *Europ. J. Pharmacol. - Mol. Sec.*, *269*, 255-264.

Rasouly, D., Matsuda, Y. and Lazarovici, P. (1995). Biochemical and pharmacological properties of K_{252} microbial alkaloids. In: *The Toxic Action of Marine and Terrestrial Alkaloids* (Blum, M.S. ed), pp. 161-190, Alaken Inc., Fort Collins, CO., USA.

Rasouly, D., Shavit D., Zuniga, R., Elejalde, R.B., Unsworth, B.R., Yayon, A., Lazarovici, P. and Lelkes, P.I. (1996). Staurosporine induces neurite outgrowth in neuronal hybrids (PC12EN) lacking NGF receptors. *J. Cell. Biochem.*, in press.

Sadot, E., Barg, J., Rasouly, D., Lazarovici, P. and Ginzburg, I. (1995). Short-term and long-term mechanisms of tau regulation in PC12 cells. *J. Cell Sci. 108*, 2857-2864.

Slack, B.E., Nitsch, R.M., Livneh, E., Kunz, G.M., Breu, J., Eldar, H. and Wurtman, R.J. (1993). Regulation by phorbol esters of amyloid precursor protein release from swiss 3T3 fibroblasts overexpressing protein kinase C_{α}. *J. Biol. Chem*, *268*, 21097-21101.

Tamaoki, T. (1991). Use and specificity of staurosporine, UCN-01 and calphostin C as protein kinase inhibitors. *Methods Enzymol.*, *201*, 340-347.

STRUCTURE AND EXPERIMENTAL USES OF ARTHROPOD VENOM PROTEINS

D. Jones*

Graduate Center for Toxicology
University of Kentucky
Lexington, Kentucky 40536-0305

INTRODUCTION

The utility and power of natural toxins as experimental probes is well established. Natural toxins have been used as tools to detect the existence of new biochemical molecules and new biochemical pathways, and to experimentally dissect the functional mechanisms by which cellular components act.

Classical examples of the use of natural toxins as experimental probes include tetradotoxin from the ovaries of puffer fish, which has been extensively used as a probe for the structures comprising its target, the sodium channel (1). Tetradotoxin has also been used as a tool to determine the site of action in axonal pathways of other neurotoxic agents (2,3), and also to immobilize organisms during physiological experiments on nonneural organs (4). Another example of such fruitful use of natural toxins is found in the fungal toxin puromycin, which has been used to dissect apart the translational pathway of protein biosynthesis (5), and as a tool to assist in characterizing the pre- vs. posttranslational action of other toxins or of natural regulatory molecules (6-8). Still other examples of useful natural toxins include nicotine, from plants, and muscarine, from fungi, that have been used to characterize the structure of acetylcholine receptors (9,10).

ARTHROPOD VENOM TOXINS

Arthropod-derived toxins are also among the list of classical examples of toxins used to probe biochemical processes. Scorpion toxins have been important in the experimental determination of the spacial arrangement of residues on the potassium channel (11), and also in the inference of properties of its quaternary structure (12). A scorpion toxin was used as a tool in the first isolation of a specific component of an insect sodium channel (13). Spider toxins have been similarly useful in characterization of calcium channels (14).

* Tel: 606-323-5412; Fax 606-323-1059; djones@pop.uky.edu

Natural Toxins II, Edited by B. R. Singh and A. T. Tu
Plenum Press, New York, 1996

Another arthropod-derived toxin that has been extensively used as a biochemical tool is honey bee melittin, which acts to disrupt cell membranes. This small protein been helpful in testing hypotheses about the basic structure of cell membranes (15), as well as being a practical tool in the lysis of cell membranes for analysis of cellular components (16).

Historically, most attention to arthropod venom toxins has been directed toward the defensive venoms of the honey bee and to the offensive, paralyzing venoms of predatory wasps, spiders and scorpions (17). Even within those groups, only in a few species have the venom components been studied in any detail, and those investigations are have primarily been on spiders and scorpions and not on venoms produced by insects. However, there exists an enormous reservoir of uninvestigated source material that is likely a veritable gold mine of pharmacological agents capable of revealing the existence of undiscovered and unsuspected biochemical molecules and pathways.

PHARMACOLOGICAL GOLD MINE IN INSECT VENOMS AND TOXINS

With respect to analysis of venoms of insect predators, the most familiar species studied are the predatory wasps. However, there are also other insect species that are likely to possess interesting toxins relevant to uses in basic research. For example, the molecules causing rapid paralysis that are used by predatory wasps (e.g. kinins) are likely not the same as those used by other kinds of wasps whose venom causes paralysis only after a period of time, sometimes days.

The largest number of arthropod species that inject regulatory and toxic agents into other organisms are those with parasitic habits. This list includes mites, ticks, mosquitos, and other blood-sucking flies, lice, blood-sucking bugs and parasitics wasps. Although the specific components injected by these parasites have been identified in only a few species, the results obtained thus far show that there are many potent agents in the injected materials that exhibit a wide range of interesting activities.

The saliva of some tick species (recently well reviewed, ref. 19) has been studied in detail, and contains histolytic enzymes, components affecting inflammation reactions, anticoagulants and antiplatlet factors, etc. The salivary components of mosquitos are becoming better characterized, and there exists in the mosquito saliva vasodialators, apyrases, amylases, and other tissue regulatory factors (19-21). The discovery of maxidilian in the saliva of biting sandflies has been very exciting, as it is the most potent vasodilator known to science (22). Analysis of its molecular structural basis for its potency should be very useful in elucidating the components of normally functioning vasodialatory pathways, and perhaps pathways not functioning due to genetic mutations that lead to clinical symptoms.

Another exciting development has been the discovery of a nitric oxide binding protein in the salivary gland of the blood-sucking bug *Rhodnius prolixus* (23). This protein has unusual binding properties that are different than those of the related protein found in humans, and comparative analysis of the differences in structure of the human vs. bug proteins should generate testable hypotheses on the structure-function properties of the natural human protein.

In contrast to the above, virtually nothing is known on the molecules injected into the host tissue by human lice. In fact, so far as this author is aware, there has yet to be published the isolation and identification of a single protein from the saliva of these insects. However, my laboratory is now engaged in a large scale analysis of, and cloning of, these proteins, and the results may further yield the discovery of interesting toxin molecules that function to redirect host cellular pathways in directions favorable to the parasite.

VENOM FROM PARASITIC WASPS

Although the above described groups of parasites of humans can be anticipated to yield new toxin molecules, by far the largest untapped reservoir awaits within the insects that are parasitic on other insects, and especially the parasitic wasps. The species parasitize an extremely vast array of host arthropods, and the venom of each different species of parasitic wasp is tailored to the different physiological templates of the hosts it parasitizes. Given the estimate that as many as several hundred thousand different species of parasitic wasps exist, the number of different venom molecules yet to be discovered is staggering. The scope of the research yet to be done can be appreciated in that the first reported isolation and cloning of a venom protein from a parasitic wasp did not appear until 1992 (24). The venom proteins of many parasitic wasps do not paralyze their host, but instead act or are necessary to suppress the host immune response (25, 26), to cause endocrine disruptions (27), to affect cellular behaviors (28, 29), or possess enzymatic properties (30).

STRUCTURE OF VENOM PROTEINS FROM PARASITIC WASPS

As a general trend, the venoms of parasitic wasps possess a much more complex array of high molecular weight proteins than the venoms of bees and predatory wasps. The structures of a number of comparatively small proteins have been reported from honey bee venom, such as melittin (31), apamin (32), and hyaluronidase (33). The venom of predatory wasps has been reported to contain small components such as kinins (17).

There is much less data available on the isolation and structure of the larger venom proteins found in nonparalysing venom of parasitic wasps. In our own studies we have analysed the venom of parasitic wasps in the subfamily Cheloninae (Braconidae), which sting the eggs of moths and then develop as larval internal parasites inside the larvae of the host caterpillar. Our studies have found some very intriguing molecules whose structure-function relationships are now being pursued. In one of our recent studies we have isolated and cloned a chitinase from the venom (30). The chitinase is present in the venom in a mature form (not as a proenzyme). Yet, the venom is stored in a chitin-lined venom reservoir, with no obvious sign of autodigestion of the storage reservoir by the chitinase. Analysis of the primary and inferred secondary structure of the chitinase shows that it does not possess a C-terminal region rich in serine and threonine, as do reported yeast chitinases (34) and chitinases that function to digest the chitin-lined insect cuticle during molting (35). Also, chitinases have a variable C-terminal domain that in yeast contains sequences similar to cellulose binding motifs (34), while in the venom chitinase the sequence of this domain is unlike those reported for any other chitinase. Thus, analysis of the the C-terminal domain in the venom chitinase may offer clues as to the function of the enzyme and how it does not digest the chitin-lining of the venom reservoir.

The venom from this parasitic wasp has yielded a 33 kDa protein with an intriguing repeating internal structure (24), and which appears to participate in suppression of the immune response of host tissues (26). The primary structure of the mature protein possesses a hydrophobic 5 kDa N-terminal sequence that contains the only cysteine residues (7 of them) in the protein. The remainder of the protein is a series of 12 hydrophillic repeats of a 14 residue core sequence. An interesting exception is that repeat 7 is truncated, and terminates in the only proline residue in the molecule. Computational secondary structure analysis predicts that many of the repeats have an amphipathic character. These analyses of the primary and secondary structure generate a model for the tertiary structure in which the hydrophobic 5 kDa N-terminus, internally cross-linked by cysteine-cysteine disulphide bonds, serves as the core of the protein, and that the amphipathic repeats wrap around this core.

This model of the tertiary structure can be tested by direct crystallization studies (which have thus far been unsuccessful due to the inability to generate crystals) or by indirect means such as epitope mapping. We are currently pursuing the second approach, in which antibodies specific for a particular protein segment are tested for the abililty to immunoprecipitate the protein in native form in solution.

The very N-terminus of the protein, shown in the sequence below, contains two adjacent aspartic acid residues, and thus may be predicted to be on the exterior of the native protein.

```
                                         *   *
Chelonus venom protein           I F S F D D L V C

Ascogaster venom protein     V S N T E N E E S M L
```

This hypothesis was tested using a polyclonal antibody generated against the N-terminal sequence shown above. The antibody was capable of binding to the 33 kDa protein and, with protein A-sepharose beads, immunoprecipitating the protein, supporting the hypothesis that this segment is on the exterior (24). Similar studies are underway with the antisera generated against other segments of the protein.

The question of mechanism of action of the protein is still elusive. The protein is stable in host tissue for 48 hrs, but then rapidly degrades as the host tissue reaches late embryonic development (36). Preliminary studies suggest that it binds to a component of the membrane fraction of host tissues. In the primary structure of the repeats of the protein a consensus motif, shown below, is evident. However, no catalytic or binding motifs are evident in the consensus repeat sequence.

repeat no.	1	2	3	4	5	6	7	8	9	10	11	12	13	14
residues	A/V	D/E	X	V/L	S	G	S	Z	D/E	Q	X	K/R	X	S/T

(Z indicates hydrophobic residue, X indicates no conserved residue).

Residues 5, 6 and 7 are an invariant Ser - Gly - Ser motif for all repeats, and by this conservation may be important in the function (or structure) of the protein.

Another approach to identifying conserved, functionally important residues is comparison of the sequence of homologs of the protein in different species. This approach has been very useful in identifying candidates for important residues in toxins from, for example, snakes (37). In the present case, polyclonal antiserum against the repeating region of the 33 kDa protein containing repeats crossreacts with proteins of ca. 42, 44 and 46 kDa in the venom of a related parasitic wasp species (*Ascogaster quadridenta*), showing that the sequence of at least one motif in the repeats has been sufficiently conserved as to cross react with anti-33 kDa protein antiserum (38). The N-terminus of two of these *Ascogaster* proteins have been obtained, and both have identical sequence and both contain two adjacent acidic amino acid residues, in conservation with the 33 kDa protein from *Chelonus* (see above). The sequence of the actual repeats in the homologs proteins from *Ascogaster* venom is currently being determined.

Incredibly tantalizing as far as function is that the most conserved motif in the repeats of the 33 kDa protein (Ser-Gly-Ser) is the also a conserved motif found in the 4 repeats of the immunoprotective surface protein of a microbial parasite (*Leishmania*) that spends part of its life cycle in the salivary glands of sandflies (39). We thus have two kinds of parasites, one a parasitic wasp that exclusively parasitizes insects, and one a microbial parasite that parasitizes both humans and sandflys, and both parasites utilize a protein containing repeats with a common conserved motif. The answer to the question of why this situation exists will undoubtably be interesting and probably illuminative of biochemical pathways not yet fully understood.

The repeats in 33 kDa venom protein from *Chelonus* and its homologs in *Ascogaster* raises the question as to why has there been selection for these proteins to have such a structure. There are several possibilities to be contemplated. One is that the repeating structure protein spawns an aggregation event and the repeating structure is necessary for the manifestation of this function. An example can be seen in the bacterial ice nucleation protein, which is composed of over one hundred repeats, each of such a length that each can exactly coordinate the binding of a water molecule. When each repeat has so bound a water molecule, the water molecules are situated so as to form a single layer that, in turn, can bind a second layer of water molecules, and so on, that results in a matrix of ice forming (40). It is conceivable that once in the host tissue, the 33 kDa protein interacts with a kind of host molecule in such a way as to serve as a locus that initiates an aggregation event.

Alternatively, there are examples of repeating sequences in proteins that serve a structural function. One example is a glue protein reported from certain mollusks, in which the repeats are involved in a cross-linking phenomenon as the glue 'sets' (41). In the case of the 33 kDa venom protein, it is not known to be involved in formation of a structure in the female wasp that synthesizes the protein, and no such structure is yet known to be formed in the vicinity of the injection site in the host.

Another possibility is that the repeating structure is related to the co-evolution between the parasite and its host. The repeating amino acid structure is also reflected at the nucleic acid level, and nucleic acid repeats are known to be sites of homologous recombination events in which the repeats can become further duplicated or lost. A comparison of the sequence of the repeats in the 33 kDa protein show that the sequences of some repeats are more closely related than they are to other repeats, supporting the proposition that the repeat region has evolved by duplication/loss of repeats or blocks of repeats (24). The existence in the *Ascogaster* venom of several homologs of the 33 kDa protein, each with identical N-termini, but different in size by increments corresponding to the size of a repeat (39) is further supportive of this hypothesis. The host is under selection pressure for mutations that escape interaction with the 33 kDa protein. The presence of the repeating region and its propensity for mutation by homologous recombination creates a mechanism by which the protein can evolve fast enough to keep pace with changes in the target in the host.

VENOM PROTEIN REPEATING STRUCTURE: A GENERAL PRINCIPLE?

The consideration arises as to whether the repeating structure of the 33 kDa protein and its homologs in parasitic wasps is a singular situation or is instead suggestive of a principle operating on the venoms of other insects and arthropods. In this regard, there is a report that a component of the black widow venom is also composed of an internal repeating structure (42). There is circumstantial evidence to suggest that the basis for the repeating structure may at least in part be related to a repeating structural feature of the target axonal protein. As mutations in the target are selected for that do not disrupt the function of the axonal protein, but which do not interact with the venom protein, the repeating structure of the venom protein itself confers evolutionary plasticity so that the spider can keep molecular pace with changes in the target.

The proposed mechanism of evolution of the 33 kDa venom protein finds a counterpart in the *Leishmania* immunoprotective surface protein, in which rapid evolution of the repeating region enables the parasite to escape the immune response of the host. A further interesting variation on this theme is also found in another agent injected into the host tissue by the parasitic wasps. There replicates in the ovary of braconid and ichneumonid wasp parasites viruses in the family Polydnaviridae, and these viruses are also injected into the host tissue along with the venom (43, 44). It has been reported that the genomic sequence encoding transcripts that are expressed in the envenomated tissue contains a repeating primary structure that would promote homologous recombination (45).

SUMMARY

In summary, the intial studies conducted thus far into the components of venoms of parasitic wasps and other arthropods have already yielded a number of interesting properties of the proteins therein. These properties have already offered the possibilities of additional principles operating in the evolution of venoms. That so many unexpected rewards have already surfaced with the relatively little experimental digging conducted thus far generates great anticipation that indeed there remains a pharmacological gold mine awaiting to be discovered in components of the other insect venoms as yet unmined by science.

ACKNOWLEDGMENTS

This work was supported, in part, by NIH GM 33995 and the University of Kentucky program in Research and Graduate Studies. The author is appreciative of the samples of *Ascogaster* wasps provided by Dr. John Brown.

REFERENCES

1. Lipkind, G.M., and Fozzard, H.A., 1994. A structural model of the tetrodotoxin and saxitoxin binding site of the Na^+ channel. Biophys. J. 66:1-13.
2. Ennis, C. and Minchin, M.C., 1993. The effect of toxin I, a K^+ channel inhibitor, on [^{3}H] noradrenaline release from rat cerebral cortex. Eur. J. Pharmacol. 248:85-88.
3. Fletchner, P.L., Fletcher, M.D., Possani, L.D., 1992. Characteristics of pancreatic exocrine secretion produced by venom from the Brazilian scorpion, *Tityus serrulatus*. Eur. J. Cell Biol. 58:259-270.
4. Nijhout, H.F. and Williams, C.M., 1974. Control of moulting and metamorphosis in the tobacco hornworm, *Manduca sexta* (L.): Cessation of juvenile hormone secretion as a trigger for pupation. J. Exp. Biol. 61:493-501.
5. Pestka, S., 1974. The use of inhibitors in studies on protein synthesis. Meth. Enzymol. 30:261-182.
6. Keightley, D.A., Lou, K.J., and Smith, W.A., 1990. Involvement of translation and transcription in insect steroidogenesis. Mol. Cell. Endocrinol. 74:229-237.
7. Mattia, E., den Blaauwen, J., van Renswoude, J., 1990. Role of protein synthesis in the accumulation of ferratin mRNA during exposure of cells to iron. Biochem. J. 267:553-555.
8. Semenkovich, C.F., Coleman, T., Goforth, R., 1993. Physiologic concentrations of glucose regulate fatty acid synthase activity in Hep G2 cells by mediating fatty acid synthase mRNA stability. J. Biol. Chem. 265:6961-6970.
9. Elgoyhen, A.B., Johnson, D.S., Boulter, J., Vetter, D.E., Heineman, S., 1994. Alpha 9: an acetylcholine receptor with novel pharmacological properties expressed in rat cochlear hair cells. Cell 79:705-715.

10. Wyatt, C.N., Peers, C. 1994. Nicotinic acetylcholine receptors in isolated type I cells of the neonatal rat carotid body. Neuroscience 54:275-281.
11. Stocker, M. and Miller, C., 1994. Electrostatic distance geometry in a K^+ channel vestibule. Proc. Natl. Acad. Sci. 91:9509-9513.
12. Russell, S.N., Overturf, K.E., and Horwitz, B., 1994. Heterotetramer formation and charybdotoxin sensitivity of two K^+ channels cloned from smooth muscle. Am. J. Physiol. 267: C1729-1733.
13. de Lima, M.E., Couraud, F., Lapied, B., Pelhate, M., Diniz, C.D. and Rochat, H., 1988. Photoaffinity labelling of scorpion toxin receptors associated with insect synaptosomal Na+ channels. Biochem. Biophys. Res. Comm. 151:187-192.
14. Adams, M.E., Mintz, I.M., Reily, M.D., Thanabal, V., Bean, B.P., 1993. Structure and properties of omega-agatoxin IVB, a new antagonist of P-type calcium channels. Mol. Pharmacol. 44:681-688.
15. Banks, B.E.C. and Shipolini, R.A., 1986. Chemistry and phramacology of honey-bee venom. IN: "Venoms of the Hymenoptera" (T. Piek, ed.), pp. 330-416
16. Smith, R.J., Friede, M.H., Scott, B.J. and von Holt, C., 1988. Isolation of nuclei from melittin destabilized cells. Anal. Biochem. 169:390-394.
17. Piek, T., 1990. Neurotoxins from venoms of the Hymenoptera — twenty five years of research in Amsterdam. Comp. Biochem. Physiol. C 96:223-233.
18. Vinson, S.B. and Iwantsch, G.F., 1980. Host regulation by insect parasitoids. Quart. Rev. Biol. 55: 143-165.
19. Champagne, D.E., Smartt, C.T., Ribeiro, J.M., James, A.A., 1995. The salivary gland-specific apyrase of the mosquito *Aedes aegypti* is a member of the 5'-nucleotide family. Proc. Natl. Acad. Sci. U.S.A. 92:694-698.
20. Grossman, G.L., James, A.A. 1993. The salivary glands of the vector mosquito, *Aedes aegypti*, express a novel member of the amylase gene family. Insect Mol. Biol. 1:223-232.
21. Champagne, D.E., Ribeiro, J.M., 1994. Sialokinin I and II: vasodialtory tachykinins from the yellow fever mosquito *Aedes aegypti*. Proc. Natl. Acad. Sci. 91:138-142.
22. Lerner, E.A. and Shoemaker, C.B., 1992. Maxadilan. Cloning and functional expression of the gene encoding this potent vasodialator. J. Biol. Chem. 267:1062-1066.
23. Ribeiro, J.M., Hazzard, J.M., Nussenzverg, R.H., Champagne, D.E., Walker, F.A., 1993. Reversible binding of nitric oxide by a salivary heme protein from a bloodsucking insect. Science 260:539-541.
24. Jones, D., Sawicki, G. and Wozniak, M., 1992. Sequence, structure and expression of a wasp venom protein with a negatively charged signal peptide and a novel repeating internal structure. J. Biol. Chem 267:14871-14878.
25. Stoltz, D.B., Belland, E.R., Lucarotti, C.J., and Mackinnon, E.A., 1988. Venom promotes uncoating in vitro and persistence in vivo of DNA from a braconid polydnavirus. J. Gen. Virol. 69:903-907.
26. Taylor, T. and Jones, D., 1989. Isolation and characterization of the 32.5 kDa protein from venom of an endoparasitic wasp. Biochim. Biophys. Acta 1035:37-43.
27. Coudron, T.A., 1991. Host-regulating factors associated with parasitic Hymenoptera. ACS Symp. Ser. 449:41-65.
28. Li, X. and Webb, B.A., 1994. Apparent functional role for a cysteine-rich polydnavirus protein in suppression of the insect cellular immune response. J. Virol. 68:7482-7489.
29. Strand, M.R. and Pech, L.L., 1995. *Microplitis demolitor* polydnavirus induces apoptosis of a specific haemocyte morphotype in *Pseudoplusia includens*. J. Gen. Virol. 76:283-291.
30. Krishnan, A., Nair, P.N. and Jones, D., 1994. Isolation, cloning, and characterization of a new chitinase stored in active form in chitin-lined venom reservoir. J. Biol. Chem. 269:20971-20976.
31. Drake, A.F. and Hider, R.C., 1979. The structure of melittin in lipid bilayer membranes. Biochim. Biophys. Acta 555:371-373.
32. Huyghues-Despointes, B.M. and Nelson, J.W. 1992. Stabilities of disulphide bond intermediates in the folding of apamin. Biochemistry 31: 1476-1483.
33. Gmachl, M. and Kreil, G., 1993. Bee venom hyaluronidase is homologous to a membrane protein of mammalian sperm. Proc. Natl. Acad. Sci. U.S.A. 90:3569-3573.
34. Kuranda, M.J. and Robbins, P.W. (1991) Chitinase is required for cell separation during the growth of *Saccaromyces cerevisiae*. J. Biol. Chem. 266:19758-19767.
35. Kramer, K.J., Corpuz, L., Choi, H.K., Muthukrishnan, S., 1993. Sequence of a cDNA and expression of the gene encoding epidermal and gut chitinases of *Manduca sexta*. Insect Biochem. Molec. Biol. 23: 691-701.
36. Leluk, J. and Jones, D., 1989. *Chelonus* sp. near *curvimaculatus* venom proteins: Analysis of their potential role and processing during development of host *Trichoplusia ni*. Arch. Insect Biochem. Physiol. 10:1-12.

37. Menez, A., Boulain, J.-C., Bouet, F., Courderc, J., Faure, G., Rousselt, A., Tremeau, O., Gatineau, E. and Fromageot, P., 1984. On the molecular mechanisms of neutralization of a cobra neurotoxin by specific antibodies. J. Physiol., Paris, 79: 196-206.
38. Jones, D. and Coudron, T., 1993. Venoms of parasitic Hymenoptera as investigatory tools. IN: Parasites and Pathogens of Insects (Beckage, N.E., Thompson, S.N. and Federici, B.A., eds.), Acad. Press, pp. 227-244.
39. Lohman, K.L., Langer, P.J. and McMahon-Pratt, D. 1990. Molecular cloning and characterization of the immunologically protective surface glycoprotein of GP46/M-2 of *Leishmania amazonensis*. Proc. Natl. Acad. Sci. U.S.A. 87:8393-8397.
40. Ruggles, J.A., Nemecek-Marshall, M. and Fall, R., 1993. Kinetics of appearance and disappearance of classes of bacterial ice nuclei support an aggregation model for ice nucleus assembly. J. Bacter. 175:7216-7221.
41. Wiliams, T., Marumo, K., Waite, J.H., Henkins, R.W., 1989. Mussell glue protein has an open conformation. Arch. Biochem. Biophys. 269:415-422.
42. Kiyatkin, N., Dulbova, I., Grishin, E., 1993. Cloning and structural analysis of alpha-latroinsectotoxin cDNA. Abundance of ankyrin-like repeats. Eur. J. Biochem. 213:121-127.
43. Summers, M.D. and Dib-Hajj, S.D., 1995. Polydnavirus-facilitated endoparasite protection against host immune defenses. Proc. Natl. Acad. Sci. U.S.A. 92:29-36.
44. Chelliah, J. and Jones, D., 1990. Biochemical and immunological studies of proteins from polydnavirus from *Chelonus* sp. near *curvimaculatus*. J. Gen. Virol. 71: 2353-2359.
45. Theilmann, D.A. and Summers, M.D. 1988. Idenfication and comparision of *Campoletis sonorensis* virus transcriptions expressed from four genomic segments in the host insects *Campoletis sonorensis* and *Heliothis virescens*. Virol. 167:329-341.

METAMORPHOSES OF A CONOTOXIN

Eliahu Zlotkin, Dalia Gordon, Iris Napchi-Shichor, and Michael Fainzilber

Department of Cell and Animal Biology
Institute of Life Sciences
The Hebrew University of Jerusalem
91904 Jerusalem, Israel

INTRODUCTORY REMARKS: NEUROPHARMACOLOGY AND CHEMICAL ECOLOGY

Venom is a mixture of various substances produced in a specific gland in the body of the venomous animal and introduced through a piercing injecting mechanism into the body of another animal in order to paralyze it or to kill it. In nature, venoms are employed by slow predatory animals in order to capture through an immediate paralysis their relatively fast prey organisms [1]. In the laboratory, however, venoms are employed as sources for the search of various pharmacological and mainly neurotoxic substances.

The conceptual and methodological combination of the above aspects of chemical ecology (venoms in nature) and neuropharmacology (venoms in the lab) has resulted in the finding of the phenomenon of animal group selective neurotoxins [2], which affect the nervous system of only a given group of organisms such as the insect selective neurotoxins derived from scorpion venom [1] and the more recently discovered mollusc selective neurotoxins derived from the venoms of molluscivorous conid snails [3]. Both groups of toxins are low molecular weight polypeptides which affect sodium conductance exclusively in the nervous systems of the respective groups of animals [4,5].

The present study reveals that the combination of the ecological and neuropharmacological approaches has resulted in additional attractive information, namely the functional diversity of a neurotoxic polypeptide.

THE DUALITY OF AN AGONIST-ANTAGONIST

Conotoxins

Peptide toxins derived from *Conus* snail venoms have in recent years proved to be among the most versatile ligands for the study of excitable systems [6]. These small peptides possess clearly defined conformations constrained by a high content of disulfide bridges, with highly variable sequences in conotoxins belonging to the same pharmacological

Natural Toxins II, Edited by B. R. Singh and A. T. Tu
Plenum Press, New York, 1996

category. So far, only one set of conotoxin probes for sodium channels has been characterized, the muscle-specific blocker μ-conotoxins [7]. Recently, we and others isolated and characterized a new conotoxin from the snail-eating species *Conus textile* [8,9]. This new toxin, designated conotoxin-TxVIA, was found to be specifically toxic to molluscs in a series of bioassays and completely inactive, in intracerebral application, to rats [9] and other vertebrates.

A detailed electrophysiological study in isolated cultured *Aplysia* neurons has revealed that conotoxin-TxVIA causes a marked prolongation of action potentials, resulting from a specific inhibition of sodium current inactivation [10]. The macroscopic effects of TxVIA on mollusc neurons are similar to those of other toxins that slow sodium channel inactivation in vertebrates, namely sea anemone and α-scorpion toxins, and the high molecular mass *Conus striatus* (CsTx) and *Goniopora* (GPT) toxins [10-13]. There is no homology in the chemical structure of TxVIA and any of these other toxins, despite their similar electrophysiological effects. It was therefore of interest to study the binding ofTxVIA and to define its interactions with other sodium channel neurotoxins.

TxVIA Binds with High Affinity to Both *Helix* and Rat Central Nervous System

Preliminary experiments revealed specific binding of ^{125}I-labeled TxVIA to both mollusc (*Helix*) central nervous system and rat brain membrane preparations (data not shown). This result was rather surprising considering the lack of toxicity of TxVIA to vertebrates, and we therefore attempted to optimize conditions for binding of TxVIA to each preparation, in order to be able to characterize and compare the binding sites. Manipulation of the ionic compositions of the binding buffers revealed that TxVIA binding is strongly dependent on the concentration of Ca^{2+} in the medium (data not shown). In subsequent experiments, we used Ca^{2+} at 5 mM, although both Mg^{2+} and Co^{2+} can be substituted with varying degrees of efficacy (data not shown). TxVIA binding is not significantly affected by substitution of choline for Na^+ in the medium (Table I), in contrast to GPT, which is strongly dependent on alkali metal cations [12].

Kinetic experiments reveal similar ligand-receptor association and dissociation rates in both rat and mollusc preparations (Fig. 1). Scatchard analyses of binding at equilibrium show that ^{125}I-labeled TxVIA binds to a single class of high affinity and low capacity sties in both *Helix* and rat membranes (Fig. 1C, Table I). There is a good correlation between the K_d values calculated from the kinetic rate constants of Fig. 1,A and B (3.4 nM for *Helix* and 2.5 nM for rat), and the K_d values obtained directly from Scatchard analysis (Table I). Voltage

Table 1. TxVIA binding to *Helix* and rat central nervous system membranes

	Helix		Rat	
Conditions	K_d nM	B_{max} pmol/mg of protein	K_d nM	B_{max} pmol/mg of protein
Na^+ binding buffer	2.8±0.4	0.96±0.08	1.9±0.3	0.67±0.06
Na^+ buffer, lysed vesicles	2.1±0.5	0.87±0.17	1.8±0.2	0.62±0.07
Choline binding buffer	3.6±1.1	1.13±0.31	2.2±0.3	0.71±0.07
Choline buffer for sodium flux			3.8±0.6	0.57±0.11

Data were taken from Scatchard analyses of ^{125}I-labeled TxVIA binding using the LIGAND program. Values are means ± S.D. of at least three separate experiments.

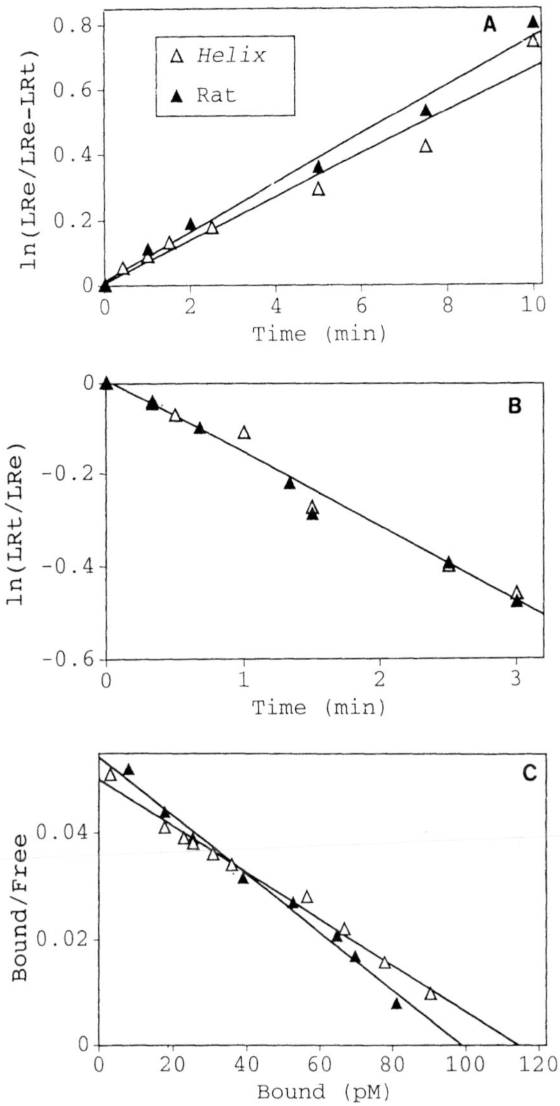

Figure 1. Characterization of TxVIA binding to *Helix* and rat neuronal membranes. **A**, association kinetics of TxVIA. Membranes were incubated with 0.2 nM ^{125}I-labeled TxVIA for the designated times, and the amount of specifically bound ^{125}I-labeled TxVIA was determined following rapid filtration. Nonspecific binding was determined in the presence of 1 μM unlabeled TxVIA and was subtracted from all data points. Data analysis was performed according to [45]. Initial association rate constants $K_{on} = 0.8 \times 10^6$ M^{-1} s^{-1} (*Helix*) and 1.1×10^6 M^{-1} s^{-1} (rat). **B**, dissociation kinetics of TxVIA. Membranes were preequilibrated with 0.2 nM ^{125}I-labeled TxVIA for 1 h and aliquots rapidly diluted 100-fold in binding buffer without toxin. Specific binding of ^{125}I-labeled TxVIA was then determined at the designated times. Initial dissociation rate constant $K_{off} = 2.7 \times 10^{-3}$ s^{-1} for both preparations. **C**, Scatchard analyses of saturation binding curves of TxVIA. Membranes were incubated with 0.2 nM ^{125}I-labeled TxVIA and the indicated concentrations of unlabeled TxVIA for 1 h and specific binding determined as described. Scatchard plot analyses were performed with the computer program LIGAND. Equilibrium binding constants determined in this experiment were as follows: *Helix*, K_d = 2.4 nM; B_{max} = 0.84 pmol/mg; rat, K_d = 1.9 nM; B_{max} = 0.72 pmol/mg; Δ, *Helix*; ▲, rat.

dependence of TxVIA binding to intact synaptosomes was examined in Na^+ buffers containing the Na^+ ionophore Gramicidin A and in osmotically lysed preparations. The presence of Gramicidin A, which completely depolarizes synaptosomes [14], did not significantly change the saturable specific binding of TxVIA in both *Helix* P2 and rat brain synaptosomes (data not shown). Scatchard analyses of biding in osmotically lysed depolarized membranes clearly show that TxVIA binding is not dependent on membrane potential in *Helix* and rat brain preparations (Table I), in contrast to the voltage dependence of all the other peptide neurotoxins (sea anemone, α-scorpion, CsTx, and GPT toxins) that affect sodium channel inactivation [15,12-13].

Competition with TxVIA Binding in *Helix* and Rat Membranes

In order to further characterize the TxVIA receptor, we tested the ability of several sodium channel neurotoxins to compete with ^{125}I-labeled TxVIA binding. As can be seen in Fig. 2A, sodium channel toxins that bind to receptor sites 1 (tetrodotoxin), 3 (sea anemone toxin II), 4 (β-scorpion toxin), and 5 (brevetoxin-PbTx-1) on sodium channels do not inhibit binding of TxVIA. On the other hand, both CsTx and GPT compete with TxVIA and completely inhibit its binding in both rat and *Helix* membranes. Interestingly the lipid-soluble channel activator veratridine, which binds to receptor site 2 on the sodium channel, also reduces specific binding of TxVIA. The isotoxin TxVIB, and also peptide fractions from additional *Conus* venoms, are able to completely inhibit TxVIA binding in both *Helix* and rat membranes (Fig. 2A and data not shown).

CsTx and GPT compete with TxVIA binding in a dose-dependent manner (Fig. 2,B and C). These competition curves show that GPT displaces TxVIA only at concentrations 2-3 orders of magnitude higher than those required of the other peptide toxins. Veratridine also inhibits TxVIA binding in a dose-dependent manner in both *Helix* and rat membranes (Fig. 2). Inhibition of TxVIA binding may result from competition on a single binding site, steric interference with binding at adjacent sites, or via a negative allosteric interaction between distant sites. We therefore attempted to clarify the mechanism(s) of inhibition of TxVIA binding by veratridine, CsTx, and GPT.

TxVIA Has No Effect on Sodium Flux in Rat Brain Synaptosomes

The data presented above show that TxVIA binds to homologous high affinity sites in mollusc central nervous system and rat brain. Therefore, the previously reported lack of toxicity of TxVIA to vertebrates [8,9] cannot be explained by a pharmacologically "non-

Figure 2. Inhibition of ^{125}I-labeled TxVIA binding by sodium channel neurotoxins. A, membranes were incubated with 0.2 nM ^{125}I-labeled TxVIA and saturating concentrations of each of the other toxins for 1 h. Nonspecific binding, determined in the presence of 1 μM TxVIA, was subtracted from all data points. Data are shown as percent inhibition of specific TxVIA binding. Toxins and concentrations used were as follows: *TTX*, tetrodotoxin (1 μM); *Ver*, veratridine (200 μM); *ATXII*, sea anemone toxin II (1 μM); *TsVII*, β-scorpion toxin (0.5 μM); *Brev*, brevetoxin-PbTx-1 (2 μM); *CsTx*, *Conus striatus* toxin (1 μM); *GPT*, *Goniopora* toxin (4 μM); *TxVIB*, conotoxin-TxVIB (1 μM); *Cn-Vt*, *Conus nigropunctatus* venom peptide fraction (0.2 mg/ml). Black bars, *Helix*; hatched bars, rat. B and C, competitive displacement curves for ^{125}I-labeled TxVIA binding from *Helix* (B) and rat (C) preparations. Neuronal membranes were incubated with 0.2 nM 125I-labeled TxVIA and increasing concentrations of the other toxins. The amount of ^{125}I-labeled TxVIA bound at each data point is expressed as a percentage of the maximal specific binding in the system without additional toxins. □, TxVIA; ▲, CsTx; O, GPT; ■, Ver. IC_{50} values were calculated using DRUG analysis in the LIGAND program and determined as follows for *Helix* and rat, respectively: TxVIA, 3.0 and 2.5 nM; CsTx, 26.4 and 3.8 nM; GPT, 0.7 and 0.5 μM; ver, 29 and 63 μM.

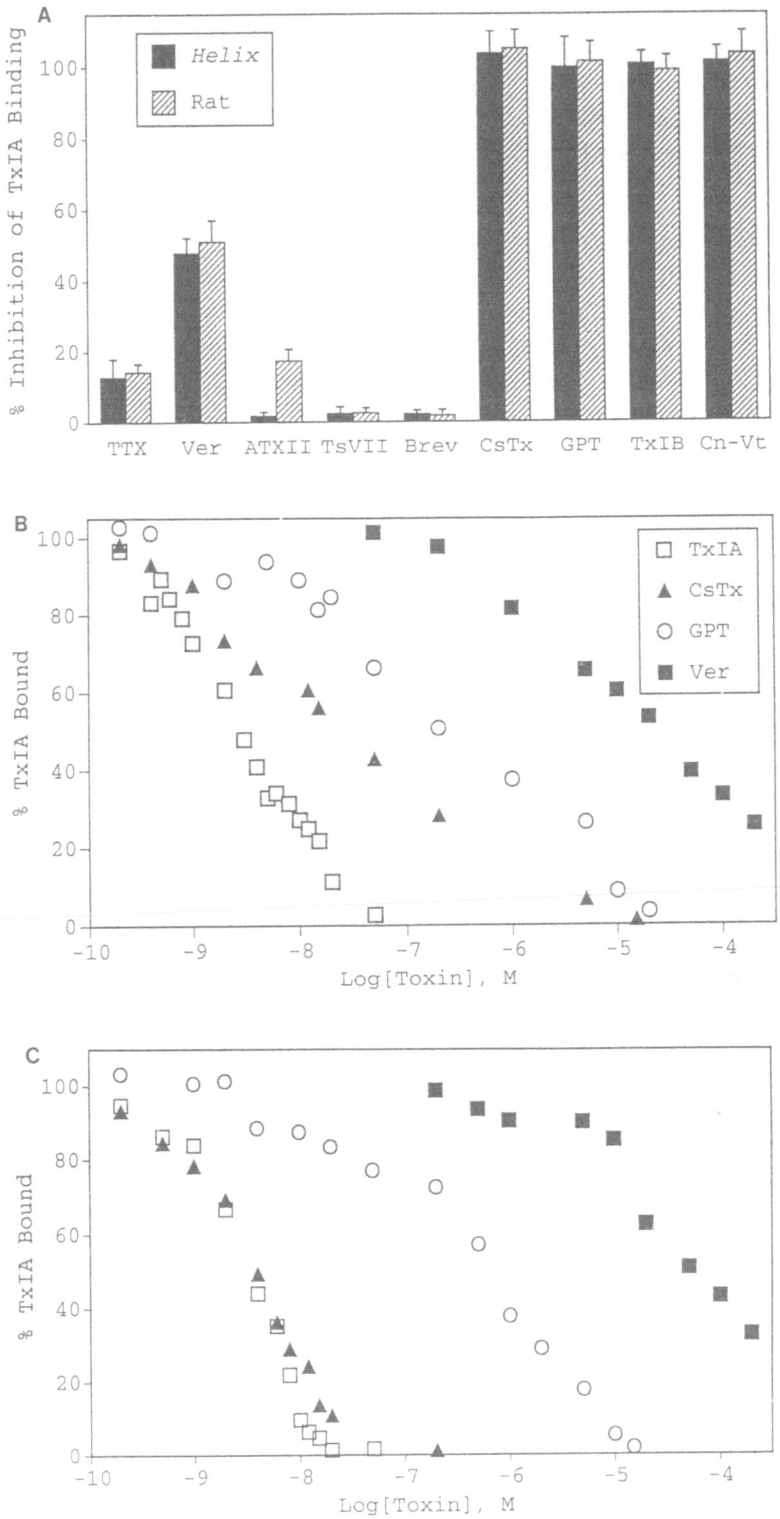
A
Helix
Rat
% Inhibition of TxIA Binding
0
20
40
60
80
100
TTX Ver ATXII TsVII Brev CsTx GPT TxIB Cn-Vt
B
TxIA
CsTx
GPT
Ver
% TxIA Bound
Log[Toxin], M
C
% TxIA Bound
Log[Toxin], M

relevant" binding. In additional experiments we demonstrated that TxVIA has no effect on *Gambusia* fish even at doses over 100 times higher than the effective paralytic dose in molluscs. Moreover, TxVIA causes no observable effects when applied directly to rat brain *in vivo* even at very high doses of up to 30 nmol/rat (see Fig. 3).

In order to examine the possibility that TxVIA might have activity in rat brain sodium channels *in vitro* we studied the effects of TxVIA and other toxins on ^{22}Na influx in rat brain synaptosomes, which is a well characterized system for the study of sodium channel neurotoxins [16]. Since the flux medium contains only 0.8 mM Mg^{2+} in divalent cations, we first verified the binding of TxVIA under these conditions. Scatchard analysis of TxVIA binding in flux medium reveals a reduced affinity (K_d = 3.8 nM) with no significant change in binding capacity (B_{max}), when compared with binding in standard TxVIA binding media (Table I). We then examined the effects of TxVIA and other sodium channel toxins on the stimulation of ^{22}Na uptake in synaptosomes. None of the toxins tested, including TxVIA, were able to initiate sodium influx (data not shown), except for the sodium channel activator veratridine. Sodium influx was found to reach maximal levels in our system at approximately 20 μM veratridine (data not shown).

The effects of the toxins on veratridine-stimulated uptake of ^{22}Na was examined in synaptosomes exposed to 2 μM veratridine. Brevetoxin, sea anemone ATXII, and CsTx all synergically increased the veratridine-stimulated flux, as expected from previous studies [17]. In contrast, TxVIA had no discernible effect on veratridine-stimulated sodium influx. Subsequently, we tested the ability of brevetoxin, ATXII, and CsTx to increase veratridine-induced flux when preincubations were carried out in the presence of 8 μM TxVIA. TxVIA inhibits almost 2-fold the ^{22}Na flux enhancement induced by CsTx but has no effect on flux enhancements induced by the other toxins. This result is in accordance with the competition of CsTx with TxVIA in binding experiments (Fig. 2).

TxVIA Protects Against CsTx-Induced Effects in Rat Brain *in Vivo*

In order to further characterize the antagonistic interaction observed *in vitro* between CsTx and TxVIA, we studied their effects in rat brain *in vivo* by intracerebroventricular injections. Rats implanted with guide cannulae were injected intracranially with toxins dissolved in artificial cerebrospinal fluid. Behavior and activity of the toxin-injected rats were

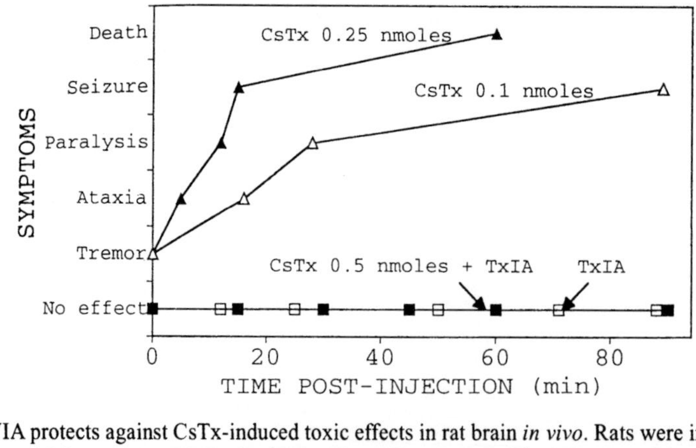

Figure 3. TxVIA protects against CsTx-induced toxic effects in rat brain *in vivo*. Rats were injected intracranially with 0.1 or 0.25 nmol/rat of CsTx (Δ,▲), or 30 nmol/rat TxVIA (□). Behavior of the injected animals was observed continuously for up to 90 min (see text). Protection was demonstrated by simultaneous injection of 0.5 nmol of CsTx with 30 nmol of TxVIA to the same animal (■).

observed for up to 4 h post-injection. Rats that did not exhibit any ill effects were subsequently injected intracranially with a lethal dose of crude *Conus* venom, in order to verify that their guide cannulae were not blocked and that their brains had indeed been exposed to previous toxin applications. As can be seen from the results summarized in Fig. 3, CsTx caused paralysis and seizures in a dose-dependent manner (0.1-0.25 nmol/rat), leading to death after 1 h at the higher dose. In contrast, TxVIA alone had no effect on the rats even when injected at a dose of 30 nmol/rat. However, upon simultaneous injection of the two toxins, 30 nmol TxVIA clearly protect against the toxic effects of 0.5 nmol CsTx (more than twice the lethal dose), indicating that TxVIA acts as an antagonist in rat brain. This protective effect is apparently absolute at the doses used, within the time frame of the experiment.

TxVIA Binds to a New Receptor Site on the Sodium Channel

The most striking result of the present study is that a mollusc sodium channel agonist acts as an antagonist at the homologous binding site on the rat brain sodium channel. As will be detailed below, we suggest that this pharmacological duality of TxVIA provides a novel selective tool, which may enable new insights into structure-function relationships in sodium channels.

Five different categories of peptide toxins (α-scorpion, sea anemone, CsTx, GPT, and TxVIA) have been shown so far to slow inactivation of sodium channels [10,12,13, 15]. Several lines of evidence suggest that the binding sites of TxVIA are distinct from those of all the other toxins that affect sodium channel inactivation. The data indicate two distinguishing features of the TxVIA binding site as follows: (1) TxVIA binding is not dependent on neuronal membrane potential (Table I), in contrast to the marked dependence of the other toxins on membrane polarization; and (2) TxVIA binding is affected via a negative allosteric interaction by the site 2 toxin veratridine (Fig. 2), as opposed to the positive cooperativity between alkaloid toxins binding at site 2 and each of the other peptide toxins that affect inactivation [12,13,15].

Competitive displacement binding experiments enable a clear discrimination between receptor site 3, which binds α-scorpion and sea anemone toxins, and the CsTx, GPT, and TxVIA binding sites [12] (Fig. 2). Additional experimental data support the distinction between the TxVIA binding site and those of CsTx and GPT. The concentration of GPT required for 50% inhibition of TxVIA binding ($IC_{50} \approx 600$ nM) is significantly higher than that of CsTx (Fig. 2) and also higher than the apparent K_d previously calculated for GPT in mouse neuroblastoma cells ($\approx$25 nM at -55 mV [12]). These data suggest that the relatively weak competition between GPT and TxVIA may result from steric hindrance between binding at different sites. The TxVIA and GPT binding sites can also be distinguished by the strong dependence of GPT on monovalent alkali cations for its activity [12], in contrast to the lack of effect of replacing Na^+ by choline in TxVIA binding media (Table I). The strong inhibition of TxVIA binding by CsTx ($IC_{50} = 3.8$ nM in rat brain, Fig. 2), and the protection by TxVIA from CsTx-induced effects both *in vitro* and *in vivo* (Fig. 3), indicates that the competition between these two toxins is mutual. Since the inhibition of TxVIA binding by CsTx is not a consequence of negative allosteric interactions, the two receptor sites (TxVIA *versus* CsTx) appear to be overlapping or sufficiently close for occupancy of one to preclude binding at the other. The marked difference between CsTx and GPT in inhibition of TxVIA binding provides additional support for the previously suggested distinction between the CsTx and GPT binding sites [13]. Altogether, the data suggest that CsTx, GPT, and TxVIA bind to adjacent sites, which may form part of a putative extracellular region of the sodium channel. It should be noted that the pharmacological similarity of TxVIA binding in mollusc and rat neuronal membranes (Table I, Figs. 1 and 2) strongly suggests that TxVIA binds to homologous sites in these preparations.

TxVIA Represents a New Category of Sodium Channel Toxins

The chemical [9], electrophysiological [10], and pharmacological (this study) data on TxVIA indicates that it belongs to a new category of sodium channel probes, hereby designated the δ-conotoxins. δ-Conotoxins are defined as small peptide toxins from *Conus* snails, which affect sodium channel inactivation through binding at a new site, distinct from other neurotoxin receptor sites affecting inactivation (see above). Conotoxin-TxVIB also belongs to this category, and the screening of other *Conus* venoms has also revealed δ-conotoxins displacing fractions (Fig. 2A). It seems most likely that additional members of this category should be obtainable from the over 500 known species of *Conus* snails.

Electrophysiological studies on δTxVIA in *Aplysia* neurons revealed that its primary effect is to slow inactivation of macroscopic sodium currents, without markedly affecting sodium channel activation kinetics [10]. It was previously suggested that one of the possible mechanisms for δTxVIA effects on *Aplysia* neurons could be the activation of a subgroup of delayed silent sodium channels [10]. The binding data does not support this possibility, since the binding capacities measured for δTxVIA in rat brain synaptosomes (≈0.6-0.7 pmol/mg, Table I) are similar to those previously determined for α-scorpion toxin in this system (0.59 pmol/mg) [14]; and the capacity for δTxVIA in the *Helix* P2 is in the same range (≈0.9-1.1 pmol/mg, Table I). It does not, therefore, seem likely that δTxVIA binds to a small subgroup of channels but rather to the main population of sodium channels that participate in generation of neuronal action potentials. The competitive and allosteric effects of other sodium channel toxins on δTxVIA binding (see above) also support this conclusion.

Interaction Between the δTxVIA Binding Site and Other Sites on the Sodium Channel

α-Scorpion and sea anemone toxins exhibit significant bidirectional positive cooperativity in binding and action with site 2 toxins [15]. Binding of CsTx was also shown to increase [^{3}H]batrachotoxin binding at site 2 [13], and GPT was shown to increase veratridine-stimulated ^{22}Na influx in N18 cells [12]. It is therefore interesting to note that our sodium flux data suggest that the allosteric interaction between veratridine and TxVIA is unidirectional, as there is no effect of δTxVIA binding on veratridine stimulation of sodium flux in rat brain synaptosomes. This aspect deserves further study. In this context, it is noteworthy that certain anticonvulsant drugs were shown to reduce batrachotoxin binding via a negative allosteric interaction [18]. Unidirectional negative allosterity was also demonstrated for the effects of TTX binding at site 1 on batrachotoxin binding [19].

The δTxVIA binding site is most likely on the extracellular side of the sodium channel. This assumption is based on the chemistry of δTxVIA as a water-soluble peptide [9], the rapidity of its effect when applied extracellulary to an isolated *Aplysia* neuron [10], the competitive inhibition of δTxVIA binding by large peptide toxins that act extracellulary (Fig. 2), and the protection by δTxVIA from CsTx-induced effects in rat brain (Fig. 3). The uniqueness of the δTxVIA binding site is accentuated in that it is the only extracellular site affecting sodium channel inactivation that undergoes negative allosteric interaction by a toxin binding at site 2.

Phyletic Differences in Sodium Channel Structure Have Functional Significance

The δTxVIA binding site may represent a conserved region of the sodium channel, since δTxVIA binds at similar high affinities to both mollusc and rat brain sodium channels.

However, a structural and/or functional difference in this region between vertebrate and mollusc neuronal sodium channels is suggested by the dramatic differences in δTxVIA activity, being an agonist in mollusc neurons and an antagonist in rat brain. The latter was demonstrated by the following: (1) its lack of toxicity in vertebrate bioassays [9] (see also above); (2) its lack of effect on ^{22}Na influx in rat brain synaptosomes; (3) its inhibition of CsTx enhancement of veratridine-induced sodium flux; and (4) its protection against CsTx-induced seizures in rat brain *in vivo* (Fig. 3). Although a previous study has reported on a chemical derivative of ATXII with reduced toxicity but full binding capability [20], this is the first reported case of a natural toxin that distinguishes between sodium channels on the basis of activity but does not discriminate on the basis of binding. Binding and activity of TxVIA might therefore be viewed as a two-step process as follows: a primary binding step, which is similar in both mollusc and rat channels, and a secondary "activation" step, which is postulated to involve conformational changes in the channel and which does not occur in the rat. For example, δTxVIA binding to mollusc sodium channels may prevent the conformational changes leading to channel inactivation from the open state but not affect these changes upon binding to the rat sodium channel.

The functional significance of phyletic differences in sodium channel interactions with δTxVIA can serve as a lead for the study of structural elements underlying different gating modes of sodium channels. The primary structures of mollusc sodium channels from the squid *Loligo* and the snail *Aplysia* have recently been elucidated [21-23]. Both channels contain domains that are highly homologous to rat brain sodium channel sequences [21,23]. Localization of the δ-conotoxin binding site on the rat and mollusc channels will provide important information on the external region in sodium channels involved in binding of at least three (δTxVIA, CsTx, GPT) of the toxins that slow channel inactivation. Intracellular regions of the sodium channel have been shown to be critical for channel inactivation [24], and mutant channels with single amino acid substitutions in intramembranal segments exhibit abnormal inactivation profiles. The α-scorpion toxin binding site has been localized on the external loops of the sodium channel between transmembrane segments S5 and S6 [25], and the other peptide toxins that affect inactivation are also thought to bind to extracellular sites on the channel (see above). However, it is not yet known how neurotoxin binding at extracellular regions of sodium channels can affect the inactivation process at intramembranal segments of the channel. Since δTxVIA discriminates between binding and function in rat brain *versus* mollusc neuronal sodium channels, it may be the probe of choice for elucidation of this phenomenon.

In conclusion, δTxVIA defines a new extracellular site affecting inactivation of sodium channels, and as such it represents a new category of pharmacological probes of sodium channels. The unusual mode of phyletic selectivity exhibited by this toxin provides a unique lead for the study of structural elements affecting the gating modes of sodium channels.

THE DUALITY OF ALLOMONE – KAIROMONE

Chemical Ecology in Marine Systems

Many behaviors in aquatic organisms are elicited by external chemical stimuli, and chemosensory signals thereby have a crucial role in marine ecosystems [26]. The aquatic environment with its theoretically infinite dilutions, and the diversity of the ubiquitous background "chemical noise" [27,28], combine to produce stringent requirements for a selective chemical signal to be recognized. Peptides have been suggested as ideal candidates for such specific cues, due to the high information content inherent in their structures, and their solubility in water [29,30]. In this study we show that conotoxin-TxVIA, a phylogenically selective venom neurotoxin, is exploited as a specific warning signal by a prey snail of the genus *Strombus*. This is the first report demonstrating that a peptide of known structure acts as an inter-species chemosensory cue in the marine environment, and presents a unique case in chemical ecology whereby the selectivity of a peptide allomone is utilized by the target prey as a specific kairomone. The specific neuroactivity of this defined chemical stimulus exemplifies the existence of intriguing parallels between the chemistry of neuronal systems and chemical communication in marine ecosystems.

Prey-Predator Interactions

Certain marine snails react to potential predators by vigorous and stereotyped escape behavior. A spectacular example of specialized escape movements is exhibited by snails of the genus *Strombus* which perform "kangaroo-like" leaping movements when confronted with potential predators [31,32]. This behavior conveys a selective advantage in enabling successful escape from a predator and is mediated by distance chemoreception [33,34]. Interestingly, the full repertoire of escape behavior is not dependent on prior experience with a predator, but stereotypically appears at a fixed stage of the early development of juvenile snails [35].

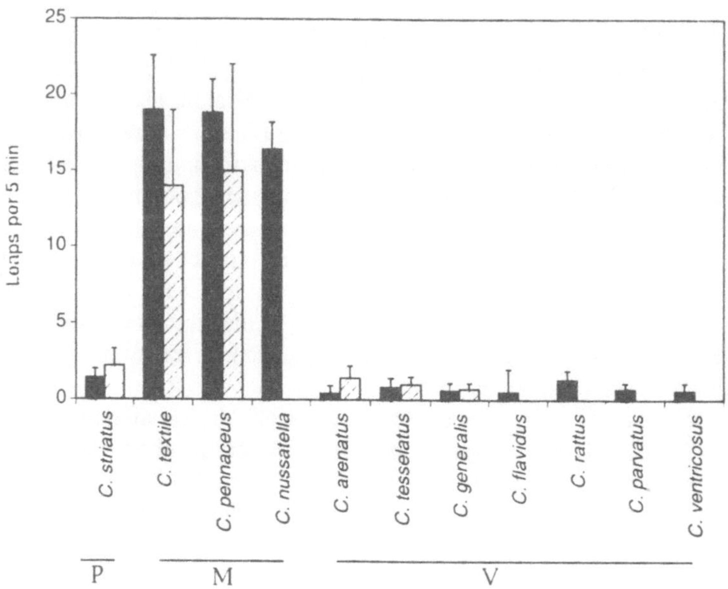

Figure 4. Adult *Strombus gibberulus* reaction to application of conditioned water (CW) from 11 species of Conidae (black bars), or to venom peptide fractions from six species of Conidae (checked bars). CW was obtained by overnight confinement of 3-5 large adult cones in a container of 1 liter clean filtered artificial sea water (SW). The G50-Vt venom peptide fraction was prepared as previously described [9]. CW was tested by placement of five strombs in the center of the container used for the overnight conditioning by the cones. Venom peptides (10 mg fraction protein in 100 μl of artificial sea water [ASW]) were tested by application adjacent to a stromb in a container of 1 liter clean ASW. Stromb reactions were quantified by counting the total number of leaps performed by each individual animal within five minutes of application of the test substance. Mean ± Std. dev. for 3 replicates of each experiment are shown. The Conidae tested represented one piscivorous species (*C. striatus*), three molluscivorous species (*C. textile, C. pennaceus, C. nussatella*), and six vermivorous species (*C. arenatus, C. tesselatus, C. generalis, C. flavidus, C. rattus, C. parvatus, C. ventricosus*).

Gonor [33] demonstrated that *Strombus* escape behavior is induced by conditioned water (CW, water where the predatory snails were immersed) from venomous *Conus* snails. Since *Conus* snails are highly specialized in their feeding ecology [36,37], we hypothesized that an ability to differentiate chemically between molluscivorous and non-molluscivorous Conidae should be selectively advantageous for *Strombus*. Accordingly, CW from 11 species of Conidae were tested for their capability to elicit escape responses in adult *Strombus gibberulus*. CW from three molluscivorous Conidae elicited a marked response in the *Strombus*, whereas CW from six vermivorous and one piscivorous species had little or no effect (Fig. 4), thus confirming that *Strombus* are able to discriminate solely on the basis of chemical stimuli between potentially dangerous and non-dangerous species belonging to the same predatory genus.

These results suggest the existence of specific chemical cues in molluscivorous Conidae. *Conus* venoms comprise a complex mix of peptides [38,6], some of which we have shown to be specifically toxic to molluscs, in correlation with the feeding specificity of the relevant snails [39,9]. We therefore wanted to find out if *Conus* venom peptides could fulfill the role of the specific alarm substances recognized by the *Strombus*. Peptide fractions from six different venoms were tested for their capability to elicit alarm responses in *Strombus*, and fractions from the two molluscivorous species tested were indeed active; although at relative high doses and in a less consistent manner than the CW (Fig. 4). A comparison between *Conus textile* CW and its venom on HPLC revealed markedly different compositions (Fig. 5), thus suggesting that although *Conus textile* venom components can elicit escape responses in *Strombus*, the primary stimulus is not venom-derived. Since *Strombus* will also react to CW conditioned for only 5 minutes with *Conus textile*, the primary stimulus seems to be a compound(s) constantly emanating from the *Conus*. However, venom components are also released to the surroundings by *Conus textile*, as "puffs" of venom released during search for prey, or while stinging prey (data not shown). We therefore examined the effects of co-application of a pure *Conus textile* toxin with CW.

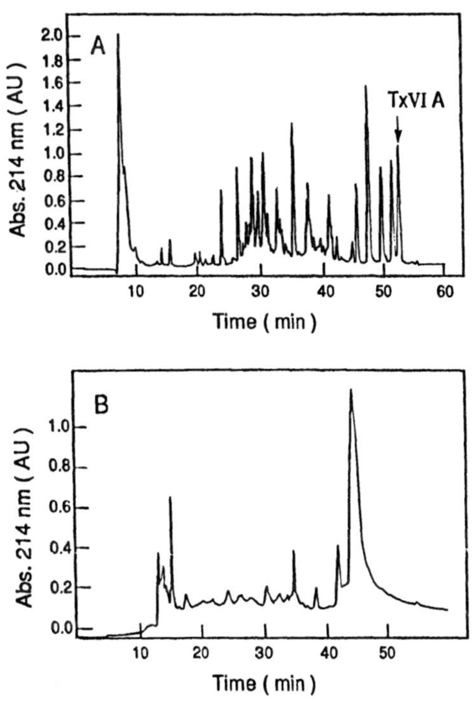

Figure 5. Anlytical HPLC separation of 300 µg venom peptides (A) and 10 ml conditioned water (B) from *Conus textile*. Details on preparation of the substances are given in the legend to Fig. 4. Separations were performed on an analytical C18 reverse phase column (Vydac wide pore, 4.6 x 250 mm, 5 µm particle size) at a flow rate of 0.5 ml/min. Substances were loaded on the column in aqueous 0.1% trifluoroacetic acid (TFA) and eluted with a linear gradient of 0-60% acetonitrile in 0.1% aqueous TFA in 0-60 minutes. On-line detection and spectral analysis was performed with a Hewlett-Packard diode array detector. The spectrum of the main peak obtained from the CW (B) is not identical to those of any of the venom derived peptides (A) that are eluted at similar times from the column (not shown). Attempts to isolate the active component(s) of *Conus textile* CW on reverse phase cartridge columns and Amicon filters were not successful, due to loss of the biological activity.

Venom Elicits Kairomonal Effects

The toxin utilized was conotoxin-TxVIA the above mentioned molluscicidal. TxVIA at doses of up to 20 nmoles did not elicit any jumping response from adult *Strombus*. However, when TxVIA was applied in the presence of a sub-threshold level of *C. textile* CW, the combination of these two factors caused a spectacular reaction in *Strombus* (Fig. 6). Marked escape responses were also observed in animals exposed to TxVIA, and afterwards to a

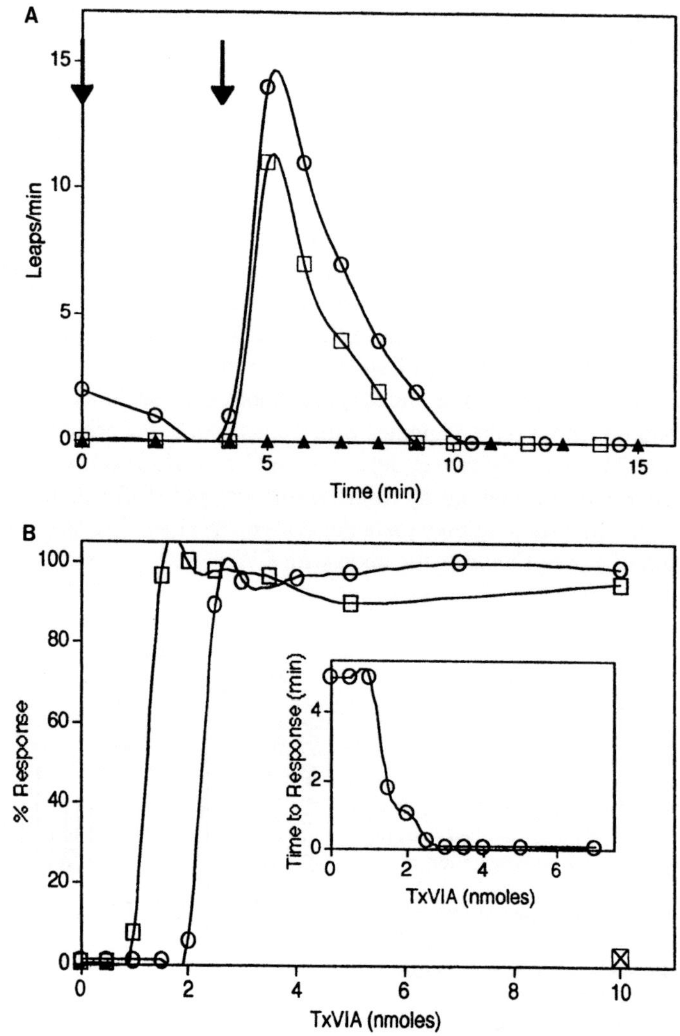

Figure 6. Synergism between a pure *C. textile* conotoxin and *C. textile* CW in eliciting *Strombus* escape response. (A) Adult *Strombus gibberulus* reaction to co-applications of 3 nmoles TxVIA with 10 ml CW. The arrows designate times of application of each substance. Squares - first arrow TxVIA, second arrow CW; circles - first arrow CW, second arrow TxVIA; triangles - first arrow *C. textile* CW, second arrow venom peptide fractions from *C arenatus, C. generalis,* or *C. tesselatus*. (B) Dose-dependence of the response. Squares - TxVIA applied together with 10 ml CW; Circles - TxVIA applied after preexposure of the animals to CW for five minutes, and their transfer to clean water. Inset shows the time period between TxVIA application and the first leap by the strombs in the second experiment. Crossed square - 10 nmoles of TxVIA incubated at 95°C for 45 minutes, and then applied to test strombs together with CW. Each data point in (A) and (B) represents the average of two replicates, except for the crossed square which represents four replicates.

sub-threshold dose of CW (Fig. 6). There was no reaction of the test *Strombus* to combinations of *C. textile* CW with peptide fractions from three other non-molluscivorous *Conus* venoms, or to application of 10 nmoles of heat-inactivated TxVIA together with CW (Fig. 6). Exposure of adult *Strombus* to sub-threshold levels of *C. textile* CW together with increasing doses of TxVIA clearly showed a dose-dependence in the triggering of the reaction (Fig. 6), with the threshold effective dose approximately 1.5 nmoles. In a second series of experiments strombs were exposed to CW, transferred to clean water, and afterwards exposed to TxVIA. Using this protocol the threshold dose of TxVIA was 2.5 nmoles (Fig. 6). Assuming that the peptide is diluted in approximately 10% of the experimental volume, the effective concentration range of TxVIA under these conditions is in the order of 0.1-0.3 μM. Interestingly this is the same concentration range as that effective in electrophysiological assays on isolated molluscan neurons [9,10].

A field experiment was carried out in order to simulate a situation whereby a "puff" of *Conus textile* toxins is released by a hunting cone in the vicinity of strombs in their natural environment. Three experimental groups of four adult *Strombus gibberulus* each were transferred to a 20 cm deep sand gully approximately 7 meters downstream of five large adult *Conus textile*. The *Strombus* were observed for 15 minutes, during which time no unusual movements were observed. After 15 minutes seven nmoles TxVIA were applied into the center of the experimental group. Immediate reactions were observed for eight strombs (three, three and two in each group), consisting of a violent burst of jumping movements which lasted for 1-2 minutes, and totaled 9-13 jumps per stromb. The movement was in a random manner, perhaps in order to confuse a predator. The other four strombs started jumping 1-1.5 minutes later, and their reaction was similar in its intensity and duration. Four minutes after application of TxVIA all the strombs had returned to normal behavior. In control experiments none of the animals reacted to applications of BSA under the same conditions.

Cooperativity Between Toxin and Kairomone

The data presented in this report suggest that a venom-derived peptide may function as a specific alarm signal for prey snails. The *Strombus* are able to distinguish between molluscivorous and non-molluscivorous Conidae (Fig. 4), their escape reaction is induced by a synergic mix of venom-derived and other chemicals released (Fig. 6 and the field experiment), is dose-dependent in a threshold "all or none" manner (Fig. 6), and is apparently specific for bioactive conformations of the venom peptide (lack of effect of heat-inactivated TxVIA, Fig. 6). Specificity of the *Strombus* escape response is further ensured by the requirement for a high concentration of a cocktail of specific toxins (Fig. 4), or a synergic combination of a toxin together with other *Conus* metabolites (Fig. 6). The ecological relevance of this phenomenon was demonstrated in a field experiment, and by observations of venom release by live cones. This is to the best of our knowledge the first documented case of a defined peptide acting as an interspecific chemosensory cue in marine ecosystems, and highlights the potential of water soluble macromolecules in fulfillment of roles that are taken up by volatile organics in the terrestrial environment. The ecological interaction described herein is unique in that a prey animal is able to "turn the tables" on its selective predator by taking advantage of the selectivity of its attack allomones as specific warning signals. This phenomenon exemplifies an emerging theme in marine chemical ecology, whereby various organisms are able to tap into existing chemical signals in the environment, in order to utilize them in a different manner from their original purpose [41]. Other examples of this theme include utilization of barnacle settlement cues as prey location odors by muricid snails [41], and the utilizaiton of algal surface molecules as metamorphosis and settlement cues by abalone larvae [42,43].

Another intriguing aspect of the current study is that it implies the existence of structural and functional similarities between chemosensory receptors and central nervous system receptors/ion channels. Such similarities might have provided *Strombus* with a preadaptation, whereby any selective chemical targeted against receptors in its nervous system might be specifically recognized by the related chemosensory receptors. In this context it is interesting to note that certain invertebrate larvae utilize neurotransmitter-like molecules (GABA or Dopa mimics) as cues for settlement and metamorphosis [43]. Furthermore neurohormone peptides can function both internally and externally (in a pheromone-like manner) to coordinate reproductive physiology and behavior in *Aplysia* [44]. Structural and/or functional similarities between external chemoreceptors and CNS channels/receptors may be widespread in marine animals, and this aspect clearly warrants further study.

CONCLUDING REMARKS: THE "DOUBLE DUALITY" OF A CONOTOXIN

TxVIA conotoxins possess a multiplicity of actions detection of which is dependent on our point of view:

(1) The peptide conotoxin TxVIA is selectively toxic to molluscs and slows sodium current inactivation in mollusc neurons, binds with high affinity to new sites on sodium channels in both mollusc and rat central nervous systems, despite its lack of toxicity to vertebrates. Furthermore, TxVIA protects from the toxic effects of Conus striatus toxin in rat brain. The TxVIA binding site differs from other neurotoxin receptor sites affecting sodium channel inactivation in that binding is not voltage-dependent and undergoes negative allosteric modulation by veratridine. TxVIA therefore represents a novel category of sodium channel probes, designated δ-conotoxins. TxVIA is shown to discriminate between sodium channels in different phyla by activity but not by binding, thus providing a lead for the study of structural elements affecting gating modes of sodium channels. In other words, from the neuropharmacological point of view, on the cellular and subcellular level the toxin reveals the duality of an agonist-antagonist.

(2) On the other hand, when observed on the organismic level from the ecological point of view the toxin reveals the duality of an allomone-kairomone. In other words, for the venomous predator the toxin serves as a tool to paralyze its prey (allomone) and for the prey the toxin supplies a warning cue (kairomone) in order to identify its predator.

Acknowledgments

This research was supported by a research Grant 90-00186 from the U.S.A.-Israel Binational Science Foundation, by a grant from the Interuniversity Fund for Ecology, Jewish National Fund, Israel and an award from the Landau fund, Mifal Hapayis, Israel. A special thanks to Dan Corcos and Solly Singer for their help in the field.

REFERENCES

1. Zlotkin, E. In: *Comparative Invertebrate Neurochemistry* (Lunt, G.G. and Olsen R.N., Eds.) Croom Helm Publ. London pp. 256-324 (1988).
2. Zlotkin, E. *Phytoparasitica*, 19, 177 (1991).
3. Fainzilber, M. PhD Thesis, Hebrew Univ. (1993).
4. Pelhate, M. and Zlotkin, E. *J. Exp. Biol.*, 97, 67 (1982).
5. Hasson, A. M.Sc. Thesis, Hebrew Univ. (1993).

6. Olivera, B.M., Rivier, J., Scott, J.K., Hillyard, D.R., and Cruz, L.J. *J. Biol. Chem.* 266, 22067 (1991).
7. Gray, W.R., Olivera, B.M., and Cruz, L.J. *Annu. Rev. Biochem.* 57, 665 (1988).
8. Hillyard, D.R., Olivera, B.M., Woodward, S., Corpuz, G.P., Gray, W.R., Ramilo, C.A., and Cruz, L.J. *Biochemistry* 28, 358 (1989).
9. Fainzilber, M., Gordon, D., Hasson, A., Spira, M.E., and Zlotkin, E. *Eur. J. Biochem.* 202, 589 (1991).
10. Hasson, A., Fainzilber, M., Gordon, D., Zlotkin, E., and Spira, M.E. *Eur. J. Neurosci.* 5, 56 (1993).
11. Gonoi, T., Hille, B., and Catterall, W.A. *J. Neurosci.* 4, 2836 (1984).
12. Gonoi, T., Ashida, K., Feller, D., Schmidt, J., Fujiwara, M., and Catterall, W.A. *Mol. Pharmacol.* 29, 347 (1986).
13. Gonoi, T., Ohizumi, Y., Kobayashi, J., Nakamura, H., and Catterall, W.A. *Mol. Pharmacol.* 32, 691 (1987).
14. Ray, R., Morrow, C.S., and Catterall, W.A. *J. Biol. Chem.* 253, 7307 (1978).
15. Catterall, W.A. *Annu. Rev. Biochem.* 55, 953 (1986).
16. Tamkun, M.M., and Catterall, W.A. *Mol. Pharmacol.* 19, 78 (1981).
17. Catterall, W.A., and Gainer, M. *Toxicon* 23,497 (1985).
18. Willow, M., and Catterall, W.A. *Mol. Pharmacol.* 22, 627 (1982).
19. Brown, G.B. *J. Neurosci.* 6, 2064 (1986).
20. Barhanin, J., Hugues, M., Schweitz, H., Vincent, J.P., and Lazdunski, M. *J. Biol. Chem.* 256, 5764 (1981).
21. Sato, C., and Matsumoto, G. *Biochem. Biophys. Res. Commun.* 186, 61 (1992a).
22. Sato, C., and Matsumoto, G. *Biochem. Biophys. Res. Commun.* 186, 1158 (1992b).
23. Dyer, J.R., Johnston, W.L., and Dunn, R.J. *Neurobiology of Aplysia Conference Abstracts*, pp. 26, Cold Spring Harbor Laboratory, Cold Spring Harbor, NY (1993).
24. Catterall, W.A. *Pharm. Rev.* 72(4) Suppl. S15 (1992).
25. Thomsen, W.J., and Catterall, W.A. *Proc. Natl. Acad. Sci. U.S.A.* 86, 10161 (1989).
26. Atema, J. In: *Sensory Biology of Aquatic Animals* (Atema, J., Fay, R.R., Popper, A.N., and Tavogla, W.N., Eds.), Springer, New York, pp. 29-56 (1988).
27. Carr, W.E.S. In: *Sensory Biology of Aquatic Animals* (Atema, J., Fay, R.R., Popper, A.N., and Tavogla, W.N., Eds.), Springer, New York, pp. 3-27 (1988).
28. Williams, J.D., Holland, K.N., Jameson, D.M., and Bruening, R.C. *J. Chem. Ecol.* 18, 2107 (1992).
29. Gurin, S., and Carr, W.E.S. *Science* 174, 293 (1971).
30. Rittschof, D., and Bonaventura, J. *J. Chem. Ecol.* 12, 1013 (1986).
31. Parker, G.H. *J. Exp. Zool.* 36, 205 (1922).
32. Gonor, J.J. *Veliger* 7, 228 (1965).
33. Gonor, J.J. *Veliger* 8, 226 (1966).
34. Kohn, A.J., and Waters, V. *Anim. Behav.* 14, 340 (1966).
35. Berg, C.J. *Am. Zool.* 12, 427 (1972).
36. Kohn, A.J. *Ecology* 4, 1041 (1966).
37. Kohn, A.J., and Nybakken, J.W. *Mar. Biol.* 29, 211 (1975).
38. Olivera, B.M., Rivier, J., Clark, C., Ramilo, C.A., Corpuz, G.P. Abogadie, F.C., Mena, E.E., Woodward, S.R., Hillyard, D.R., and Cruz, L.J. *Science* 249, 257 (1990).
39. Fainzilber, M., and Zlotkin, E. *Toxicon* 30, 465 (1992).
40. Fainzilber, M., Kofman, O., Zlotkin, E., and Gordon, D. *J. Biol. Chem.* 269, 2574 (1994).
41. Rittschof, D. *Amer. Malac. Bull.* Spec. Ed. 1, 111 (1965).
42. Morse, D.E. *Bull. Mar. Sci.* 37, 697 (1985).
43. Morse, D.E. *Bull. Mar. Sci.* 46, 465 (1990).
44. Painter, S.D. *Biol. Bull.* 183, 165 (1992).
45. Weiland, G.A. and Molinoff, P.B. *Life Sci.* 29, 313 (1981).

PURIFICATION AND CHARACTERIZATION OF NERVE GROWTH FACTORS (NGFs) FROM THE SNAKE VENOMS

Kyozo Hayashi,[1] Seiji Inoue,[2] and Kiyoshi Ikeda[2]

[1] Department of Molecular Biology
Gifu Pharmaceutical University
Mitahora-Higashi, Gifu 502, Japan
[2] Department of Biochemistry
Osaka University of Pharmaceutical Sciences
Matsubara, Osaka 580, Japan

INTRODUCTION

In general, *Elapidae* venoms contain low molecular weight proteins, such as neurotoxins and cytotoxins, and are poor in enzymes, while *Viperidae* and *Crotalidae* venoms contain higher molecular weight proteins and are rich in enzymes.

One of the most interesting features of snake venoms is the high content of NGFs. NGF is a polypeptide hormone which participates in the development and maintenance of several tissues of neural crest origin including the sympathetic nervous system.

Recently, it is suggested that NGF is an endogenous trophic factor for the cholinergic neurons which are considered to be important for memory and other cognitive processes (Hefti, 1986; Knusel, *et al.*, 1991). The loss of the cells consisting of cholinergic neurons has been emphasized for patients with Alzheimer's disease (Coyle, *et al.*, 1983).

Although NGFs are reported to present in *Elapidae*, *Viperidae*, and *Crotalidae* venoms in rich, there is no clear explanations for the physiological role of the NGF richness in snake venoms. Levi-Montalcini has suggested that a highly specific neurotrophic molecule such as NGF is utilized by the reptiles as a carrier of other neurotoxins devoid of specific receptors in the central and peripheral nervous systems (Levi-Montalcini, 1987).

Amino acid sequences of mammalian NGFs have already been determined. However, until the nucleotide sequence of NGF cDNA cloned from the venom gland of the Thailand cobra *Naja naja siamensis* was determined (Selby *et al.*, 1987), a tentative sequence of the Indian cobra *Naja naja* NGF (Hogue-Angeletti, *et al.*, 1976), based on the partial sequence and amino acid compositions of the peptides which had been obtained by enzymatic digestion of the NGF, was the only information on the amino acid sequences of snake NGF available so far.

Natural Toxins II, Edited by B. R. Singh and A. T. Tu
Plenum Press, New York, 1996

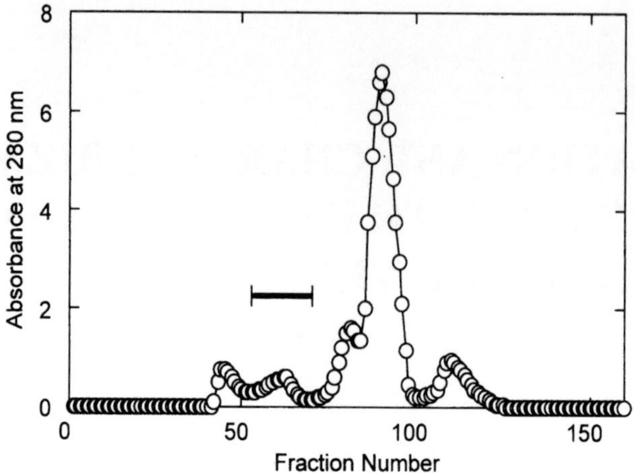

Figure 1. Gel filtration of *Naja naja atra* venom on a Sephadex G-50 column.

Elapidae NGF

We first tried to isolate and determine the amino acid sequence of *Naja naja atra* NGF (Oda, T., *et al.*, 1989).

The venom was applied to a Sephadex G-50 column (Fig. 1), and the fractions containing NGF were then applied to a CM-Sephadex C-50 column and the adsorbed proteins were eluted with an acetate buffer, pH 6.5. The eluate containing NGF thus obtained was then applied to a Mono-S column (Fig. 2). The NGF fraction was further purified by reversed-phase HPLC on a TSK-phenyl column.

During the course of purification, NGF in the fraction was immunologically detected using its cross-reactivity with anti-NGF antiserum against *Naja naja* NGF. The biological activity of NGF toward chicken embryo dorsal root ganglia was estimated by the method

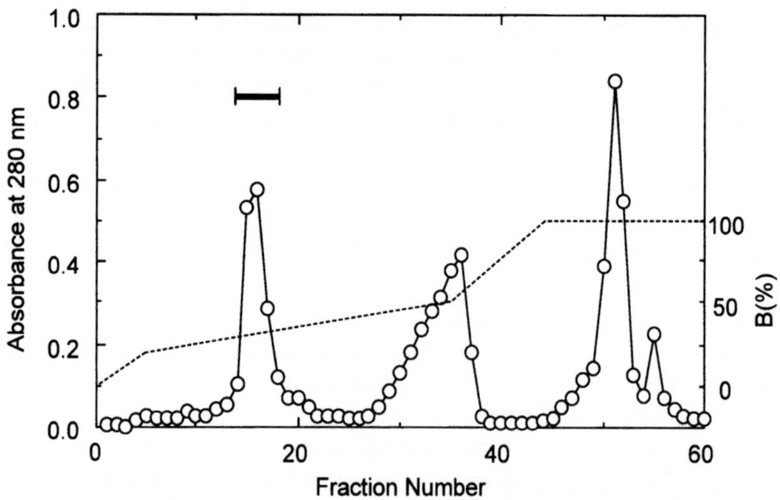

Figure 2. Mono S column chromatography of the fraction obtained from the CM-Sephadex C-50 column chromatography. The proteins were eluted by a gradient from solution A, 0.05 M sodium acetate buffer (pH 5.0), to solution B, 0.5 M sodium acetate buffer (pH 6.0).

Table 1. The amino acid compositions of snake venom NGFs. Values in parentheses were taken from the sequence

Amino acid	*Naja naja atra* NGF		*Naja naja siamensis* NGF		*Naja naja* NGF		*Vipera russelli russelli* NGF		*Crotalus adamanteus* NGF	
Asp	15.6	(16)	16.1	(16)	16.9	(17)	20.0	(20)	19.3	(20)
Thr	12.1	(13)	12.1	(13)	12.9	(14)	10.4	(11)	10.6	(10)
Ser	7.17	(8)	7.06	(8)	7.34	(8)	7.30	(7)	6.57	(6)
Glu	10.1	(10)	9.90	(10)	9.09	(9)	7.00	(6)	6.46	(6)
Pro	3.84	(4)	3.90	(4)	3.57	(4)	4.08	(4)	3.37	(4)
Gly	6.09	(6)	6.28	(6)	6.87	(6)	6.27	(6)	5.82	(6)
Ala	5.02	(5)	5.12	(5)	5.11	(5)	6.56	(6)	6.59	(6)
Val	9.15	(10)	9.58	(10)	9.51	(10)	14.2	(15)	11.6	(14)
Met	2.07	(2)	1.84	(2)	1.73	(2)	2.95	(3)	1.68	(2)
Ile	5.51	(6)	5.62	(6)	5.60	(6)	3.97	(4)	5.24	(7)
Leu	3.07	(3)	3.11	(3)	3.28	(3)	2.44	(2)	2.01	(2)
Tyr	3.03	(3)	2.86	(3)	2.69	(3)	2.93	(3)	3.45	(4)
Phe	3.78	(4)	3.95	(4)	3.86	(4)	5.77	(6)	4.89	(4)
Lys	9.19	(10)	9.84	(10)	8.81	(9)	6.22	(6)	6.16	(6)
His	4.03	(4)	3.87	(4)	3.71	(4)	3.09	(3)	2.54	(3)
Arg	3.01	(3)	2.98	(3)	3.04	(3)	7.11	(7)	7.60	(8)
Cys	5.31	(6)	5.03	(6)	5.24	(6)	5.79	(6)	N.D.	(6)
Trp	N.D.	(3)	N.D.	(3)	N.D.	(3)	N.D.	(3)	N.D.	(3)
$GlcNH_2$	0.00	–	0.00	–	0.00	–	2.11	+	N.D.	+
Total		116		116		116		117		117

a) Cys was determined as pyridylethylcysteine.

b) Trp was not determined (N.D.).

c) Glucosamine was detected by amino acid analysis (+).

described by Varon *et al.* (Varon *et al.*, 1972). The homogeneity was confirmed by isoelectric focussing in the presence of 8M urea and SDS-PAGE. Finally, about 3 mg of the purified NGF was obtained from 1 g of the lyophilized venom.

The amino acid composition (Table 1) of the purified *Naja naja atra* NGF was in good agreement with the composition deduced from the sequence determined.

The total amino acid sequence of this NGF was determined by a sequencer equipped with an on-line PTH-analyzer. The summary of determination of the sequence is given in the Fig. 2 and this is the first report of the complete amino acid sequence of snake nerve growth factor.

The sequence of *Naja naja atra* NGF was found to be composed of 116 amino acid residues. This sequence was identical to that deduced from the nucleotide sequence of NGF cDNA from the Thailand cobra *Naja naja siamensis* (Selby *et al.*, 1987).

In order to compare the amino acid sequence with those of other snake NGFs, we purified and determined the sequences of NGFs from the *Naja naja siamensis* and *Naja naja* venoms (Inoue, S., *et al.*, 1991). These cobra NGFs were purified by the same procedures as those for *Naja naja atra* NGF (Fig. 1).

Figure 4 shows the determined sequences of these cobra NGFs. Previously, Hogue-Angeletti *et al.* reported a tentative sequence of *Naja naja* NGF which had been based on the partial sequences and amino acid compositions of the peptides obtained from the enzymatic digest of the NGF (Hogue-Angeletti *et al.*, 1976). However, the sequence reported by Hogue-Angeletti *et al.* differed in 29 amino acid residues from that determined by our study. The stippled residues of *Naja naja* NGF in Fig. 3 indicate the different residues.

The amino acid sequence of *Naja naja siamensis* NGF determined by our study was identical to that of *Naja naja atra* NGF. Only two amino acid replacements were observed

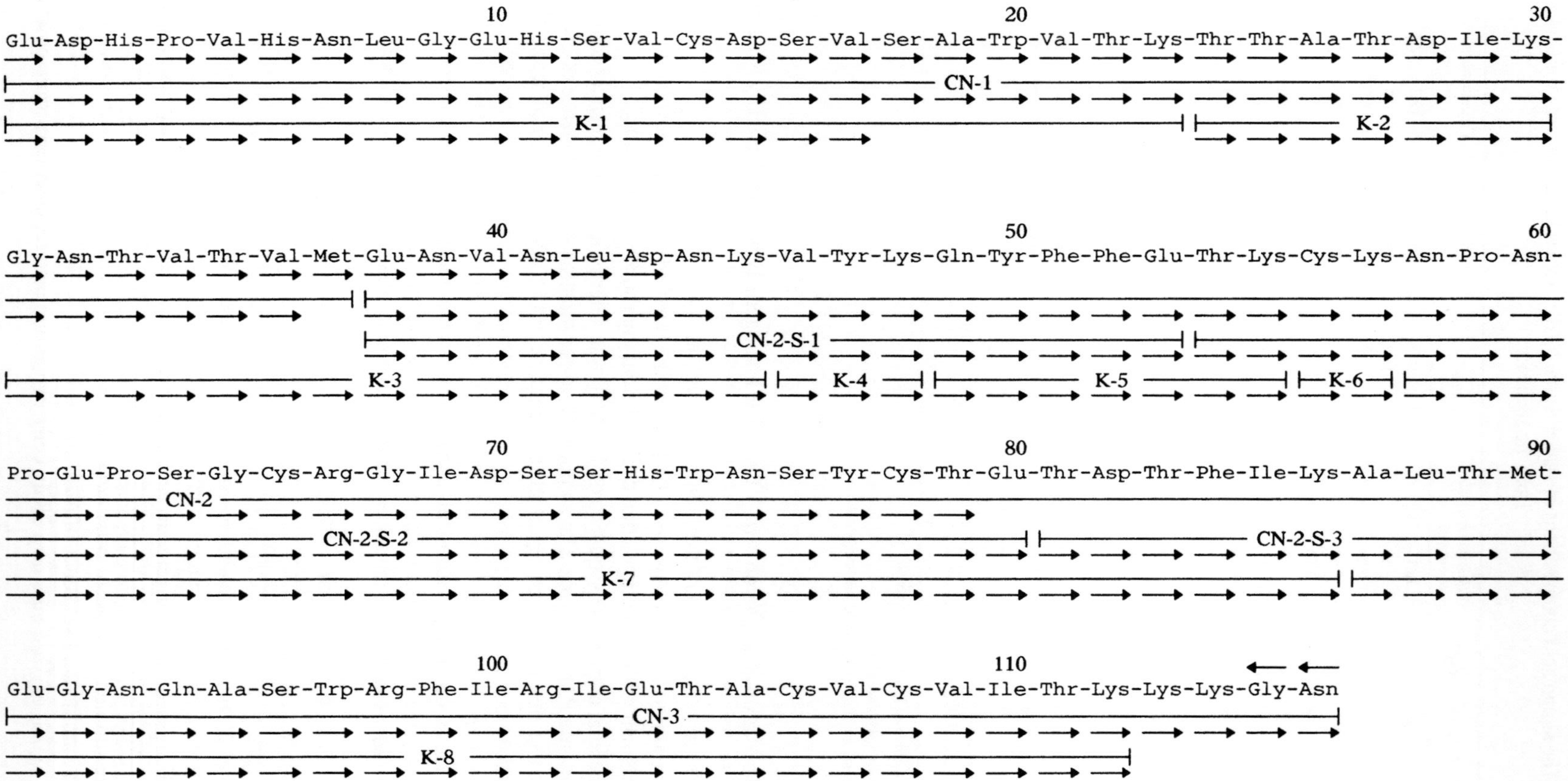

Figure 3. Summary of the sequence studies on *Naja naja atra* NGF.

	1 ... 10 ... 20 ... 30
(1) *Naja naja* NGF	- E D H P V H N L G E H S V C D S V S A W V T - K T T A T D I K
(2) *Naja naja siamensis* NGF	- E D H P V H N L G E H S V C D S V S A W V T - K T T A T D I K
(3) *Naja naja atra* NGF	- E D H P V H N L G E H S V C D S V S A W V T - K T T A T D I K
(4) Mouse NGF	S S T H P V F H M G E F S V C D S V S V W V G D K T T A T D I K

	40 ... 50 ... 60 ... 70
(1)	G N T V T V M E N V N L D N K V Y K Q Y F F E T K C K N P N P E P S G C R G I D S S H
(2)	G N T V T V M E N V N L D N K V Y K Q Y F F E T K C K N P N P E P S G C R G I D S S H
(3)	G N T V T V M E N V N L D N K V Y K Q Y F F E T K C K N P N P E P S G C R G I D S S H
(4)	G K E V T V L A E V N I N N S V F R Q Y F F E T K C R A S N P V E S G C R G I D S K H

	80 ... 90 ... 100 ... 110 ... 116
(1)	W N S Y C T E T D T F I K A L T M E G N Q A S W R F I R I D T A C V C V I T K K T G N
(2)	W N S Y C T E T D T F I K A L T M E G N Q A S W R F I R I E T A C V C V I T K K K G N
(3)	W N S Y C T E T D T F I K A L T M E G N Q A S W R F I R I E T A C V C V I T K K K G N
(4)	W N S Y C T T T H T F V K A L T T D E K Q A A W R F I R I D T A C V C V L S R K A T R

Figure 4. Comparison of the amino acid sequences of three cobra NGFs and mouse NGF. The boxed residues indicated the amino acid residues that differ from those for *N. naja* NGF.

between the sequences of *Naja naja* and *Naja naja siamensis* NGFs. However, this two amino acid replacements in cobra NGFs had no effect on the biological activity.

Viperidae NGF

Unlike cobra and mammalian NGFs, *Vipera* venom NGF had been reported to be a glycoprotein, but the purification had been incomplete. So, we tried to purify the NGF from the venom. As a result, we succeeded in purifing the NGF by a methods simpler than these reported methods (Koyama, *et al.*, 1992).

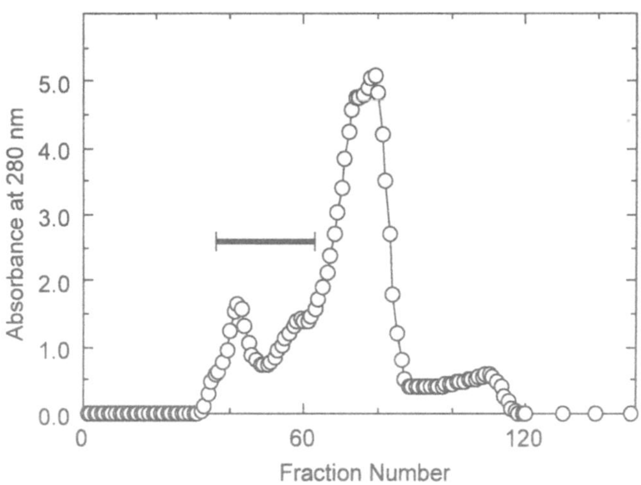

Figure 5. Gel filtration of *Vipera russelli russelli* venom on a Sephadex G-50 column with 1% acetic acid.

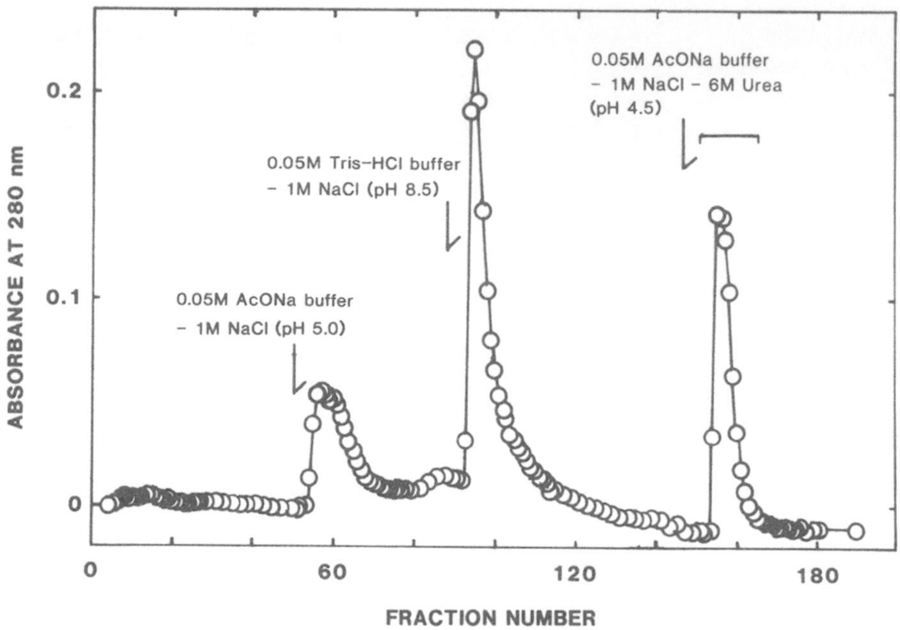

Figure 6. Blue Sepharose CL-6B column chromatography of the sample obtained from S-Sepharose column chromatography.

The *Vipera russelli* venom was first fractionated on a Sephadex G-50 column (Fig. 5), and the fractions exhibiting NGF activity were applied to a S-Sepharose column. The fractions containing the NGF activity were then applied to a Blue-Sepharose CL-6B column (Fig. 6). The NGF was adsorbed so tightly on this column that the elution needed a buffer containing 6 M urea and 0.5 M NaCl. The eluates containing NGF were dialyzed against to a phosphate buffer, pH 5.0. The NGF, thus purified, gave a single peak on reversed-phase HPLC pattern. Finally, 2.6 mg of the purified NGF was obtained from 1 g of the lyophilized venom.

As is shown in Fig. 7, the purified NGF showed a blurred protein band (apparent molecular mass of 17.5 kDa) on SDS-PAGE (*lane* A). When it had been treated with

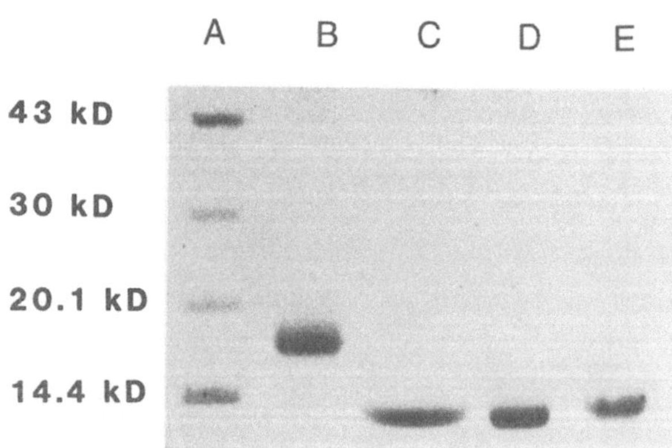

Figure 7. SDS-PAGE of *Vipera russelli russelli* NGF.

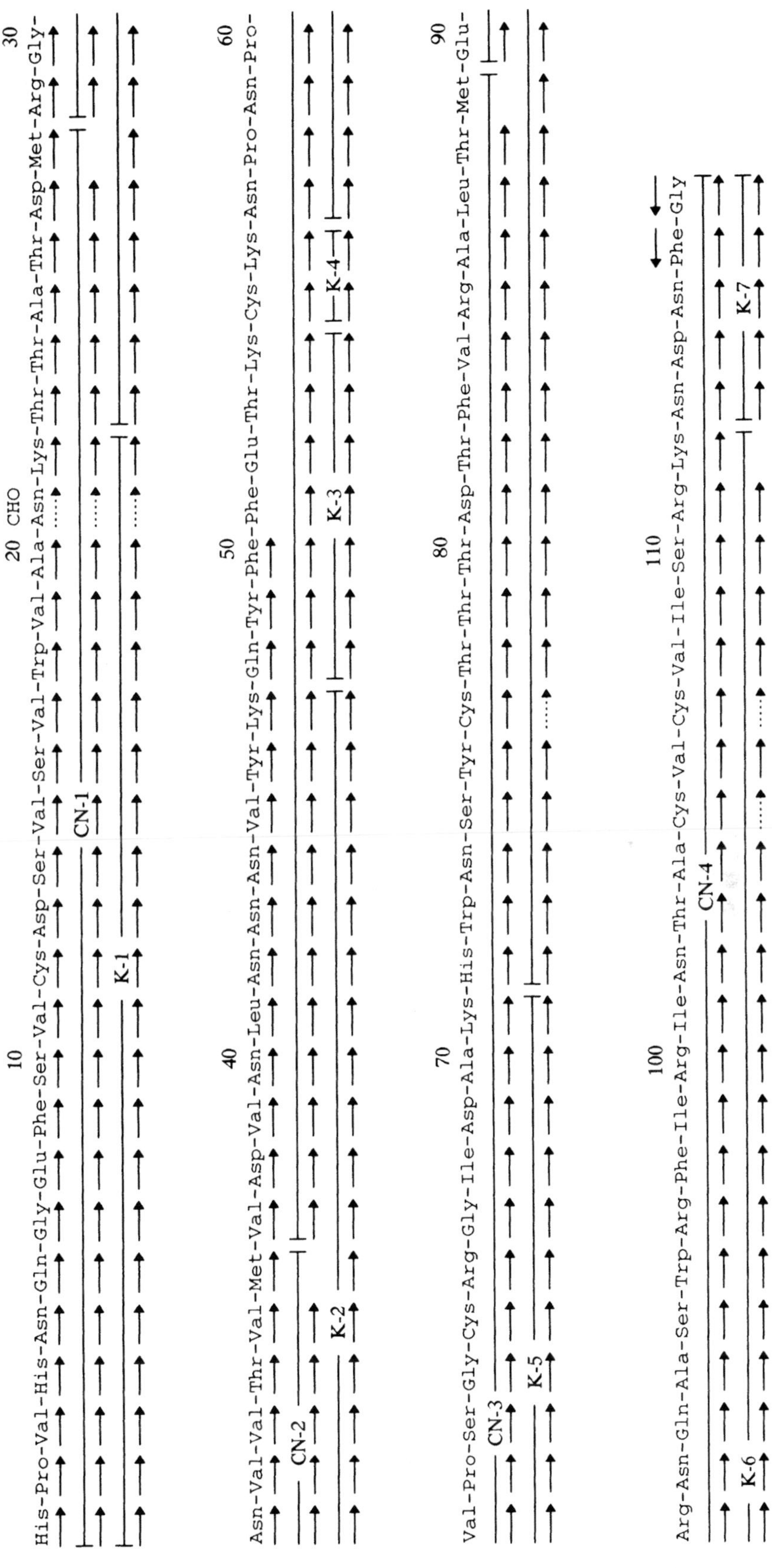

Figure 8. Summary of the sequence studies on *Vipera russelli russelli* NGF.

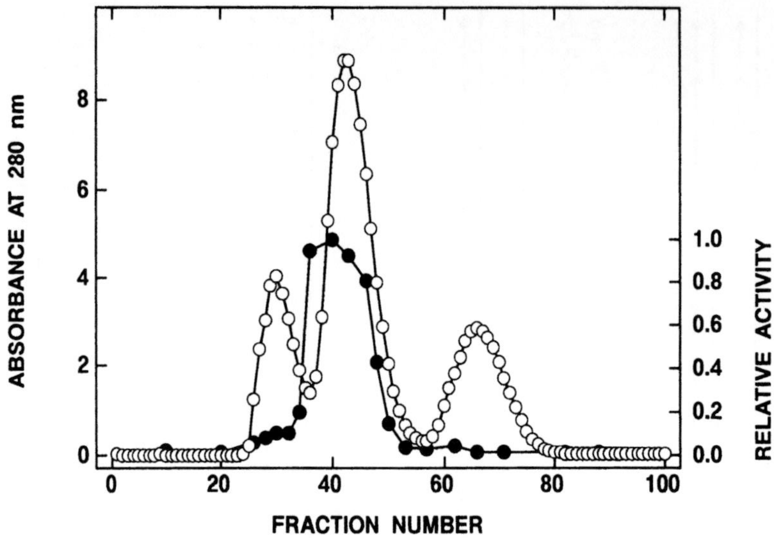

Figure 9. Gel filtration of *Crotalus adamanteus* venom on a Sephadex G-100 column.

N-glycanase, the apparent molecular mass was reduced to 13.5 kDa (*lane* B), which was similar to those of cobra NGF (*lane* C) and mouse NGF (*lane* D). This result suggests that *Vipera russelli* NGF is an *N*-linked glycoprotein.

The amino acid composition of *Vipera russelli* NGF is shown in Table 1. 2.11 moles of glucosamine per mole of the NGF were found.

As is shown in Fig. 8, *Vipera russelli russelli* NGF was completely sequenced and found to be composed of 117 amino acid residues with one residue, Asn 21, being *N*-linked glycosylated.

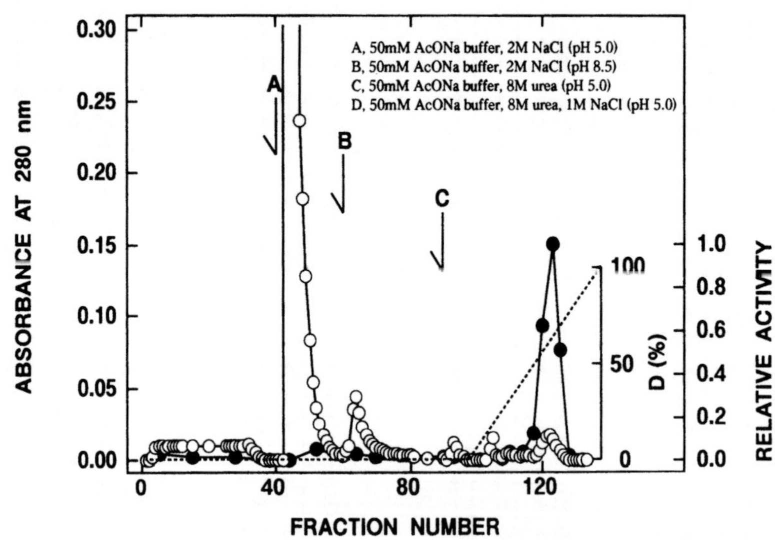

Figure 10. Blue Sepharose CL-6B column chromatography of the sample obtained from the DEAE cellulose column chromatography.

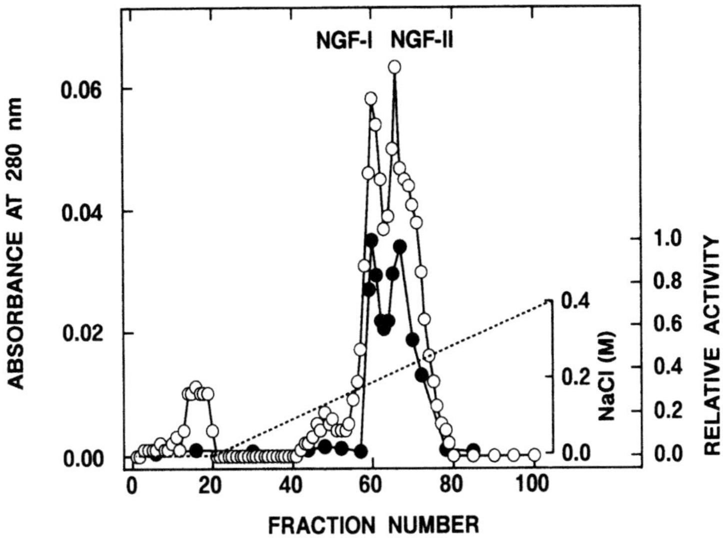

Figure 11. Mono S column chromatography of the sample obtained from the Blue Sepharose CL-6B column chromatography.

Crotalidae NGF

Recently, we also purified and determined the amino acid sequence of *Crotalus adamanteus* NGF (Horie, *et al.*, 1994).

Figure 9 shows the result of gel filtration of the venom. The fractions exhibiting NGF activity were applied to a DEAE cellulose column, and the unbound fraction was then applied to a Blue-Sepharose CL-6B column. As shown in Fig. 10, the NGF was adsorbed so tightly to the column that its elution required a buffer containing 8 M urea and NaCl. As was the case for the purification of *Vipera russelli* NGF, the Blue-Sepharose column chromatography was also very effective for the purification of this NGF, since the NGF fraction obtained in this step was practically pure as far as judged from the results of reversed-phase HPLC and SDS-PAGE. About 5 mg of NGF fraction was obtained from 1 g of the lyophilized venom.

However, when the obtained NGF fraction was chromatographed on a Mono S column (Fig. 11), the NGF activity was separated into two fractions (NGF-I and NGF-II). From the sequence analysis of the respective fractions, NGF-I was found to be an intact NGF, and NGF-II lacked four amino acid residues at the *N*-terminus. The *N*-terminal four amino acid residues were apparently not responsible for the biological activity.

In order to determine whether the NGF-II was originally included in the venom or was derived by proteolytic degradation of NGF during the course of purification, a sample of the venom was applied directly onto a Blue-Sepharose column. The NGF preparation thus obtained was found to contain NGF-II in almost the same ratio as that in the previous preparation, indicating that the NGF-II existed originally in the venom.

The purified *Crotalus adamanteus* NGF was found to be a glycoprotein whose apparent molecular mass was estimated to be about 16.5 kDa by SDS-PAGE.

As shown in Fig. 12, *Crotalus adamanteus* NGF was completely sequenced and found to be composed of 117 amino acid residues. The N-glycosylation site, Asn-23, was at the same position as that for *Vipera russelli russelli* NGF.

Figure 12. Summary of the sequence studies on *Crotalus adamanteus* NGF.

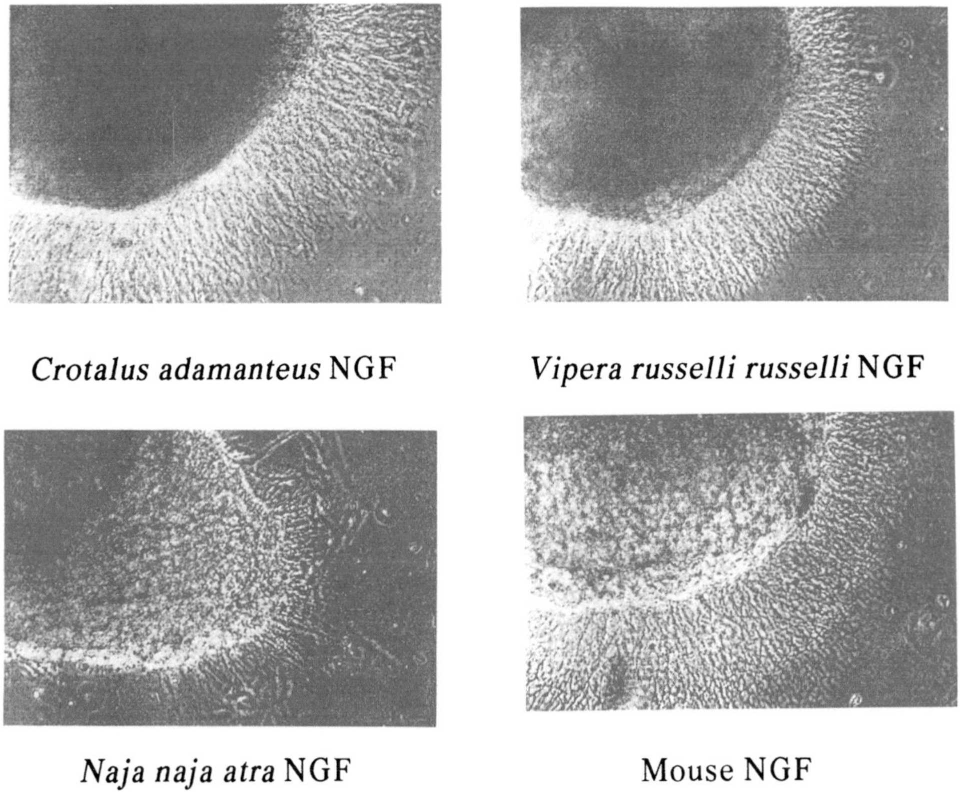

Figure 13. Comparisons of biological activities of NGFs toward chicken embryo dorsal root ganglia.

BIOLOGICAL ACTIVITY OF SNAKE VENOM NGFs

The biological activities of purified NGF or NGF preparations can be measured in vitro by its ability to support the survival and neurite outgrowth of cultured neurons. Chicken dorsal root ganglia (Varon *et al.*, 1972) or PC12 cells (Greene, 1977) have been most frequently used.

When the biological activities of snake venom NGFs were assayed by using PC12 cells, all the snake venom NGFs showed the same biological activity as mouse NGF. Activities of these NGFs toward PC12 cells are also indistinguishable each other.

However, a difference in biological activity was observed among the NGFs when the chicken dorsal root ganglia were used as targetting cells. As shown in Fig. 13, cobra NGFs have lower biological activities than mouse NGF with respect to the increase in length of neurites that grow from chicken embryo dorsal root ganglia at an optimal concentration of 10 ng/ml. On the contrary, the purified *Vipera* and *Crotalus* NGFs shows almost the same biological activity as mouse NGF toward chicken embryo dorsal root ganglia at the concentration of 10 ng/ml. Since all the NGFs elicit its maximal neurite outgrowth from embrionic dorsal root ganglia at a concentration of 10 ng/ml, the snake venom NGFs might have affinities similar to that of mouse NGF for the NGF receptor on the cells from chicken dorsal root ganglia. The difference between cobra NGFs and the other NGFs in increasing the length of neurite outgrowth might be a reflection of the different responces to the intracellular event after the binding of NGF to its receptor on the cell membrane.

```
                                   10        20        30        40        50
(1)  Naja naja atra           -EDHPVHNLGEHSVCDSVSAWVT-KTTATDIKGNTVTVMENVNLDNKVYKEY
(2)  Naja naja                -EDHPVHNLGEHSVCDSVSAWVT-KTTATDIKGNTVTVMENVNLDNKVYKEY
(3)  Crotalus adamanteus      -EDHPVHKRGEYSVCDSVNVWVANKTIATDLRGNIVTVMVDVNINNNVYKQY
(4)  Vipera russelli russelli ---HPVHNQGEFSVCDSVSVWVANKTTATDMRGNVVTVMVDVNLNNNVYKQY
(5)  Chicken                  -TAHPVLHRGEFSVCDSVSMWVGDKTTATDIKGKEVTVLGEVNINNNVFKQY
(6)  Mouse                    SSTHPVFHMGEFSVCDSVSVWVGDKTTATDIKGKEVTVLAEVNINNSVFRQY
(7)  Rat                      SSTHPVFQMGEFSVCDSVSVWVGDKTTATDIKGNEVTVLGEVNINNSVFKQY
(8)  Guinea Pig               SSTHPVFHMGEFSVCDSVSVWVADKTTATDIKGKEVTVLAEVNVNNNVFKQY
(9)  Bovine                   SSSHPVFHRGEFSVCDSISVWVGDKTTATDIKGKEVMVLGEVNINNSVFKQY
(10) Human                    SSSHPIFHRGEFSVCDSVSVWVGDKTTATDIKGKEVMVLGEVNINNSVFKQY

               60        70        80        90        100       110       120
(1)  FFETKCKNPNPEPSGCRGIDSSHWNSYCTETDTFIKALTMEGNQASWRFIRIETACVCVITKKKGN--
(2)  FFETKCKNPNPEPSGCRGIDSSHWNSYCTETDTFIKALTMEGNQASWRFIRIDTACVCVITKKTGN--
(3)  FFETKCRNPNPVPSGCRGIDARHWNSYCTTTDTFVKALTMEGNQASWRFIRIDTACVCVISRKNGN--
(4)  FFETKCKNPNPVPSGCRGIDAKHWNSYCTTTDTFVRALTMERNQASWRFIRINTACVCVISRKNDNFG
(5)  FFETKCRDPRPVSSGCRGIDAKHWNSYCTTTHTFVKALTMEGKQAAWRFIRIDTACVCVLSRKSGRP-
(6)  FFETKCRASNPVESGCRGIDSKHWNSYCTTTHTFVKALTTDEKQAAWRFIRIDTACVCVLSRKATRRG
(7)  FFETKCRARNPVESGCRGIDSKHWNSYCTTTHTFVKALTTQDRQAAWRFIRIDTACVCVLTRKAPRRG
(8)  FFETKCRDPSPVDSGCRGIDSKHWNSYCTTTHTFVKALTTANKQAAWRFIRIDTACVCVLNRKAARRG
(9)  FFETKCRDPNPVDSGCRGIDAKHWNSYCTTTHTFVKALTMDGKQAAWRFIRIDTACVCVLSRKTGQRA
(10) FFETKCRDPNPVDSGCRGIDSKHWNSYCTTTHTFVKALTMDGKQAAWRFIRIDTACVCVLSRKAVRRA
```

Figure 14. Comparison of the amino acid sequences of NGFs. The stippled boxes show the residues different from those of the other sequences.

SEQUENCE COMPARISON OF SNAKE VENOM NGFs

Figure 14 shows the comparison of the amino acid sequences of snake venom NGFs determined in the present study with those of mammalian NGFs. 65 amino acid residues are completely conserved among all the sequences. The residues, Cys (15, 58, 68, 80, 108, and 110), Trp (21, 76, 99), Tyr 52, His 75, Thr 91, Asp (16, 30, 72), Glu 55, Lys 25, and Arg (69, 100, and 103), which had been shown to be essential for the stability of the protein or subunit interactions, were conserved (Bradshaw, *et al.*, 1994).

Snake venom NGFs have some characteristic residues, His 7, Met 39, Tyr 49, Asn 60, Pro 65, Asp 84, Asn 95, Ser 98, and Ile 112. Of these residues, Met 39, Tyr 49, Asp 84, Ser 98, and Ile 112 are conserved as Leu, Phe, His, Ala, and Leu, respectively, in all the mammalian NGFs. In particular, the amino acid replacement of His 84 of the mammalian NGFs to Asp 84 in the snake venom NGFs is of great interest. His 84 has been shown to be a part of the *Trk* A (high-affinity receptor of NGF) binding region in NGF molecule (Ibanez *et al.*, 1993). The replacement of His 84 by Asp in the snake venom NGFs might compensate for the amino acid replacement at the other position of the snake venom NGFs in the receptor binding.

As mentioned above, the biological activities of cobra NGFs are lower than those of *Vipera* and *Crotalus* NGFs and those of mammalian NGFs. The sequences of cobra NGFs had a deletion at the residue 24 and had some unique residues, Asp 45, Glu 51, Glu 64, Ser 74, Glu 82, Ile 87, and Lys 114, which were conserved as Phe, Asn, Val, Lys, Thr, Val, and Arg, respectively, in all of the other sequences. Some of these residues might be responsible for the difference in their biological activities toward chicken dorsal root ganglia.

In spite of the high sequence homology among snake venom NGFs, the biological activities of *Viper russelli russelli* and *Crotalus adamanteus* NGFs were almost the same as

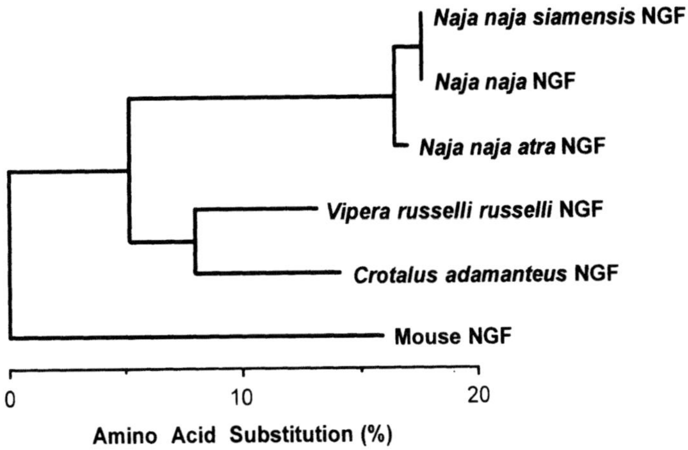

Figure 15. A phylogenetic tree of snake venom NGFs.

that of mouse NGF. This result suggests that the glycosylation at Asn 24 might be responsible for the biological activity.

Figure 15 shows the most probable molecular phylogenetic tree, constructed according to the neighbor-joining method (Saitou and Nei, 1987) on the basis of the matrix showing amino acid differences among NGFs shown in Fig. 14.

The tree is in general agreement with the consensus evolutionary relationship among the venomous snakes.

Unlike the cobra NGFs and mammalian NGFs, *Vipera* and *Crotalus* NGFs are glycoproteins. The conservation of the *N*-glycosylation site (Asn-21 in *Vipera russelli* NGF and Asn-23 in *Crotalus adamanteus* NGF) suggest that the common ancestral NGF had the same *N*-glycosylation site before the divergence of the *Viperidae* and *Crotalidae* families. Therefore, most of the snake venom NGFs from the both families must be glycoproteins. In order to elucidate this possibility, it is important to purify NGFs from the other snake venoms and to determine their amino acid sequence.

We are now studying to clarify the carbohydrate structure and the biological role of the sugar moiety of *Vipera* and *Crotalus adamanteus* NGFs.

REFERENCES

Bradshaw, R.A., Murry-Rust, J., Ibanez, C.F., McDonald, N.Q., Lapatto, R., and Blundell, T.L., 1994, Nerve growth factor: structure/function relationships, *Protein Sci.*, 3:1901-1913.

Coyle, J., Price, D.L., and Delpng, M.R., 1983, Alzheimer's disease: a disorder of cortical cholinergic innervation, *Science*, 219:1184-1190.

Greene, L.A., 1977, A quantitative bioassay for nerve growth factor (NGF) activity employing a clonal pheochromocytoma cell line, *Brain Res.*, 133:350-353.

Hefti, F., 1986, Nerve growth factor promotes survival of septal cholinergic neurons after fimbrial transections, *J. Neurosci.*, 6:2155-2162.

Hogue-Angeletti, R.A., Frazier, W.A., Jacobs, J.W., Niall, H.D., and Bradshaw, R.A., 1976, Purification, characterization, and partial amino acid sequence of nerve growth factor from cobra venom, *Biochemistry*, 15:26-34.

Horie, S., Inoue, S., Ikeda, K., and Hayashi, K., 1994, Purification and amino acid sequence of nerve growth factor from *Crotalus adamanteus* venom, *J. Nat. Toxins*, 3:165-176.

Ibanez, C.F., Ilag, L.L., Murry-Rust, J., Persson, H., 1993, An extended surface of binding to Trk tyrosine kinase receptors in NGF and BDNF allows the engineering of a multifunctional pan-neurotrophin, *EMBO J.*, 12:2281-2293.

Inoue, S., Oda, T., Koyama, J., Ikeda, K., and Hayashi, K., 1991, Amino acid sequences of nerve growth factors derived from cobra venoms, *FEBS Lett.*, 279:38-40.

Knusel, B., Winslow, J.W., Rosenthal, A., Burton, L.E., Seid, D.P., Nikolics, K., and Hefti, F., 1991, Promotion of central cholinergic and dopaminergic neuron differentiation by brain-derived neurotropic factor but not neurotropin 3, *Proc. Natl. Acad. Sci. USA.*, 88:961-965.

Koyama, J., Inoue, S., Ikeda, K., and Hayashi, K., 1992, Purification and amino-acid sequence of a nerve growth factor from the venom of *Vipera russelli russelli*, *Biochim. Biophys. Acta*, 1160:287-292.

Levi-Montalcini, R., 1987, The nerve growth gactor: thirty-five years later, *EMBO J.* 6:1145-1154.

Oda, T., Ohta, M., Inoue, S., Ikeda, K., Furukawa, S., and Hayashi, K., 1989, Amino acid sequence of nerve growth factor purified from the venom of the Formosan cobra *Naja naja atra*, *Biochem. Int.*, 19, 909-917.

Saitou, N., and Nei, M., 1987, The neighbor-joining method: a new method for reconstructing phylogenetic tree, *Mol. Biol. Evol.*, 4:406-425.

Selby, M.J., Edward, R.H., and Rutter, W.J., 1987, Cobra nerve growth factor: structure and evolutionary comparison, *J. Neurosci. Res.*, 18:293-298.

Varon, S., Nomura, J., Perez-Polo, J.R., Shooter, E.M., Kennedy, Jr.J.P., 1972, The isolation and assay of the nerve growth factor proteins, *Methods Neurochem.*, 3:203-229.

35

SNAKE VENOMS AS PROBES TO STUDY THE KINETICS OF FORMATION AND ARCHITECTURE OF FIBRIN NETWORK STRUCTURE

A. Azhar,[1] F. S. Ausat,[1] F. Ahmad,[1] C. H. Nair,[2] and D. P. Dhall[2]

[1] Coagulation and Haemostasis Research Unit
Department of Biochemistry
University of Karachi
Karachi-75270, Pakistan
[2] Vascular and Thrombosis Research Unit
Woden Valley Hospital
P.O. Box 11, Woden
ACT 2606, Australia

INTRODUCTION

Although it is widely recognized that snake venoms have both haemostatic and neurotoxic proteins, chemical mode of action of most of these proteins is not clear. Often these proteins have a procoagulant or anticoagulant activity (Denson, 1969; Rosing *et al*, 1988). The procoagulant action may be due to one or more actions including conversion of factor X to activated factor Xa, the direct conversion of prothrombin to thrombin, or the conversion of fibrinogen to fibrin by means of a thrombin-like action (Gaffney, 1977). Since many snake venoms have been shown to possess enzymes which can cleave off one or other fibrinopeptide preferentially or exclusively, these venoms may prove excellent probes for studying the effects of cleavage of each fibrinopeptide separately. Moreover, some of the venom enzymes have been shown to activate Factor XIII and are likely to affect the ultimate structure of fibrin networks by catalyzing formation of covalent cross-linkings (Pirkle *et al*, 1986). Furthermore, many snake venom enzymes possess proteolytic activities of sufficiently narrow specificity that allows their use in investigation of a wide variety of structure-function relationship of fibrinogen and fibrin. The current investigation deals with the effects of crude venoms from snakes, *Oxyuranus scutellatus scutellatus* and *Pseudonaja texitilis* on fibrin networks developed in human plasma.

Natural Toxins II, Edited by B. R. Singh and A. T. Tu
Plenum Press, New York, 1996

MATERIALS AND METHODS

Crude venoms collected from *Oxyuranus scutellatus scutellatus* (Taipan snake) and *Pseudonaja texitilis* (common brown snake) were snap frozen and stored at -70°C for further use. Appropriate concentrations of snake venoms, equivalent in coagulant activity to 1 uml^{-1} bovine thrombin with respect to thrombin clotting time of normal plasma, were used for the development of networks. Networks were developed with snake venoms (12 ugml^{-1} of *O. scutellatus* venom or 4 ugml^{-1} of *P. texitilis* venom) in platelet-free plasma obtained from healthy donors and mass-length ratio (μ_T), permeability (τ), compaction, and susceptibility to lysis with streptokinase were studied according to Nair *et al* (1991). The network characteristics were compared with bovine thrombin induced networks.

Development of Networks

Blood was collected from healthy volunteers using atraumatic venepuncture from the antecubital vein and mixed with 3.8% (w/v) trisodium citrate in a ratio of 9:1. Except for experiments on fibrinolysis, 35 KIUml^{-1} of Trasylol (aprotinin) (Bayer, Germany) was added to the blood to inhibit fibrinolytic activity. The blood was centrifuged at 2,400 × g for 10 minutes to obtain platelet-poor plasma (PPP), which was recentrifuged at 4000 × g for five minutes to obtain essentially platelet-free plasma (PFP), with a platelet count of less than 3000μl^{-1}. Networks were developed at 22 ± 2°C by the addition of bovine thrombin (Parke Davis, U.S.A.) and $CaCl_2$, in the final concentrations of 1 uml^{-1} and 25 mM respectively, or respective concentration of venom and 25 mM $CaCl_2$, in plasma pre-mixed with a trace of ^{125}I-labelled fibrinogen (Amersham, U.K.).

Kinetics of Network Development

Kinetics of network development was followed at 800 nm according to the method described by Hantgan and Hermans (1979). The increase in optical density was recorded, a

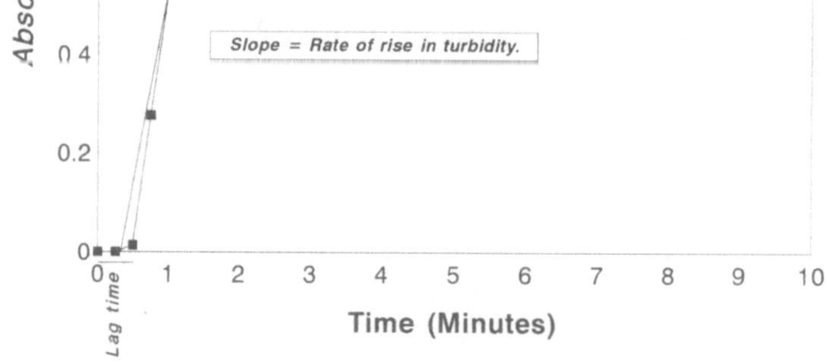

Figure 1. Kinetics of network development at 800 nm. Slope of the line represents the rate of rise in turbidity and the distance from origin to X-intercept the lag time.

tangent was drawn through the 0.25 absorbance point and extrapolated to 0 intensity. The slope of this line defined the rate of rise in turbidity ($ODsec^{-1}$). The distance from origin to the intercept of this line and 0 intensity defined the lag time (Figure 1). The lag time represents the phase prior to visible fibre growth.

Turbidimetric Technique

0.9 ml of platelet-free plasma was pipetted into cuvettes of 10 mm path length. 0.1 ml of an appropriate mixture of thrombin and $CaCl_2$ or venom and $CaCl_2$ was added, and the solution was stirred with a thin glass stirrer. Networks were left to develop at room temperature (22 ± 2°C). After the optical density had plateaued, turbidity (optical density × $\ln_{10}$) was recorded at wavelengths of 608, 650, 700, 750, and 800 nm, using unclotted plasma in the reference cell. Then $C/T\lambda^3$ was plotted as a function of $1/\lambda^2$, where T is the turbidity (optical density × $\ln_{10}$), λ is the wavelength and C is the concentration of fibrin in networks in $mgml^{-1}$. The intercept of this plot, A, was used to calculate the average mass-length ratio, μ_T, according to the equation (Nair *et al.*, 1991):

$$\mu_T = (10/1.48A) \times 10^{12} \text{ daltonscm}^{-1}$$

Permeation Technique

Fibrin networks were formed in glass tubes, which had been pre-etched by immersion in 1% hydrofluoric acid for 48 hours. Each tube was 100 mm in length and 3 mm in internal diameter and the individual internal diamensions of each tube were carefully measured. Clots were formed in tubes with the bottom end sealed with two layers of Parafilm (American Can Co., Greenwich, U.S.A.). Networks were allowed to develop for a time corresponding to that required for the turbidity to plateau by adding appropriate thrombin-$CaCl_2$ mixture or venom-$CaCl_2$ mixture to 0.8 ml of platelet-free plasma. The tube containing the clots were then attached, using plastic tubing, to horizontally mounted 1 ml pipettes on retort stands. The networks were then perfused with a Tris-HCl saline buffer (pH 7.35, ionic strength 0.153 M) at a pressure head of not more than 150 mm of water. The permeability or the Darcy constant, τ, of the fibrin networks were calculated from the equation (Carr *et al.* 1977):

$$\tau = Qh\eta/FtP$$

Where,
Q is the volume flow through the network in time t,
η is the viscosity of the perfusion buffer,
h is the length of the clot,
F is its cross-section, and
P is the applied pressure.

Compaction

Compaction is based on the method of Dhall *et al* (1976). Fibrin networks were formed in 1.5 ml Eppendroff microcentrifuge tubes by mixing 0.1 ml of an appropriate thrombin $CaCl_2$ or venom-$CaCl_2$ mixture with 0.9 ml of plasma or fibrinogen solution, pre-sprayed with a lecithin-based aerosol spray to render the surface nonadhering. The clots were centrifuged at 8,000 x g for 45 seconds in the Eppendroff microcentrifuge, Model 3200 (Germany), after they had been allowed to clot for a period of time corresponding to that required to reach a plateau in the turbidity curve. The volume of fluid expelled from networks was measured with a 1 ml syringe, and expressed as a percentage of the initial volume.

Fibrinolysis

Fibrinolysis was followed by measuring release of radioactivity from ^{125}I-prelabelled plasma, and was expressed as a percent of initial fibrin content of networks. Networks were washed with Tris-HCl saline buffer (pH 7.35, ionic strength 0.153 M) for 24 hours with three changes of buffer and placed in tubes containing equal volumes of plasma and Tris-HCl saline buffer. The tubes were then rotated in an elliptical rotator at 37°C and fibrinolysis was induced with streptokinase (KabiVitrium, Stockholm, Sweden). The release of ^{125}I-fibrin was monitored as an index of the rate of lysis.

Fibrinogen Concentration in Plasma

Fibrinogen concentration in plasma was determined according to the modified method of Ratnoff and Menzies (1951).

Statistical Analyses

Statistical analyses were performed using Minitab (Minitab, Inc., PA, U.S.A.) statistical software. Student's t test was used to determine significance levels.

RESULTS

Networks developed with *O. scutellatus* showed an extended lag phase (Figures 2). The lengthening of lag phase represents a delay in the release of fibrinopeptides from the fibrinogen molecule. Unlike *O. scutellatus*, lag phase in the clotting curve in networks developed with *P. texitilis* venom was not different from that in bovine thrombin networks

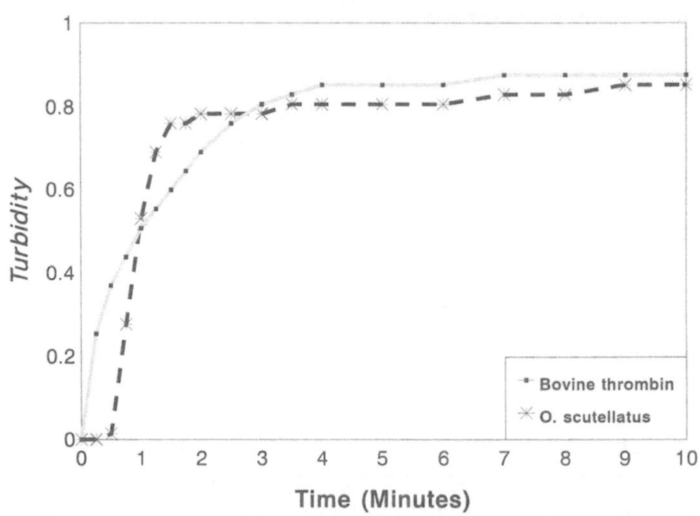

Figure 2. Turbidity curve in networks developed with bovine thombin (1uml^{-1}) or *O. scutellatus* venom (12 ugml^{-1}). Turbidity is absorbance at 800 nm × ln_{10}.

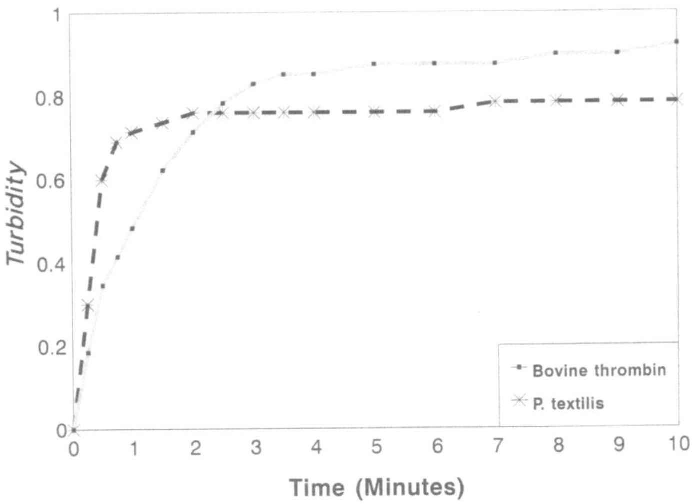

Figure 3. Turbidity curve in networks developed with bovine thombin (1uml^{-1}) or *P. texitilis* venom (4 ugml^{-1}). Turbidity is absorbance at 800 nm × $\ln_{10}$.

(Figure 3). The rate of rise in turbidity in networks developed with *P. texitilis* and in those developed with *O. scutellatus* were significantly higher than that of networks developed with bovine thrombin ($p < 0.001$). The rapid increase in rate of rise in turbidity in networks developed with *O. scutellatus* and *P. texitilis* venom induced networks suggests a rapid generation, more extensive protofibril network and faster polymerization of fibrin monomers after the release of fibrinopeptides, compared to those in networks developed with bovine thrombin. Once the equilibrium phase is reached, there was a slow steady increase in turbidity.

Figure 5 shows comparisons of characteristics in networks developed either with bovine thrombin or with snake venoms. The results are expressed as a percentage of the

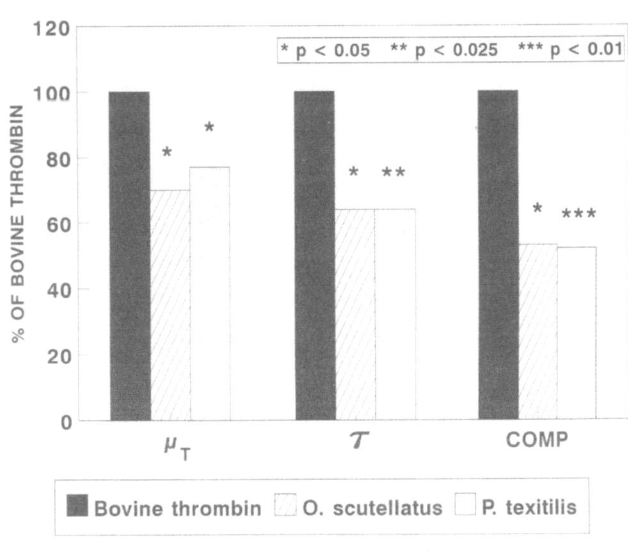

Figure 4. Comparison of network characteristics developed with bovine thombin (1uml^{-1}), *O. scutellatus* venom (12 ugml^{-1}) and *P. texitilis* venom (4 ugml^{-1}). Results are expressed as percent of respective index of bovine thrombin induced networks.

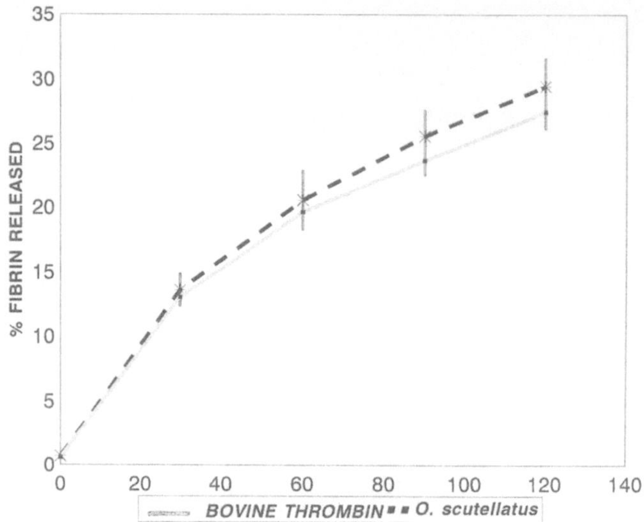

Figure 5. Susceptibility to lysis with streptokinase (2,500 $IUml^{-1}$) in networks developed with bovine thombin or *O. scutellatus* venom. The fibrin released has been expressed as a percent of initial fibrin content of networks.

respective index in bovine thrombin networks. Networks developed with *O. scutellatus* and *P. texitilis* venom had significantly lower μ_T, permeability (τ), and compaction, when compared to those developed with thrombin. Networks developed with venoms were therefore made up of thinner fibres, had decreased permeability and greater tensile strength. It seems that a higher initial monomer concentration in the protofibril network and an enhanced fibrin polymerization has resulted in thinner fibres. Another factor contributing to thinner fibrin fibres may be enhanced end-to-end polymerization, which results in decreased thickness of fibres as well as an increase in the number of fibres in the minor network.

Susceptibility to lysis by streptokinase of networks developed with snake venoms is compared with that of bovine thrombin induced networks in Figures 5 and 6. Statistically

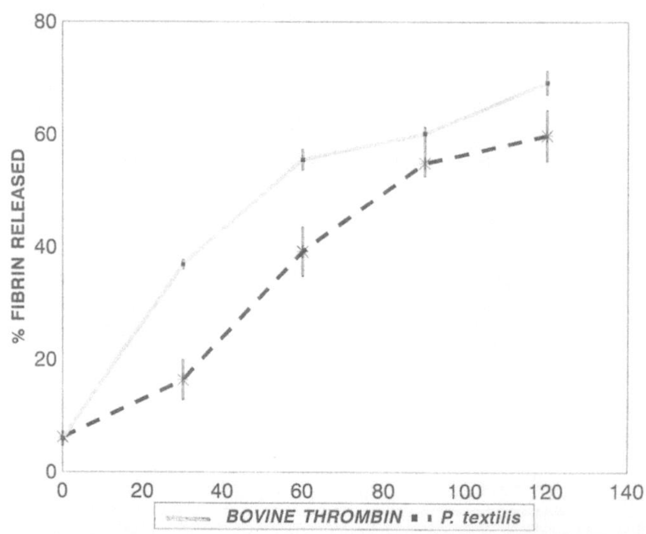

Figure 6. Susceptibility to lysis with streptokinase (2,500 $IUml^{-1}$) of networks developed with bovine thombin or *P. texitilis* venom. The fibrin released has been expressed as a percent of initial fibrin content of networks.

significant difference was not found between the rate of lysis of networks developed with *O. scutellatus* and those developed with bovine thrombin (Figure 5). Networks developed with *P. texitilis* venom, on the other hand, show greater resistance to lysis when compared to networks developed with bovine thrombin (Figure 6).

DISCUSSION

It is well established that fibrin monomer generation, polymerization, and fibrin assembly are affected by a number of proteins in plasma (Nair *et al*, 1990). The study of the kinetics of growth of networks shows that final network characteristics are determined by events preceding the appearance of fibrin fibres (Nair *et al*, 1986). There are three phases of network development: a lag phase, a phase of rapidly increasing turbidity, and an equilibrium phase. The lag phase corresponds to the time required for release of fibrinopeptides, for generation of fibrin monomers and the initial aggregation of the monomers to a network of protofibrils. Once the thickness of the fibres exceeds the wavelength of the incident light, turbidity begins to rise. The main increase in turbidity during the phase of rapidly rising turbidity is from the growth of thickness of fibres (Wolfe and Waugh, 1981). If the amount of fibrin in the protofibrils at the end of the lag phase is reduced and the conversion of fibrinogen to fibrin remains unaffected, a larger amount of fibrin is incorporated into relatively fewer but thicker fibres. An alteration in enzyme kinetics can lead to such a situation which can lead to networks with significantly thicker fibres. An increase in the concentration of fibrin monomers in the protofibril networks, on the other hand, may lead to a situation where fibrin is mainly incorporated into a greater number of thinner fibres. While there was no difference in the lag phase in networks developed with bovine thrombin and *P. texitilis* venom, a significant difference in the rate of rise in networks developed with bovine thrombin that in networks developed with venoms was observed. The final turbidity was lower in networks developed with *O. scutellatus* or *P. texitilis*. Therefore, despite the difference in the lag phase in networks developed with venoms, the difference in final turbidity can be attributed to the difference in rate of rise in turbidity which suggests a higher initial concentration of fibrin monomers in the protofibril networks and an increased polymerization and their incorporation into the network. It is interesting to note that the final structure of networks developed with both venoms is similar. Although the venoms described here seem to act through thrombin generation, the final concentration of thrombin may have differed significantly between the systems, resulting in significant difference in kinetics of network formation. Increasing concentration of thrombin have been shown to reduce permeability and fibre thickness of networks (Shah *et al*, 1987). Activation of other factors in coagulation cascade by the components of venoms may also affect the final architecture of the networks. The differences in structure appear to be determined by the amount of thrombin generated and differences in the concentation of fibrin monomers and their rate of polymerization in the networks.

O. scutellatus venom is known for its procoagulant properties, derived primarily through prothrombin activation which requires phospholipids and $CaCl_2$ for its action (Gaffney, 1977; Govers-Riemslang *et al*, 1988; Marshall & Herrmann, 1983; Rosing et al, 1988). It also shares many properties with Factor Xa and is known to contain a Factor Va-like protein. Inhibition of activity of this protein by some of the inhibitors of Xa suggest that it might belong to the class of serine proteases (Govers-Riemslang *et al*, 1988; Rosing *et al*, 1988). The characteristics of networks developed with *O. scutellatus* venom showed thinner fibrin fibres, decreased permeability, and increased tensile strength attributable to thrombin-like activity and Factor Xa and Va-like activities which collectively seem to enhance thrombin generation and promote early incorporation of fibrin into protofibrils. This venom also induced a small but insignificant increase in the degree of susceptibility to lysis by streptokinase confirming observations by Marshall and Herrmann (1983). The presence of fibrinolytic activity in venoms been suggested to act as anticoagulant, thereby facilitating distribution of the venom throughout the circulation (Markland Jr., 1988).

P. texitilis venom has also been reported to contain a prothrombin activator which has its own Factor Va-like cofactor (Govers-Riemslang *et al*, 1988; Marshall & Herrmann, 1983). It is interesting to note that unlike networks induced by *O. scutellatus* venom, the lag phase in networks developed with *P. texitilis* was not prolonged. Despite this dissimilarity the characteristics of networks were very similar. The thrombin-like and Va-like activities in both suggests that similar features of the networks are probably from similar components in these venoms. However, Factor V or X, when added to either plasma or to purified fibrin networks do not alter fibrin networks developed with bovine thrombin (Sullivan, unpublisehed observations). There was a significant increase in resistance to lysis in networks developed with *P. texitilis* venom possibly from an antifibrinolytic component in the crude venom. The structural changes in the networks may also lead to an increased resistance to lysis. This observation is in agreement with Marshall and Herrmann (1983) who could not show any fibrinolytic activity in the plasma clots developed with *P. texitilis* over a period of 24 hours.

Investigations described in this study have highlighted the suitability of snake venoms in investigating the kinetics and development of plasma networks. The crude venoms investigated here have been shown to modify network properties through altering the generation of thrombin and through affecting the concentration and polymerization of fibrin monomers. The use of purified components of snake venoms in investigation of structure-function relationship of fibrinogen and fibrin is an area of future research.

ACKNOWLEDGEMENTS

Authors wish to thank Assoc. Professor A. N. Whitaker, University of Queensland, Department of Medicine, Princess Alexandra Hospital, Queensland, Australia for providing crude venoms.

REFERENCES

Carr, M.E. Jr., Shen L.L., Hermans, J., 1977, Mass length ratio of fibrin fibres from gel permeation and light scattering. Biopolymers, 16:1.

Denson, K.W.E., 1969, Coagulant and anticoagulant actions of snake venoms. Toxicon, 7:5.

Dhall, T.Z., Bryce W.A.J., and Dhall, D.P., 1976, Effects of dextran on the molecular structure and tensilr behaviour of human fibrinogen. Thromb Haemostas, 35:737.

Gaffney, P.J., 1977, International committee communications. Report of the task force on standards for thrombin and thrombin like enzymes. Thromb Haemostas, 38:562.

Govers-Riemslag, J.W.P., Speijer, H., Zwaal, R.F.A., Rosing, J., 1988 Purification and characterization of the prothrombin activator from *Oxyuranus scutellatus* (Taipan snake). Markland, Jr. F.S. (Eds.), Marcel Dekker Inc., Newyork, Basel, pp 41.

Hantgan, R.R., and Hermans, J., 1979, Assembly of fibrin: A light scattering study. J Biol Chem, 254:11272.

Marshall, L.R., Herrmann, R.P., 1983, Coagulant and anticoagulant actions of australian snake venoms. Thromb Haemostas, 50:707.

Markland, Jr. F.S., 1988, Fibrin(ogen)olytic enzymes from snake venoms. In: 'Heamostasis and animal venoms', Pirkle H. & Markland Jr. F.S. (Eds.), Marcel Dekker Inc., Newyork, Basel, pp 149.

Nair, C.H., Azhar, A., and Dhall, D.P., 1991, Studies on fibrin network structure in human plasma: Part One: Methods for clinical application. Thromb Res, 64:455.

Nair, C.H., Azhar, A., Dhall, D.P., 1990, The effects of some plasma proteins on fibrin network structure. Blood Coag Fibrinolysis, 1: 469.

Nair, C.H., Shah, G.A., Dhall, D.P., 1986 Effects of temperature, pH and ionic strength and composition on fibrin network structure and its development. Thromb Res, 42: 809.

Pirkle, H., Vukasin, P., Theodor, I.D.A., and Simmons, G., 1986, Thickness of fibrin fibers induced by venom enzymes: Influence of secondry enzymatic action. In: 'Fibrinogen and its derivatives', G. Muller-Berghaus *et al*, (Eds.), Elsevier science publishers B.V. (Biomedical Division), pp 87.

Ratnoff, O.D., Menzie, C., 1951, A new method for the determination of fibrinogenin small samples of plasma. J .Lab.Clin.Med, 37: 316.

Rosing, J., Zwaal, R.F.A., and Tans, G., 1988, Snake venom prothrombin activators. In: 'Haemostasis and animal venoms', Pirckle, H. and Markland Jr., F.S. (Eds.), Marcel Dekker Inc., New York, Basel, pp 3.

Shah, G.A., Nair, C.H., Dhall, D.P., 1987, Comparison of fibrin networks in plasma and fibrinogen solution. Thromb Res, 45:257.

Wolfe, J. K., Waugh, D.F., 1981, Relations between enzymatic and association reactions in the development of bovine fibrin clot structure. Arch Biochem Biophys, 211:125.

FIBROLASE, AN ACTIVE THROMBOLYTIC ENZYME IN ARTERIAL AND VENOUS THROMBOSIS MODEL SYSTEMS

F. S. Markland

University of Southern California, School of Medicine
Cancer Research Laboratory #106
1303 N. Mission Rd.
Los Angeles, CA 90033

INTRODUCTION

Pharmacologic dissolution of an established thrombus has become an accepted therapeutic approach for many patients who develop thrombotic occlusive disease (1). Intravenous infusion of plasminogen activators, including recombinant tissue plasminogen activator (rt-PA), urokinase (UK), streptokinase (SK), and anisoylated plasminogen streptokinase-activator complex (APSAC), is effective in restoring blood flow in occluded arteries and veins (1,2). SK, APSAC, and UK activate circulating plasminogen as well as fibrin-bound plasminogen within the thrombus. The widespread systemic activation of the fibrinolytic system, leads to the depletion of α_2-antiplasmin (α_2-AP), and generation of free plasmin that degrades several plasma proteins, including fibrinogen, and factors V and VIII. On the other hand, rt-PA and recombinant single-chain urokinase-type plasminogen activator (scu-PA), activate plasminogen preferentially on the fibrin surface, where the fibrin associated plasmin is protected from rapid inactivation by α_2-AP. Despite the availability of fibrin-specific thrombolytic agents, currently available therapy has a number of important limitations. A significant percentage (25-30%) of patients with acute myocardial infarction are resistant to reperfusion within 90 minutes despite the use of the most potent thrombolytic agents or combinations (3) and systemic fibrinogenolysis with accompanying bleeding is encountered frequently (4). Further, 10-30% of patients experience acute coronary reocclusion following thrombolytic therapy (5). There is also a small but significant risk of neurological complications (6) including stroke (7) and intracranial hemorrhage (8). Additionally, there is concern about the rapidly acting plasma inhibitor of t-PA, PAI-1 (9), which is significantly increased in myocardial infarction (10), and may, with other factors, predispose patients to reinfarction (11).

The limited efficacy and potentially life-threatening side effects of currently available thrombolytic agents remain a problem. Investigators have attempted to overcome these problems by enhancing thrombolytic activity and improving targeting to the clot. Thus, a new generation of thrombolytic agents have evolved including mutant forms of scu-PA and

Natural Toxins II, Edited by B. R. Singh and A. T. Tu
Plenum Press, New York, 1996

t-PA, hybrids between t-PA and scu-PA or between plasmin and t-PA, and synergistic combinations of t-PA and scu-PA (12). Despite these modifications, all current thrombolytic agents depend upon the generation of plasmin through activation of plasminogen.

Snake venoms, particularly those from North American pit vipers, contain direct-acting fibrinolytic proteinases (13). Fibrolase is the fibrinolytic enzyme from southern copperhead venom (14). Fibrolase is a non-glycosylated metalloproteinase with a molecular weight of 23,000; it contains one mole of zinc per mole of protein (14). The enzyme has an isoelectric point of approximately pH 6.8. The amino acid sequence of the enzyme has been determined (15). The enzyme belongs to the major family of metalloproteinases known as the metzincins (16) and is a member of the large subfamily of snake venom metalloproteinases called the adamalysin subfamily. This subfamily was named for the first member of the family whose three-dimensional structure was determined, adamalysin II from eastern diamondback rattlesnake venom (17). The three-dimensional structure of adamalysin II reveals that the zinc binding site is minimally composed of three histidine residues which are identically located in the amino acid sequence of fibrolase. Further, there is a methionine-turn, which forms a hydrophobic base for the three active site histidine residues, that is in an identical location, relative to the zinc-binding site, in the two enzymes (18).

Fibrolase has been shown to degrade fibrin clots made from purified fibrinogen or from blood plasma (14). Fibrolase represents one of the few highly effective fibrinolytic enzymes that is not a plasminogen activator; it acts directly on fibrin and is not inhibited by serine proteinase inhibitors (SERPINS) that interfere with PA-based thrombolytic agents (14). Our studies demonstrated conclusively that fibrolase is a direct-acting fibrinolytic enzyme that does not activate plasminogen; it acts by a completely different mechanism than the plasminogen activators (19,20). Degradation of purified human fibrin (and fibrinogen) by the fibrinolytic enzyme was examined using SDS-PAGE analysis (19,20). These studies revealed that fibrolase degrades the α-chain of fibrin and fibrinogen most rapidly. There is also total degradation of the β-chain, although at a somewhat slower rate. Little, if any, degradation of the γ-chain or the γ-γ dimer occured. The major site of fibrolase cleavage in the α-chain occurs between Lys413-Leu414 (19). These findings clearly demonstrate that fibrolase acts directly on fibrin/fibrinogen and does not require any blood-borne intermediates in this reaction.

The protein C anticoagulant pathway plays a critical role in the regulation of coagulation (21). Our studies revealed that fibrolase does not activate protein C as shown by lack of generation of activated protein C chromogenic substrate hydrolyzing activity and by lack of prolongation of partial thromboplastin time by protein C following incubation with fibrolase (19).

The present report outlines studies with fibrolase in several different animal thrombosis model systems. Infusion of fibrolase proximal to an occlusive thrombus produces rapid and effective thrombolysis in carotid (22) and renal arterial, and iliac venous thrombosis model systems (23). Recently the recombinant form of fibrolase has been purified from a yeast expression system (24). It appears to be identical in all respects to the natural enzyme and has been used successfully in the carotid arterial thrombosis model system (22). Both natural and recombinant forms of the enzyme have effective thrombolytic activity in the different animal models employed; there are no observable side effects nor toxicity, and minimal or no observable hemorrhaging.

METHODOLOGY

Purification of Natural Fibrolase

A four-step open column procedure was originally developed for purification of fibrolase from crude snake venom (14). Recently, we developed a rapid two-step, analytical scale, high performance liquid chromatography (HPLC) method for purification that utilizes hydrophobic interaction (HIC) HPLC and hydroxyapatite (HAP) HPLC (25). However, for large scale enzyme purification (10 grams of venom) by the HPLC method, a final step has been added to insure homogeneity (Mono Q HPLC). The enzyme purified by these methods was homogenous by denaturing acrylamide gel electrophoresis (SDS-PAGE), reverse phase (RP)-HPLC, HAP-HPLC, and isoelectric focusing (14). Peak integration following RP-HPLC of crude southern copperhead venom indicates that fibrolase comprises 8-9% of the total protein in the venom. Fibrolase has been shown by immobilized pH gradient isoelectric focusing (26), sequence analysis (15), capillary electrophoresis (27), and cation exchange HPLC (28), to exist in two isoforms with identical enzymatic activities. The isoforms can be separated semi-preparatively by weak cation exchange HPLC (carboxymethyl, Synchropak CM300, SynChrom, Inc., Lafayette, IN), with recoveries of close to 80% activity. For the *in vivo* model studies it is not necessary to separate the isoforms, but for crystallization of the protein this is a crucial step (27).

Purification of Recombinant Fibrolase

In collaboration with Drs. P. Valenzuela and P. Riquelme from Chiron Corp., Emeryville, CA, fibrolase has been cloned and the recombinant protein expressed from a yeast system. Investigators at Chiron isolated southern copperhead snake venom gland mRNA, constructed and screened cDNA libraries, and prepared a full length fibrolase cDNA. Using proprietary expression vectors, secretion of fibrolase from yeast (*S. cerevisiae*) was achieved via a plasmid-mediated α-factor leader-fibrolase expression system under transcriptional control of the inducible ADH/GAP (alcohol dehydrogenase/glyceraldehyde phosphate dehydrogenase) promoter. Fibrolase yields of 30-60 mg/L were obtained using a proteinase deficient yeast strain in a defined medium containing zinc. The enzyme (hereafter referred to as r-fibrolase) was purified from the yeast broth by a procedure that provides the enzyme in good yield and high purity. Briefly, yeast broth was concentrated, diafiltered and quickly diluted with an equal volume of 8 M urea containing 0.5 mM $ZnCl_2$ to allow full activation of r-fibrolase. This step was necessary due to the low pH of the yeast culture medium which did not allow full zinc binding to the enzyme. The active enzyme was purified by a four-step procedure including Q-Sepharose ion exchange chromatography, phenyl Sepharose hydrophobic interaction chromatography, and Sephacryl S-100 gel permeation chromatography. Again we employed cation exchange HPLC (using the CM300 column) to resolve isoforms of r-fibrolase that were presumably generated by the low pH of the yeast culture medium (24). The first isoform of r-fibrolase eluting from the HPLC column, which was identical to the early eluting isoform of natural fibrolase, was used for the animal model studies (22).

Animal Model Studies

Initial *in vivo* studies using the rabbit were conducted at the Angiographic Research Facility at the VA Medical Center, La Jolla, in collaboration with Dr. Joseph J. Bookstein, University of California, San Diego. More recently we employed a canine reoccluding carotid arterial thrombosis model in collaboration with Dr. Benedict R. Lucchesi, University of Michigan Medical School.

Rabbit acute renal arterial thrombosis model. In vivo evaluation of purified fibrolase was initially performed, in collaboration with Joseph Bookstein, using an acute rabbit renal arterial thrombosis model. Fresh thrombi were made by adding bovine thrombin to rabbit blood. The clot was allowed to set and retract at room temperature for one hour. Rabbits were anesthetized and each kidney was catheterized under fluoroscopic guidance via a femoral artery. Control arteriograms were performed. Retracted clot, approximately 0.25 ml, was then injected into each renal artery via the arterial catheters. Complete arterial occlusion was confirmed fluoroscopically. The animal was then heparinized. Five minutes later, a repeat arteriogram was obtained for documentation. Within 10 minutes after the post-embolic arteriogram, infusion of the fibrinolytic enzyme, at the site of the thrombus, was begun on one side; the other kidney served as control. On the test side, the enzyme (1 mg/ml) was infused at the rate of 0.1 ml/min for a period of 30-60 minutes or longer. Arteriograms were repeated at intervals of 30 minutes, and the rate of lysis on the two sides was compared.

Rabbit Subacute Iliac Venous Thrombosis Model. For these studies, again in collaboration with Joseph Bookstein, subacute clots were formed in the iliac veins of Flemish giant rabbits and fibrolase was infused at the site of the thrombus on one side, with the contralateral thrombus serving as control. To establish this model, rabbits were anesthetized and an occluding spring coil device was selectively introduced retrograde from the jugular vein into each iliac vein. The device became embedded in the vessel wall inducing an occluding thrombus within a few hours. A single venogram was recorded to verify positioning of the device. After 48 hours the rabbits were studied. A control venogram was recorded using both femoral veins to confirm thrombus formation. The largest thrombus was selected as the drug test side, the smaller served as the internal control. Rabbits were heparinized by continuous infusion through the peripheral ear vein. Parathrombic infusion of venom enzyme on one side was via a catheter inserted through the femoral vein. Venographic studies were performed at 30 min intervals to monitor clot lysis. On the test side, the enzyme was infused at the rate of 0.1 ml/min, using a concentration of 1 mg/ml of fibrolase, for a period of 30-120 minutes. Arteriograms were repeated at intervals of 30 minutes, and the rate of lysis on the two sides was compared (23).

Canine Reoccluding Carotid Arterial Thrombosis Model. Recently, in collaboration with Benedict Lucchesi, we have examined activity of recombinant fibrolase in a canine reoccluding carotid arterial thrombosis model (22). Male mongrel dogs were used for the study. Each dog was anesthetized, intubated, and ventilated with room air under positive pressure. Both common carotid arteries and the right internal jugular vein were exposed. A catheter was inserted into the jugular vein for blood sampling and administration of the test drug. Arterial blood pressure was monitored from the cannulated femoral artery with the use of a blood pressure transducer. Heart rate was recorded throughout the experimental protocol from the standard limb lead II of the ECG. A doppler flow probe was placed on each common carotid artery proximal to both the point of insertion of an intraarterial electrode and a mechanical constrictor. A mechanical constrictor is adjusted to control vessel circumference and produce a regional stenosis so that the pulsatile flow pattern is reduced by 25-30% without altering mean flow. Blood flow velocity in each carotid vessel was monitored continuously. Electrolytic injury to the intimal surface of each carotid artery was accomplished with the use of an intravascular electrode. Each intraarterial electrode is connected to the anode of a dual channel stimulator. The cathode is attached to a distant subcutaneous site. The current delivered to each vessel is maintained at 300 μA. The anodal electrode is positioned to have the uninsulated portion of the electrode in contact with the endothelial surface of the vessel.

The anodal current is applied to the intimal surface of the carotid artery for a maximum period of 4 hours or is terminated 30 min after blood flow in each vessel remains stable at zero flow velocity to verify having achieved formation of a stable occlusive thrombus. Three experimental groups of five or six animals per group were studied. Group A consisted of animals in which the right carotid artery underwent electrolytic injury and thrombosis followed by the intraarterial administration of APSAC, 0.1 U/kg, immediately proximal to the occlusive lesion. Group B was similar to the previous group, but in addition animals were treated with a single intravenous dose of 0.8 mg/kg of the GPIIb/IIIa receptor antagonist, monoclonal antibody 7E3 F(ab')2 (29). Group C had both carotid arteries subjected to electrolytic injury leading to occlusive thrombus formation. Both arteries are instrumented in an identical manner and vessel wall injury is induced simultaneously in each carotid artery. Fibrolase (4 mg/kg, in a volume of 3 ml) was infused over 5 min proximal to the thrombus in the left carotid artery only. Physiological saline (identical rate as fibrolase infusion) was simultaneously infused proximal to the thrombus in the right carotid artery and this vessel was used as the control vessel to allow each dog to serve as its own control. If lysis and reperfusion was achieved, 0.8 mg/kg of the 7E3 F(ab')2 was administered intravenously as in the previously described Group B.

Reperfusion is defined as the restoration of carotid artery blood flow velocity to 20% of baseline values. Patency is defined as measurable carotid artery blood flow velocity. Blood pressure, heart rate and carotid artery flow velocity were monitored for 2 hours after achieving successful thrombolysis.

Venous blood (20 ml) was withdrawn for platelet studies from the jugular cannula into a plastic syringe containing sodium citrate anticoagulant at baseline, 5 minutes after thrombolysis by APSAC or fibrolase, 30 min after completion of the infusion of 7E3, and 2 hours after infusion of 7E3. Concentrations of red blood cells, white blood cells, hemoglobin, platelets and the hematocrit were determined. *Ex-vivo* platelet aggregation was determined by established turbidometric methods with a four-channel aggregometer. Aggregation is induced with arachidonic acid (0.65 mM and 0.325 mM) and ADP (20 μM and 5 μM). To assess the anticoagulation state of the animals, activated partial thromboplastin time (aPTT) was determined.

At the conclusion of the study, each vessel segment was ligated and removed without disturbing the intravascular thrombus. The vessel segment was opened along its length and the intact thrombus mass was lifted off the intimal surface of the vessel. The weight of the thrombus was then determined. After respective vessel segments were obtained, each dog was euthanized and a post-mortem examination was performed.

RESULTS

Rabbit Acute Renal Arterial Thrombosis Model

In the rabbit renal arterial thrombosis model, rapid lysis occurred on the side of infusion of fibrolase with reperfusion being achieved in approximately 30 min, a rate as fast as that achieved with plasminogen activators (Figure 1). Toxicity was tested by infusing fibrolase into two additional rabbits in the absence of arterial thrombus. The left carotid artery in each of the two rabbits was operatively exposed, and cannulated with a needle, through which arterial pressure was recorded. Needle electrodes were placed on three limbs enabling continuous EKG recording. Monitoring in these rabbits showed no significant electrocardiographic or manometric changes following infusion of the venom enzyme into normal rabbit kidneys. In one additional rabbit, fibrolase was infused into a normal kidney for 35 min to evaluate possible tissue injury related to the venom enzyme. Bilateral arteriograms were performed before and after infusion and two days thereafter. Blood urea nitrogen and creatinine were determined and kidneys were examined histologically two days after fibrolase infusion. No toxicity was observed angiographically, and no histologic or functional evidence of injury was apparent.

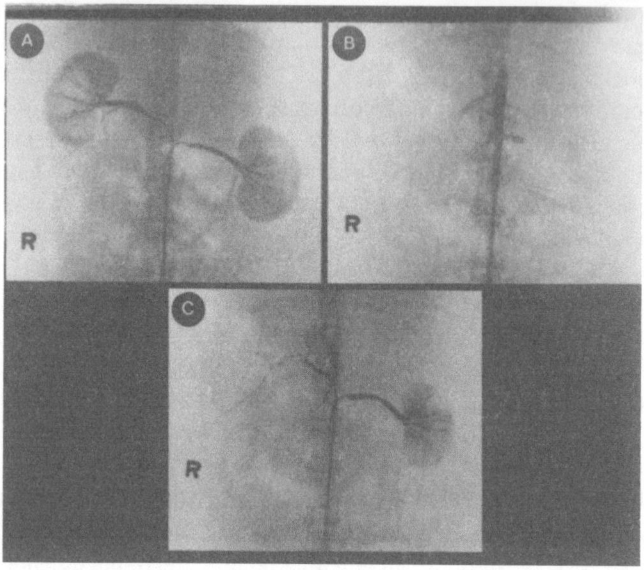

Figure 1. Thrombolysis produced by natural fibrolase in the rabbit acute renal arterial thrombosis model. The arteriograms show: A, contrast medium injected through the catheters into both normal renal arteries; B, arterial occlusion after introducing thrombus in both renal arteries; C, after infusion of fibrolase into the occluded left renal artery there is almost complete clearance of the clot at 30 min. Note the lack of clearance in the non-infused (right) kidney. After 4 hours the right renal artery remained occluded, but was cleared after a 20 min infusion of fibrolase directly into the right renal artery.

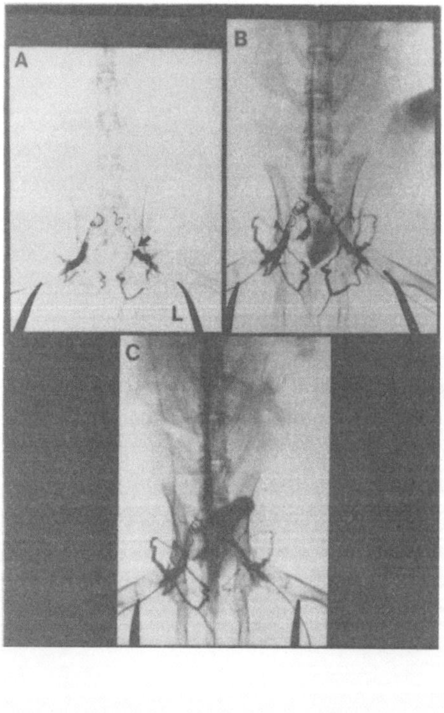

Figure 2. Thrombolytic effectiveness of natural fibrolase in a subacute iliac venous thrombosis model in the rabbit. Occluding spring coils were introduced retrograde from the jugular vein into both iliac veins. After 48 hours cutdowns on each iliac vein were performed below the coils and a catheter was introduced. Standard heparin infusion was initiated via the ear vein. A control angiogram was taken. Infusion of venom enzyme was then begun and repeat venograms were taken to document the extent of lysis. A, venogram at 0 time; B, following 1 hour of fibrolase infusion; C, following 2 hours of enzyme infusion.

These model studies revealed that selective application of highly purified fibrolase produced rapid thrombolysis. The autologous thrombus lysed promptly whereas there was no detectable lysis of the contralateral thrombus. The lack of physiologic alterations attributable to fibrolase and the absence of demonstrable histologic changes, further suggest the favorable therapeutic potential of the enzyme. These studies indicate that the venom fibrinolytic enzyme appears to function independently of the native fibrinolytic system and offers promise as a safe and effective agent for selective thrombolysis.

Rabbit Subacute Iliac Venous Thrombosis Model

In the iliac venous thrombosis model system six animals were studied with good lysis observed venographically in five (23). Figure 2 presents representative venograms illustrating the effectiveness of thrombolysis induced by selective application of natural fibrolase. No toxicity, little or no hemorrhaging, and no evidence of other side effects were observed in the animals studied. These studies again demonstrate the favorable therapeutic potential of native fibrolase.

Canine Reoccluding Carotid Arterial Thrombosis Model

We examined the thrombolytic activity of r-fibrolase using an experimental arterial thrombosis model capable of exhibiting thrombolysis and reocclusion. A total of 17 dogs were allocated among three groups. Group A animals (15±2 kg, n=6) received APSAC. Group B (13±1 kg, n=6) animals received APSAC plus 7E3 F(ab')2 monoclonal antibody, 5 minutes after achieving clot lysis. Group C (9±1 kg, n=5) animals were administered r-fibrolase plus 7E3 F(ab')2 monoclonal antibody, 5 minutes after thrombolysis.

Group A: APSAC-Induced Thrombolysis. The 6 animals in Group A developed occlusive arterial thrombi in 133±12 min. The influence of APSAC alone immediately proximal to the occlusive lesion lysed the thrombus in each of the 6 animals within a mean time of 17±3 min. Despite the efficacy of APSAC in achieving thrombolysis, reocclusion of the carotid artery occured in all animals within an additional 40±5 min. Blood flow velocity throughout the experimental protocol is summarized (Figure 3A). The mean weight of the thrombi removed from the occluded carotid arteries was 62±9 mg.

Group B: APSAC-Induceed Thrombolysis Plus 7E3 F(ab')2 Monoclonal Antibody. The mean time to carotid artery occlusion for the 6 animals in Group B was 108±13 min. The occlusive carotid arterial thrombi in Group B animals treated with APSAC lysed within 26±8 min. After clot lysis was achieved, 7E3 F(ab')2 was administered and the carotid artery remained patent in each of the 6 animals for the remainder of the experimental protocol. Blood flow velocity throughout the experimental protocol is summarized (Figure 3A). The carotid artery thrombi removed at the end of the study had a mean weight of 30±8 mg.

Group C: R-Fibrolase-Induced Thrombolysis Plus 7E3 F(ab')2 Monoclonal Antibody. Group C animals were subjected to thrombus formation in the right and left carotid arteries. The administration of 0.9% NaCl (3.0 ml over 5 min) proximal to the thrombus in the right carotid artery failed to achieve clot lysis in each of the 5 animals. The administration of r-fibrolase proximal to the thrombus in the left carotid artery achieved thrombolysis in all animals within a mean time of 6±1 min. The subsequent administration of 7E3 F(ab')2 maintained left carotid artery patency in four of the five animals so treated. One animal rethrombosed despite the presence of the 7E3 F(ab')2. The results for Group C are summarized graphically (Figure 3B) which compares the blood flow velocity in the left carotid arteries (infused with r-fibrolase plus 7E3) versus the right carotid arteries (infused with

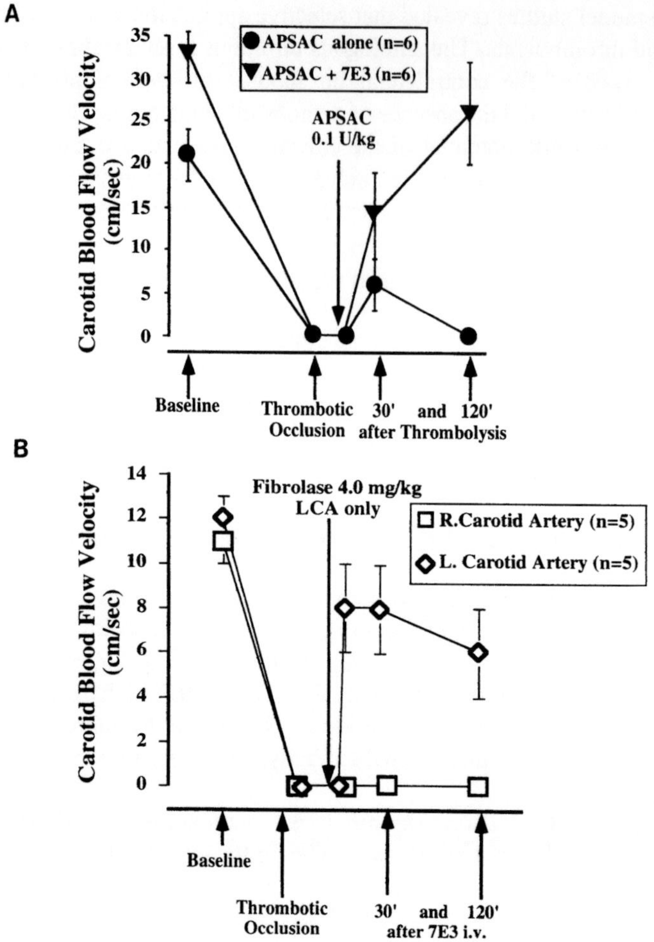

Figure 3. Carotid blood flow velocity in treated and control groups. A, carotid blood flow velocity in groups treated either with APSAC alone or APSAC plus 7E3. Thirty min after thrombotic occlusion of the vessel, APSAC was infused immediately proximal to the thrombus. Five min after restoration of the left carotid artery blood flow velocity, the 7E3 F(ab')2 monocloanl antibody was administered intravenously to only one of the groups receiving APSAC. B, carotid blood flow velocity in both the right and left carotid arteries after occlusive thrombus formation. The right carotid artery served as the control vessel and remained occluded throughout the course of the experimental procedure. Thirty min after occlusion of the left carotid artery, r-fibrolase was infused immediately proximal to the occlusive thrombus. Clot lysis was achieved in each of the five animals in the group. Five min after restoration of blood flow in the left carotid artery, the 7E3 antibody was administered intravenously. Blood flow velocity was maintained in the left carotid artery in four of the five vessels that received the combined treatment.

saline). r-Fibrolase resulted in the restoration of blood flow in the left carotid arteries (to about 70% of the pre-injury state), whereas the right carotid arteries remained occluded after saline treatment. There was a significant decrease (40%) in the weight of the residual thrombi in the left carotid arteries of the animals treated with r-fibrolase plus 7E3 compared to the weight of thrombi retrieved from the right carotid arteries treated with saline (23±7 mg *vs* 38±6 mg). Since the mean body weight in Group C animals, and thus the relative lumen size of the respective carotid arteries, was less than that of the other two groups, a comparison among groups of respective carotid flow velocity and residual thrombus weights was not appropriate.

Effects of r-Fibrolase Upon Physiologic and Hematologic Parameters. With respect to the physiologic and hematologic parameters measured in the experimental protocol, the data indicated that there were no changes in the circulating number of erythrocytes or platelets after treatment with r-fibrolase and 7E3. The hemoglobin and hematocrit values remain relatively constant during the observation period. There was an increase in the total white cell count in blood samples removed 120 min after the administration of r-fibrolase but the significance of this, if any, is not known at present. The mean arterial blood pressure and the heart rate were not altered after the administration of r-fibrolase. Towards the end of the experimental protocol, however, mean arterial blood pressure declined very slightly from the initial control value. The relationship of the decrease in mean arterial pressure to the administration of r-fibrolase is not readily apparent.

The administration of r-fibrolase (4 mg/kg), immediately proximal to the occlusive carotid artery thrombus, did not alter the aPTT from its baseline value of 111±6 sec vs 113±18 sec after r-fibrolase. Platelet aggregation in response to either arachidonic acid or ADP was decreased by approximately 60%. Complete inhibition of *ex vivo* platelet aggregation resulted following the administration of the 7E3 F(ab')2 monoclonal antibody.

Lack of Hemorrhagic Effect of r-Fibrolase. The chest was opened in all r-fibrolase treated dogs and the pericardium exposed for evidence of excessive blood infiltration. The appearance of the heart and surrounding pericardium appeared normal in each dog. The intestines, liver, and kidneys were examined and found to be normal in appearance. There was no evidence of local or internal hemorrhaging at any site in any of the dogs which received r-fibrolase.

DISCUSSION

Despite the development of more specific thrombolytic agents, currently available therapy has a number of important limitations (3-11). Investigators have attempted to overcome these problems by enhancing thrombolytic activity and improving targeting to the thrombus. Thus a new generation of thrombolytic agents have evolved, but all thrombolytic agents still depend upon the generation of plasmin.

In view of the potential limitations of plasminogen activator-based thrombolytic therapy, alternative agents having potent thrombolytic activity on their own, or that may act synergistically with plasminogen activators, need to be evaluated. In this regard investigations are presented here that are aimed at examining a mechanism of thrombolytic therapy different from presently available therapies. North American pit viper venoms contain direct-acting fibrinolytic metalloproteinases. Fibrolase, the fibrinolytic enzyme from southern copperhead venom, represents a highly effective thrombolytic agent that is not a plasminogen activator. The fibrinolytic action of fibrolase, a metalloproteinase, results from a direct proteolytic action on fibrin that is not susceptible to inhibition by SERPINS known to interfere with PA-based thrombolytic agents. Fibrolase acts rapidly and directly on the thrombus with minimal alterations to the hemostatic system.

The present results demonstrate that natural fibrolase exhibits effective *in vivo* thrombolytic activity in several animal thrombosis model systems. Infusion of the enzyme proximal to an occlusive thrombus induced rapid and specific thrombolysis in rabbit renal arterial and iliac venous thrombosis model systems. No evidence of hemorrhaging or alterations to the hemostatic system were observed in these studies. Additionally, no toxicity was observed and no angiographic, histologic, or functional evidence of side effects were obtained. The enzyme rapidly lysed 48 hr aged thrombi in the venous thrombosis model. This suggests that one of the primary mechanisms of thrombus resistance to PA-based agents, aging of the thrombus, might not be applicable with fibrolase. Thus, these studies revealed that highly purified natural fibrolase, in the presence of low dose heparin, produced rapid and effective thrombolysis. Although untoward side effects or toxicity were not apparent, one must withhold judgement on this until the enzyme can be evaluated more completely under conditions where it is administered in a fully effective systemic dose.

The primary intent of the study using the canine carotid arterial thrombosis model was to determine if the recombinant enzyme could achieve thrombolysis in an experimental animal model in which a platelet-rich, occlusive arterial thrombus, develops in response to deep arterial wall injury. We found that fibrolase produced rapid thrombolysis. The experimental model is similar to one originally developed by Lucchesi in the coronary artery (30) and subsequently adapted to the carotid artery to minimize experimental loss due to ventricular fibrillation (31). The use of the carotid has the additional advantage of greater yield of experimental data, and when necessary allows the animal to serve as its own internal control, thereby facilitating the interpretation of results and simplifying statistical evaluation. This was an important consideration in the present study where r-fibrolase was in limited supply thus accounting for the need to employ relatively small animals and to administer the enzyme by local injection immediately proximal to the occlusive platelet-rich arterial thrombus.

This report is not intended to compare fibrolase to other thrombolytic agents or to suggest that the enzyme possesses a therapeutic advantage over current therapeutic interventions used for thrombolysis. Since fibrolase is a fibrin(ogen)-specific enzyme and acts independently of the need for activation of plasminogen, it was considered important to assess its *in vivo* capacity to induce thrombolysis in experimental models in which other known thrombolytic agents would achieve a comparable effect in lysing occlusive arterial or venous thrombi. In the carotid arterial model APSAC resulted in successful thrombolysis and restoration of carotid artery blood flow in each of the animals to which it was administered. However, rethrombosis followed the use of APSAC within a relatively short time without subsequent adjunctive therapy. r-Fibrolase induced rapid thrombolysis, but was also not able to prevent rethrombosis in the reoccluding carotid arterial thrombosis model system without adjunctive antiplatelet therapy. Therefore we combined the enzyme with an antiplatelet agent, the fibrinogen receptor antagonist 7E3 (29). This combination induced rapid and sustained thrombolysis and inhibited platelet function without alterations in physiologic or hematologic parameters including the aPTT.

The amino acid sequence of fibrolase exhibits approximately 60% identity to several venom hemorrhagic metalloproteinases (15). However, fibrolase possesses no *in vitro* hemorrhagic activity (14). The lack of *in vivo* hemorrhagic activity is apparent with the recombinant enzyme in the carotid arterial thrombosis model studies. A single cDNA clone was used in the yeast expression system, therefore only a single gene product, r-fibrolase, could be produced. Thus, since only a single protein was infused in the canine model studies, there would be no chance of a contaminating hemorrhagic proteinase. Hemorrhagic activity, if observed in the canine model studies, would have to be an inherent activity of fibrolase. Under these conditions there was complete absence of internal bleeding as observed following gross examination of each of the dogs which received the combination of r-fibrolase and the antiplatelet antibody.

Although fibrolase enters the general circulation after dissolution of the venous or arterial thrombus, there is minimal lysis of the contralateral thrombus presumably due to slow inactivation of the enzyme by α_2-macroglobulin. Previous *in vitro* investigations have shown that fibrolase is inactivated in a stoichiometric manner by α_2-macroglobulin (14). Despite possible inactivation by α_2-macroglobulin, fibrolase has been used successfully following systemic administration. Ahmed *et al.* (32) showed that fibrolase effectively dissolved femoral arterial thrombi in rabbits following a single intravenous bolus injection of the enzyme (3.7 mg/kg). Thrombi were formed by injecting thrombin into a ligated section of the femoral artery. Blood flow was reestablished within 61±10 min in the three animals treated with fibrolase. This time was not much different from that with streptokinase (0.08 mg/kg). Fibrolase did not alter heart rate or blood pressure during the course of thrombolysis. Thus it was concluded that fibrolase is an efficacious thrombolytic agent following intravenous bolus administration.

The present investigations, although limited to the local application of fibrolase, demonstrate that under these conditions the enzyme lyses venous or arterial thrombi rapidly and with no observable systemic or hematologic side effects. In these thrombosis model systems the enzyme, either alone or in combination with antiplatelet therapy, offers a unique, safe, and specific mechanism for clot dissolution and may prove useful as a clinically effective alternative to, or for use in synergistic combination with, presently used thrombolytic agents.

ACKNOWLEDGEMENTS

The authors would like to acknowledge Drs. Patricio Riquelme and Pablo Valenzuela, Chiron Corporation, Emeryville, CA, for providing r-fibrolase. This work was supported in part by NIH Grant HL 31389. Figure 3 was reproduced with permission of the American Heart Association from F.S. Markland, G.S. Friedrichs, S.R. Pewitt and B.R. Lucchesi, 1994, Thrombolytic Effects of Recombinant Fibrolase or APSAC in a Canine Model of Carotid Artery Thrombosis, Circulation 90:2448-2456.

REFERENCES

1. Abel, H., 1992, Thrombolysis: the logical approach for the treatment of vascular occlusions, Acta Cardiology 47:287-295.
2. Lijnen, H.R. and Collen, D., 1991, Strategies for the improvement of thrombolytic agents, Thrombos. Haemostas. 66:88-110.
3. Collen, D, 1990, Coronary thrombolysis: streptokinase or recombinant tissue-type plasminogen activator, Ann. Intern. Med. 112:529-538.
4. Lawrence, P.F. and Goodman, G.R., 1992, Thrombolytic therapy, Surg. Clin. North Am. 72:899-918.
5. Rapaport, E., 1991, Thrombolysis, anticoagulation, and reocclusion, Am. J. Cardiol. 68:17E-22E.
6. Caramelli, P., Mutarelli, F.G., Caramelli, B., Tranchesi, B., Pileggi, F. and Scaff, M., 1992, Neurological complications after thrombolytic treatment for anti myocardial infarction: emphasis on unprecedented manifestations, Acta Neurologica Scand. 85:331-333.
7. Maggioni, A.P., Franzosi, M.G., Santoro, F., White, H., Vandewerf, F. and Tognoni, G., 1992, The risk of stroke in patients with acute myocardial infarction after thrombolytic and antithrombotic treatment. New Engl. J. Med. 327:1-6.
8. Kase, C.S., Pessin, M.S., Zivin, J.A., delZoppo, G.J., Furlan, A.J., Buckley, J.W., Snipes, R.G. and LittleJohn, J.K., 1992, Intracranial hemorrhage after coronary thrombolysis with tissue plasminogen activator, Am. J. Med. 92:384-390.
9. Kruithof, E.K.O., 1988, Inhibitors of plasminogen activators, In: Tissue-Type Plasminogen Activator (t-PA): Physiological and Clinical Aspects (C. Kluft, Ed.), CRC Press, Inc., Boca Raton, pp 189-210.
10. Gram, J., Kluft, C. and Jespersen, J., 1987, Depression of tissue plasminogen activator (t-PA) activity and rise of t-PA inhibition and acute phase reactants in blood of patients with acute myocardial infaction (AMI), Thrombos. Haemostas. 58: 817-821.
11. Vaughan, D.E., Declerck, P.J., Van Houtte, E., DeMol, M. and Collen, D., 1992, Reactivated recombinant plasminogen activator inhibitor-1 (rPAI-1) effectively prevents thrombolysis *in vivo*, Thrombos. Haemostas. 68:60-63.
12. Collen, D., 1994, Development of new fibrinolytic agents, In: Haemostasis and Thrombosis (A.L. Bloom, C.D. Forbes, D.P. Thomas and E.G.D. Tuddenham, Eds.), 3rd Ed., Churchill Livingstone, New York, pp 625-637.

13. Markland, F.S., 1988, Fibrinogenolytic enzymes from snake venoms, In: Hemostasis and Animal Venoms (H. Pirkle and F.S. Markland, Eds.), Marcel Dekker, Inc., New York, pp. 149-172.
14. Guan, A.L., Retzios, A.D., Henderson, G.N. and Markland, F.S., 1991, Purification and characterization of a fibrinolytic enzyme from venom of the southern copperhead snake (*Agkistrodon contortrix contortrix*), Arch. Biochem. Biophys. 289:197-207.
15. Randolf, A., Chamberlain, S.H., Chu, C., Retzios, A.D., Markland, F.S. and Masiarz, F.R., 1992, Amino acid sequence of fibrolase, a direct-acting fibrinolytic enzyme from *Agkistrodon contortrix contortrix* venom, Protein Science 1:590-600.
16. Bode, W., Gomis-Ruth, F-X. and Stockler, W., 1993, Astacins, serralysins, snake venom and matrix metalloproteinases exhibit identical zinc-binding environments (HEXXHXXGXXH and Met-turn) and topologies and should be grouped into a common family, the 'metzincins', FEBS Letters 331:134-140.
17. Gomis-Ruth, F-X., Kress, L.F. and Bode, W., 1993, First structure of a snake venom metalloproteinase: a prototype for matrix metalloproteinases/collagenases, EMBO J. 12:4151-4157.
18. Gomis-Ruth, F-X., Kress, L.F., Kellerman, J., Mayr, I., Lee, X., Huber, R. and Bode, W., 1994, Refined 2.0 A^0 x-ray crystal structure of the snake venom zinc-endopeptidase adamalysin II. Primary and tertiary structure determination, refinement, molecular structure and comparison with astacin, collagenase and thermolysin, J. Mol. Biol. 239:513-544.
19. Retzios, A.D. & Markland, F.S., 1988, A direct-acting fibrinolytic enzyme from the venom of *Agkistrodon contortrix contortrix:* Effects on various components of the human blood coagulation and fibrinolytic systems, Thrombos. Res. 52:541-552.
20. Ahmed, N.K., Tennant, K.D., Markland, F.S. and Lacz, J.P., 1990, Biochemical characteristics of fibrolase, a fibrinolytic protease from snake venom, Haemostas. 20:147-154.
21. Esmon, C.T., 1983, Protein C: biochemistry, physiology and clinical implication, Blood 62:1155-1158.
22. Markland, F.S., Friedrichs, G.S., Pewitt, S.R. and Lucchesi, B.R., 1994, Thrombolytic effects of recombinant fibrolase or APSAC in a canine model of carotid artery thrombosis. Circulation 90:2448-2456.
23. Markland, F.S., Bookstein, J.J., and Machado, T., 1989, *In vivo* fibrinolytic activity of fibrolase, a metalloproteinase from the venom of *Agkistrodon contortrix contortrix*, Thrombos. Haemostas. 62:121.
24. Loayza, S.L., Trikha, M., Markland, F.S., Riquelme, P. and Kuo, J., 1994, Resolution of isoforms of natural and recombinant fibrolase, the fibrinolytic enzyme from *Agkistrodon contortrix contortrix* snake venom, and comparison of their EDTA sensitivities, J Chromatog. B 662:227-243.
25. Retzios, A.D., and Markland, F.S., 1990, HPLC-based two-step purification of fibrinolytic enzymes from the venoms of *Agkistrodon contortrix contortrix* and *Agkistrodon piscivorus conantii*, Prot. Express. Purific. 1:33-39.
26. Markland, F.S. and Guan, A.L., 1986, The Use of IPG to purify a fibrinolytic snake venom enzyme, Protides of the Biological Fluids 34:807-810.
27. Markland, F.S., Morris, S., Deschamps, J.R. and Ward, K.B., 1993, Resolution of isoforms of natural and recombinant fibrinolytic snake venom enzyme using high performance capillary electrophoresis, J. Liquid Chromatog.16:2189-2201.
28. Trikha, M., Schmitmeier, S. and Markland, F.S., 1994, Purification and characterization of fibrolase isoforms from venom of individual southern copperhead (*Agkistrodon contortrix contortrix*) snakes, Toxicon 32:1521-1531.
29. Coller, B.S., 1985, A new murine monoclonal antibody reports an activation-dependent change in the conformation and/or microenvironment of platelet GPIIb/IIIa complex, J. Clin. Invest. 76:101-108.
30. Romson, J.L., Haack, D.W. and Lucchesi, B.R., 1980, Electrical induction of coronary artery thrombosis in the ambulatory canine: a model for *in vivo* evaluation of anti-thrombotic agents, Thromb. Res. 17:841-853.
31. Rote, W.E., Mu, D-X. and Lucchesi, B.R., 1993, Thromboxane antagonism in experimental canine carotid artery thrombosis, Stroke 24:820-828.
32. Ahmed, N.K., Gaddis, R.R., Tennant, K.D. and Laez, J.P., 1990, Biological and thrombolytic properties of fibrolase: a new fibrinolytic protease from snake venom, Haemostas. 20:334-340.

37

MASS SPECTROMETRIC INVESTIGATIONS ON PROTEINACEOUS TOXINS AND ANTIBODIES

T. Krishnamurthy,* M. Prabhakaran, and S. R. Long

U.S. Army Edgewood RD&E Center
Aberdeen Proving Ground, Maryland 21010-5423

INTRODUCTION

Recent development in biological mass spectrometry has contributed immensely to the growth in diverse fields such as molecular biology, biochemistry, pharmacology, toxinology, immunology and biotechnology.[1-11] The mass spectrometric techniques provide versatility in ionization techniques and modes of detection along with femtomolar sensitivity, accuracy and reproducibility. This has enabled solving assorted complex structures of biopolymers with ease.[1-19] In general, the mass spectrometric techniques have been demonstrated to be powerful tools for the characterization and detection of larger as well as smaller chemical and biological molecules.[1-19]

Fast atom bombardment, liquid-SIMS (secondary ion mass spectrometry), electrospray (ESI), and matrix assisted laser desorption (MALDI) ionization modes have been applied successfully for the investigations of biomolecules.[1-19] However, ESI and MALDI are the two most frequently adopted techniques for investigations of biopolymers[1-18] Details involving the principles and application of all of these techniques can be found elsewhere.[1-18] The samples may be introduced either directly or after liquid chromatographic separation. All of the above techniques, with the exception of MALDI, have been adopted for the LC/MS experiments.[1-19] Although most of the reported LC/MS investigations involved the electrospray ionization of the molecules, continuous flow-FAB ionization techniques have also been found useful.[19,20]

During the electrospray ionization, each of the basic sites are protonated and hence instead of formation of a singly charged protonated molecular ion, a series of multiply charged ions are formed.[10,11,21] In the case of large biomolecules with molecular masses exceeding 100 kDa, the m/z values of these ions in most instances fall between m/z 300-1800 and hence could be measured using mass spectrometers with limited mass range (<2,000). Measurement of mass/charge (m/z) ratios of the multiply charged ions of a large biomolecule,

* The author to whom the correspondence should be addressed.

Natural Toxins II, Edited by B. R. Singh and A. T. Tu
Plenum Press, New York, 1996

followed by deconvolution of the measured m/z values of the ions, lead to the accurate measurement of the molecular mass of the biomolecule. Molecular masses exceeding 100 kDa can be measured using an instrument with mass range ordinarily limited to 2 kDa or less.[10,11,21] In all other modes of ionization in most instances, singly charged ions are primarily formed. However, doubly and triply charged ions of basic molecules may also be observed during the FAB and MALDI ionizations. Hence, mass spectrometers with higher mass ranges are required to measure the molecular masses of larger molecular ions generated during these ionization processes.

The molecular masses of ions from either a single component or of complex mixtures may be measured with either low or high resolution conditions. High resolution mass measurements were carried out using double focussing instruments containing magnetic and electrostatic sectors for focussing each ion based on its mass and energy and thus contributing to the increased resolution of the ions. Scanning quadrupole or time-of-flight mass spectrometers are for the low resolution mass measurements. In most cases, time-of-flight (TOF) mass spectrometers are used to analyze the ions generated during the laser desorption (MALDI) of larger biomolecules.[6,9-11] The ions migrate along the drift region of the TOF mass spectrometer and their m/z values are measured based on the time taken for the migration to the detector. The mass range is unlimited during the TOF measurements and the molecular masses of biomolecules exceeding 1000 kDa can easily be measured though with limited accuracy.[6,9-11] The ions generated from the ESI or MALDI of molecules could also be measured with higher resolution using a double focussing mass spectrometer, in the latter case using an array detector. [11,22] Most of the reported ESI investigations have been conducted using quadrupole mass spectrometers.[11,22]

Alternatively, smaller molecules with molecular masses less than 6 kDa may also be ionized under fast atom bombardment (FAB) or continuous flow-FAB (frit-FAB) and detected with high resolution.[19,20] This approach enables the ionization and detection of smaller hydrophilic molecules in the presence of hydrophobic molecules, which is essential for the investigation of complex mixtures.[19,20]

The molecular masses of biomolecules, regardless of the difference in their polarity and size, may thus be determined using electrospray, MALDI and continuous flow-FAB ionization methods. For structural characterization of the molecules, specific protonated molecular ions may be selected and subjected to collisionally induced dissociation (CID) at elevated kinetic energies by neutral atoms such as helium, argon or xenon. The resulting daughter (fragment) ions may then be recorded in their corresponding daughter (MS/MS) spectra.[2-4,11,18,19] The CID experiments are carried out using tandem mass spectrometers only. The instrumentation contains two mass spectrometers in tandem with a collision cell between them.[23] Even though triple quadrupole and four sector magnetic tandem mass spectrometers are commonly used for detecting CID products, hybrid tandem mass spectrometers have also been used in a limited manner.[23] Both low and high energy collision experiments are possible. The former is carried out using either triple quadrupole or hybrid instrumentation and the latter in high resolution four sector tandem mass spectrometers.[18,23] After collisionally induced dissociation, spectra of daughter or precursor ions or neutral losses may be selectively recorded. Daughter spectral data of large and small molecules are widely used for their structural characterization as well as detection and analyses.

Under CID conditions, the precursor ions of the peptide(s) dissociate to form various types of both N-terminus and C-terminus ions.[3,24] Hence, amino acid sequences of a peptide can be derived from the corresponding daughter spectral (MS/MS) data.[3,24] Larger proteins may be sequenced by combining tandem mass spectral (MS/MS) techniques along with enzymatic degradations and a limited amount of Edman degradation procedures.[3,10] The mass spectrometric techniques have also been applied for determining the number of cysteines present in a protein and establishing the positions of disulfide bridges present in the molecule.[25] Post translational modifications occurring during the biotechnological processes have also been established using mass spectrometry.[2,10,11] In addition to the proteins, characterization of other biopolymers such as glycoproteins, oligosaccharides, oligonucleotides, glycolipids and phospholipids has also been possible by this approach.[10,11] Such investigations were carried out using both low and high resolution tandem mass spectrometers.

Thus, regardless of modes of detection, the MS and MS/MS techniques are widely applied for the investigation of large biomolecules as well as small chemical molecules. The methods can be applied for the measurements of their molecular masses as well as their analysis and structural characterization.

The majority of individual toxins isolated from complex venom mixtures have been found to be proteinaceous.[17,26-35] Mass spectrometric techniques have been successfully adopted for investigations of some proteinaceous toxins.[17,26-34] This study was carried out to explore the potential universal application of MS and MS/MS techniques for the investigations of proteinaceous toxins. During this investigation various types of toxins originating from different sources were selected. Detection limits under most instances were established to be in the femtomolar range. Amino acid sequences of smaller cyclic peptides were established, with minor chemical modifications, by the tandem mass spectrometric method.[26] Four disulfide bridges present in a cardiotoxin were located using only a few nanomoles of the intact peptide.[25] Allied techniques such as automated Edman degradation, HPLC and electrophoretic techniques are also essential for establishing the N-terminus residues and for sample preparations and purifications.

EXPERIMENTAL

The electrospray and continuous flow-FAB experiments were carried out on a Finnigan-MAT TSQ700 triple quadrupole low resolution tandem mass spectrometer and a JEOL four sector HX110/HX110 magnetic high resolution tandem mass spectrometer, respectively. The electrospray ionization source was manufactured by Analytica of Branford, CT. The MALDI experiments were done on a Vestec-2000 time-of-flight mass spectrometer equipped with a nitrogen laser. Chromatographic purifications of standards were performed on Applied Biosystems 140B microbore HPLC systems.

Toxin and antibody standards were purchased from Calbiochem (San Diego, CA), Bachem (King of Prussia, PA) and Sigma Chemicals (San Diego, CA). Spectra-grade solvents, MS standards and calibrants were purchased from Fischer Scientific (Fairlawn, NJ). The protein isolated from Clostridium botulinum was provided by Prof. Bal Ram Singh of University of Massachusetts, Dartmouth, MA as part of a collaborative project. Likewise, the hemorrhagic protein from Agkistrodon acutus was supplied by Prof. Antony Tu of Colorado State University, Fort Collins, CO.

Stock solutions of toxins were prepared at a concentration of 1nmole/μl, using 1-5% aqueous acetic acid and stored at -40°C in small aliquots until use. Standard solutions at a concentration of 1-10 pmole/μl were prepared by diluting the stock solution prior to use. Dilute standards were discarded after 48 hours.

Electrospray Ionization - MS and MS/MS Analysis

A syringe (100 μl) filled with a 1/1 mixture of 1-10% aqueous acetic acid and methanol was placed in a Harvard syringe pump. The solution was pumped at a speed of 1-2 μl/min into a blank fused silica capillary column (25 cm x 320 μm) and carried into the electrospray ionization source via a rheodyne 8125 injector with a 5 μl sample loop. Sample solution was either infused at a constant rate of 1μl/min or injected into the injector and carried into the ionization source at a rate of 1-2 μl/min. Liquid nitrogen was used as the drying gas, maintaining the temperature of the heater at 180°C. A potential of 3200-3600 V was applied through the needle of the electrospray assembly, inducing the ionization. The multiply charged ions were detected by scanning the mass spectrometer at a range of either m/z 300-1800 or 2000-4000 in 5-6 seconds. Tuning of the instrument and the mass calibrations were performed using a mixture of a tetrapeptide (MRFA) and myoglobin (10 pmole/μl) solution.

For MS/MS measurements, the triply charged molecular ions of conontokin (Glu-5)-G were mass selected at MS1 and allowed to undergo collisionally induced dissociations at the collision cell using argon as the collision gas. The optimum CID conditions of precursor ions were experimentally determined. The precursor ions were dissociated by maintaining the collision gas pressure and energy at 4.0 mTorr and -15ev, respectively. The product ion (MS/MS) spectrum was recorded by scanning the Q3(MS2) quadrupoles from m/z 50 to1400 in 4 seconds.

MALDI-MS Analysis

Sinapinic acid solution (9 μl; 10 mg/ml) in a 2:1 mixture of 0.1% aqueous trifluoro acetic acid:acetonitrile was mixed with sample solution (1 μl; 2-10 pmole/ul) and vortexed for a few seconds. The mixture (1 μl; 0.2-1.0 pmole) was transferred to a pin in the sample holder (autosampler) and allowed to air dry. Similarly, all twenty four pins located in each of the autosampler were loaded with samples and calibrant. The sample holder was loaded into the mass spectrometer and each pin in turn was selected for analysis. The sample was monitored using a video camera and a selected area was exposed to the nitrogen laser (337 nm). The pin was rotated and the signal was monitored. The area giving maximum signal was selected and bombarded with the laser and the mass/time data was acquired. The mass spectrum was obtained by the automatic conversion of the acquired data into the corresponding mass/charge files. Mass calibration was performed using angiotensin, myoglobin or bovine serum albumin depending upon the molecular masses of the analytes. Mass assignments were made from the specific calibration file generated for individual autosamplers.

Continuous Flow Frit-FAB -MS and -MS/MS Analysis

A syringe (1 ml) filled with 1.5% glycerol in a 1/1 mixture of deionized water and methanol was placed in Harvard syringe pump. The solution was pumped at a speed of 2 μl/min into a blank fused silica capillary column (35cm, 60μm i.d.) and carried into the continuous flow frit-FAB probe via a rheodyne 8125 injector with 20 μl sample loop. Once the constant flow and pressure was attained, the matrix ions were monitored to verify the stability of the fluid flow and ionization conditions. Sample solution (1 μl) was injected. The generated ions were detected by scanning MS1 from m/z 400-3000 at scan rates set up for specific calibration files and the high resolution (2,000) mass spectra in profile mode were generated.

For tandem mass spectrometric (MS/MS) analysis, the MS1 (mass spectrometer 1) was set to pass only the ^{12}Carbon peak of the selected precursor ion. The collision cell was floated at 8kV to provide the collision energy of 2kV. Xenon was used as the collision gas and the collision gas pressure was adjusted to lower the precursor ion intensity by 75%. The generated daughter ions were detected by a JEOL ADS-11 array detector by setting the dispersion at 10%. The dead time of 0.3 seconds was used to stabilize the magnetic field in between the magnetic and electrostatic steps. The preset daughter ions were monitored at preset intervals (35 segments for m/z 69-1500) and recorded. The number of exposures (8) per scan and the number of scans (4) per segment were increased to improve the sensitivity of detection. The data, recorded using spectra calc software, was automatically transferred to the complement data system and processed.

Other specific chemical and enzymatic degradation or derivatization procedures are described elsewhere in detail.[25,26,34]

RESULTS AND DISCUSSION

1. Measurement of Molecular Masses and Detection

A. Electrospray Ionization. All samples were purified over a reverse phase RP300 (C_8) narrowbore column to remove any inorganic buffers and other ionizable impurities. Experiments were carried out either by infusing toxin solutions into the electrospray ionization source via a blank fused silica capillary column or injecting known amounts into the blank column and eluting into the ESI source. The percentage of acetic acid in 50/50 methanol was adjusted between 1-5% based on the size, hydrophilicity, and solubility of analyzed molecules. Regardless of the nature of the analyte, the detection limit for each of the analyzed toxins was similar regardless of the mode of introduction into the ESI source. The minimum analyzed amounts ranged between 100-500 femtomoles in most instances. Based on the recorded mass spectra, the minimum detectable limits could be extrapolated to be an order of magnitude less than the analyzed amounts. The measured and calculated molecular masses of several toxins are listed in Table 1.

a. Toxins. The ESI-mass spectrum of ω-conotoxin GVIA (Figure 1a) indicates the molecular ions with two, three and four charges. This toxic peptide contains 3 basic residues, one lysine and two arginines, and the N-terminus amino group. Hence, there are four potential positions in GVIA for protonation during the ESI process. The observed multiply charged ions in the mass spectrum are in agreement with the projected profile based on the amino acid (aa) sequences of the molecule. On deconvolution of the observed ions, the molecular mass of the singly charged protonated molecular ion was calculated to be 3037.6 ± 0.8 Da (Figure 1b). This is in good agreement with the value (3038.4 Da) calculated from the amino acid sequences of the intact peptide.

The hepatotoxic peptides isolated from blue-green algae (cyanobacteria; microcystins) are cyclic peptides.[26,34-38] The microcystins[38] are seven membered peptides with an unusual aromatic amino acid and mixture of D- and L- amino acids.[26,34 - 38] Despite the presence of one or two basic residues in the molecule, these peptides are sparingly soluble in water. Nodularin, a pentapeptide, was also isolated from cyanobacteria (Nodularia spumigina). These cyclic peptides had low ionization efficiencies in comparison with other proteinaceous toxins investigated during this study. The deconvoluted ESI spectrum of a mixture (Figure 2) containing microcystin-LR, microcystin-RR, microcystin-YR and nodularin was obtained by infusing a solution containing each of the components at a concentration of 2 pmole/μl at a rate of 1 μl/min. The single scan spectrum derived from 133 femtomole indicated singly charged protonated molecular ions of all of the peptides. The nodularin ionized better under ESI conditions than the rest (Figure 2).

The bee venom peptide, apamin[39], was also analyzed under similar conditions. The four basic residues (one lysine, two arginine and one histidine) along with the N-terminus amino group are potential sites for protonation. The m/z of the molecular ions with 2-4 charges are found (Figure 3a). The singly charged ion was not observed since its m/z value exceeded that of the scanned range. The observed molecular mass (2,026.8 Da) was in good agreement with

Table 1. Molecular masses of proteinaceous toxins (Daltons)

Toxins	Calculated	Electrospray-MS*	MALDI-MS
Joro Spider Toxin	565.7	564.9	563.8
NSTX-3 Spider Neurotoxin[51]	664.8	664.7 ± 0.1	663.2
Nodularin Toxin	824.2	824.6	823.2
Microcystin-LR	994.2	994.6	995.0
Microcystin-RR	1,038.2	1,038.2	1,033.6
Microcystin-YR	1,045.2	1,045.0	1,046.5
Conopressin S	1,003.2	1,002.2 ± 0.4	1,004.1
Conantokin G	2,264.2	—	2,264.9**
[Glu-5]Conantokin G	2,044.2	2,043.3 ± 0.1	2,041.2
μ-Conotoxin GIIIB	2,641.3	2,641.3 ± 1.4	2,642.7
ω-Conotoxin GVIA	3,038.4	3,037.6 ± 0.8	3,036.7
Naja Naja atra Phospholipase A2	—	13,257.7 ± 0.3(Major)	13,226.6
		13,026.1 ± 0.5(Minor)	19,855.8
NN atra Cardiotoxin I	6,693.1	6,693.0 ± 1.0	6,677.8
NN atra Cardiotoxin II	6,742.1	6,742.2 ± 0.5	6,749.0
NN atra Cardiotoxin III	6,739.4	6,739.1 ± 0.2	6,723.3
NN atra α-toxin	—	6,787.0 ± 0.2(Major)	
		6,885.7 ± 2.0(Minor)	
NN nigricolis Phospholipase A2	—	13,257.7 ± 0.3(Major)	13,344.0
		13,354.0 ± 0.9(Minor)	20,102.0
NN nigricolis Cardiotoxin	6,817.5	6,816.9 ± 0.4	6,806.9
α-Cobratoxin[35]	7,821.1	7,820.6 ± 0.7	7,812.2
α-Dendrotoxin[52]	‘7,800	7,048.3 ± 0.5	7,022.0

*Deconvoluted data

**Ions due to the loss of 1-5 COOH groups were more pronounced.

the calculated value (2027.3 Da) and the observed error in the mass measurement was 0.02%. The sample consumed for the single scan spectrum was 133 femtomoles.

Cardiotoxin IV isolated from Naja Naja atra[40] was also subjected to ESI-MS analysis. There are 12 sites in the intact molecule which could be protonated during the electrospray

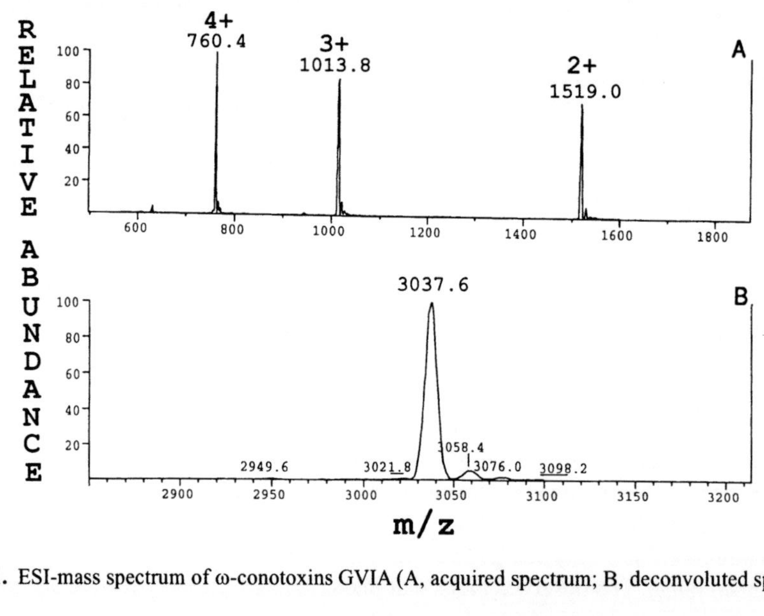

Figure 1. ESI-mass spectrum of ω-conotoxins GVIA (A, acquired spectrum; B, deconvoluted spectrum).

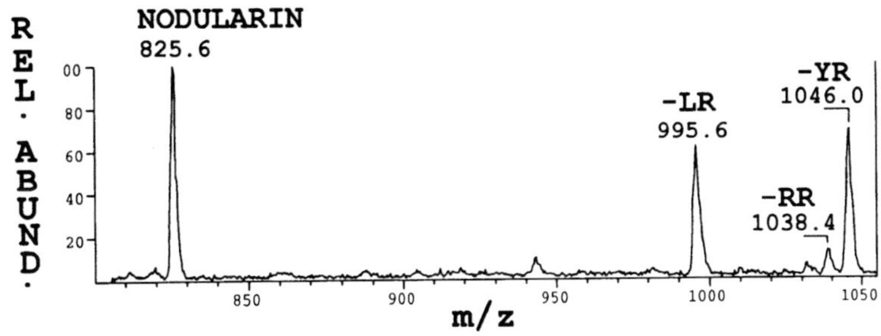

Figure 2. ESI-mass spectrum of blue-green Algal hepatotoxins (-LR, Microcystin-LR; -RR, Microcystin-RR; -YR, Microcystin-YR).

ionization. The ESI spectra were recorded by infusing (Figure 4a, 133 fmole) and by injecting (Figure 4b, 500 fmole) the sample. The molecular mass obtained on deconvoluting the multiply charged ions of the cardiotoxin was measured to be 6785.2 ± 0.8 Da, which was in agreement with the calculated value (6785.0 Da). The observed error was less than 0.003%. Similar results were observed for cardiotoxins I, II, III from the same species (Table 1).

A post-synaptic (α-bungarotoxin) and a pre-synaptic (β-bungarotoxin, phospholipase-A2) neurotoxins isolated from Bungarus multicinctus[35,41] were ionized separately and their corresponding ESI spectra were recorded. The results are shown in Figure 5. The molecular mass of α-bungarotoxin was measured (7,983.4 ± 0.6 Da) with minor (0.08%) error in comparison with the calculated value (7,990.4 Da). The measured molecular mass of β-bungarotoxin was (20,658.8 ±1.5 Da).

The ESI-mass spectrum (Figure 6) of another pre-synaptic neurotoxin (Mojave Toxin; phospholipase-A2) isolated from Crotalus snake[42] was also recorded. The measured

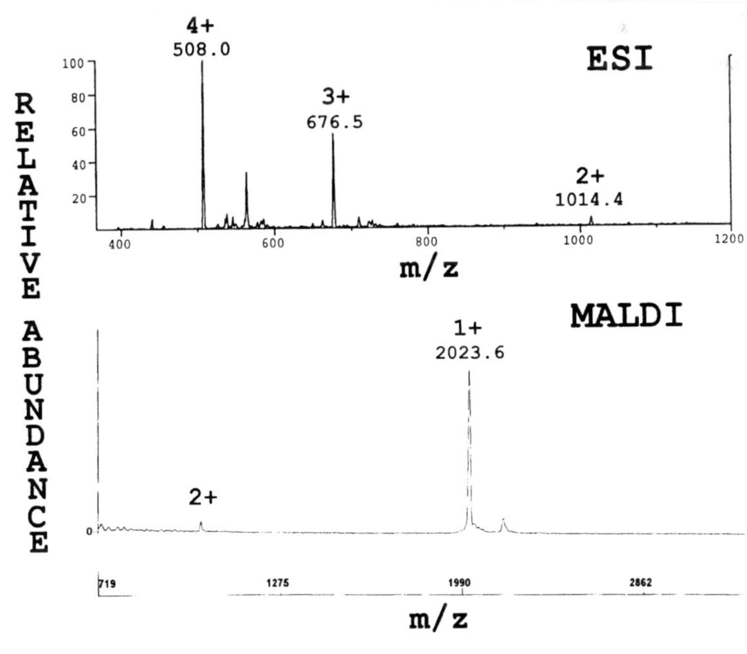

Figure 3. ESI- and MALDI- mass spectra of Apamin (α-cyano-4-hydroxycinnamic acid matrix was used for MALDI experiments).

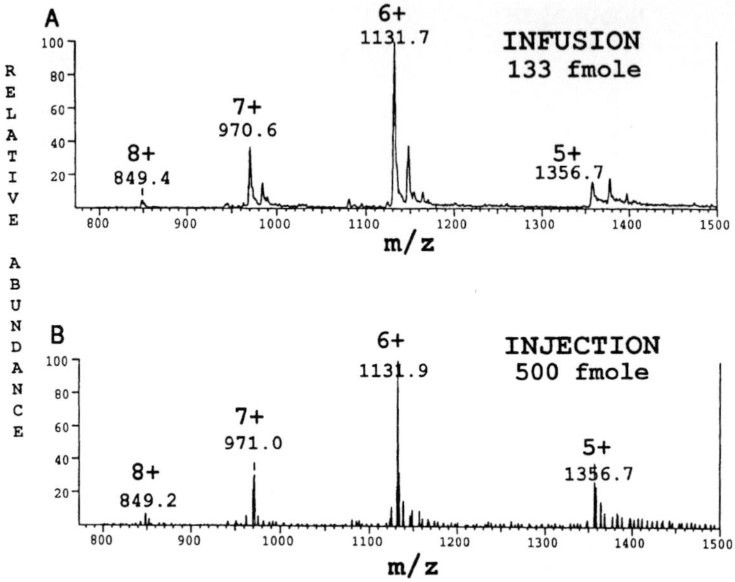

Figure 4. ESI-mass spectrum of Naja Naja atra cadiotoxin IV (A, singe scan; B, cumulative data).

molecular mass was 14,185.8±1.1 Da in comparison with the calculated value of 14,187.3. The calculated percent error was 0.01%.

A mixture containing a post-synaptic neurotoxin (Conotoxin GI & MI), a sodium channel blocker (μ-Conotoxin GIIIB), and a calcium channel blocker (ω-Conotoxin GVIA) isolated from Conus geographus[43] along with a post-synaptic neurotoxin from Conus magus[44] (Conotoxin MI) was analyzed without prior chromatographic separation. The observed multiply charged ions for all of the four components were deconvoluted automatically. The deconvoluted spectrum (Figure 7) showed the molecular masses of all toxins very distinctly. The measured values for α-Conotoxin GI (1,437.7 Da), α-Conotoxin MI (1,494.1), μ-conotoxin GIIIB (2,641.5 Da), and Conotoxin GVIA(3,035.5 Da) were in agreement with the calculated values. This clearly demonstrates that the ESI-MS method could also be adopted for mixture analysis without prior chromatographic separations. Ionic suppression was not noticed during the ESI analysis of this mixture. The sample consumed during this process was 500 femtomoles of each of the components.

Similar results were observed when a mixture of Erabutoxin A[35] (6,837.8 Da), Erabutoxin B[35] (6,860.8 Da), and a cardiotoxin[35] (6,817.5 Da) and α-toxin (~6,800 Da) from Naja Naja nigricolis was analyzed by the ESI-MS technique. The measured values shown in the deconvoluted spectrum of the mixture (Figure 8) provide a clear indication that the ESI-MS method can be applied to the analysis of mixtures containing components with only slight variations in molecular masses.

The ESI mass spectrum of a chromatographically purified unknown hemorrhagic toxin[45] isolated from Agkistrodon acutus is shown in Figure 9a. The observed heterogeneity indicates that the molecule could be a glycoprotein. The measured molecular mass was 22,879 Da. Further investigation for the purification and characterization of the molecule is underway.

b. Antibodies. Extensive desalting of the sample was required prior to the electrospray ionization of the antibodies. Desalting of the samples was attempted by ultrafiltra-

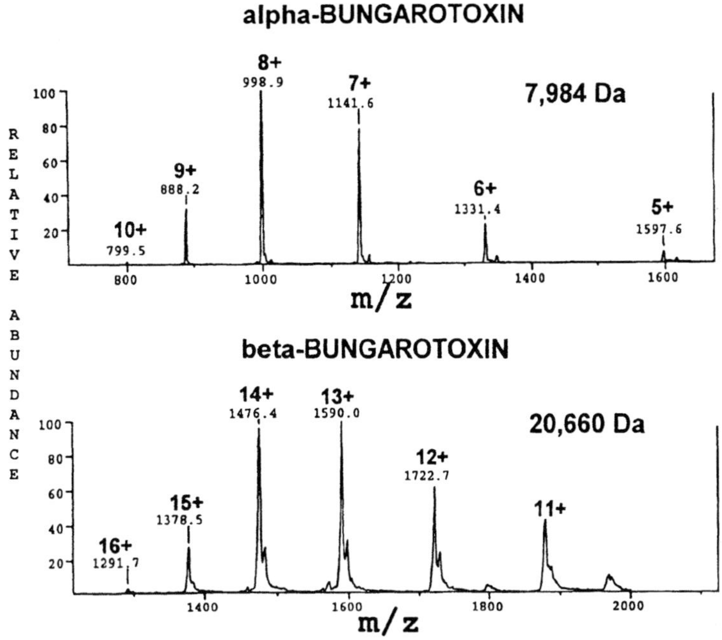

Figure 5. ESI-mass spectrum of α- & β-Bungarotoxins (acquired spectra; single scan data; sample infusion).

tion using Pharmacia PD-10 columns, AMICON microcon concentrator tubes with the cut off at 50 kDa, and dialysis using membranes from spectra/por with a cut off at 50 kDa. The samples were also purified by chromatographing over a reverse phase C_4 HPLC column. Samples prepared by ultrafiltration using Microcon tubes and chromatographic purification were found best suited for the ionization of antibodies. More rigorous and systematic approaches for optimizing the desalting procedures are underway.

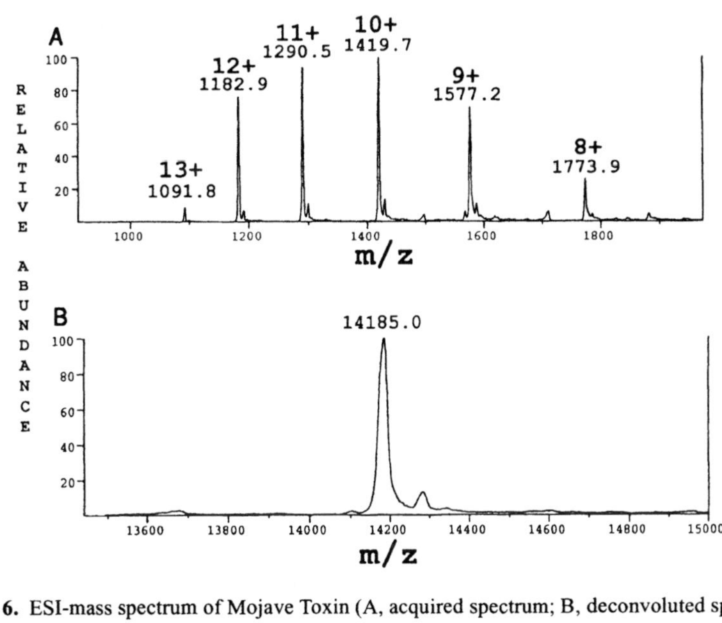

Figure 6. ESI-mass spectrum of Mojave Toxin (A, acquired spectrum; B, deconvoluted spectrum).

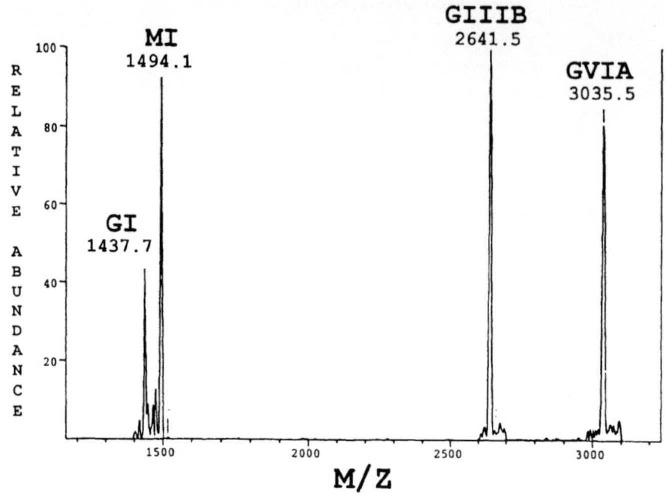

Figure 7. Deconvoluted ESI-mass spectrum of a mixture of conotoxins.

Multiply charged ions for the antibodies have m/z values exceeding 2,000 Da. Hence, tuning of the mass spectrometer for the high mass range (2-4 kDa) was required. The mass calibrations were done using myoglobin as the calibrant. Mass assignments in the high mass range were verified using bovine serum albumin (66,504 Da) and transferrin apo-human plasma (79,561 Da). Two commercially acquired IgGs, human α-antitrypsin and human α-acid glycoprotein, were selected for this investigation. Antibody solutions were made up in 10% acetic acid in 50/50 water/methanol mixture and infused directly into the ESI source and ionized. However, dissolving the samples in 0.2% formic acid produced better results. Use of sheath flow (2-methoxyethanol or methanol) and sheath gas did not improve the

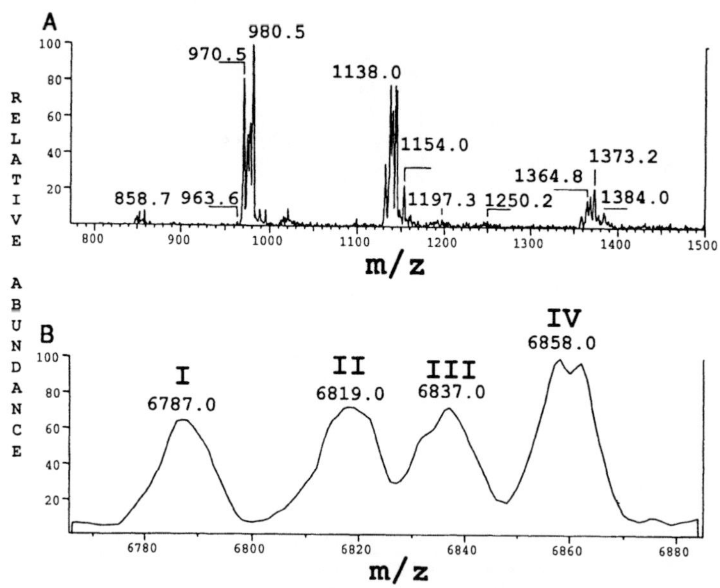

Figure 8. ESI-mass spectrum of snake venom polypeptides (A, acquired spectrum; B, deconvoluted spectrum; I, Naja Naja atra α-toxin; II, Naja Naja atra cardiotoxin; III, Erabutoxin A; IV, Erabutoxin B).

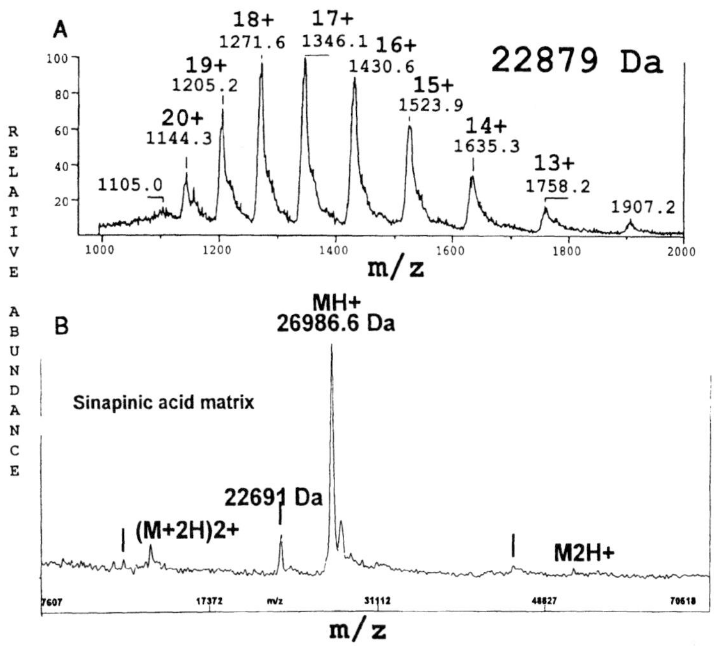

Figure 9. Mass spectra of hemorrhagic toxin from Agkistrodon acutus (A, acquired ESI-mass spectrum; B, MALDI-mass spectrum).

ionization under electrospray conditions. The single scan ESI-spectra of the infused antibody solutions were recorded (Figure 10). Molecular masses of the human α-antitrypsin and human α-acid glycoprotein were calculated, on deconvolution of the corresponding observed multiply charged ions, to be 148,286 Da and 149,435 Da, respectively.

B. Matrix Assisted Laser Desorption Ionization (MALDI).[6,12,13] a. Toxins. Sample solutions (1 μl; 2-10 pmole/μl) were treated with matrix solution (9 μl) and vortexed. A 1 μl aliquot of each of the resulting sample solutions containing 0.2 - 1.0 pmole/μl was loaded on the individual pins of the sample holder and allowed to air dry slowly and form fine crystals. All experiments were conducted using either sinapinic acid(SA) or α-cyano-4-hydroxycinnamic acid (ACHCA) as the matrix materials. Individual samples were irradiated with nitrogen laser (337 nm) and the resulting ions were detected using a time-of-flight mass spectrometer. Mass assignments for the ions observed in the corresponding time-of-flight (TOF) mass spectra were made using the mass calibration created by using external calibrants for individual sample holders. Peptides such as angiotensin (1,296 Da), proteins such as myoglobin (16,952 Da) and bovine serum albumin (66,504 Da) were used as external calibrants. Based on the observed ion counts and signal to noise ratio (s/n) during the MALDI analysis of these toxins and antibodies (0.2 - 1.0 pmole), their minimum detectable limits are projected to be 50 - 200 femtomoles or less for biomolecules with molecular masses ranging from 650 Da to 150 kDa. Buffers do not totally inhibit the ionization of these molecules under MALDI conditions in contrast with the ESI process. However, removal of them from the samples enhances the ionization yield.

Blue-green algal hepatotoxic peptides such as nodularin, microcystin-LR, -RR, and -YR [26,36,37] were analyzed under MALDI-TOF-MS conditions using ACHCA as the matrix. Only singly charged protonated molecular ions were observed. The corresponding molecular masses were measured with better than 0.1% error (Table 1). When a mixture containing 200

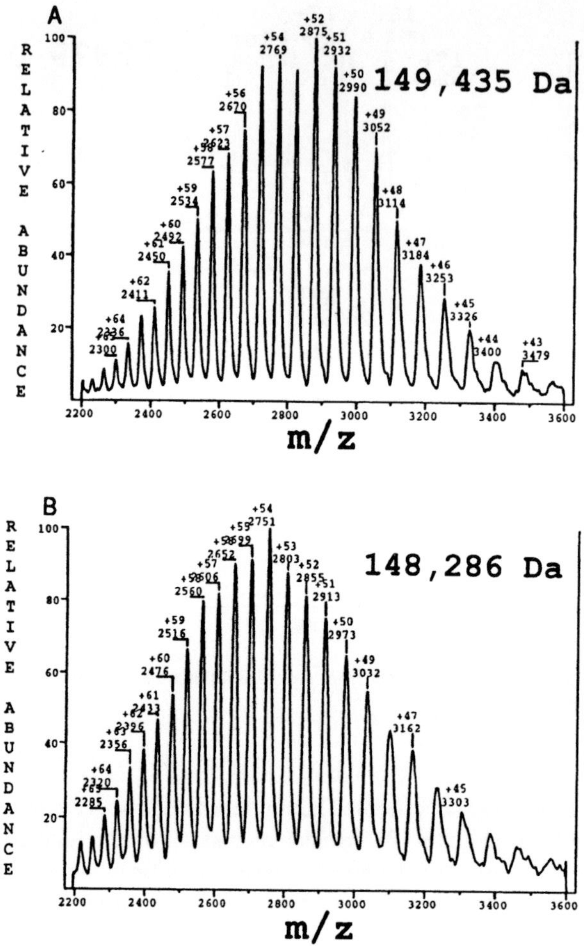

Figure 10. ESI-mass spectra of antibodies. (infusion; A, human α-acidglycoprotein; B, human α-antitrypsin).

fmoles of each component was analyzed, the nodularin was detected with maximum sensitivity. Observed sensitivity for these cyclic peptides during MALDI analysis was better than during ESI analysis. Apamin[39], the bee venom component, also was detected (2,022.6 Da; 0.2% error) with similar sensitivity under MALDI conditions using ACHCA as the matrix. Both singly and doubly charged molecular ions were observed (Figure 3b). The accuracy can be improved using the internal calibrant during the mass measurements.

The MALDI-TOF-MS method of detection is ideal for mixture analysis. A complex mixture containing nine toxic and biologically active conus peptides[43,44] (Figure 11), 0.5 pmoles of each component, was mixed with ACHCA and subjected to MALDI-TOF analysis. Singly charged molecular ions of α-, μ-, and ω- conotoxins, along with that of conopressins were observed with excellent sensitivity. Conantokin G has five γ-carboxyglutamic acid residues [43,44,46] known to be heterogeneous due to the instability of the γ-carboxy residues. This is very clearly established in the TOF mass spectrum (Figure 11). There are at least five ions observed for the molecule, each formed by decomposition of one or more γ-carboxy groups to its corresponding decarboxylated residues (glutamic acid). ESI-MS analysis of Conantokin G was not possible at 0.5 picomole level which is perhaps due to the heterogeneity of the molecule. The corresponding [Glu5] isomer, [Glu5]conantokin G displayed only

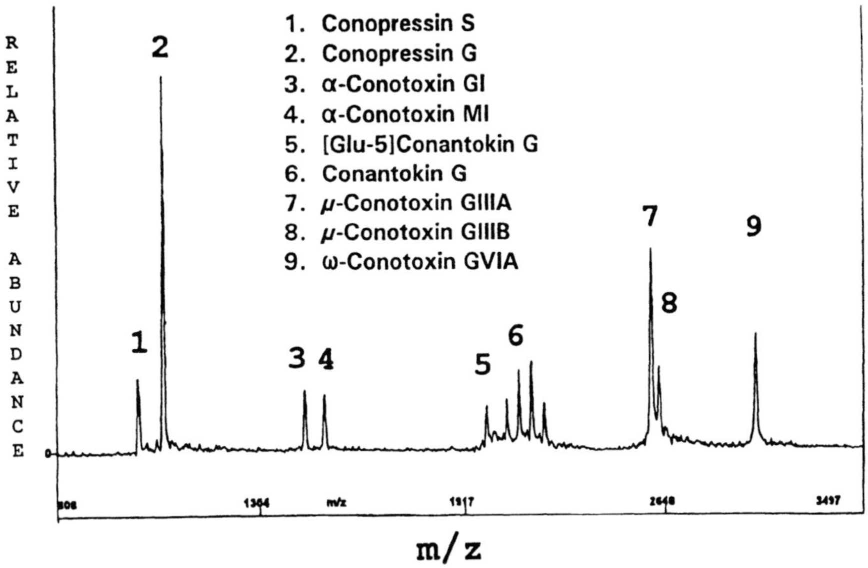

Figure 11. MALDI-mass spectrum of conotoxin mixture (α-cyano-54-hydroxycinnamic acid matrix).

a singly charged protonated molecular ion. Each of these components not only has different biological activity but also possesses various levels of hydrophilicity. Hence, such a mixture cannot be analyzed under FAB conditions. ESI-MS analysis of this mixture was also not possible. Different polarities of the components as well as formation of multiply charged ions for individual components with similar m/z values might have also contributed to the inability to detect and analyze the individual components present in the mixture under ESI conditions.

Toxic polypeptides originating from various snake venoms were also analyzed by the MALDI-TOF-MS method. Their corresponding molecular masses were measured with reasonable accuracy (Table 1). When post-synaptic neurotoxic (α-bungarotoxin, α-cobratoxin, Naja Naja nigricolis α-toxin) and cytotoxic (cardiotoxins I, II, III and IV from Naja Naja atra [35], and cardiotoxin from Naja naja nigricolis) polypeptides[35] were ionized using ACHCA as the matrix, molecular ions with single, double and triple charges were observed. However, in a matrix of sinapinic acid, in addition to the multiply charged ions for monomer, ions due to dimer, tetramer and octamer were also observed. Other species such as trimer, pentamer, hexamer or heptamer were not detected (Figure 12). Erabutoxins A and B under MALDI conditions, with sinapinic acid as the matrix, exhibited ions due to the corresponding dimers only. Whereas, the α-dendrotoxin,[53] which blocks the potassium channels exhibited ions due to dimer and tetramers when ionized using sinapinic acid as well as ACHCA. Pre-synaptic neurotoxins, such as β-bungarotoxin (Figure 13), mojave toxin and phospholipase from Naja Naja atra, could only be effectively ionized using sinapinic acid as the matrix. In addition to the singly, doubly and triply charged ions of the monomer, ions due to dimer, trimer, tetramer, pentamer and hexamer were also observed. A fraction with hemorrhagic activity originating from Agkistrodon acutus also formed ions due to monomer, dimer and tetramer but not trimer (Figure 9b).

All of the above observations indicate that single chained polypeptides such as pre-synaptic neurotoxins, cardiotoxins and the hemorrhagic toxin form molecular aggregates in a specific manner. Conversely, the process seems to be non-specific among pre-synaptic neurotoxins with an α and β chain. The former phenomenon was observed during the ESI

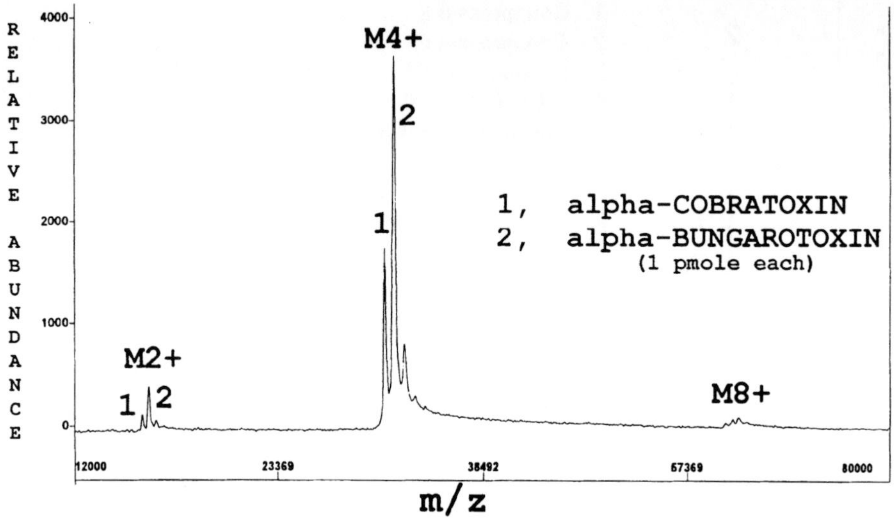

Figure 12. MALDI mass spectra of α-Cobratoxin and α-Bungarotoxin (Sinapinic acid matrix).

investigations of concanavalin A.[47] The molecular aggregation is projected to be taking place in solution, not in the gas phase.[47] Based on the available data of these toxic proteins, the mechanism of the formation of the molecular aggregates cannot be explained. Further investigations to understand this phenomenon are underway.

The MALDI process was also applied for analyzing the protein isolated from Clostridium botulinum (Figure 14). A crude proteinaceous extract from Brucella suis after ultrafiltration over a PD-10 column, was also analyzed using sinapinic acid matrix. Several ions were observed with negligible background noise which indicated the potential application of the methodology in distinguishing the bacterial products using specific biomarkers. This should also have enormous applications in the health related fields.

b. Antibodies. Antibody raised against cholera was purified over PD-10 columns and the standard solution (10 pmole/μl) was prepared in 5% aqueous acetic acid. It was mixed with sinapinic acid and the concentration of the antibody present in the analyzed sample was

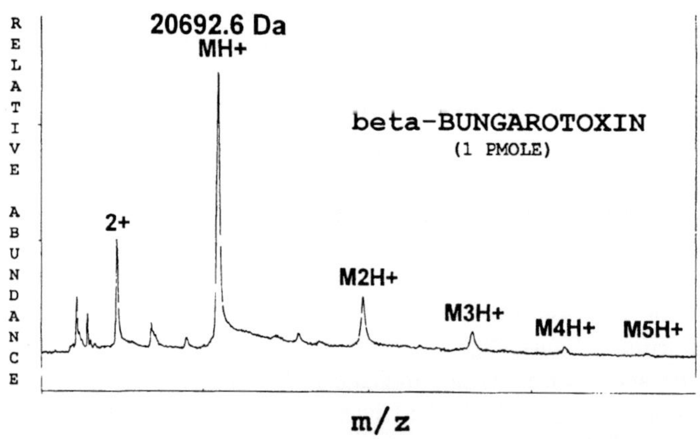

Figure 13. MALDI-mass spectrum of β-Bungarotoxin (Sinapinic acid matrix).

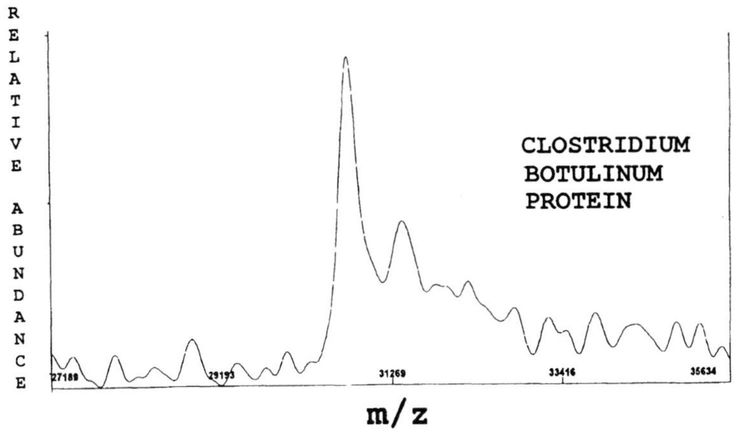

Figure 14. MALDI mass spectrum of a Clostridium botulinum protein (Sinapinic acid matrix) MH+, 30724.9 Da.

adjusted to be 1 pmole. The MALDI spectrum of goat anti-cholera (Figure 15) showed only singly, doubly and triply charged protonated molecular ions and the molecular mass was measured as 146,144 Da.

C. Continuous Flow(Frit)-FAB Ionization (High Resolution Mass Measurements).[19,20] Ionization of hydrophilic molecules under the conventional FAB conditions always posed a challenge, especially in the presence of hydrophobic molecules. The observed sensitivity in most instances was poor and the minimum detectable amounts were between 0.1 - 1.0 nanomole. In addition, direct on-line sample introduction of the effluent from a HPLC column was not possible. All these potential problems could be resolved using the continuous flow-FAB ionization.[19] This technique has been highly useful for the ionization of extremely hydrophilic and thermally labile molecules.[20] Continuous flow-FAB involves introducing the carrier solution continuously into the ionization source and bombarding the fluid with energized xenon or cesium ions.[20] The sample is introduced either as part of the carrier solution or injecting into the carrier solution at specific intervals.[19] One of the effective ways of introduction of the carrier solution as well as the sample into the source was demonstrated to be infusing the solution through a porous frit.[19,20] Minimum detectable limits are comparable with the values observed during the electrospray ionization.

We performed continuous flow frit-FAB experiments in the JEOL HX110-110 high resolution tandem mass spectrometer with QPD array detector.[19] Combination of frit-FAB ionization along with the measurement of ions at high resolution provides an excellent tool for identifying molecules of similar molecular masses with ease. Most of the known conotoxins and related conus peptides are extremely hydrophilic and/or possess great adsorptive properties. Figure 16 indicates the comparison of the results obtained during the ionization of conopressin G under static or conventional and continuous flow-FAB conditions. The mass measurement was carried out at a mass resolution of 2000. The observed signal (Figure 16) will be enhanced by at least an order of magnitude when the resolution is decreased. The observed sensitivities under frit-FAB high resolution mass measurements are thus comparable with those observed during the electrospray ionization and low resolution mass measurements. Some of the hydrophilic conotoxins could be analyzed at 0.5 - 1.0 picomole level under frit-FAB high resolution MS conditions.

In addition to the mass measurements, the frit-FAB ionization was also adopted while developing a new methodology for the reduction of the disulfide bond and derivatization of

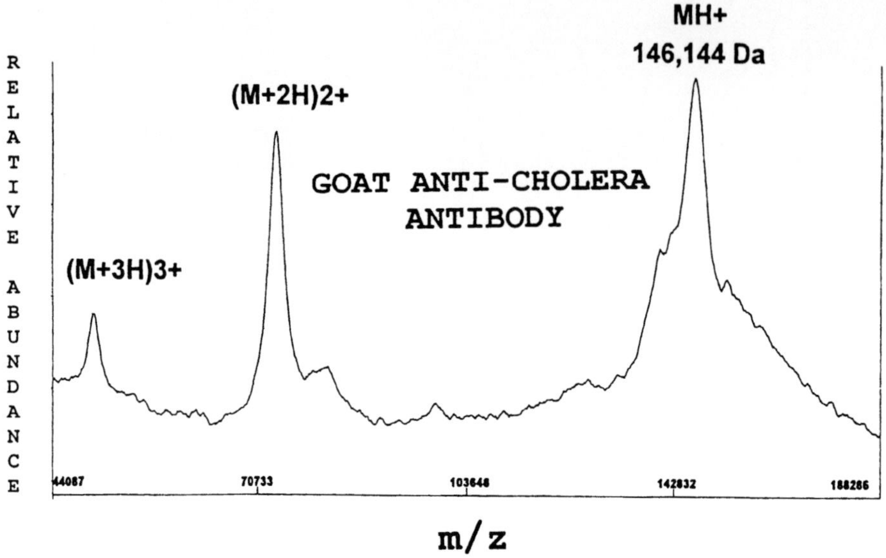

Figure 15. MALDI mass spectrum of anti-cholera antibody (Sinapinic acid matrix).

the thiol group. The published accounts report the use of dithiothreital and tris(2-carboxyethyl)phospine for reducing the disulfide bridges.[48] However, we observed that mixing of the intact peptide with thioglycerol and dilute ammonium hydroxide accomplished the reduction of the disulfide bridges almost instantaneously. The projected mass difference between the intact and reduced conopressins is 2 Daltons. The reduction process was followed by merely monitoring the ions after mixing the peptides and reagents on the FAB

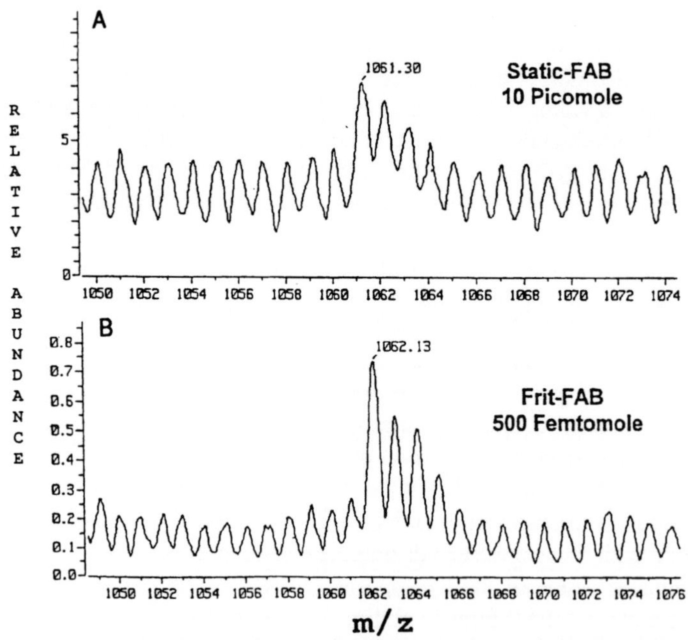

Figure 16. Fast atom bombardment mass spectra of conopressin G (A, Static-FAB; B, continuous flow-FAB).

probe tip and introducing into the ionization source. Alternatively, the reaction mixture was also introduced into the Frit-FAB source by injecting into a short (10") blank fused silica capillary column (Figure 17). The reduced form could be maintained, without undergoing oxidation of the thiol group, in solution flushed with nitrogen for at least 24 hours. The ethylpyridyl derivatives of the peptides were prepared by addition of vinylpyridine directly to the reduction product, soon after mixing the intact peptide with thioglycerol and dilute aqueous ammonium hydroxide. The ethylpyridyl derivatives were obtained in excellent yields on adopting this new procedure(Figure 17). The high resolution profile mass spectral data of the ethylpyridyl derivative (Figure 17c) indicated the product from the reduced conopressins in the naturally occuring amide forms, as well as the corresponding acid forms. Even though the reduced amide and the acid forms differ only by one Da, they could be distinguished based on the Carbon$_{12}$ (C_{12}) and C_{13} peak ratios observed during the high resolution mass measurements (Figure 17c).

The application of the Frit-FAB ionization along with high resolution mass measurements can be utilized not only for the accurate mass measurements but also for monitoring reactions and reaction products.

2. Primary Structures of Proteinaceous Toxins

A. Sequencing of Polypeptides. The methods used for ionizing biological molecules such as proteinaceous toxins induce softer ionization of the molecules, forming only either the singly charged or a series of multiply charged protonated molecular ions. Ions formed by the degradation of the parent molecule, as found in the electron impact ionization mass spectra, are completely absent during softer ionization processes. Hence, in order to characterize the peptides or proteins, degradation of the molecular ions needs to be accomplished by other means. It is usually carried out by selecting the parent ions of choice, subjecting them to collisionally induced dissociation, and acquiring the daughter ion (MS/MS or

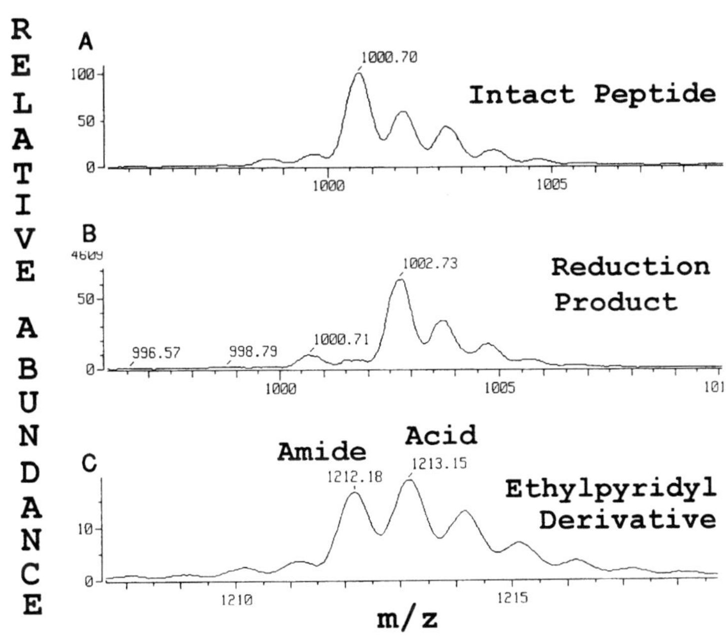

Figure 17. Frit-FAB mass spectra of intact (A), reduced (B), and Ethylpyridyl © Conopressin S.

product ion) spectrum.[3,10,23,24] The collisions may be induced by providing either low (10-200 eV) or high (2-10 KeV) energies.[24] The product ions are detected by simple triple quadrupole or four sector magnetic tandem mass spectrometers.[23,24] On few occasions, the processes are also carried out in hybrid tandem mass spectrometers.[23,24]

The daughter ions of a peptide are formed by cleavage at and around the amide linkages between two amino acid moieties leading to the formation of ions, generated by the cleavage from the N-terminus as well as the C-terminus. The measured m/z values of these daughter ions gave valuable information for deriving the amino acid sequences of peptides.[3,10,23,24] Details concerning the types of cleavages occurring in peptides and the nomenclature of the generated daughter ions are discussed elsewhere.[49] Normally, there are three types of ions formed during the CID process.[2,3,10,24] Immonium ions of amino acids as well as the ions formed by cleavages of bonds around the amide linkages are generated under both low and high energy collisions.[10,24] However, the side chain or the ring cleavages are accomplished only under high energy CID conditions.[10,24] The latter type of ions are useful in identifying the isobaric amino acids such as leucine and isoleucine.[24]

We had earlier deduced the total amino acid sequences of some unknown blue-green algal toxic peptides utilizing tandem mass spectrometric techniques in combination with simple chemical modifications and derivatization procedures.[26] The cyclic peptides were hydrolyzed using trifluoroacetic acid. The linear hydrolytic products and their corresponding methyl esters were subjected to FAB ionization and the individual types of ions were analyzed using CID processes. The sequences of the individual peptides were derived from the corresponding recorded MS/MS spectra.[26] Some of the peptides were found to contain Asp and dehydroalanine which had not been observed previously.[26,36,37]

When multiply charged precursor or parent ions formed during the electrospray ionization are subjected to CID processes, both singly as well as multiply charged daughter ions are formed. The interpretation of the data from these experiments becomes more complex. However, the total sequence could be derived and at times complementary information may also be obtained during the process.[11] The ESI-MS/MS spectrum of [Glu-3,4,7,10,14]-Conantokin G (Figure 18) with the molecular mass of 2,044.2 Da could successfully be sequenced by subjecting the doubly charged ion to CID processes. The basic residues Lys and Arg are located closer to the C-terminus, promoting the formation of the C-terminus daughter ions. The doubly charged "y" ions provided the total sequence of the molecule. In addition, the singly charged "y" ions for up to ten residues are also present. Thus, the sequence of a peptide which exceeded the accurate mass range of the mass spectrometer (2 kDa) could be obtained from the doubly charged daughter ions.

The continuous flow-FAB ionization is superior to the ESI process during the structural characterization of proteins and other biopolymers. In most instances, regardless of the number of basic moieties present in the peptides, mostly the singly charged molecular ions, occasionally along with doubly charged ions, are observed. Hence, the singly charged species required for the CID experiments are obtained preferentially. We have combined this valuable tool along with array detection and high resolution mass measurements of daughter ions in order to determine the primary structures of some toxic conus peptides. Conopressin G was reduced (Figure 17b) and derivatized with vinylpyridine (Figure 17c) using the newly developed procedure. The singly charged protonated molecular ions of the reduction product and the ethylpyridyl derivative were subjected to CID in the presence of xenon at a collision energy of 2 KeV. The CID spectra of the reduction product and the derivative are shown in Figure 19 and 20, respectively. The sequence ions in the MS/MS spectra enabled not only determination of the total amino acid sequence of the molecule but also distinguished the third residue from the N-terminus to be isoleucine. The wa_7 and wb_7 ions distinguished the residue to be isoleucine since leucine would form only a single "w" ion with the mass difference of 45 from the corresponding "z" ion and not two "w" ions with 15 and 29 amu

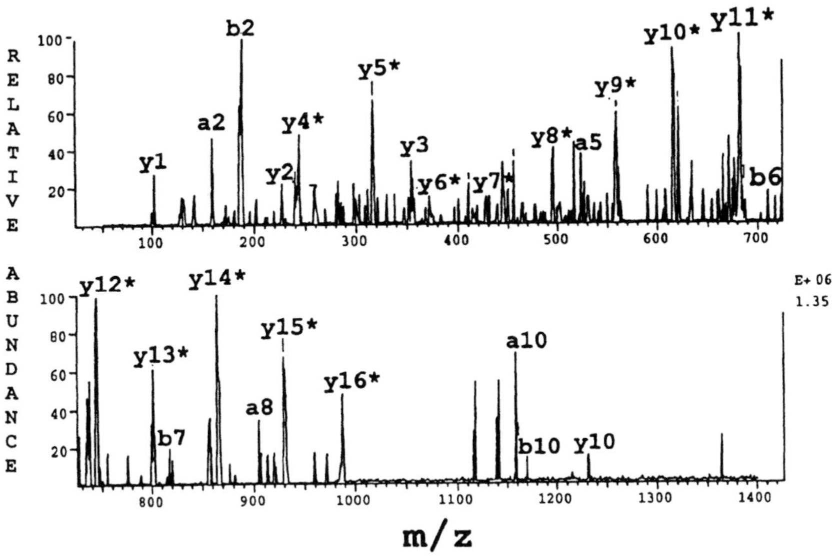

Figure 18. ESI-MS/MS spectrum of [Glu5]-Conantokin G. [GEEELQENQELIREKSN] (*. doubly charged daughter ions.

from the corresponding "z" ion. Such a distinction cannot be accomplished from the corresponding low energy collision spectrum.

The CID processes of peptides are routinely carried out for molecules with molecular mass of 3 kDa or less. During the sequencing of larger polypeptides or proteins, the intact peptide is subjected to enzymatic degradation using at least two different enzymes. The molecular masses of the degradation products are measured and their corresponding CID spectra are recorded. The derived sequences of various enzymatic fragments along with the N-terminus sequences obtained by the automated Edman deg-

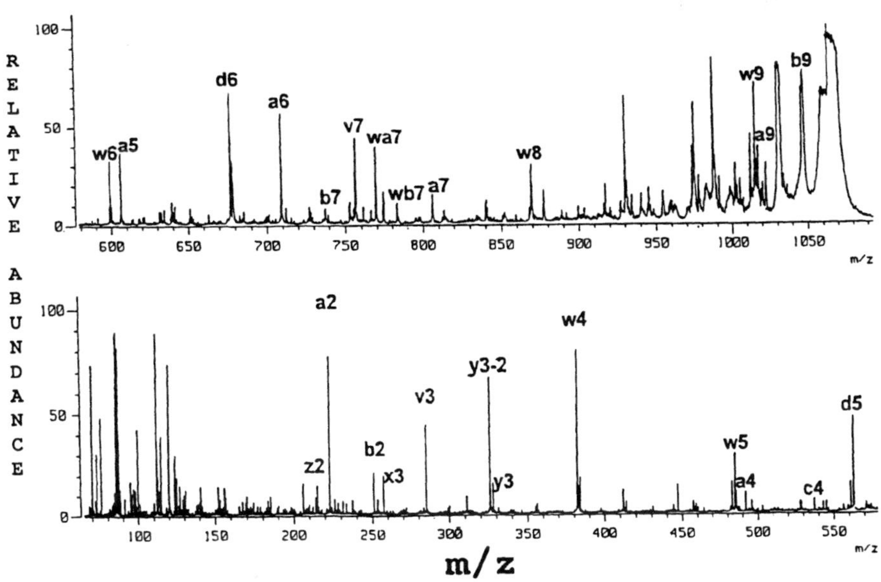

Figure 19. Continuous flow-Frit-FAB-CID spectrum of reduced conopression G. M + H = 1064.52.

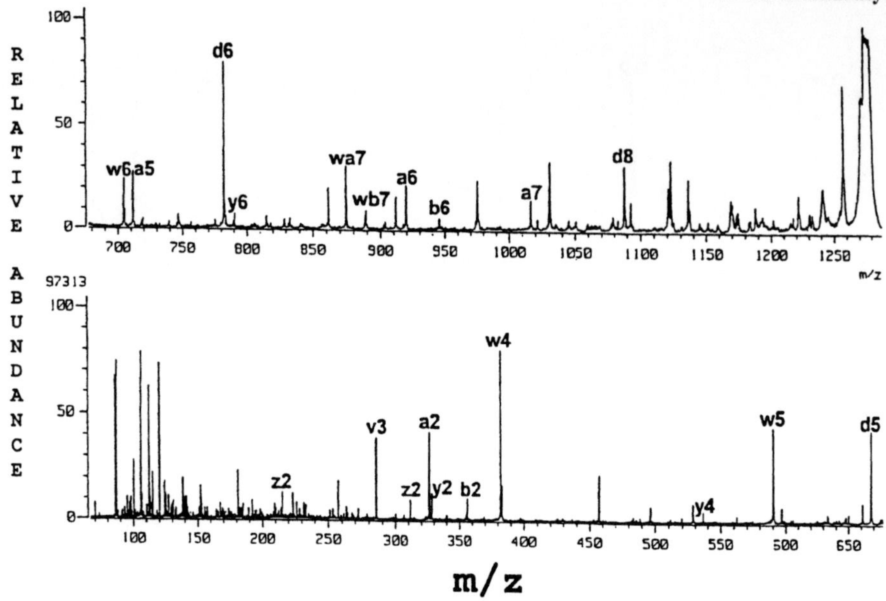

Figure 20. Continuous flow Frit-FAB-CID spectrum of Ethylpyridyl derivative of Conopressin G. M + H = 1274.64.

radation process are used to compute the total amino acid sequences of the larger protein.[3,10,11,24] Figure 21 displays the FAB-MS/MS spectrum of a tryptic digest fragment of an unknown snake venom component. The derived sequence indicated that the fragment must be the C-terminus fragment.

B. Chirality of Amino Acids in Toxic Peptides. Blue-green algal toxic peptides (microcystins) have been known to contain both D- and L- amino acids. Chiralities of the individual amino acids, obtained by the acid hydrolysis of the intact peptide, cannot be obtained chromatograpically since the enantiomers elute with identical retention times. However, on derivatization with another optically active derivatizing agent, the resulting diastereomers originating from a D- and L- isomer of the same amino acid would elute from a chromatographic column at different times. The derivative from a L-amino acid eluted earlier than the corresponding D- isomer. Based on this principle, we determined the chiralities of amino acids present in some unknown microcystins.[34] Both D- and L- isomers of amino acids, detected in the investigated peptides, were derivatized with 1-fluoro-2,4-dinitrophenyl-5-L-alanine-amide (Marfey's reagent).[34] The retention times of the diastereomers over a reverse phase C_{18} column and their corresponding mass spectra were obtained by the LC/Thermospray-MS[34] analysis of the derivatives. The most abundant ions for individual amino acid derivatives were selected for the selected ion monitoring. The hydrolysates of microcystins were derivatized with Marfey's reagent and separated over the C_{18} column and introduced directly into the thermospray ionization source. The observed m/z values of the ions along with their retention data were sufficient to determine the chirality of the individual amino acids present in the toxic peptides.[34] Figure 22 illustrates the LC/MS analysis of the diastereomeric derivatives obtained from the hydrolysates of microcystin-LR isolated from the bloom found in the Akers Lake, Norway. It was found to contain D-Ala, D-Glu, *erythro* D-β-methyl-Asp, L-Leu and L-Arg based on the LC/MS data generated for the FDAA derivatives of microcystin-LR hydrolysates.[34] This procedure was found to be more sensitive than the previously known methods.[25,50]

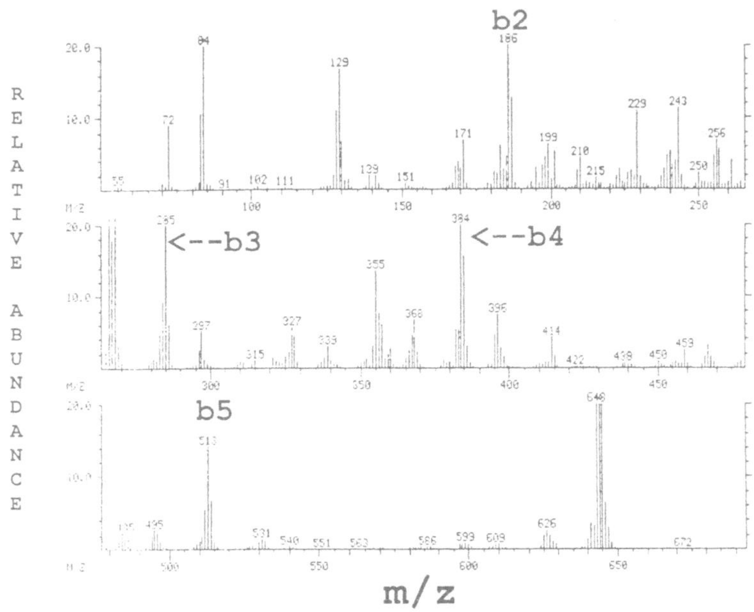

Figure 21. FAB-MS/MS spectrum of a tryptic digest peptide (GKVVEL).

C. Location of Disulfide Bridges in Toxins. The amino acid sequences of a cardiotoxin (Figure 23) isolated from Naja Naja atra was determined by the automated Edman degradation method.[40] The locations of the disulfide bridges were previously unknown.[25] All four disulfide bridges present in the molecule were determined by our innovative approach as follows.[25] The intact peptide was subjected to tryptic digestion and the products were chromatographically separated. The molecular masses of all the fragments were generated and analyzed. One of the fragments contained a branched peptide with one of the disulfide bridges intact. Based on the measured molecular mass of the fragment, the disulfide bridge was identified to be between C-3 and C-21. Another tryptic fragment, containing all three

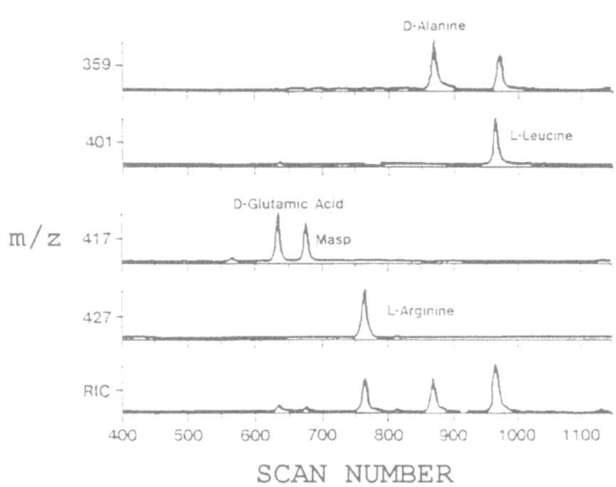

Figure 22. Thermospray mass spectrum of microcystin-LR hydrolysates.

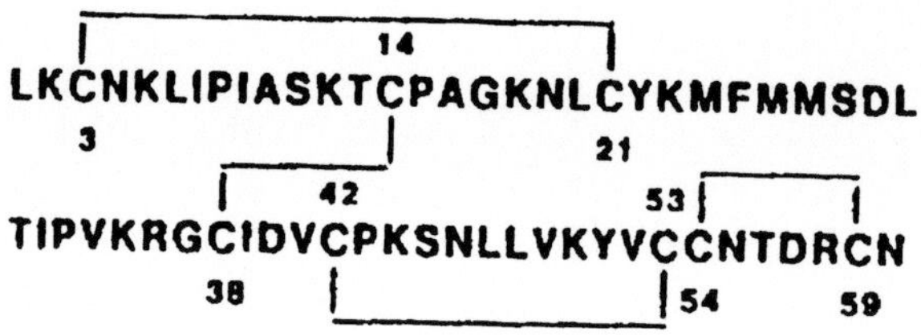

Figure 23. Disulfide linkages in a Naja Naja atra cardiotoxin.

disulfide bridges intact, was subjected to Asp-N digestion and the fragments were identified by ESI-MS data. This process led to the identification of another disulfide bridge between C-14 and C-38. The Asp-N product containing the other two intact disulfide bridges required either enzymatic or chemical cleavages to identify the linkages. However, the adjacent cysteines in positions 53 and 54 and absence of enzymatically cleavable sites in the residue complicated the process.[25] Careful review of the sequences of the fragment revealed the possibility of accomplishing the goal by subjecting the residue through three cycles of Edman sequencing process followed by the mass measurement of the products. A new method was developed and applied for this scheme. The Asp-N product loaded on the PVDF membrane was subjected to three cycles of automated Edman degradation and the residue was desorbed from the membrane. The measured molecular mass of the only detected fragment (821.5 Da) could only correspond to the heptapeptide with the disulfide bridge between C-54 and C-59. The fourth linkage was thus assigned by default to between C-42 and C-53. Thus, using only a few nanomoles of the intact peptide, we established the location of all four disulfide bridges (Figure 23) present in a tightly bound toxic polypeptide.[25]

CONCLUSIONS

Electrospray, matrix assisted laser desorption, and continuous flow Frit-FAB ionization methods along with mass spectrometric methods of detection have been found to be valuable tools for the investigation of biomolecules including larger proteinaceous toxins and antibodies. Measurement of molecular masses, exceeding 100 kDa can be routinely accomplished using only 0.1-1.0 picomole of the intact peptides. Primary structures of proteinaceous toxins can be derived by utilizing collisionally induced dissociations of molecular ions and tandem mass spectrometric methods. Allied techniques such as Edman degradation, chemical and enzymatic degradation, derivatization, and HPLC techniques are used to facilitate the characterization of biomolecules by mass spectrometric techniques. Specific toxins or other biomolecules can be detected very efficiently when present in either a complex mixture or sample matrices. The procedures are sensitive, selective, time and cost effective, accurate and reproducible.

REFERENCES

1. Matsuo, T., Caprioli, R. M., Gross, M.L., and Seyama, Y., Eds., 1994, In Mass Spectrometry, Present and Future; Wiley, New York.
2. Burlingame, A.L., Boyd, R.K., and Gaskell, S.J., 1994, In Mass Spectrometry, Anal. Chem., 66, 634R-683R.
3. Hunt, D.F., Shabanowitz, J., Michel, H., and Engelhart, V.H., 1993, In Methods in Protein Sequence Analysis; Imhori, K., Saklyama, F., Eds., Plenum Press, New York, pp 127-133

4. Fenselau, C., Ed., 1994, In Mass Spectrometry for the Characterization to Microorganisms; ACS Symposium Series 541; American Chemical Society, Washington, D.C.
5. Desiderio, D.M., Ed., 1994, In Mass Spectrometry: Clinical and Biomedical Applications, Plenum Press, New York, Vol., 2.
6. Cotter, R.J., Ed., 1994, In Time of Flight Mass Spectrometry, ACS Symposium Series, 549, American Chemical Society, Washington, D.C.
7. Niessen, W.M.A., van der Greef, J., 1992, In Liquid Chromatography-Mass Spectrometry, Mercel Dekker, New York.
8. Settineri, C.A., Burlingame, A.L., 1995, In Carbohydrate Analysis: High Performance Liquid Chromatography and Capillary Electrophoresis, El Rassi, Z., Ed., Elsevier Science Publishers, Amsterdam, The Netherlands.
9. Emary, W.B., Hand, O.W., Cooks, R.G., 1995, In Lasers and Mass spectrometry, Lubman, D.M., Ed., Oxford University Press, New York, 1990
10. McCloskey, J.A., Ed., 1990 Mass Spectrometry, Methods in Enzymology, Volume 193, Academic Press, New York
11. 1993, Journal of the American Society for Mass Spectrometry, Special Issue on Electrospray Ionization & Applications to Peptide & Protein Chemistry, Volume 4.
12. Cotter, R.J., 1992, In Time-of-flight Mass Spectrometry for the Structural Analysis of Biological Molecules, Anal.Chem., 64, 1027A-1039A
13. Cotter, R.J., 1991, In Matrix-Assisted Laser Mass Spectrometry of Biopolymers, Anal. Chem., 63, 1193A-1203A.
14. Huberty, M.C., Vath, J.E., Yu, W., Martin, S.A., 1993, Site-specific Carbohydrate Identification in Recombinant Proteins using MALDI-TOF MS, In Anal. Chem., 65, 2791-2800.
15. Pfeiffer,G., Dabrowski, U., Dabrowski, J., Strim, S., Strube, K.H., and Geyer, R., 1992, Carbohydrate Structure of a thrombin-like Serine Protease from Agkistrodon rhodostoma, In Eur. J. Biochem., 205, 961-978
16. Xiang, F., and Beavis, R.C., 1993, A Method to Increase contaminant Tolerance in Protein-Matrix-assisted Laser Desorption/Ionization by the Fabrication of Thin Protein-doped Polycrystalline films, In Rapid Commun. In Mass Spectrom., 199-204
17. Xhang, Y., Chen N., Wang, L., Yang, B., Qian X., and Ma L., 1994, Separation and identification of Proteins obtained from Agkistrodon acutus snake Venom by Capillary Zone electrophoresis and laser Desorption/Ionization Mass Monitoring, Biomedical chromatography, 8, 148-150.
18. Covey T., Huang, E.C., and Henion, J.D., 1991, Stuctural Characterization of Protein Tryptic Peptides via Liquid Chromatography/Mass spectrometry and Collision-Induced Dissociation of Their Doubly charged Molecular Ions, In Anal. Chem., 63, 1193-1200.
19. Kenny, P.T., and Orlando, R., 1992, Tandem mass Spectrometric Analysis of Peptides at the Femtomole Level, in Anal. Chem., 64, 957-960.
20. Caprioli, R.M., 1990, In Continuous-Flow fast Atom Bomardment Mass Spectrometry, John Wiley, New York.
21. Fenn, J.B., Mann, M., Meng, C.K., Wong, S.F., and Whitehouse, C.M., 1990, Electrospray Ionization - Principles and Practice, In Mass Spectrom.Rev., 9, 37-70
22. Fabris, D., Kelly, M., Murphy, C., Wu, Z., and Fenselau, C., 1993, High-Energy Collision-Induced Dissociation of Multiply Charged Polypeptides Produced by electrospray, In J.Amer.Soc.Mass Spectrom., 4, 652-661
23. McLafferty, Ed., 1983, In Tandem Mass Spectrometry, John Wiley, New York
24. Biemann, K., 1990, Sequencing of Peptides by Mass Spectrometry and High Energy Collision-Induced Dissociation, In Ref 10, pp 455-479.
25. Krishnamurthy, T., Hauer, C.R., Prabhakaran, M., Freedy, J.G., and Hayashi, K., 1994, Identification of Disulfide Bridges in a Cardiotoxic Peptide by Electrospray Ionization, In Biol. Mass Spectrom., 23, 719-726.
26. Krishnamurthy, T. Szafraniec, L., Hunt, D.F., Shabanowitz, J., Yates, III, J.R., Hauer, C.R., Carmichael, W.W., Skulberg, O., Codd, G.A., and Missler, S., 1989, in Structural Characterization of Toxic Cyclic Peptides from Blue-green Algae by Tandem Mass Spectrometry, In Proc.Natl.Acad.Sci.USA, 86, 770-774.
27. Kaiser, I.I., Griffin, P.R., Aird, S.D., Hudiburg, S., Shabanowitz, J., Francis, B., John, T.R., Hunt, D.F., and Odell, G.V., 1994, Primary Structures of Proteins from the Venom of Mexican Red Knee Terantula, In Toxicon, 32, 1083-93

28. Ren, K., Bannon, J.D., Pancholi, V., Cheung, A.L., Robbins, J.C., Fischetti, V.A., and Zabriskie, J.B., 1994, Characterization and Biological Properties of a New Staphylococcal Exotoxin, In J. Exp. Med., 180, 1675-83.
29. Slos, P., Speck, D., Accart, N., Kolbe, H.V., Schubnel, D., Bouchon, B., Bischoff, R., and Kieny, M.P., 1994, Recombinant Cholera Toxin B Subunit in E. Coli: High-level Secretion, Purification, and Characterization, In Protein Expression Purif., 5, 518-26
30. Perkins, J.R., Tomer, K.B., 1994, Capilllary Electrophoresis/Electrospray Mass Spectrometry using a High-Performance magnetic Secttor Mass Spectrometer, In Anal.Chem., 66, 2835-40
31. Monks, S.A., Gould, A.R., Lumley, P.E., Alewood, P.F., Kem, W.R., Goss, N.H., and Norton, R.S., 1994, Limited Proteolysis Study of Structure-function Relationships in ShI, a Polypeptide Neurotoxin from a Sea Anemone, In Biochim.Biophys.Acta, 1207, 93-101.
32. Garcia, M.L., Garcia-calvo, M., Hidalgo, P., Lee, A., and MacKinnon, R., 1994, Purification and Characterization of Three Inhibitors of Voltage-Dependent K+ Channels from Leiurus Quinquestriatus Va. Hebraeus Venom, In Biochemistry, 33, 6834-9.
33. Gowda, D.C., Jackson, C.M., Hensley, P., and Davidson, E.A., 1994, Facor X-activating glycoprotein of Russell's Viper Venom Polypeptide Compositions and Characterization of the Carbohydrate Moieties, In J. Biol. Chem., 269, 10644-50.
34. Krishnamurthy, T., 1994, Chiralities in Microcystins, In J. Amer.Soc. For Mass Spectrom., 5, 724-30.
35. Dufton, M.J., and Hider, H.C., 1983, Conformational Properties of the Neurotoxins and Cytotoxins isolated from Elapid Snake Venoms, In CRC Crit.Rev. Biochem., 14, 113-171.
36. Sivonen, K., Namikoshi, M., Evans, W.E., Fardig, M., Carmichael, W.W., and Reinhart, K.L.,1992, Three New Microcystins, Cyclic Heptapeptide Hepatotoxins, from Nostoc. Sp Strain 152, In Chem.Res.Toxicol. 5, 464-469.
37. Namikoshi, M., Reinhart, K.L., Saka, R., Slotts, R.R., Dahlem, A.M., Beasley, V.R., Carmichael, W.W., and Evans, W.R.,1992, Identification of 12 Hepatotoxins from a Homer Lake Bloom of the Cyanobacteria Microcystis aeruginosa, Microcystis viridis, and Microcystis wesenbergii: Nine New Microcystins, In J.Org.Chem., 57, 866-872.
38. Carmichael, W.W., Krishnamurthy, T., Beasley, V.A., Min-Juan, Y., Bunner, D.L., Moore, J.E., Eloff, J.N., Reinhart, K.L., Falconer, L., Runnegar, M., Corham, P., Skulberg, O.M., Harada, K., and Watanabi, M., Naming of Cyclic Heptapetide Toxins of Cyanobacteria (Blue-green Algae), In Toxicon, 26, 971-73.
39. Chau, M.H., and Nelson, J.W., 1992, Cooperative Disulfide Bond Formation in Apamin, In Biochemistry, 31, 4445-4450. .
40. Chiou, S., Raynor, R.L., Bin, Z., Chambers, T.C., and Kuo, J.F.,1993, Cobra Venom Cardiotoxin (Cytotoxin) Isoforms and Neurotoxin: Comparative Potency of Protein Kinase C Inhibitors and Cancer Cell Cytotoxicity and Modes of Enzyme Inhibition, In Biochemistry, 32, 2062.
41. Tu, A., Ed., 1977, In Venoms: Chemistry & Molecular Biology, Wiley Interscience, New York.
42. Aird, S.D., Kruggel, W.G., and Kaiser, I.I., 1990, Amino Acid Sequence of the Basic Subunit of Mojave Toxin from the Venom of the Mojave Rattlesnake (Crotalus scutulatus), In Toxicon, 28, 669-673.
43. Olivera, B.M., Rivier, J., Scott, J.K., Hillyard, D.R., and Cruz, L.J., 1991, Conotoxins, In J.Biol.chem., 266, 22067-22070.
44. Olivera, B.M., 1987, Neuronal Calcium channel Antagonists. Discrimination between Calcium Channel Subtypes using ω-Conotoxin from Conus magus Venom, In Biochemistry, 26, 2086-209.
45. Tu, A., Colarado State University, Fort Collins, CO, USA, Personal Communication.
46. Olivera, B.M., Rivier, R., Clark, C., Railo, C.A., Corpuz, G.P., Abogadie, Fe C., Mena, E.E., Woodward, S.R., Hillyard, D.R., and Cruz, L.J., Diversity of Conus neuropeptides, In Science, 249, 257-263.
47. Loo, J.A., Ogorzalek, R.R., and Andrewa, P.C., 1993, Primary to Quarternary Protein Structure Determination with Electrospray Ionization and Magnetic Sector Mass Spectrometry, In Org.Mass Spectrom., 28, 1640-1649.
48. Fisher, W.H., Rivier, J.E., and Craig, A.G., 1993, In Situ Reduction Suitable for Matrix-assisted Laser Desorption/Ionization and Liquid Secondary Ionization using Tris)2-carboxyethyl)phosphine, In Rapid Comm. In Mass Spectrom., 7, 225-228.
49. Roepstrorff, P., 1984, Proposal for a Comman Nomenclature for Sequence Ions in Mass Spectra of Peptides, Biomed. Mass Spectrom., 11, 601.
50. Botes, D.P., Tuinman, A.A., Wessels, P.L., Viljoen, C.C., Kruger, H., Williams, D.H., Santikarn, S., Smith, R.J., and Hammonds, S., 1984, The Structure of Cyanoginosin-LA, a Cyclic Heptapeptide Toxin from the Cyanobacterium Microsystis aeruginosa, In J.Chem.Soc.Perkin Trans., 2311-2318
51. Teshima, T., Wakamiya, T., Arakami, Y., Nakajima, T., kawai, N., and Shiba, T., 1987, Total synthesis of ADDA, the Unique C_{20} Amino Acid of Cyanobacterial Hepatotoxins, In Tetrahedron Letters, 29, 3509-3510

52. Muniz, Z.M., Pracej, D.N., and Dolly, J.O., 1992, Characterization of Monoclonal Antibodies against Voltage-Dependent K^+ Channels Raised Using α-Dendrotoxin Acceptors Purified from Bovine Brain, In Biochemistry, 31, 12297-12303.
53. Sahara, Y., Robinson, H.P., Miwa, A., and Kawai, N., 1991,A Voltage-clamp Study of the Effects of Joro Spider Toxin and Zic on Excitatory synaptic Transmission in CA1 Pyramidal Cells of the Quinea Pig Hippocampal Slice, In Neurosci.Res., 10, 200-210.

38

DETECTION OF THE STAPHYLOCOCCAL TOXINS

M. S. Bergdoll

Food Research Institute
University of Wisconsin
1925 Willow Drive
Madison, Wisconsin 53706

INTRODUCTION

The staphylococci are one of the most common of microorganisms and are common inhabitants of most animals. They probably have been around almost since the beginning of time, but were not recognized as important organisms until 1880 when Ogston (1880) demonstrated that they caused many types of purulent infections in humans. Rosenbach (1884) gave them the name *Staphylococcus*. These organisms formed both yellow and white colonies which were given the names *Staphylococcus aureus* and *Staphylococcus albus*, but later classification by pigmentation was discontinued and all were called *S. aureus*. A second species, *S. epidermidis*, was added in 1908 by Winslow and Winslow (1908); the major difference in the two species was that the latter species did not produce coagulase. It was not until 1974 that a third species, *S. saprophyticus*, was added, and from that time forward the addition of new species proceeded on a regular basis until at this writing 31 species, along with several subspecies, have been identified.

Classification of the Coagulase-Positive Staphylococci

The majority of the new species have been non-producers of coagulase, so-called coagulase-negative species (CNS). *S. aureus* remained the sole species that produced coagulase although Hajek and Marsalek (1973) had subdivided this species into biotypes according to the animals from which the coagulase-positive staphylococci were isolated (Table 1). In general, any staphylococci that produced either coagulase or thermonuclease (TNase), were considered to be *S. aureus*. This species was the most important one as it was the major cause of staphylococcal diseases and infections. Also, it was the most frequent producer of the toxins implicated as the cause of some of the staphylococcal diseases, such as the enterotoxins are the cause of staphylococcal food poisoning.

The taxonomists were not satisfied with the status quo and reclassified some of the biotypes of *S. aureus* as separate species. Thus *S. aureus* biotype E (isolated from dogs)

Natural Toxins II, Edited by B. R. Singh and A. T. Tu
Plenum Press, New York, 1996

Table 1. Biotypes of *S. aureus* strains[a]

Biotype	Animal
A	Man
B	Swine
	Poultry
C	Cattle
	Sheep
	Goats
D	Hares
E	Dogs
	Horses
F	Pigeons

[a]Hajek and Marsalek, 1973

became *S. intermedius* (Hajek, 1976) and biotype F (isolated from pigs) became *S. hyicus* (Devriese et al., 1978). More recently another coagulase-positive species was added, *S. delphini* (Varaldo et al., 1988), isolated from dolphins. Classification of the coagulase-positive staphylococci into species is given in Table 2. Normally *S. aureus* is pigmented, but this is not an identifying characteristic because some strains may be non-pigmented. The only difference in coagulase production is that only about 45% of *S. hyicus* are coagulase-positive, with a positive reaction requiring 24 hours (Devriese et al., 1978). However, it is possible for some strains to be negative for the identifying properties, such as some *S. aureus* strains may be coagulase-negative. Only *S. delphini* is TNase-negative. Anaerobic fermentation of mannitol is observed only with *S. aureus*; also, acetoin production is limited to *S. aureus*. Because these properties are not 100%, it may be difficult to specifically classify some strains. In a recent study of human staphylococcal strains in Brazil, three strains fit the criterion for *S. intermedius*, and one fit the requirements for *S. hyicus*; however, these species are seldom if ever isolated from humans. Phage typing and comparison to strains isolated from other sites in the same individuals indicated these strains to be *S. aureus*. Anytime a strain is isolated from a human that fits some characteristics of *S. aureus*, it should be examined for enterotoxin production.

Transference of Staphylococci from Animals to Humans

S. aureus is a common inhabitant of animals and a major question is can animal *S. aureus* be transferred to humans and become colonized. There is little evidence for this, as no study has been conducted to determine whether such transfer can occur. It is of interest that the toxic shock syndrome toxin (TSST) produced by human *S. aureus* is also produced

Table 2. Characterization of coagulase-positive staphylococcal species

Property	aureus	intermedius	hyicus	delphini
Pigment	+	—	—	—
Coagulase	+	+	±	+
TNase	+	+	+	—
Mannitol(aer)	+	±	—	+
Mannitol(an)	+	—	—	—
Hemolysins	+	±	—	+
Clump. factor	+	±	—	—
Acetoin prod.	+	—	—	—

by *S. aureus* strains isolated from bovine animals; however, the TSST produced by strains isolated from sheep and goats is not identical to the human TSST (Ho et al.,1989), although it reacts with the antibody to the human TSST.

The other coagulase-positive species are seldom if ever isolated from humans, but is it possible for such transfer to occur. Two studies of veterinary students who worked with dogs was carried out to determine whether *S. intermedius* could be transferred to humans and become colonized (Talon et al., 1989a, 1989b). Only one student became colonized with an *S. intermedius* strain, indicating that transfer from animal to humans is a rare occurrence. Of greater significance, however, is the fact that in one study three individuals had skin infections due to *S. intermedius* (Talon et al., 1989) and in the other study this species was isolated from humans with infected canine-inflicted bite wounds (Talon et al, 1989a).

THE STAPHYLOCOCCAL ENTEROTOXINS

The staphylococcal enterotoxins are relatively low molecular weight proteins, 27,000 to 29,000 daltons. The enterotoxins that have been identified are classified as enterotoxins A (SEA), B (SEB), C_1 (SEC_1), C_2 (SEC_2), C_3 (SEC_3), D (SED), and E (SEE) (Bergdoll,1979). One additional enterotoxin has been identified (enterotoxin H) (Su & Wong, 1994), with others still unidentified. This enterotoxin was involved in a food poisoning outbreak in Brazil (Bergdoll,1994). The unidentified enterotoxins are relatively unimportant because only about five percent of staphylococcal food poisoning outbreaks are due to the unidentified enterotoxins. Of the identified enterotoxins, the enterotoxin C's are very closely related and can be identified by their cross-reactions with antibodies prepared against any one of the SEC's (Reiser et al.,1984). The other enterotoxins are identified by antibodies specific for each of the enterotoxins, although cross-reactions between SEB and the SEC's (Lee et al.,1980) and SEA and SEE (Lee et al., 1978) do exist. Monoclonal antibodies have been prepared that are useful in the sandwich ELISA for the detection of all of the identified enterotoxins.

ENTEROTOXIN PRODUCTION BY THE STAPHYLOCOCCI

Very little work has been done on the newer coagulase-positive species, hence, it is not known how their specific properties, such as production of toxins and other biologically active substances, compare to those of *S. aureus*. This may be of little importance because the other coagulase-positive species are seldom isolated from humans, the major source of staphylococci implicated in food poisoning. However, if transfer to humans of species, such as *S. intermedius*, should occur and these individuals were to handle foods, it might be possible for the foods to be contaminated with the organisms. If such were to occur, food poisoning could result because *S. intermedius* strains can produce enterotoxin (Table 3; Hirooka et al., 1988). One such outbreak has been reported (Anon., 1992). If the source of the organism is important, further testing would be in order.

No studies have been made on the transference of *S. hyicus* from pigs to humans, although this may be less likely because people's contact with pigs is not as intimate as it is with dogs. If studies reveal that *S. hyicus* and *S. delphini* can produce enterotoxin, it would indicate that any coagulase- or TNase-positive staphylococci should be checked for enterotoxin production.

Any animal enterotoxigenic strain can be involved in food poisoning as animal products can be contaminated with animal *S. aureus* and if the right conditions exists the staphylococci will grow and produce enterotoxin. The major opportunity for this is in dairy

Table 3. Enterotoxin production by staphylococci from dogs[a]

Number	Enterotoxin	%
81/254		15.5
14/81	A	17.3
2/81	B	2.5
59/81	C	72.8
13/81	D	16.0
2/81	E	2.5

[a]Hirooka et al., 1988

products, such as raw milk from mastitic animals. There have been outbreaks from milk from mastitic animals and from cheese made from unpasteurized milk. Also, there have been outbreaks from sheep cheese, which could be related to enterotoxigenic staphylococci from the sheep. There is always the possibility that staphylococci could be transferred to food from pets kept in the confines of the home; however, there is no proof of this ever happening, but then most food poisonings that occur in the home are never investigated.

Determination of Source of Enterotoxigenic Staphylococci

Aside from the subdivision of *S. aureus* into biotypes, it is further subdivided into strains. It is difficult to determine whether a staphylococci isolated form one source is identical to one isolated from another source. One method of doing this is phage typing, for example, to determine the source of the staphylococci causing food poisoning. If the staphylococci isolated from the food has an identical phage pattern to one from a food handler they are considered identical or the same strain. Phage typing is not easily done and many strains are never subjected to it.

PRODUCTION OF ENTEROTOXIN

The staphylococci grow under all types of conditions, but not all of these are favorable for toxin production. Although staphylococci will grow under anaerobic conditions, growth is much slower with little or no toxin produced. Growth is much faster and proceeds to a higher level when adequate O_2 is available. For example, staphylococci will grow in a still culture such as 10 ml of medium in a test tube, but very little toxin will be produced. Growth and toxin production can be enhanced by shaking the tube, the more vigorous the shaking the better the results (Kato et el., 1966). However, the depth of the layer of medium will

Table 4. Concentration of NAK medium on enterotoxin A production (μg/ml)[a,b]

Concentration	Volume in 250 ml flasks		
%	15 ml	50 ml	75 ml
2	2.4	3.1	2.6
3	5.2	4.5	2.9
4	9.7	4.1	2.8
5	8.6	3.9	2.7
6	8.6	3.6	2.5

[a]Kato et al., 1966
[b]Incubated for 24 hours at 37°C on gyrotory shaker at 280 rpm.

Table 5. Aeration effect on toxic shock syndrome toxin production[a]

Type	Growth x 10^9	TSST (μg/ml)
Tube[b]	6.5	15.3
Sparger[c]	16.0	120.0
Shake flask	11.0	11.3

[a]Wong and Bergdoll, 1990
[b]100 cc air and 100 cc CO_2/min.
[c]20 cc air and 5 cc CO_2/min.

affect the aeration by the shaking. In general, the thinner the layer of fluid in a flask the more effective the shaking will be (Table 4). How the air is delivered to the medium is important also, for example, merely bubbling air through a stirred medium is not nearly as effective as breaking the air bubbles into fine particles with a sparger (Wong and Bergdoll, 1990; Table 5).

Growth of the staphylococci on a membrane over an agar-medium stimulates good toxin production (Robbins et al.,1974; Fig. 1). It is assumed that under these conditions that all organisms are adequately exposed to the air to provide for optimum toxin production. Also, increasing the surface area in a flask such as placing the medium in a sac placed in the flask, along with shaking, greatly increases toxin production (Robbins et al.,1974).

The medium is also important for toxin production. The staphylococci will grow in many different media, but not always is there good toxin production. There are many different types of pancreatic digests of casein, but only special types are suitable for toxin production. However, the addition of yeast extract aids in making them suitable for toxin production (Wong and Bergdoll,1990). Also, brain heart infusion (BHI) broth supplemented with yeast extract provides a good medium for production of the enterotoxins.

DETECTION OF THE ENTEROTOXINS

All methods for detection of the enterotoxins are based on the use of antibodies prepared against the enterotoxins. Most of the antibodies in use have been prepared in rabbits using the individual purified enterotoxins (Robbins & Bergdoll,1984). These polyclonal antibodies react with the enterotoxins in gels to give precipitin reactions which are highly

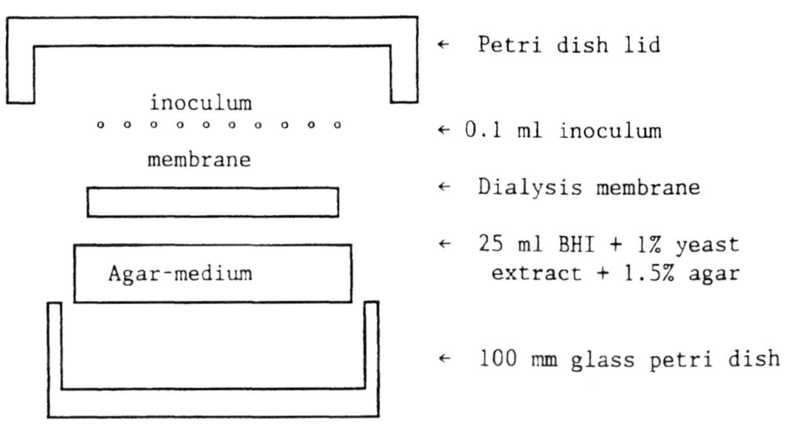

Figure 1. An expanded view of a 100 x 15 mm petri dish and contents used in the membrane-over-agar method.

specific. Protein A, a common metabolite of *S. aureus*, reacts non-specifically with IgG and can interfere with the detection of the enterotoxins. This interference can be eliminated by producing the antibodies in sheep because protein A does not react with sheep IgG. Monoclonal antibodies to all of the enterotoxins have been produced, but they cannot be used in gels because their reactions with the enterotoxins do not result in the formation of precipitates.

Detection of Enterotoxin Production by Staphylococcal Strains

Gel Diffusion Methods. Many types of gel reactions have been used in the detection of the enterotoxins, the most common ones being some form of either the Ouchterlony gel plate or the microslide. These methods have been used widely in the determination of the enterotoxigenicity of staphylococcal strains. The modification of the Ouchterlony gel plate test that is used in the Food Research Institute and recommended to others is the optimum sensitivity plate (OSP) method (Robbins et al., 1974; Fig. 2). It is easy to use and in conjunction with production of the enterotoxins by the membrane-over-agar or sac culture methods (Robbins et al.,1974) is of adequate sensitivity for detection of most enterotoxigenic

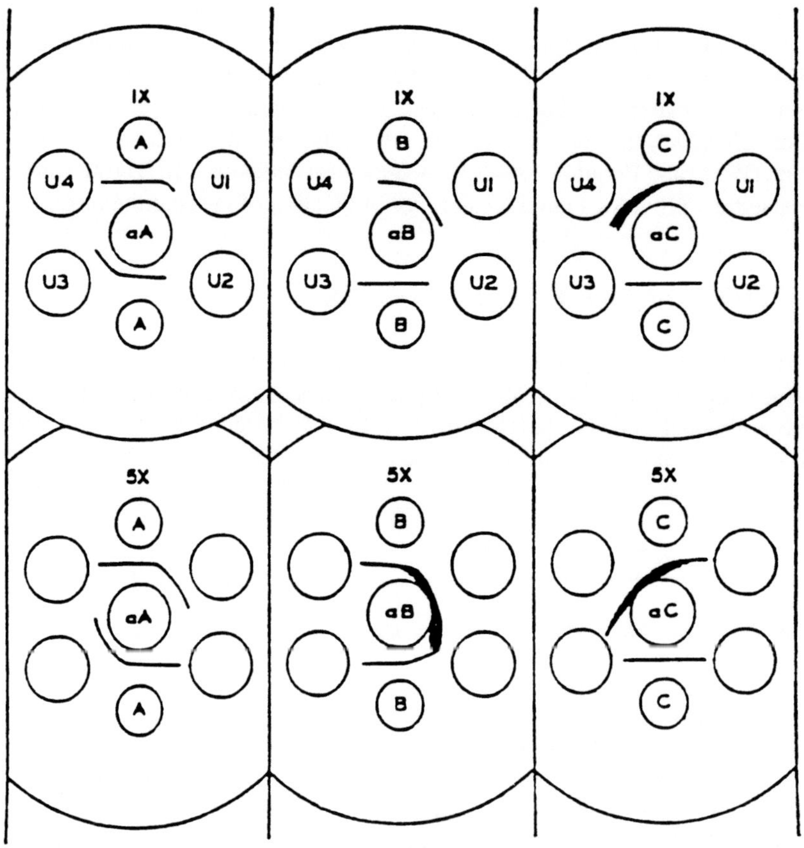

Figure 2. Enterotoxin analysis by the Optimum Sensitivity Plate (OSP) method. A, B, C = SEA, SEB, SEC, respectively (4 mg/ml). aA, aB, aC = anti-SEA, anti-SEB, anti-SEC, respectively. Unknown U1 contains approximately 0.5 mg A and 6 mg B/ml; unknown U2 contains no A, B, or C; unknown U3 contains approximately 2 mg a/ml; unknown U4 contains approximately 16 mg C/ml. 1X = unconcentrated unknowns' 5X = 5-fold concentrated unknowns.

Table 6. Low enterotoxin A producers[a,b]

S. aureus	Shake	Semi-solid	Sac	Membrane
1507	—	+ (S)[c]	+	+
1508	—	+	+	+
17	—	+ (HS)[d]	+	+ (H)[e]
19	—	+ (HS)	+	+ (H)
F-68	—	+ (S)	+ (VL)[f]	+

[a]Robbins et al., 1974
[b]Brain heart infusion broth medium
[c]After staining
[d]Hook after staining
[e]Hook
[f]Very low

staphylococci (Table 6). The normal sensitivity is 0.5 μg/ml but can be increased to 0.1 μg/ml by a 5-fold concentration of the staphylococcal culture supernatant fluids. Over the years any strain that produced enough toxin by maximum production methods to be detectable by gel diffusion methods was considered a toxin-producing strain.

The microslide is used by some investigators but care is needed in preparing the slides, even so, the results are often difficult to interpret. Many things can go wrong with this method and experience is very important in using it successfully (Casman & Bennett,1965).

Sensitive Detection Methods. Detection of toxin production by use of the more sensitive detection methods such as reversed passive latex agglutination (RPLA) (Igarashi et al., 1986) and enzyme-linked immunosorbent assay (ELISA) indicated that some strains produced only 10 to 20 ng/ml, which was not detectable by the OSP method. This was confirmed when 100-fold concentrates of culture supernatant fluids from five strains that tested positive for SEA by ELISA were positive by the OSP method. Examination of 110 strains that were positive when administered to monkeys, but were negative for any of the identified enterotoxins by MOA-OSP, were examined for enterotoxin production by ELISA (Kokan and Bergdoll, 1987). Twenty-six strains were found to be positive with most of them producing enterotoxin D (SED) (Table 7). Some of the strains were isolated from food poisoning outbreaks. The amount of toxin produced appeared to be 10-15 ng/ml of culture supernatant fluid.

The importance of this low production was investigated to determine what amount of enterotoxin production is significant in food poisoning. The production of 10 to 20 ng of enterotoxin/ml may be of significance because only 100 to 200 ng of SEA was shown to be necessary to produce food poisoning (Evenson, et al., 1987; Table 8) with the amount present

Table 7. *S. aureus* negative for enterotoxin production by OSP[a]

Source	Monkey positive	ELISA positive
Food poisoning	38	10
Foods	25	4
Fish (raw)	11	6
Nares (human)	26	4
Nares (horse)	4	1
Miscellaneous	6	1

[a]Kokan and Bergdoll, 1987

Table 8. Enterotoxin A analysis in chocolate milk[a]

	Enterotoxin A	
Sample	ng/ml	ng/half pint
1	0.40	94
2	0.73	172
3	0.48	113
4	0.63	149
5	0.78	184
6	0.65	153
Ave.	0.61	144

[a]Evenson et al., 1988

in the vehicle, chocolate milk, being 0.50 to 0.75 ng/ml. Although the amount of enterotoxin produced by the membrane-over-agar method is 5-10 times that produced in shake flasks or even possibly in foods, if growth is sufficient, 10^7, 1 to 2 ng of enterotoxin/g of food will be produced (Pereira et al., 1991; Table 9). This would be adequate to result in staphylococcal food poisoning in sensitive individuals. It is important, therefore, that standard procedures be developed for the examination of staphylococcal strains for enterotoxin production; currently a research project is underway to develop a standard method for examination of staphylococcal strains for production of low amounts of enterotoxin.

It can be noted that the production of toxins is not one of the properties on which the classifications are based, as production of the toxins cuts across species, even to include an occasional CNS. Production of the toxins by any of the CNS is considered rare, if indeed actual; however, there is one report that coagulase-negative staphylococci from several different species do produce one or more of the identified toxins (Valle et al.,1990). This is a debatable question because one of the sensitive detection methods, ELISA, was used to assay for toxin production. The amount of toxin produced by these strains was less than 10 ng/ml of culture supernatant fluid. The fact that the ELISA methods are sensitive to less than 1 ng/ml of culture supernatant fluid, it is questionable whether a strain producing less than 10 ng/ml can be considered toxin-positive. It is an open question as to how much toxin a strain should produce to be labeled a toxin producer, as this has not been defined. Some of the strains are under study to determine if the toxin production reported can be verified and whether these strains can produce enterotoxin in foods.

Table 9. Production of enterotoxin D by *S. aureus* strain 542[a]

		SED produced in 24 h (ng)	
Method	cfu/g	ELISA	OSP
Shake flask		17	nd
Sac culture		106	nd
Cream			
25°	2.4×10^7	2.1	nd
30°	6.2×10^7	3.1	nd
Ham			
25°	2.1×10^7	nd	nd
30°	1.2×10^9	8.2	nd

[a]Pereira et al., 1991

DETECTION OF ENTEROTOXIN IN FOODS

Detection of enterotoxin in foods requires sensitive methods, such as the ELISA and RPLA methods. The quantity of enterotoxin present in foods involved in food poisoning outbreaks may vary considerably, from less than 1 ng/g to greater than 50 ng/g. Usually, little difficulty is encountered in detecting enterotoxin in foods involved in food poisoning outbreaks, however, outbreaks do occur in which the amount of enterotoxin is less than 1 ng/g, such as the case with the chocolate milk (Table 8). In such cases, the enterotoxin can be detected only by the most sensitive methods. Another situation in which it is essential to use a very sensitive method is in determining the safety of a food for consumption. In this situation it is necessary to show that no enterotoxin is present by the most sensitive methods available.

The ELISA Methods

The ELISA methods were applied to the detection of the enterotoxins in foods soon after they were originally developed for the detection of other proteins. Essentially all ELISA methods are of the sandwich type in which the antibody is treated with the unknown sample before the antibody-enterotoxin complex is treated with the enzyme-antibody conjugate. This type is preferred because the amount of enzyme and, thus, the color developed from the enzyme-substrate reaction, is directly proportional to the amount of enterotoxin present in the unknown sample. This eliminates the need for the highly purified enterotoxins as crude or only partially purified enterotoxin is needed for preparation of a standard curve.

The majority of users of ELISA methods use microtiter plates to which the antibodies are attached. The large number of wells in a microtiter plate provide for doing several samples at one time, although there may not be uniformity in all of the wells, particularly those around the edge of the plate. A plate reader is necessary to record the results which adds to the cost. One kit is available commercially, RADISCREEN (R-biopharm GmbH, Darmstadt, Germany), that utilizes microtiter plates (Park et al.,1994).

An alternate procedure, the use of polystyrene balls to which the antibodies are attached, has been developed (Stiffler-Rosenberg & Fey, 1978; Freed et al., 1982). A kit utilizing the ball method is available commercially (Fey, et al., 1983). The ball method is more cumbersome because each ball must be handled separately. The main advantage is that a relatively large volume of the unknown sample can be used, thus, increasing the amount of enterotoxin adsorbed per sample. This makes possible the use of 1 ml volumes of substrate so that the color developed can be read in a simple colorimeter, an instrument that most laboratories would have available. The sensitivity of the ELISA methods is between 0.5 to 1.0 ng/g of food (Table 10). Those who have used it have found it to be a very good method

Table 10. Detection of enterotoxins in foods by ELISA[a]

Food	Enterotoxin	Amount added ng/g	Amount detected ng/g
Cheese	D	0.63	0.15
	E	0.63	0.38
Cheese food	A	0.63	0.36
Genoa sausage	A	0.63	0.36
Potato salad	B	0.63	0.18
Ham	A	0.63	0.34
Milk	A	0.63	0.63

[a]Freed et al., 1982

Table 11. Detection of enterotoxins (SE) in foods from outbreaks[a]

Food	*S. Aureus* count	SE by strain	SE detected in food		
			Gel-diff.	ELISA kit	RPLA kit
Ham	1.5 x 10^9	A		A	A
Smoked bacon	1.0 x 10^9	A,D	A,D	A,D	A,D
Pork	6.0 x 10^9	B	B	B	B
Lasgne, dried	2.0 x 10^8	A	nd[b]	A	A
Corned beef	4.0 x 10^7	A,B	A	A,B	A,B
Salmon, canned	3.5 x 10^6	A	nd	A	A
Beef rolls	4.0 x 10^6	A,D		A	nd[d]
Chicken	1.0 x 10^6	C		C	nd
Sheep cheese	nd			A	nsa[c]
Meat pies	1.0 x 10^6	A		nd	nd
Corned beef	8.5 x 10^4	A		nd	nd

[a]Wieneke and Gilbert, 1987
[b]nd, not detectable
[c]nsa, non-specific agglutination
[d]Extract not tested for D

for detecting enterotoxin in foods (Wieneke & Gilbert, 1987; Table 11). Although the test can be completed in one day, recommendations are that the antibody-coated balls be shaken with 20 ml of food extract overnight to obtain the highest sensitivity.

A third type of ELISA method for enterotoxin detection in foods makes use of dip sticks that have wells with nitrocellulose paper in the bottoms containing the specific antibodies to the enterotoxins, each well containing an antibody to a different enterotoxin (Fig. 3). Only one test is necessary per sample because all of the enterotoxins can be tested for with the same dip stick (Transia, 8 Rue Saint Jean de Dieu, 69007 Lyon, France). The monoclonal antibodies used were developed in the Bergdoll laboratory (Thompson, et al., 1984; Thompson et al.,1986). The sensitivity of the method was less than that of the ball method and showed non-specific reaction with the SEC detection (Wieneke, 1991;Table 12). The nonspecific reaction may have been due to the use of polyclonal antibody for the second SEC antibody.

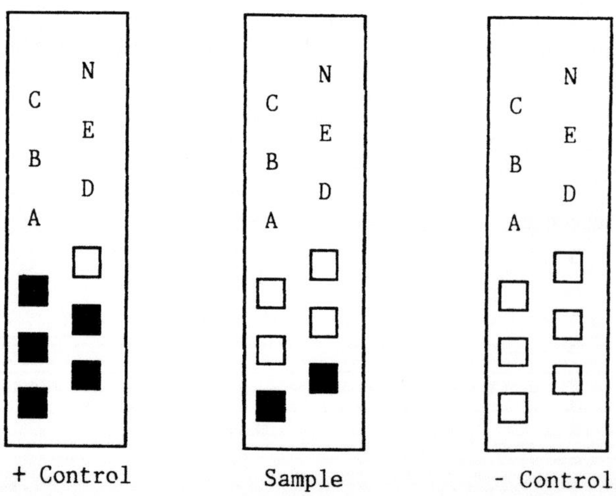

Figure 3. The dipstick ELISA. A, B, C, D, E = SEA, SEB, SEC, SED, SEE, respectively. N = normal rabbit serum. Positive control = sample containing all of the enterotoxins; unknown sample contains A and D; Negative control = buffer containing no enterotoxins.

Table 12. Detection of enterotoxin in foods from outbreaks[a]

		ELISA		
Food	RPLA	BALL	DIP-ST	TUBE
Ham	A	A	A,C[b]	+
Lasagne	A	A	—[c]	+
Chicken	D	D	D	+
Pork	A	A	A	+
Rice	A	A	A	+
Chicken	—	A	A	+
Sandwich	—	A	—	—

[a]Wieneke, 1991
[b]Non-specific reaction
[c]Not detectable

Two enzymes are commonly used in the ELISA methods, horseradish peroxidase or alkaline-phosphatase, with little difference in the sensitivity of the two. The yellow color produced by the alkaline-phosphatase may be more difficult to see visually at low levels than the blue or green color produced by the horseradish peroxidase. The major problem with the horseradish peroxidase is that an oxidizing agent is necessary for color development, whereas a substrate tablet is all that is needed for the alkaline-phosphatase color development. Alkaline-phosphatase is used in the ball kit, whereas horseradish peroxidase is utilized in the other ELISA kits. Alkaline-phosphatase is preferable because there is no problem with the substrate, whereas with the horseradish peroxidase the oxidizing agent is subject to deterioration.

The sensitivity of the ELISA methods are equivalent to <0.5 to 1.0 ng of enterotoxin per gram of food, with the dip stick kit somewhat less sensitive than the other two kits. If necessary greater sensitivity can be obtained by some type of concentration of the extract, although this is not generally required for examination of foods implicated in food poisoning outbreaks.

The RPLA Method

In the RPLA method, the antibody-coated latex particles agglutinate when brought in contact with the enterotoxins. The method utilizes microtiter plates, with the latex particles settling to the bottom of the wells in a dot if no enterotoxin is present in the solution added. One problem is that an occasional food extract will result in non-specific agglutination, which masks any positive reaction from the presence of enterotoxin (Weineke & Gilbert,

Table 13. Enterotoxin detection in foods from outbreaks by RPLA[a]

Food	Enterotoxin	Amount (ng)
Rice ball	A	40
Rice ball	A	45
Omelet	A	4
Rice ball	A	8
	D	16
Omelet	A	64
Rice ball	A	10

[a]Igarashi et al., 1985

1987). An RPLA kit produced in Japan is available commercially through Oxoid (Oxoid Limited, Wade Road, Basingstoke, Hants RG24 0OW, England). The method is adequately sensitive for the detection of enterotoxin in most foods that are implicated in food poisoning outbreaks (Igarashi, et al., 1985; Table 13); however, it may be inadequate for detection of the small amounts of enterotoxin that sometimes is present (Table 8). This method is the sensitive method of choice for the examination of staphylococcal strains for the production of enterotoxin.

Screening Methods

In some instances it may not be necessary to determine the type of enterotoxin, only if enterotoxin is present; for example, in examining the marketability of suspect foods. The food would not be marketable if any enterotoxin were present. For this purpose, including all of the enterotoxins in one test saves time as only one is needed. However, this would not save time in examining foods implicated in food poisoning out-breaks, because identification of the type of enterotoxin is valuable in tracing the source of the contamination. Two ELISA kits that include all of the enterotoxins in one test are now available commercially. One kit utilizes small tubes coated with monoclonal antibodies to all of the enterotoxins (Transia, 8 Rue Saint Jean de Dieu, 69007 Lyon, France). The method is sensitive to 0.2 ng/ml of extract; a bright blue color is developed with this concentration of enterotoxins. Although this is a screening test, if the enterotoxins were present at these low levels it would be difficult to analyze for individual enterotoxins because the other methods are less sensitive. It would be especially difficult if more than one enterotoxin were present.

The other ELISA kit utilizes microtiter plates and has a sensitivity of at least 2 to 24 ng/ml (Park et al., 1992) (Bioenterprises Pty. Ltd., Roseville, New South Wales, Australia). Although the sensitivity is adequate for most foods involved in food poisoning outbreaks, the SEA in the chocolate milk would not have been detectable. This is critical, particularly in the case of testing suspect foods whose marketability are contingent on the absence of enterotoxin. The tube kit mentioned above should provide adequate sensitivity for the screening of suspect foods, but one report indicates otherwise (Wieneke, 1991; Table 12).

ENTEROTOXIN EXTRACTION FROM FOODS

The extraction method is relatively simple as usually no concentration is necessary for the sensitive detection methods (Freed et al.,1982; Table 14). Usually only about 50% of the enterotoxin is recoverable from the foods (Fig. 4). It is necessary to centrifuge the food extracts at relatively high speed which may be a problem for some laboratories, particularly those in the developing countries. This could be a particular problem if the RPLA method is used because the extract must be completely clear to avoid interference with the agglutination. In this case, it is necessary for some extracts to be filtered through a non-protein adsorbing filter. Preparation of the food extracts for use in the ball method

Table 14. Food extraction for detection of enterotoxin by ELISA[a]

1. Grind to homogenous slurry with 1-1.5 ml of fluid/g food.
2. Adjust pH to 4.5; centrifuge.
3. Adjust supernatant fluid to pH 7.5; centrifuge if necessary.
4. Extract with $CHCl_3$ if fat interferes; centrifuge and filter.

[a]Freed et al., 1982

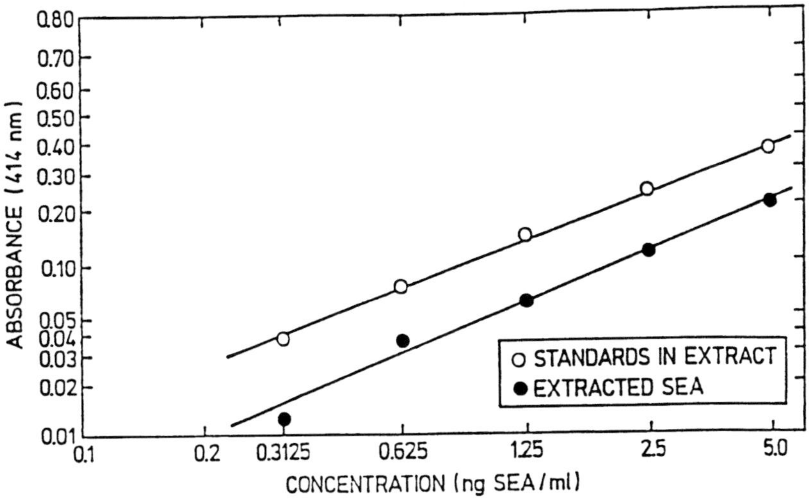

Figure 4. Recovery of enterotoxin from foods by simple extraction. Standards in extract: SEA added to extract from ham free of enterotoxin. Extracted SEA: SEA added to ham at time of extraction.

requires less treatment as cloudiness will not interfere with the take-up of the enterotoxin by the antibodies adsorbed to the surface of the balls. For example, milk can be used without any treatment.

CONCLUSIONS

The current procedures are adequate for detection of enterotoxin in foods and it is possible to do them in one day. There is always the request that the time for doing an analysis be shortened, but in reality, there is no great need to obtain results in less than the time now required. We have come a long way in improving the methods for enterotoxin detection in foods, but it is likely that further improvements will be made in the future.

REFERENCES

Anon. 1992. Outbreak that did not seem right spread like butter. *Food Prot. Report*, July-August.

Bergdoll, M. S. 1979. "Staphylococcal intoxications" in Food-borne infections and intoxications, 2nd edition, H. Reimann and F. L. Bryan (Eds.), Academic Press, New York, NY, pp. 443-494.

Bergdoll, M. S. 1994. Unpublished data.

Casman. E. P., and Bennett, R. W. 1965. Detection of staphylococcal enterotoxin in food. *Appl. Microbiol.* 13:181-189.

Devriese, L. A., Hajek, V., Oeding, P., Meyer, S. A., and Schleifer, K. H. 1978. *S. hyicus* (sompolinsky 1953) comb. nov. and *S. hyicus* subsp. *chromogenes* subsp. Nov. *Intl. J. Syst. Bacteriol.* 28: 482-490.

Evenson, M. L., Hinds, W. M., Bernstein, R. S., and Bergdoll,M. S. 1987. Estimation of human dose of staphylococcal enterotoxin A from a large outbreak of food poisoning involving chocolate milk. *Int. J. Food Microbiol.* 7:311-316.

Fey, H., and Pfister, H. 1983. A diagnostic kit for the detection of staphylococcal enterotoxins (SET) A, B, C and D (SEA, SEB, SEC, SED) in Immunoenzymatic techniques, S. Avrameas et al.(Eds.), Elsevier/North-Holland Publishing Co., Amsterdam, pp. 345-348.

Freed, R. C., Evenson, M. L., Reiser, R. F. and Bergdoll, M. S. 1982. Enzyme-linked immunosorbent assay for detection of staphylococcal enterotoxins in foods. *Appl. Environ. Microbiol.* 44:1349-1355.

Hajek, V. 1976. *Staphylococcus intermedius*, a new species isolated from animals. *Int. J. Syst. Bacteriol.* 26:401-408.

Hajek, V., and Marsalek, E. 1973. The occurrence of enterotoxigenic *Staphylococcus aureus* strains in hosts of different animal species. *Zbl. Bakt. I. Abt. A* 223: 63-68.

Hirooka, E. Y., Müller, E. E., Freitas, J. C., Vicente, E., Yoshimoto, Y., and Bergdoll, M. S. 1988. Enterotoxigenicity of *Staphylococcus intermedius* of canine origin. *Int. J. Food Microbiol.* 7:185-191.

Ho, G., Campbell, W. H., and Carlson, E. 1989. Ovine-associated *Staphylococcus aureus* protein with immunochemical similarity to toxic shock syndrome toxin 1. *J. Clin. Microbiol.* 27:210-212.

Igarashi, H., Shingaki, M., Fujikawa, H., Ushioda, H., and Terama, T. 1985. Detection of staphylococcal enterotoxins in food poisoning outbreaks by reversed passive latex agglutination in The staphylococci, J. Jeljaszwicz (Ed.) *Zbl. Bakt. Suppl.* 14, Gustav Fischer Verlag, Stuttgart, New York, pp. 255-257.

Igarashi, H., Fujikawa, H., Shingaki, M., and Bergdoll, M. S. 1986. Latex agglutination test for staphylococcal toxic shock syndrome toxin 1. *J. Clin. Microbiol.* 23:509-512.

Kato, E., Khan, M., Kujovich, L., and Bergdoll, M. S. 1966. Production of enterotoxin A. *Appl. Microbiol.* 14:966-972.

Kokan, N. P., and Bergdoll, M. S. 1987. Detection of low-enterotoxin-producing *Staphylococcus aureus* strains. *Appl. Environ. Microbiol.* 53:2675-2676.

Lee, A. C.-M., Robbins, R. N., and Bergdoll, M. S. 1978. Isolation of specific and common antibodies to staphylococcal enterotoxins A and E by affinity chromatography. *Infect. Immun.* 21:387-391.

Lee, A. C.-M., Robbins, R. N., and Bergdoll, M. S. 1980. Isolation of specific and common antibodies to staphylococcal enterotoxins B, C_1 and C_2. *Infect. Immun.* 27:432-434.

Ogston, A. 1880. Uber Abscesse. *Arch. Klin. Chir.* 25:588-600.

Park, C. E., Ahkatar, M., and Rayman, M. K. 1992. Nonspecific reactions of a commercial enzyme-linked immunosorbent assay kit (TECRA) for detection of staphylococcal enterotoxins in foods. *Appl. Environ. Microbiol.* 58:2509-2512.

Park, C. E., Akhtar, M., and Rayman, M. K. 1994. Evaluation of a commercial enzyme immunoassay kit (RIDASCREEN) for detection of staphylococcal enterotoxins A, B, C, D, and E in foods. *Appl. Environ. Microbiol.* 60:677-681.

Pereira, J. L., Salzberg, S. P., and Bergdoll, M. S. 1991. Production of staphylococcal enterotoxin D in foods by low-producing staphylococci. *Int. J. Food Microbiol.* 14:19-26.

Reiser, R. F., Robbins, R. N., Noleto, A. L., Khoe, G. P., and Bergdoll, M. S. 1984. Identification, purification, and some physicochemical properties of staphylococcal enterotoxin C_3. *Infect. Immun.* 45:625-630.

Robbins, R. N., and Bergdoll, M. S. 1984. Production of rabbit antisera to the staphylococcal enterotoxins. *J. Food Protect.* 47:172-176.

Robbins, R., Gould, S., and Bergdoll, M. 1974. Detecting the enterotoxigenicity of *Staphylococcus aureus* strains. *Appl. Microbiol.* 28:946-950.

Rosenbach, F. J. 1884. Mikroorganismen bei den undinfectionskrankheiten des Menschen. J. F. Bergmann, Wiesbaden, Germany.

Stiffler-Rosenberg, G., and Fey, H. 1978. Simple assay for staphylococcal enterotoxins A, B, and C: modification of enzyme-linked immunosorbent assay. *J. Clin. Microbiol.* 8:473-479.

Su, Y-c., and Wong, A. C. L. 1994. Unpublished data.

Talan, D. A., Staatz, D., Staatz, A., and Overturf, G. D. 1989. Frequency of *Staphylococcus intermedius* as human nasopharyngeal flora. *J. Clin. Microbiol.* 27:2393.

Talan, D. A., Staatz, D., Staatz, A., Goldstein, E. J. C., Singer, K. and Overturf, G. D. 1989. *Staphylococcus intermedius* in canine gingiva and canine-inflicted would infections: a newly recognized zoonotic pathogen. *J. Clin. Microbiol.* 27:78-81.

Thompson, N. E., Ketterhagen, M. J., and Bergdoll, M. S. 1984. Momoclonal antibodies to staphylococcal enterotoxins B and C: Cross-reactivity and localization of epitopes on tryptic fragments. *Infect. Immun.* 45:281-285.

Thompson, N. E., Razdan, M., Kunstmann, G., Aschenbach, J. M., Evenson, M. L., and Bergdoll, M. S. 1986. Detection of staphylococcal enterotoxins by enzyme-linked immunosorbent assays and radioimmunoassays: comparison of monoclonal and polyclonal antibody systems. *Appl. Environ. Microbiol.* 51:885-890.

Valle, J., Gomez-Lucia, E., Priz, S., Goyache, J., Orden, J. A., and Vadillo, S. 1990. Enterotoxin production by staphylococci isolated from healthy goats. *Appl. Environ. Microbiol.* 56:1323-1326.

Varaldo, P. E., Kilper-Balz, R., Biavasco, F., Satta, G., and H. Schleifer, K. H. 1988. *Staphylococcus delphini* sp. nov., a coagulase-positive species isolated from dolphins. *Int. J. System. Bacteriol.* 38:436-439.

Wieneke, A. A., and Gilbert, R. J. 1987. Comparison of four methods for the detection of staphylococcal enterotoxin in foods from outbreaks of food poisoning. *Int. J. Food Microbiol.* 4:135-143.

Wieneke, A. A. 1991. Comparison of four kits for the detection of staphylococcal enterotoxin in foods from outbreaks of food poisoning. *Int. J. Food Microbiol.* 14:305-312.

Winslow, C-E. A., and Winslow, A. R. 1908. The systematic relationships of the *Coccaceae*. John Wiley, New York.

Wong, A. C. L., and Bergdoll, M. S. 1990. Effect of environmental conditions on production of toxic shock syndrome toxin 1 by *Staphylococcus aureus*. *Infect. Immun.* 58:1026-1029.

39

DETECTION AND IDENTIFICATION OF *Clostridium botulinum* NEUROTOXINS

Charles L. Hatheway[1] and Joseph L. Ferreira[2]

[1] Centers for Disease Control and Prevention
National Center for Infectious Diseases
Division of Bacterial and Mycotic Diseases
1600 Clifton Road, N.E., Atlanta, Georgia 30333
[2] U.S. Food and Drug Administration
Southeastern Regional Laboratories
60 - 8th Street, N.E., Atlanta, Georgia 30309

INTRODUCTION

Botulism is a serious neuroparalytic illness that affects humans and various domestic and wild animal and avian species. It is due to the neurotoxic effect of a toxin produced by the anaerobic bacterium *Clostridium botulinum*. Botulism is most commonly known as a foodborne intoxication of humans; it also can result from growth of the toxigenic organism in a wound or, in the case of infant botulism, from colonization of the intestinal tract.

Detection and identification of the botulism neurotoxin have been essential for diagnosis of the illness and for identifying the causative food. When van Ermengem showed the lethality for animals of the ham that caused the Ellezelles botulism outbreak in 1895 (73), the bioassay naturally became the standard test for botulism neurotoxin. The mouse is very sensitive, and the mouse LD_{50} determined by i.p. injection became the quantitative unit.

Kempner surmised from previous reports on tetanus and diphtheria toxins that botulism toxin is immunogenic and in 1897 prepared the first botulism antitoxin that neutralized the biologic activity of the toxin (41). Antibodies against the toxin have provided means for specific identification of the toxin ever since. Serologic differentiation of toxins from strains isolated from two different outbreaks of botulism was noted by Leuchs (43) in 1910. At present, seven serologic types, designated by letters A through G, have been recognized. International Standards for *Clostridium botulinum* antitoxin were prepared against types A, B, C, D and E and have been distributed through the World Health Organization (WHO) (5). The standard for type F was added subsequently. Although reference antitoxins are available for identifying type G neurotoxin (e.g., from CDC), no international antitoxin standard exists. Positive identification of the neurotoxins is made by specific neutralization of the biologic activity by the International Standard,

Natural Toxins II, Edited by B. R. Singh and A. T. Tu
Plenum Press, New York, 1996

Table 1. Clostridia that produce botulism neurotoxin

Species	Characteristic	Toxin type
Clostridium botulinum I	Proteolytic	A B F
Clostridium botulinum II	Nonproteolytic	E B F
Clostridium botulinum III	*C. novyi*-like	C D
Clostridium argentinense	Asaccharolytic	G
Clostridium baratii	Lecithinase	F
Clostridium butyricum		E

or by a reference antitoxin (other than in the case of type G) that has been standardized against the International Standard. The International Unit (IU) for each toxin type is defined by the neutralizing potency of each antitoxin standard. Originally, the antitoxins were calibrated so that 1 IU neutralized 10,000 mouse LD_{50} of test toxin for toxin types A, B, C and D, and 1,000 LD_{50} for type E (5). This quantitative value cannot be relied upon since it varies with toxins from different strains, and even with different preparations of toxin from the same strain (34). The importance of Antitoxin Standards for identifying the toxin type is borne out by the confusion between toxin types C and D between European investigators on the one hand, and South African and North American investigators on the other (56). The problems could not be solved by relating to toxigenic reference cultures; the critical cultures had lost their toxigenicity. Reference diagnostic antitoxins calibrated in comparison to the WHO Standards are currently available from the Centers for Disease Control and Prevention.

Six phenotypically and genetically diverse groups of clostridia that can produce botulism neurotoxins have been recognized (Table 1). Three of them retain the name *Clostridium botulinum*, although their phenotypic and genetic diversities warrant individual species designations. A separate name, *C. argentinense* has been proposed for organisms that produce type G toxin (67). Rare neurotoxigenic strains of *C. butyricum* (1) and *C. baratii* (30) have been implicated as causes of type E and type F botulism, respectively in humans.

TESTS FOR BOTULISM NEUROTOXINS

Tests for detection and quantitation of botulism neurotoxins are used for the following purposes:

- Investigation of botulism
- Identification of *Clostridium botulinum* and other botulinogenic organisms
- Studies on food safety
- Basic studies on neurotoxins
- Purification of neurotoxins
- Preparation and standardization of therapeutic neurotoxin
- Preparation of immunizing toxoid
- Measurement and standardizing of antitoxin

The various methods used to detect botulism toxin are listed in Table 2. Some of the assays for toxin can be modified for detection and measurement of antitoxin. Although detection of toxin is often the primary purpose of the test, sometimes it is secondary to identification of toxigenic organisms. In the investigation of botulism, detection of substantial amounts of toxin in the suspected food is confirmatory of the diagnosis. This is also true for detection of any demonstrable amount in the patient's blood or feces.

Table 2. Methods for detection of botulism neurotoxin and determining neurotoxigenicity of bacterial strains

TOXIN DETECTION	
Bioassays in mice	
	i.p. challenge, lethality endpoint.
	i.v. challenge, time to death.
	i.m. challenge, local muscle paralysis.
Immunoassays	
	Flocculation
	Passive latex agglutination
	Immunodiffusion
	Radioimmunoassay
	Enzyme-linked immunosorbent assay
	Enzyme-linked clotting assay
	Biosensor
TOXIN GENE DETECTION	
Gene probe	
Polymerase chain reaction	

Isolation of the toxigenic organism from the patient's feces in the absence of toxin is also strong confirmatory evidence. Determining toxigenicity is necessary for identifying the organism because there are no phenotypic characteristics unique to toxigenic organisms, and there are phenotypically identical nontoxigenic organisms corresponding to each toxigenic organism. For identification of organisms, a qualitative assay that detects the toxin and identifies it by toxin type is sufficient. Recently, tests for identifying the structural gene in the organisms by gene probes or by the polymerase chain reaction have been proposed for this purpose.

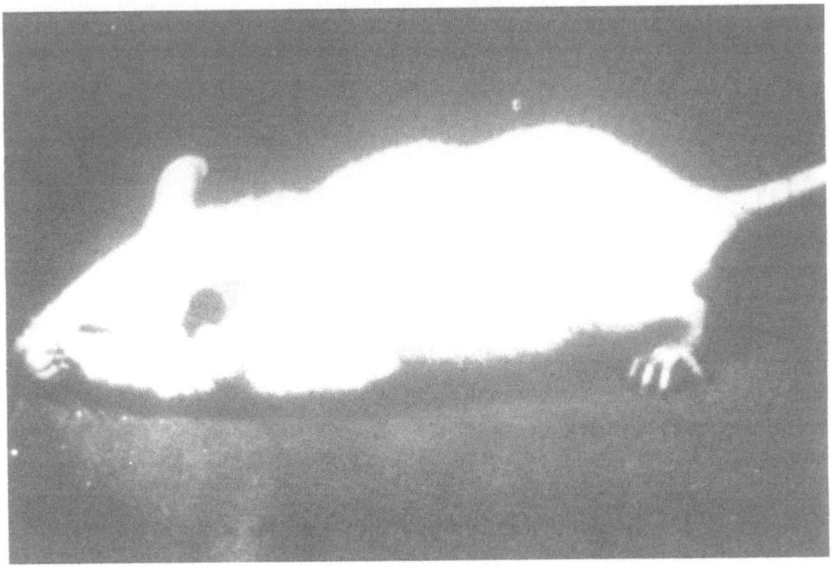

Figure 1. Mouse with botulism.

BIOASSAYS

The small size of the mouse and its high sensitivity to all types of botulism neurotoxin have made it ideal for assaying biologic activity. The usual test is performed by intraperitoneal injection (31,39,63). The ability to observe the classic symptoms of the illness when working with unknown specimens enhances its specificity (Figure 1).

Signs of botulism in a mouse include generalized weakness progressing to severe paralysis; atrophy of the abdominal area (after i.p. injection), giving a "wasp-like" appearance to the body; ruffled fur; and, especially, shallowness of breathing which will progress to labored breathing and final respiratory failure. Toxin type is clearly determined by neutralization of biologic activity (protection of mice) with type-specific reference antitoxins. This is usually done by adding 1 unit of antitoxin to a sample to be tested and, after this "in vitro neutralization," injecting the mixture into mice (31). Another protocol requires pretreatment of separate groups of mice by i.p. injection with 1 unit of the several antitoxins under consideration (e.g., A, B, E, and polyvalent), and 1 h later injecting 2 mice from each group with a sample of the test material (39). One modification of a bioassay for detection and identification of type C toxin used groups of mice actively immunized with type C toxoid (45).

A stable reference toxin derived from a preparation of crystalline type A neurotoxin was proposed for standardizing botulism toxin bioassays performed in different laboratories under diverse conditions (63). Standardized procedures for using the bioassay for detecting botulism neurotoxin in foods and identifying toxigenicity and toxin type of the organisms were established on the basis of the results of a collaborative study (40).

Toxin is quantitated by lethality endpoint titration, and the toxin unit most commonly used is the median lethal dose (LD_{50}). One LD_{50} of the crystalline type A toxin standard for a 20 g mouse was determined as 0.043 ng (63). The crystals consist of complexes of neurotoxin and nontoxic components. The LD_{50} for chromatographically purified type A botulism neurotoxin has been calculated as approximately 6 picograms (32,60). Sometimes lethality is referred to in terms of minimum lethal dose (MLD), which is the amount that causes all injected animals to die, rather than 50%. The LD_{50}, not being an all or none endpoint can be interpolated and is therefore the more reliable unit. However, for accurate determination of the LD_{50}, close incremental dilutions and 6 to 10 mice per dilution are needed. Antitoxins are quantitated by their neutralization of a test dose of toxin (50% endpoint) in relation to the performance of an antitoxin standard in a simultaneous parallel titration (33).

When large numbers of samples have to be evaluated, such as fractions in a column chromatography purification study, the lethality endpoint calculations require an excessive number of mice because a series of dilutions of each sample have to be injected into a number of mice. This problem has been alleviated to some extent by using intravenous injection (tail vein) of a single dilution of a preparation into several mice and determining the time to death (3,61). The LD_{50}/ml of the sample is calculated from the linear relationship between LD_{50} and time to death determined for a standard toxin preparation.

Intramuscular injection of 0.1 ml of toxin into the gastrocnemius muscle of mice with the observation of local paralysis as the endpoint has been studied as an alternative method by Sugiyama et al. (68). The 50% paralytic endpoint dose (ED_{50}) was between 0.06 and 0.09 i.p. LD_{50} for all toxin types A-F. Pearce et al. (54) used both the lethality and muscle paralysis assays to compare the potency of two commercial preparations of type A botulism neurotoxin marketed for clinical use in treatment of neuromuscular disorders. They found disparities between the relationship between the activities for each preparation; 0.2 lethal units of the

American product caused a paralytic effect of 1.2, while 0.6 lethal units of the British product caused a paralytic effect of 1.07.

FLOCCULATION, PRECIPITATION, AGGLUTINATION AND PASSIVE HEMAGGLUTINATION REACTIONS

Because of the complexity of bioassays, the expenses for animals and facilities, and ethical considerations, many investigators have sought in vitro methods for detection and quantitation of botulism toxins and antitoxins. One of the earliest was the Ramon flocculation test (57), in which the reaction between a toxin and its homologous antitoxin standard was visualized by a precipitation reaction. This was advantageous in vaccine production because the toxoid, which cannot be measured by bioassay can be quantitated in a comparable manner as the toxin by this test. The flocculation unit (Lf) was used as the antigenic unit employed in the formulation of botulism toxoids used for human immunization (7).

Passive Hemagglutination (PHA) and Reversed Passive Hemagglutination (RPHA)

In 1966, Johnson et al. (38) devised test systems using formalinized sheep red blood cells (SBRCs) to which botulism toxins (PHA) or antitoxins (RPHA) were attached by bis-diazotized benzidine linkage. Toxin-coated SRBCs would agglutinate in the presence of type-specific antitoxin. The agglutination could be inhibited by adding test samples containing toxin of the proper type to the antitoxin before adding the toxin-coated SRBCs. The reactions were inhibited by 43, 39, and 245 LD_{50} of type A, B, and E toxins, respectively. RPHA tests were capable of detecting 0.75-1.3 LD_{50} of type A, 2.3 of type B, and 150 of type E toxins. Tests using toxins or antitoxins adsorbed to bentonite particles were much less sensitive than the HA tests. Cross-reactions between type A and B toxin preparations occurred in the PHA and RPHA tests because antibodies in the reagents reacted to non-neurotoxin antigenic components common to both types. These could be eliminated, according to Johnson et al. (38), by adding toxoid of the heterologous type to each reaction mixture.

Evancho et al. (12) later attempted to standardize the RPHA using purified type A toxin and toxoid. They found no differences in RPHA results using SRBCs coated with antitoxins produced in response to crystalline toxin or toxin partially purified by ethanol fractionation. The tests were able to detect type A toxin at a minimum concentration of 27 LD_{50}/ml, and no cross reactions were observed with culture filtrates containing toxins of types B, C, D, E or F.

Sakaguchi et al. (62) noted cross-reactions in RPHA employing antibodies produced against crystalline type A and partially purified type B toxins. The cross-reactions involved nontoxic components of the antigen preparations, which they demonstrated by immunodiffusion studies. They improved the RPHA test by using affinity-purified antibodies, which were specific for the type A and B neurotoxins. Tests employing SRBCs coated with these antibody preparations detected type A toxin at a concentration of 8-89 LD_{50}/ml and type B at 10-59 LD_{50}/ml. No cross-reactions between the toxin types occurred. Differences in the sensitivities of the tests based on LD_{50} reported by the various investigators may reflect differences in specific activities of the antigen preparations rather than real sensitivities with regard to actual antigen protein mass.

Reversed Passive Latex Agglutination (RPLA)

Another development of this type of test was the reversed passive latex agglutination as described by Horiguchi et al. for types A, B and E toxins (35). They reported a minimum detectable concentration of 3.9 ng/ml for type A toxin, comparable to about 390 LD_{50} of the highly purified preparation.

Agar Gel Diffusion Methods

Agar diffusion methods included capillary immunodiffusion (46), Ouchterlony (74), and toxic colony overlay detection. A simple gel immunodiffusion agar procedure was used by Ferreira et al. to detect toxin-producing colonies of *C. botulinum* type A (20) and was later expanded to include type B, E, and F toxins surrounding colonial growth (21). Colonies growing on thin layers of agar were overlayered with gel-diffusion agar containing specific antitoxins. Precipitin rings formed around the colonies that produced toxin.

Detection of Toxin on Fixed Preparations with Labeled Antibodies

Another serologic test for rapid identification of toxigenic *C. botulinum* type A colonies using an enzyme detection system was developed by Smith and Hamdy (66). Fixed smears of cells on glass slides from isolated agar colonies were incubated with type A antitoxin conjugated to horseradish peroxidase; this was followed by the addition of a diaminobenzidine substrate. Microscopic observation at 1,000X showed that the toxin-producing cells had a reddish-brown color. Those lacking type A toxin (*C. sporogenes* or *C. botulinum* type B) were a faint brown color, due possibly to cross-reactivity with nontoxin antigens in these organisms. Goodnough et al., (28) developed an immunoblot technique for detecting toxins from toxic colonies on agar plates. Type A, B, or E botulism neurotoxins in and around the colonies were transferred to nitrocellulose membrane filters. The immobilized toxins were detected using specific rabbit antitoxins, followed by anti-rabbit antibody conjugated to alkaline phosphatase. An insoluble colored precipitate was deposited at the site of the antigen-antibody-enzyme complex.

Although most fluorescent antibody applications for *C. botulinum* were directed toward identifying organisms by means of their cellular antigens (26,27), Inukai and Riemann (37) used this technique for detecting intracellular type E botulinal toxin. Type E strains were fixed onto glass slides, and toxin was detected with absorbed horse antibody followed by rabbit anti-horse fluorescein-labeled antibody and fluorescent microscopy. They found that non-toxin-producing type E strains were negative but toxigenic type E strains were positive. Toxigenic cultures were clearly stained after 6 h incubation, at which time they contained less than 100 MLD/ml.

RADIOIMMUNOASSAY

Because of the high sensitivity of radioimmunoassay (RIA) for detection of other antigens, a competitive binding technique was adapted for type A botulism neurotoxin by Boroff and Shu-Chen (4). Toxin being measured competed with ^{131}I-labeled toxin for binding with a specific antitoxin. Although the sensitivity was reported in the abstract as 100 mouse MLD, the data showed a sensitive range between 400 and 3000 MLD, and obvious reduction of label in the immune precipitates was seen only with samples containing 1000-2000 MLD of test toxin. Since their reaction was performed in a total volume of 0.5 ml, the concentration would have to be more than 2 times the minimal detectable amount. Habermann and Bernáth used RIA for measuring botulism antibodies in serum of persons immunized with botulism toxoid (29). They found a good correspondence between radioimmunologic measurements and toxin-neutralizing antibodies. Because of the development of enzyme immunosorbent assay (ELISA) techniques, which were more sensitive and did not require the use of radioisotopes, the RIA approach lost favor.

MICROTITER PLATE ENZYME-LINKED IMMUNOASSAYS (ELISA)

Many of the early ELISA methods devised for botulism neurotoxin detection, like most of the in vitro tests,suffered from a lack of specificity, due to impurities in the antigen preparation used to produce the antitoxins. More purified toxins are now available for the production of better quality antitoxins. The most sensitive ELISA protocols use an indirect assay sometimes referred to as the sandwich assay. In the basic procedure, a specific antitoxin is first adsorbed to the surface of the wells of a plastic plate. The toxin added to the wells is then bound by these antibodies and detected with a second antitoxin which is conjugated to an enzyme or other labeling molecule. The amount of label is measured by supplying the enzyme substrate, which is converted to a colored product that is measured colorimetrically. Some ELISA protocols use a polyclonal antitoxin on one side of the sandwich and a monoclonal on the other side. Other assays use the same antibody for both sides but label the antibody the second time it is used. A modification of the sandwich assay is the double sandwich ELISA, which employs a third antibody that is conjugated to an enzyme and is directed against the second antitoxin; it is an anti-antibody such as rabbit anti-horse IgG. The steps in a typical application of this assay for botulism toxin are shown in Figure 2.

Antibodies may be conjugated to enzymes such as alkaline phosphatase or horseradish peroxidase, or to other tags such as fluorescein or biotin (55). The fluorescein is measured by antifluorescein antibodies coupled with an enzyme, and biotin is measured by an avidin-enzyme conjugate, which binds with high affinity to biotin. Michalik et al.(47) used human albumin to tag their second antitoxin, and then used an anti-albumin antibody conjugated to peroxidase for detection of botulism toxins A and B. They reported a sensitivity for this method of 100 and 300 MLD for toxin types A and B, respectively.

ELISA amplification techniques are sometimes used to generate 10-fold increases in absorbance values (55,64). An amplification system supplied by Gibco BRL (Gaithersburg, MD; cat. no. 9589SA), uses alkaline phosphatase in the antibody conjugate to generate NADH from NADPH (Figure 3). The NADH is used as substrate in a secondary cyclic enzyme reaction in which a tetrazolium salt is reduced to a formazan dye, an intensely colored product that is measured spectrophotometrically.

Step 1 WELL~Goat Ab

Step 2 WELL~Goat Ab~Toxin

Step 3 WELL~Goat Ab~Toxin~Rab Ab

Step 4 WELL~Goat Ab~Toxin~Rab Ab~Goat Anti-Rab I
AlkPhos

Step 5 Add Substrate
Substrate + AlkPhos → COLOR

Figure 2. Double sandwich ELISA. Scheme for a typical adaptation for detecting botulism neurotoxin.

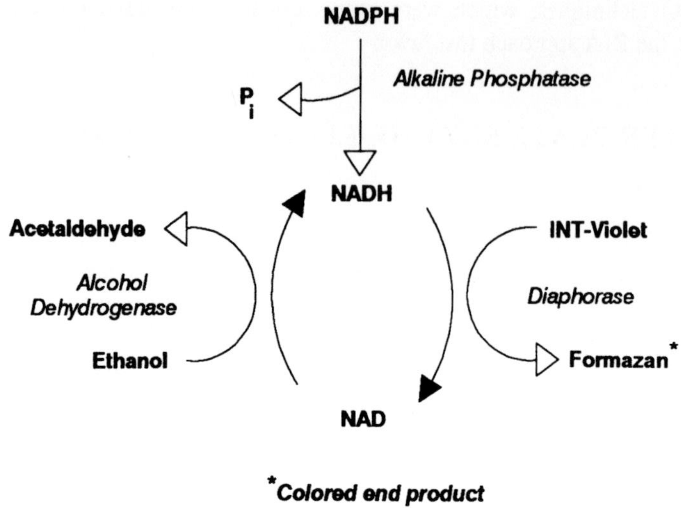

Figure 3. Enzymatic system for amplification of ELISA reactions. Reproduced from instructions supplied with product, cat. no. 9589SA; figure used with permission of Gibco BRL/Life Technologies, Inc.

Developmental Studies

Notermans et al. have reviewed the early developments of the ELISA for botulism neurotoxin (51,52). Most of the interest has been with toxin types A, B and E because these are the types that cause almost all of the human botulism. In 1978 Notermans et al. (50) used polystyrene tubes with polyclonal horse and rabbit antitoxin to detect type A toxin with a sensitivity of 50-100 LD_{50}. In 1979 Kozaki et al. (42) reported that a double sandwich ELISA technique improved the sensitivity from about 5,000 LD_{50} to about 400 LD_{50} for type B botulism toxin. When EDTA was added to toxic samples at a concentration of 5 mM just before incubation in the ELISA the ELISA extinction values were observed to rise twofold. Notermans et al. (48) used a combination of rabbit and horse antitoxins for type E toxin detection with a sensitivity of about 80 LD_{50}. Huhtanen et al. (36) compared the ELISA plate system with the mouse bioassay for detecting types A and B botulism toxins. Their ELISA, which used a combination of goat and rabbit polyclonal antibodies, was generally 8-32 times less sensitive than the mouse bioassay. They found that trypsin treatment of the toxic materials from some type A and B strains *decreased* the sensitivity of the ELISA but *increased* neurotoxin activity for the bioassay.

Monoclonal antibodies and amplification systems contributed to improved specificity and sensitivity of the ELISA. Ferreira et al. used a combination of monoclonal and polyclonal antibodies to detect type A toxin at a sensitivity of approximately 10 MLD/ml (16). Shone et al. (64) used monoclonal antibodies and polyclonal guinea pig IgG to detect type A toxin using an amplified ELISA system and obtained a sensitivity of 5-10 LD_{50}/ml, approaching the sensitivity of the bioassay. Gibson et al., used a modification of this same assay for detecting type A botulism toxin in culture and pork (25). The monoclonal antibody was used to immobilize the toxin and a guinea pig polyclonal antitoxin was used as the second antibody. They reported a sensitivity of approximately 33 LD_{50} /ml. They also evaluated a monoclonal antibody system for type B *C. botulinum* in pure culture and in a model system for detecting toxin in cured meat (24). The detection level was about 20 LD_{50}/ml. Ransom et al. (58) used polyclonal goat and rabbit antisera and the amplification system described by Shone et al. (64) for detecting type A and type E toxins. While the sensitivity appears excellent (approximately 20 [~2 LD_{50}] pg/ml), only the type A assay appears to be toxin specific. In a different approach, Potter et al. (55) used a single trivalent antibody product to capture and detect types A, B, and E botulism toxins. The second antibody reagent was biotinylated and the biotin was detected using an alkaline phosphatase-conjugated streptavidin and an ELISA amplification system. The detection levels for this system were 9 LD_{50} for type A and <1 LD_{50} for types B and E.

In regard to other toxin types, Notermans et al. (49) described an ELISA for detecting types C and D botulism toxins. The type C ELISA detected type C toxin at 500 LD_{50} but also detected type D toxin. The type D ELISA detected type D toxin at 100 LD_{50} and also detected type C toxin. Lewis et al. (44) developed an ELISA method for detecting type G botulism toxin. This double sandwich method used a combination of goat and rabbit antitoxins. A mouse toxicity of 16 LD_{50} correlated with an optical density of 0.32 in the ELISA in one type G strain, and in a second strain, 1 LD_{50} was 0.34 by ELISA. Ferreira et al. used a combination of monoclonal and polyclonal antibodies to detect type F toxins each at a sensitivity of approximately 10 MLD/ml (19).

Diagnostic Application

Dezfulian et al. (8) evaluated ELISA for detecting type A and B toxins in fecal samples from infants for diagnosis of infant botulism. The ELISA gave positive results for all 22 infant specimens that had been confirmed as toxic by bioassay, and for 5 specimens confirmed by culture but not by bioassay. Positive ELISA results were obtained with specimens not confirmed by conventional means; these were possibly false positives. Rubin et al. reported detecting of antibodies reactive to botulism toxin preparations in serum samples from two infants who had recovered from infant botulism (59).

Botulism Antitoxin Detection

Siegel (65) compared an ELISA with the mouse bioassay for determination of type A, B, and E antitoxin titers in humans. She found poor correlation between these two methods

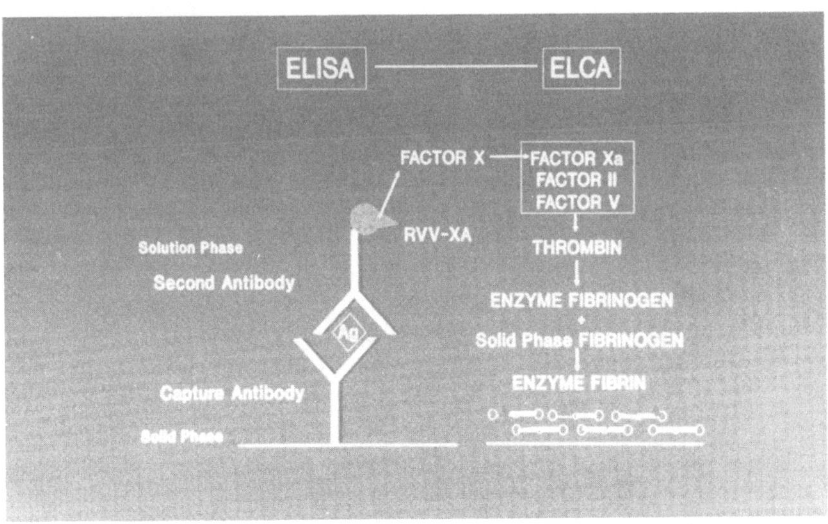

Figure 4. Schematic depiction of the ELCA; used with permission of the authors and publisher (10).

for serum samples taken early in a series of immunizations. The correlation was higher for hyperimmune serum samples.

ENZYME-LINKED CLOTTING ASSAY (ELCA)

Doellgast et al. (10) developed a novel form of the ELISA called the enzyme-linked clotting assay (ELCA) system that is extremely sensitive. A schematic representation of the test is shown in Figure 4. The test can be divided into two stages. First, toxins are bound in an ELISA microtiter plate previously coated with a capture antitoxin. A second antitoxin labeled with a coagulation activating enzyme isolated from Russell's viper venom (RVV-XA) is then added to form a sandwich complex on the plate. In the second stage, this complex is detected by providing all of the components of the blood clotting system including alkaline phosphatase-labeled fibrinogen. The thrombin which is formed in the cascade reaction

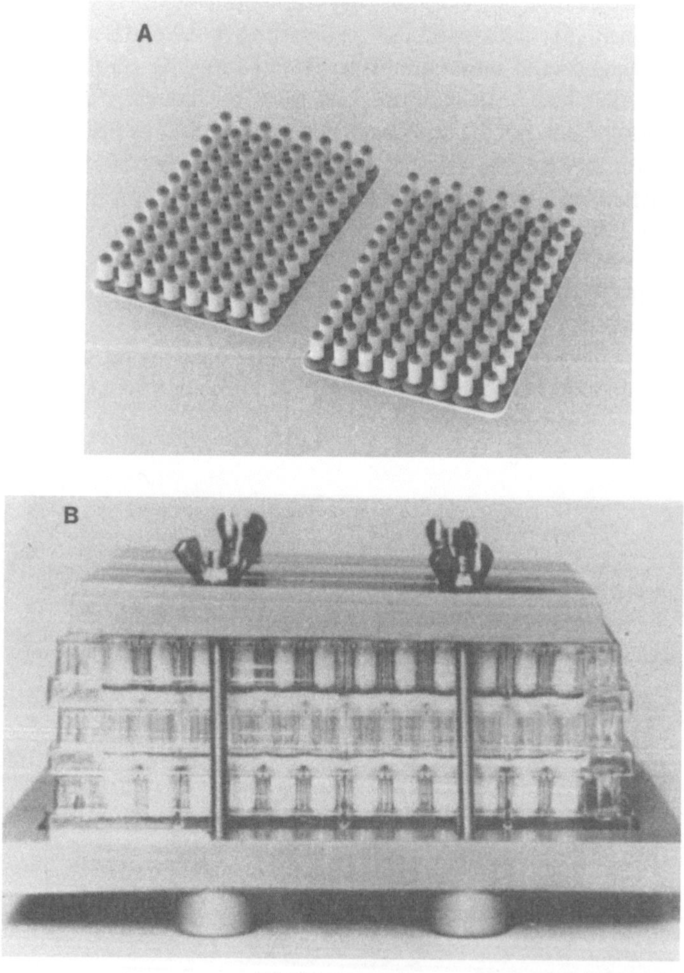

Figure 5. Apparatus for incorporation of alkaline phosphatase into solid-phase fibrin in the ELCA. (A) Silicone rubber nubs. (B) Nubs inserted into microtiter plate wells. (Courtesy G.J. Doellgast and ELCATech).

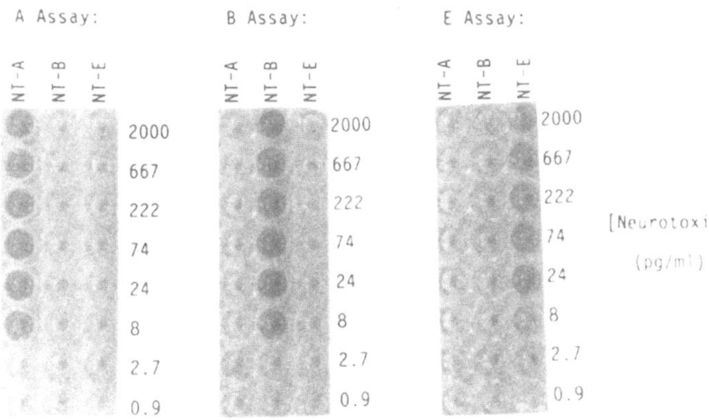

Figure 6. Results of type A, B, and E botulism neurotoxin assays using the ELCA; sensitivity and specificity for each type can be seen. Photograph from reference 10, used with permission of the authors and the publisher.

initiated by RVV-XA incorporates labeled fibrinogen into fibrin, which forms on a solid-phase surface, i.e., a set of silicone rubber nubs (Figure 5) molded to fit precisely into the 96 wells of the microtiter plate. The nubs contain unlabeled fibrinogen before insertion into the wells, and in the wells that contain thrombin, labeled fibrinogen from the liquid phase will be incorporated into fibrin that is produced on the nubs. The label incorporated into the nubs is then assayed by transferring the nubs into a new microtiterplate whose wells contain enzyme substrate; the reaction is measured photometrically as in the conventional ELISA.

The reactions are, in general, all or none, and quantitation is obtained by determining the highest dilution yielding a positive reaction. The assay can detect as little as 5-10 pg/ml for types A, B, and E (Figure 6), and the data presented show no cross-reactions against heterologous toxin types. The sensitivity equals or exceeds that of the mouse bioassay.

This method has been modified more recently using a solution-phase complexing of the toxin to either biotin-labeled antitoxin or chicken antitoxin prior to binding of the complex to the solid-phase plate (9). Doellgast et al. (11) have also devised an adaptation of the ELISA/ELCA for detecting human antibodies to botulism toxins types A, B, and E.

BIOSENSOR AND FIBEROPTIC METHODS

Optic sensor techniques employing antibodies have been developed for recognition and quantitative determination of biologic molecules (2). A separate antibody-coated fiber is used for each determination. The determination is done in a single step, using a single reagent, a fluorescent-labeled solution-phase antibody; no washing is needed since only the fluorescence on the surface of the optical fiber is measured. By using fiberoptics, the presence of analytes and their concentration can be determined within 10 min. A fiber optic method for detecting toxin in solution was shown by Ogert et al.(53) to detect type A toxin at concentrations as low as 5 ng/ml. This system utilizes specific antitoxin covalently attached to the surface of a fiber. The toxin is bound by the immobilized antitoxin, and a second antitoxin conjugated to rhodamine is added. The fluorescent complex is excited by laser light and measured by a photodiode. Further developments in the use of this technology for detection of botulism neurotoxins are reported in the poster presentation program of this symposium by Dr. B. R. Singh.

Table 3. Sensitivity of neurotoxin assays[a]

Method	Toxin type A	Toxin type B	Toxin type E	References
Mouse Bioassay	2	2	2	(definition of LD_{50})[b]
RPHA	1.6-60	4.6-60	150	Johnson et al., 1966 Sakaguchi et al., 1974
RPLA	390			Horiguchi et al., 1984
Capillary Tube	100	100	100	Mestrandrea, 1974
Immunodiffusion		370	560	Vermilya et al., 1968
Fluorescence			100	Inukai et al., 1968
Radioimmunoassay	400-1000			Boroff & Shu-chen, 1973
Biosensor	200			Ogert et al., 1992
ELISA				
Review of methods through 1982	100-1000	100-1000	100-1000	Notermans et al., 1982 (51)
MAb[c] amplification	5-10			Shone et al., 1985
Mab amplification; cultures & pork	33	20		Gibson et al., 1987, 1988
Polyclonal Ab, no amplification; diverse strains tested	2-64	0.1-32		Huhtanen et al., 1992
Polyclonal Ab, amplification; reacts w/HA	2		2	Ransom et al., 1993
Trivalent ABE syst.; no type identification; biotin/avidin detection	9	<1	<1	Potter et al., 1993
ELCA				
Clotting system amplification	<1	<1	<1	Doellgast et al., 1993, 1994

[a]Sensitivities expressed as minimum detectable amount in mouse i.p. LD_{50}/ml.
[b]At 2 LD_{50}/ml, 50% of mice injected with 0.5ml will die.
[c]MAb, monoclonal antibody.

SENSITIVITY OF ASSAYS FOR BOTULISM NEUROTOXIN

The sensitivities of the various assays expressed as mouse i.p. LD_{50} for botulism neurotoxin are shown in Table 3. The minimum for the bioassay should be 2 LD_{50}/ml because 0.5 ml is customarily injected, and 1 LD_{50} would result in deaths in half of the injected mice. The results with in vitro tests can be deceptive because they can show reactions between antibodies and antigens unrelated to biologically active toxin. The reactions may occur with denatured toxin that has no biological activity, or with antigens other than toxin. These problems may account for the apparent high sensitivity reported for some of the RPHA studies (38).

The sensitivity of the early ELISA methods (51) was in the same range as previous in vitro methods, but increased greatly with the incorporation of amplifier systems (58,64). Specificity has also improved with the preparation of antitoxins against highly purified toxins and use of monoclonal antibodies (64). The reports on the ELCA method describe meticulous preparation of high quality reagents and an exceptionally effective amplification which utilizes the Russell's viper venom factor X activator to initiate the clotting cascade (9,10). Its sensitivity appears to equal or exceed that of the mouse.

The biosensor-fiberoptic approach to detection of botulism neurotoxin promises the most rapid results, but at this early stage in its development, its sensitivity is much less than that of the bioassay and the microtiter assays.

MOLECULAR BIOLOGY TECHNIQUES

Molecular techniques can be used to identify genes for toxins and establish the ability of the organism to produce the toxin. If reliable, the results can obviate the need for toxin assays currently required for identifying the organism. The results are qualitative and do not identify the toxin itself nor quantitate it.

The nucleotide sequences for tetanus toxin and botulism neurotoxin genes types A-F have been published. Recently, researchers have used this information and new techniques for the detection of these genes in neurotoxin-producing clostridia. These methods include the polymerase chain reaction (PCR) and DNA hybridization detection assays using toxin gene probes. Wernars and Notermans presented preliminary data on identification of *C. botulinum* type B using a DNA probe to detect the neurotoxin gene (75). Bacterial colonies were lysed on nylon membranes and tested with a ^{32}P-labeled probe. In most other reports, toxin gene-specific probes were used for identification of PCR-generated toxin gene fragments.

Polymerase Chain Reaction

In 1992, Ferreira et al. reported a PCR method for the detection of type A toxigenic organisms (17). The PCR-amplified fragment from one strain was labeled with digoxigenin and used as a DNA probe to confirm that the PCR fragments amplified from other type A strains were similar in sequence. The same probe was also useful for detecting of toxin gene sequences in chromosomal DNA preparations. The detection sensitivity of the PCR for vegetative cells in culture was 1 cell/10 μl. This method was extended for the detection of type A organisms in foods (14), although it was useful only for the detection of toxigenic vegetative cells. The PCR was negative when spores were used instead of cells. Szabo et al., (69) devised a PCR method and a radiolabeled oligonucleotide probe for the detection of the type B toxin gene from proteolytic and nonproteolytic botulism organisms. Subsequently they reported on the detection of botulinum neurotoxin types A to E using the PCR and specific primer pairs for each toxin gene type (70). They also described an alkaline lysis procedure that released DNA from spores as well as vegetative cells for use in the PCR. They studied the use of PCR for detecting type A, B, and E toxigenic organisms in food, soil and infant feces (71) and applied the method to the investigation of a case of equine botulism (72).

Ferreira et al. described a PCR method for detecting toxigenic type F proteolytic and nonproteolytic *C. botulinum* as well as the partial nucleotide sequence for the toxin gene of the proteolytic type F Langeland strain (18). A multiplex PCR assay was reported by Ferreira and Hamdy (15) to detect types A and E in a single PCR tube and types B and F in a second tube.

Campbell et al. described the use of the PCR and degenerate primers to amplify toxin gene fragments from types B, E, and F (6). Oligonucleotide probes specific for toxin gene types B, E, or F were labeled with fluorescein and detected using an enhanced chemiluminescence kit. In a similar study, Fach et al. (13) also used degenerate PCR primers to amplify fragments from toxin genes A-G. An internal type-specific oligonucleotide probe was used to determine the toxin gene type.

Table 4. PCR detection of type A, B, and E neurotoxin genes in 205 strains of *Clostridium botulinum* and 5 strains of *Clostridium butyricum*[a]

		Positive in PCR		
Type of strains	No. of strains	A	B	E
A	79	79	43	0
B	77	1	77	0
E	49	0	0	46
C. butyricum (toxigenic)	5	0	0	5

[a]From Franciosa et al., 1994 (22).

Franciosa et al. studied the effectiveness of the PCR for the detection of type A, B, and E toxin genes in 209 strains of *C. botulinum* and 29 other *Clostridium* spp. (23). They found that the PCR was sensitive and specific for the detection of these toxin genes (Table 4). All organisms that produced type A, B, or E neurotoxins (including 5 toxigenic strains of *C. butyricum*) yielded an amplified DNA fragment consistent with the specific primer promoted PCR. Three type E strains that did not yield a product were found to be nontoxigenic by bioassay. Surprisingly, the authors also discovered that 43 of the 79 type A strains examined also contained a type B toxin gene; only one of them produced biologically active type B toxin. Specific digoxigenin-labeled DNA probes were used to verify the toxin gene types, and the presence of toxin in the cultures was determined with the bioassay and ELISA. More recent studies in our laboratories (22) showed that reverse transcription coupled with the PCR can demonstrate the presence of type B toxin gene m-RNA in some of the PCR-positive type A strains.

DISCUSSION

Although the interest in the botulism neurotoxins is keen, the progress in devising reliable in vitro tests has been slow. This is because i) botulism is a rare disease and thus, few clinical laboratories need the test; ii) because of the low demand, there is no incentive for commercial development; iii) the neurotoxin is potent biologically and the sensitivity needed to detect it by physicochemical means is very demanding; iv) the existence of 7 different types requires a battery of reagents to detect them all; v) with the known diversity of botulism neurotoxins, we can never be sure that we might be missing types heretofore unknown without the use of a biological detection system.

As noted, there have been numerous publications on the diverse methods proposed for detection, identification and quantitation of the botulism neurotoxins. Detection of the toxin genes by PCR or DNA probes can serve well for identifying toxigenic organisms, but would have questionable application for establishing the presence of or quantifying toxin. The most promising practical tests for toxin are enzyme-antibody based assays such as the ELISA or ELCA. However, at present there are no sources of reagents that allow the rest of the scientific world to use the published protocols. Therefore, each laboratory must try to develop its own, and everyone's immunoassay remains a private experimental procedure at this time. The conventional bioassay, with the available diagnostic reagents (distributed primarily, if not exclusively through CDC) and the International Standards for *Clostridium botulinum* Antitoxins that have been available through WHO, remains by default the common reliable test for botulism neurotoxin. For this reason, the U.S. Food and Drug Administration in collaboration with the Centers for Disease Control and Prevention is attempting to develop reagents that can be used with a common protocol. The reagents are expected to be made available to qualified investigators through one of these collaborating governmental agencies.

REFERENCES

1. Aureli, P., Fenicia, L., Pasolini, B., Gianfranceschi, M., McCroskey, L.M., and Hatheway, C.L., 1986, Two cases of type E infant botulism caused by neurotoxigenic *Clostridium butyricum* in Italy, *J. Infect. Dis.* 154:201-206.
2. Bluestone, B.I., Craig, M., Slovacek, R., Stundtner, L., Urciuoli, C., Walczak, I., and Luderer, A., Evanescent wave immunosensors for clinical diagnosis, *in*: "Biosensors with Fiberoptics," Wise, D.L., and Wingard, L.B., Jr., eds., Humana Press, Clifton, New Jersey (1991).
3. Boroff, D.A., and Fleck, U., 1966, Statistical analysis of a rapid in vitro method for the titration of the toxin of *Clostridium botulinum*, *J. Bacteriol.* 92:1580-1581.
4. Boroff, D.A., and Shu-Chen, G., 1973, Radio-immunoassay for type A toxin of *Clostridium botulinum*, *Appl. Microbiol.* 25:545-549.
5. Bowmer, E.J., 1963, Preparation and assay of the International Standards for *Clostridium botulinum* types A, B, C, D and E antitoxins, *Bull. Wld. Hlth. Org.* 29:701-709.
6. Campbell, K.D., Collins, M.D., and East, A.K., 1993, Gene probes for identification of the botulinal neurotoxin gene and specific identification of neurotoxin types B, E, and F, *J. Clin. Microbiol.* 31:2255-2262.
7. Cardella, M.A., Botulinum Toxoids, *in*: "Botulism: Proceedings of a Symposium," Lewis, K.H., and Cassel, K., Jr., eds., U.S. Department of Health, Education, and Welfare, Public Health Service (PHS Publ. No 999 FP-1), Cincinnati, Ohio (1964).
8. Dezfulian, M., Hatheway, C.L., Yolken, R.H., and Bartlett, J.G., 1984, Enzyme-linked immunosorbent assay for detection of *Clostridium botulinum* type A and type B toxins in stool samples from infants with botulism, *J. Clin. Microbiol.* 20:379-383.
9. Doellgast, G.J., Beard, G.A., Bottoms, J.D., Cheng, T., Roh, B.H., Roman, M.G., Hall, P.A., and Triscott, M.X., 1994, Enzyme-linked immunosorbent assay and enzyme-linked coagulation assay for detection of *Clostridium botulinum* neurotoxins A, B, and E and solution-phase complexes with dual-label antibodies, *J. Clin. Microbiol.* 32:105-111.
10. Doellgast, G.J., Triscott, M.X., Beard, G.A., Bottoms, J.B., Cheng, T., Roh, B.H., Roman, M.G., Hall, P.A., and Brown, J.E., 1993, Sensitive enzyme-linked immunosorbent assay for detection of *CLostridium botulinum* neurotoxins A, B, and E using signal amplification via enzyme-linked coagulation assay, *J. Clin. Microbiol.* 31:2402-2409.
11. Doellgast, G.J., Triscott, M.X., Beard, G.A., and Bottoms, J.D., 1994, Enzyme-linked immunosorbent assay - enzyme-linked coagulation assay for detection of antibodies to *Clostridium botulinum* neurotoxins A, B, and E and solution-phase complexes, *J. Clin. Microbiol.* 32:851-853.
12. Evancho, G.M., Ashton, D.H., and Briskey, E.J., 1973, A standardized reversed passive hemagglutination technique for the determination of botulinum toxin, *J. Food Sci.* 38:764-767.
13. Fach, P., Gibert, M., Griffais, R., Guillou, J.P., and Popoff, M.R., 1995, PCR and gene probe identification of botulinum neurotoxin A- B- E- F- and G-producing *Clostridium* spp. and evaluation in food samples, *Appl. Environ. Microbiol.* 61:389-392.
14. Ferreira, J.L., Baumstark, B.R., Hamdy, M.K., and McCay, S.G., 1993, Polymerase chain reaction for the detection of *Clostridium botulinum* type A in foods, *J. Food Protect.* 56:18-20.
15. Ferreira, J.L., and Hamdy, M.K., 1995, Detection of botulinal toxin genes: types A and E or B and F using the multiplex polymerase chain reaction, *J Rapid Meth. Automat. Microbiol.* 3:177-183.
16. Ferreira, J.L., Hamdy, M.K., Herd, A.L., McCay, S.G., and Zapatka, F.A., 1987, Monoclonal antibody for the detection of *Clostridium botulinum* type A toxin, *Mol. Cell. Probes* 1:337-345.
17. Ferreira, J.L., Hamdy, M.K., McCay, S.G., and Baumstark, B.R., 1992, An improved assay for identification of type A *Clostridium botulinum* using the polymerase chain reaction, *J Rapid Meth. Automat. Microbiol.* 1:29-39.
18. Ferreira, J.L., Hamdy, M.K., McCay, S.G., Hemphill, M., Kirma, N., and Baumstark, B.R., 1994, Detection of *Clostridium botulinum* type F using the polymerase chain reaction, *Mol. Cell. Probes* 8:365-373.
19. Ferreira, J.L., Hamdy, M.K., McCay, S.G., and Zapatka, F.A., 1990, Monoclonal antibody to type F *Clostridium botulinum* toxin, *Appl. Environ. Microbiol.* 56:808-811.

20. Ferreira, J.L., Hamdy, M.K., Zapatka, F.A., and Hebert, W.O., 1981, Immunodiffusion method for detection of *Clostridium botulinum* type A, *Appl. Environ. Microbiol.* 42:1057-1061.
21. Ferreira, J.L., Hamdy, M.K., Zapatka, F.A., and Hebert, W.O., 1983, Immunodiffusion method for detection of *Clostridium botulinum* types A, B, E and F, *J. Food Safe.* 5:87-94.
22. Ferreira, J.L., Hatheway, C.L., Johnson, E.A., and Collins, M.D., Unpublished collaborative studies on neurotoxin genes of *Clostridium botulinum*.
23. Franciosa, G., Ferreira, J.L., and Hatheway, C.L., 1994, Detection of type A, B, and E botulism neurotoxin genes in *Clostridium botulinum* and other *Clostridium* species by PCR: evidence of unexpressed type B toxin genes in type A toxigenic organisms, *J. Clin. Microbiol.* 32:1911-1917.
24. Gibson, A.M., Modi, N.K., Roberts, T.A., Hambleton, P., and Melling, J., 1988, Evaluation of a monoclonal antibody-based immunoassay for detecting type B *Clostridium botulinum* toxin produced in pure culture and an inoculated model cured meat system, *J. Appl. Bacteriol.* 64:285-291.
25. Gibson, A.M., Modi, N.K., Roberts, T.A., and Shone, C.C., 1987, Evaluation of a monoclonal antibody-based immunoassay for detecting type A *Clostridium botulinum* toxin produced in pure culture and an inoculated model cured meat system, *J. Appl. Bacteriol.* 63:217-226.
26. Glasby, C., and Hatheway, C.L., 1983, Fluorescent-antibody reagents for the identification of *Clostridium botulinum*, *J. Clin. Microbiol.* 18:1378-1383.
27. Glasby, C., and Hatheway, C.L., 1984, Evaluation of fluorescent-antibody tests as a means of confirming infant botulism, *J. Clin. Microbiol.* 20:1209-1212.
28. Goodnough, M.C., Hammer, B., Sugiyama, H., and Johnson, E.A., 1993, Colony immunoblot assay of botulinal toxin, *Appl. Environ. Microbiol.* 59:2339-2342.
29. Habermann, E., and Bernath, S., 1975, Preparation, measurement and possible use of human antitoxin against *Cl. botulinum* A, B, and E toxins, *Med. Microbiol. Immunol.* 161:203-210.
30. Hall, J.D., McCroskey, L.M., Pincomb, B.J., and Hatheway, C.L., 1985, Isolation of an organism resembling *Clostridium barati* which produces type F botulinal toxin from an infant with botulism, *J. Clin. Microbiol.* 21:654-655.
31. Hatheway, C.L., Botulism, *in*: "Laboratory Diagnosis of Infectious Diseases: Principles and Practice," Balows, A., Hausler, W.H., Jr., Ohashi, M., and Turano, A., eds., Springer-Verlag, New York (1988).
32. Hatheway, C.L., 1990, Toxigenic clostridia, *Clin. Microbiol. Rev.* 3:67-98.
33. Hatheway, C.L., and Dang, C., Immunogenicity of the neurotoxins of *Clostridium botulinum*, *in*: "Therapy with botulinum toxin," Jankovic, J., and Hallett, M., eds., Marcel Dekker, Inc., New York (1994).
34. Hatheway, C.L., Ferreira, M.C., and McCroskey, L.M., Evaluation of various factors in the mouse toxicity test for identification of botulinal toxin. Abstr., Ann. Mtg., Am. Soc. Microbiol., Las Vegas, 1985.
35. Horiguchi, Y., Kozaki, S., and Sakaguchi, G., 1984, Determination of *Clostridium botulinum* toxin by reversed passive latex agglutination, *Jpn. J. Vet. Sci.* 46:487-491.
36. Huhtanen, C.N., Whiting, R.C., Miller, A.J., and Call, J.E., 1992, Qualitative correlation of the mouse neurotoxin and enzyme-linked immunoassay for determining *Clostridium botulinum* types A and B toxins, *J. Food Protect.* 12:119-127.
37. Inukai, Y., and Riemann, H., 1968, Detection of intracellular botulinum E toxin by fluorescent antibody technique, *Jap. J. Vet. Res.* 16:39-43.
38. Johnson, H.M., Brenner, K., Angelotti, R., and Hall, H.E., 1966, Serological studies of types A, B, and E botulinal toxins by passive hemagglutination and bentonite flocculation, *J. Bacteriol.* 91:967-973.
39. Kautter, D.A., Lynt, R.K., and Solomon, H.M., *Clostridium botulinum*, *in*: "Bacteriological Analytical Manual," U.S. Food and Drug Administration, Washington, D.C. (1984).
40. Kautter, D.A., and Solomon, H.M., 1977, Collaborative study of a method for the detection of *Clostridium botulinum* and its toxins in foods, *J. Assoc. Off. Anal. Chem.* 60:541-545.
41. Kempner, W., 1897, Weiterer Beitrag zur Lehre von der Fleischvergiftung. Das Antitoxin des Botulismus, *Ztschr. Hyg. Infektionskh.* 26:481-500.
42. Kozaki, S., Dufrenne, J., Hagenaars, A.M., and Notermans, S., 1979, Enzyme-linked immunosorbent assay (ELISA) for detection of *Clostridium botulinum* type B toxin, *Jpn. J. Med. Sci. Biol.* 32:199-205.
43. Leuchs, J., 1910, Beitraege zur Kenntnis des Toxins und Antitoxins des *Bacillus botulinus*, *Ztschr. Hyg. Infektskh.* 65:55-84.
44. Lewis, G.E.J., Kulinski, S.S., Reichard, D.W., and Metzger, J.F., 1981, detection of *Clostridium botulinum* type G toxin by enzyme-linked immunosorbent assay, *Appl. Environ. Microbiol.* 42:1018-1022.
45. Lüthgen, W., 1972, Eine modifikation des Mäusetestes zum Nachweis der Botulismus C Intoxikation des Geflügels, *Berl. Münch. Tierärzl. Wochenschr.* 85:107-110.
46. Mestrandrea, L.W., 1974, Rapid detection of Clostridium botulinum toxin by capillary tube diffusion, *Appl. Microbiol.* 27:1017-1023.

47. Michalik, M., Grzybowski, J., Ligieza, J., and Reiss, J., 1986, Enzyme-linked immunosorbent assay (ELISA) for the detection and differentiation of *Clostridium botulinum* toxins type A and B, *J. Immunol. Meth.* 93:225-230.
48. Notermans, S., Dufrenne, J., and Kozaki, S., 1979, Enzyme-linked immunosorbent assay for detection of *Clostridium botulinum* type E toxin, *Appl. Environ. Microbiol.* 37:1173-1175.
49. Notermans, S., Dufrenne, J., and Kozaki, S., 1982, The relation between toxicity and toxin-related-antigen contents of *Clostridium botulinum* types C and D cultures as determined by mouse bioassay and ELISA, *Jpn. J. Med. Sci. Biol.* 35:203-211.
50. Notermans, S., DuFrenne, J., and Van Schothorst, M., 1978, Enzyme-linked immunosorbent assay for detection of *Clostridium botulinum* toxin type A, *Jpn. J. Med. Sci. Biol.* 31:81-85.
51. Notermans, S., Hagenaars, A.M., and Kozaki, S., 1982, The enzyme-linked immunosorbent assay (ELISA) for the detection and determination of *Clostridium botulinum* toxins A, B, and E, *Methods Enzymol.* 84:223-238.
52. Notermans, S., and Nagel, J., Assays for botulinum and tetanus toxins, *in*: "Botulinum Neurotoxin and Tetanus Toxin," Simpson, L.L., ed., Academic Press, Inc., San Diego (1989).
53. Ogert, R.A., Brown, J.E., Singh, B.R., Shriver-Lake, L.C., and Ligler, F.S., 1992, Detection of *Clostridium botulinum* toxin A using a fiber optic-based biosensor, *Anal. Biochem.* 205:306-312.
54. Pearce, L.B., Borodic, G.E., First, E.R., and MacCallum, R.D., 1994, Measurement of botulinum toxin activity: evaluation of the lethality assay, *Toxicol. Appl. Pharmacol.* 128:69-77.
55. Potter, M.D., Meng, J., and Kimsey, P., 1993, An ELISA for detection of botulinal toxin types A, B, and E in inoculated food samples, *J. Food Protect.* 56:856-861.
56. Prevot, A.R., 1953, Rapport d'introduction du President du Sous-Comite *Clostridium* pour l'unification de la nomenclature des types toxigeniques de *C. botulinum*, *Int. Bull. Bacteriol. Nomenclature* 3:120-123.
57. Ramon, G., 1924, Sur la toxine et sur l'anatoxine diphtheriques; pourvoir floculant et propriétés immunisantes, *Ann. Inst. Pasteur* 38:1-10.
58. Ransom, G.M., Lee, W.H., Elliot, E.L., and Lattuada, C.P., Enzyme-linked immunosorbent assays (ELISAs) to detect botulinum toxins using high titer rabbit antisera, *in*: "Botulinum and Tetanus Neurotoxins," DasGupta, B.R., ed., Plenum Press, New York (1993).
59. Rubin, L.G., Dezfulian, M., and Yolkin, R.H., 1982, Serum antibody response to *Clostridium botulinum* toxin in infant botulism, *J. Clin. Microbiol.* 16:770-771.
60. Sakaguchi, G., 1983, *Clostridium botulinum* toxins, *Pharmacol. Ther.* 19:165-194.
61. Sakaguchi, G., and Sakaguchi, S., 1968, Rapid bioassay for *Clostridium botulinum* type E toxins by intravenous injection into mice, *Jpn. J. Med. Sci. Biol.* 21:369-378.
62. Sakaguchi, G., Sakaguchi, S., Kozaki, S., Sugii, S., and Ohishi, I., 1974, Cross reaction in reversed passive hemagglutination between *Clostridium botulinum* type A and B toxins and its avoidance by the use of antitoxic component immunoglobulin isolated by affinity chromatography, *Jpn. J. Med. Sci. Biol.* 27:161-172.
63. Schantz, E.J., and Kautter, D.A., 1978, Standardized assay for *Clostridium botulinum* toxins, *J. Assoc. Off. Anal. Chem.* 61:96-99.
64. Shone, C., Wilton-Smith, P., Appleton, N., Hambleton, P., Modi, N., Gatley, S., and Melling, J., 1985, Monoclonal antibody-based immunoassay for type A *Clostridium botulinum* toxin is comparable to the mouse bioassay, *Appl. Environ. Microbiol.* 50:63-67.
65. Siegel, L.S., 1988, Human response to botulinum pentavalent (ABCDE) toxoid determined by a neutralization test and by an enzyme-linked immunosorbent assay, *J. Clin. Microbiol.* 26:2351-2356.
66. Smith, C., and Hamdy, M.K., 1992, Immunoenzyme for rapid screening of *Clostridium botulinum* type A cells, *J Rapid Meth. Automat. Microbiol.* 1:149-163.
67. Suen, J.C., Hatheway, C.L., Steigerwalt, A.G., and Brenner, D.J., 1988, *Clostridium argentinense*, sp. nov: a genetically homogenous group composed of all strains of *Clostridium botulinum* toxin type G and some nontoxigenic strains previously identified as *Clostridium subterminale* or *Clostridium hastiforme*, *Int. J. Syst. Bacteriol.* 38:375-381.
68. Sugiyama, H., Brenner, S.L., and DasGupta, B.R., 1975, Detection of *Clostridium botulinum* toxin by local paralysis elicited with intramuscular challenge, *Appl. Microbiol.* 30:420-423.
69. Szabo, E.A., Pemberton, J.M., and Desmarchelier, P.M., 1992, Specific detection of *Clostridium botulinum* type B by using the polymerase chain reaction, *Appl. Environ. Microbiol.* 58:418-420.
70. Szabo, E.A., Pemberton, J.M., and DesMarchelier, P.M., 1993, Detection of the genes encoding botulinum neurotoxin types A to E by the polymerase chain reaction, *Appl. Environ. Microbiol.* 59:3011-3020.
71. Szabo, E.A., Pemberton, J.M., Gibson, A.M., Eyles, M.J., and Desmarchelier, P.M., 1994, Polymerase chain reaction for detection of *Clostridium botulinum* types A, B, and E in food, soil and infant feces, *J. Appl. Bacteriol.* 76:539-545.

72. Szabo, E.A., Pemberton, J.M., Gibson, A.M., Thomas, R.J., Pascoe, R.R., and Desmarchelier, P.M., 1994, Application of PCR to a clinical and environmental investigation of a case of equine botulism, *J. Clin. Microbiol.* 32:1986-1991.
73. Van Ermengem, E., 1897, Ueber einen neuen anaeroben Bacillus and seine Beziehungen zum Botulismus, *Ztschr. Hyg. Infektionskh.* 26:1-56.
74. Vermilyea, B., Walker, H.W., and Ayers, J.C., 1968, Detection of botulinal toxins by immunodiffusion, *Appl. Microbiol.* 16:21-24.
75. Wernars, K., and Notermans, S., Gene probes for detection of food-borne pathogens, *in*: "Gene probes for bacteria," Macario, A.J.L., and de Macario, E.C., eds., Academic Press, Inc., San Diego (1990).

40

DETECTION OF BOTULINUM NEUROTOXINS USING OPTICAL FIBER-BASED BIOSENSOR

Bal Ram Singh and Melissa A. Silvia

Department of Chemistry
University of Massachusetts Dartmouth
North Dartmouth, Massachusetts 02747

INTRODUCTION

Botulinum neurotoxins, produced by the anaerobic bacterium *Clostridium botulinum*, are the most toxic poisons known to man. The neurotoxins are food poisons. Once ingested, the neurotoxin is absorbed through the intestinal mucosal layer into the blood stream. It acts at the neuromuscular junction to inhibit the release of acetylcholine (a neurotransmitter) from nerve endings (Simpson, 1989). The result is the dreaded botulism disease, which is manifested by flaccid muscle paralysis.

Toxins produced by different strains of *Clostridium botulinum* have been classified into seven different serotypes, namely types A, B, C, D, E, F, and G. These neurotoxins are large proteins with molecular weights of 150 kilodaltons (kDa), and are synthesized as single chain polypeptides. The biologically active form of the neurotoxins consists of a dichain which is produced by endogenous or exogenous proteolytic cleavage (nicking) of the single chain polypeptide (DasGupta, 1989). This dichain form is composed of a light chain having a molecular weight of 50 kDa which is covalently linked by at least one disulfide bond to a 100 kDa heavy chain. The heavy chain plays a role in the binding of the toxin to peripheral synapses, and the light chain is associated with the intracellular activity of blocking acetylcholine release.

All the seven serotypes of botulinum neurotoxin are produced as complexes (Sakaguchi, 1983). These complexes consist of the 150 kDa toxin portion and a non-toxin portion of varying molecular size and composition. Depending on the serotype, the non-toxin portion may contain up to 7 proteins as in types A and B, or only one protein as in type E botulinum (Ohishi et al., 1977; Singh et al., 1995a). We refer the proteins comprising the non-toxic portion of the complex as neurotoxin associated proteins (NAPs), and the protein found in type E botulinum neurotoxin complex as neurotoxin binding protein (NBP; Singh et al., 1995a). Presently, the role and importance of the NAPs in the activity of botulinum toxins are not fully understood. However, it is clear that these complexing proteins, which do not dissociate from the toxin until intestinal adsorption of the toxin occurs, serve to protect the

Natural Toxins II, Edited by B. R. Singh and A. T. Tu
Plenum Press, New York, 1996

neurotoxin. NAPs preserve the toxic activity of the neurotoxin by protecting it from heat, acidity, and proteases present in the stomach. In this way, NAPs can protect the purified toxins both in the environment and in the digestive system upon ingestion. It has been asserted that the presence of and protection provided by these NBPs make botulinum toxins potent food as well as potential agents in biological warfare poisons (Ohishi et al., 1977; 1980; Singh et al., 1995b). This idea is supported by the fact that tetanus toxin which has 35% sequence homology with botulinum neurotoxins are not produced as complexes, and consequently, are not poisons found in foods or considered as biological warfare agents (Ember, 1991; Singh et al., 1995b).

Clostridium botulinum is ubiquitous, in both terrestrial and aquatic environments, and is particularly common in some geographical areas. Its occurrence in foods, though occasional, could result in death upon ingestion of even very low levels of the toxin. Despite well-validated methods of controlling the bacterial growth, foodborne outbreaks of botulism still occur from time to time. Foodborne botulism may result after faulty processing or more commonly after recontamination and/or temperature abuse of an adequately processed product (Gibson et al., 1987). *Clostridium botulinum* is a major concern for processors of slightly acidic foods, since the ingestion of even pre-formed toxin may cause severe illness which is difficult to treat and recovery of those afflicted is slow. In addition to foodborne botulism, infant botulism is the most common form of human botulism in the US with approximately 75-100 hospitalized cases a year (cf. Arnon, 1993). A crucial step in developing preventive measures against these toxins is the availability of a technique that offers the rapid and sensitive detection of their presence in the environment and in foodstuffs.

Until recently, there have been only two primary techniques available for the detection of botulinum neurotoxins. The first of these, which is the most widely accepted and sensitive technique for the detection of botulinum neurotoxins in serum and food extracts, is the mouse bioassay (Sakaguchi, 1983). Although the mouse bioassay is the most sensitive method, with the ability to detect less than 5 mouse 50% lethal doses (MLD_{50}s)/mL, the assay takes up to four days to complete and requires a large number of mice if the toxin is to be quantified. In addition, the mouse toxicity results are not in themselves specific; specificity is imparted only by carrying out parallel toxin neutralization tests with homologous antisera (Shone et al., 1985). Furthermore, as future modifications are made to botulinum neurotoxins relative to their use as therapeutic agents, quantification relative to its toxicity to mice may not be possible. Thus, despite the apparent sensitivity offered by the mouse bioassay, its use as a routine detection technique for botulinum neurotoxins is not only impractical, but also may be obsolete in some areas of research.

The second technique which is commonly used, is Enzyme Linked Immunosorbent Assays (ELISA). Numerous radioimmunoassays and various ELISAs including amplified ELISA and Enzyme Linked Coagulation Assay (ELCA) systems have been developed for the detection of these toxins (Ashton et al., 1985; 1988; Dezfulian et al., 1984; Doellgast et al., 1993; Gibson et al., 1987; Michalik et al., 1986; Notermans et al., 1982; Potter et al., 1993). Detection limits reported as a result of the many ELISA experiments involving these toxins range around 100 MLD_{50} units/mL; and very recently, ELCA and amplified ELISA techniques have reported detection of 1-10 MLD_{50} units/mL of types A, B and E neurotoxin (Doelgast et al., 1993; Potter et al., 1993). While enzyme-linked immunosorbent assays have been shown to rival the mouse bioassay in terms of sensitivity and are superior in that they are readily made selective by the use of specific antibodies, the technique still requires analysis times of 24 hours or longer.

Recently, a novel approach for the detection of *Clostridium botulinum* neurotoxins was developed (Ogert et al., 1992). This technique involved a two-step sandwich immunoassay utilizing an evanescent wave of light to excite fluorescently labeled antibodies bound to type A toxin captured on the surface of antibody immobilized fibers. Despite the many advantages offered by the fiberoptic immunosensor, the detectable level of type A toxin was 5 ng/mL (approximately 200 MLD_{50} units/mL). However, the work of Ogert et al. (1992) demonstrated that the use of various antibody systems in the generation of the sandwich complex resulted in different levels of detectability of type A toxin.

In this research, we have investigated detection of the complex molecule of type E *Clostridium botulinum* rather than just the purified neurotoxin portion. Type E botulinum neurotoxin is produced in the form of a complex that is referred to as the M complex, and consists of a non-toxic component (the NBP; Singh et al., 1995a) as well as the neurotoxin component. The M toxin is composed of the 150 kDa protein (neurotoxin or 7S protein) and a 120 kDa NBP (Singh et al., 1995a). Since the NBP component of the complex does not separate from the toxin in the environment (Ohishi et al., 1977; 1980), it makes more sense to develop a detection scheme for the complex rather than just the purified neurotoxin.. The nature of this neurotoxin demonstrates the applicability of the immunosenor in several areas of research, in biomedical fields due to recent implication of therapeutic uses of the toxin, in the food and agriculture industries because botulinum is a food poison and in military applications because of its threat as a potential agent in biological warfare.

MATERIALS AND METHODS

Instrumentation

The fiberoptic biosensor used in the detection of type E botulinum complex consists of five main components: the sensor (7 subcomponents), the detector (3 subcomponents), the lock-in amplifier, the laser power supply and a lap-top computer. A schematic representation of all of the biosensor's components and subcomponents with illustration of the laser light path and the returning fluorescent signal is shown in Figure 1.

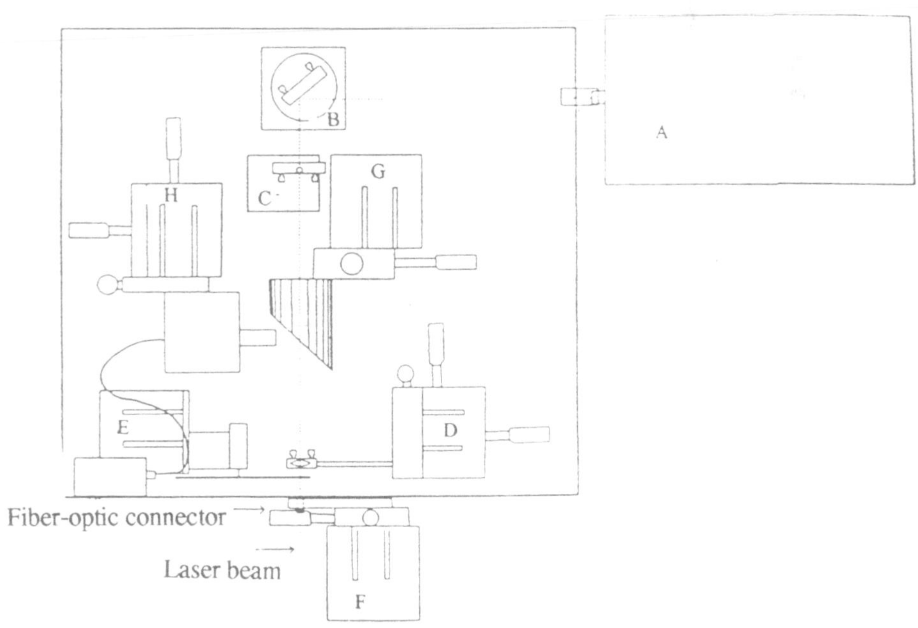

Figure 1. Schematic representation of all of the fiberoptic immunosensor's component.

The following lists and describes the function of the seven subcomponents which make up the sensor:

- A. *Laser head*: An Omnichrome Model 532 air-cooled single line argon ion gas laser operating at 514.5 nm.
- B. *Mirror*: A pyrex mirror which deflects the laser beam 90 degrees.
- C. *Line Filter*: A notch laser line filter with a it's filter centered at 514.5 nm for the rejection of unwanted plasma lines (wavelengths) from the laser.
- D. *Objective Lens*: An objective lens which focuses the excitation light into the optical fiber and collimates the returning fluorescent signal.
- E. *Chopper*: The chopper reduces background noise from scattered excitation light by modulating the excitation light out of the DC noise of the sensor.
- F. *Fiber Holder*: A fiber holder which connects the optical fibers to the sensor and allows for optimum alignment of the fiber with the incident laser beam.
- G. *Paraboloid*: The off-axis paraboloid collects the collimated fluorescent light from the objective lens and focuses it onto the detector.

The detector is made up of three different subcomponents:

- 1. *Aperture*: The aperture functions to lower background signal by rejecting most of the uncollimated light incident on the paraboloid.
- 2. *Emission Filter*: Filters out unwanted excitation light (green light) while allowing the fluorescent light (red light) into the detector.
- 3. *Photodiode*: Detects the fluorescent signal and sends an electrical current proportional to the incident light to the lock-in amplifier.

The lock-in-amplifier used with this biosensor is a Stanford Research Corporation Model SR510. The function of the lock-in amplifier is to demodulate the fluorescent signal and provide a digital output of the signal voltage both to the photodiode load resistor and to the laptop computer. The laptop computer is a Gridcase Model 1530 loaded with Stanford Research Systems software which runs the acquisition, controls the lock-in-amplifier, and collects and displays the test results.

Affinity Purification of Polyclonal Antibodies

Polyclonal antibodies purified from horse serum that had been hyperimmunized with *C. botulinum* type E complex toxoid, were obtained from Dr. J. E. Brown, US Army through the Naval Research Laboratory (NRL) in Washington, DC. These anti-toxin E antibodies were subsequently affinity purified against type E complex using the method described by Ogert et al. (1992), except for the fact that we used Affigel 15 as the column matrix (Singh et al., 1989) instead of activated sepharose 4B. The objective was to enhance biosensor detectability for type E complex, using affinity purified antibodies as both the immobilized capture antibodies and the labeled antibodies for generation of a fluorescent signal.

Fluorescent Labeling of Proteins

Antibodies were labeled with an extrinsic fluorescent probe for the generation of a fluorescent signal during biosensor detection. The fluorescent probe used was tretramethyl-rhodamine-5-isothiocyanate (TRITC) purchased from Molecular Probes, Inc., Eugene, Oregon and stored in a desiccator at less than 0 °C prior to use.

A major consideration in choosing an appropriate ratio of TRITC molecules per protein molecule was based on the possibility of the TRITC binding at a site near to the antibody-antigen binding site. Therefore, in order to preserve binding specificity of the antibody, as well as for fluorescent quenching concerns, a low to moderate degree of labeling is preferred. Based on ELISA results, TRITC-labeling did not significantly alter the binding of IgG to the toxin (data not shown).

The procedure for TRITC labeling of proteins was based on the methods described in Ogert et al. (1992) with a few modifications. In general, an amount of TRITC was dissolved in 200 microliters of dimethylformamide (DMF) and added to 100 mL of a 0.05 M borate-buffered saline (BBS) solution, pH 9.3 containing 0.04 M NaCl. The solution was stirred to a homogeneous mixture by use of a magnetic stir bar and stir plate. The protein solution to be labeled was added to a clean dialysis bag and dialyzed in the TRITC/BBS solution at 4 °C for a pre-determined length of time. After conjugation had occurred, the dialysis bag containing the labeled protein was removed from the TRITC/BBS solution and dialyzed in 500 mL of PBS at 4 °C. The dialysis buffer was replaced with clean PBS every hour for a total of three hours before being left to dialyze overnight. This procedure was necessary in order to remove any free TRITC molecules that may be present in the dialysis bag but not covalently bound to an IgG molecule.

There are two variables that affect the conjugation of TRITC to a protein, the amount of TRITC used, and the time allowed for the reaction to occur. Prior to labeling of affinity purified antibodies, the amount of TRITC and the time needed to achieve a labeling ratio between 1.5 and 1 TRITC labels per protein were determined. The TRITC labeled protein solutions used in all biosensor experiments were prepared at a protein concentration of 5 μg/mL in PBS containing 2.0 mg/mL Bovine Serum Albumin (BSA). The BSA was used as a blocking agent to block non-specific binding sites on the fiber.

Preparation of Optical Fibers

Optical fibers were provided by NRL and all fiber tapering was conducted at the laboratory in Washington, DC. Fibers were made from step-index plastic clad silica optical fiber (200 m core, Quartz et Silica, Quartz Products) with a black plastic ferrule on the proximal end. The proximal face of the fiber is polished after insertion into the ferrule. The total length of the fiber is approximately 1 meter with only 10 centimeters of the entire fiber being used for assay purposes. At the distal end of the fiber (last 12 cm), the cladding was stripped with a razor blade and the fiber core was tapered by computer-controlled immersion into hydrofluoric acid.

There were five steps involved in the preparation of the optical fibers: tapering, cleaning, silanization, crosslinker attachment, and covalent binding of antibodies. The procedures used were based on methods outlined previously (Bhatia et al., 1989; Ogert et al., 1992). The goal was to immobilize affinity purified antibodies on the surface of the fibers for the capturing of complex molecules in solution and ultimately detection using the biosensor.

The fluorescence emission generated upon the binding of a fluorophore to the surface of the optical fibers occurs in higher order modes. The refractive index of the cladding being higher than that of the buffer solution means that the clad portion of the fiber is not able to support as many modes as the immersed portion of the fiber. The fluorescent light propagating in the higher order modes is, therefore, preferentially lost as they enter the clad portion of the fiber. A mode will be cut off when its field does not decay outside of the core. In a geometric sense this means that the field associated with this mode is not totally internally reflected at the core cladding interface. Rather, they are partially transmitted and there is continuous radiation loss. Therefore, a reduction in the radius of the unclad portion of the fiber through tapering limits the number of modes that the unclad portion of the fiber may carry and improves coupling of the returning fluorescent signal. Fiber tapering, also serves to increase the intensity of the evanescent wave at the surface of the fiber and thereby increases the potential for the bound molecules to fluoresce. Although fiberoptic immunosen-

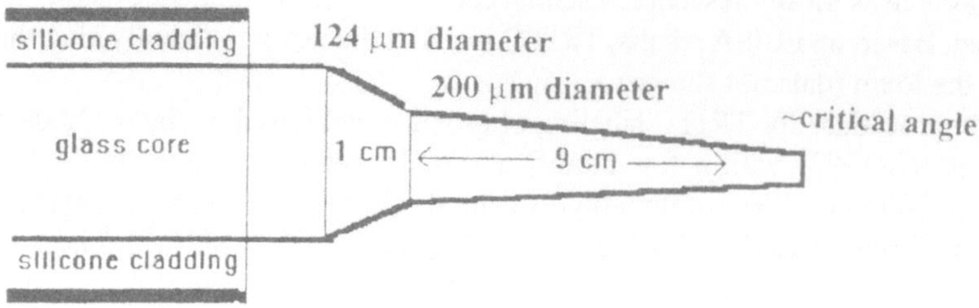

Figure 2. Schematic diagram of a tapered optical fiber.

sor allows for an ultrasensitive detection of a variety of biomolecules, the percentage of fluorescent coupling is rather low and in our system is approximately 4%.

During the process of tapering, the fiber diameter was reduced from 200 to 124 μm over approximately 1 cm and then very slowly over the remaining 9 cm to about 75 μm diameter. The fiber was tapered such that the angle of incidence at the end of the fiber nearly reached the critical angle. A schematic representation of a tapered optical fiber is shown in Figure 2.

Biosensor Studies

The detection of type E *C. botulinum* complex using the fiberoptic biosensor was accomplished using a two-step sandwich immunoassay. The assay involved (1) the incubation of antibody bound fibers in a solution of type E complex followed by (2) signal collection upon incubation of the resulting complex bound fibers in a fluorescently labeled antibody solution. The procedures and fundamentals involved are discussed below in detail.

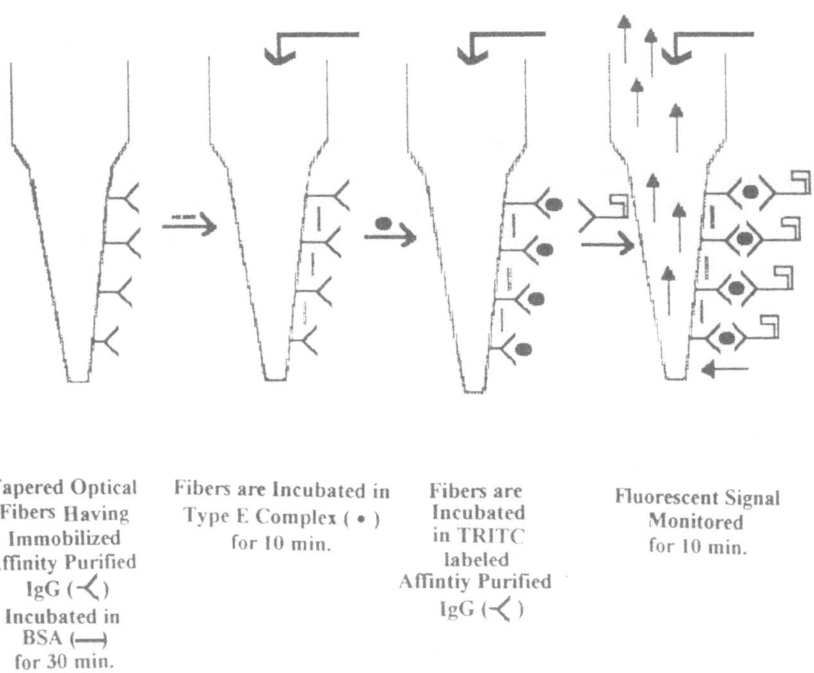

Figure 3. Two-step sandwich immunoassay used in the detection of *Clostridium botulinum* neurotoxins.

The two step sandwich immunoassay (Fig. 3) used in the detection of botulinum type E complex was conducted as follows. In each of the experiments, fibers having covalently attached affinity purified IgG were incubated in 5 mL-pipets containing 20 mg/mL BSA in PBS solution for 30 min. This was done to block the non-specific sites on the surface of the fiber and thereby decrease the background signal. Background signal was then collected by measuring the fluorescent signal produced upon incubation of BSA blocked fibers in a 5.0 µg/mL TRITC-labeled affinity purified IgG solution. In each of the experiments, a positive control was also analyzed to ensure that the crosslinking procedure was successful. The positive control involved the incubation of BSA blocked fibers in a 2 mL-pipet containing 50 µg/mL of anti-horse IgG for ten min followed by signal monitoring in the affinity purified IgG-TRITC solution. The detection of a fluorescent signal above background was indicative of antibody-antigen binding, and therefore, successful crosslinking of the antibody to the fibers. Fibers used to detect type E complex were first incubated in BSA for 30 min followed by 10 min of incubation in varying concentrations of the complex contained in a 2 mL-pipet. Binding was then monitored by signal collection upon incubation in the affinity purified IgG-TRITC solution.

RESULTS AND DISCUSSION

An analytical technique for rapid and continuous monitoring of botulinum neurotoxins was developed. The methods involved the use of affinity purified antibodies in the detection of type E complex using a fiberoptic immunosensor. Experiment was conducted to investigate the detection of type E complex at a concentration intermediate between 100 pg/mL and 500 pg/mL. The resulting binding curves from this experiment are shown in Figure 4. While the signal curve for the 100 mg/ml type E complex and for the negative control (type A botulinum neurotoxin) showed negative slope, the curves for 300, 500 and

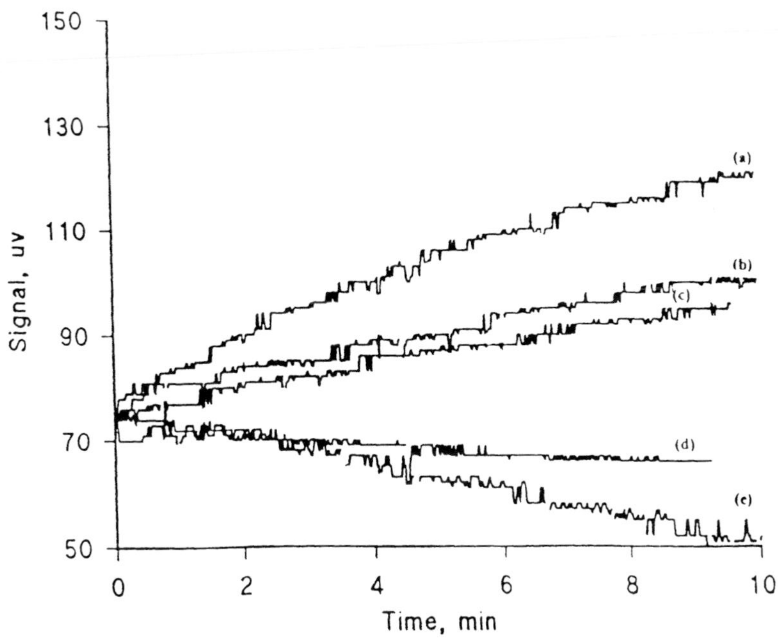

Figure 4. Binding of 1000 pg/mL (a), 500 pg/mL (b), 300 pg/mL (c), 100 pg/mL (d) type E complex, and 10 ng/mL type A toxin = negative control (e) to fibers having immobilized affinity purified IgG.

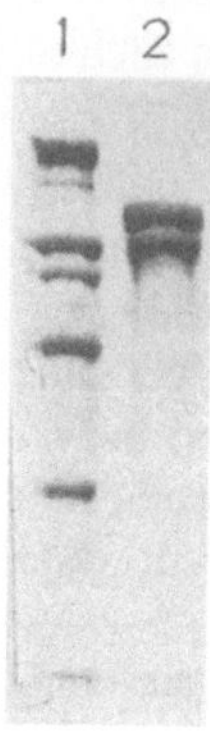

Figure 5. Gel electrophoresis of type E *Clostridium botulinum* complex. Lane 1 = molecular weight standards. Lane 2= type E complex: 1st band = 150 kDa toxin band, 2nd band = 120 kDa NBP band.

1000 pg/ml type complex showed considerable positive slope. The data for 300 pg/ml type E complex was reproducible although sometimes it did not show significant positive slope. We believe the negative slope of 100 pg/ml as well as variation in the slope is due the fiber to fiber variability, rather than unreliability of the detection. We are currently designing ways to correct this problem. Results from this last experiment offer the evidence, that if functioning properly, the immunosensor may be sensitive enough to detect a concentration of 300 pg/mL, or at least a concentration of type E complex less than 500 pg/mL may be possible, and additional work is warranted.

Previous work by Ogert et al. (1992), demonstrating the detection of 5 ng/mL of type A botulinum neurotoxin, combined with the many advantages offered by using such an

Table 1. Available Techniques for the Detection of *Clostridium botulinum* neurotoxins

Technique	Detection Limit (pg or pg/mL)	Detection Limit[e] (MLD_{50} units)	Total Analysis Time
ELISA[a]	100-1000 pg/mL (type A)	10-100 MLD_{50} units/mL	24 hours
ELISA[b]	40 pg (type A)	9 MLD_{50} units	30 hours
	20 pg (type B)	< 1 MLD_{50} units	
	35 pg (type E)	< 1 MLD_{50} units	
ELCA[c]	8 pg/mL (type A)	1.2 MLD_{50} units/mL	24 hours
	8 pg/mL (type B)	< 1 MLD_{50} units/mL	
	8 pg/mL (type E)	< 1 MLD_{50} units/mL	
FIS[d]	5 ng/mL (type A)[e]	1000 MLD_{50} units/mL	50 minutes
	150 pg/mL (type E)[f]	15 MLD_{50} units/mL	

[a]Conventional enzyme linked immunosorbent assay detection limits are based on those reported in the study conducted by Shone et al. (1985)
[b]Amplified enzyme linked immunosorbent assay detection limits are based on those reported in the study conducted by Potter et al. (1993).
[c]Enzyme linked coagulation assay detection limits are based on those reported in the study conducted by Doellgast et al. (1993).
[d]Fiberoptic immunosensor detection limits based on reported MLD_{50}s and MLD_{50}units/mL of 5 pg and 5 pg/mL, 23 pg and 10 pg/mL, and 37 pg and 10 pg/mL for types A, B and E ,respectively (Doelgast et al., 1993; Potter et al., 1993).
[e]Based on the immunosensor study conducted by Ogert et al. (1992).
[f]Based on the detection limit asserted for the toxic component of type E botulinum determined in the present study.

immunsensor, prompted efforts to increase the sensitivity of this immunoassay. The present study, which utilized affinity purified polyclonal antibodies as both the immobilized capture antibodies and the fluorescently labeled antibodies, has demonstrated the definitive detection of 500 pg/mL type E botulinum complex, and offers evidence for the detection of concentrations as low as 300 pg/mL. Since the complex molecule of type E botulinum neurotoxin is composed of approximately 50% toxin and 50% neurotoxin binding proteins (Singh et al., 1995a; Fig. 5), a 300 pg/mL concentration of complex would correspond to a detection of approximately 150 pg/mL of the actual toxic component of type E botulinum. Such detection offers comparable sensitivity to other method of neurotoxin assay with considerable advantage in the time required for analysis. Table 1 lists the methods currently available for the detection of botulinum neurotoxins, their detection limits, and the time required for analysis.

CONCLUSION AND FUTURE PROSPECTS

Although it appears that the fiberoptic immunosensor may be less sensitive than amplified ELISA and ELCA assays in the detection of botulinum neurotoxins (Table 1), the immunosensor still offers the detection of 15 MLD_{50} units/mL in less than a minute (time corresponding to actual signal collection). This combined with the immunosensor's ability for remote and continuous monitoring, makes it the preferable method when rapid detection is needed, or for environmental monitoring in the case of biological warfare. Moreover, the ability of the immunosensor to detect 15 MLD_{50} units/mL is comparable to the mouse bioassay which has a sensitivity of 5 MLD_{50} units/mL and the increased sensitivity offered by the other two methods is insignificant when considering the toxicity of the neurotoxin.

Furthermore, research efforts described herein, have resulted in at least a ten-fold improvement in the detection of *Clostridium botulinum* neurotoxin using a fiberoptic immunosensor. Additionally, the increased sensitivity of the assay by using affinity purified polyclonal antibodies as both the immobilized capture antibodies and the fluorescently labeled antibodies suggests that the investigation into additional assay systems may be worthwhile.

Some of the inherent difficulties associated with the use of this fluorescence-based immunosensor include fiber to fiber variability in the signal, cost of the instrument as well as fibers, and availability of portable commercial instrument. Still, further characterization of the instrument and the methods surrounding the immunoassay are necessary in order to increase its potential application as a sophisticated analytical technique. Relevant to the detection of botulinum neurotoxins, binding in a variety of biological and environmental matrices needs to be investigated. Also, to increase the applicability of the biosensor as a technique that offers reliable quantitative results, a method for internal standardization should be investigated. The fiberoptic immunosensor in its present state is best described as a technique that allows for the selective detection and monitoring of botulinum neurotoxins at concentrations at or above 300 pg/mL. However, this technique would provide only semi-quantitative results as far as the determination of unknown concentration of the toxin is concerned.

REFERENCES

Arnon, S. S. (1993) Clinical trial of human botulism immune globulin. In: Botulinum and Tetanus Neurotoxins: Neurotransmission and Biomedical Aspects (B. R. DasGupta, ed.), Plenum Press, New York. pp. 477-482.

Ashton, A. C., Growther, J. S., Dolly, J. D., 1985, A sensitive and useful radioimmunoassay for neurtoxin and it's hemagglutinin complex *for Clostridium botulinum. Toxicon* 23: 235-246.

Bhatia, S. K., Shriver-Lake, L. C., Prior, K. J., Geoger, J., Calvert, J. M., Bredehurst, R., Ligler, F. S., 1989, Use of thiol-terminal silanes and heterobifunctional crossliners for immobilization of antibodies on silica surfaces. *Anal. Biochem.* 178: 408-413.

DasGupta, B. R., 1989, The structure of botulinum neurotoxin. In: Botulinum Neurotoxin and Tetanus Toxin (Simpson, L. L., ed.), Academic Press, SanDiego. pp. 53-67.

Dezfulian, M., Hatheway, C. L., Yolken R. H., 1984, Enzyme-linked immunosorbent assay for detection of Clostridium botulinum type A and B toxins in stool samples of infants with botulism. *J.Clin. Microbiol.* 20: 379-388.

Doellgast, G. J., Triscott, M. X., Beard, G. A., Bottoms, J. D., Cheng, T., Roh, B. H., Roman, M. G., Hall, P. A., and Brown, J. E., 1993, Sensitive Enzyme-Linked Immunosorbent Assay for Detection of *Clostridium botulinum* Neurotoxins A,B, and E Using Signal Amplification via Enzyme-Linked Coagulation Assay. *J. Clin. Microbiol.* 31: 2402-2409.

Ember, L., 1991, Chemical weapons: plan proposed to destroy Iraqi arms. *Chem. Eng. News*. 69:6.

Fujii, N., Kouichi, K., Yokosawa, N., Yashiki, T., Tsuzuki, K., and Oguma, K., 1993, The complete nucleotide sequence of the gene encoding the nontoxin component of *Clostridium botulinum* type E progenitor toxin. *J. Gen. Microbiol.* 139: 79-86.

Gibson, A.M., Modi, N.K., Roberts T.A, et al., 1988, Evaluation of a monoclonal antibody based immunoassay for detecting type B *Clostridium botulinum* produced in pure culture and an inoculated cured meat system. *J. Appl. Bacteriol.* 64: 285-291.

Gibson, A. M., Modi, N. K., Roberts, T. A., Shone, C. C., Hambleton, P., and Melling, J., 1987, Evaluation of a monoclonal antibody-based immunoasssay for detecting type A *Clostridium botulinum* toxin produced in pure culture and an innoculated model cured meat system. *J. Appl. Bacteriol.* 63: 217-226.

Michalik, M.,Grezbowski, J., Ligieza J., 1986, Enzyme-linked immunosorbent assay (ELISA) for the detection and differentiation of *Clostridium botulinum* toxin type A and B. *J. Immun. Meth.* 93: 225-230.

Notermans, S., Hagenaars, M., Kozaki, S., 1982, The enzyme-linked immunosorbent assay (ELISA) for the detection and determination of *Clostridium botulinum* A, B and E. *Meth. Enzymol.* 84: 223-238.

Ohishi, I., Sugii, S., and Sakuchi, G., 1977, Oral toxicities of *Clostrdium botulinum* toxins in response to molecular size. *Infect. Immun.* 16: 107-109.

Ohishi, I., Sugii, S., and Sakuchi, G., 1980, Oral toxicities of *Clostridium botulinum* type C and D toxins of different molecular size. *Infect. Immun. 28*: 303-309.

Ogert, R. A., Brown, E., Singh, B. R., Shriver-Lake L. C., and Ligler F. S., 1992, Detection of Clostridium Botulinum Toxin A Using a Fiber Optic-Based Biosensor. *Anal. Biochem.* 205: 1-7 .

Potter, M. D., Meng, J., and Kimsey, P., 1993, An ELISA for Detection of Botulinal Toxin Types A, B, and E in Inoculated Food Samples. *J. Food Prot.* 56: 856-861.

Sakaguchi, G., 1983, *Clostridium botulinum* toxins. *Pharmac. Ther.* 19: 164-194.

Shone, C., Witon-Smith, P., Appleton, N., Hambleton, P., Modi, N., Gatley, S., and Melling, J., 1985, Monoclonal Antibody-Based Immunoassay for Type A *Clostridium botulinum* Toxin is Comparable to the Mouse Bioassay. *Appl. & Env. MicroBiol.* 50: 63-67.

Simpson, L. L. , 1989, Botulinum Neurotoxins and Tetanus Toxin. Academic Press, San Diego.

Singh, B. R., Chai, Y. G., Robertson, D. T. and Song, P. S., 1989, A photoreversible phytochrome affinity column chromatography for putative phytochrome receptor studies. *J. Biochem. Biophys. Meth.* 18: 105-112.

Singh, B. R., Foley, J. and Lafontaine, C., 1995a, Physico-chemical characterization of the botulinum neurotoxin binding protein from type E botulinum producing *Clostridium botulinum. J. Prot. Chem.* 14: 7-18.

Singh, B. R., Li, B. and Read, D., 1995b, Botulinum versus tetanus neurotoxins: why is botulinum neurotoxin a food poison but not tetanus? *Toxicon* 33:1541-1547.

COMPARATIVE STUDIES OF ANTISERA AGAINST DIFFERENT TOXINS

N. Nascimento, P. J. Spencer, R. A. de Paula, H. F. Andrade, Jr., and J. R. Rogero

Supervisão de Radiobiologia
Coordenadoria de Bioengenharia - IPEN-CNEN/S.P.
Caixa postal 11049
05422-970, São Paulo, S.P., Brazil

INTRODUCTION

Snake venom main components are proteins and peptides, many of them presenting enzymatic and/or toxic activities. Neutralization of these activities is one of the goals of envenomation treatment and this is usually achieved by mean of serotherapy. Snake venom antisera are produced in horses and, due to high toxicity of the immunogens, about 10% of the animals die after prime injection. Previous works of our group have shown gamma rays to attenuate snake venom toxicity without affecting its immunological properties, inducing an enhanced humoral and cellular immune response without the deleterious effects of native venom [1] [2] [3]. The venom attenuation induced by irradiation is a result of interaction of the venom components with the free radicals formed by water radiolysis, mainly hydroxyl radical and hidrated electron. These radicals act removing hidrogens, breaking S-S bonds or promoting deamination. As a result of the interaction of these radicals and others lesser known with the venom components, larger molecular weight aggregates are formed, with few if any toxic or enzymatic activity but presenting epitopes that induce a protective response when the irradiated venom is employed as immunogen [3]. These facts point toward venom irradiation as a solution to improve sera production. de Paula *et al.*[4], when immunizing rabbits with irradiated crotoxin, South American rattlesnake's main toxin, obtained polyclonal antibodies specific for the *Crotalus* genus and with high neutralizing capacity. These immunoglobulins proved to be useful as diagnostic and therapeutic agents. *C.d.terrificus* neurotoxic venom is composed by few protein fractions, mainly crotoxin, while the Brazilian most prevalent snake on what refers to accidents (85%), *Bothrops sp*, venom has proteolytic and coagulant activities, with several proteins and fractions involved, with phospholipase activity and minor neurotoxicity, serotherapy being the main treatment. These activities generate intense local signs in animals used in antisera production, with losses and low efficiency. In the present work, we detoxified *Bothrops jararaca* venom by mean of gamma radiation, immunized rabbits with the obtained toxoid and investigated the

Natural Toxins II, Edited by B. R. Singh and A. T. Tu
Plenum Press, New York, 1996

raised antisera efficiency against toxicity, cross-reactivity with other snake venoms from the Elapid and Crotalid group and neutralization of phospholipase A_2 activity which, as suggested by other authors [5], is an indicator of antisera potency.

MATERIAL AND METHODS

Venom Irradiation

Bothrops jararaca whole venom, purchased from Instituto Butantan, was dissolved in 150mM NaCl to a 2mg/ml final concentration and irradiated at room temperature and atmosphere with 2kGy of ^{60}Co radiation, at a dose rate of 480Gy/h in a Gammacell 220 (Atomic Energy of Canada Ltd, Canada) device.

Toxicity

Toxic activity was determined in mice by lethal dose 50% (LD50) and calculated by the Spearman-Karber method as preconized by WHO [6].

Immunization

Three month old male New Zealand rabbits were immunized with irradiated *B. jararaca* venom. Prime injection was of 1mg in complete Freund adjuvant. The second was of the same venom amount in incomplete adjuvant and the booster consisted of 1mg toxoid in saline solution.

Enzyme Linked Immuno Sorbent Assay (ELISA)

96 wells microplates were coated with 10μg/ml of either *B. jararaca, B. neuwiedi, B. cotiara, B. alternatus, B. moojeni, B. jararacussu, L. muta, M. frontalis, C. d. terrificus, C. d. cumanensis* and *C.d. terrificus* PLA_2 standard (Sigma Chemical Co, St. Louis, U.S.A.). Serial dilutions of the sera were assayed against those venoms. Reaction was developed by 1:2000 horse radish peroxidase coupled goat anti-rabbit IgG (Sigma), 0.02% OPD (orto-phenyldiamine) and 0.02% hidrogen peroxyde. The reaction was interrupted with 200mM citric acid. Optical densities were determined at 450nm in microplate reader (Dynatech MR5000). Negative controls consisted of rabbits serum collected before immunization. For comparative cross reaction between each venom, multiple sera dilution was performed, revealing that the best serum dilution to achieve 0.5 O.D. in ELISA was 1/2000, when reacting with autologous venom. The quantitative crossreactivity was expressed as percentage of autologous venom.

IgG Purification

IgGs were purified from whole antisera by affinity chromatography in Protein-A Sepharose medium (Pharmacia, Uppsala, Sweden), according to a low salt schedule [7].

Western Blot

30μg of the same venoms as above were submitted to SDS-PAGE (15% acrylamide) under non-reducing conditions and transferred to nitrocellulose membrane (0.45μm pore; Sigma). After blocking with 1% BSA, the membrane was allowed to react with 20μg of the purified IgG for 2 hours at room temperature and incubated with peroxydase conjugated anti-rabbit IgG. The reacting bands were revealed with incubation with diaminobenzidine 1mg/ml and H_2O_2 0.01% in citrate buffer. Reaction was intensified using cobalt chloride as enhancer. The proteins remaining in the overloaded separating gel were stained with Coomassie Brilliant Blue 250-R (Sigma).

Phospholipase A_2 Activity

This enzymatic activity was assayed by indirect hemolysis as described by Guttierrez *et al* [5] with a slight modification: instead of sheep erythrocytes, mice red blood cells were employed. Previous experiments have shown that this little change had no effect on the assay sensitivity [8]. 0.3μg of each venom were incubated with either phosphate buffered saline, pure or 1:10 diluted antiserum and applied to 2mm wells punched on a glass plate covered with a mixture of 1.2% fresh washed mice erythrocytes, 1.2% of 1:4 diluted egg yolk and 10mM calcium chloride in 0.8% agarose medium. Phosphate buffered saline was used as negative control. The diameter of the hemolysis haloes were measured after 24 hours incubation at 37°C.

RESULTS

Irradiation resulted in a 6.5 folds toxicity attenuation (table 1).

On what refers to immunization, antibodies titers, assayed by ELISA were similar to those obtained when immunizing the animals with native venom (data not shown). However, local signs following injection,when present, were discrete, suggesting once again attenuation of venom activity. Immunoenzymatic assay indicates similar immunoreactivity for all venoms assayed excepting *Micrurus* venom which presented lower reactivity.

Table 1. Toxicity of native and irradiated samples as determined by Guarnieri [2]

Samples	LD50 (mg/kg)	Relative toxicity
Native venom	1.94 (1.52-2.48)	1
Irradiated venom	12.55 (9.82-16.04)	6.5

Table 2. Quantitative crossreaction between B.jararaca irradiated venom antiserum and other snake venoms, as determined in ELISA

Venoms	% of Autologous venom
B. jararaca	100
B. neuwiedi	98
B. cotiara	97
B. alternatus	99
B. moojeni	96
B. jararacussu	96
L. muta	94
M. frontalis	30
C. d. terrificus	70
C. d. cumanensis	97

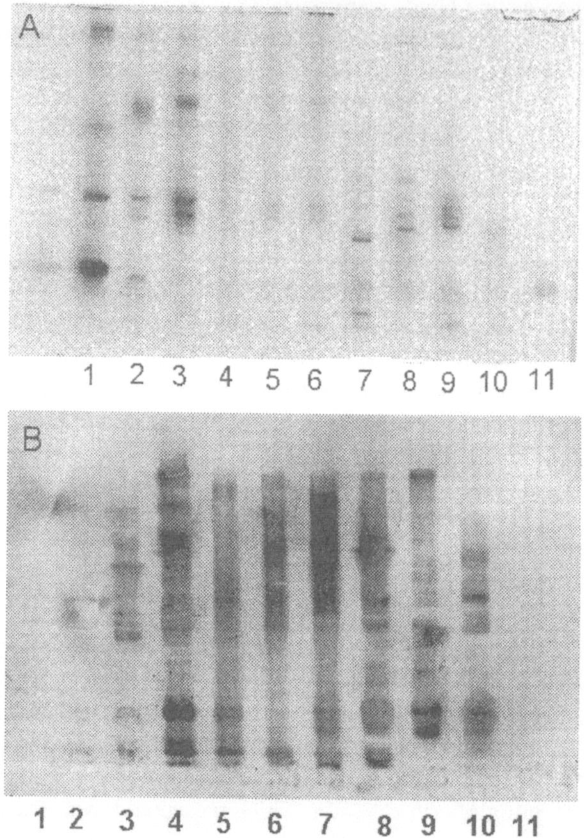

Figure 1. A. SDS-Page of the Assayed Venoms. B. Western Blot of the Assayed Venoms. Lane 1= C.d.terrificus, 2- C.d.cumanensis, 3- B.jararaca, 4- B.neuwiedi, 5- B.cotiara, 6-B.alternatus, 7-B.moojeni, 8-B.jararacussu, 9-L.muta, 10-M.frontalis, 11-C.d.terrificus Phospholipase A_2.

Table 3. PLA_2 Activity Neutralization

	Hemolysis halo (cm)		
Venoms	Pure Ab	1:10 diluted Ab	PBS
B. jararaca	0	0.80	1
B. neuwiedi	0	0	0
B. cotiara	0	0	0
B. alternatus	0	0	0
B. moojeni	0.65	0.75	0.80
B. jararacussu	1.10	1.20	1.20
L. muta	0.40	0.60	1
M. frontalis	2	2	2
C. d. terrificus	1.10	1.20	1.30
C. d. cumanensis	0	0	0
PLA_2 (standard)	1.10	1.20	1.20

Western blot data (Figs. 1a & 1b) indicate a similar immunoreactivity of all bothropic venoms, excepting *B. jararacussu* which presented lower detectable reactivity in the low molecular weight region at the end of the membrane. This might be due to species-specific myotoxins which may not cross-react since *B. jararaca* venom does not contain such compounds, although these myotoxins are structurally close to phospholipase A_2 with low molecular weight [9].

When assaying PLA_2 activity neutralization (table 3), the antisera neutralized efficiently the autologous venom, while venoms from other species of the same genus were only partially neutralized. Other bothropic venoms (*B. neuwiedi, B. cotiara and B. alternatus*) presented no detectable PLA_2 activity although previous standardization assays demonstrated this assay to be quite sensitive *C.d. terrificus* venom was not neutralized by the antibodies tested, neither did standard phospholipase or elapidic venom. Curiously, *L. muta* venom was almost neutralized.

DISCUSSION

The detoxification induced by gamma rays in Bothrops venom was less efficient as than the one described for *C.d.terrificus* venom. This fact could be ascribed to the diversity of fractions present in the bothropic venom, associated with a less specific activities of some of the components. Other explanation is the presence of small peptides and proteins that could protect and compete with the main components for the free radical generated by radiation. The toxoid, despite its relatively high toxicity, was able to induce an antisera in rabbits that presents a good protective activity against the original venom. ELISA and Western Blot analysis indicate quantitative and qualitative high level of cross-reactivity of all the venoms assayed against this antibody, suggesting shared antigenic determinants in most venoms, as expected by the close relationship of the tested snakes. Phylogenetically more distant snake venoms presented less cross reactivity but almost all presented antigens recognized in the molecular weight range of PLA_2 enzymes. When the direct inhibitory action of this antisera was tested against this enzymatic activity of each venom, it efficiency was not equal or proportional, suggesting that these conserved cross reacting epitopes are not important for the enzyme activity. Irradiation preserves epitopes that could induce neutralizing antibodies that may be shared by *L.muta* PLA_2, which was almost completely neutralized by the antisera, but others venoms presented the original activity. As suggested by other authors [10] the toxic and the enzymatic site of toxic phospholipases A_2 may be located on different sites of the enzyme molecule and we can not speculate whether the bound antibodies neutralize or not the enzyme toxicity. Further experiments should be done with purified enzymes in order to confirm these findings. As a whole, the irradiation of bothropic venom is an attenuation method that could induce better antisera, that could help to solve the chronic problems of antiophidic sera production.

ACKNOWLEDGEMENTS

This work was supported by Conselho Nacional de Pesquisa e Desnvolvimento (CNPq).

REFERENCES

1. MURATA, Y - *Efeitos da radiação gama no veneno de Crotalus durissus terrificus.* São Paulo, 1988. (Dissertação de mestrado, Instituto de Pesquisas Energéticas e Nucleares).

2. GUARNIERI, M.C. - *Estudo dos efeitos da radiação gama de 60Co nas propriedades bioquímicas, biológicas e imunológicas do veneno de Bothrops jararaca.* São Paulo, 1992. (tese de doutorado, Instituto de Pesquisas Energéticas e Nucleares).
3. NASCIMENTO, N. - *Estudo comparativo entre crotoxina nativa e irradiada. Aspectos bioquímicos e farmacológicos.* São Paulo, 1991. (Dissertação de mestrado, Instituto de Pesquisas Energéticas e Nucleares).
4. de Paula, R.A.; Nascimento, N.; Spencer, P.J.; Andriani, E.P.; Sanalios, R.B. & Rogero, J.R - Reatividade dos anti-soros produzidos contra venenos atenuados por radiação gama frente a diferentes venenos. IX Reunião Anual da Federação de Sociedades de Biologia Experimental (FESBE., 1994.
5. GUTTIÉRREZ, J.M.; AVILA, C.; ROJAS, E. & CERDAS, L. - An alternative *in vitro* method for testing the potency of polyvalent antivenom produced in Costa Rica. *Toxicon* **26**: 411-413, 1988.
6. WORLD HEALTH ORGANIZATION - *Progress in characterization of venoms and standardization of antivenoms.* Geneva, 1981.
7. HARLOW, E. & LANE, D. Eds: Antibodies: A Laboratory Manual. Cold Spring Harbor Labs ed.
8. Spencer, P.J.; Andriani, E.P.; Nascimento, N.; DE Paula, R.A.; Sanalios, R.B. & Rogero, J.R. - Improvement of an inexpensive method for phospholipase A_2 activity determination. IX Reunião da Sociedade Brasileira de Bioquímica e Biologia Molecular, 1994.
9. DOS-SANTOS, M.C.; GONÇALVES, L.R.C.; FORTES-DIAS, C.L.; CURY, Y.; GUTIÉRREZ, J.M. & FURTADO, M.F. - A eficácia do antiveneno botrópico-crotálico na neutralização das principais atividades do veneno de *Bothrops jararacussu. Rev. Inst. Med. Trop. São Paulo* **34**: 77-83, 1992.
10. KINI, R.M. & EVANS, H.J. - A model to explain the pharmacological effects of snake venom phospholipases A_2. *Toxicon* **27**: 613-635, 1989.

42

NEW APPROACHES IN ANTIVENOM THERAPY

V. Choumet,[1] F. Audebert,[1] G. Rivière,[1] M. Sorkine,[1] M. Urtizberea,[2] A. Sabouraud,[2] J.-M. Scherrmann,[2] and C. Bon[1]

[1] Unité des Venins
Institut Pasteur
25, rue du Dr Roux
75724 Paris Cedex 15, France
[2] Unité INSERM U26
Hôpital Fernand Widal
200 rue du Faubourg Saint Denis
75010 Paris, France

INTRODUCTION

Snake bites are a major public health problem throughout the world, particularly in tropical countries where mortality and morbidity rates are very high. The specific treatment of ophidian envenomation is serotherapy. It consists in the empirical administration of large amounts of specific antivenoms. Although serotherapy has been discovered a century ago, the mechanism by which antibodies of antivenoms neutralize venom proteins *in vivo* is still poorly understood. In order to rationalize antivenom therapy, clinicians should be able to know as quickly as possible the severity of the envenomation, to assess the appropriateness of serotherapy and to adapt the dose of serum to inject to the gravity of envenoming. Moreover, some parameters of antivenom administration such as the delay in administration after snake bites, and the route of injection and the type of antibodies to be used, are dependent on the understanding of the kinetics of venom and antivenom. Such investigations have been performed in the case of envenomations by European vipers in France. An ELISA was developed to determine venom concentrations in biological samples of patients bitten by vipers in an attempt to establish a correlation between venom levels and clinical symptoms. The kinetics of *Vipera aspis* venom was studied using experimental envenomations in rabbits. The effect of antivenom administration on the toxicokinetics of venom was also determined.

Natural Toxins II, Edited by B. R. Singh and A. T. Tu
Plenum Press, New York, 1996

MATERIALS AND METHODS

Sandwich ELISA was performed with Fab'2 purified from IPSER Europe antivenom, acccording to Audebert *et al.*, (1992; 1993; 1994a). Iodination of venom proteins was performed with [^{125}I]iodine using iodogen following the recommendations of the manufacturer. The labelling yield of venom was 99 % (200 µCi/mg) and the radioactivity was precipitable by trichloroacetic acid was 94 %. Toxicokinetic studies were performed as described by Audebert *et al.* (1994b). To study the effect of antivenom injection on *Vipera aspis* venom toxicokinetics, five rabbits were injected intramuscularly with 700 µg/kg of *Vipera aspis* venom and 7 h later received an intravenous injection of half a dose (2.5 ml, 125 mg) of IPSER Europe serums diluted with 2.5 ml saline. Venom concentrations in plasma were measured by ELISA and by counting the radioactivity of the TCA-precipitable fraction. The total area under the plasma concentration-time curve ($AUC0\text{-}\infty$) was calculated by the trapezoïdal rule from zero to the last experimental time (96 h) and to infinity by extrapolation using C_{96}/β, C_{96} being the plasma concentration at 96 h and β the terminal slope. The terminal half-life ($t_{1/2}\beta$) was calculated as $0.693/\beta$.

RESULTS AND DISCUSSION

A prospective enquiry was conducted in 1990 and 1991 in France by the Unité des Venins in order to collect epidemiological, clinical and biological data from hospitals. *Vipera aspis* venom antigens were quantified in urine and blood samples by a sandwich enzyme linked immunosorbant assay.

One hundred fifty charts were analyzed, from patients presenting documented viper bites. Oedema was the most frequent local sign (120), it appeared within the first 2 h, regularly developing to reach a maximum between 48 h and 72 h after the bite. Systemic signs rapidly appeared during the first 3 h. They consisted in vomiting and/or diarrhoea (37 cases), slight or severe hypotension (25 cases), shock (4 cases) and bleeding (2 cases). A relationship was observed between these systemic signs and the extent of oedema, which

Table 1. Clinical gradation and plasma venom concentrations of envenomations by European vipers

Grade	Name	Characteristics/symptoms	Mean values of plasma venom concentration (enquiry of 1991)
0	No envenomation	Fang marks	1 ± 0.3[(a)]
		No oedema or local reaction	n = 8
1	Minimal	Local oedema around the bite area envenomation	5 ± 1.8
		No systemic symptoms	n = 12
2	Moderate envenomation	Regional oedema involving major part of an extremity	32 ± 7
		Moderate systemic symptoms (slight hypotension, vomiting, diarrhoea)	n = 14
3	Severe envenomation	Extensive oedema spreading towards the trunk	126 ± 50
		Severe systemic symptoms (prolonged hypotension, shock, bleeding)	n = 4

(a) ng of *Vipera aspis* venum per ml of plasma

allowed the establishment of a grading scale (Table 1). Because this classification is intended as a therapeutic guide, the final evolution was not taken into account, as in other grading scales (Reid, 1976; Persson and Irestedt, 1981). Thirty nine patients were not envenomed [grade 0]. Envenomations were minimal for 67 cases [grade 1], moderate in 42 patients [grade 2] and severe in 6 cases [grade 3].

To examine whether the clinical gradation is related to the amount of inoculated venom, ELISA was used to quantify venom concentrations in serum samples collected from bitten patients during the first hours of their hospitalization. In the enquiry of 1990, biological samples were collected during the first 12 hours after the bite. For patients classified in grades 0 and 1, venom concentrations were not statistically different from those obtained for the control group (non envenomed persons). Patients presenting severe envenomation [grade 3] were clearly distinguished from others. In the enquiry of 1991, a more sensitive ELISA was used to quantify venom concentrations in samples collected during the first 4 h after the bite. A clear distinction between minimal, moderate and severe envenomations was established. As in the enquiry of 1990, for patients classified in grades 0 or 1, the venom concentrations were not statistically different (Table 1). On the other hand, patients classified in grade 2 and in grade 3 showed significantly higher venom concentrations than those calculated in grades 0 and 1 ($p \leq 0,001$). Furthermore, the values obtained from patients in grade 3 were statistically different from those obtained for patients in grade 2 ($p \leq 0,004$) (Table 1). A good correlation was thus observed between the venom levels in blood samples collected during the first hours of hospitalization and the clinical signs of envenomation. Furthermore, a clear venom concentration threshold of 15 ng/ml was established above which patients might be moderately or seriously envenomed. The ELISA appeared as a useful and predictive tool for the management of envenomed patients, especially in the early evaluation of the severity of envenomation, before symptoms reached their maximal development. Samples collected during the course of hospitalization were used to study the kinetics of venom levels in the plasma of patients. As shown in Table 2, venom antigens could be detected rapidly after the bite and their apparent half life was 8 h whatever the grade, suggesting that the kinetics of the venom did not depend on its concentration. This also indicated that the venom had disappeared 48 h after envenomation. However, this study did not allow an accurate determination of venom kinetic parameters, since only few samples were available and the amount of venom inoculated by the snake was unknown. The toxicokinetics of *Vipera aspis* venom was thus studied after experimental envenomations in rabbits.

The level of venom antigens in plasma was studied after intravenous injection of 250 $\mu g.kg^{-1}$ of unlabelled *Vipera aspis* venom. The venom level in plasma was shown to follow a biexponential decline and became undetectable after 96 h (Figure 1A). The pharmacokinetic parameters are shown in Table 3. The terminal half-life of the venom is about 12 h, indicating that most of the injected venom is eliminated 3 days after administration. This is in good agreement with the toxicokinetic study of *Vipera russelli* venom which showed a half-life of 8.9 h (Maung-Maung-Thwin *et al.*, 1988). The volume of distribution was found

Table 2. Kinetic parameters in patients envenomed by European vipers. T min is the time elapsed between the bite and the first sampling. Cmax is the maximal concentration detected. $T_{1/2}$ is the half-life calculated as $0.693/\beta$

Grade	Tmin (h)	Cmax (ng/ml)	$T_{1/2}\beta$ (h)
II	0.5 to 5	30.8 ± 28	8.3 ± 2.5
III	2 to 3	118 ± 105	8.6 ± 2.6

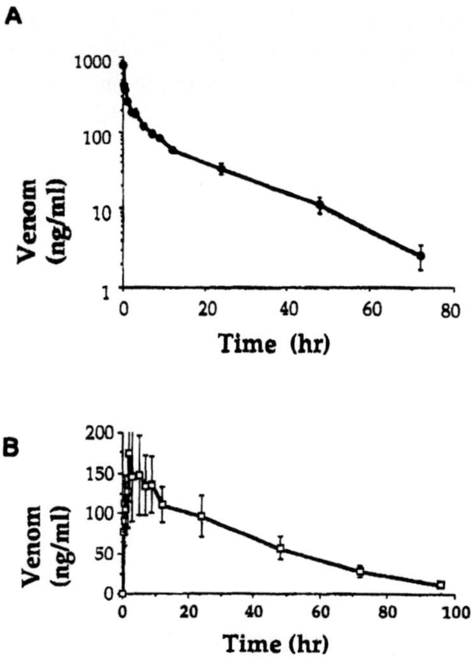

Figure 1. Pharmacokinetics of *Vipera aspis* venom after intravenous (A) and intramuscular (B) administration. Rabbits were injected with *Vipera aspis* venom via an intravenous route with 250 μg.kg^{-1} (A) or via an intramuscular route with 700 μg.kg^{-1} (B). Plasma samples were collected at the indicated times and assayed by ELISA. Venom concentrations are expressed in nanograms per milliliter as mean ± s.e.m (n = 5).

to be larger than the blood volume, indicating that the venom largely distributes in the extravascular compartment.

The toxicokinetics of *Vipera aspis* venom was studied after intramuscular injection of three doses (300, 500 and 700 μg.kg^{-1}). The time course of plasma venom concentration after administration of 700 μg.kg^{-1} of venom is shown in Figure 1B. Venom levels in plasma increased within the few hours after the venom administration. This has also been observed in cases of human envenomations by early quantifications of plasma venom concentrations using an ELISA (Table 2). Maximal values proportional to the injected dose were reached after 1.5 to 5 h. After 24 h, the plasma venom concentrations followed a monoexponential decline, with a half-life of 32.5 h (Table 3). In the range of the injected doses, the areas under the curves and the maximal concentrations were proportional to the dose of injected venom. The kinetics of venom absorption was studied using the deconvolution method. Absorption was rapid during the first 24 h and continued at a slower rate over the subsequent 72 h. This slow absorption process contributed to the maintenance of high venom concentrations in the plasma for a long period after intramuscular injection and explained the apparent terminal

Table 3. Kinetic parameters for *Vipera aspis* venom after intravenous and intramuscular injections

Injection		AUCo-∞								
Route	Dose (mg.kg^{-1})	ng.ml^{-1} hr^{-1}.kg^{-1}	Cmax ng.ml^{-1}	Tmax. (hr)	F (%)	T1/2 α (hr)	T1/2β (hr)	Vdss (l.kg^{-1})	Cls (l.hr^{-1}.kg^{-1})	Fu (%)
I.V.	250	1040±90	—	—	—	0.71±0.09	12±1	1.2±0.04 (0.7)[c]	0.084±0.006 (0.072)[c]	1.4
I.M.	700	2470±460	160	2	63±17	—	29[b]±2	—	—	2.6

[a] The values are the means ± S.E.M (n = 5)

[b] Apparent terminal half life

[c] Numbers in parentheses are the volume of distribution and the systemic clearance for iodinated venom

AUC: area under the curve; F: bioavailability; T 1/2α: distribution halflife; T 1/2β: terminal halflife; Vdss: volume of distribution; Cls:systemic clearance; Fu: fraction of venom in urine.

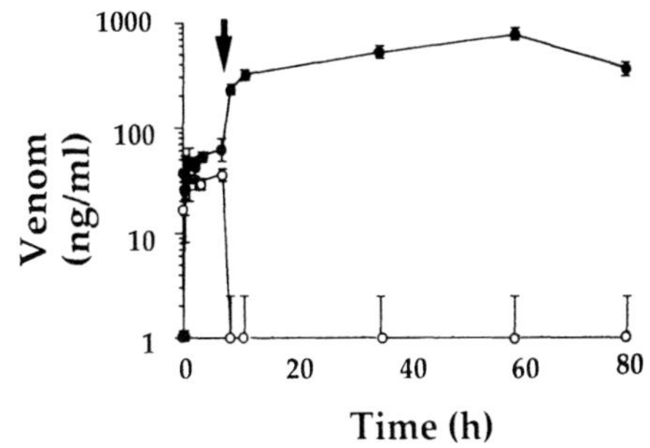

Figure 2. Effect of antivenom administration on plasma disposition of *Vipera aspis* venom in experimentally envenomed rabbits. Five rabbits were intramuscularly injected with 700 μg.kg^{-1} of ^{125}I-labelled *Vipera aspis* venom. Seven hours later, they were intravenously injected with 2.5 ml of Ipser Europe serum diluted with 2.5 ml of saline. Plasma samples were analyzed by ELISA for their content in free antigens (○) and by counting radioactivity for their total concentration of antigens (●).

half-life which was 3-fold higher than that measured after intravenous injection. Thus, venom disposition was limited by this complex absorption process, with absorption and disposition occurring simultaneously. This observation is in good correlation with the studies performed after subcutaneous injection of *Bothrops jararaca* venom in mice and with a clinical study of accidental envenomations with *Calloselasma rhodostoma* (Barral-Netto *et al.*, 1990; Ho *et al.*, 1986). Sixty three percent of the venom injected intramuscularly reached the blood circulation.

The effect of antivenom administration on the kinetics of *Vipera aspis* envenomation was also examined. Radiolabelled *Vipera aspis* venom was injected intramuscularly to mimic the route of administration in case of accidental envenomations. The use of a radiolabelled venom allowed the quantification of plasma venom components free and bound to antivenom antibodies. Free venom proteins were detected by ELISA. The plasma concentration time profile of venom in antivenom-treated animals is shown in Figure 2. The plasma concentration curves measured by counting radioactivity or by ELISA superimposed before the administration of the antivenom. After intravenous injection of 125 mg of IPSER Europe serum, the total venom concentrations in plasma rapidly increased more than ten-fold and remained elevated during three days, whereas the plasma levels of free venom antigens measured by ELISA rapidly decreased and remained undetectable for the same period of time. Antivenom Fab'2 administration therefore results in the immunocomplexation of venom proteins in the vascular compartment and in the plasma redistribution of venom antigens from the extravascular compartment to the vascular compartment.

The phenomenon of plasma sequestration and redistribution has been observed for digitalis, colchicine and phencyclidine after Fab infusion (Smith *et al.*, 1976; Sabouraud *et al.*, 1992; Valentine *et al.*, 1994). Such experiments could also be used to determine the optimal route for serum administration, to test the efficiency of a precocious or a delayed serotherapy, to define the dose of antivenom for optimal efficacy and to compare the relative effectiveness of different preparations of antivenoms. Studies similar to those carried out in the case of viper envenomations could be performed in the case of envenomations by other snakes or by scorpion stings and would help to rationalize antivenom treatment.

REFERENCES

Audebert, F., Sorkine, M. and Bon, C. (1992) Envenoming by viper bites in France: clinical gradation and biological quantification by ELISA. *Toxicon* 30, 599-609.

Audebert, F., Grosselet, O., Sabouraud, A. and Bon, C. (1993) Quantitation of venom antigens from European vipers in human serum or urine by ELISA. *J. Anal. Toxicol.* 17, 236-240.

Audebert, F., Sorkine, M., Robbe-Vincent, A. and Bon, C. (1994a) Viper bites in France: clinical and biological evaluation; kinetics of envenomations. *Hum. Exp. Toxicol.* 13, 683-688.

Audebert, F., Urtizberea, M., Sabouraud, A., Scherrmann, J.-M. and Bon, C. (1994b) Pharmacokinetics of *Vipera aspis* venom after experimental envenomation in rabbits. *J. Pharmacol. Exp. Ther.* 268, 1512.

Barral-Netto, M., Schriefer, A., Vinhas, V. and Almeida, A. R. (1990) Enzyme-liked immunosorbent assay for the detection of *Bothrops jararaca* venom. *Toxicon* 28, 1053-1061.

Ho, M., Warrell, D. A., Looareesuwan, S., Phillips, R. E., Chanthavanich, P., Karbwang, J., Supanaranond, W., Viravan, C., Hutton, R. A. and Vejcho, S. (1986) Clinical significance of venom antigen levels in patients envenomed by the Malayan pit viper (*Calloselasma rhodostoma*). *Am. J. Trop. Med. Hyg.* 35, 579-587.

Maung-Maung-Thwin, Khin-Mee-Mee, Mi-Mi-Kyin and Thein-Than (1988) Kinetics of envenomation with Russell's viper (*Vipera russelli*) venom and of antivenom use in mice. *Toxicon* 26, 373-378.

Persson, H. and Irested I. (1981) A study of 136 cases of adder bites treated in Swedish hospitals during one year. *Acta Medica Scandinavica* 210, 433-439.

Reid, H. A. (1976) Adder bites in Britain *Brit. Med.* J. 2, 153.

Sabouraud, A., Urtizberea, M., Cano, N., Grandgeorge, M., Rouzioux, J.-M. and Scherrmann, J.-M. (1992) Colchicine-specific Fab fragments alter colchicine disposition in rabbits. *J. Pharmacol. Exp. Ther.* 260, 1214-1219.

Smith, T. W., Haber, E., Yeatman, L. and Butler, V. P. (1976) Reversal of advanced digoxin intoxication with Fab fragments of digoxin specific antibodies. *N. Engl. J. Med.* 294, 797-800.

Valentine, J. L., Arnold, L. W. and Owens, M. (1994) Antiphencyclidine monoclonal Fab fragments markedly alter phencyclidine pharmacokinetics in rats. *J. Pharmacol. Exp. Ther.* 269, 1079-1085.

DISTRIBUTION OF DOMOIC ACID IN SEAWEEDS OCCURRING IN KAGOSHIMA, SOUTHERN JAPAN

T. Noguchi[1] and O. Arakawa[2]

[1] Laboratory of Food Hygiene
Faculty of Fisheries
Nagasaki University
Nagasaki 852, Japan
[2] Laboratory of Marine Botany and Environmental Science
Faculty of Fisheries
Kagoshima University
Kagoshima 890, Japan

INTRODUCTION

The anthelminthic principle in the red alga *Chondria armata* was isolated and designated "domoic acid" by Daigo[1]. Later, the sale of this amino acid, as well as L-kainic acid, as an anthelminthic, was prohibited. Since then, not much attention had been paid to domoic acid except as an excitatory amino acid. However, in 1987 a food poisoning incident due to the ingestion of cultured mussels *Mytilus edulis* occurred in Eastern Canada[2]. The main symptoms of one hundred seven persons poisoned were gastrointestinal and neurological disorders. Severely affected patients suffered from memory loss ("amnesic shellfish poison"; ASP), a special characteristic of this poisoning. Four victims died. The causative agent was shown to be the excitatory amino acid domoic acid, which was present in high concentration in the digestive gland of the cultured mussel. It appeared that the domoic acid came from the diatom *Nitzschia pungens* forma *multiseries* which occurred as a "red tide" and on which the mussel fed.

In Japan, red tides due to *N. pungens* have occurred. However, thus far there have been no reports of ASP due to the ingestion of shellfish. Since 1991, ASP screening of cultured bivalves and of diatoms has been carried out in Japan. Domoic acid has not been detected in industrially important shellfish from 1991 to 1994, nor in diatoms, except for a *N. pungens* sample (0.01 pg of domoic acid per cell) collected from a red tide which occurred in Hiroshima Bay in August 1994. On the other hand, large amounts of domoic acid were detected in the red alga *C. armata* occurring in Kagoshima Prefecture, Southern Japan.

This paper deals with the distribution of domoic acid in seaweeds occurring in Kagoshima Pref. and with its possible origin.

Natural Toxins II, Edited by B. R. Singh and A. T. Tu
Plenum Press, New York, 1996

MATERIALS AND METHODS

Chondria armata

Samples were collected in Kagoshima Pref. at Tokunoshima, in June 1993, at Yakushima, in June, July, December 1994, and at Hanasezaki, in August 1993 and August 1994. Fresh samples, except those from Tokunoshima, were frozen and transported to the Laboratory of Marine Biochemistry, University of Tokyo. Tokunoshima samples were first dried, then sent to Tokyo.

Other Red Algae

Samples of *Coelothrix irregularis* were collected at Yakushima, in July, December 1993, and January 1994. Samples of *Jania capillacea* were collected at Yakushima, in late July 1993, and at Arasaki, Kanagawa Pref., in August 1993. Samples of *Digenea simplex* were collected at Ishigaki Island, Okinawa Pref., in April 1993 to March 1994.

Green Alga

Samples of *Boodlea coacta* were collected at Yakushima, in January 1994.

Crab

The xanthid crab *Atergatis floridus* was collected at Yakushima, in January 1994.

HPLC Analysis of Domoic Acid

To 1-2 g samples of seaweeds an equal volume of 0.1 N HCl was added, and the mixture heated in a water bath for 5 min, then filtered through filter paper. The filtrate was applied to a Sep-pak C18-cartridge (Waters, Milford, MA, USA). Ten μl of eluate was submitted to HPLC analysis for domoic acid on a Tosoh HPLC equipped with a Finepack SIL 18-T-5 column (4.6 x 250 mm), using a mixture of acetonitrile, heptafluorobutyric acid, water (12 : 0.43 : 87.57) as the eluant at a rate of 1.0 ml per min. The domoic acid in the eluate was monitored at 242 nm with a Tosoh UV 8011 monitor. The quantity of domoic acid was calculated using a calibration curve of amount of authentic domoic acid (Sigma, St. Louis, MO, USA) *vs* absorbance at 242 nm, as follows:

$$\text{Domoic acid (ppm)} = (5.2 \times \text{Peak area} \times 10^{-4} - 2.6)\,(\text{Dilution} \times 2)/\,\text{Sample volume } (\mu l)$$

Mouse Bioassay

Domoic acid was assayed using ddY strain mice (body weight 18-20 g, male) by intraperitoneal administration. At a 40 μg dosage of authentic domoic acid, the mouse showed a typical "scratching syndrome", where it repeatedly scratches its abdominal parts with its hind limbs and then moves around for 45 min after injection, but does not die. At a dose of 80 μg, mice died in 15 min. Because of its low sensitivity, the mouse bioassay was used only to check high levels of domoic acid.

Organisms Adhering to Raw *Chondria armata*

Immediately after fresh *C. armata* samples were collected in late July 1993 at Yakushima and in August 1993 and August 1994 at Hanasezaki, the adhering fouling organisms were removed from the live seaweed by stroking it repeatedly with a writing brush dipped in seawater. The washed seaweed, fouling organisms, and the filtered seawater wash were submitted to HPLC analysis for domoic acid.

Microorganisms Isolated from *Chondria armata*

Cells and culture broths of 25 microorganisms isolated by Dr. K. Ohwada, Ocean Research Institute, University of Tokyo, from *C. armata* collected in January 1993 at Yakushima were submitted to HPLC analysis for domoic acid.

RESULTS AND DISCUSSION

Domoic Acid of Red Algae

Chondria armata. The domoic acid content was investigated in seaweed from 3 areas in Kagoshima Pref. The Tokunoshima sample (dried) collected in June 1993 showed a very high score of 4,300 ppm. Dried *C. armata* from this area was formerly collected and

Table 1. Content of domoic acid in seaweeds

Algae	Place of collection	Date of collection	Domoic acid (ppm)
Red algae			
Chondria armata	Tokunoshima, Kagoshima	Jun., '93	4,300 (dry)
	Yakushima, Kagoshima	Jun., '93	550 - 648 (wet)
		Jul., '93 (twice)	52 - 928 (wet)
		Dec., '93	994 (wet)
		Jan., '94	924 (wet)
	Hanasezaki, Kagoshima	Aug., '93	381 (wet)
		Aug., '94	201 (wet)
Jania capillacea	Yakushima	Jul., '93	38 - 82 (wet)
		Dec., '93	709 (wet)
		Jan., '94	358 (wet)
Coelothrix irregularis	Yakushima	Jul., '93	52 - 405 (wet)
		Dec., '93	591 (wet)
		Jan., '94	775 (wet)
	Ishigaki Is., Okinawa	Jun., '93	14 (wet)
		Dec., '93	6 (wet)
		Jan., '94	<1 (wet)
		Feb., '94	1 (wet)
Amphiroa spp.	Yakushima	Jul., '93	37 (wet)
	Arasaki, Kanagawa	Aug., '93	<1 (wet)
Digenea simplex	Ishigaki Is.	Apr., '93 ~ Feb., '94	<1 (wet)
Green algae			
Boodlea coacta	Yakushima	Jan., '94	5 (wet)

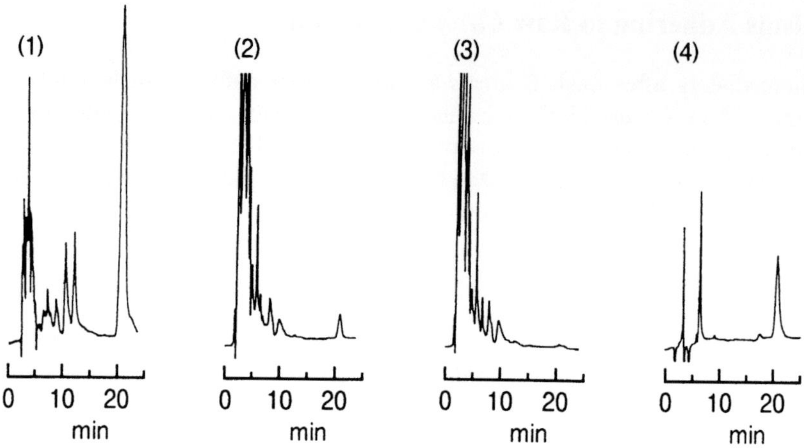

Figure 1. HPLC analysis for domoic acid of red algae: (1) *Chondria armata* (994 ppm), (2) *Coelothrix irregularis* (6 ppm), (3) *Digenea simplex* (<1 ppm), (4) standard of domoic acid.

used as an anthelminthic. However, the domoic acid content in this seaweed had never been determined. Five Yakushima samplings were carried out from June 1993 to January 1994. The results of the HPLC analysis for domoic acid are shown in Table 1 and Figure 1. The domoic acid content in this seaweed remained near 1000 ppm (wet) for all seasons except for a dip in early July 1993. The Yakushima samples were collected from many tide pools. There was little variation in domoic acid between tide pool locations. At Hanasezaki, *C. armata* was collected in August 1993 and August 1994. The domoic acid contents were 381 and 201 ppm, respectively, lower than the scores of the Yakushima samples.

Jania capillacea. This seaweed is always found entwined with *C. armata* in the tide pools at Yakushima. Its domoic acid content was 38-709 ppm, as high as that of *C. armata* during early July 1993 to January 1994 (Table 1). This seaweed has never been reported to contain domoic acid. It is not certain whether or not the origin of the domoic acid in this seaweed is connected with its presence in *C. armata*.

Coelothrix irregularis. This seaweed is also found entwined around the body as well as the root of *C. armata* in the tide pools of Yakushima. It showed a high concentration (52-775 ppm) of domoic acid, as high as those of *C. armata* and *J. capillacea* from Yakushima. However, at Ishigaki Island, Okinawa Pref., 940 km from Yakushima, this seaweed showed a far lower score (0.3-14 ppm) of domoic acid than Yakushima samples. This seaweed at Ishigaki Island also occurs entwined around the body of a red alga. A regional variation in the domoic acid content of *C. irregularis* is clearly apparent. The origin of domoic acid in this seaweed is not certain.

Amphiroa spp. This seaweed was collected in late July 1993 from the same tide pools as *C. armata* at Yakushima. It, however, does not become entwined with *C. armata*. This seaweed showed a moderate concentration (37 ppm) of domoic acid, while a sample collected at Arasaki, Kanagawa Pref., contained less than 1 ppm of domoic acid. Regional variations in domoic acid content in *Amphiroa* spp. samples was apparent.

Digenea simplex. This seaweed is known to contain a high concentration (1940 ppm) of kainic acid, an anthelminthic and excitatory amino acid related to domoic acid. Samples of this seaweed collected from Ishigaki Island, Okinawa Pref. contained less than 1 ppm domoic acid.

Green Algae

Boodlea coacta. This seaweed was collected from the same tide pools as the red algae at Yakushima and is found not entwined around other algae. This green alga showed a slight concentration (5 ppm) of domoic acid.

From the above results on algae, *C. armata* from all areas of collection in Kagoshima Pref. showed a high concentration of domoic acid. *J. capillacea* and *C. irregularis* from Yakushima, where they are found entwined around *C. armata*, also showed almost as high scores of domoic acid as *C. armata*. The red alga *Amphiroa* spp. from Yakushima, which is not found entwined with *C. armata* in the same tide pools, showed a moderate content of domoic acid. However, *C. irregularis* and *Amphiroa* spp., collected in areas other than Kagoshima Pref. and free of *C. armata*, generally showed a far lower content of domoic acid. The red alga *D. simplex* collected away from Kagoshima Pref. did not show any domoic acid. The green alga *B. coacta* from Yakushima showed a slight content of domoic acid. It would appear that certain conditions at Kagoshima Pref. may be favoring the production of domoic acid in the red algae.

Microorganisms Obtained from *C. armata*

Fresh *C. armata* collected in July 1993 at Yakushima showed 928 ppm of domoic acid. With samples washed repeatedly with a brush dipped in seawater, the domoic acid content of the washed seaweed decreased to 277 ppm, while the seawater wash contained 600 ppm of domoic acid. Some diatoms were found in the wash. Fresh *C. armata* collected at Hanasezaki in August 1993 and August 1994 were treated in a similar fashion. A native sample showed 381 ppm of domoic acid before washing and 328 ppm after washing. In the filtrate from the wash water domoic acid was barely detected, while the wash water residue showed 22.8 ppm domoic acid. In another sample, 201 ppm domoic acid was present before washing and 132 ppm after washing. The wash water filtrate had 7 ppm domoic acid while the residue contained 30-50 ppm. The residues consisted of several diatoms and other particulate matter. From these results, the domoic acid in *C. armata* collected at Yakushima may be partly derived from fouling organisms. However, this does not seem to be the case with the Hanasezaki samples. Further investigation is necessary on this matter.

Of the 25 microorganisms isolated from *C. armata*, no domoic acid (less than 1 ppm) was found in any of the cells or culture broths. It appears that none of these is involved in the origin of domoic acid in this seaweed.

The diatom *N. pungens* (4,000,000 cells) collected from Hiroshima Bay, Hiroshima Pref., by Dr. T. Uchida, Nansei National Fisheries Research Institute, contained about 0.01 pg of domoic acid per cell. This yield is very low when compared with the 0.2 pg per cell found in *N. pungens* forma *multiseries* from Canada. This would suggest that the domoic acid in the red algae of Japan arises from a route different from the *N. pungens* forma *multiseries* route in Canadian mussel.

Crab

The xanthid crab *Atergatis floridus* is known to contain paralytic shellfish poison (PSP) or tetrodotoxin. The viscera of specimens collected from tide pools at Yakushima in late July 1993 showed 6.0 Mouse Units of PSP per gram by the mouse assay and a content of 10 ppm of domoic acid. Since the crab feeds on seaweeds, it is suggested that the domoic acid may have originated from the food web.

ACKNOWLEDGMENTS

The authors express sincere appreciation to Dr M. Ikawa, University of New Hampshire, for reviewing this manuscript.

REFERENCES

1. Daigo, K., 1959, Studies on the constituents of *Chondria armata*, II. Isolation of an anthelmintical constituent, *J. Jpn. Pharm. Assoc.* 79:353-356.
2. Wright, J. L. C., Boyd, R. K., de Freitas, A. S. W., Falk, M., Foxall, R. A., Jamieson, W. D., Laycock, M. V., McCulloch, A. W., Mclnness, A. G., Odense, P., Pathak, V. P., Quilliam, M. A., Ragan, M. A., Sim, P. G., Thipault, P., and Walter, J. A., 1989, Identification of domoic acid, a neuroexcitatory amino acid, in toxic mussels from eastern Prince Edward Island, *Can. J. Chem.* 67:481-490.

INDEX

MIX
Papier aus verantwortungsvollen Quellen
Paper from responsible sources
FSC® C105338
FSC
www.fsc.org

If you have any concerns about our products,
you can contact us on
ProductSafety@springernature.com

In case Publisher is established outside the EU,
the EU authorized representative is:
Springer Nature Customer Service Center GmbH
Europaplatz 3, 69115 Heidelberg, Germany

Printed by Libri Plureos GmbH
in Hamburg, Germany